A BASIC ATLAS OF RADIO-WAVE PROPAGATION

A BASIC ATLAS OF RADIO-WAVE PROPAGATION

SHIGEKAZU SHIBUYA
Yokohama, Japan

Translated by
HARUO ISHIZUKA
Japan

A WILEY-INTERSCIENCE PUBLICATION
JOHN WILEY & SONS
New York • Chichester • Brisbane • Toronto • Singapore

A BASIC ATLAS OF RADIO-WAVE PROPAGATION SYSTEMS by Shigekazu Shibuya
Copyright © 1983 by Corona Publishing Co., Ltd.
English translation rights arranged with Corona Publishing Co., Ltd.

Copyright © 1987 by John Wiley & Sons, Inc.

All rights reserved. Published simultaneously in Canada.

Reproduction or translation of any part of this work beyond that permitted by Section 107 or 108 of the 1976 United States Copyright Act without the permission of the copyright owner is unlawful. Requests for permission or further information should be addressed to the Permissions Department, John Wiley & Sons, Inc.

Library of Congress Cataloging in Publication Data:

Shibuya, Shigekazu, 1925–
A basic atlas of radio-wave propagation.

"A Wiley-Interscience publication"
1. Radio wave propagation. I. Title.
TK6553.S49 1986 621′.3841′1 86-1658
ISBN 0-471-88183-X

Printed in the United States of America

10 9 8 7 6 5 4 3 2 1

FOREWORD

Advances in electronics and communication technology are rapidly moving the world toward a globally integrated service communication network of telephone, telegraph, and image signals that includes digital data processing. Among the contributors to the phenomenon are the technology of satellite communication, broadcasting, mobile radio, radiolocation, remote control, measurement, electronic switching, and very-broad-band transmission. In all of these areas, radio communication plays an important role.

The state of the art of artificial satellites has reached a level of sophistication that permits exploration of interplanetary space. This would not be possible without corresponding advances in radio communication technology.

Achievements in radio astronomy during the past few years have disclosed a new view of the universe to replace the one supported by traditional optical astronomy.

Refining radio communication technology will require perfecting and developing systems engineering capable of application to specific areas and a total system.

The author approaches systems engineering through a review of radio-wave propagation using the aforementioned areas. The many figures contained in this book add to our understanding of radio-wave propagation.

Publication of this book is timely and should be valuable to those engaged in radio facilities and systems projects and the study of radio-wave propagation.

SHIGERU YONEZAWA

PREFACE

Radio waves continuously reach the earth from the sun and other galaxies. However, these radio waves are not artificially controlled, unlike those created by man. It is the artificially controlled radio waves that will be explored in this book.

Because the phenomenon of propagation is an essential characteristic of radio waves, any research on radio waves is meaningless without a discussion of propagation.

Recent years have seen sophisticated and significant changes in radio engineering. Vastly important has been the tremendous progress in solid-state electronic circuitry. This progress has been responsible for the production of radio equipment that is greatly reduced in size and weight and is both more reliable and easier to maintain.

Moreover, advances in radio-wave engineering knowledge may require that the work done by radio engineers be subdivided into two specialities: research and development, and systems engineering. Sharing between these two groups will be essential.

A potential result of this split will be the creation of a new dawn for radio engineering; two research avenues will be explored at the same time: radiation and propagation.

This text has been written with this new era in mind. I have reviewed the processes of radio-wave propagation used during the last 30 years, eliminating errors and adding useful descriptions of phenomena previously overlooked. Another of my goals has been to try to understand the phenomenon of radio-wave propagation under cloud cover.

I have tried to avoid difficult equations and have used many figures and graphs to help clarify the written material. These figures also help to illuminate various phenomena and concepts, and indicate applicable parameters.

The problems presented here can also be explored using a computer. The figures will be useful for verifying the accuracy of these computer programs. A word of caution: accuracy of a specific value at a certain point will not always guarantee the accuracy of values at points near that point.

Upgraded and digitalized microwave transmission systems (including satellite communication systems) are being demanded today. In a digital radio system it is very important to suppress the reflection interference propagation path (see Chapter 2) by decreasing the propagation echo ratio and the path-length difference between direct and reflected waves and by minimizing frequency-selective fading.

Achievements in Japanese microwave technology are the result of three decades of research and development, production, installation, operation, and maintenance. I shall be very grateful if this book is able to contribute to the development of microwave engineering in any country in the world.

SHIGEKAZU SHIBUYA

Yokohama, Japan

ACKNOWLEDGMENTS

It is my great pleasure to have this book published by John Wiley & Sons, New York, in cooperation with Corona Publishing Co., Ltd., and the translator, Mr. Haruo Ishizuka.

Publication of the English version of this book will take seven years from inception to publication. Therefore, I want to thank especially the translator and the editors, all of whom tackled a difficult and long-term challenge.

CONTENTS

A BASIC ATLAS OF RADIO-WAVE PROPAGATION

INTRODUCTION

HOW TO USE *A BASIC ATLAS OF RADIO-WAVE PROPAGATION*

Chapter 0 presents basic concepts of the physics of radio propagation which are used in the chapters that follow.

Detailed descriptions of propagation are given in Chapters 1 through 6. The first three chapters, Chapter 1, Free-Space Propagation Path, Chapter 2, Reflection Interference Propagation Path; and Chapter 3, Diffraction Propagation Path, will probably be referred to frequently by readers. Although the theories found in Chapter 1 are simplified, they form the foundation for the other chapters.

Chapter 4, Troposcatter Propagation Path, and Chapter 5, Absorption Propagation Path, are hypothetical because much of the material is based on empirical data.

Chapter 6, Passive-Relay Propagation Path, may have more use in the near future.

Chapter 7 summarizes noise and the signal-to-noise ratio.

Chapter 8 presents systems engineering procedures and discusses fading in terms of causes, classifications, statistical processing, estimation, propagation design, performance analysis, and evaluation.

Chapter 9 provides the astronomy- and geography-related information needed by propagation engineers and gives application graphs with comprehensive explanations suitable even for beginners.

Each chapter presents and explains theories, and gives analog graphs, nomographs, or tables. Chapters 1 to 8 also give examples, which supplement the written information and provide readers with the opportunity to connect theory with practical applications. Chapter 8 presents many examples, arranged systematically for the reader's convenience.

It is advisable not to skip the notes included in each chapter because they provide logical justification, proofs, exceptions, and restrictions.

A chart index is given at the beginning of the chart section for each chapter. For main topic items, use the index.

0

BASIC CONCEPTS

0.1 ABBREVIATIONS AND IMPORTANT VALUES

The abbreviations, specific symbols, frequently used measuring units, and common constants employed in propagation design are defined in the following sections.

A. Specific Symbols

This book uses approximately 500 symbols for propagation calculations, each of which uses English and Greek letters, numbers, astronomical graphic symbols, and specialized characters.

Greek letters and astronomical graphic symbols unfamiliar to the reader are found in Fig. 1. Among the astronomical symbols used, the sun is denoted by ⊙, the moon by ☾, Mars by ♂, and Venus by ♀. Figure 1*b* lists all of the astronomical symbols used in this book. Due to recent advances in radio astronomy and interplanetary exploration, the possibility of earth-to-planet communication has increased. Therefore all graphic symbols shown will also have distance and luminosity included.

B. Numerical Multiplier Prefix

The numerical multiplier prefix is used to facilitate the expression of very large or small numbers containing many zeros in decimal notation. Figure 2*a* provides typical examples. The prefix abbreviation, such as k, M, or p, always precedes the corresponding unit symbol, as in km, MHz, or pF.

More general notation applicable for any range of numerical values is also used. Computer outputs that include scientific calculation are printed in specific form.

The following methods of notation have been used in this atlas:

Decimal Notation	Prefix Notation	Power Notation	Example of Computer Printout
6,370,000 m	6.37 Mm	6.37×10^{6} m	6.37 E 06 M
142,800,000,000 $	142.8 G$	$1.428 \times 10^{11}$$	1.428E 11$
0.000 325 W	325 μW	3.25×10^{-4} W	3.25 E −04 W
1,200,000	—	1.2×10^{6}	1.2 E +06

C. Common Measuring Units

Names, symbols, and magnitudes of measuring units commonly used are also found in Fig. 2*a*.

D. Common Constants for Propagation Design

Figure 2*b* lists the numerical values of the constants used frequently in this book along with common logarithms, natural logarithms, and reciprocal numbers.

1. In propagation calculations, dB (decibel) $10 \log_{10}$ is widely used. Where $\log_{10}$ will not be confused with the natural logarithm $\log_e$, $\log_{10}$ may be written simply as log. The abbreviation ln may be used for the natural logarithm $\log_e$.
2. The decibel value of a reciprocal number has the same absolute value as that of the original number, but is of opposite sign.
3. Ka is a frequently used value.

$$K = \frac{\text{effective earth radius}}{\text{actual earth radius}}$$

where K (effective earth radius factor) $= 4/3$ for atmosphere with standard refractivity.

Since the actual earth radius a is 6370 km, for $K = 4/3$, the value $Ka = 8500$ km is used for approximate calculations.

4. Boltzmann's constant k is used to calculate thermal noise, the signal-to-noise (S/N) ratio, and threshold input power.

NOTE. In Chapter 7, K is used instead of k. K is used when there is no possibility of confusion. K also represents the Kelvin temperature.

5. For normal temperature T_0, a range of 288 to 300 K (15 to 27°C) is used; the difference between the lowest and highest temperatures when expressed in decibels is only 0.2 dB. For example, $T_0 = 290$ K or $T_0 = 300$ K is used to simplify calculations.

0.2 ELECTROMAGNETIC WAVES

In 1861 James Clerk Maxwell, the Scottish physicist, showed that light is an electromagnetic wave. Then the German physicist Heinrich Hertz succeeded in generating electromagnetic waves in 1888. In 1895 Wilhelm Röntgen discovered X rays, and Guglielmo Marconi invented a system of wireless telegraphy in 1897. The remarkable advances in electronics engineering that followed have had an enormous impact on communications.

A. Classification of Electromagnetic Waves

In Fig. 3 electromagnetic waves are subdivided into six categories according to frequency, namely, radio waves, infrared rays, visible rays (light), ultraviolet rays, X rays, and γ (gamma) rays.

In theory the lowest limit of electromagnetic waves is set at nearly zero, the highest limit at infinity, and infinity is included within the γ rays so as to cover the entire frequency spectrum (overlap is inevitable).

The velocity of electromagnetic waves is equal to the velocity of light c ($\approx 3 \times 10^8$ m/s). Therefore the wavelength λ (the distance traveled within a period of unit oscillation) is given by

$$\lambda = c/f = 3 \times 10^8/f \quad \text{(m)} \qquad \text{(free-space wavelength)}$$

where f is the frequency in hertz. Figure 3 provides scales N and L using the relations $f = 3 \times 10^N$ Hz and $\lambda = 10^L$ m.

Radio Waves

Radio waves are defined by the radio regulation of the International Telecommunication Union (ITU) as "electromagnetic waves of frequencies arbitrarily lower than 3000 GHz, propagated in space without artificial guide," where 3000 GHz corresponds to 3×10^{12} Hz and $\lambda = 0.1$ mm. In the frequency spectrum where N ranges from 3 to 11, eight abbreviations, VLF to EHF, are given for each decimal range. The higher-frequency region of radio waves, $11 < N \leq 12$, overlaps with the infrared region. Details of these abbreviations are given in Fig. 4. Occupied bandwidth in Fig. 4 may be used for reference to measure a transmittable amount of information.

The bandwidth of HF on lower frequency bands is only slightly more than 10% greater than the VHF bandwidth. Therefore development of the higher frequency region is unavoidable for an information based society.

Infrared Rays

The spectrum spanned by infrared rays extends from the extreme end of red visible rays (typical wavelengths 0.75, 0.77, and 0.81 μm) to near the EHF band of radio waves (typical wavelengths 0.4, 1, and 5 mm). It should be noted that although radio waves are generally classified by frequency, infrared and shorter wavelength electromagnetic waves are classified by wavelength to facilitate numerical expression. Units ordinarily used are:

μm (10^{-6} m, micrometers) for infrared rays.
μm, nm (10^{-9} m, nanometers).
Å (10^{-10} m, angstroms) for light rays.
nm, Å for ultraviolet rays.
Å for X rays.
Å, pm (10^{-12} m, picometers) for γ rays.

Visible Rays (Light)

Electromagnetic waves with wavelengths ranging from 0.38 to 0.77 μm (3800 to 7700 Å) can be seen by the human eye.

Ultraviolet Rays

They are electromagnetic waves whose wavelengths are shorter than that of light, that is, from approximately 3800 (or 3970) to approximately 100 Å.

X Rays

X rays are electromagnetic waves with wavelengths of 100 to 0.01 Å. The shorter the X-ray wavelength, the denser the material that may be penetrated. Accordingly, shorter X rays may be called hard X rays and longer ones are called soft X rays.

γ Rays

γ rays are, specifically, electromagnetic waves ranging from 1 to 0.002 Å, which emanate from radioactive material. However, they can also indicate any electromagnetic waves with wavelengths of 1 Å or shorter, regardless of origin. Since this range overlaps that of X rays, artificially generated hard X rays may be called γ rays.

B. Electromagnetic Field in Free Space

Electromagnetic waves in free space are pure TEM (transverse electromagnetic) waves whose electric and magnetic field directions are both perpendicular to the direction of propagation, as shown in Fig. 5*a*.

States of oscillation at two different points separated by one wavelength (λ) or its integral multiple ($n\lambda$) along the direction of propagation are assumed to be equal and are called in phase. By contrast, the states at two different points separated by $\lambda/2$ or its odd-number multiple $(2n - 1)\lambda/2$ are assumed to be equal in amplitude of oscillation but opposite in polarity to one another.

The position of a point on the waveform with respect to the start of the cycle is called the phase. The phase may be expressed in angles (radians, degrees), distance, wavelength ratio, or time or period ratio. However, graphic representation by angle or wavelength ratio is used most frequently. Figure 5*a* shows various methods of representation.

For example, if the separation between two points is r, the phase difference (angle) φ is given by the relation $\lambda = 2\pi$ rad (= 360°),

$$\varphi \text{ (rad)} = 2\pi r/\lambda$$

$$\varphi \text{ (°)} = 360r/\lambda$$

When the result exceeds 2π rad or 360°, deduct $2n\pi$ rad or $n \times 360°$ ($n = 1, 2, \ldots$) from it.

Since wave motion is considered to be a series of repetitions of a cycle of sinusoidal oscillations corresponding to a wavelength, any phase may be substituted for the corresponding phase within a specified cycle.

In TEM waves, the electric field E and the magnetic field M are spatially orthogonal to one another, but in phase (their crests or null points coincide in time), and their magnitudes are proportional to each other. Therefore the state of an electromagnetic wave is determined if either the electric or the magnetic field is known.

The electric field strength E (electric potential gradient per meter in the direction of oscillation, V/m, mV/m, μV/m) is normally used to express the magnitude of an electromagnetic wave.

C. Electromagnetic Waves within an Enclosed Space

When electromagnetic waves are enclosed within a space whose volume is small compared with the wavelength, such as the spaces of a waveguide or coaxial cable, they assume specific modes of the electromagnetic field.

Figure 5*b* shows useful modes that contain a longitudinal component of either the electric or the magnetic field. The only exception is the coaxial cable (TEM mode).

TE Mode (Transverse Electric Mode)

In the TE mode the electric field vector is everywhere perpendicular to the direction of propagation, and a longitudinal magnetic component exists.

Examples of various modes, such as TE_{01} and TE_{11} of the circular waveguide type and TE_{10} of the rectangular waveguide type, are shown in Fig. 5*b*. The TE mode in the figure can be recognized because the field configuration on a plane including the waveguide axis indicates the magnetic field by a dashed line in the longitudinal direction.

TM Mode (Transverse Magnetic Mode)

The TM mode is similar to the TE mode, but E is replaced by H and "electric" by "magnetic." An example of TM_{01} for a circular waveguide is also shown in Fig. 5*b*. The field configuration of the cross-sectional plane including the waveguide axis shows the electric field E by a solid line.

REMARK. No dashed or solid line is shown in the field configuration for a coaxial cable. This is a feature of the TEM mode.

Meaning of Subscripts

The electromagnetic field mode differs depending upon the spacing (measured in wavelength) between side walls and the method of excitation even when the waveguide shapes are the same. Two-digit subscripts indicate the mode, and the rule indexing the subscripts differs for the circular waveguide and the rectangular waveguide as follows (for electric field configuration on a cross-sectional plane perpendicular to the waveguide axis):

1. Circular waveguide.
   a = number of planes perpendicular to the electric field
   b = number of cylinders on which the electric field becomes zero (including inner surface)
2. Rectangular waveguide.
   a = number of a half-cycle in electric field distribution over the longer side
   b = number of a half-cycle in electric field distribution over the shorter side

The subscripts are given in the order ab. For example, the TE_{01} mode for the circular waveguide has no plane perpendicular to the electric field ($a = 0$), and only one cylinder (inner surface of waveguide) of zero electric field strength ($b = 1$).

The TM_{01} mode is identical to the TE_{01} mode when the electric field and the magnetic field replace one another.

Cutoff Frequency and Effects of Higher Modes

The wavelength capable of being transmitted is limited, depending on the modes and physical size of the waveguide. An electromagnetic wave with a frequency whose wavelength is greater than the cutoff wavelength cannot propagate within the waveguide. However, when the size (expressed in wavelength ratio) of a waveguide is increased or there is a bending, deformation or discontinuation, the higher order modes are generated, resulting in a higher transmission loss.

Although no cutoff wavelength exists in coaxial cables, if the spacing between inner and outer conductors exceeds approximately $\lambda/4$, the higher order modes are apt to generate. Consequently a smaller diameter coaxial cable has to be used in a higher microwave frequency band, but with higher ohmic and dielectric loss. For this reason use of the waveguide feeder is preferable in the higher microwave frequency region.

0.3 WAVE FRONT AND DIRECTIVITY

A. Wave Front

The set of points at which the waves that simultaneously left a wave source arrive after the same time period is called the wave front. Figure 6 illustrates typical types of wave fronts. All points on the same wave front are in phase, and all antennas of the same type when placed over the same wave front generate receiving outputs that are all in phase.

Spherical Wave

Waves emitted from a single wave source (ideally an isotropic antenna) within an isotropic medium (ideally an infinitely extended vacuum space) form spherical wave fronts.

Although a half-wave dipole, a dipole array, a parabolic antenna, etc., can never form a spherical wave in the neighborhood of an antenna, wave fronts in the far field may be treated as plane waves within a small portion of a sphere. The field strength varies according to the angle from the antenna toward a position on the sphere; this feature is called directivity.

Cylindrical Wave

The emission of waves from an infinitely extended linear wave source placed within an isotropic space forms a cylindrical wave front as a result of the total effect of the infinite number of point wave sources arranged linearly.

Although there is no actual antenna that can be treated as an infinitely long wave source, the electric field in the neighborhood of a single linear antenna (including a multistage dipole) or a diffraction field obtained behind a knife-edge more than several times the width of the first Fresnel zone radius, may be treated as a cylindrical wave.

However, the wave front at a great distance from the antenna or one formed through diffraction by a narrow knife-edge becomes nearly a spherical wave, and a small portion of the cylindrical wave is considered a plane wave.

Plane Wave

A wave source at an infinitely long distance or a wave source of an infinitely extended flat plane forms a plane wave front. However, in practice, if the plane wave front is approximated by a set of points in which mutual phase differences do not exceed a certain value, such as $\lambda/16$ or $\lambda/24$, a portion of the wave front of spherical waves or cylindrical waves may be treated as a plane wave.

Aperture-type antennas, such as a parabolic antenna or a horn antenna, produce the plane wave on the aperture plane by aligning the phases.

B. Directivity

The angular performance of antenna radiation measured at a point on a distant spherical wave front for a transmitting antenna, or the reception of a plane wave by a receiving antenna is called directivity.

Directivity is indicated by showing the performance for all directions of a solid angle. It is represented by directivity patterns on two different planes perpendicular to one another (e.g., one in the vertical plane and the other in the horizontal plane or, for linear polarization, the E plane and the H plane) and includes the antenna. Directivities in the other directions may be interpolated, measured when necessary, or computed theoretically (requiring complicated calculations). Figure 6 shows two examples of directivity.

Radiation Axis

The main radiation axis is the direction of the maximum strength of radiation (the maximum reception in the case of a receiving antenna).

For planar antennas to generate plane waves and for linear array antennas fed in phase, the direction perpendicular to the linear element or the aperture plane coincides with the radiation axis.

Antenna gain normally refers to the value on the main radiation axis. Values in other directions are expressed as gain at angle θ in the vertical (or horizontal) plane.

Directivity Coefficient D_θ

The directivity of an antenna is expressed by the directivity coefficient D_θ at an angle θ. D_θ is a ratio of the electric field strengths, a ratio of the receiver input voltages for a receiving antenna (in antilogarithms) in the direction of the main radiation axis and at an angle θ, or a ratio expressed in 20 log (dB).

Gain G

Gain G is the effectiveness of a given antenna in relation to an isotropic antenna. Antenna gain is the ratio in decibels of the electric field strengths obtained from a great distance when transmitted both through the antenna being considered and through an isotropic antenna.

For a receiving antenna, the ratio of signal power values obtained at the receiver input terminals is used.

The power ratio in antilogarithms is converted into decibels applying 10 log.

Half-Power Width and Main Beam Width

To describe directivity sharpness or the high gain feature of an antenna, the half-power width or the main beam width is used.

1. The half-power width $\theta_{1/2}$ is the angular spacing between two points on the main lobe, at which the directivity coefficient is equal to 0.707 ($= 1/\sqrt{2}$) on both sides of the radiation axis.
2. The main beam width θ_B is the angular spacing between the two closest points of the minimum directivity coefficient on both sides of the radiation axis of the antenna radiation pattern. For antennas with an aperture side length that is sufficiently long in comparison with the wavelength, the approximate θ_B value is given by

$$\theta_B \simeq 2.4\theta_{1/2}$$

Because the actual value varies depending on the power distribution of aperture and phase disarray, more precise values should be determined by actual measurement. It should be noted that since propagation is reciprocal, an antenna gives the same directivity and gain for transmission and reception.

C. Requisites for Approximation by Plane Wave

The conditions for approximating a spherical wave by a plane wave are determined by the distance from the wave source and the tolerable phase difference λ/u (or $2\pi/u$), where u is a tolerance factor, as shown in Figs. 19 and 20 and as described below.

Point Wave Source (Fig. 19)

Let l denote the radius of the sphere, r $(= n\lambda)$ the radius of a circle obtained by cutting a portion of the sphere by a plane, and L $(= kr)$ the distance between the point wave source and the center of the circular plane. Assuming $L \gg r$, then the difference Δ $(= \lambda/u)$ between path lengths from the point wave source to the center of the circle and to the circumference is given by

$$\Delta = l - L \simeq r^2/2L$$

Substituting λ for Δ, L, and r,

$$\lambda/u = n^2\lambda^2/2kn\lambda$$

we have

$$u = 2k/n$$

$$k = un/2$$

The resulting graph, which gives the relationship between k, u, and n (λ is not included), can be used for any frequency.

If a phase difference value of $\lambda/16$ is a practical criterion for a plane wave, then the relation between n and k is given by line $\lambda/16$ in Fig. 19. The area which can be approximated by the plane (equivalent to a circle of $r = n\lambda$) and the distance from the wave source $L = kr$ are then determined.

Actual Wave Source (Ordinary Antenna) (Fig. 20)

We denote by $2r_1$ the wave source length (the segment of a linear antenna, length in an arbitrary direction on the aperture plane for a planar antenna, and diameter D_ϕ for a parabola). L $(= kr)$ is the distance between the center of line $2r_1$ and the point on a line $2r_2$ which is perpendicular to L and lies in a plane with segments $2r_1$ and L such that segment $2r_2$ is parallel to segment $2r_1$, as shown in Fig. 20.

Maximum phase difference occurs for path L. (Although one end of segment $2r_1$ is connected with one end of segment $2r_2$, these two endpoints are located on different sides of segment L.)

We then determine the condition that limits the difference $l - L$ within the tolerable value for approximation to the plane wave.

Figures 19 and 20 demonstrate that the relation given for the point wave source can be utilized when $r = r_1 + r_2$.

Figure 20 presents a nomograph defining the relationship between wavelength λ, $r\ (= r_1 + r_2)$, L, and the tolerable phase difference coefficient u ($\Delta = \lambda/u$, $u = \lambda/\Delta$) to be used for propagation design.

Determination of the plane wave is particularly important to the measurement of antenna directivity, the determination of multiantenna spacing for suppressing reflected waves, and the determination of antenna spacing for space diversity. The value of u is selected to be $\geq$ 32, 24, 16, or 12, according to our purpose.

Moreover, results of performance measurements on antenna directivity, gain, and cross-polarization discrimination are not reliable if the criteria for plane waves are not satisfied, the propagation path produces reflections (e.g., ground reflection), or the propagation medium is not homogeneous and isotropic.

The measurement of large-aperture antennas, especially those used by earth stations for satellite communications and by radio telescopes, requires a very large separation L, a higher value of u, and a free-space propagation medium. Consequently radio waves from a geostationary satellite or celestial body are used in the measurement.

0.4 POLARIZATION

A. Types of Polarization

The direction of oscillation of an electric field is always perpendicular to the direction of propagation. For an electromagnetic wave, if its oscillation occurs only within a plane containing the direction of propagation, it is called plane polarized or linearly polarized. This is so because the locus of oscillation of the electric field vector within a plane perpendicular to the direction of propagation forms a straight line. On the other hand, when the locus of the tip of an electric field vector forms an ellipse or a circle, the electromagnetic wave is called an elliptically polarized or circularly polarized wave.

An arbitrarily polarized wave can be analyzed as two mutually orthogonal plane polarized waves. Particular cases are the linearly polarized wave and the circularly polarized wave.

Figure 7 illustrates typical processes for generating various kinds of polarized waves using two polarized waves E_x and E_y of different amplitude ratios and relative phases. Although the magnetic field H is not shown in the figure to avoid confusion, there are magnetic fields with amplitudes proportional to and in phase with E_x and E_y, but orthogonal to the corresponding electric field vectors.

B. Linearly Polarized Wave (Figs. 7 and 8)

When there is only E_x or E_y, or E_x and E_y are in phase or inverse to each other, the result is a linearly or plane polarized wave. In other words, the x axis or the y axis coincides with the direction of polarization, and hence no electric field component is produced on the plane perpendicular to the polarization plane. On the component plane at angle θ to the polarization plane, an electric field proportional to $\cos\theta$ is produced. Accordingly, if the electric field axis of a receiving antenna is tilted by θ from the polarization plane, the receiving voltage will be $\cos\theta$ times the value expected when the axis coincides with the polarization plane. However, if the axis is perpendicular to the polarization plane, the receiving voltage will be zero ($= \cos\pi/2$).

Horizontally Polarized Wave

Radio waves whose electric field vector is parallel to the earth surface are called horizontally polarized waves. These radio waves radiate from a horizontal antenna. If the direction of wave propagation is away from the earth surface, the polarization plane cannot be horizontal. However, if the direction of the electric field vector is horizontal, the radio wave is received normally with a horizontal antenna whose radiation axis is set in the direction of the radio wave.

Vertically Polarized Wave

Radio waves whose electric field vector is vertical to the earth surface are called vertically polarized waves. If the direction of propagation of the radio wave is toward the earth surface, the direction of the electric field cannot be vertical to the earth, even if the polarization plane is vertical. In this condition the voltage induced in a receiving vertical antenna is lower in proportion to $\cos\varepsilon$, where ε is the slant angle. In this case the receiving antenna should be tilted correctly to coincide with the direction of the electric field vector. A radio wave whose polarization plane is vertical may be called a vertically polarized wave.

Obliquely Polarized Wave

A radio wave whose polarization plane is neither vertical nor horizontal is called an obliquely polarized wave. Normal reception of this wave requires the antenna to be erected so that the antenna's electric field axis coincides with the electric field vector of the radio wave.

An obliquely polarized wave is considered to be a combination of mutually in-phase or opposite phase horizontally polarized wave components and vertically plane polarized wave components. Thus any horizontal or vertical antenna is able to receive the wave at some level of tilting, depending on the tilt angle. Consequently it is used when the types of receiving antennas actually employed are unknown, or both vertical and horizontal receiving antennas are in use. It is used also for mobile communication where the altitudes of the mobile station are unstable. The most useful of the obliquely polarized waves is the wave polarized obliquely at 45°, the midpoint between vertical and horizontal polarization. There are two kinds of polarization: one tilted 45° to the right, and the other tilted 45° to the left. In either kind, the magnitudes of the electric field strengths of both vertical and horizontal components is $1/\sqrt{2}$ ($= \cos 45°$, or -3 dB) of the electric field strength of the combined wave. Therefore any of the horizontal or vertical receiving antennas can be used, provided gain reduction by 3 dB is permitted.

When the antenna is tilted 45° correctly to the right or left, there is no gain reduction. However, it should be noted that if the antenna is tilted 45° to the wrong side, i.e., to the left or right, the reception becomes zero since the polarization plane of the wave approaches a right angle to the electric field axis of the receiving antenna.

C. Elliptically and Circularly Polarized Waves (Figs. 7 and 8)

When component waves on the x and y planes are not in phase or in reverse phase and exhibit an intermediate phase difference, the locus of the electric field vector tip forms an ellipsoid. These waves are called elliptically polarized waves.

If the amplitudes of both component waves are equal ($E_x = E_y$) and the phase difference is $\pm\pi/2$ $(=\lambda/4)$, the resultant locus forms a circle, and the wave is appropriately called a circularly polarized wave.

A circularly polarized wave is considered to be a special case of an elliptically polarized wave.

Elliptical (circular) polarization is classified into right-hand polarization and left-hand polarization. In general, elliptical polarization may be classified further as right-hand circular polarization and left-hand circular polarization. However, deletion of one of these produces circular polarization.

Consequently a right-hand elliptically polarized wave can be received satisfactorily by an antenna designed for right-hand circular polarization, while the output of the same wave received through the antenna for left-hand circular polarization will also be obtained.

A circularly polarized wave can only be received by an antenna of matching circular polarization. In practice, the elliptically polarized wave is rarely used for transmission; usually the circularly polarized waves are employed.

Elliptically polarized waves become a matter of great concern when an emitted circularly polarized wave becomes elliptically polarized. This occurs when reversely polarized wave components are generated during such propagational processes as reflection of HF waves by the ionosphere, irregular reflection by man-made structures such as buildings, and scattering by rain drops in the VHF band or above.

A circularly polarized wave can be divided into two mutually orthogonal, transverse components of equal amplitude. Thus it can provide constant and good reception by an antenna for linearly polarized waves which has been installed transversely to the direction of propagation at any angle, such as vertical, horizontal, or oblique.

Accordingly, for mobile stations such as on aircraft and radiosonde, interruption of communication may be minimized and reliability improved when a circularly polarized antenna is employed on one side and a linearly polarized antenna on the other.

D. Definition of Right-Hand and Left-Hand Rotation

It would be difficult to mistake vertical polarization for horizontal polarization. However, care should be taken to distinguish between the left hand and the right hand in oblique polarization as well as in elliptical or circular polarization because of the following:

The definition differs by country and field of science and may change with time. "Different" means "opposite" in practice.

No preference is given for either, that is, left or right, by definition.

Any definition may be adopted if it is used consistently in a specific field or in a number of fields.

Even in a single field, such as optics or radio engineering, the definition may differ by country.

An erroneous interpretation may cause a serious effect. In oblique–oblique or circular–circular configurations, if any one of the transmitting and receiving stations uses an antenna with incorrect polarization, no signal can be received. However, no problem occurs if both stations happen to have installed the wrong antennas.

According to recent developments in international communication, particularly satellite communication, it was feared that a misinterpretation due to a different definition might result in incorrect facility configuration. As a result, a report with the following proposals was adopted by CCIR in 1963.

As regards the right-hand and left-hand rotations in the circular or elliptical polarization employed in a radio system, it is necessary that they be defined clearly to prevent misunderstanding. The following definitions are given to eliminate ambiguity.

1. Right-hand (clockwise) polarized wave. An elliptically or circularly polarized wave in which the electric field vector observed in any fixed plane normal to the direction of propagation, while looking in the direction of propagation, rotates with time in a right-hand or clockwise direction.

NOTE. For a circularly polarized wave, the tip of the electric field vector travels along the left-hand helix.

2. Left-hand (counterclockwise) polarized wave. An elliptically or circularly polarized wave in which the electric field vector observed in any fixed plane normal to the direction of propagation, while looking in the direction of propagation, rotates with time in a left-hand or counterclockwise direction.

NOTE. For a circularly polarized wave, the tip of the electric field vector travels along the right-hand helix.

This book will use the preceding definitions. The same definitions may apply to oblique polarization.

1. Right-hand polarization. When viewing the wave front from the wave source rather than a point in front of the wave front, the direction of the electric field vector slants to the right against the perpendicular.
2. Left-hand polarization. The same definition applies, but the direction of the electric field vector slants to the left against the perpendicular.

E. Nature of Polarization and Its Application

Polarization Rotation Caused During Propagation

In general an antenna is designed for a specified polarization. However, if there is a change of polarization characteristics in propagation (so-called polarization rotation), the type and direction of the receiving antenna should be determined accordingly.

If the polarization rotation is unpredictable or unstable, normal reception will become difficult.

In ionospheric propagation in the HF or lower frequency bands, the radio waves emitted as linearly polarized waves generate a polarization rotation to a part of the waves during passage through the ionosphere. They return to the receiving point on the ground in the form of an elliptically polarized wave. Hence radio waves transmitted by a horizontal polarization antenna can be received by vertical polarization receiving antenna.

In tropospheric scattering encountered in VHF or higher frequency bands, no such significant rotation of polarization occurs, except for a specific reflection.

In tropospheric scattering or in scatter and absorption propagation of 10 to 15 GHz or higher due to rain drops, the generation of some opposite polarization components is observed, but not enough to prevent propagation by normal polarization.

Relation between Normal and Opposite Polarization and Cross-Polarization

The desired polarization is called normal polarization and the undesired polarization, opposite polarization or cross-polarization. Examples are given in the following:

Linear polarization.

Vertical polarization–horizontal polarization.

Right-hand 45° polarization–left-hand 45° polarization (in general both polarization planes are perpendicular to one another).

Circular polarization.

Right-hand circular polarization–left-hand circular polarization (in general, the elliptically polarized waves are indicated where the major to minor axis ratio is equal but the rotation direction is opposite).

The relationship between linear and circular polarization is not referred to as cross-polarization.

Antenna Cross-Polarization

In general an actual antenna gives a slight gain to the cross-polarized wave due to engineering and structural restrictions.

The directivity gain pattern obtained over the entire direction on a representative plane for cross-polarization with respect to the maximum gain for the normal polarization is called antenna cross-polarization discrimination. It is an important factor in determining the antenna performance.

Special care should be taken in measuring the performance to ensure that the antenna counterpart employed in the test has sufficiently good cross-polarization discrimination characteristics and generates no elliptical polarization component for both linear and circular polarization. It is also necessary to minimize the effect caused by reflection from intermediate ground objects on the propagation path used for the test.

Use of Cross-Polarization

Since the antenna gain to cross-polarization is small and the polarization rotation produced during propagation generally is ignored, such characteristics can be applied in the system design.

Cross-Polarization Antenna. A reduction in the number of antennas to be installed may generate outstanding economical savings in the supporting structures.

The sharing of an antenna by a number of radio frequencies is an old practice enhanced by the fact that combination and separation can readily be achieved by means of a branching filter.

The cross-polarization antenna has been developed to ease the density of frequency sharing by adding cross-polarized frequencies primarily in the microwave frequency bands.

A combination of and separation into the normal and opposite polarization waves (vertical and horizontal, right-hand 45° and left-hand 45°, or right-hand circular and left-hand circular) occurs between the antenna feed point and the aperture utilizing a single aperture in common as follows.

1. Cross transmission and reception. The transmission and the reception RF channel groups are assigned two frequencies with different polarizations. By assuming the same configuration for the facing station, a two-way (upper and lower or east–west and west–east) link may be established by two antennas.
2. Cross RF channel group. This group is used when the specified frequency band cannot accommodate all the necessary RF channels. It uses separate antennas for transmission and reception, and each antenna accommodates two RF channel groups employing two different polarizations. Here a greater number of RF channels can be accommodated.

Prevention of Interference by Cross-Polarization. When co-channel interference is unavoidable because of terrain conditions or unexpected interference is caused by another system, relocation of the station is usually difficult and some other solution must be sought.

The cross-polarization method may be the first choice as a solution. In this method the interfering side, or the side suffering from the interference, changes the polarization to hold the cross-polarization between the desired and the interfering frequencies. If both the desired and the interfering frequencies use the same linear polarization, the polarization of either the desired or the interfering frequencies may be rotated by 90° about the antenna axis.

In the case of circular polarization, the interference wave may contain an elliptical polarization component. This component reduces the effect of suppressing interference depending upon the direction of interference, thereby requiring modification of the system to secure cross-polarization between two linearly polarized frequencies.

F. Polarization Rotation by Reflection

Reflection from the ground surface is called forward reflection because the reflected waves propagate toward the direction of the facing station.

On the other hand, the reflection employed in a primary radar system is called backward reflection since the reflected waves proceed backward.

Also included in the backward reflection is the reflection from a nearby object in an urban area, such as the reflection from a building, which causes interference with the reception in the counter direction system.

Effects of the reflection upon polarization differ greatly between forward reflection and backward reflection. Also, it must be noted that a metallic reflector exhibits different characteristics from an ordinary terrain object as shown in the following.

Linear Polarization

In the case of reflection from a plane that is parallel with or perpendicular to the polarization plane, no polarization rotation occurs since the phase shift takes place within the original polarization plane. This is the case where the vertically or horizontally polarized waves are reflected from the ground surface or the vertical sidewall of a building.

An oblique plane may cause a reflection associated with rotation of the polarization plane depending upon the slope angle and the electrical characteristics of the reflection plane.

In the case of metallic reflectors, the direction of polarization of the wave after reflection can be determined geometrically. (The direction of polarization before reflection is indicated by an arrow on the plane and the direction of the arrow viewed from the receiving point agrees with the expected direction of polarization.)

Circular Polarization

It is not so easy to determine circular polarization because the component waves must be analyzed taking into account both their amplitude and their phase. Figure 8*b* illustrates the concept of three types of reflection.

Metallic Plane Reflection. Independent of the angle of incidence, a circularly polarized wave reverses its polarization upon reflection. Accordingly when reflections are repeated an odd number of times, the receiving antenna must be paired with the transmitting antenna with its polarization opposite to that of the receiving antenna.

Backward Reflection from Geographical Objects. A circularly polarized wave becomes a countercircularly polarized wave if the reflecting object is smooth and sufficiently large for the wavelength, and the ray of incidence is perpendicular to the reflection plane. Otherwise it becomes a counterelliptical polarization wave.

Forward Reflection from Ordinary Ground. The effect differs according to the grazing angle (complement of the angle of incidence) ψ. A circularly polarized wave becomes a circularly or elliptically polarized wave if ψ is smaller than the Brewster angle ψ_B, but a counterelliptically polarized wave if ψ is greater than ψ_B. The value of ψ_B, which is a function of the wavelength and the dielectric constant and conductivity of the earth surface, is approximately 4 to 25° at frequencies above the

UHF band. For most of the terrestrial radio link system, including an over-the-water propagation path, $\psi < 3°$, the reflection does not produce counterpolarization, and the original circular polarization is approximately maintained. In the case of a perfect dielectric ground, circular polarization becomes horizontal polarization at $\psi = \psi_B$, where the vertical component of the circular polarization vanishes. The typical example is dry land.

It has been proposed that the use of circular polarization can eliminate the reflected wave through formation of a strong counterpolarized reflected wave from the sea and on snowy or icy earth surfaces. However, this is not true for the ordinary range of ψ at frequencies in the VHF band or above. Therefore suppression of reflection over a flat propagation path is still an important problem in propagation path engineering.

0.5 REFERENCE ANTENNA

A. Antenna Directivity and Radiation Characteristics

In true free space the propagation path is to be treated as free space for any type of antenna with any radiation pattern. However, in the ordinary terrestrial propagation path, transmission performances are affected not only by conditions of topography and atmosphere, but also by radio-wave energy illumination from a transmitting antenna and reception of incoming waves by a receiving antenna. Hence the antenna polarization and radiation pattern should not be disregarded.

For example, the propagation path to or from an earth station with a large-aperture antenna can be treated as a free space even when a sea surface, plain, etc., exists in the surrounding area. This is because the high directivity of the antenna eliminates the influence of interference waves arriving from any direction other than the satellite.

In general, an analysis of propagation characteristics should include the study of propagation through isotropic antennas as a first step, and the prediction of propagation characteristics taking into account the directivity of an actual antenna as a second step.

The first step may be omitted depending on the specific antenna directivity, such as in a satellite communication propagation path.

The second step may be carried out by modifying the result of the first step by taking into account the effect of antenna directivity, or it may need a specific analysis due to complicated correlations.

B. Reference Antenna

An antenna that is to be the basis for the theory and design of practical antennas and that is also to be used as a reference in performance measurements is called a reference antenna. Figure 9 shows various parameters of three typical reference antennas.

For theoretical studies in VHF and higher frequency bands, in principle, the isotropic antenna is used; for auxiliary means of measurement, the half-wave dipole is employed. The Hertzian dipole is used in theoretical studies in HF and lower frequency bands.

Isotropic Antenna

An antenna that sends or receives energy equally in all directions is called an isotropic antenna. Since it corresponds to a hypothetical point source, it may be called a point source, a point wave source, an omnidirectional antenna, or a perfectly isotropic antenna.

1. Gain. $G = 1$ (= 0 dB), to be a reference for all antennas and to be uniform in all directions.
2. Effective aperture area. $A_i = \lambda^2/4\pi$, to be dependent only on wavelength.
3. Radiation pattern. To form a circle on any plane, including the antenna.

Half-Wave Dipole

A $\lambda/2$ linear antenna with a feed point in the center is called a half-wave dipole. Its impedance is approximately 73 ohms. The half-wave dipole is considered to be convenient, its theoretical analysis easy, and the theoretically predicted performance can be almost realized in practice.

1. Gain. $G = 1.64$ (2.15 dB) in the direction of maximum directivity.
2. Effective aperture areas. $A_e \approx 3.3\lambda^2/8\pi$.
3. Radiation pattern. E plane (polarized plane) has a figure-eight-shaped pattern with its maximum radiation in the direction normal to the line. The H plane (plane on which the magnetic field vector varies) is a (nondirectional) circle. The E plane includes the line element, while the H plane is perpendicular to the E plane.

Hertzian Dipole (Hertzian Doublet)

Two infinitely small conducting balls are placed in a plane with an infinitely small distance between them. The plane that contains these two balls is the E plane, while the plane at a right angle to the E plane is the H plane. The gain and the directivity of the dipole are similar to those of a linear antenna that is shorter than a half-wavelength (doublet or infinitesimal dipole). Hertzian dipoles are discussed no further since they do not have practical applications for VHF or higher frequency bands.

REMARKS

1. The concepts of the isotropic antenna and the Hertzian dipole are imaginary and theoretical and therefore are not realizable.
2. Various types of linear antennas based on the half-wave dipole are used in VHF and UHF bands below 1 GHz. Do not confuse the gain referred to the half-wave dipole with that referred to the isotropic antenna. The latter may be called the isotropic gain or absolute gain.

3. The relation between the isotropic gain G and the relative gain G_d is

$$G = 1.64G_d \quad \text{(antilogarithms)}$$

$$[G] = [G_d] + 2.15 \quad \text{(dB)}$$

0.6 RADIO-WAVE INTENSITY

The radio-wave intensity of an arbitrary point in a propagation space may be expressed as electric field strength, magnetic field strength, power passing through a unit area, and power received through a reference antenna placed at a point of interest (Fig. 10). Among these, the electric field strength and the magnetic field strength are in phase and proportional to one another. Therefore the electric field strength usually is employed since it is physically more comprehensive.

A. Electric Field Strength

The electric potential gradient per meter in the direction of the electric field vector is called field intensity or field strength. The units are as follows.

1. Antilogarithm a.

$$\left.\begin{array}{ll} \text{V/m} & \\ \text{mV/m} & (= 10^{-3}\ \text{V/m}) \\ \mu\text{V/m} & (= 10^{-6}\ \text{V/m}) \end{array}\right\} a = 10^{b/20}$$

2. dB values b.

$$\left.\begin{array}{ll} \text{dBV/m} & \text{(decibels referred to 1 V/m as 0 dB)} \\ \text{dBmV/m} & \text{(decibels referred to 1 mV/m as 0 dB)} \\ \text{dB}\mu\text{V/m} & \text{(decibels referred to 1 }\mu\text{V/m as 0 dB)} \end{array}\right\} b = 20 \log a$$

B. Flux Density

Power that passes through an area of 1 m^2 on the wavefront in unit time is called flux density or Poynting power. The units are as follows.

1. Antilogarithm a.

$$\left.\begin{array}{ll} \text{W/m}^2 & \\ \text{mW/m}^2 & (= 10^{-3}\ \text{W/m}^2) \\ \mu\text{W/m}^2 & (= 10^{-6}\ \text{W/m}^2) \\ \text{pW/m}^2 & (= 10^{-12}\ \text{W/m}^2) \end{array}\right\} a = 10^{b/10}$$

2. dB values b.

$$\left.\begin{array}{ll} \text{dBW/m}^2 & \text{(decibels referred to 1 W/m}^2\text{ as 0 dB)} \\ \text{dBmW/m}^2 & \text{(decibels referred to 1 mW/m}^2\text{ as 0 dB)} \\ \text{dB}\mu\text{W/m}^2 & \text{(decibels referred to 1 }\mu\text{W/m}^2\text{ as 0 dB)} \\ \text{dBpW/m}^2 & \text{(decibels referred to 1 pW/m}^2\text{ as 0 dB)} \end{array}\right\} b = 10 \log a$$

C. Isotropic Input Power

The power received (absorbed power) by an isotropic antenna is called isotropic input power. The units are as follows.

1. Antilogarithm a.

$$\left.\begin{array}{ll} \text{W} & \\ \text{mW} & (10^{-3}\text{ W}) \\ \mu\text{W} & (10^{-6}\text{ W}) \\ \text{pW} & (10^{-12}\text{ W}) \end{array}\right\} a = 10^{b/10}$$

2. dB values b.

$$\left.\begin{array}{ll} \text{dBW} & \text{(decibels referred to 1 W as 0 dB)} \\ \text{dBmW} & \text{(decibels referred to 1 mW as 0 dB)} \\ \text{dB}\mu\text{W} & \text{(decibels referred to 1 }\mu\text{W as 0 dB)} \\ \text{dBpW} & \text{(decibels referred to 1pW as 0 dB)} \end{array}\right\} b = 10 \log a$$

D. Relationship between Various Methods of Expression

The three kinds of expressions mentioned in Sections 0.6 A to C are mutually convertible as follows. ([ ] show dB values and λ is expressed in meters.)

1. The field intensity E and the flux density P_u are given by

$$E \to P_u \begin{cases} P_u = E^2/120\pi \\ P_u\ (\text{dB}) = [E] - 25.76 \end{cases}$$

$$P_u \to E \begin{cases} E = \sqrt{P_u \cdot 120\pi} \\ E\ (\text{dB}) = [P_u] + 25.76 \end{cases}$$

2. The field intensity E and the basic input P_i are given by

$$E \to P_i \begin{cases} P_i = \lambda^2 E^2/480\pi^2 \\ P_i\ (\text{dB}) = [E] + 20 \log \lambda - 36.76 \end{cases}$$

$$P_i \to E \begin{cases} E = (\pi/\lambda)\sqrt{P_i \cdot 480} \\ E\ (\text{dB}) = [P_i] - 20 \log \lambda + 36.75 \end{cases}$$

3. The flux density P_u and the basic input P_i are given by

$$P_u \to P_i \begin{cases} P_i = \lambda^2 P_u / 4\pi \\ P_i\ (\mathrm{dB}) = [P_u] + 20 \log \lambda - 10.99 \end{cases}$$

$$P_i \to P_u \begin{cases} P_u = 4\pi P_i / \lambda^2 \\ P_u\ (\mathrm{dB}) = [P_i] - 20 \log \lambda^2 + 10.99 \end{cases}$$

REMARK. The following table shows the relationship between various units requiring no modification of the equations.

E	P_u	P_i
V/m, dBV/m	W/m^2, dBW/m^2	W, dBW
mV/m, dBmV/m	μW/m^2, dBμW/m^2	μW, dBμW
μV/m, dBμV/m	pW/m^2, dBpW/m^2	pW, dBpW

The units shown in the same horizontal row may be used without modification of equations.

E. Received Voltage and Receiving Input Power

In an aperture antenna such as a parabolic antenna, the available power that is the maximum matched power is used as the receiving input power. However, where a linear antenna based on a half-wave dipole is employed, an open voltage (induced voltage) at the feed point may be referred to as the received voltage V_t for engineering design purposes.

Receiving Input Power P_r

To determine P_r, use of the flux density P_u or the basic input P_i at the receiving point is convenient. When the effective aperture area of the receiving antenna A_e ($= A\eta$, A being the actual aperture area and η the aperture efficiency) and the gain G (referred to the isotropic antenna) are assumed, P_r is calculated as

$$P_r = P_u A_e \quad (= [P_u] + 10 \log A_e) \quad (\mathrm{dB})$$

$$P_r = P_i G \quad (= [P_i] + 10 \log G) \quad (\mathrm{dB})$$

Receiving Voltage V_t

Calculation of V_t is easily carried out using the field intensity E and the gain relative to that for a half-wave dipole (power ratio),

$$V_t = E\sqrt{G_d}\, l_e = E G_d (\lambda/\pi)$$

$$[V_t] = [E] + [G_d] + 20 \log \lambda - 9.94 \quad (\mathrm{dB})$$

where l_e is equal to λ/π, the effective length of the half-wave dipole.

F. Field Intensity Conversion Graph

For the convenience of propagation design, the relation between flux density P_u (dBm), isotropic input power P_i (dBm), and input voltage to the half-wave dipole V_{td} (dBμV) is shown in Fig. 21 to cover the field intensity range $E = -60$ to $+180$ (dBμV/m). Figure 22 is used as an index to find the appropriate figure.

0.7 PROPAGATION LOSS

Such expressions as field intensity, flux density, and isotropic input power are convenient for stating a local condition in a propagation path. However, they are not considered to be complete because these measures vary according to transmitting conditions (the transmitter output and antenna gain in the direction of the receiving point under consideration).

To supplement these expressions, an imaginary transmission loss between isotropic antennas at the transmitting and receiving points under consideration is defined. This is called propagation loss or basic transmission loss (Fig. 11).

A. Free-Space Propagation Loss Γ_0

Assuming that a power P is radiated through an isotropic antenna in free space, the flux density $P_u = P_{u0}$ (flux density in free space) at a distance D is given by

$$P_{u0} = P/4\pi D^2$$

which is the radiated power per surface area of a sphere with radius D. When it is received by an isotropic antenna, the isotropic input power in free space P_{i0}, taking into account that the effective aperture area of the isotropic antenna is $A_i = \lambda^2/4\pi$, is given by

$$P_{i0} = P_{u0}A_i = P(\lambda/4\pi D)^2$$

Hence the free-space propagation loss Γ_0 is given by

$$\Gamma_0 = P/P_{i0} = (4\pi D/\lambda)^2 \quad \text{(antilogarithm)}$$

$$\Gamma_0 = 20 \log D\ (\text{km}) - 20 \log \lambda\ (\text{cm}) + 121.98 \quad (\text{dB})$$

B. Propagation Loss Γ

For ordinary propagation paths the propagation loss is defined as

$$\Gamma = P/P_i$$

where P is the power radiated by an isotropic antenna, and P_i is the isotropic input power.

Rewriting this,

$$\Gamma = \underbrace{(P/P_{i0})}_{\Gamma_0} \quad \underbrace{(P_{i0}/P_i)}_{\Gamma_p} \quad \text{(additional propagation loss)}$$

That is,

$$\Gamma = \Gamma_0 \Gamma_p \quad \text{(antilogarithm)}$$

$$\Gamma = [\Gamma_0] + [\Gamma_p] \quad \text{(dB)}$$

If it is a free-space propagation path,

$$P_i = P_{i0}$$

which results in $\Gamma_p = 1$ (or 0 dB).

C. Additional Propagation Loss Γ_p

This dimensionless quantity is used widely to indicate a degree of deviation of the propagation path under consideration from the corresponding free-space propagation path. It should be noted that Γ and Γ_0 are also dimensionless quantities. Variation of Γ_p with time leads to fading.

When such a variation is treated statistically, an increment (in decibels) of the propagation loss at an arbitrary time percentage in the cumulative probability, over the median propagation loss, is called fading evaluation loss F_e.

The statistical values obtained when the receiving point is moved around within a certain area, instead of the statistical values in the time domain, may be called location factor or terrain factor. In other words, analysis of a propagation path may be equivalent to estimating quantitatively the variation pattern (in terms of mean value, median, deviation, etc.) of the additional propagation loss with changing location and time.

0.8 RADIO TRANSMISSION LOSS

A. Transmission Units

Denoting the voltages and currents at both ends A and B of a transmission system by V_1, V_2, I_1 and I_2, the ratio of apparent powers $V_1 I_1$ and $V_2 I_2$ is called the transfer constant θ and is expressed in nepers or decibels as

$$\theta \text{ (Np)} = \tfrac{1}{2}\ln(V_1 I_1 / V_2 I_2)$$

$$\theta \text{ (dB)} = 10\log(V_1 I_1 / V_2 I_2)$$

where ln is the natural logarithm $\log_e$, Np is the power ratio of natural logarithm multiplied by $\frac{1}{2}$ and dB is the power ratio of the common logarithm multiplied by 10.

The neper and the decibel may be called transmission units. The neper is used widely in transmission network theory, while the decibel is used almost totally in radio-wave propagation theory, with a few exceptions.

Conversion between these two units is made as follows:

$$1 \text{ Np} = 8.686 \text{ dB}$$

$$1 \text{ dB} = 0.1151 \text{ Np}$$

B. Transmission Loss

Although there are a number of definitions (working attenuation, image attenuation, insertion transfer, and transmission transfer) for the transfer constant of the network, depending on impedance matching conditions, we define the transfer constants (power ratios) between

Transmitting and receiving isotropic antennas.
Transmitting antenna input and receiving antenna output.
Transmitter output and receiver input.

These transfer constants are called propagation loss, antenna-to-antenna transmission loss, and span loss (Fig. 12).

Propagation Loss Γ (or Basic Transmission Loss L_b)

The transmission loss caused when an isotropic antenna is employed at each end of the path is called propagation loss because the loss is intrinsic to the propagation path. The equation is given in Section 0.7B.

Antenna-to-Antenna Transmission Loss L_a

The transmission loss between the transmitting and receiving antenna feed points is given by

$$L_a = P_1/P_2 \quad \text{(antilogarithm)}$$

$$[L_a] = [P_1] - [P_2] \quad \text{(dB)}$$

where P_1 is the radiation power and P_2 is the available power at the receiving antenna output. (If the transmitting antenna has an ohmic loss, the ohmic loss is included in the antenna gain.)

When the transmitting and receiving antenna gains (the power ratio referred to the isotropic antenna) in the directions of the facing antennas are G_1 and G_2, respectively,

$$L_a = \Gamma/(G_1G_2/L_c) \quad \text{(antilogarithm)}$$

$$[L_a] = [\Gamma] - [G_1] - [G_2] + [L_c] \quad \text{(dB)}$$

where L_c is the antenna-to-medium coupling loss. L_c may be ignored except in the troposcatter system (refer to Chapter 4). It is equal to 1 or 0 dB for free space and approximately 1 for all the propagation paths except the troposcatter path.

$G_c = G_1G_2/L_c$ is called overall antenna gain or path antenna gain, and using this relation,

$$L_a = \Gamma/G_c \quad \text{(antilogarithm)}$$

$$[L_a] = [\Gamma] - [G_c] \quad \text{(dB)}$$

Span Loss L_s

The transmission loss between the transmitter output terminal and the receiver input terminals is called span loss,

$$L_s = P_t/P_r \quad \text{(antilogarithm)}$$

$$[L_s] = [P_t] - [P_r] \quad \text{(dB)}$$

where P_t is the transmitter output and P_r is the receiver input (the available power at the output of the receiving feeder system).

Denoting the transmitting feeder system loss (in power ratio) by L_{f1} and the receiving feeder system loss by L_{f2},

$$L_s = \Gamma(L_{f1}L_{f2})/G_1G_2/L_c \quad \text{(antilogarithm)}$$

$$[L_s] = [\Gamma] + [L_{f1}] + [L_{f2}] - [G_1] - [G_2] + [L_c] \quad \text{(dB)}$$

REMARK. In general the term antenna system means antenna plus feeder system (including the feeder and the branching filter).

0.9 REFRACTION OF RADIO WAVES IN TROPOSPHERE

A. Troposphere

The density of the atmosphere containing dry air, water vapor, clouds, rain drops, sleet, sandy dust, etc., surrounding the earth becomes greater or smaller as the altitude decreases or increases. While the temperature decreases with altitude up to approximately 10 to 12 km above the earth, it remains almost constant up to approximately 20 km; then it increases with altitude, decreases again, and finally increases to approximately 1500 K at an altitude between 400 and 700 km.

With a further increase in altitude, the temperature again decreases toward the temperature of the universe (estimated as 1.7 K to 6 K).

As shown in Fig. 13, the portion of the atmosphere that extends from the earth surface up to an altitude where the temperature-decreasing tendency stops is called the troposphere. The upper boundary of the troposphere is called tropopause. The altitude of the tropopause varies somewhat according to latitude, region, and season.

Figure 13 shows the height distribution of the atmospheric pressure and the temperature for the standard atmosphere specified by the International Civil Aviation Organization (ICAO). The upper limit of the troposphere is assumed at 11 km above the earth.

For most communication systems, except the satellite communication system, the radio propagation paths established in the VHF or higher frequency bands are confined within the troposphere. Consequently it is necessary to study the effects of the vertical and horizontal structures of atmospheric refractivity within the troposphere.

The standard radio atmosphere used in radio-wave propagation indicates a tropospheric atmosphere with the vertical structure of radio refractivity corresponding to $K = 4/3$ (refer to CCIR Recommendation 310-2), which is different from that defined in Fig. 13.

B. Expression of Atmospheric Refractivity

The measures n, m, N, and M used to express the vertical refractivity gradient of the standard and the nonstandard radio atmosphere are defined in Fig. 14 as follows. (Refer also to CCIR Recommendations 310-2, 369-1, and 453.)

1. Refractive index n,

$$n = c/v$$

where c is the velocity of electromagnetic waves in a vacuum and v is the velocity of radio waves in the air under consideration.

2. Refractivity N,

$$N = (n - 1) \times 10^6$$

where N is a function of the absolute temperature of the atmosphere T (Kelvins) the atmospheric presure p (mbar), and the vapor pressure e (mbar), expressed by

$$N = \frac{77.6}{T}\left(p + 4810\frac{e}{T}\right)$$

3. The modified refractive index m the sum of the refractive index n of the air at a given height above sea level and the ratio of this height h to the radius of the earth a,

$$m = n + h/a$$

4. Refractive modulus M,

$$M = (m - 1) \times 10^6$$

$$M = N + (h/a) \times 10^6$$

REMARK. N and M are the quantities obtained by subtracting 1 from n and m, respectively, and multiplying by 10^6 because n and m are values very close to 1 ($n = 1.000278$, $m = 1.000356$, for example) and considered inconvenient. The values given in the example can be expressed by $N = 278$ and $M = 356$. Usually these are written as N.U. for N unit and M.U. for M unit.

C. *M* Curves and Definition of *K*

A curve showing the vertical distribution of M is called an M curve or M profile. Typical M curves are shown in the upper half of Fig. 14. The following descriptions are given based on CCIR Recommendation 310-2.

Classification of Radio Refractivity by the Shapes of M Curves

1. Standard refraction. Reference for various types of M curves. At the standard refraction, the vertical gradient of M, dM/dh, is constant at 0.12 M.U./m. This value indicates the average condition of the M distribution of the earth and is used.
2. Subrefraction. This occurs when the vertical gradient of M is positive and greater than that for the standard refraction. Such a condition occurs when the lapse rate of the refractive index (n, N) is smaller than that for the standard refraction. When the atmosphere is homogeneous, the gradient is

$$dM/dh = 10^6/a = 0.157 \text{ M.U./m}$$

   corresponding to $K = 1$.
3. Superrefraction. This occurs when the vertical gradient of M is smaller than that for the standard refraction or the lapse rate of n or N is outstanding ($K > 4/3$). In such a condition the radio ray is bent downward, tending to improve visibility.
4. Atmospheric radio duct. When the M curve contains a portion of a negative gradient, a duct is formed, as shown in the figure. Thus the radio wave energy is confined between two horizontal boundaries (the lower boundary may be the earth surface) if the radio frequency is sufficiently high (generally VHF or higher, depending upon the thickness and the intensity of the duct). Such a propagation mode is called duct propagation. Within such a duct, irregular refraction and reflection take place between boundaries, and mutual interferences among a number of thus produced waves traveling over different paths occur. This results in a great variation in field intensity accompanied by extraordinarily low or high field intensity. When the receiving point is outside the duct, considerable attenuation-type fading may occur in addition to the

multipath interference. The atmospheric radio duct may be called radio duct or M duct.

Effective Earth Radius Coefficient K

The coefficient K is used widely.

1. Definition of K. When the M curve slope is constant (corresponding to a straight line), the radio ray takes an arcing path. In connection with a radio line-of-sight distance (distance to the radio horizon), the ratio of an effective earth radius for which propagation can be rectilinear to the actual earth radius (average radius $a = 6.37 \times 10^6$ m) is called simply effective earth radius coefficient, or K. Thus

$$K = R/a, \qquad R = Ka$$

2. The equation between the atmospheric refractive index n and K is

$$K = 1/[1 + (a/n)dn/dh]$$

3. For standard K ($K = 4/3$),

$$dn/dh = -0.039 \times 10^6 \qquad \text{(see note)}$$

[or n is extremely close to 1. The equation given in item 2 becomes

$$K = 1/[1 + 6.37 \times 10^6(-0.039 \times 10^{-6})] \approx 1.33 \approx 4/3$$

Thus standard $K = 4/3$.

NOTE. $n = 1 + N \times 10^{-6} = 1 + M \times 10^{-6} - h/a$ and $dn/dh = (dM/dh) \times 10^{-6} - 1/a$. While $dM/dh = 0.118$/m, which is the standard refraction (≈ 0.12 M.U./m), we have $dn/dh = 0.118 \times 10^{-6} - 1/(6.37 \times 10^6) \approx -0.039 \times 10^{-6}$.

D. K and Path Profile Chart

In the propagation design for the VHF or higher frequency band it is most important in system analysis to find the spatial relationship between the radio ray and various ground-based objects. If this is not done, an engineering design is not feasible. However, if the curved radio ray is to be drawn on a diagram as it is, with appropriate scales for ease of analysis, both the earth profile arc and the radio ray arc form parabolic or more complicated curves. These curves require an enormous amount of work to prepare a path profile chart on which the radio ray and the earth profile are drawn.

The period from the introduction of VHF radio communication until the end of World War II saw many radio engineers frustrated by such time-consuming work. The introduction of the K concept and the establishment of a path profile chart utilizing K were epochal achievements.

The path profile with distance on the horizontal axis and height above sea level modified by the effective earth surface corresponding to K has the following features.

1. A radio ray traveling between any two points on the chart becomes a straight line.
2. Profiles of terrain undulation and man-made structures on the earth are easily recorded over the geoid (the reference level for zero meters above sea level).
3. Any part of the path profile chart paper may be used, provided that the horizontal distance and the vertical height are plotted correctly. Therefore a single sheet of path profile chart paper may accommodate more than one path profile.
4. Scales on the horizontal and vertical axes may be modified for convenience of application, provided the vertical scale is multiplied by n^2 when the horizontal scale is multiplied by n.

The standard path profile paper is prepared for $K = 4/3$. For specific purposes, path profile paper for $K = 1.00$ (corresponding to the actual earth) or $K = 1.15$ (corresponding to the refractivity of the light ray) may be used. However, it is not recommended that path profile paper be used for $K < 1$ and $K > 2$ because of the nature of K and for other reasons mentioned later. Figure 18 shows three kinds of standard path profile chart paper to be chosen according to the path length and height.

Remarks. The path profile chart paper is prepared in the following steps.

1. Determine the paper size.
2. Prepare the rectangular coordinate for the total path length D and the maximum height above sea level H.
3. Draw the geoid on the rectangular coordinate according to the equation

$$x = d_1 d_2 / 2Ka \simeq d_1 d_2 / 17, \qquad \text{for } K = 4/3$$

   where x is the rise of the geoid, d_1 is the distance to an arbitrary point from site 1, and $d_2 = D - d_1$.
4. Draw parallel curves to the geoid from 0 m above sea level up to H m.

E. Variation and Definable Range of K

The value of K varies as the vertical profile of the atmospheric refractive index changes. It is important to determine the allowable range of variation of K in the propagation design, because it defines the relation between the radio path and the earth profile. Although several data on the statistical distribution have been reported in Japan, in other countries, and by the CCIR, no reliable figure has been reported except estimates of its average tendency.

It should be noted that giving those reports too much credit may mislead those concerned with site selection or fault detection investigation. Supplemental descriptions of K with regard to its nature and application follow.

Essential Nature of K and Its Definable Range

K is an effective coefficient introduced to facilitate the propagation path design by substituting a straight line for the curved radio ray under the condition that the gradient of the M line (instead of the M curve) be uniform over the entire propagation path.

Even when the M curve contains small irregularities at individual points, application of K is possible if the irregularities are averaged and can be approximated by a straight line over the entire path. When a sound bending or radio duct is within or close to the first Fresnel zone of the radio ray, K should not be introduced in the propagation design.

The definable range of K is considered to be between 1 and 2. If K is definable, the propagation is called normal propagation; if it is not, it is called abnormal or anomalous propagation. The path profile paper is applicable only for the range of normal propagation ($1 \leq K \leq 2$).

Application of K in Propagation Design

1. Line-of-sight and mountain diffraction propagation path.
   (a) The standard value of 4/3 is used for electrical and geometrical analysis of the average propagation condition.
   (b) Estimates of type K (interference K) and type T (tangential ridge K) fading, transmitting vertical angle, etc., are carried out for the definable range of K ($1 \leq K \leq 2$).
   (c) Any direct or indirect effect caused by abnormal refraction, etc., is considered to have been included in the concept of fading phenomenon and in the estimate of the statistical values.
2. Spherical-earth diffraction propagation path. The procedures outlined in item 1(a) and (b) apply.

REMARKS. Although values of K ranging from 1 to 20 are sometimes found in equations or nomographs in traditional theories, they are all imaginary and temporary values and are useless in practice.

3. Tropospheric propagation path. The standard value of 4/3 is used in principle as a value of K for electrical and geometrical analysis to estimate long-term medians of radio-wave field intensity indicating the average propagation condition. However, a value between 1 and 2 (e.g., $K = 1.5$, $Ka = 8600$ km) may be used according to differences in local radio meteorology.
4. Satellite communication propagation path. Most of the propagation path is outside the troposphere and the elevation angle to the satellite is so great (more than 3 to 5° in practice) that the concept K is not applicable.

Further Remarks Concerning K

1. An apparent K estimated from the height pattern, the angle of incidence, and the statistical value of the variation of the receiving field is sometimes called effective K. However, K itself is an effective concept, as mentioned before. Therefore the

term effective K inevitably makes it more ambiguous. The effective K is apt to extend the range of variation of K because of uncertainties, such as effects due to ducts or birefringence and measuring error. Consequently, it is very difficult to detect true K.

2. Effective K was proposed in early VHF system development to simplify system design where the reliability is quite low (95 to 99%) and type K fading is dominant. Therefore such a statistical value may not be applied to a high-quality system (better than 99.9% reliability) or to estimate K in an interference-free propagation path.

3. Such confusion is found even among some researchers and specialists who, according to their reports, determine the effective K as K. Hence it is desirable not to use such a misleading expression as effective K, but to use instead the term apparent K for those K that are accompanied by a statement of measurement conditions.

4. Some reports (CCIR Recommendation 338-1 and related papers) propose that the values of K for cumulative time rates of 99.9 and 0.1% are 0.6 to 0.8 and ∞ to -4, respectively. Here ∞ means that the effective earth surface becomes a plane, and negative K should indicate a concave surface. If true, routing design and suppression of intersystem interference would become practically impossible. This is a typical example of using apparent K instead of K.

5. Lower values of K, such as 2/3 and 0.8, are used as design parameters or in the path profile in Japan and some other countries. However, there seem to be no sound reasons for applying these values. It is desirable not to use these values because adherence to them restricts flexibility in selecting station sites. This leads to a doubling of labor in economical system design.

6. Some of the figures in this book contain curves for $K < 1$ and $K > 2$. However, these are for reference and comparison only.

0.10 GROUND REFLECTION

A. Ground and Ground Reflection (Fig. 15)

Classification of Ground

Ground in radio propagation means the earth surface, that is, vegetation, villages, towns, etc., over the earth surface. The ground may be classified into seas, lakes, rivers, swamps, low plains, plateaus, deserts, hilly land, mountains, frozen soil, snow and ice fields, urban areas, etc. The ground may also be classified according to the conditions of vegetation and man-made processing into ferny land, steppes, savanna, prairie, bush, forests, woods, jungle, paddy fields, farmland, meadows, plantations, villages, residential areas, high-rise building areas, industrial areas, airports, harbor areas, etc.

Ground Reflection

The main effects of the ground upon propagation in the VHF or higher frequency bands are caused by an obstruction (producing a diffraction loss) and a reflection (interfering with the direct wave). The physical characteristics (in terms of size, extent, roughness, shape, relative position, etc.) of the ground have a significantly greater effect than its electrical characteristics such as conductivity and permittivity. A flat ground surface produces a sound radio-wave reflection, frequently causing harmful interference with the direct waves.

However, our experience shows that the average canceling-out effect between direct and reflected waves is not very harmful in maintaining the necessary system performance, except for the case of sound reflection.

However, a strong ground-based duct that occurs persistently near the ground surface causing reflection increases markedly the probability of deep fading (G type plus D type). This effect intensifies rapidly with the propagation echo ratio ρ_θ.

Consequently it is most important in the interference propagation path design to estimate ρ_θ and to determine various parameters in order to minimize ρ_θ so that it can be ignored. (Refer also to Chapter 8 for fading.)

The reflection coefficient that is intrinsic to the propagation path terrain is called the effective reflection coefficient ρ_e, and the one that includes the antenna directivity effect is called the propagation echo ratio ρ_θ.

B. Effective Reflection Coefficient ρ_e

Definition

When a radio wave is transmitted through an isotropic antenna, the ratio of the received electric field vectors of the reflected wave to those of the direct wave is called the effective reflection coefficient,

$$\dot{\rho}_e = \rho_e e^{-j\Theta} = \dot{E}_r/\dot{E}_d$$

where ρ_e = absolute value of $\dot{\rho}_e$
Θ = phase of ρ_e
$\dot{E}_r$ = electric field vector of reflected wave
$\dot{E}_d$ = electric field vector of direct wave

Composition of $\dot{\rho}_e$

Absolute Value ρ_e. Methods used to determine ρ_e differ as shown in Fig. 15.

$\rho_e = RD_v$ (smooth ground, flat or sphere)

$\rho_e = S$ (rough ground)

$\rho_e = RD_vZ$ (smooth ground plus ridge)

$\rho_e = S$ (rough ground plus ridge, $Z > S$)

$\rho_e = SZ$ (rough ground plus ridge, $Z \leq S$)

$\rho_e < 0.1$ (mountainous area, undulating hilly land, multiple ridges)

where ridge = shielding ridge against reflected ray
R = flat-plane reflection coefficient
D_v = spherical-ground divergence factor
S = rough-ground reflection coefficient
Z = diffraction coefficient of shielding ridge

Phase Θ. It is difficult to determine the phase Θ of ρ_e except for reflection over smooth ground,

$$\Theta = \delta + \phi_R \qquad \text{(smooth ground, flat plane, sphere)}$$

$$\Theta = \text{uncertain} \qquad \text{(smooth ground plus ridge)}$$

$$\Theta = \text{uncertain} \qquad \text{(rough ground, including path with a ridge)}$$

where δ = path-length phase difference (phase delay of reflected wave from directed wave)
ϕ_R = phase angle of reflection coefficient

The equation that applies when a ridge covers smooth ground can be expressed mathematically as

$$\Theta = \delta + \phi_R + \phi_Z$$

where δ = path-length phase difference when there is no ridge
ϕ_Z = phase component of ridge diffraction coefficient (corresponding to phase difference at receiving point between cases with and without ridges)

If the ridge cannot be regarded as a perfect knife-edge, estimation of ϕ_z is not possible. Also, the accuracy of measurement of the phase difference is worsened because the reflection point moves between the propagation paths with and without the ridge. As a result, the phase Θ remains uncertain.

As regards Θ, its specific value is not important, but this is not the case for its range of variation. For terrestrial propagation paths normally

$$\phi_R \approx \pi$$

In the case of rough ground it is assumed that the sum of the estimated value δ_s corresponding to δ and the approximate value of ϕ_s (equivalent phase of rough surface reflection coefficient) equals $\delta + \pi$,

$$\Theta \approx \delta_s + \phi_s = \delta + \pi$$

A more practical method for estimating Θ is to measure the pitch in the height pattern and also the separation between the minimum point and the presumed antenna mounting level, and then make a determination based on measured data collected over a very long period. However, it is only possible to estimate an approximate value.

C. Propagation Echo Ratio ρ_θ

In the UHF and higher frequency bands a high-directivity antenna is realizable economically. Therefore the reflected wave can be suppressed between the transmitting antenna input and the receiving antenna output even when the effective reflection coefficient ρ_e intrinsic to that propagation path is large.

The ratio of the reflected-wave voltage component $\dot{V}_r$ to the direct-wave voltage component at the antenna terminal is expressed by $\dot{\rho}_\theta$ and is called propagation echo ratio,

$$\dot{\rho}_\theta = \rho_\theta e^{-j\Theta} = \dot{V}_r/\dot{V}_d$$

$$\rho_\theta = \rho_e D_{\theta 1} D_{\theta 2} = \rho_e D_{\theta T}$$

where ρ_θ = absolute value of $\dot{\rho}_\theta$

Θ = phase of ρ_θ

$D_{\theta 1}, D_{\theta 2}$ = directivity coefficients in direction of reflected wave when the directivity coefficient in direction of direct wave is assumed to be 1

$D_{\theta T}$ = overall directivity effect, $= D_{\theta 1} D_{\theta 2}$

0.11 VARIOUS TYPES OF PROPAGATION

Except for the intersatellite communication system, the actual propagation path that may be assumed to be free space (an unobstructed and isotropic vacuum space) is limited to only a particular case satisfying specific conditions. This is due to the presence of the tropospheric atmosphere and the earth surface.

The condition of radio propagation may be discussed in terms of the characteristics and of the effect of various nonfree-space factors between transmitting and receiving points. Also, it is possible to show a number of individual patterns and combinations of these patterns. Such a categorized pattern is called propagation type.

A. Physical Classification of Propagation Waves (Fig. 16 top)

Direct Wave

When the path is assumed to be the line of sight between transmitting and receiving points, most of the energy radiated from the transmitting antenna reaches the receiving point directly along a radio-wave ray (shortest-time path determined by Snell's law). Such a radio wave is called a direct wave.

Reflected Wave

The ground provides good reflection for radio waves and man-made buildings may also produce effective reflections if their dimensions are greater than the wavelength of the radio wave. In the HF or lower frequency bands the ionosphere may reflect the radio wave back to earth.

Communication systems utilizing the reflected wave are found in primary radar and passive repeater systems employing a metallic reflector.

Refracted Wave

The atmosphere over the earth's surface is not homogeneous. Its density decreases with altitude and the refractivity approaches unity (the value for vacuum space). Consequently a radio wave radiated from a ground-based antenna reaches the receiving point by traveling along an arc bent slightly downward (a curvature radius that is four times the earth radius under standard atmospheric condition, corresponding to $K = 4/3$).

Departures from the standard atmospheric condition may lead to subrefraction, superrefraction, and duct (see Fig. 14) as a result of refraction specific to each of these phenomena.

Diffracted Wave

When the radio path between transmitting and receiving points is obstructed by mountains, buildings, etc., the field intensity obtained at the receiving point is considered to be produced by waves diffracted by such obstacles.

For radio waves of longer wavelength (lower frequency), the diffraction is greater. Therefore diffracted waves are used extensively in the VHF and lower frequency bands.

Surface Wave

Radio-wave components propagating over the earth's surface can be placed in the following three categories: 1. a component propagating over a conductive body surface (Sommerfeld's surface wave); 2. a component, along a conductive wire enclosed by a dielectric sheath (Goubau's surface wave); and 3. a component, over the earth's surface (ground wave). The ground wave exhibits less propagation loss, and accordingly, higher field intensity at lower frequencies for vertical polarization and if traveling along a path over land rather than over sea (with greater conductivity and dielectric constant).

The frequency range where the surface wave is applicable is up to the HF band, and above approximately 50 MHz the other modes of propagation waves become dominant.

Scatter Wave

Scatter waves are produced by irregularities of the refractive index in a portion of the troposphere, partial refraction or reflection in the ionosphere (ionized layer), irregular reflection by a rough ground surface, and scattering by rain droplets at microwave frequencies above 10 to 15 GHz.

The most important among these is tropospheric scatter for radio waves of approximately 100 MHz or higher at transhorizon distances. If the transhorizon portion of the distance exceeds 50 to 100 km, a stronger field intensity than that in the diffraction mode is obtained.

The ionospheric scatter occurs in frequencies up to 80 to 100 MHz. However, this is of little value for practical applications. The scattering by rain drops generates a polarization distortion (generation of a counterpolarization component) in addition to an attenuation at a frequency higher than 10 GHz.

B. Propagation Path Types and Applicable Frequency Bands

Among the various types of propagation-wave components mentioned in the preceding, several types play a role as transmission media.

In this respect, propagation paths are classified as shown in Fig. 16, where the path type is named after the prevailing propagation-wave component. The applicable frequency bands are also indicated.

Type I—Free-Space Propagation Path

This is by far the most preferred type of propagation paths. Such an almost ideal propagation path is found between satellite and satellite or between satellite and earth (when the elevation angle from the earth station is large). There is no restriction in applicable frequency except for MF or lower frequencies, where the antenna dimension and the earth surface are the limiting factors.

Terrestrial systems using VHF or higher frequencies, where the effective reflection coefficient is low and a good line of sight is secured (diffraction loss is negligible) as in the case of a mountainous region, are considered to belong to this type.

Type II—Reflected-Wave Interference Propagation Path

This may be called simply interference propagation path. The field intensity at the receiving point is the sum of the electric field vectors of the direct and ground-reflected waves. In the MF and lower frequency bands, the reflected wave cancels the direct wave because of 180° phase rotation at reflection. This makes this type of path inapplicable.

In the HF band the ionospheric wave (ionized-layer wave) is essential in determining the receiving signal level. Therefore this type of path is applicable to the VHF and higher frequency bands. However, in the upper region of the SHF or higher frequency bands absorption attenuation by precipitation, water vapor, and oxygen becomes dominant. On the other hand, the reflected wave becomes less significant because the effective reflection coefficient ρ_e decreases and the antenna directivity $D_{\theta T}$ becomes more effective. This makes this type of path inapplicable to such a frequency band.

Type III—Diffraction Propagation Path

In lower frequency regions the diffraction wave passing over the earth curvature or man-made structures must be considered; in the VHF, UHF, and SHF bands the diffraction wave passing over mountains must be considered.

For frequencies of approximately 10 GHz or higher, the diffraction loss becomes enormous in magnitude and there is significant attenuation due to precipitation. Therefore this path is not applicable at these frequencies.

This type of propagation path is used only when no other solution is available. It is found often in broadcasting systems such as television and VHF or UHF radio in order to cover many unspecified receivers. (Some receivers use this type inadvertently.)

Type IV—Absorption-Type Propagation Path

This path is to be considered but has no application for frequencies of approximately 10 GHz or higher because the scattering and absorption attenuation by rainfall, clouds, fog, water vapor, and oxygen are intensified and limit the applicable propagation path length. (This band is subject to examination but is not applicable. Accordingly, the portion is designated by the hatched area in Fig. 16.)

Type V—Tropospheric Scatter Propagation Path

At frequencies above approximately 100 MHz the radio-wave components received at a great transhorizon distance are composed mainly of the tropospheric scatter waves because the diffraction loss is excessive and the radio waves penetrate the ionized layer.

However, at frequencies higher than 8 to 10 GHz this type of path is not applicable because of dominant absorption attenuation by rainfall.

Type VI—Passive Propagation Path

This is a propagation path that contains repeating systems such as plane reflectors, back-to-back connected antennas, radio prisms, and so on, which change the radio-wave direction. A path of greater length is called main section, while a path of shorter length is called passive additional section. The passive additional section may often be treated as a free-space propagation path.

The passive propagation path is used in frequency bands of approximately 2 GHz or higher because the shorter the wavelength, the lower the total loss.

Type VII—Ground-Wave Propagation Path

This path is useful in the HF or lower frequency bands as mentioned in Section 0.11A. Typical applications are found in VLF super-long-distance submarine communication and MF standard broadcasting service. In the lower region of the VHF band only the system using vertical polarization, propagation over water with low antennas, needs to be examined.

Type VIII—Ionospheric-Wave Propagation Path

Here we examine the HF and lower frequency bands, where refraction, reflection, and scattering in the ionosphere are utilized for transmission. However, prevailing absorption and attenuation close to 1 MHz and daytime observation are excluded in practice. In the LF and VLF bands the ground wave may become dominant.

Type IX—Hybrid Propagation Path

There are a number of possible different combinations. However, our discussion is limited to some representative types.

III + VII + VIII Type. This mixed-type path is caused by the diffraction wave, the ground wave, and the ionospheric wave. It is apt to produce severe fading due to mutual interference. In the VHF and higher frequency bands types VII and VIII vanish to form type III.

II + III Type. This mixed-type path is caused by the reflected and diffracted waves such as an oversea ridge diffraction propagation path. It is encountered in the SHF and higher frequency bands.

III + V Type. This mixed-type path is caused by the diffracted wave and the tropospheric scatter wave. It is a transhorizon propagation path passing over the earth curvature with the transhorizon portion of much shorter length (50 to 60 km or less). Heavy fading occurs by mutual interference between the spheric diffraction wave and the tropospheric scatter wave.

II + IV Type. A reflected-wave interference propagation path using a frequency of approximately 10 GHz or higher is an example of this type. However, in this frequency region it may be treated as a type IV in practice because of sharp antenna directivity.

C. Propagation Path Types in VHF to SHF bands

Typical examples of the propagation path types for VHF to SHF bands are shown in Fig. 17. The features of these types are given in the right-hand side of the figure. For details refer to the respective chapters.

REMARKS

1. In practice, for a path classified as a free-space propagation path (Fig. 17*a*), the type is determined by the tolerable reflection and diffraction. To determine the effect by reflection or diffraction, refer to Chapter 2 or 3.
2. The reflected-wave interference propagation path includes a slanted plane as shown in Fig. 17*b*. Although the path over the spherical-ground surface can be treated using the parameter *Ka*, it is not very easy to solve the problem involving the slanted plane whose geometrical configuration may change in a variety of ways. The following suggests a practical approach for such a path.
   (a) Assume a flat plane that includes the slanted plane and analyze it utilizing the path profile.
   (b) On the path profile, determine a reference plane passing through the estimated reflection area on the slanted plane. Then obtain the heights h_1 and h_2 of the transmitting and receiving antennas measured from the reference plane.
   (c) Determine a reflection point in the same manner as with a horizontal flat plane, using h_1' and h_2'. When the determined reflection point falls within the estimated area of the slanted plane, it is concluded that the reflection over the slanted plane exists, otherwise no slanted plane reflection exists.

Chart Index,
Basic Matters

Greek Letters
Notations

Capital	Small	Latin name	Principal quantities	
Α	α	alpha	α, $\alpha_{\odot}$	Direct-ray vertical angle, Sun's right ascension
Β	β	beta	β, β_{12}	Reflected-ray vertical angle, Azimuth
Γ	γ	gamma	Γ, Γ_0, Γ_p, γ_{x0}	Propagation loss, Atmospheric attenuation factor
Δ	δ	delta	Δ, Δf, δ	Path-length difference, Bandwidth, Phase difference
Ε	ε	epsilon	ε, ε_c	Phase angle, Complex dielectric constant
Ζ	ζ	zeta	—	
Η	η	eta	η, η_ϕ	Antenna aperture efficiency, Performance efficiency
Θ	θ	theta	Θ, θ	Phase angle, Transfer angle, Adjacent angle
Ι	ι	iota	—	
Κ	κ	kappa	—	
Λ	λ	lambda	λ	Wavelength
Μ	μ	mu	μ	Permeability, Micro-(10^{-6})
Ν	ν	nu	—	
Ξ	ξ	xi	ξ	Center angle of effective earth
Ο	o	omikron	—	
Π	π ϖ	pi	π	Ratio of circumference to diameter of circle
Ρ	ρ	rho	ρ_e, ρ_θ	Effective reflection coefficient, Propagation echo ratio
Σ	σ ς	sigma	Σ, σ	Sum, Conductivity, Standard deviation
Τ	τ	tau	τ	Time period
Υ	υ	upsilon	—	
Φ	ϕ φ	phi	ϕ, ϕ_R, φ_Δ	Angle of incidence, Phase angle of reflection coefficient
Χ	χ	chi	χ	Fading parameter
Ψ	ψ	psi	ψ	Grazing angle
Ω	ω	omega	ω	Solid angle, Angular velocity

Letters employed in this book $\left\{ \begin{matrix} \alpha, \beta, \gamma, \delta, \varepsilon, \eta, \theta, \lambda, \mu, \xi, \pi, \rho, \sigma, \tau, \phi, \varphi, \chi, \psi, \omega \\ \Gamma, \Delta, \Theta, \Sigma \end{matrix} \right\}$

Astronomical Symbols

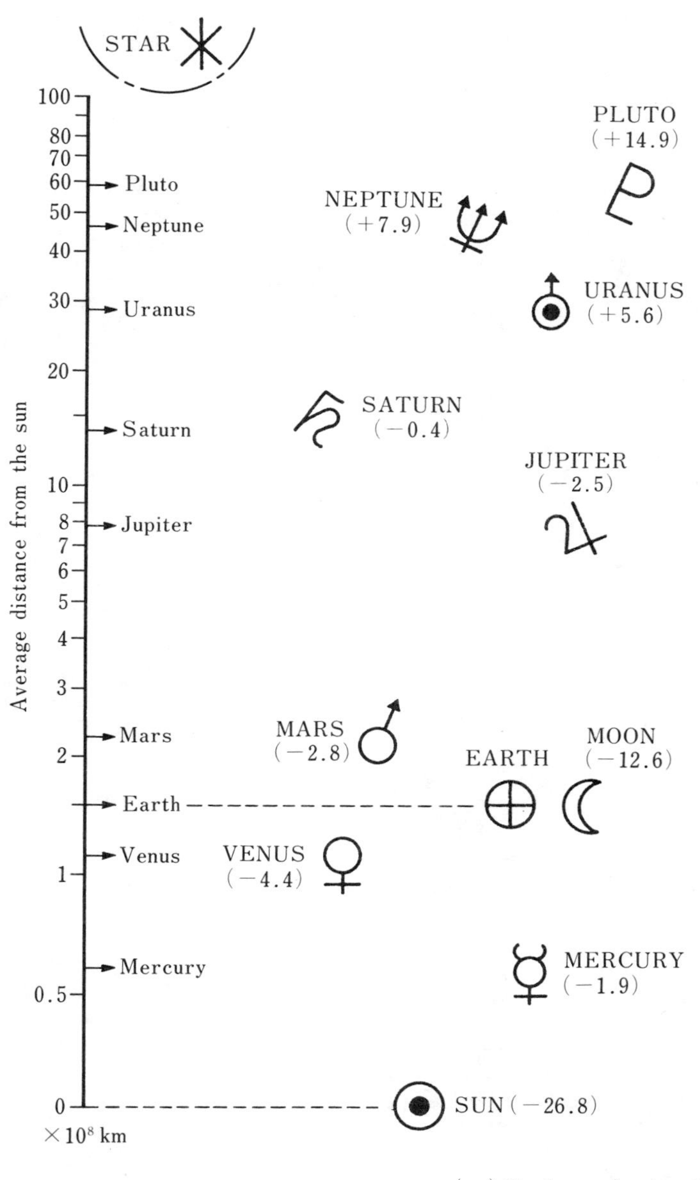

( ) Maximum luminosity

Designation of Numerical Values

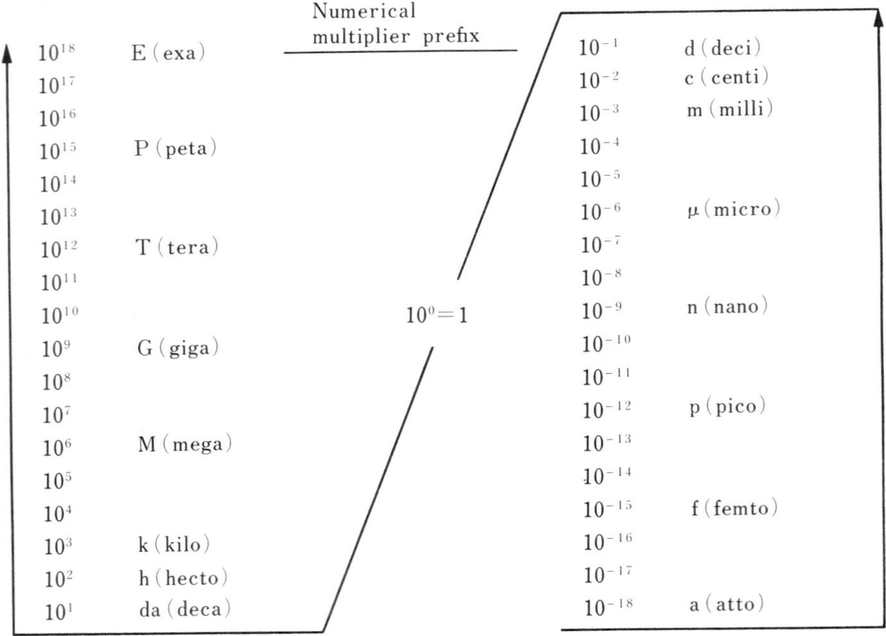

Numerical multiplier prefix
10^{18} E (exa)
10^{17}
10^{16}
10^{15} P (peta)
10^{14}
10^{13}
10^{12} T (tera)
10^{11}
10^{10}
10^{9} G (giga)
10^{8}
10^{7}
10^{6} M (mega)
10^{5}
10^{4}
10^{3} k (kilo)
10^{2} h (hecto)
10^{1} da (deca)
$10^{0}=1$
10^{-1} d (deci)
10^{-2} c (centi)
10^{-3} m (milli)
10^{-4}
10^{-5}
10^{-6} μ (micro)
10^{-7}
10^{-8}
10^{-9} n (nano)
10^{-10}
10^{-11}
10^{-12} p (pico)
10^{-13}
10^{-14}
10^{-15} f (femto)
10^{-16}
10^{-17}
10^{-18} a (atto)

Common measuring units

Quantity	Unit	Abbreviation	Equivalent to	Quantity	Unit	Abbreviation	Equivalent to
Length	Angstrom	Å	10^{-10} m	Mass	Carat	ct,car	200 mg
	Micron	μm	10^{-6} m		Kilogram	kg	——
	Meter	m	——		Ounce	oz	28.350 g
	Nautical mile	n mi	1 852 m		Pound	lb	453.592 g
	Inch	in	2.54 cm	Angle	Second	″	$\pi/648\,000$ rad
	Foot	ft	0.3048 m		Minute	′	$\pi/10\,800$ rad
	Yard	yd	0.9144 m		Degree	°	$\pi/180$ rad
	Mile	mi	1 609.34 m		Radian	rad	57.2958°
Area	Square meter	m^2	——	Force	Dyne	dyn	10^{-5} N
	Are	a	100 m^2		Newton	N	1 kg·m/s^2
	Square yard	yd^2	0.8361 m^2	Pressure	Bar	bar	750.06 mmHg
	Acre	ac	40.469 a		Atmospheric pressure	atm	1.01325 bar
	Square mile	mi^2	2.59 km^2		Mercury column height	mmHg	760 mmHg = 1 atm
Volume	Liter	l	0.001 m^3	Work	Erg	erg	10^{-7} J
	Cubic meter	m^3	1 000 l		Joule	J	1 N · m
	Cubic yard	yd^3	0.7645 m^3		Watt-hour	Wh	3 600 J

Useful Numerical Values for Propagation Design

Number	Value N	$10 \log_{10} N$ (dB)	$\log_e N$	Reciprocal $1/N$	Remarks
π	3.141 6	4.971 5	1.144 7	0.318 3	Ratio of the circumference of a circle to its diameter
π^2	9.869 6	9.943 0	2.289 5	0.101 3	
$\pi/4$	0.785 4	−1.049 1	−0.241 6	1.273 2	
$\pi/2$	1.570 8	1.961 2	0.451 6	0.636 6	
$3\pi/4$	2.356 2	3.722 1	0.857 1	0.424 4	
2π	6.283 2	7.981 8	1.837 9	0.159 2	
$\sqrt{\pi}$	1.772 5	2.485 8	0.572 4	0.564 2	
$\sqrt{2\pi}$	2.506 6	3.990 9	0.918 9	0.398 9	
$\sqrt{\pi/2}$	1.253 3	0.980 6	0.225 8	0.797 9	
e	2.718 28	4.342 9	1	0.367 9	Base of natural logarithm
e^2	7.389 06	8.685 9	2	0.135 3	
$\sqrt{e}$	1.648 72	2.171 5	0.5	0.606 5	
$\log_{10} e = M$	0.434 3	−3.622 2	−0.834 0	2.302 6	$\log_{10} N = \log_e N \times M$
	1.01	0.043 2	0.010 0	0.990 1	Widely used figures (for logarithmic calculation)
	1.02	0.086 0	0.019 8	0.980 4	
	1.03	0.128 4	0.029 6	0.970 9	
	1.05	0.211 9	0.048 8	0.952 4	
	1.1	0.413 9	0.095 3	0.909 1	
	1.2	0.791 8	0.182 3	0.833 3	
	1.3	1.139 4	0.262 4	0.769 2	
	1.7	2.304 5	0.530 6	0.588 2	
	2.	3.010 3	0.693 1	0.5	
	3.	4.771 2	1.098 6	0.333 3	
	5.	6.989 7	1.609 4	0.2	
	7.	8.451 0	1.945 9	0.142 9	
$\sqrt{2}$	1.414 2	1.505 2	0.346 6	0.707 1	
$\sqrt{3}$	1.732 1	2.385 6	0.549 3	0.577 4	
$\sqrt{5}$	2.236 1	3.494 9	0.804 7	0.447 2	
$\sqrt{7}$	2.645 8	4.225 5	0.973 0	0.378 0	
$\sqrt{8}$	2.828 4	4.515 5	1.039 7	0.353 6	
$\sqrt{10}$	3.162 3	5.0	1.151 3	0.316 2	
$\sin 1''$	$0.484\,8 \times 10^{-5}$	−53.144 4	−12.236 9	$2.062\,7 \times 10^{5}$	For astronomical survey
Ka	$\doteqdot 8.5 \times 10^{3}$	39.29	9.048	1.176×10^{-4}	$K = 4/3$ (Standard value)
$1/(2Ka)$	5.88×10^{-5}	−42.31	−9.741	1.7×10^{4}	$a = 6\,370$ km
k	1.38×10^{-23}	−228.60 (dBW)	—	7.246×10^{22}	Boltzmann's constant Joule/ K
$k \cdot T_0$	4.14×10^{-21}	−203.83 (dBW)	—	2.415×10^{20}	$T_0 = 300$ K $\doteqdot 27$ °C
$k \cdot T_0$	$\doteqdot 4.00 \times 10^{-21}$	−203.98 $\doteqdot$ −204 (dBW)	—	$\doteqdot 2.50 \times 10^{20}$	$T_0 = 290$ K $\doteqdot 17$ °C

1) Ratio of the circumference of a circle to its diameter : $\pi = 3.141\,592\,65$
Base of natural logarithm : $e = 2.718\,281\,83$

2) Boltzmann's constant : $k = 1.380\,662 \times 10^{-23}$ J/K $\approx 1.38 \times 10^{-23}$ J/K

3) Velocity of light in vacuum : $c = 2.997\,924\,58 \times 10^{8}$ m/s $\approx 3 \times 10^{8}$ m/s

Classification of Electromagnetic Waves

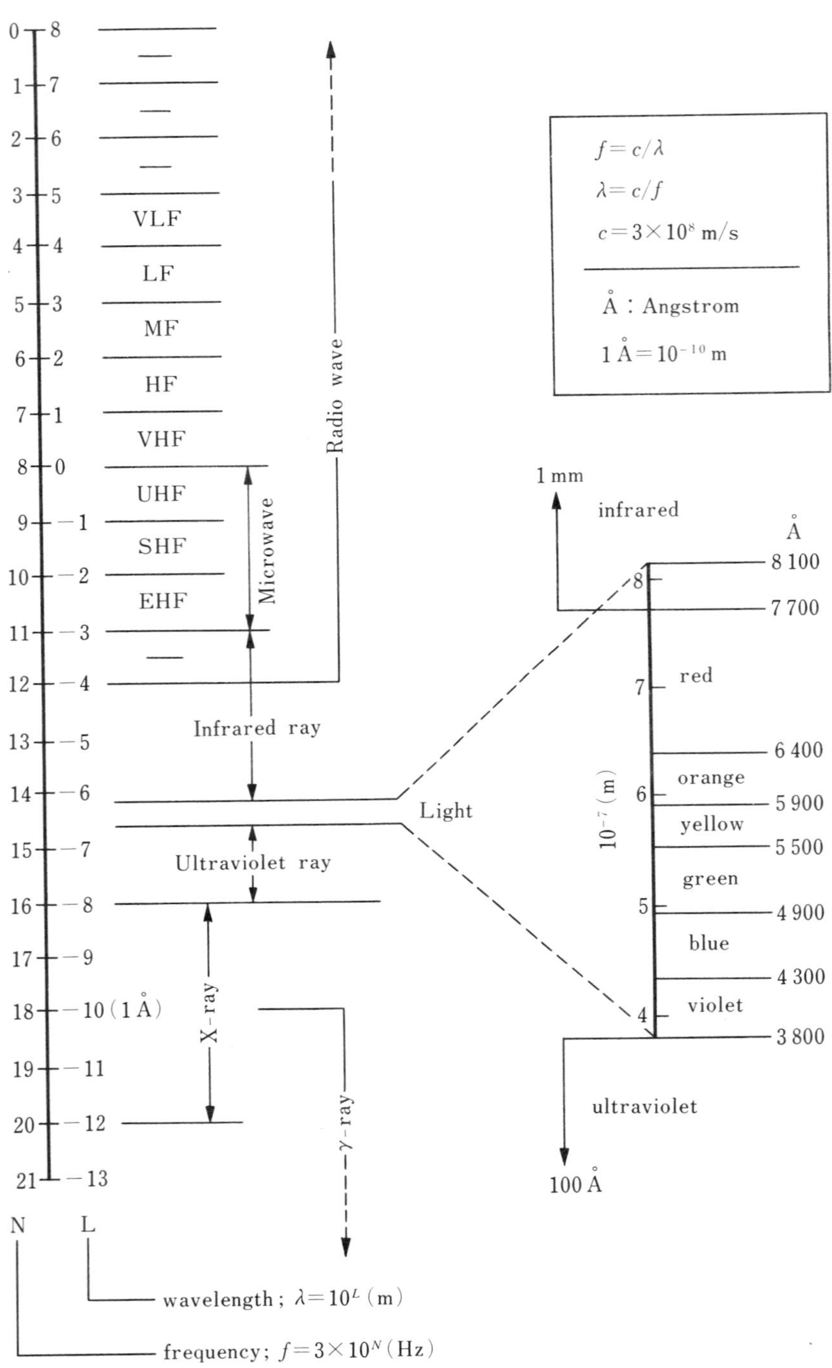

Nomenclature of Radio Waves (Hertzian Waves)

Hz=Hertz

Band No.	Symbols	Frequency range	Wavelength range	Occupied bandwidth	(Ratio)	Designation
4	VLF	3~30	100~10	2.7×10	10^{-4}	Very Low Freq. (Myriametric wave)
5	LF	30~300 kHz	10~1 km	2.7×10^{2} kHz	10^{-3}	Low Freq. (Kilometric wave)
6	MF	300~3 000	1~0.1	2.7×10^{3}	10^{-2}	Medium Freq. (Hectometric wave)
7	HF	3~30	100~10	2.7×10	10^{-1}	High Freq. (Decametric wave)
8	VHF	30~300 MHz	10~1 m	2.7×10^{2} MHz	1	Very High Freq. (Metric wave)
9	UHF	300~3 000	1~0.1	2.7×10^{3}	10	Ultra High Freq. (Decimetric wave)
10	SHF	3~30	100~10	2.7×10	10^{2}	Super High Freq. (Centimetric wave)
11	EHF	30~300 GHz	10~1 mm	2.7×10^{2} GHz	10^{3}	Extremely High Freq. (Millimetric wave)
12	——	300~3 000	1~0.1	2.7×10^{3}	10^{4}	—— (Decimillimetric wave)

↑ Normalized by the VHF bandwidth

Note : The indicated frequency range
Excludes the lower end frequency but includes the upper end frequency.
May be expressed by $0.3\times10^{B}\sim3\times10^{B}$ (Hz) where B is the frequency-band number.

Electromagnetic Waves in Free Space

Propagation Modes and Phases

Electromagmetic waves in free space (TEM mode)

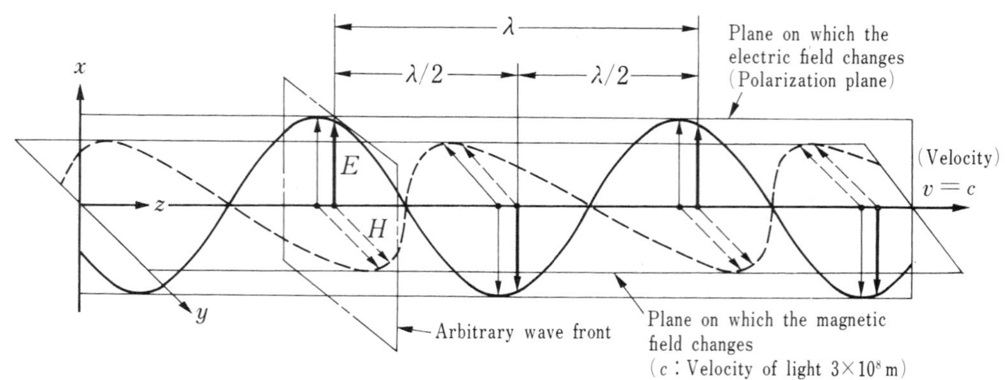

Presentation of phase

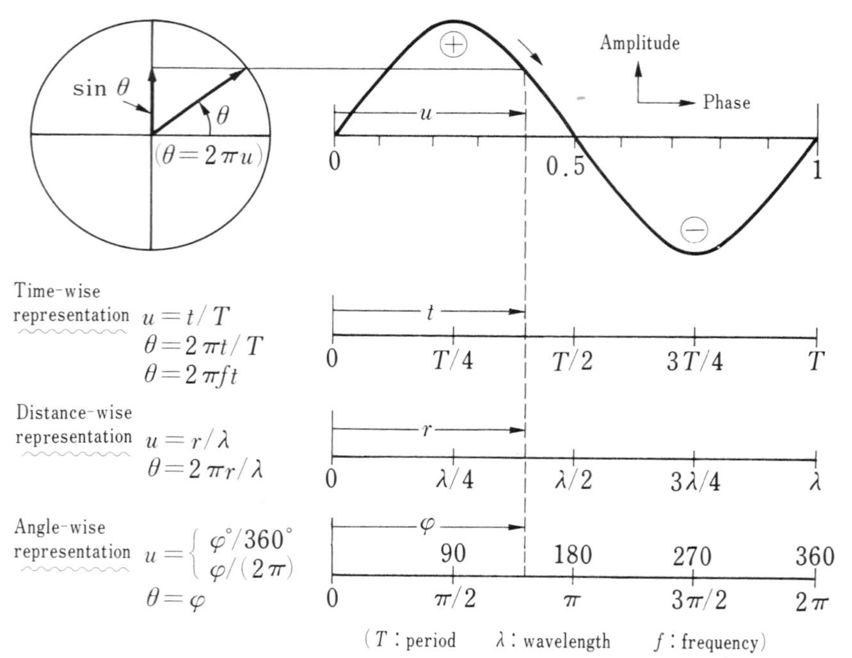

Electromagnetic Waves in Waveguides

Type	Mode	Field component / cutoff frequency	Transverse cross section	Longitudinal cross section
Coaxial cable	TEM	E_r H_ϕ —	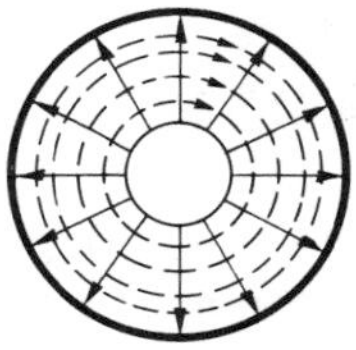	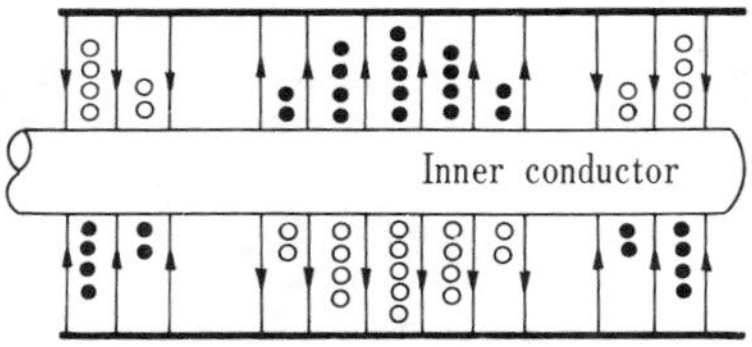 Inner conductor
Circular waveguide	TM_{01}	E_z, E_r H_ϕ 2.61 a (a : Inner diameter)		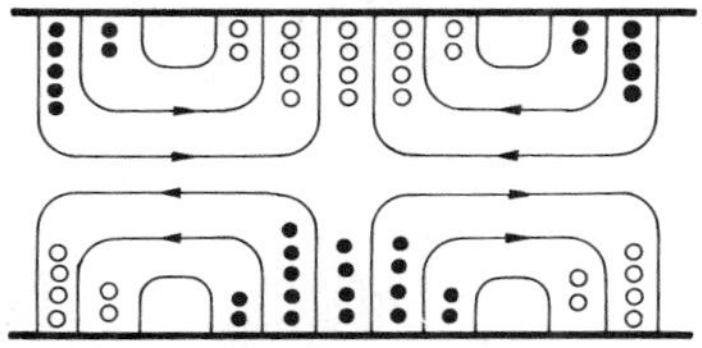
	TE_{01}	E_ϕ H_z, H_r 1.64 a (a : Inner diameter)	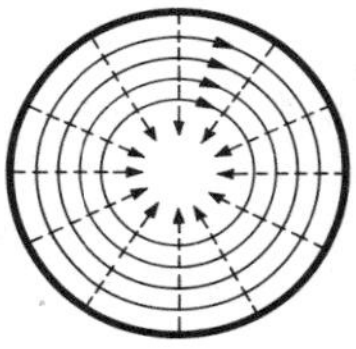	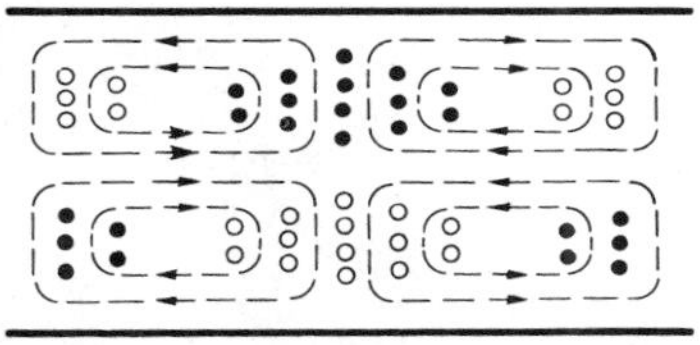
	TE_{11}	E_r, E_ϕ H_z, H_r, H_ϕ 3.41 a (a : Inner diameter)	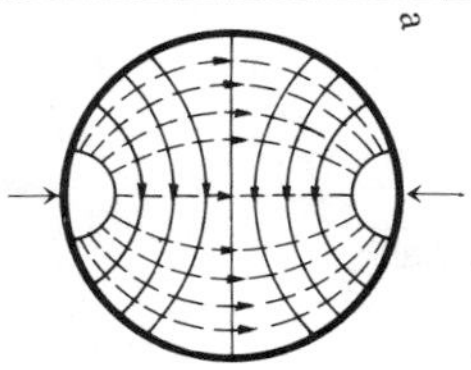 a	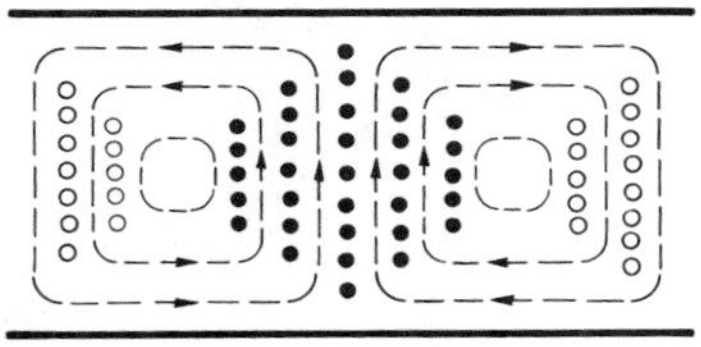
Rectangular waveguide	TE_{10}	E_y H_z, H_x 2 a (a : Rectangular waveguide)	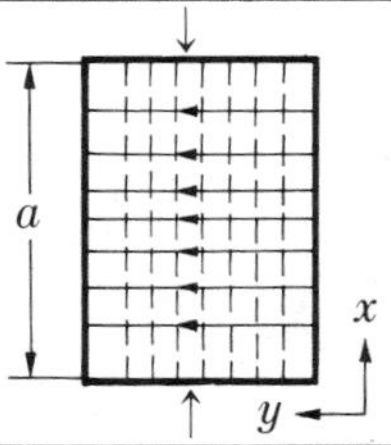 a x y	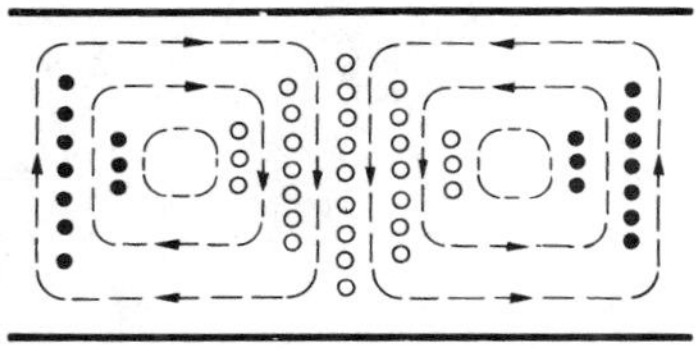
Remark		ϕ : Denoting the field present in a direction tangential to the concentric circle r : Denoting the field present in a radial direction	Electric field Magnetic field	● ○ Perpendicular to the paper but opposite in direction

0
6

Wave Front and Directivity

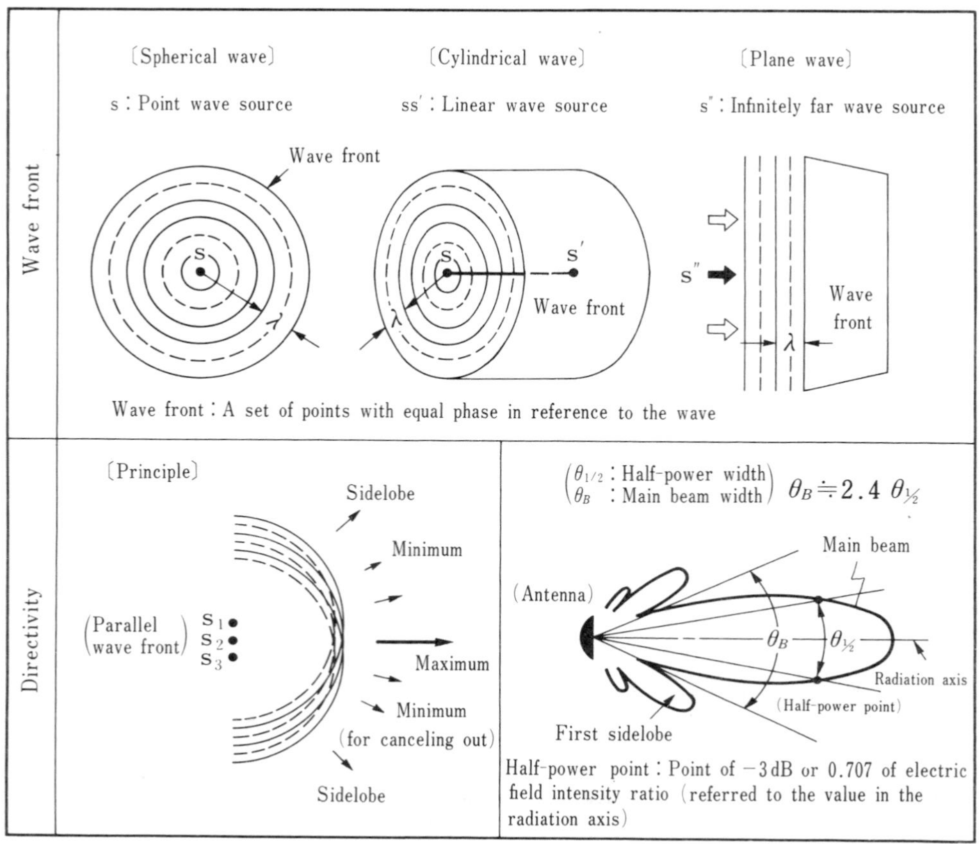

Wave front
〔Spherical wave〕
s : Point wave source
〔Cylindrical wave〕
ss' : Linear wave source
〔Plane wave〕
s'' : Infinitely far wave source
Wave front
s
λ
s
s'
Wave front
λ
s''
Wave front
λ
Wave front : A set of points with equal phase in reference to the wave
Directivity
〔Principle〕
Sidelobe
Minimum
(Parallel wave front)
s1
s2
s3
Maximum
Minimum
(for canceling out)
Sidelobe
(θ1/2 : Half-power width
θB : Main beam width)
θB ≒ 2.4 θ½
Main beam
(Antenna)
θB
θ½
Radiation axis
(Half-power point)
First sidelobe
Half-power point : Point of −3 dB or 0.707 of electric field intensity ratio (referred to the value in the radiation axis)

Vectorial Summation of Polarized Wave

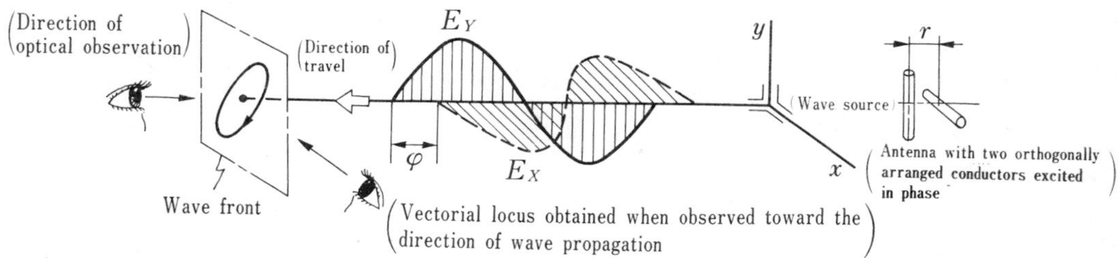

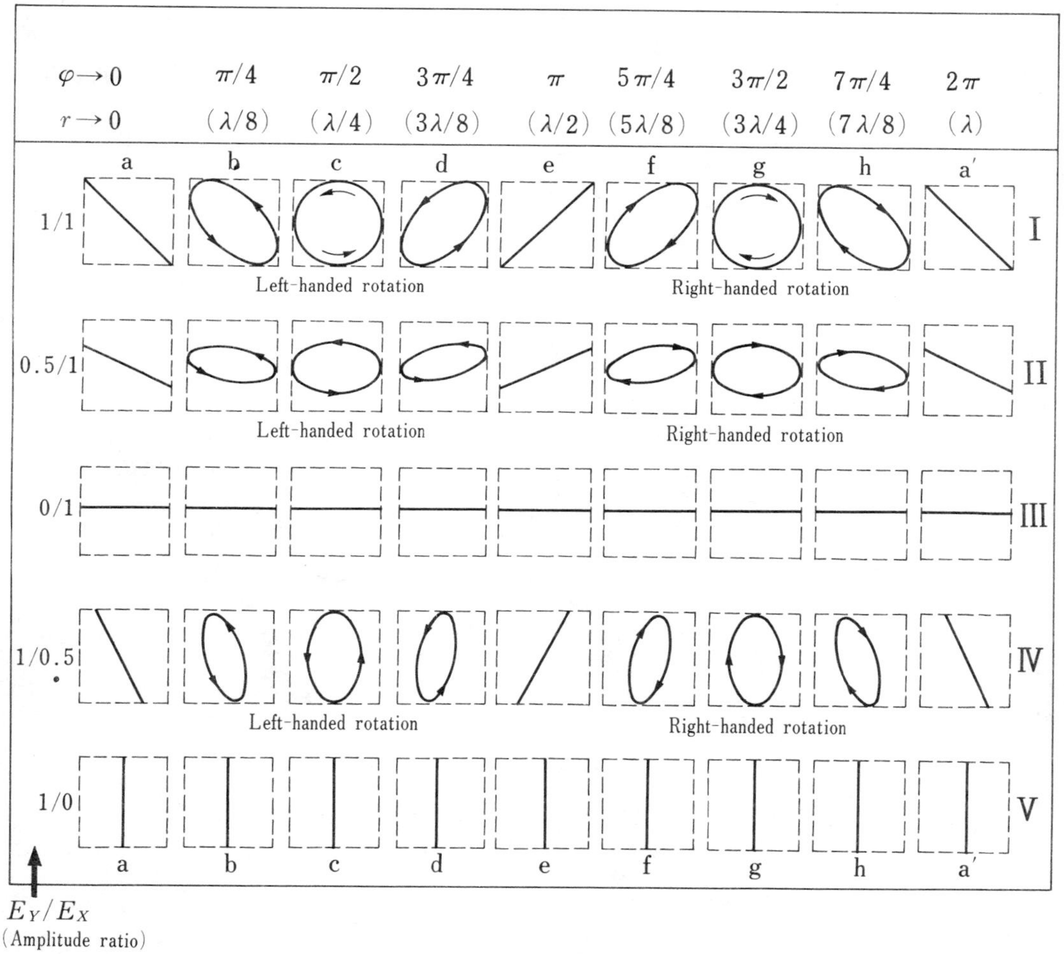

$\varphi \to 0$	$\pi/4$	$\pi/2$	$3\pi/4$	π	$5\pi/4$	$3\pi/2$	$7\pi/4$	2π	
$r \to 0$	($\lambda/8$)	($\lambda/4$)	($3\lambda/8$)	($\lambda/2$)	($5\lambda/8$)	($3\lambda/4$)	($7\lambda/8$)	(λ)	
	a	b	c	d	e	f	g	h	a′
1/1		Left-handed rotation				Right-handed rotation			I
0.5/1		Left-handed rotation				Right-handed rotation			II
0/1									III
1/0.5		Left-handed rotation				Right-handed rotation			IV
1/0									V
	a	b	c	d	e	f	g	h	a′

E_Y/E_X
(Amplitude ratio)

Linearly Polarized Waves (Plane Polarized Waves)

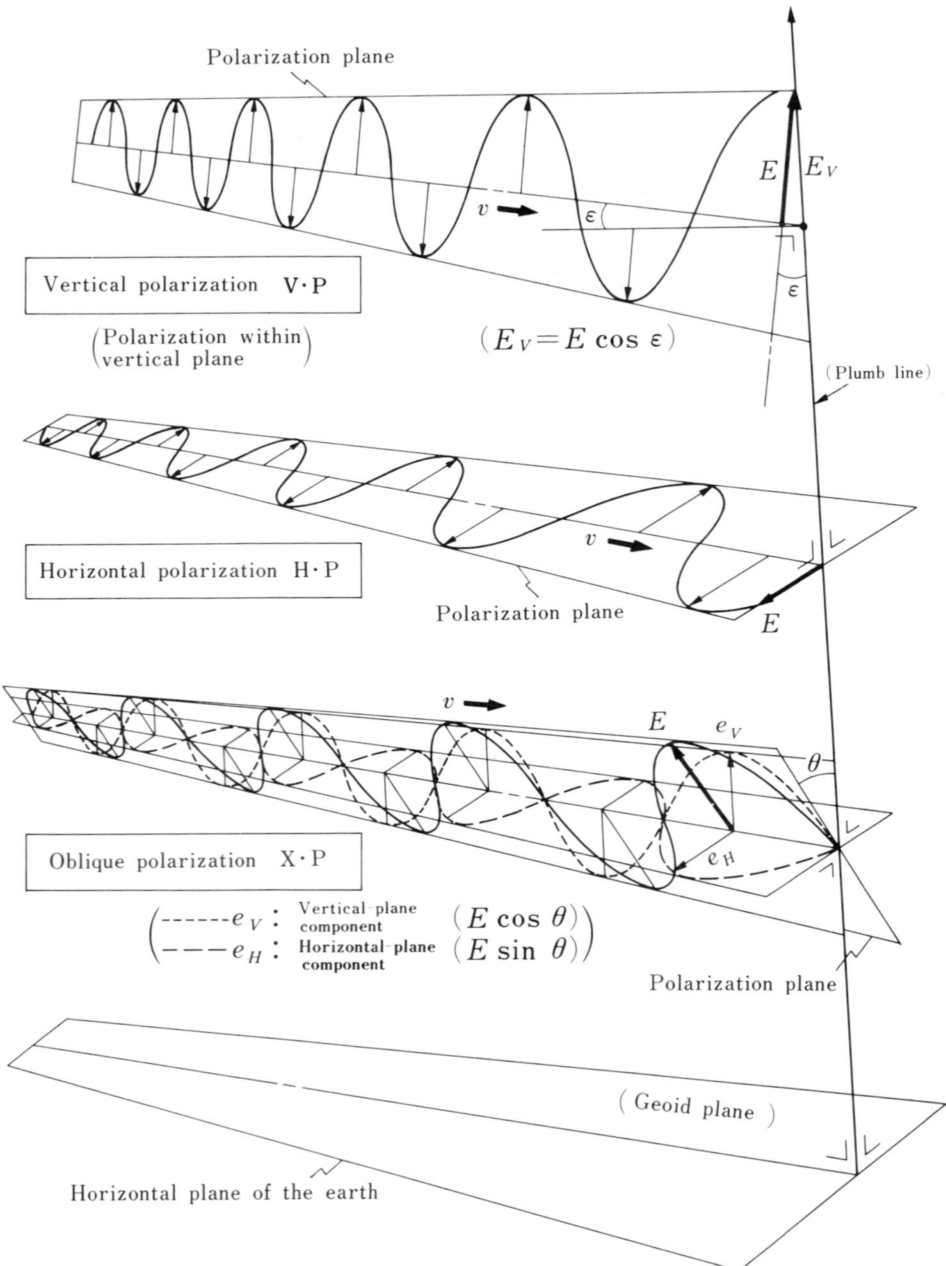

Circularly and Elliptically Polarized Waves

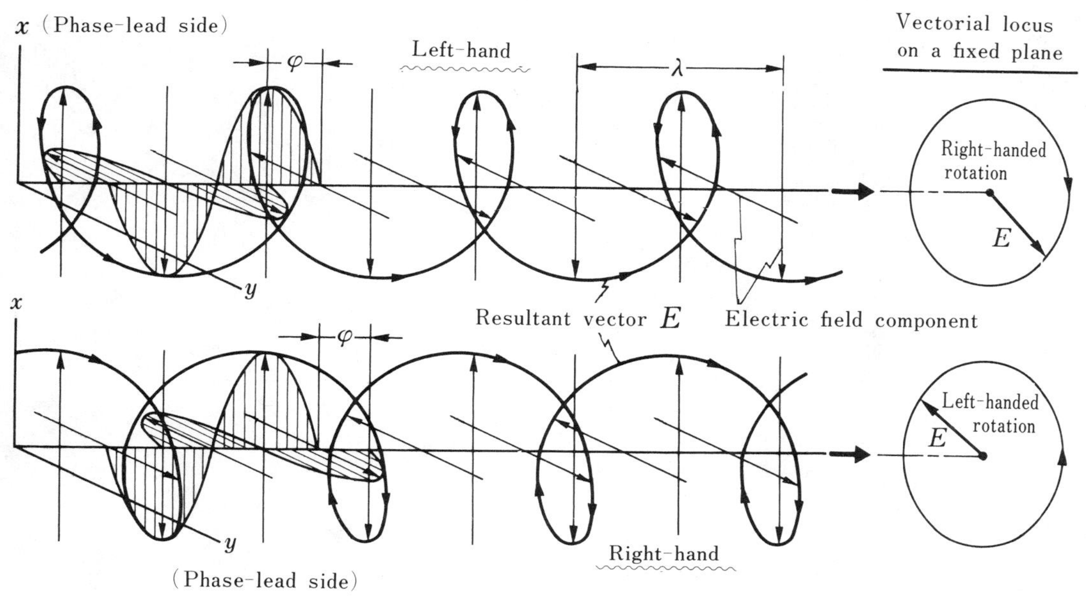

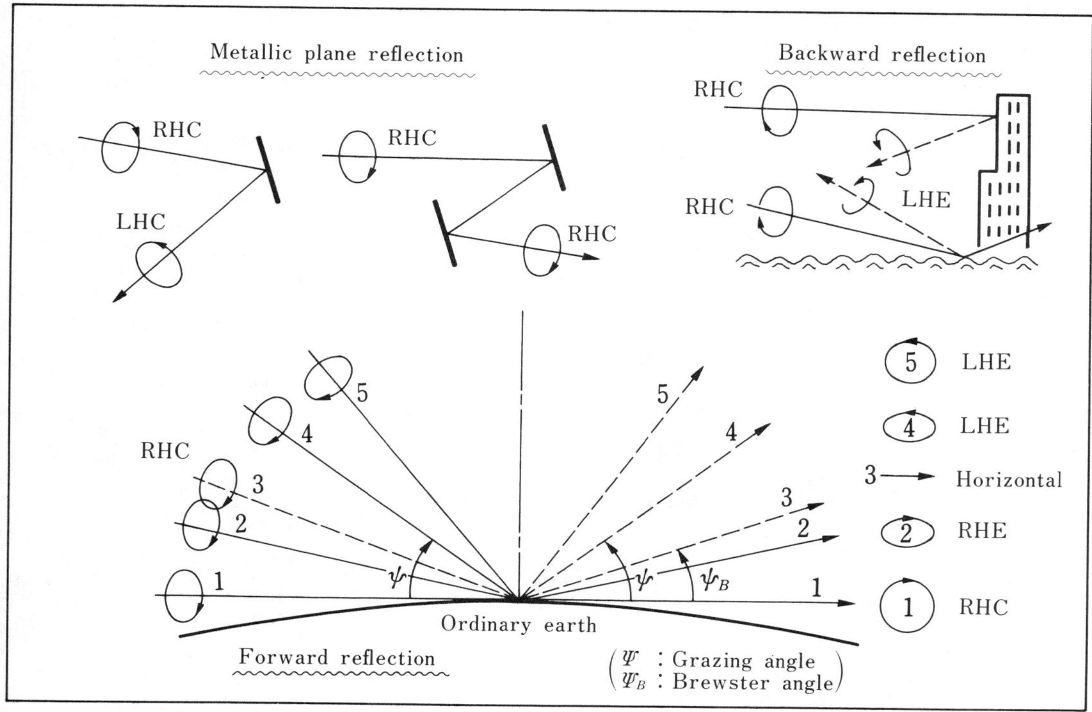

Note : RHC (Right-handed circular)
LHC (Left-handed circular)
RHE (Right-handed elliptical)
LHE (Left-handed elliptical)

Reference Antennas

		Type	Gain G	Effective aperture area A_e	Directivity pattern	
					E-plane	H-plane
Typical reference antennas	Isotropic antenna	Point wave source	1 (0 dB)	$\lambda^2/(4\pi) = A_i$	Circular shaped	Circular shaped
	Hertz dipole	$\pm q$, $\mp q$; Main axis of radiation	1.5 (1.76 dB)	$3\lambda^2/(8\pi)$	8-shaped	Circular shaped
	Half-wave dipole	Main axis of radiation; $\lambda/2$; i	1.64 (2.15 dB)	$3.3\lambda^2/(8\pi)$	8-shaped	Circular shaped
		Note	Power ratio	$G(\lambda^2/4\pi)$	—	Nondirectional

* The reference antenna adopted in this book is the isotropic antenna in principle and the half-wave dipole where necessary.

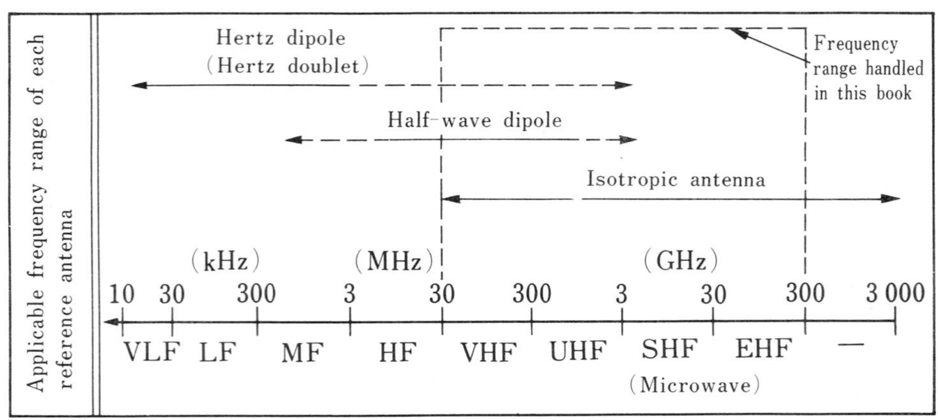

Radio-Wave Intensity

		Electric field strength	Flux density	Basic input power (Isotropic receiving input)	
Definition		Electric field, 1 m, E, Magnetic field Gradient of electric potential in the direction of oscillation of electric field	P_u, 1 m^2 Energy passing through an area of 1 m^2 on the wavefront in a unit time period	$(A_i=\lambda^2/(4\pi))$, P_i Available power received by an isotropic antenna	
Symbol		E	P_u	P_i	
Normally employed units	Antilog	μV/m, mV/m, V/m	pW/m^2, μW/m^2, mW/m^2, W/m^2	pW, μW, mW, W	*1
	dB	dB μV/m dB mV/m dB V/m (20 log E)	dB pW/m^2 dB μW/m^2 dB mW/m^2 dB W/m^2 (10 log P_u)	dB pW dB μW dB mW dB W (10 log P_i)	
Conceptional parameters and applications		l_e : Effective length of half-wave dipole	A_e : Effective aperture area	G : Gain (Power ratio)	
		$V_t=E(l_e\sqrt{G_d})$ └-- Received voltage	$P_r=A_e\cdot P_u$ └- Available receiving power	$P_r=G\cdot P_i$ └-- Available receiving power	*3
Relationship		$E=\sqrt{120\pi\cdot P_u}$	$P_u=E^2/(120\pi)$		*2
			$P_u=P_i(4\pi/\lambda^2)$	$P_i=P_u(\lambda^2/(4\pi))$	
		$E=(\pi/\lambda)\sqrt{480\cdot P_i}$		$P_i=E^2\cdot\lambda^2/(480\pi^2)$	

* 1 dBμV/m and dBμV may be abbreviated as dBμ and likewise dBmW, as dBm.
* 2 Equations for conversion between parameters on radio wave intensity at a point within a propagation space.
* 3 G_d : Gain with reference to that for $\lambda/2$ dipole.
  G (with reference to the isotropic antenna) $= G_d + 2.15$ (dB)

Propagation Loss

Propagation loss Γ

Γ

A P P_i B

D

The transmission loss (or basic transmission loss) between two points A and B under consideration, each provided with an isotropic antenna, is given by

$$\Gamma = 10 \log (P/P_i) \quad \text{(dB)}$$

where,

- presence of a radio-wave line-of-sight is not essential
- any propagation mode (free space, interference, diffraction, scattering, etc) can be assumed.

= Free space propagation loss Γ_0 + Additional propagation loss Γ_P (dB)

Propagation loss generated when the path between two points A and B is under free space condition. In this case, the loss is given by a function of the wavelength λ and the distance D.

$$\Gamma_0 = 20 \log (4 \pi D/\lambda) \text{ (dB)}$$

The excessive propagation loss over Γ_0 for the free space propagation.

$$\Gamma_P = \Gamma - \Gamma_0 \text{ (dB)}$$

Radio Transmission Loss

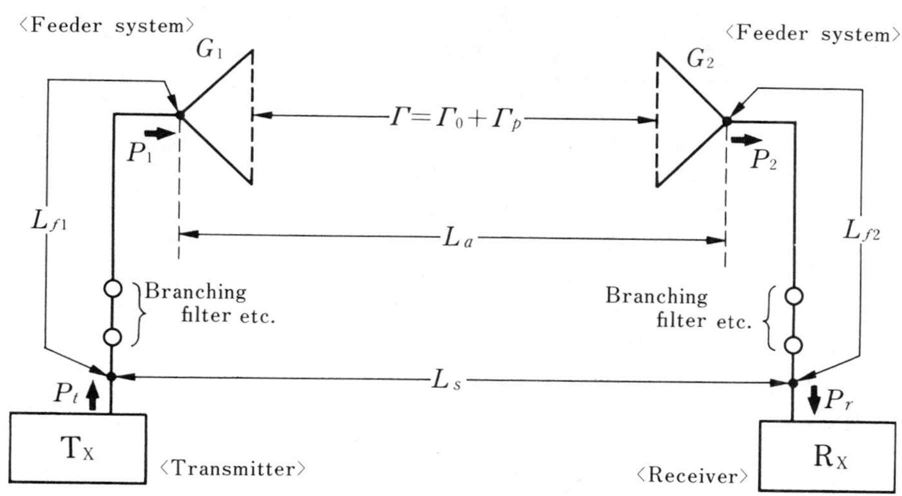

Γ_0 : Free space propagation loss

Γ_p : Additional propagation loss (in excess of attenuation in free space)

Γ : Propagation loss ($\Gamma = \Gamma_0 + \Gamma_p$) dB

L_a : Transmission loss between antennas ($L_a = \Gamma - G_c$) dB

L_s : Span loss ($L_s = P_t - P_r$) dB

G_1 : Transmitting antenna gain

G_2 : Receiving antenna gain

L_{f1} : Transmitting feeder system loss

L_{f2} : Receiving feeder system loss

P_1 : Transmitting antenna input (Transmitting power)

P_2 : Receiving antenna output (Available power)

P_t : Transmitter output

P_r : Receiver input

G_c : Antenna overall gain ($G_1 + G_2 - L_c$) dB

L_c : Antenna-to-medium coupling loss ($L_c = 0$ dB in free space)

Note : The isotropic antenna is assumed here as the reference antenna pertaining to Γ_0, Γ, G and G_2.
The antenna system implies the antenna plus the associated feeder system.
Propagation loss may be called the basic transmission loss.

0
13

Atmospheric Temperature and Pressure in Troposphere

ICAO Standard Atmosphere

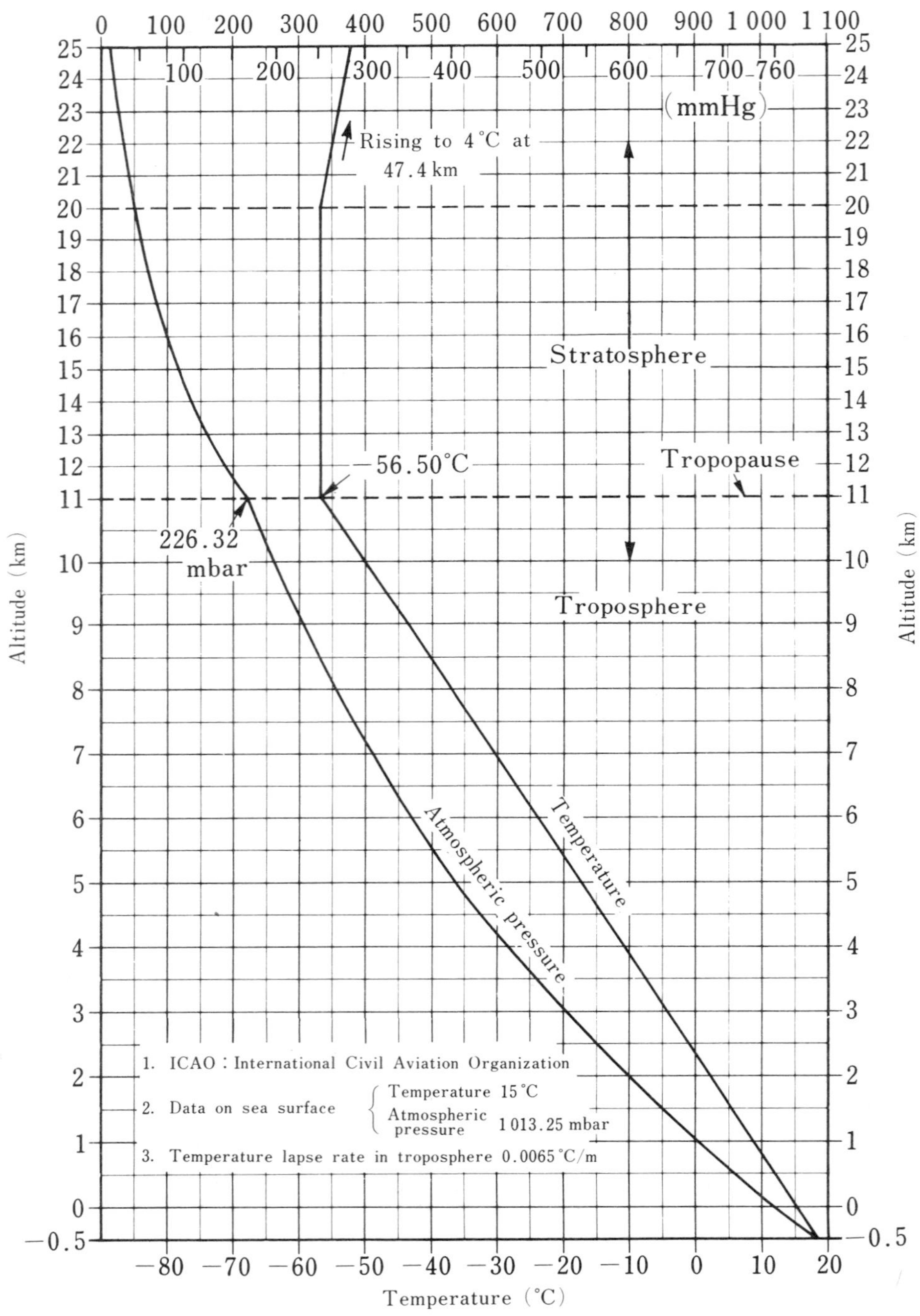

Atmospheric Refraction

M curve and duct

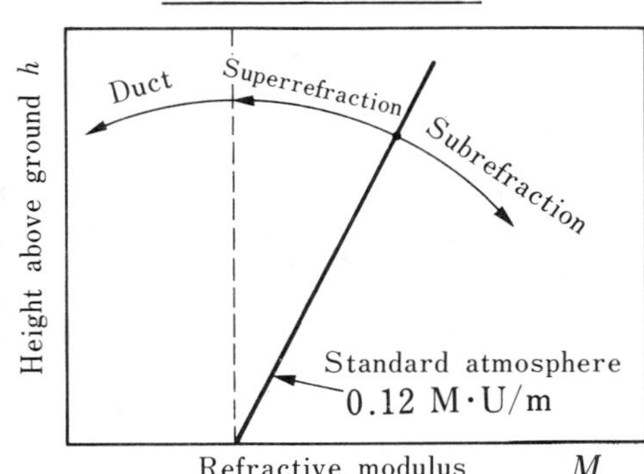

p : Atmospheric pressure (mbar)
e : Water vapor pressure (mbar)
T : Absolute temperature (K)
h : Height above ground (m)
a : Earth radius (6.37×10^6 m)
M·U : 10^{-6}

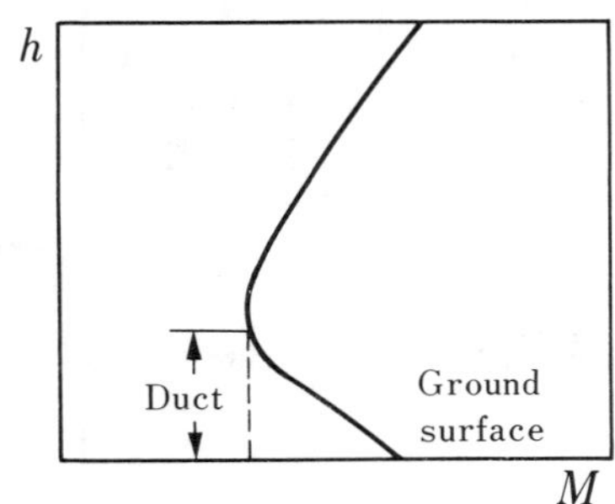

Surface duct

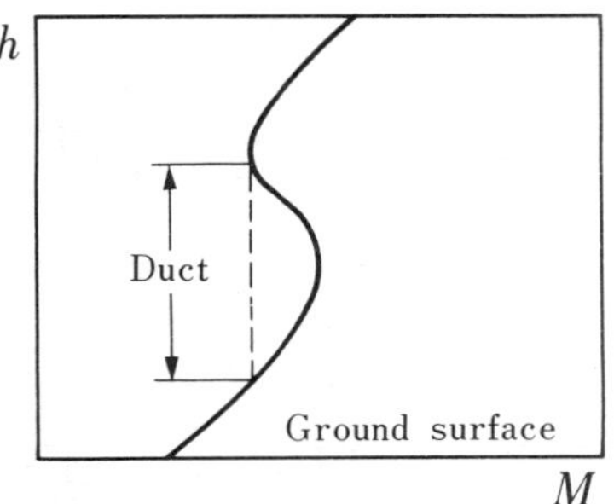

Elevated duct

Basic formulas

Refractive modulus	$M = N + \left(\frac{h}{a}\right) \cdot 10^6$
Refractivity	$N = \frac{77.6}{T}\left(p + 4\,810\,\frac{e}{T}\right)$
Refractive coefficient	$n = 1 + N \cdot 10^{-6}$
Coefficient of effective-earth radius	$K = 1 \Big/ \left(1 + \frac{a}{n} \cdot \frac{dn}{dh}\right)$

(1) K = Effective earth radius/Mean earth radius

(2) The K is defined only for the linear portion of the M curve averaged over entire propagation path.

(3) Vertical structures of the refractivity are expressed by M or N.

Ground Reflection

Effective Reflection Coefficient ρ_e and Propagation Echo Ratio ρ_θ

Smooth flat ground

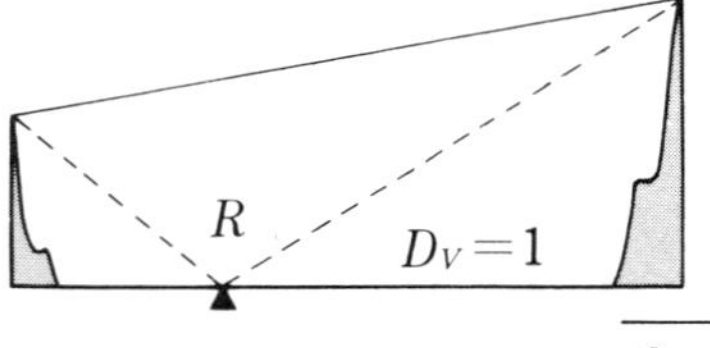

Smooth spherical ground

$R \cdot D_V$

(a)

$$\rho_e = R \cdot D_V$$

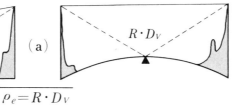

Ordinary rough ground

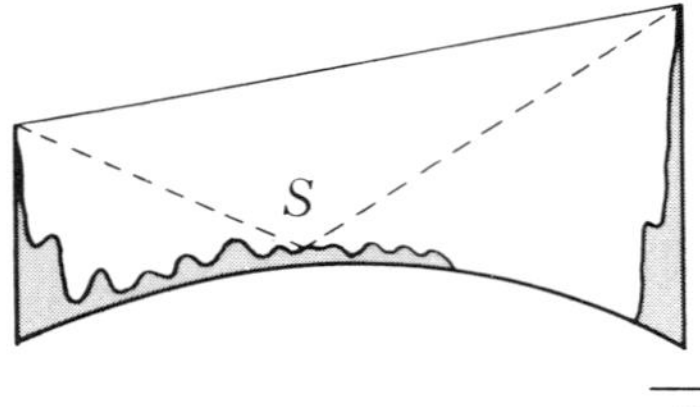

(b)

Mountainous land

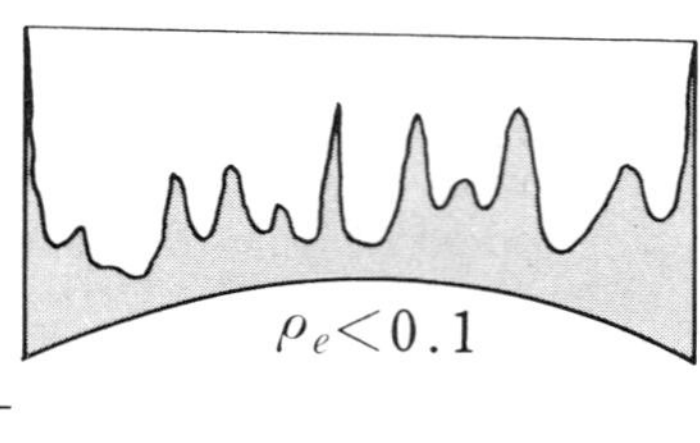

$$\rho_e = S$$

Smooth ground with ridge

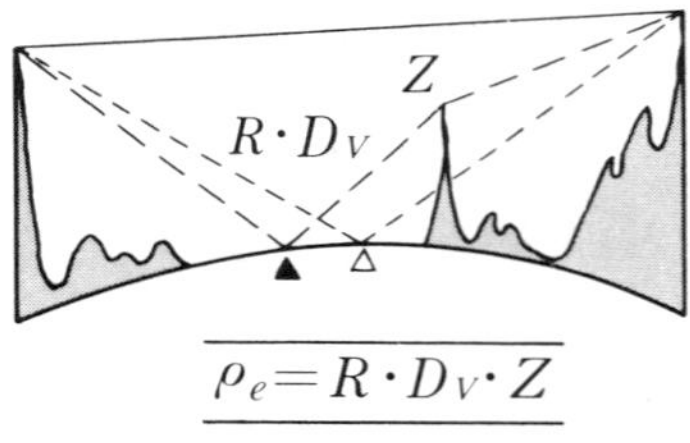

(c)

Rough ground with ridge

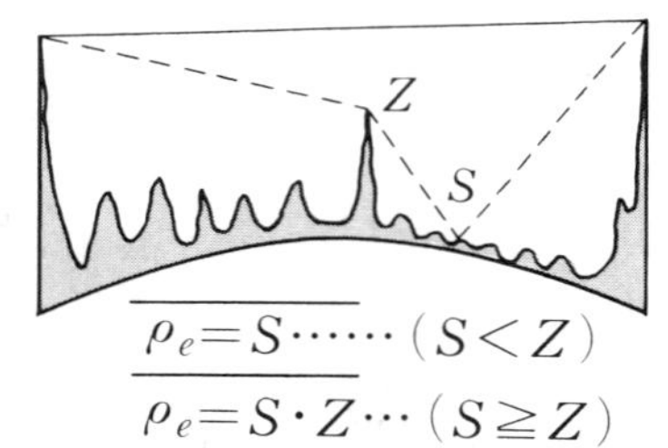

$$\rho_e = R \cdot D_V \cdot Z$$

$$\rho_e = S \cdots\cdots (S<Z)$$

$$\rho_e = S \cdot Z \cdots (S \geqq Z)$$

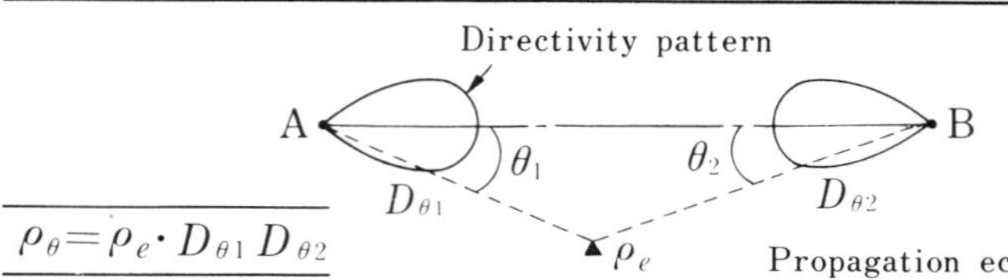

$$\rho_\theta = \rho_e \cdot D_{\theta 1} D_{\theta 2}$$

1) R : Plane reflection coefficient. D_V : Spherical ground divergence factor. S : Rough ground reflection coefficient. Z : Ridge diffraction coefficient. $D_{\theta 1}$, $D_{\theta 2}$: Directivity coefficients of reflected rays of antennas A and B (Unity for direct rays).

2) Refer to Chapter 2 for criteria on rough and smooth ground and for how to estimate R, D_V, S, Z, $D_{\theta 1}$, and $D_{\theta 2}$.

Propagation Mode

Physical classification of propagation waves

Propagation wave component	Propagation in non-ionized layer	Propagation in ionized layer
Direct wave	Free space wave	——
Reflected wave	Ground reflection wave, reflected wave through passive reflector	Ionized-layer reflected wave
Refracted wave	Standard refracted wave, Subrefracted wave, Superrefracted wave, Duct wave	Ionized-layer refracted wave
Diffracted wave	Mountain diffracted wave, spherical earth diffracted wave	——
Surface wave	Surface wave	——
Scatter wave	Troposcatter wave, precipitation-scatter wave	Ionized-layer scatter wave

Propagation path types and applicable frequency ranges

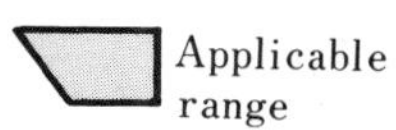

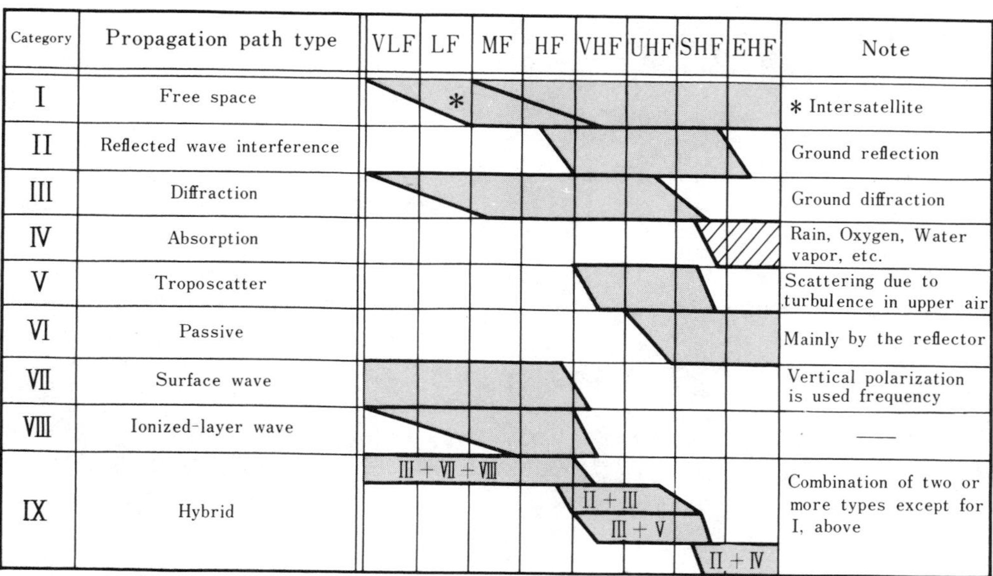

Category	Propagation path type	Note
I	Free space	* Intersatellite
II	Reflected wave interference	Ground reflection
III	Diffraction	Ground diffraction
IV	Absorption	Rain, Oxygen, Water vapor, etc.
V	Troposcatter	Scattering due to turbulence in upper air
VI	Passive	Mainly by the reflector
VII	Surface wave	Vertical polarization is used frequency
VIII	Ionized-layer wave	——
IX	Hybrid	Combination of two or more types except for I, above

Propagation Path Types (1), VHF to SHF Bands

[A] Free space propagation path (Refer also to Chapter 1)

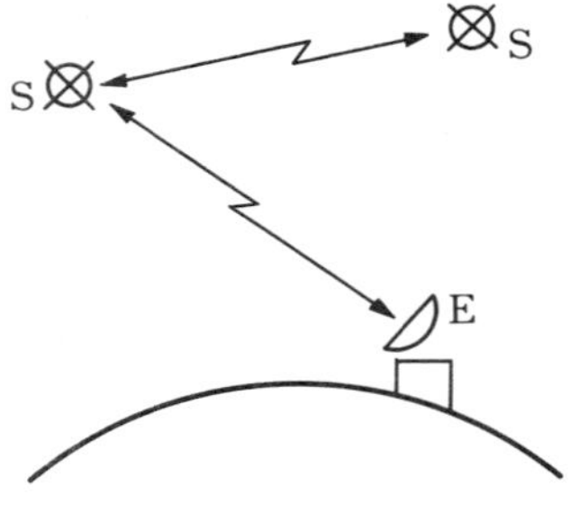

(a) Space-space type
Space-earth type

- S ; Space (Satellite or aircraft) station
  E ; Earth (Terrestrial) station

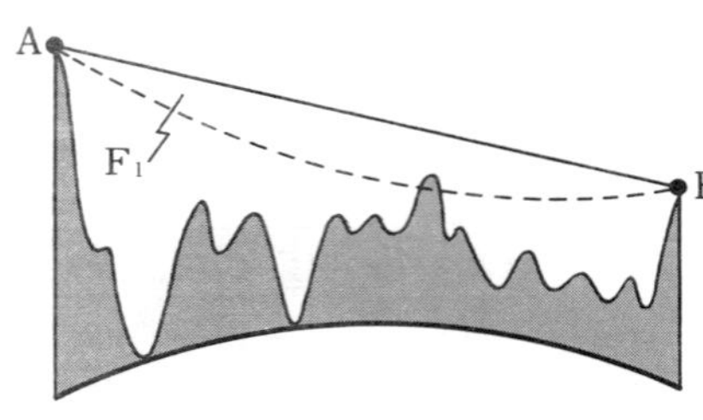

(b) Mountain propagation type

- There is no obstruction loss to the direct wave and the ground reflected wave may be ignored.
- F_1; First Fresnel zone

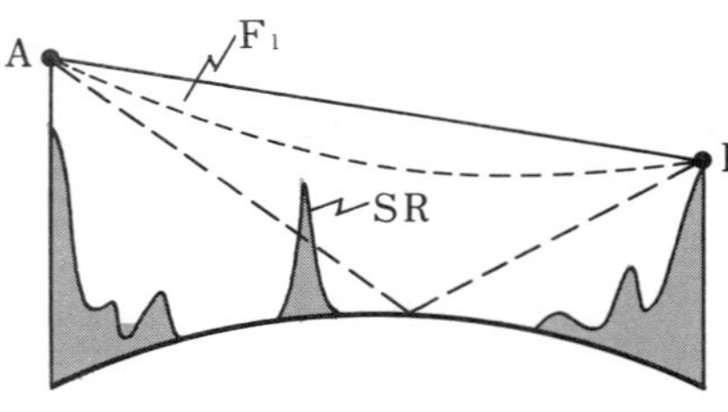

(c) Screening-the-reflected-wave type

- There is no obstruction loss to the direct wave and a would-be reflection surface area is screened by an island or cape.

Note : Effects due to attenuation by precipitation and to abnormal refraction in the atmosphere are described separately in Chapters 5 and 8.

Propagation Path Types (2), VHF to SHF Bands

[B] Reflected-wave-interference propagation path (Refer also to Chapter 2)

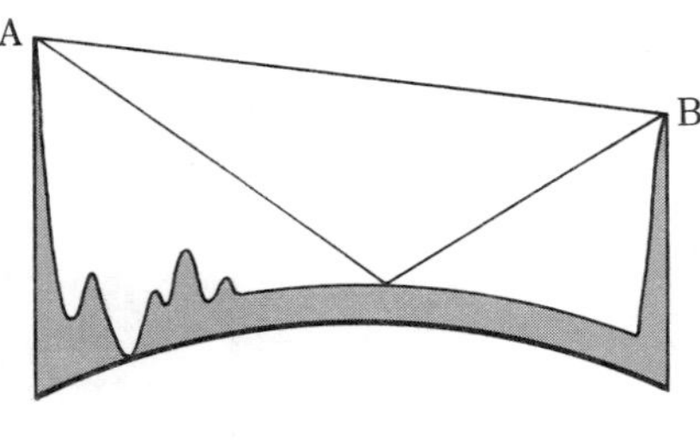

(a) Normal reflection type

Path with a reflection from the sea or lake water surface or flat ground surface

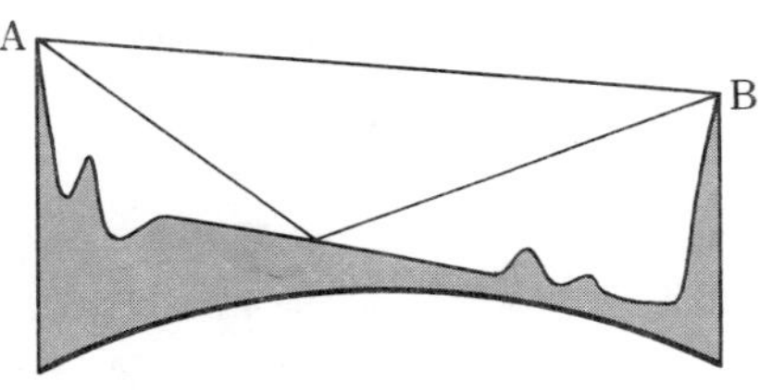

(b) Slant reflection type

Path with a reflection from the slant area as in mountain slope or fanlike land

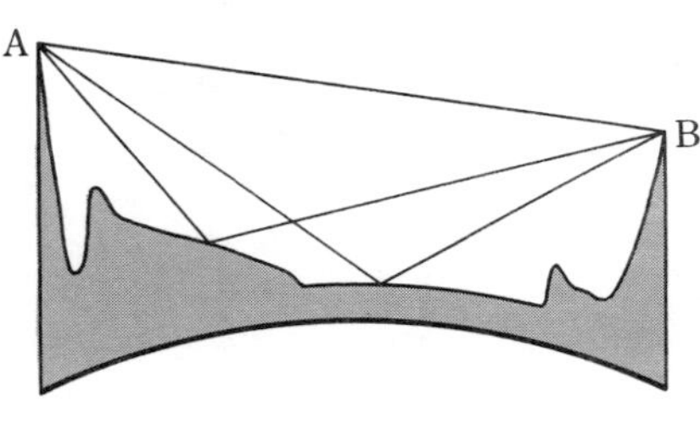

(c) Multipath reflection type

It is very rare to experience the case caused by more than two rays though cases with one direct and two reflected rays or two direct and two reflected rays have sometimes occurred.

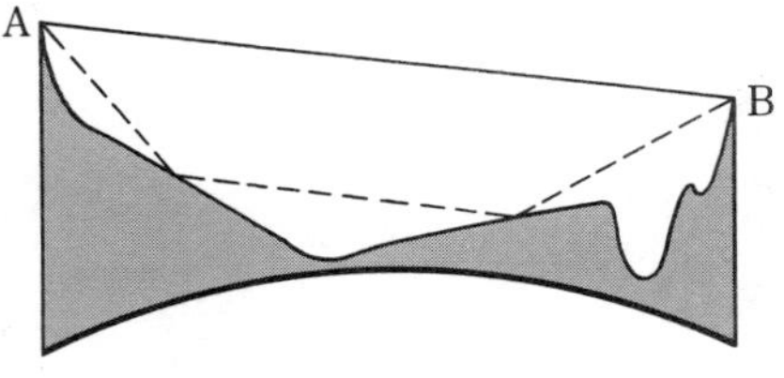

(d) Multiple reflection type

It very rarely occurs and can be ignored even when it occurs because the expected propagation echo ratio is very small.

Propagation Path Types (3), VHF to SHF Bands

[C] Diffraction propagation path (Refer also to Chapter 3)

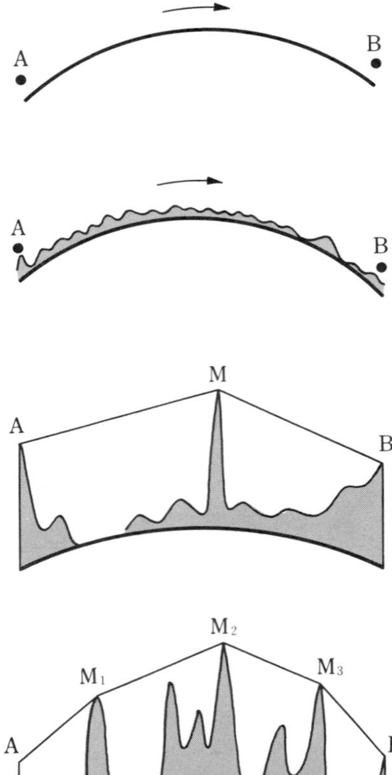

(a) Smooth spherical ground diffraction type

A typical example is the transhorizon over-seawater propagation path

(b) Rough ground diffraction type

Transhorizon overland propagation path

(c) Single ridge diffraction type

The single ridge may be a mountain or a man-made structure such as a building.

(d) Multi-ridge diffraction type

Diffraction loss will increase with the number of ridges. It is very rare to encounter this type of path with more than 3-fold ridges.

Propagation Path Types (4), VHF to SHF Bands

[D] Tropospheric scatter (T.S.) propagation path (Refer also to Chapter 4)

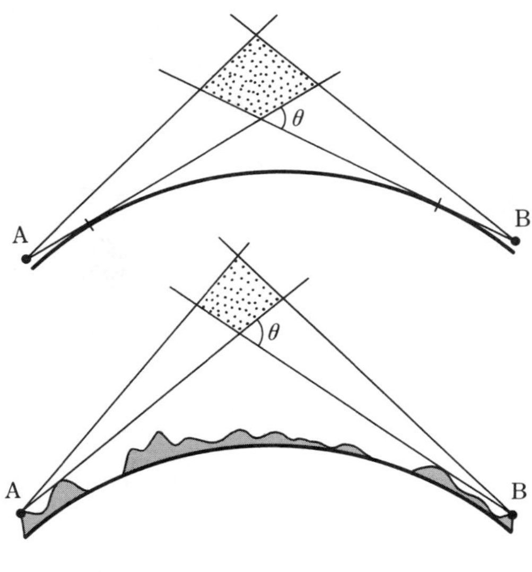

(a) T.S. type over smooth spherical ground

Transhorizon great distance propagation path passing over such smooth ground as seawater and desert area

(b) T.S. type over ordinary ground

Propagation loss increases with θ

Rough spherical surface diffraction wave may become predominant depending on conditions.

[E] Passive relay propagation path (Refer also to Chapter 6)

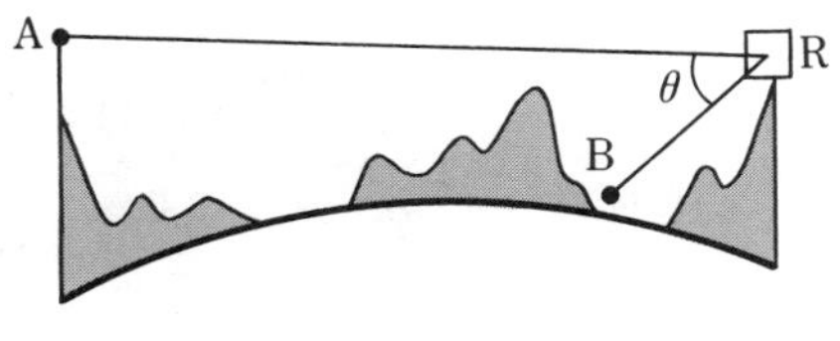

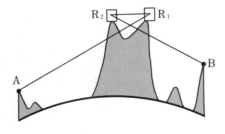

(a) Single reflector type

The smaller θ, the greater the efficiency. The double reflector type is used when $\theta \geqq 120^\circ$.

(b) Double reflector type

Spacing between reflectors R_1 and R_2 should be minimized as far as possible.

When the spacing between R_1 and R_2 is too great, the path should be treated as two-stage passive relay type resulting in greater overall propagation loss.

Propagation Path Types (5), VHF to SHF Bands

[F] Hybrid propagation path (Refer also to Chapters 2, 3 and 4)

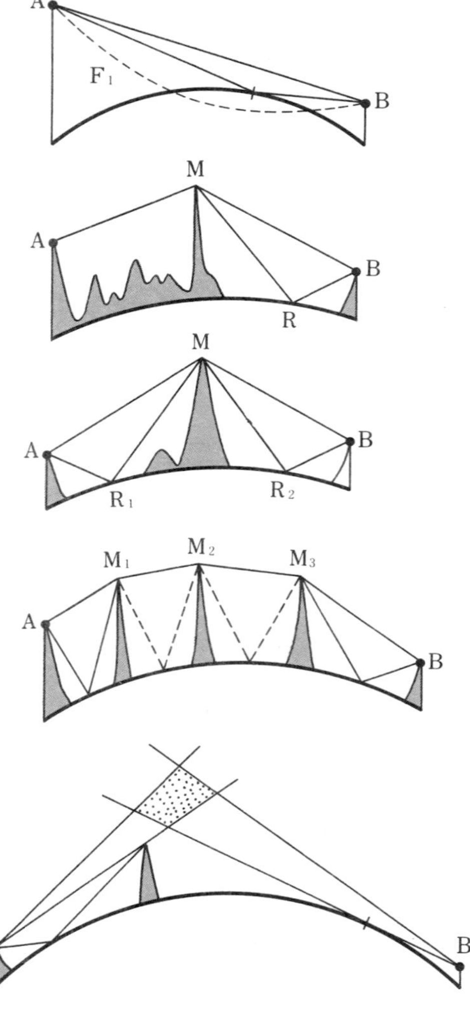

(a) Tangential path type

The line-of-sight path where a part of the first Fresnel zone (F_1) is obstructed by the spherical earth presents the hybrid propagation path generating diffraction and reflected waves.

(b) Ridge-diffraction-with-a-single-section-of-reflection type

The receiving field is analyzed into AMB and AMRB path components.

(c) Ridge-diffraction-with-reflections-on-both-sides type

The receiving field is analyzed into four (4) path components, i. e., AMB, AR_1MB, AMR_2B and AR_1MR_2B. The AR_1MR_2B component is normally negligible.

(d) Multi-ridge diffraction with reflection type

In principle, inter-ridge reflection may be ignored, analyze it in a similar manner as in (b) or (c) above.

(e) Tropospheric scatter diffraction type

Whether it is the diffraction type or the troposcatter type is determined by which is dominant, i.e., total effect by ground reflection, ridge diffraction and spherical earth diffraction or the effect by tropospheric scattering.
The path type may frequently change depending upon day or night and upon season.

Path Profile (1)

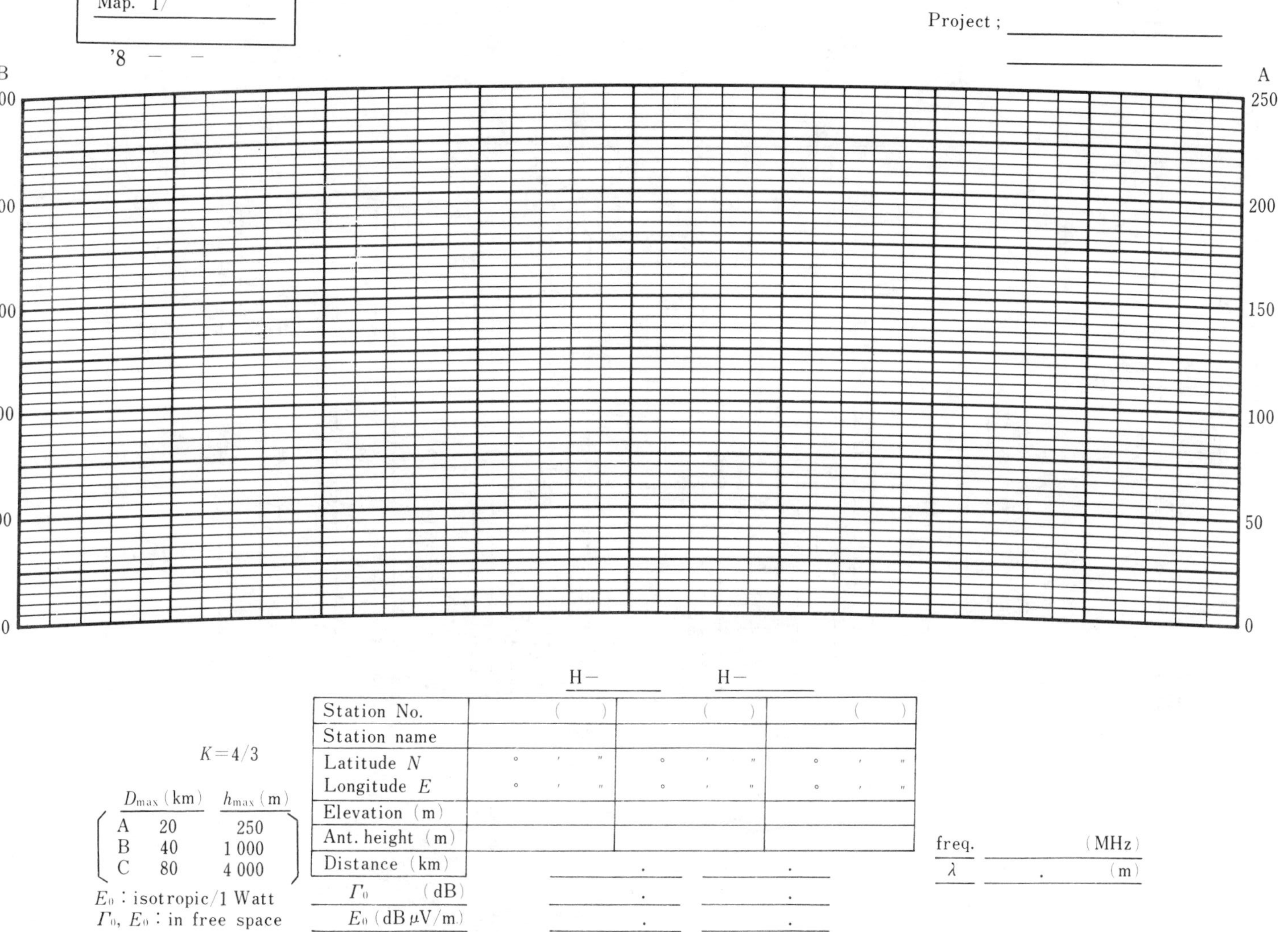
Map. 1/
'8 – –
Project ;
B
A
1 000
800
600
400
200
0
250
200
150
100
50
0
H–
H–
K=4/3
D_{max} (km) h_{max} (m)
A 20 250
B 40 1 000
C 80 4 000
E_0 : isotropic/1 Watt
Γ_0, E_0 : in free space
Station No.
Station name
Latitude N
Longitude E
Elevation (m)
Ant. height (m)
Distance (km)
Γ_0 (dB)
E_0 (dB μV/m)
freq. (MHz)
λ (m)

Path Profile (2)

Map. 1/

'8 – –

Project :

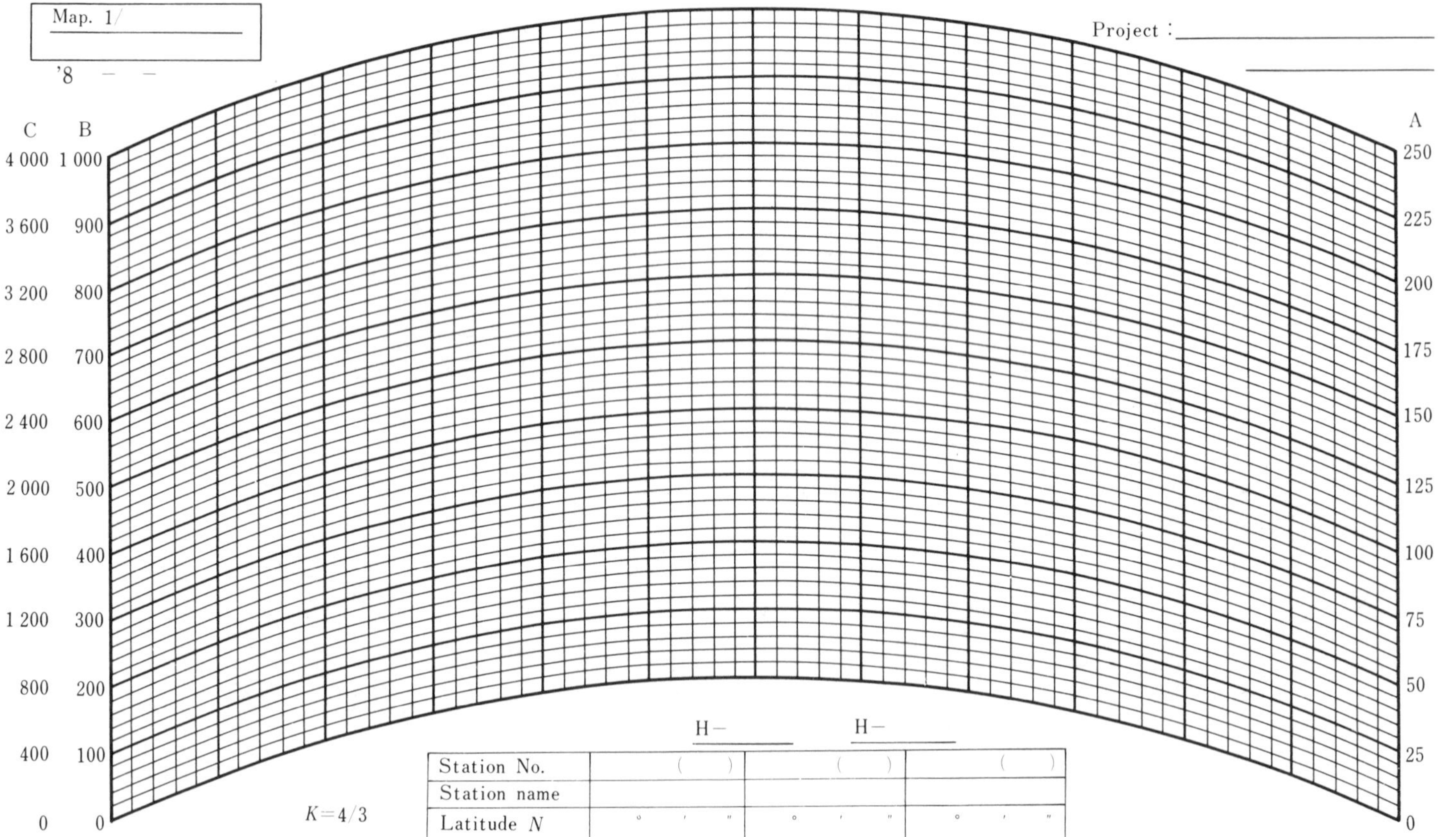

C	B	A
4 000	1 000	250
3 600	900	225
3 200	800	200
2 800	700	175
2 400	600	150
2 000	500	125
1 600	400	100
1 200	300	75
800	200	50
400	100	25
0	0	0

$K = 4/3$

	D_{max} (km)	h_{max} (m)
A	60	250
B	120	1 000
C	240	4 000

E_0 : isotropic/1 Watt

Γ_0, E_0 : in free space

	H–	H–	
Station No.	( )	( )	( )
Station name			
Latitude N / Longitude E	° ′ ″ / ° ′ ″	° ′ ″ / ° ′ ″	° ′ ″ / ° ′ ″
Elevation (m)			
Ant. height (m)			
Distance (km)	.	.	
Γ_0 (dB)	.	.	
E_0 (dBμV/m)	.	.	

freq. (MHz)

λ . (m)

0

18c

Path Profile (3)

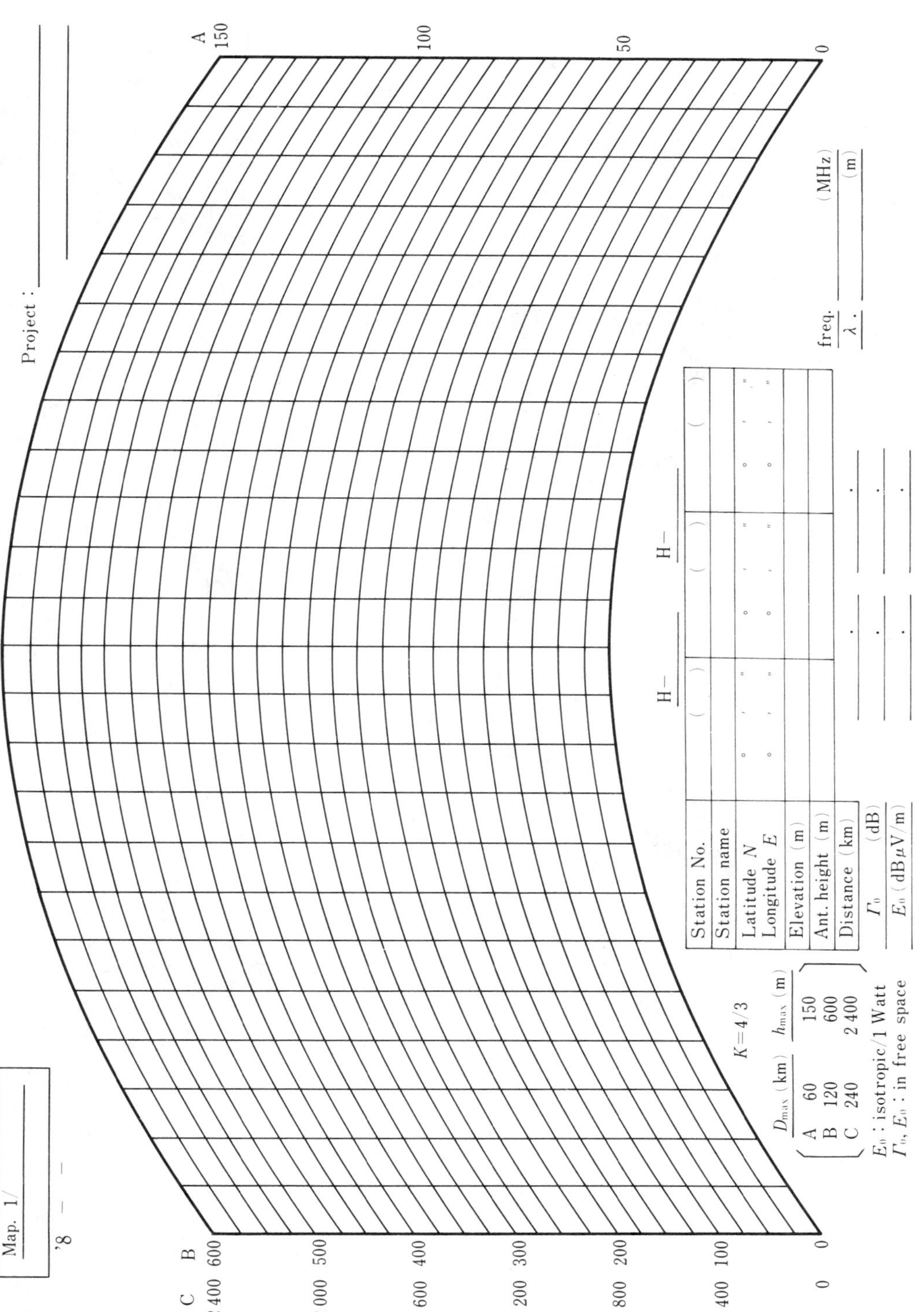

Map. 1/
'8 – –
Project :
A
150
100
50
0
C B
2 400 600
2 000 500
1 600 400
1 200 300
800 200
400 100
0 0
K=4/3
D_{max} (km) h_{max} (m)
A 60 150
B 120 600
C 240 2 400
E_0 : isotropic/1 Watt
Γ_0, E_0 : in free space
Station No.
Station name
Latitude N
Longitude E
Elevation (m)
Ant. height (m)
Distance (km)
Γ_0 (dB)
E_0 (dBμV/m)
H–
H–
freq. (MHz)
λ. (m)

Criteria for Approximation of Spherical Wave by Plane Wave

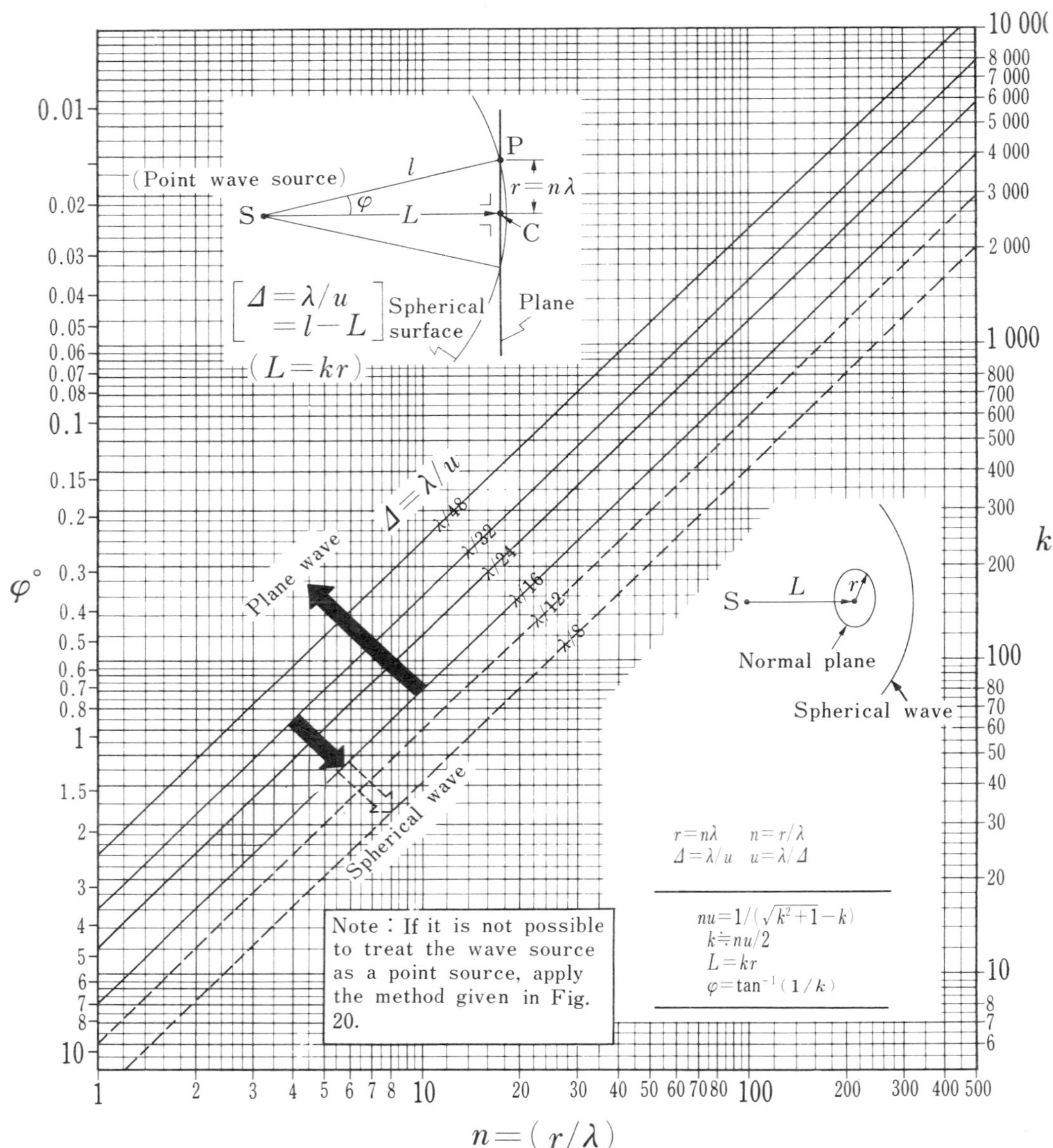

Nomograph to Define Plane Wave (Application)

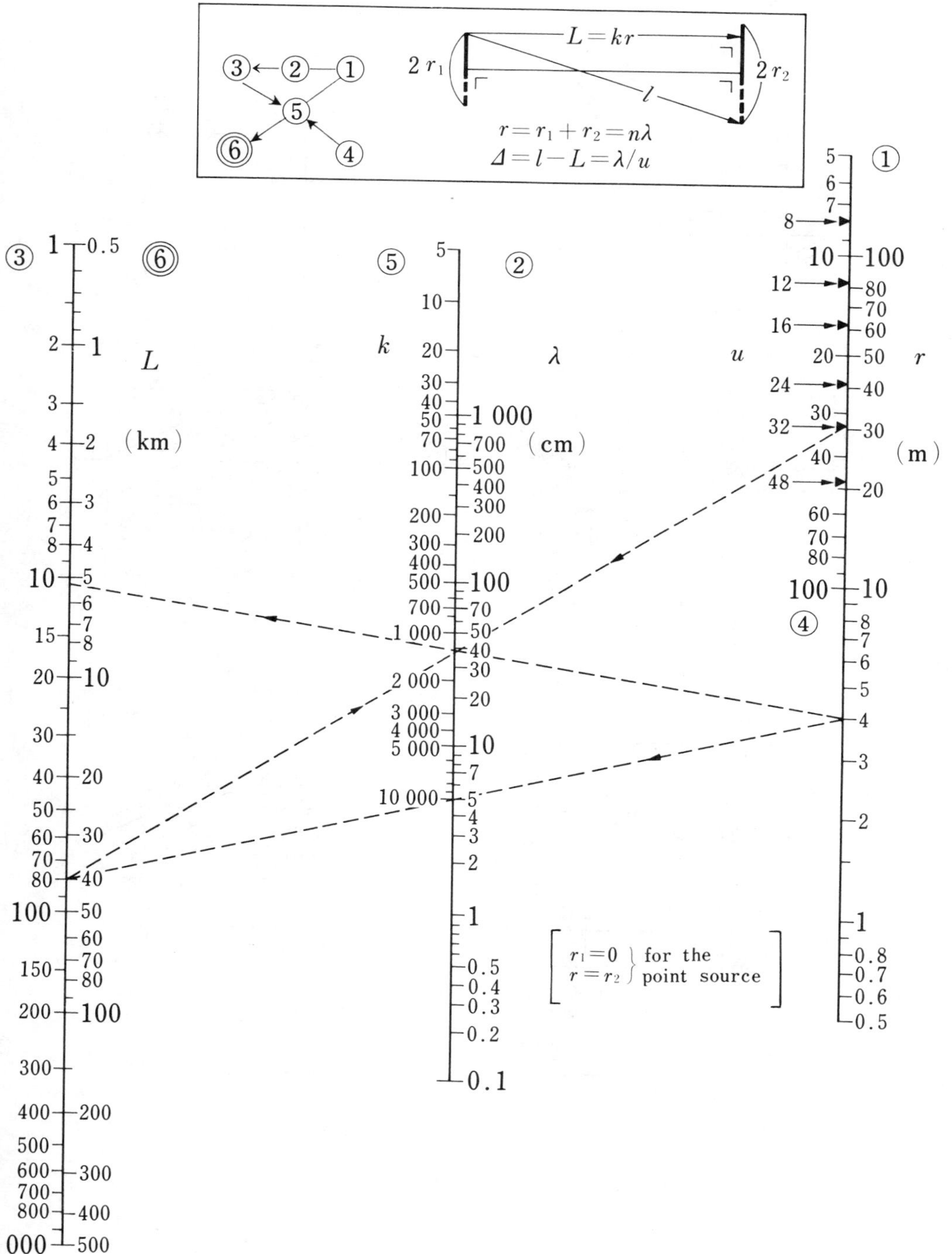

Graph for Conversion of Field Intensity (1)

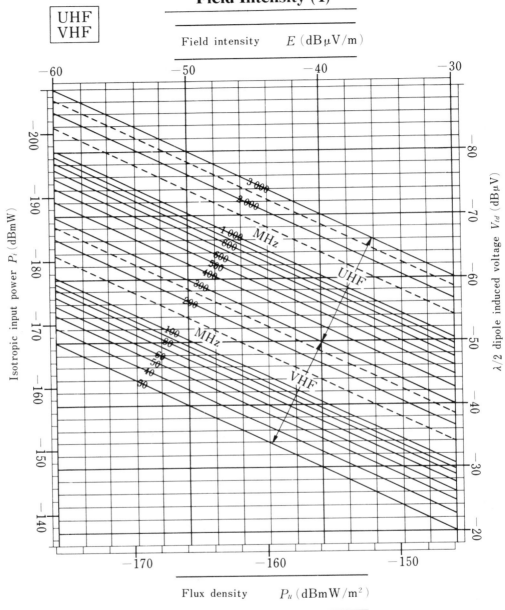

UHF
VHF

Field intensity E (dBμV/m)

Flux density P_u (dBmW/m²)

Graph for Conversion of Field Intensity (2)

EHF
SHF

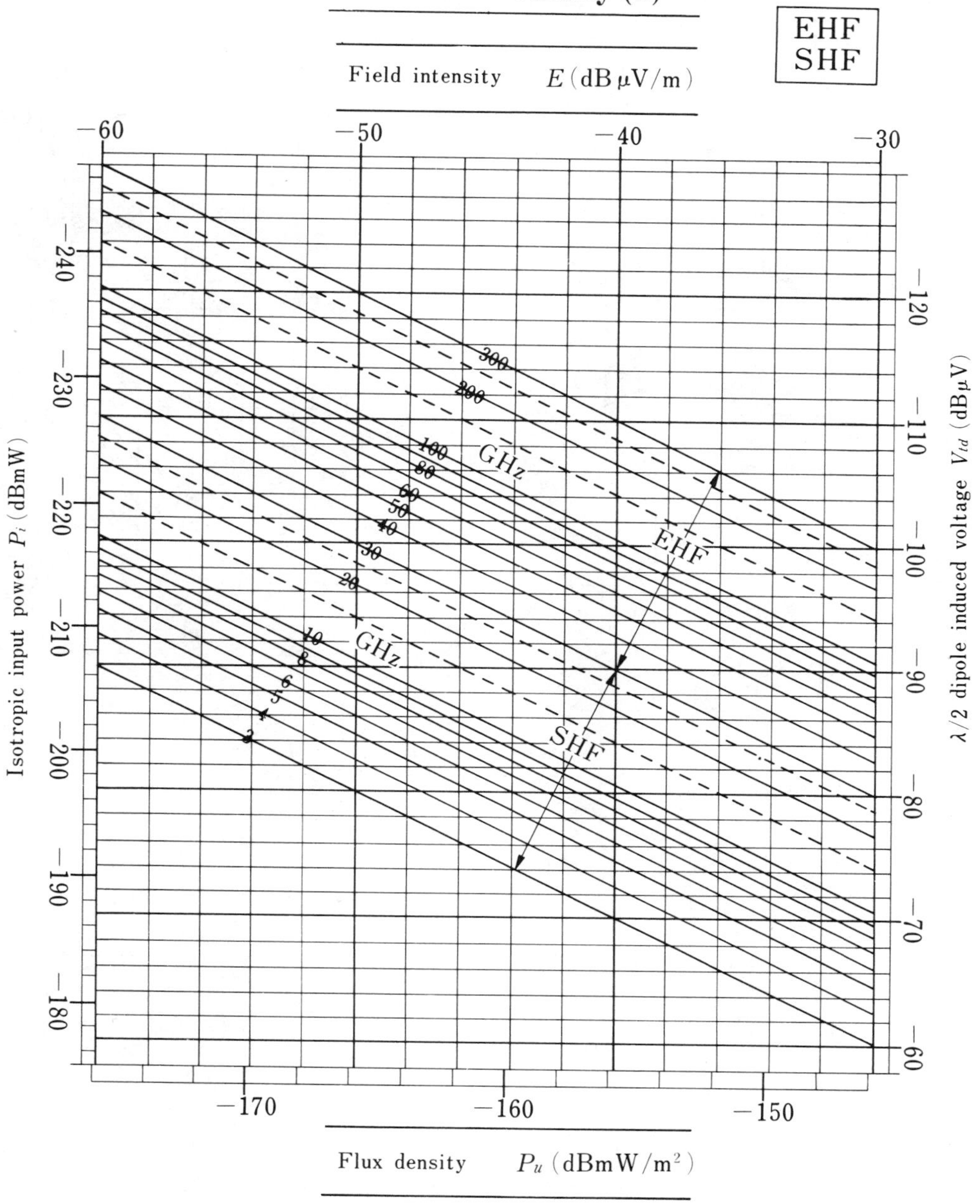

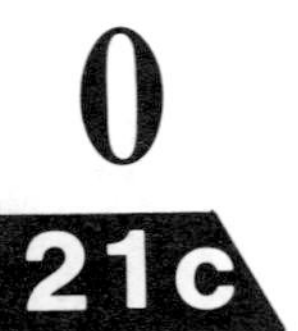

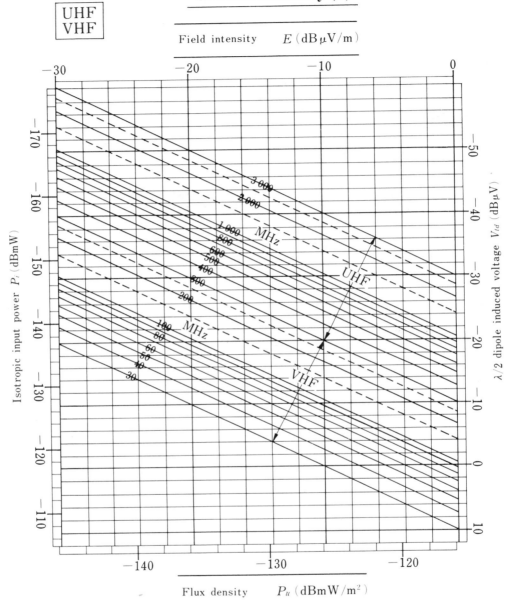

Graph for Conversion of Field Intensity (3)

UHF
VHF

Graph for Conversion of Field Intensity (4)

EHF
SHF

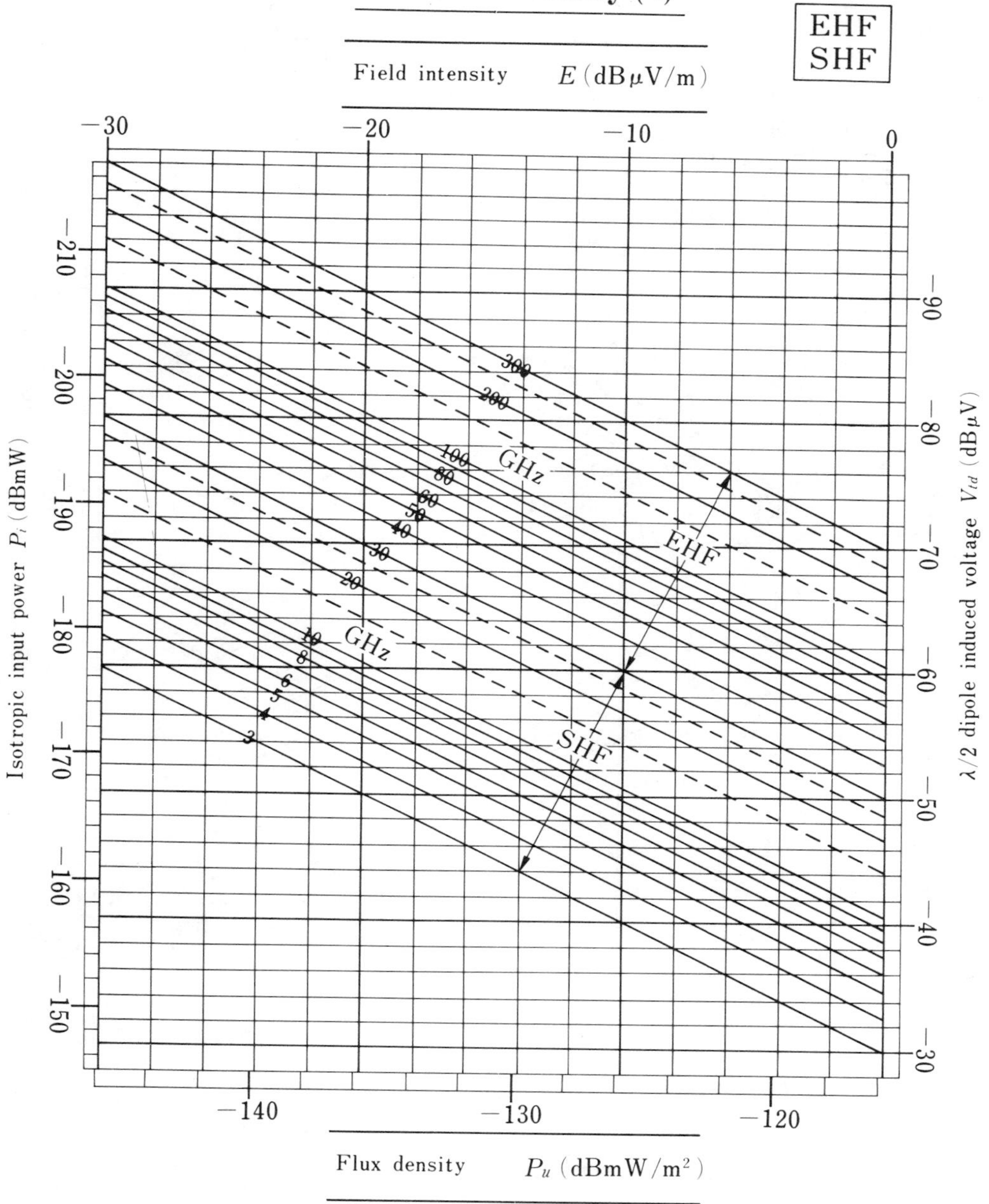

Graph for Conversion of Field Intensity (5)

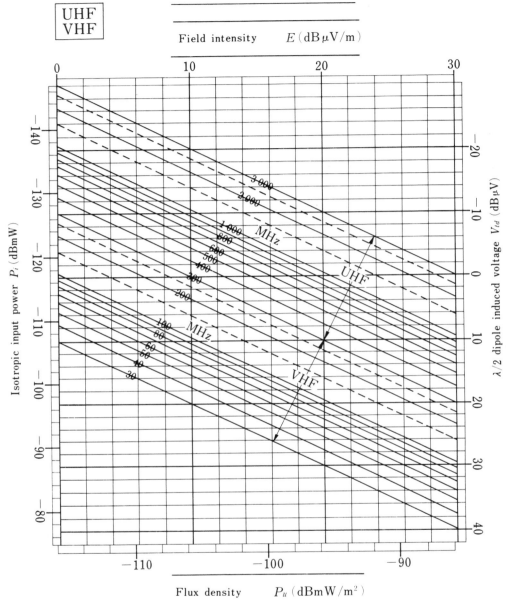

UHF
VHF

Field intensity E (dBμV/m)

0 10 20 30

Isotropic input power P_i (dBmW)

−140 −130 −120 −110 −100 −90 −80

$\lambda/2$ dipole induced voltage V_{id} (dBμV)

−20 −10 0 10 20 30 40

3 000
2 000
1 000
800
600
500
400
300
200
MHz
UHF

100
80
60
50
40
30
MHz
VHF

−110 −100 −90

Flux density P_u (dBmW/m²)

Graph for Conversion of Field Intensity (6)

EHF
SHF

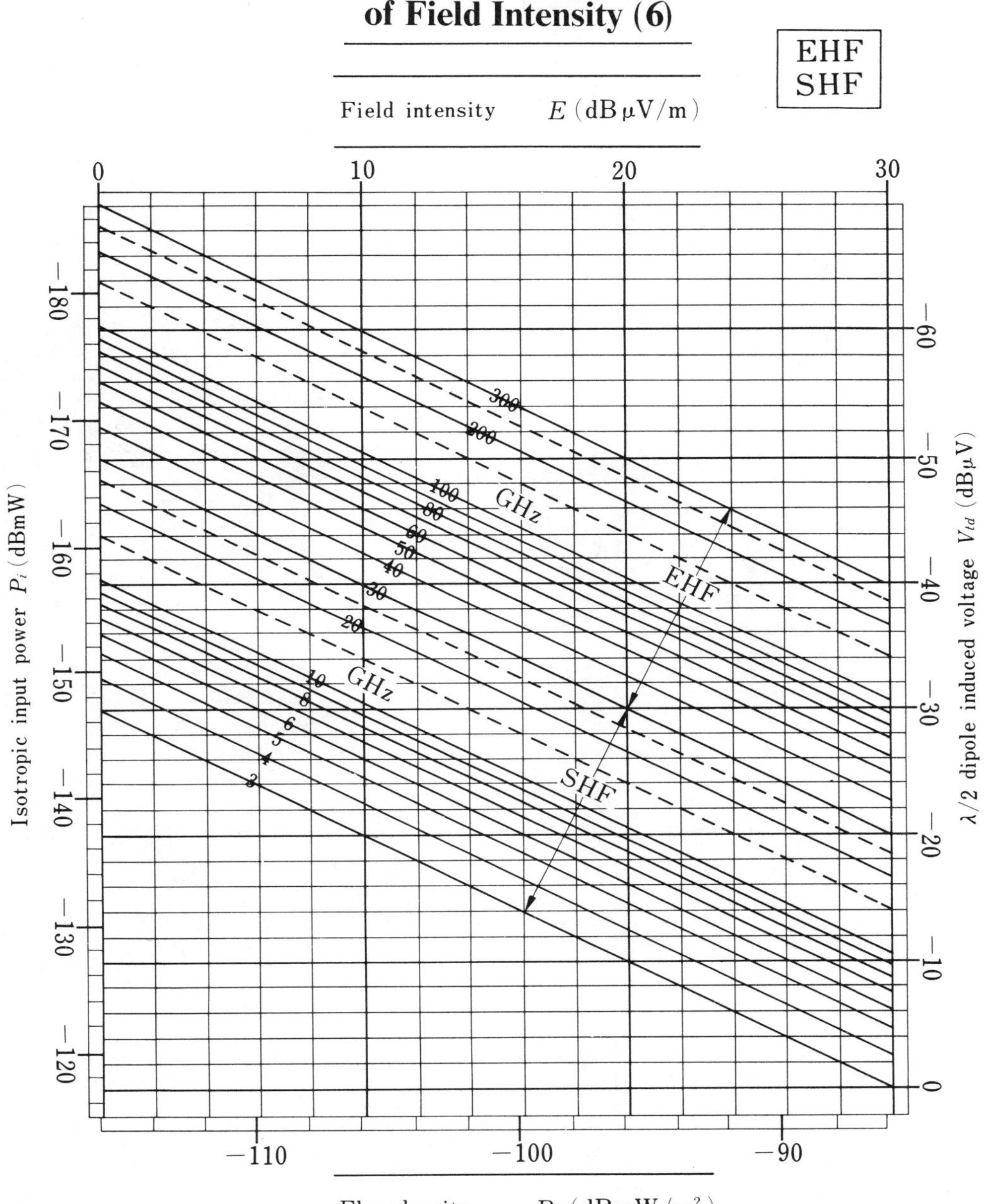

Graph for Conversion of Field Intensity (7)

UHF
VHF

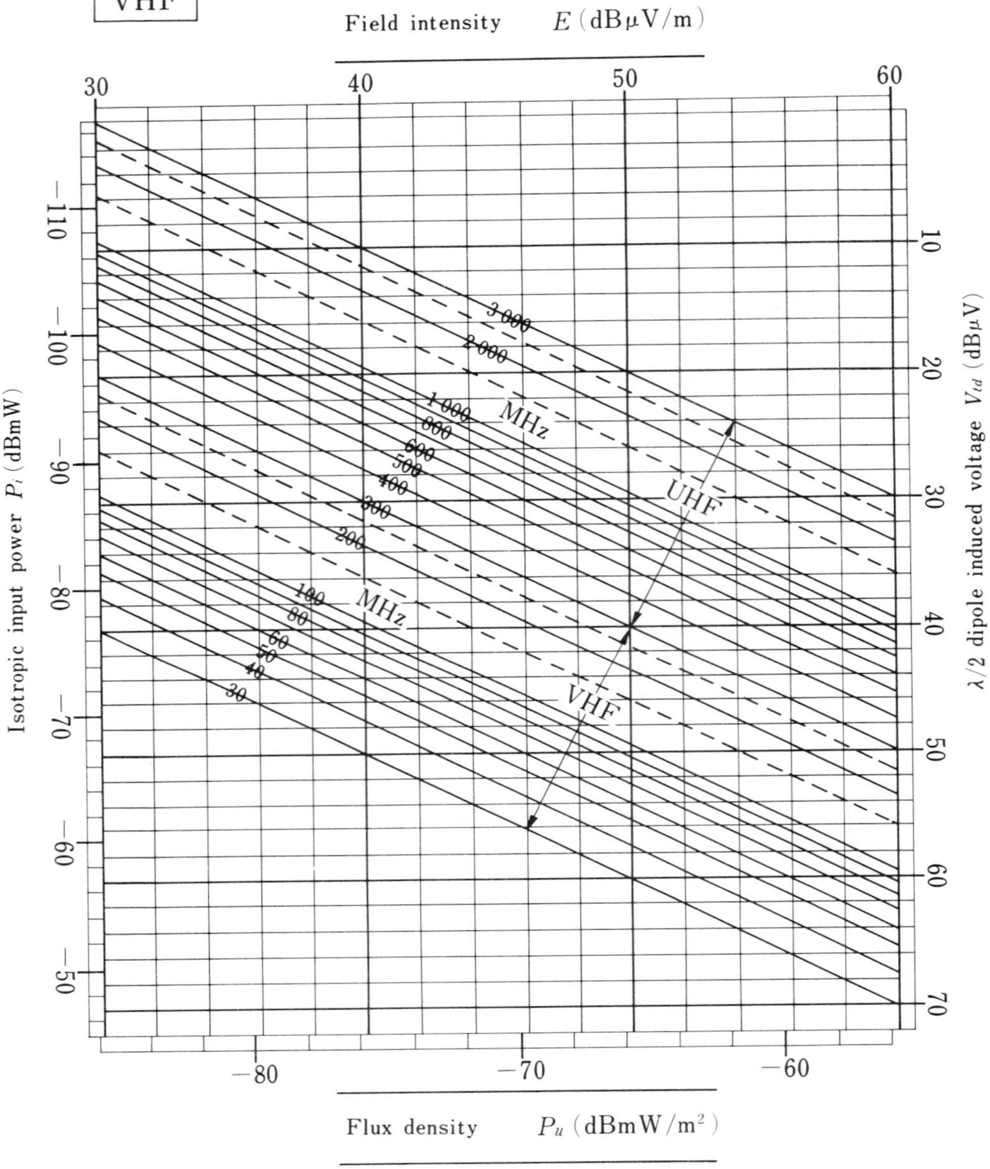

Field intensity E (dBμV/m)

30 40 50 60

Isotropic input power P_i (dBmW)

−110 −100 −90 −80 −70 −60 −50

$\lambda/2$ dipole induced voltage V_{ld} (dBμV)

10 20 30 40 50 60 70

3 000
2 000
1 000
800
600
500
400
300
200
MHz
UHF

100
80
60
50
40
30
MHz
VHF

−80 −70 −60

Flux density P_u (dBmW/m²)

Graph for Conversion of Field Intensity (8)

EHF
SHF

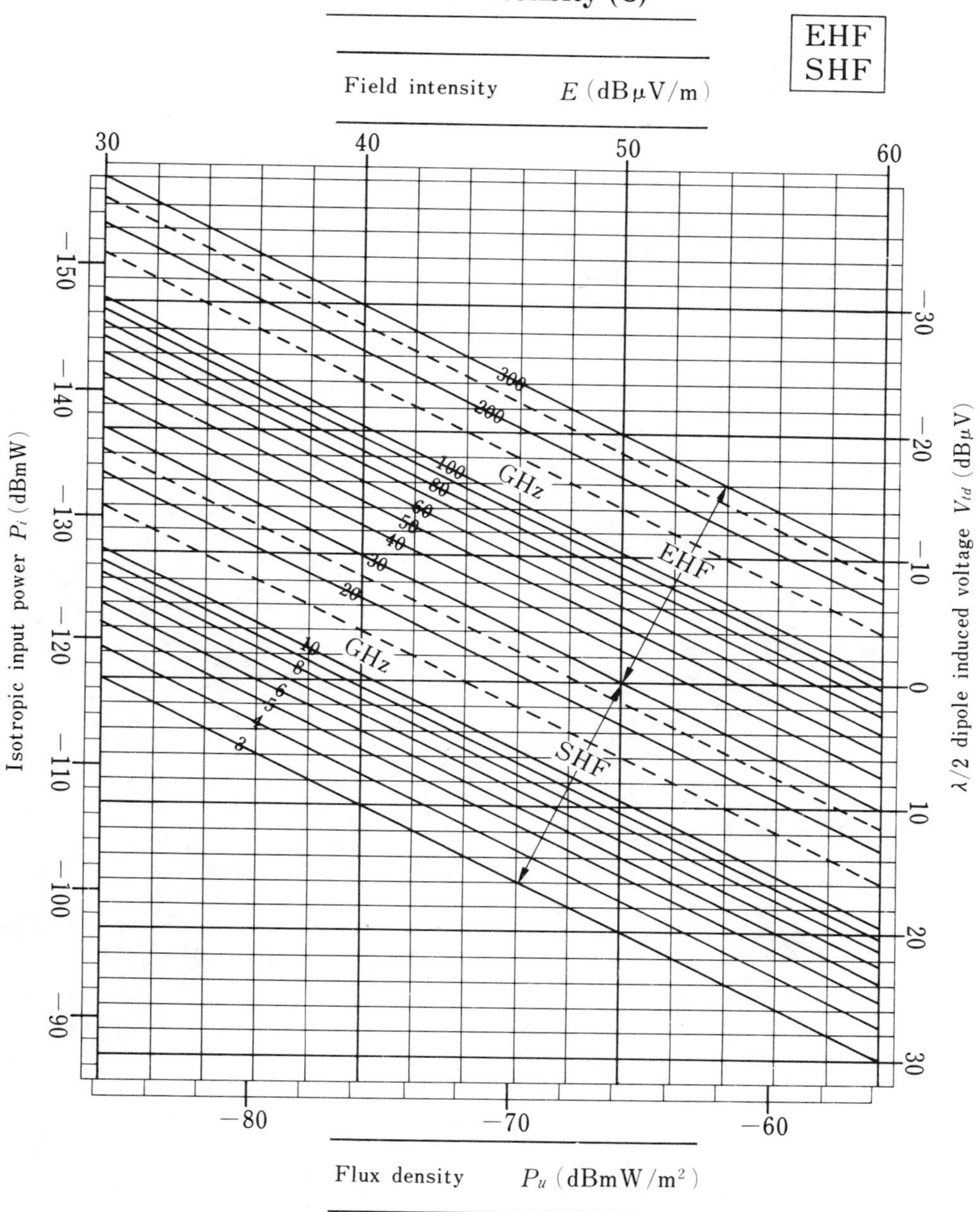

Graph for Conversion of Field Intensity (9)

UHF
VHF

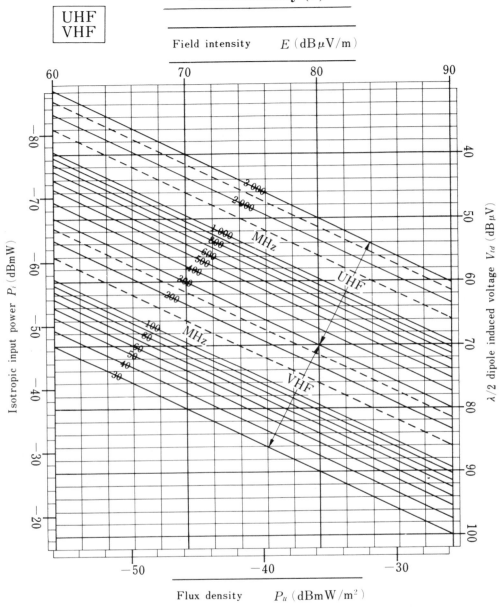

Field intensity E (dBμV/m)

60 70 80 90

Isotropic input power P_i (dBmW)

−80 −70 −60 −50 −40 −30 −20

λ/2 dipole induced voltage V_{ld} (dBμV)

40 50 60 70 80 90 100

3 000
2 000
1 000
800
600
500
400
300
200
MHz
UHF

100
80
60
50
40
30
MHz
VHF

−50 −40 −30

Flux density P_u (dBmW/m²)

Graph for Conversion of Field Intensity (10)

EHF
SHF

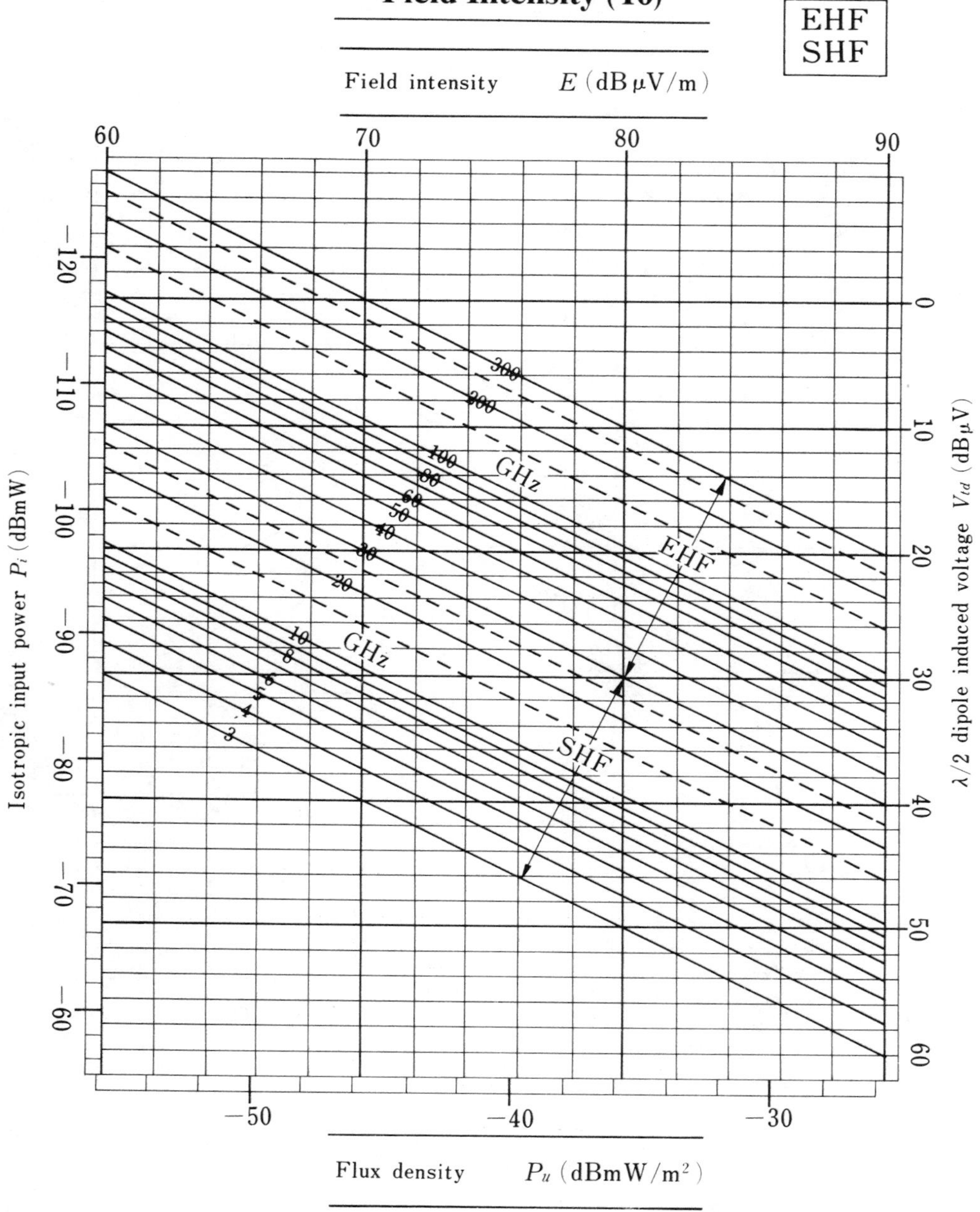

Graph for Conversion of Field Intensity (11)

UHF
VHF

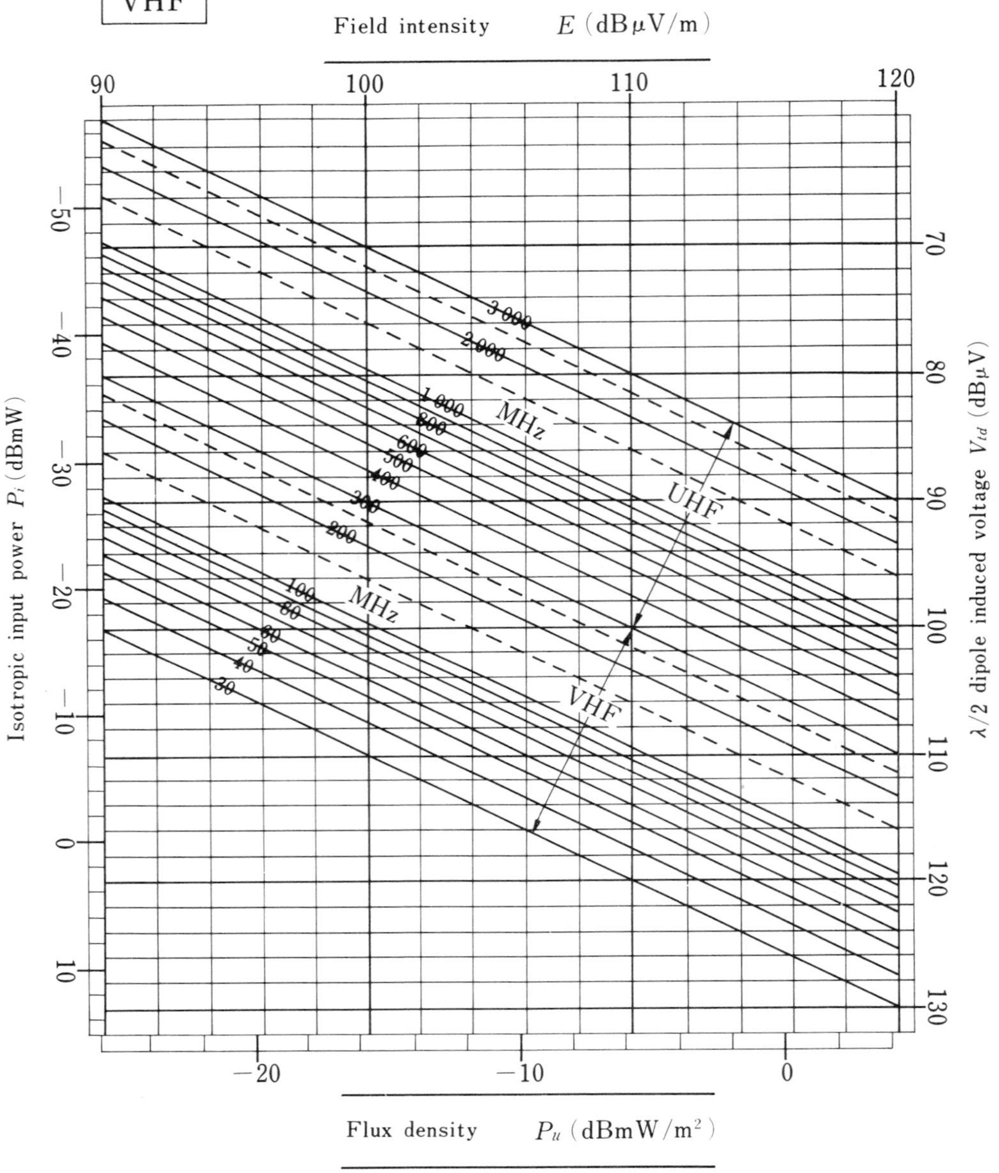

Graph for Conversion of Field Intensity (12)

EHF
SHF

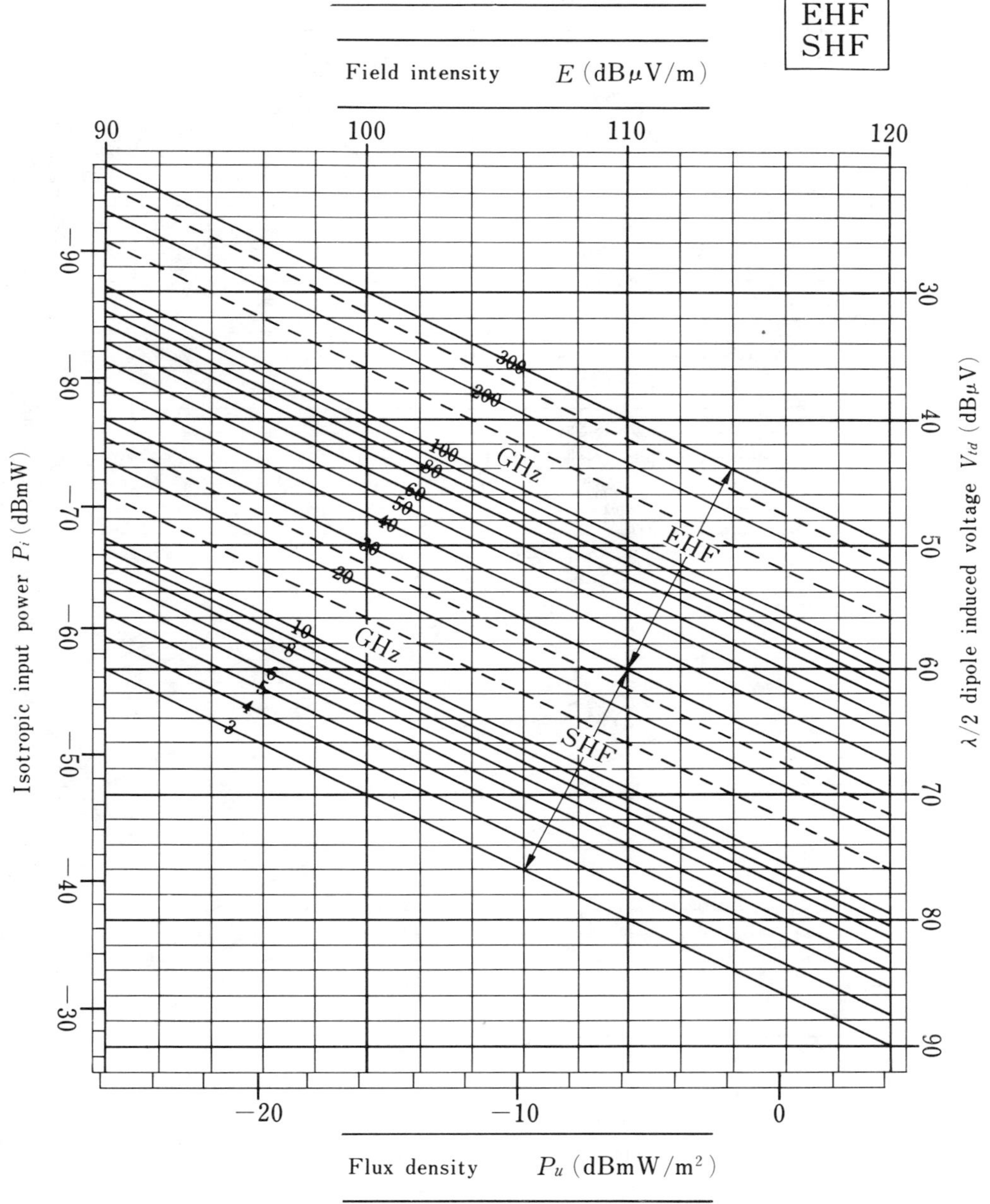

UHF
VHF

Graph for Conversion of Field Intensity (13)

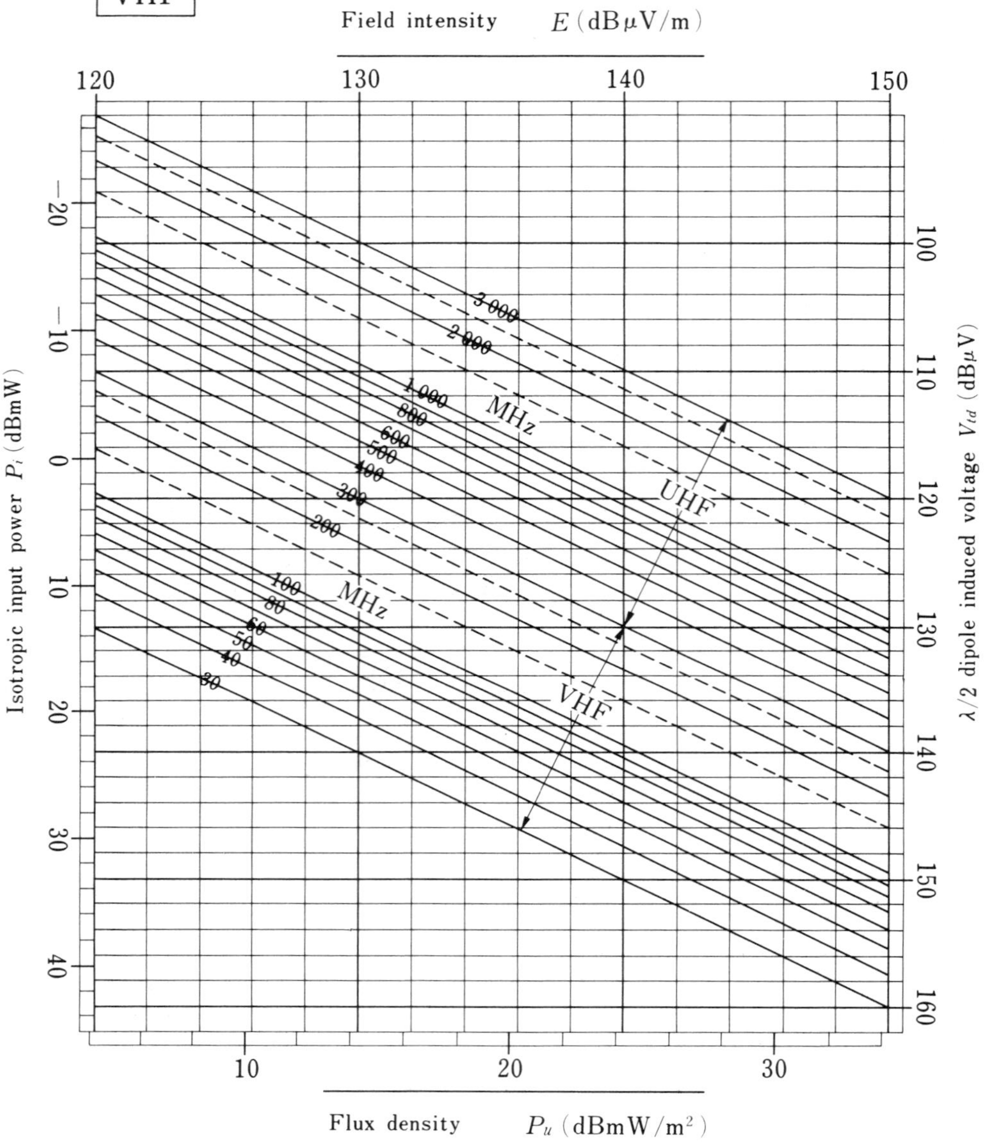

Field intensity E (dBμV/m)

120 130 140 150

Isotropic input power P_i (dBmW)

λ/2 dipole induced voltage V_{ld} (dBμV)

3 000
2 000
1 000
800
600
500
400
300
200
MHz
UHF
100
80
60
50
40
30
MHz
VHF

−20 −10 0 10 20 30 40

100 110 120 130 140 150 160

10 20 30

Flux density P_u (dBmW/m^2)

Graph for Conversion of Field Intensity (14)

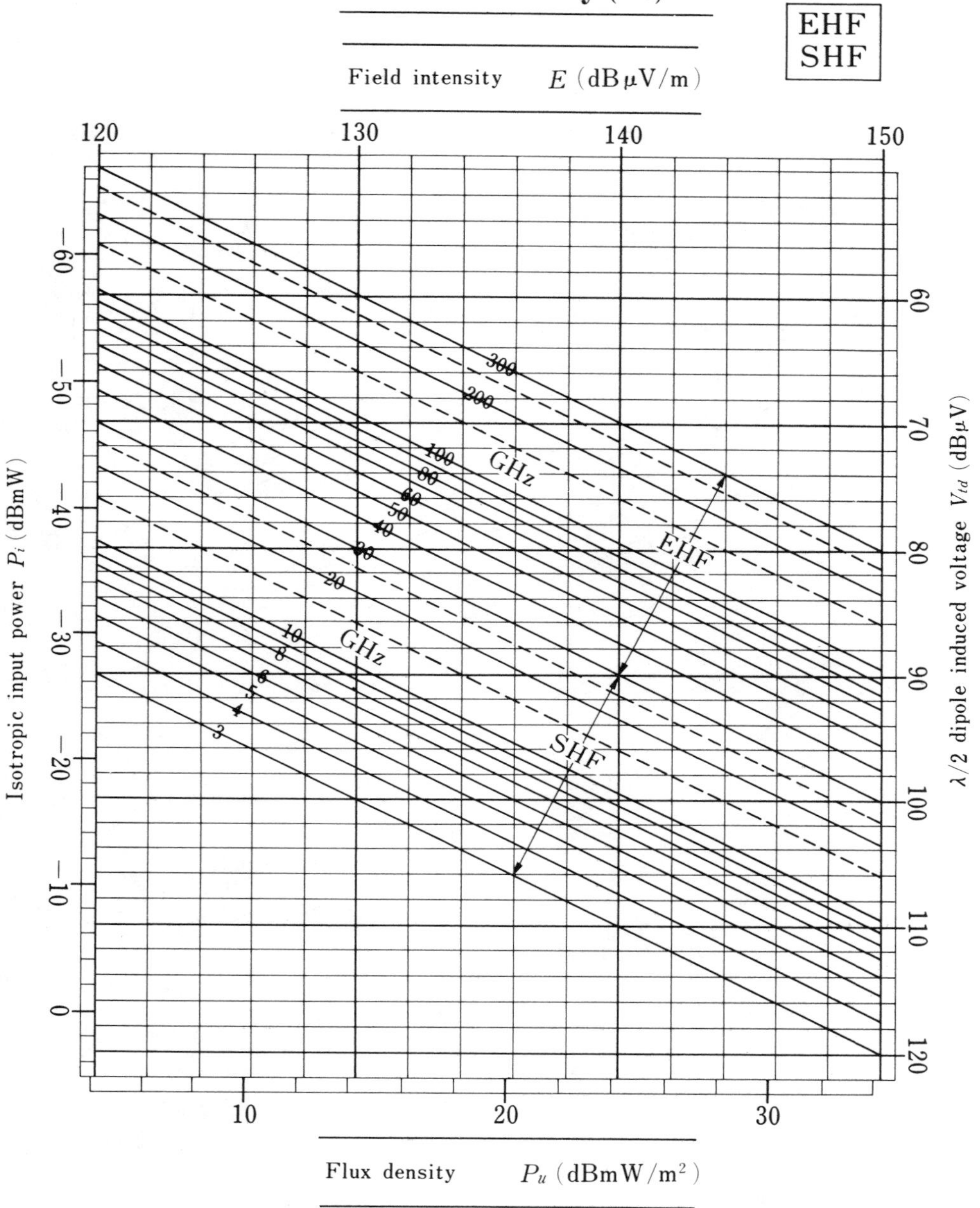

Graph for Conversion of Field Intensity (15)

UHF
VHF

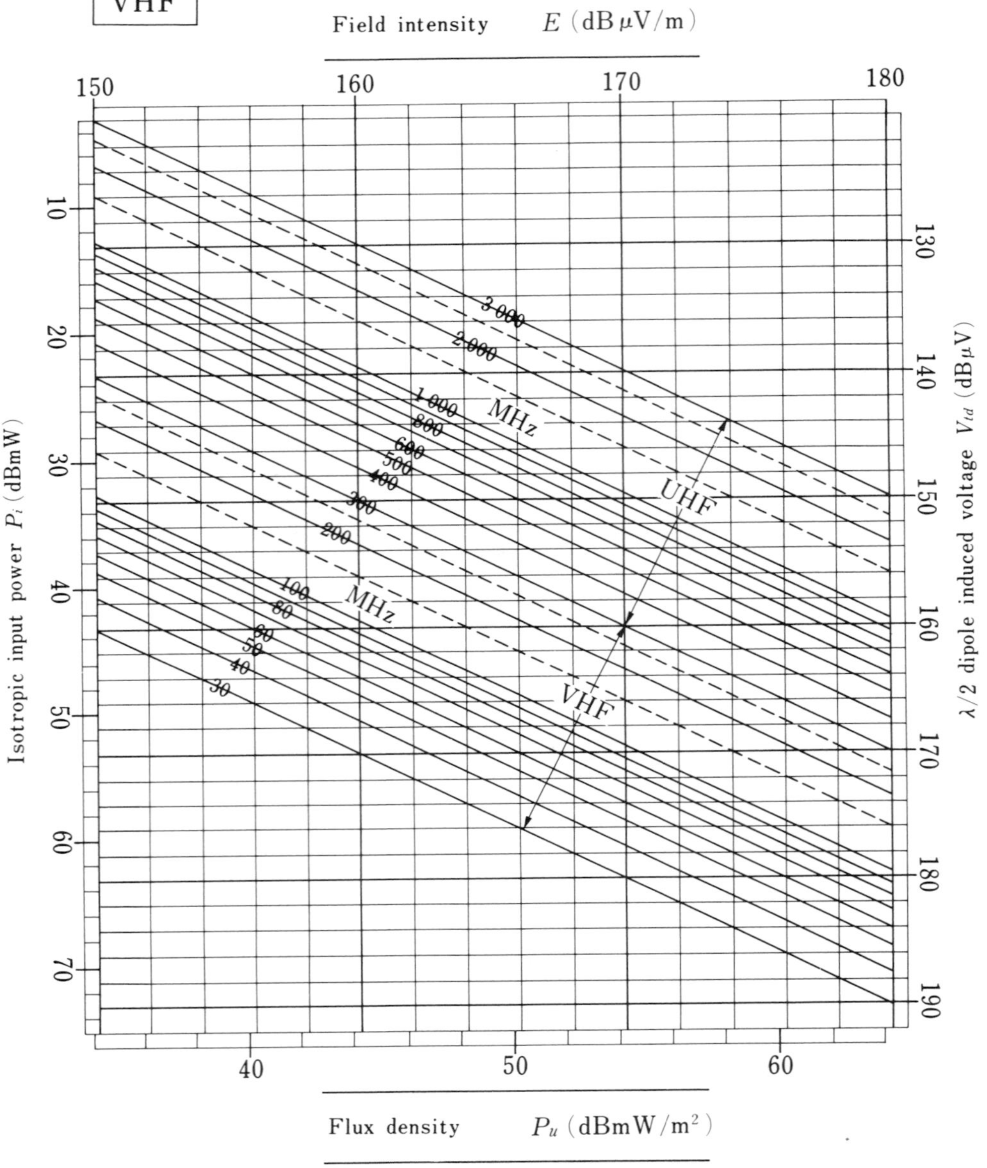

Graph for Conversion of Field Intensity (16)

EHF
SHF

Field intensity E (dBμV/m)

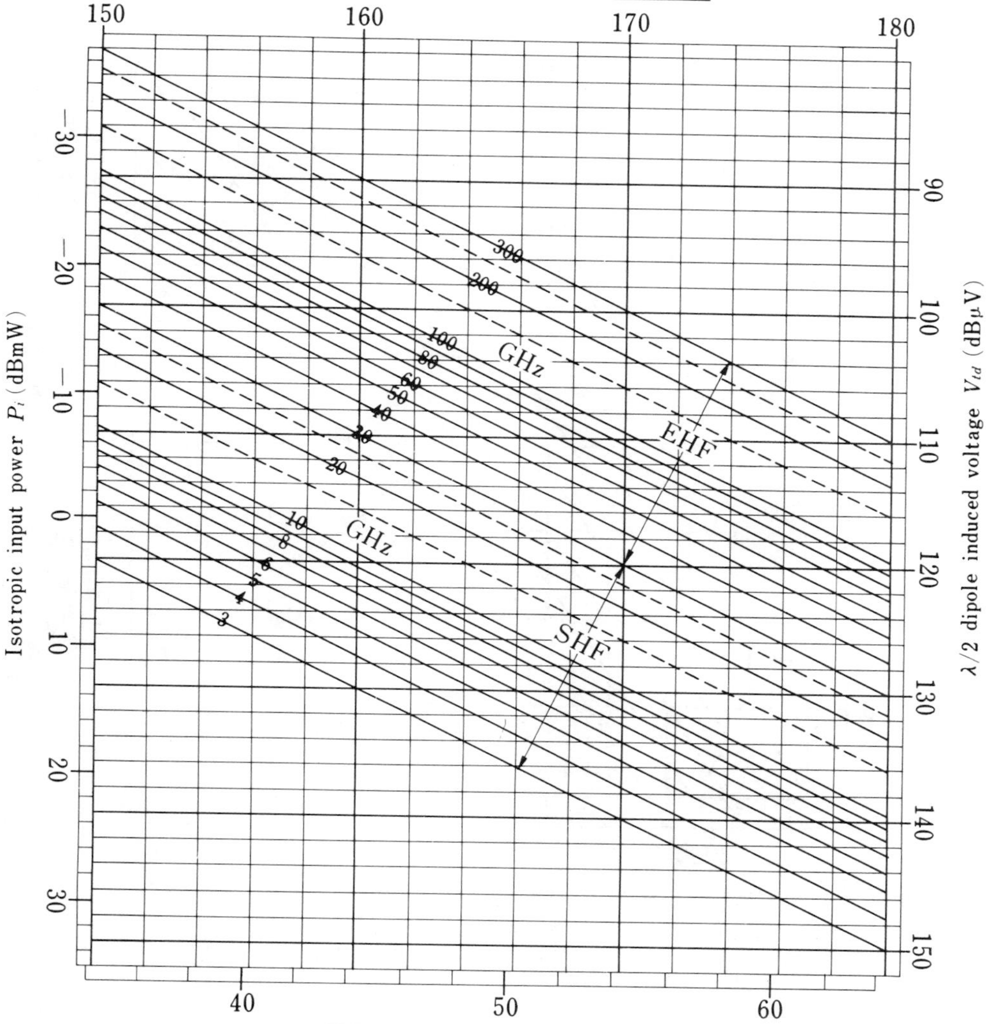

150 160 170 180

Isotropic input power P_i (dBmW)

−30 −20 −10 0 10 20 30

$\lambda/2$ dipole induced voltage V_{ld} (dBμV)

90 100 110 120 130 140 150

300 200 100 80 60 50 40 30 20 GHz

EHF

10 8 6 5 4 3 GHz

SHF

40 50 60

Flux density P_u (dBmW/m²)

Index for Field Intensity Conversion Graphs

Auxiliary Index

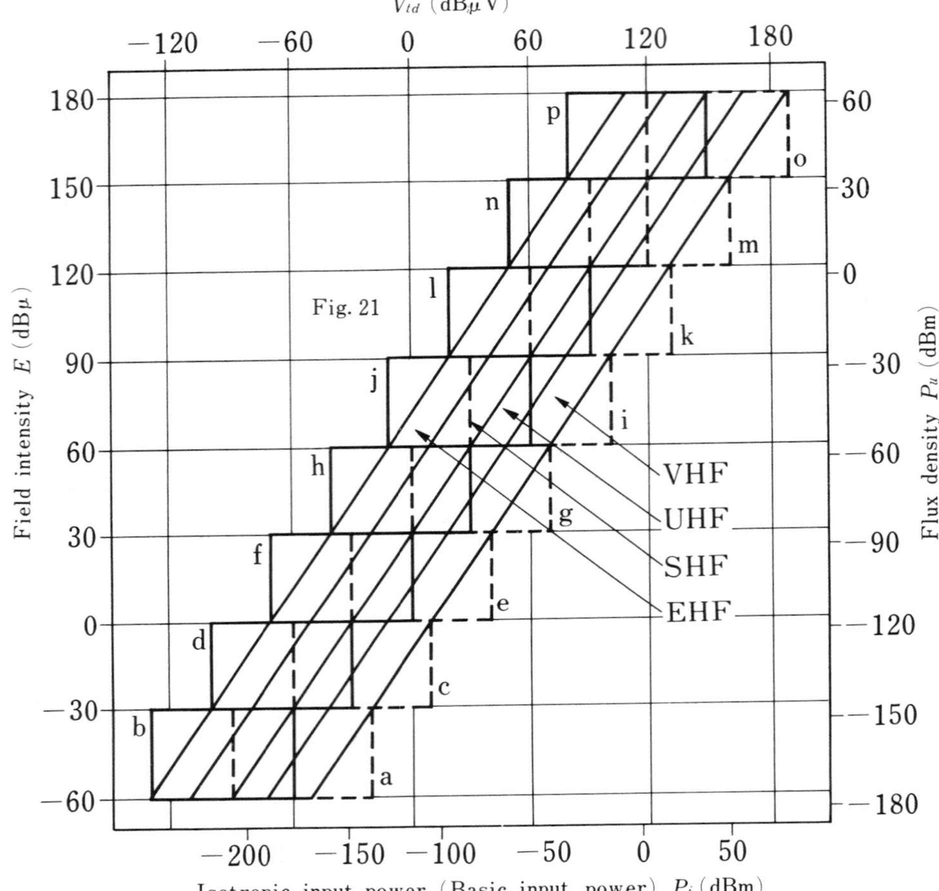

P_u (dBm) $= E$ (dBμ) -115.76
V_{td} (dBμ) $= E$ (dBμ) $+20 \log \lambda_{cm} - 49.94$
P_i (dBm) $= E$ (dBμ) $+20 \log \lambda_{cm} - 166.75$

1

FREE-SPACE PROPAGATION PATH

1.1 CONDITIONS FOR FREE-SPACE PROPAGATION PATH

A homogeneous and isotropic space that generates electromagnetic energy loss but not refraction, reflection, diffraction, absorption, or scattering is called free space.

Although an infinite vacuum is the ideal free space, in practical applications the space between artificial communication satellites can be considered free space.

Radio communications between two points on the ground are affected more or less by the atmosphere, meteorological phenomena, and topographical conditions. However, any propagation path that satisfies the following conditions may be considered as having propagation characteristics similar to those in free space.

1. Diffraction. If electromagnetic visibility is good, diffraction loss caused by intermediate terrain can be ignored.
2. Reflection. The effective reflection coefficient over the radio path is so small that interference caused by reflected waves is not harmful.
3. Refraction. When the refractive index gradient at any atmospheric altitude is approximately uniform over the entire radio path, the standard refraction ($K = 4/3$) may be assumed.
4. Absorption. Absorption and the resultant attenuation due to rainfall, fog, clouds, smog, oxygen, etc., may be ignored.

Criteria defining when actual conditions are significant may differ depending on the radio system performance or applications. However, except for specific cases, a practical determination of a free-space propagation path may be made according to Fig. 2.

Any discussion of the characteristics of other types of radio paths, such as reflected-wave interference propagation path, diffraction propagation path, or propagation path undergoing absorptive attenuation due to rain or snow, should include a comparison with the corresponding free-space propagation path. Accordingly, the first step in propagation path design is to determine the various parameters in free-space propagation, such as free-space propagation loss and free-space electric field strength.

1.2 BASIC PARAMETERS OF FREE SPACE

A. Free-Space Flux Density

An antenna that radiates power uniformly in all directions (i.e., is perfectly omnidirectional) is called an isotropic antenna. If power P is radiated from an isotropic antenna placed in free space, as shown in Fig. 1, radiated energy is distributed uniformly over the spherical surface with radius D and Poynting's power P_{u0} (which is the power flowing through a unit area on the sphere in unit time),

$$P_{u0} = P/4\pi D^2 \qquad \text{(free-space flux density, isotropic antenna radiation)} \tag{1.1}$$

where P_{u0} = free-space flux density
P = power radiated from isotropic antenna
D = distance
$4\pi D^2$ = total surface area of a sphere with radius D

B. Free-Space Basic Input Power

Assume that an isotropic antenna is placed at the point where a radio wave with flux density determined by Eq. (1.1) is received. Letting P_u denote the flux density and A_e the effective aperture area (or absorption area) of the receiving antenna, then the available power P_a is given by

$$P_a = P_u A_e \tag{1.2}$$

The absorption area of an isotropic antenna A_i as a function of wavelength λ is

$$A_i = \lambda^2/4\pi \tag{1.3}$$

Hence the available power P_a, hereafter called basic input power P_i (isotropic receiving input power), is

$$P_i = P_u A_i = P_u(\lambda^2/4\pi) \qquad \text{(basic input power, general case)} \tag{1.4}$$

The basic input power in the free space P_{i0} (isotropic antenna radiation) is

$$\boxed{P_{i0} = P_{u0}(\lambda^2/4\pi)}$$

Using Eq. (1.1), we obtain

$$\boxed{P_{i0} = P(\lambda/4\pi D)^2 \qquad \text{(free-space basic input power between isotropic antennas)}} \tag{1.5}$$

C. Free-Space Propagation Loss

The transmission loss between isotropic antennas placed in free space is called free-space propagation loss or basic transmission loss, and is denoted by Γ_0. Transmission loss, a ratio of the transmitting power to the receiving power, is defined here more specifically as the ratio of the transmitting power at the transmitting isotropic antenna to the available power at the receiving isotropic antenna. Then using Eq. (1.5),

$$\boxed{\Gamma_0 = P/P_{i0} = (4\pi D/\lambda)^2 \qquad \text{(free-space propagation loss)}} \tag{1.6}$$

Γ_0 is among the basic parameters frequently used in the theory and design of radio systems. Figures 7 to 10 are used to estimate Γ_0.

D. Free-Space Electric Field Strength

In electric circuits having a defined relationship between voltage, resistance, and power, there is also a relationship between the electric field strength E (hereafter called field strength), the characteristic impedance, and the flux density P_u,

$$P_u = E^2/120\pi \tag{1.7}$$

where 120π corresponds to the resistance (equivalent to R in $P = V^2/R$) and represents impedance in free space.

If the free-space values of P_u and E are expressed by P_{u0} and E_0, respectively, then

$$P_{u0} = E_0^2/120\pi \tag{1.8}$$

$$E_0 = \sqrt{120\pi P_{u0}} \tag{1.9}$$

Since $P_{u0} = P/4\pi D^2$ in Eq. (1.1), then

$$E_0 = \sqrt{120\pi P/4\pi D^2}$$

Hence

$$E_0 = \sqrt{30}\sqrt{P}/D \qquad \text{(free-space field strength, isotropic antenna radiation)} \tag{1.10}$$

Given a half-wave dipole power gain of 1.64 (2.15 dB) for the isotropic antenna, the free-space field strength obtained by the half-wave dipole radiation E_0' is given by

$$E_0' = \sqrt{1.64}\,E_0 = \sqrt{1.64}\sqrt{30}\sqrt{P}/D$$

Thus

$$E_0' = 7\sqrt{P}/D \qquad \text{(free-space field strength, half-wave dipole radiation)} \tag{1.11}$$

1.3 FREE-SPACE TRANSMISSION BY ACTUAL ANTENNAS

A. Antenna Gain versus Transmission Loss

When transmitting and receiving antennas have isotropic antenna gains of G_1 and G_2, respectively, flux density, transmission loss, and field strength are estimated (in antilogarithms rather than dB values),

$$\begin{array}{ll} \text{flux density } P_u & G_1 \text{ times} \\ \text{transmission loss } L & 1/G_1G_2 \text{ times} \\ \text{field strength } E & \sqrt{G_1} \text{ times} \end{array} \tag{1.12}$$

The propagation loss is denoted by Γ, and the transmission loss L is found by

$$L = \Gamma/G_1G_2 \qquad \text{(transmission loss)} \tag{1.13}$$

The free-space transmission loss L_0 is

$$L_0 = \Gamma_0/G_1G_2 \qquad \text{(free-space transmission loss)} \tag{1.14}$$

B. Effective Antenna Aperture Area versus Transmission Loss

The gain G (relative to that for the isotropic antenna) of an antenna with an effective aperture area A_e is expressed by

$$G = A_e/A_i = 4\pi A_e/\lambda^2 \tag{1.15}$$

where A_i is the absorption area of an isotropic antenna,

$$A_i = \lambda^2/4\pi.$$

When the transmitting and the receiving antennas have effective aperture areas A_{e1} and A_{e2}, respectively, the transmission loss L is given by

$$L = \Gamma/G_1G_2 \tag{1.16}$$

and

$$L = \Gamma/(4\pi A_{e1}/\lambda^2)4\pi A_{e2}/\lambda^2$$
$$= \Gamma(\lambda^2/4\pi)^2/A_{e1}A_{e2} \tag{1.17}$$

For free-space propagation,

$$\Gamma_0 = (4\pi D/\lambda)^2$$

Therefore

$$L_0 = (4\pi D/\lambda)^2(\lambda^2/4\pi)^2/A_{e1}A_{e2}$$

This simplifies to

$$L_0 = (\lambda D)^2/A_{e1}A_{e2} \quad \text{(free-space transmission loss between actual antennas)} \tag{1.18}$$

C. Transmission Loss between Unit-Area Antennas

If $A_{e1} = 1$ and $A_{e2} = 1$ in Eq. (1.18),

$$L_{uf} = (\lambda D)^2 \quad \text{(free-space transmission loss between unit-area antennas)} \tag{1.19}$$

where L_{uf} = transmission loss between unit-area antennas installed in free space
λ = wavelength
D = distance

Figures 11 and 12, which explain Eq. (1.19), allow quick estimation of the transmission loss expected for an arbitrary effective aperture area.

When the size (the effective aperture area) of an antenna to be mounted on a tower or building roof is limited, the transmission loss will be inversely proportional to the square of the frequency. This is shown by Eqs. (1.18) and (1.19).

If we consider two systems, one using 200 MHz in the VHF band, the other 8 GHz in the SHF band, then the transmission loss ratio of the two systems is

$$(200/8000)^2 = 1/1600$$

This reduction creates an improvement of 32 dB, allowing a reduction in transmission power without jeopardizing the necessary buffer, and permitting broad-band high-quality transmission of television or multichannel communications.

1.4 VOLTAGE RECEIVED BY A LINEAR ANTENNA (INDUCED VOLTAGE AND OPEN-CIRCUIT VOLTAGE AT THE FEEDING POINT)

To estimate the received voltage at a linear antenna used for a VHF or UHF band, the field strength E is used. The received voltage at a half-wave dipole V_{td} is

$$V_{td} = El_e \qquad \text{(received voltage at a half-wave dipole)} \tag{1.20}$$

In Eq. (1.20) l_e, the effective length of a half-wave dipole, is a function of the wavelength,

$$l_e = \lambda/\pi \qquad \text{(effective length of a half-wave dipole)} \tag{1.21}$$

Denoting by G_d the receiving antenna's relative gain (to a half-wave dipole), the received voltage V_t is $V_t = V_{td}\sqrt{G_d}$, and

$$V_t = El_e\sqrt{G_d} \qquad \text{(received voltage at an actual antenna)} \tag{1.22}$$

Figure 6 is a graph of l_e.

1.5 APERTURE AREA AND GAIN OF AN ANTENNA

The performance of an antenna is generally determined by how much radio-wave energy is absorbed from the space in which the antenna is installed. For example, the reception of available power P_a (matched receiving power) through a receiving antenna installed at a flux density P_u per square meter is equal to the antenna absorption of the total radio-wave energy passing through the area,

$$A_e = P_a/P_u \quad (\text{m}^2) \tag{1.23}$$

where A_e is the effective aperture area.

The concept of effective aperture area is easily understood for parabola, horn, and other types of antennas having a physical aperture area A; the ratio A_e to A is called the aperture efficiency η,

$$\eta = A_e/A \leq 1 \qquad \text{(aperture efficiency)} \tag{1.24}$$

The effective aperture area, an indicator of high-quality design and construction and fabrication accuracy, is important not only for isotropic antennas, but also for half-wave dipole, yagi, and other antennas.

As described previously, the effective aperture area of an isotropic antenna A_i is $\lambda^2/4\pi$. The antenna gain G is the ratio of this antenna's available power to the isotropic antenna's available power and is equivalent to the effective aperture area ratio, that is,

$$\boxed{G = A_e/A_i = 4\pi A_e/\lambda^2} \tag{1.25}$$

as illustrated in Fig. 13. Figure 14 shows how to estimate the effective aperture area A_e when the diamenter D_ϕ of the parabola and the aperture efficiency are known.

REMARK. The following formulas provide an estimate of parabola gain;

$$G = 4\pi A_e/\lambda^2 = 4\pi\left(\eta\pi D_\phi^2/4\right)/\lambda^2 = \eta\left(\pi D_\phi/\lambda\right)^2$$

Hence

$$G(\text{dB}) = 20\log(\pi D_\phi/\lambda) + 10\log\eta$$

or

$$G(\text{dB}) \doteqdot 20\log D_\phi(m) - 20\log\lambda(\text{cm}) + 10\log\eta + 50$$

1.6 CORRECTION TO FIELD STRENGTH AND FLUX DENSITY BY RADIATED POWER

Since charts for the free-space field strength and the flux density have been drawn relative to 1 watt of radiated power from the isotropic antenna, a correction is necessary for a directional antenna whose gain is not unity (0 dB). If the radiated power is P (watts) and the transmitting antenna gain in the direction of the receiving point is G (relative to the isotropic antenna), then the free-space values E_0' and P_{u0}' are

$$\boxed{\begin{aligned} E_0' &= E_0\sqrt{PG} \\ P_{u0}' &= P_{u0}PG \end{aligned} \quad \text{(free-space values, general case)}} \tag{1.26}$$

The correction factor P_e (dB) is given as

$$\boxed{P_e = 10\log PG \quad (\text{dB}) \qquad (\text{correction})} \tag{1.27}$$

or

$$P_e = 10\log P + 10\log G \quad (\text{dB}) \tag{1.28}$$

where P_e is effective isotropically radiated power (e.i.r.p.).

REMARK. Traditionally the effective radiated power (e.r.p.) has frequently been defined relative to the half-wave dipole. Therefore effective *isotropically* radiated power is distinguished by inserting an i, that is, e.i.r.p.

Figure 4 is used to determine the dB correction factor P_e (dB) from an antilogarithm. If the antenna gain G expressed in decibels is known, P_e is obtained by adding G to the radiated power expressed as a decibel value.

NOTE. When a radiated power in decibels is known, the e.i.r.p. in decibels for a half-wave dipole is obtained by adding 2.15 dB (1.64 in power ratio or 1.28 in voltage ratio).

Figure 5 is used to convert the field strength between antilogarithm and decibel values.

1.7 SUMMARY OF BASIC FORMULAS FOR RADIO-WAVE FIELD STRENGTH AND TRANSMISSION LOSS

	Isotropic Antenna	Actual Antenna	dB
Free-space propagation path			
Field strength	$E_0 = \sqrt{30}\sqrt{P}/D$	$E_0\sqrt{G_1}$	20 log
Flux density	$P_{u0} = P/4\pi D^2$	$P_{u0}G_1$	10 log
Basic input power	$P_{i0} = P\lambda^2/(4\pi D)^2$	$P_{i0}G_1$	10 log
Propagation loss	$\Gamma_0 = (4\pi D/\lambda)^2$	—	10 log
Transmission loss	$L_0 = (4\pi D/\lambda)^2$	Γ_0/G_1G_2, $L_{uf}/A_{e1}A_{e2}$	10 log
Ordinary propagation path			
Field strength	$E_0/\sqrt{\Gamma_p}$	$E_0\sqrt{G_1/\Gamma_p'}$	20 log
Flux density	P_{u0}/Γ_p	$P_{u0}G_1/\Gamma_p'$	10 log
Basic input power	P_{t0}/Γ_p	$P_{t0}G_1/\Gamma_p'$	10 log
Propagation loss	$\Gamma_0\Gamma_p$	—	
Transmission loss	$\Gamma_0\Gamma_p$	$\Gamma_0\Gamma_p''/G_1G_2$, $L_{uf}\Gamma_p''/A_{e1}A_{e2}$	10 log

where λ = wavelength
D = distance
P = radiation power
G_1, G_2 = transmitting or receiving antenna gain
A_{e1}, A_{e2} = effective aperture area of transmitting or receiving antenna

Propagation loss is the basic transmission loss, that is, the transmission loss between isotropic antennas, $\Gamma = \Gamma_0\Gamma_p$. Γ_p is an additional propagation loss (loss in excess of the free-space value); Γ_p' includes the effect of the directivity of the transmitting antenna, and Γ_p'' includes the effects of the directivities of both the transmitting and the receiving antennas. These are described in later chapters.

1.8 NOMOGRAPH FOR SIMPLIFIED CALCULATIONS

If an error on the order of 1 dB is allowed, the use of nomographs may improve calculation efficiency:

Figure 15*a*	Nomograph for free-space field strength (substitute for Figs. 3 and 4)
Figure 15*b*	Nomograph for free-space propagation loss (substitute for Fig. 7)
Figure 15*c*	Nomograph for transmission loss between unit aperture area antennas (substitute for Fig. 12)
Figure 15*d*	Nomograph for conversion between field strength or flux density and induced voltage or received power

EXAMPLES

Example 1.1

A television broadcast by a geostationary satellite over the Pacific Ocean (36,000 km above the equator) is made at 12 GHz and 0.2 kW. Calculate the expected flux density in the Tokyo area, assuming that the distance between the satellite and Tokyo is 38,000 km, the gain of the antenna directed toward Tokyo is 32 dB, and a free-space propagation path is available.

Solution

1. Calculate the free-space flux density.
2. Chart index → auxiliary index → Fig. 3*c*. Free-space flux density P_{u0} at $D = 38{,}000$ km,

$$P_{u0} = -162.6 \text{ dBW/m}^2$$

3. Correction factor P_e to e.i.r.p. (Fig. 4),

$$\begin{aligned} P = 0.2 \text{ kW} &\rightarrow 23 \text{ dBW} \\ G_t &= \underline{+32 \text{ dB}} \\ P_e &= 55 \text{ dBW} \end{aligned}$$

4. Flux density P_u at receiving point,

$$\begin{aligned} P_u &= P_{u0} + P_e \\ &= -162.6 + 55 = -107.6 \text{ dBW/m}^2 \\ &= -77.6 \text{ dBmW/m}^2 \end{aligned}$$

Note. 36,000 km and 12 GHz are not necessary for the solution.

Example 1.2

The installation plan of a radar station for air navigation on top of a mountain places it next to an existing 4-GHz microwave repeater station; interference is possible. Estimate the input level of the interference (at the receiving antenna output), making the following assumptions: frequency is 5 GHz; peak power is 1.4 MW; direct distance is 720 m; angle between main lobe axis of antenna for microwave repeater and direction to radar station is 7.5°; isotropic antenna gain to radar station is +4 dB; radar antenna gain is 45 dB; and scanning is done in all directions.

Solution

1. Calculate the free-space propagation loss, then the level of the receiving interference.
2. Chart index → auxiliary index → Fig. 9*a*. Propagation loss $\Gamma_0 = 103.6$ dB for $D = 720$ m and $f = 5$ GHz.
3. Level of interference. $P_1 = 1.4$ MW = 61.5 dBW (from Fig. 4). Then

$$P_2 = P_1 + G_1 + G_2 - \Gamma_0$$
$$= 61.5 + 45 + 4 - 103.6 = +6.9 \text{ dBW } (\simeq 5 \text{ watts}) \text{ peak value}$$

4. The resultant interference is so strong that it is necessary to insert a rejection filter for 5 GHz in the receiving system or take countermeasures, including changing the site of the microwave repeater station.

Example 1.3

Since Mt. Fuji (3776 m) and Mt. Tateyama (3015 m) have good visibility and the reflected waves are effectively screened by smaller mountains, a free-space propagation path exists. Check if the use of a portable VHF system is possible for a mountain rescue operation in the hop, making the following assumptions: frequency $f = 150$ MHz; transmit output $P = 0.5$ watt; antenna gain (relative to gain of half-wave dipole) $G_1, G_2 = -2$ dB each (whip type); required induced voltage $V_{req} = 10$ dBμV; and distance $D = 170$ km.

Solution

1. Calculate the induced voltage and compare it with the requirement. Determine the related items on the chart index, that is, field strength and voltage induced on half-wave dipole.
2. Field strength,

$$E_0 = 30.2 \text{ dB}\mu\text{V/m} \qquad (\text{Fig. } 3b)$$

3. Correction factor P_e to effective isotropically radiated power (e.i.r.p.) (Fig. 4),

Output power P	-3 dBW
Antenna gain (relative to gain of a half-wave dipole) G_1	-2 dB
Isotropic gain of a half-wave dipole	$+2.1$ dB
	$P_e = -2.9$ dB

4. Field strength E_0 at receiving point,

$$E_0' = E_0 + P_e$$
$$= 30.2 - 2.9 = 27.3 \text{ dB}\mu\text{V/m}$$

5. Induced voltage V_i. Effective length of a half-wave dipole (150 MHz) $l_e = -4$ dB (Fig. 6). Then

$$V_t = E_0' + l_e + G_2$$
$$= 27.3 - 4 - 2 = 21.3 \text{ dB}\mu\text{V}$$

6. It is feasible since the induced voltage is sufficiently higher than the required 10 dBμV.

Example 1.4

A broadcast is to be sent from each stadium holding a national athletic meet to a television tower using mobile radio relay equipment. Assuming the transmitting power $P_t = 320$ mW, the parabola's diameter $D_\phi = 1.5$ m at the mobile station side and 1.0 m at the tower side, and the aperture efficiency $\eta = 60\%$, calculate the margin for the input power M (dB) for the required receiving input of -40 dBm at 4 GHz, 8 GHz, and 12 GHz. Maximum distance is 22 km and the propagation path is in free space.

Solution A (Method Using Free-Space Propagation Loss Γ_0 and Antenna Gains G_1 and G_2)

1. Γ_0 for the distance of 22 km (chart index → auxiliary index → Fig. 9c),

	4 GHz	8 GHz	12 GHz
$\Gamma_0 =$	131.3	137.4	140.9

2. Antenna gains G_1 and G_2 (chart index A → auxiliary index → Fig. 13),

D_ϕ	A_e
1.5 m	1.06 m^2
1.0 m	0.47 m^2
	$\eta = 0.6$

	4 GHz	8 GHz	12 GHz
$G_1 =$	33.7	39.8	43.3
$G_2 =$	30.2	36.2	39.8
$G_c = G_1 + G_2 =$	63.9	76.0	83.1

3. Transmission loss between antennas $L_0 = \Gamma_0 - G_c$ (dB),

	4 GHz	8 GHz	12 GHz
$L_0 =$	67.4	61.4	57.8

4. Received power $P_r = P_t - L_0$ and $M = P_r - (-40)$ dB,

	4 GHz	8 GHz	12 GHz
$P_r =$	−42.4	−36.4	−32.8
$M =$	−2.4	+3.6	+7.2

Solution B (Method Using Unit-Area Antennas)

1. L_{uf} for distance of 22 km (chart index → Fig. 12),

	4 GHz	8 GHz	12 GHz
$L_{uf} =$	64.3	58.3	54.8

2. Effective area ratio A_e (Fig. 11),

$D_\phi(\eta = 0.6)$	A_e (dB)
1.5 m	$A_{e1} = +0.3$
1.0 m	$A_{e2} = -3.3$
	$A_e = A_{e1} + A_{e1} = -3.0$

3. Transmission loss L_0, received power P_r and margin M,

	4 GHz	8 GHz	12 GHz
$L_0 = L_{uf} + A_e$	67.3	61.3	57.8
$P_r =$	−42.3	−36.3	−32.8
$M =$	−2.3	+3.7	+7.2

The calculations are the same as for Solution A.4.

REMARKS

1. Because received power at 4 GHz is below the requirement, an increase of transmitting power or antenna gain is necessary.
2. Attenuation due to rainfall is great at 12 GHz. However, this is not a problem since the rain will cause the meet to be canceled.
3. Solution B is regarded as easier and more reliable than solution A. The margin M is proportional to $20 \log f$ since L_{uf} is proportional to $20 \log \lambda$. Therefore determination of M at a single frequency ratio simplifies the calculation.

Chart Index,
Free-Space Propagation Path

Concept of Free-Space Propagation

[1] Diffusion of radio wave energy in free space

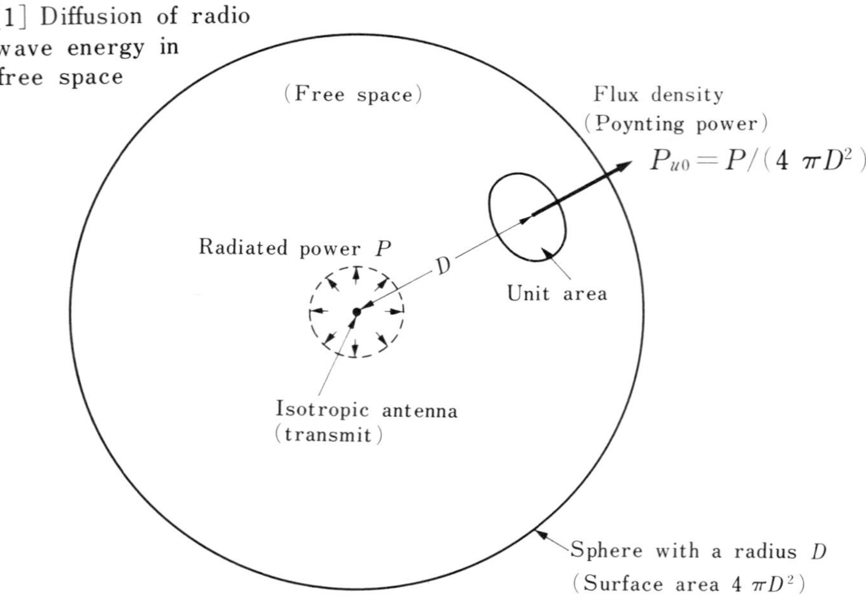

[2] Reception by isotropic antenna

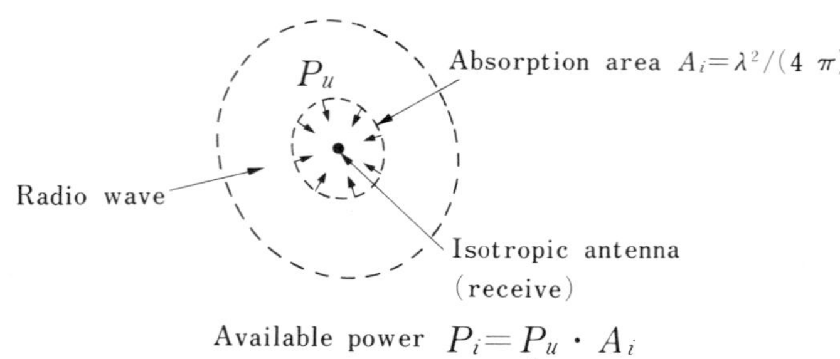

[3] Free space propagation loss

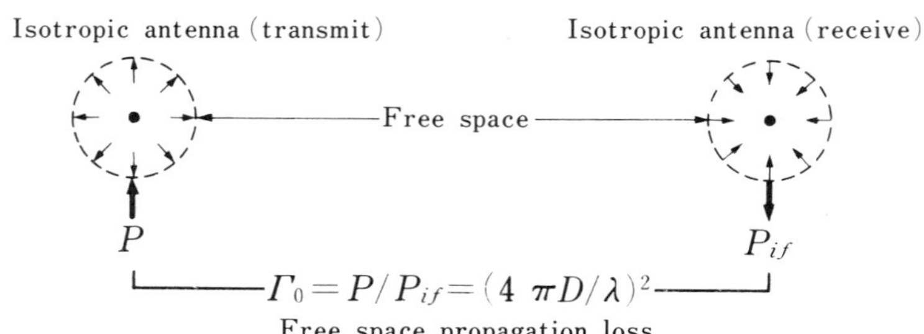

Judgment on Applicability of Free-Space Propagation

Propagation wave	Judgment	Free space			Non-free space	
	Application category	A	B	C	D	E
	Additional propagation loss	1 dB or lower	2 dB or lower	3 dB or lower	6 dB or lower	6 dB or higher
Reflection	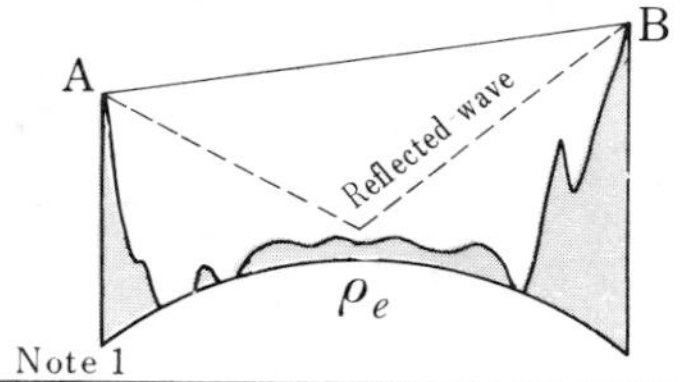 A, B, Reflected wave, ρ_e; Note 1	Equivalent reflection coefficient ρ_e				〔Interference propagation〕
		0.1 (−20dB) or lower	0.2 (−14dB) or lower	0.3 (−10dB) or lower	0.5 (−6dB) or lower	0.5 (−6 dB) or higher
Diffraction	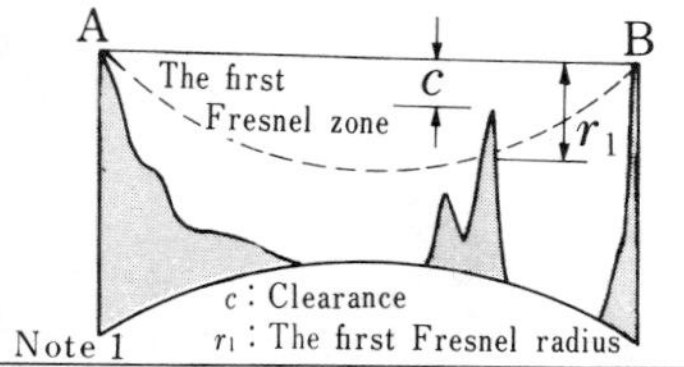 A, B, The first Fresnel zone, c, r_1; c : Clearance; r_1 : The first Fresnel radius; Note 1	Clearance parameter $u_c = c/r_1$				out of sight
		2/3 or larger	1/2 or larger	1/3 or larger	0 or larger	0 or smaller
Refraction	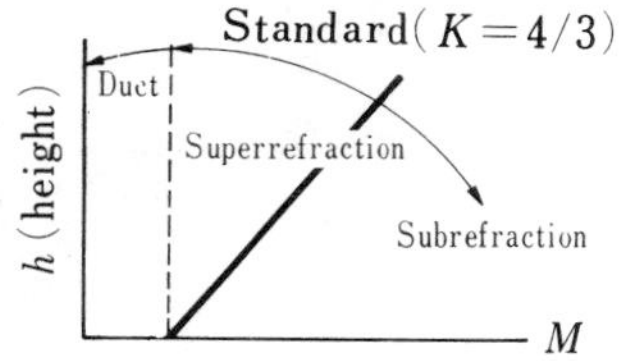 Standard ($K=4/3$), Duct, Superrefraction, Subrefraction, h (height), M	M-curve's inclination is assumed to be constant and remains as it is for the standard atmosphere ($K=4/3$). Influences by superrefraction, subrefraction and duct will be treated separately as "refraction fading."			Note 2	—
Absorption	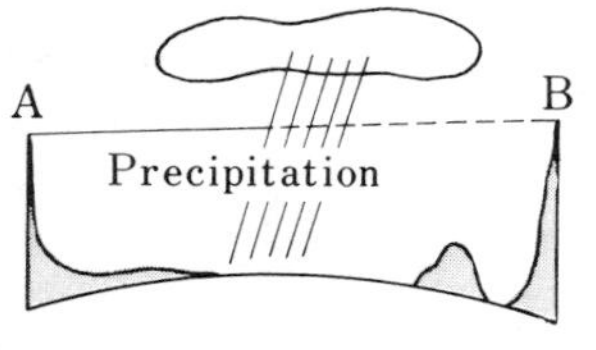 A, B, Precipitation	It can be neglected at or below 8 GHz. Attenuation generated at 8 to 10 GHz (long distance) and at 10 GHz or higher frequency will be treated separately as "absorption fading."			Note 2	—
Scattering	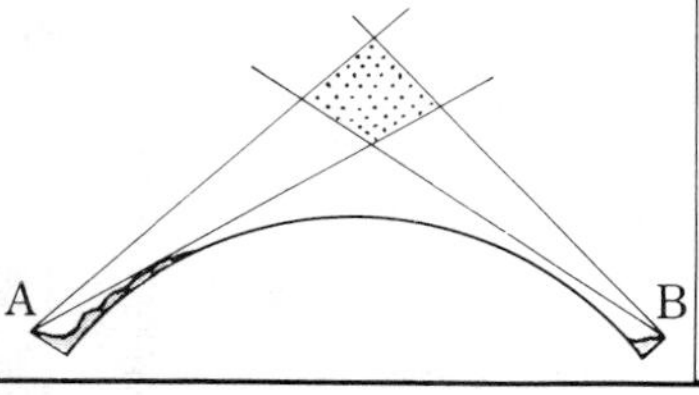 A, B	Both ionospheric-scatter wave (below 100 MHz only) and tropo-scatter wave (VHF to SHF) are extremely weak compared with the free-space wave.			Note 2	Type of propagation path is assumed as shown in the left

Note 1 : Refer to relating charts in Chapters 2 and 3 for effective reflection coefficient and clearance parameters (diffraction parameters).
When suppression effect of reflected waves by antenna directivity is expected, "propagation echo ratio $\rho_\theta (=\rho_e \cdot D_{\theta 1} \cdot D_{\theta 2})$" instead of the effective reflection coefficient ρ_e is used.

Note 2 : These are for reference only. These atmospheric phenomena will be discussed separately in other chapters.

Free-Space Field Strength and Flux Density

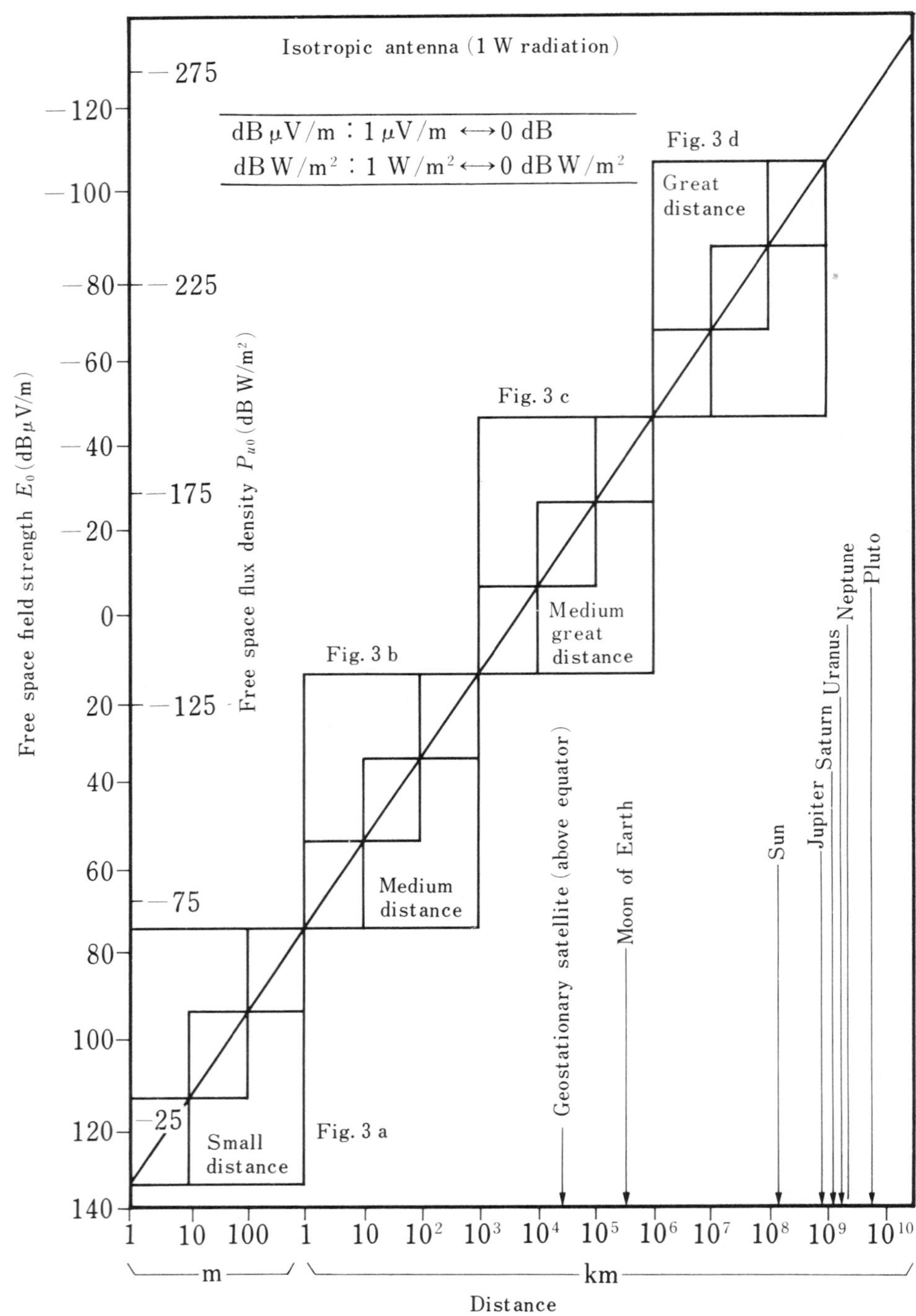

Free-Space Field Strength and Flux Density (1)

Isotropic antenna
(1 watt radiated)

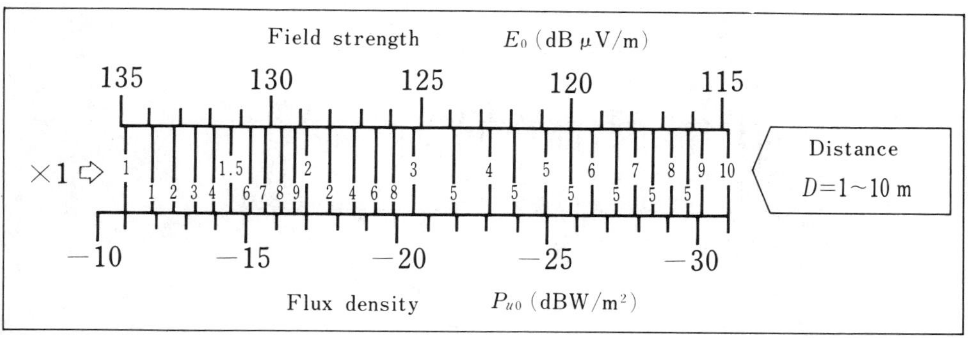

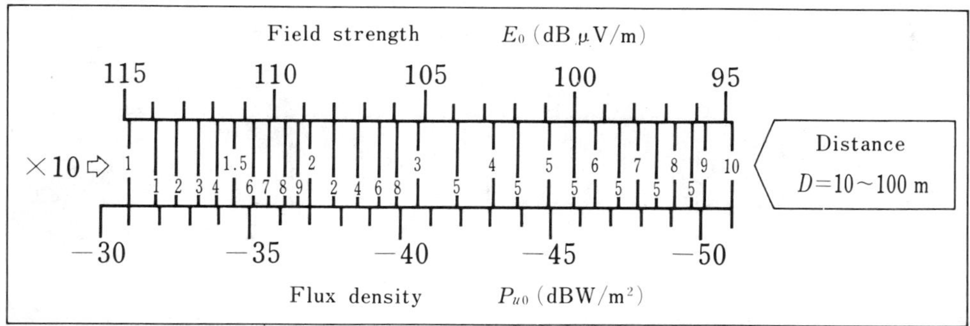

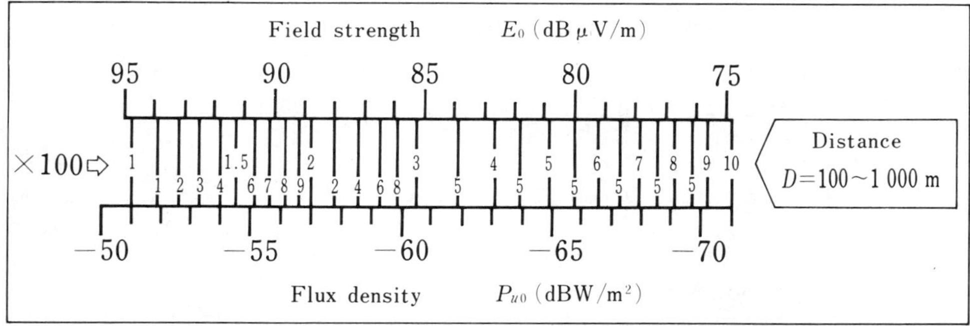

(1) Add+2.15 dB to the indicated value in case of a halfwave dipole (1 watt radiation).
(2) For ordinary condition of radiation, add the correction factor to be found on Fig. 4.

Free-Space Field Strength and Flux Density (2)

Isotropic antenna
(1 watt radiated)

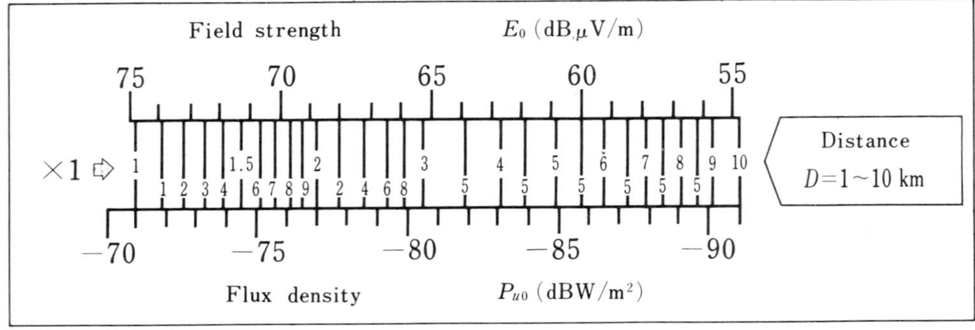

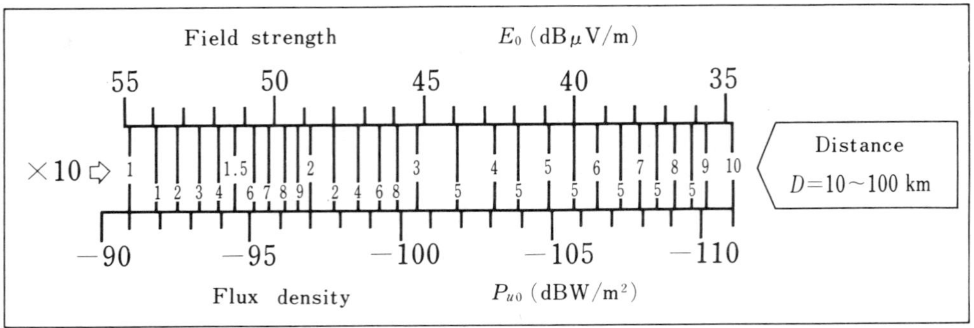

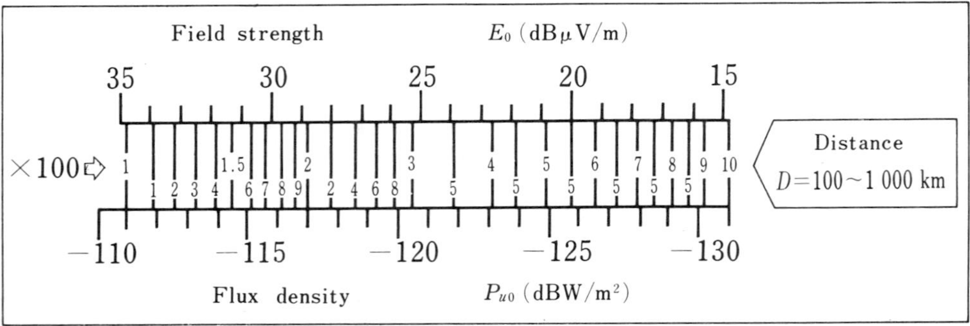

(1) Add+ 2.15 dB to the indicated value in case of a halfwave dipole (1 watt radiation).

(2) For ordinary radiation condition, add the correction factor found in Fig. 4.

Free-Space Field Strength and Flux Density (3)

Isotropic antenna
(1 watt radiated)

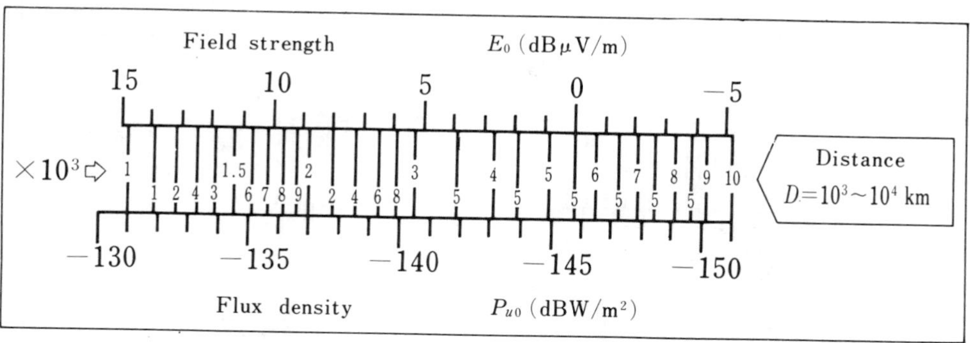

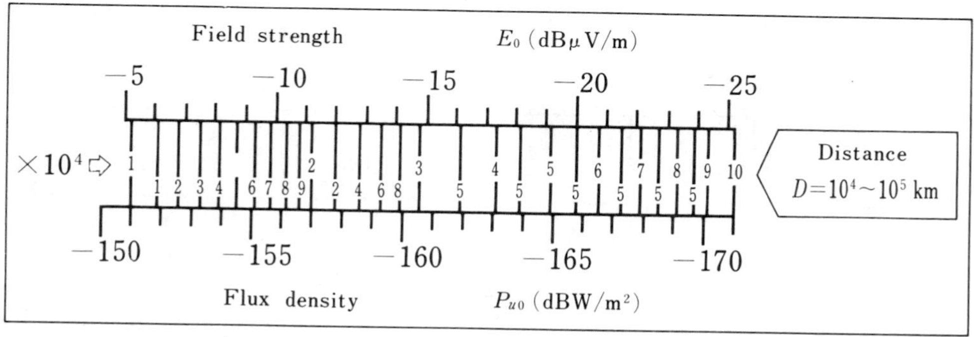

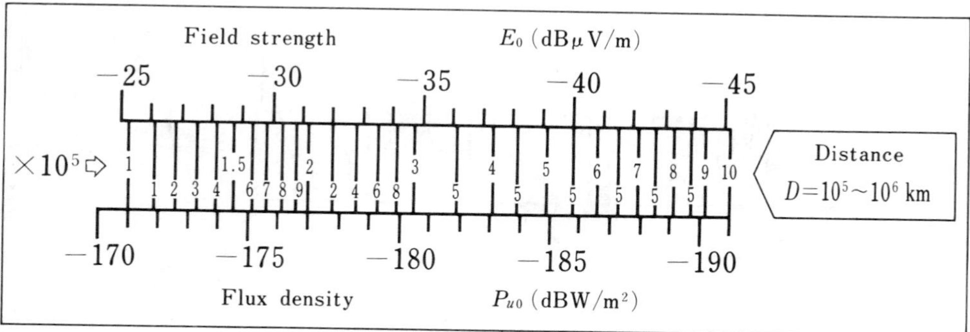

(1) Add+2.15 dB to the indicated value in case of a halfwave dipole (1 watt radiation).

(2) For ordinary radiation condition, add the correction factor found in Fig. 4.

Free-Space Field Strength and Flux Density (4)
For Great Distance

Isotropic antenna
(1 watt radiated))

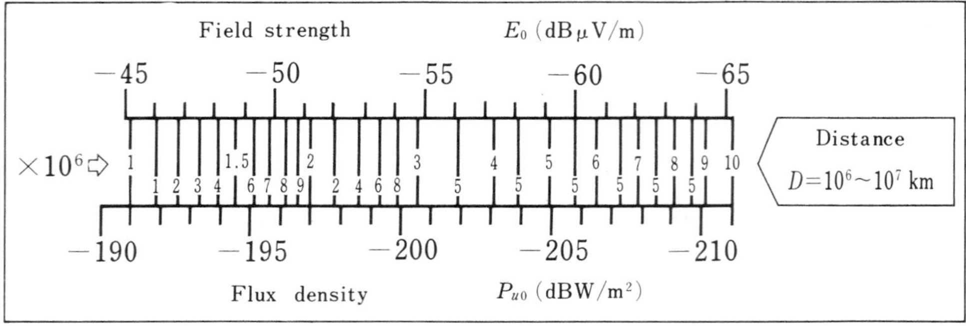

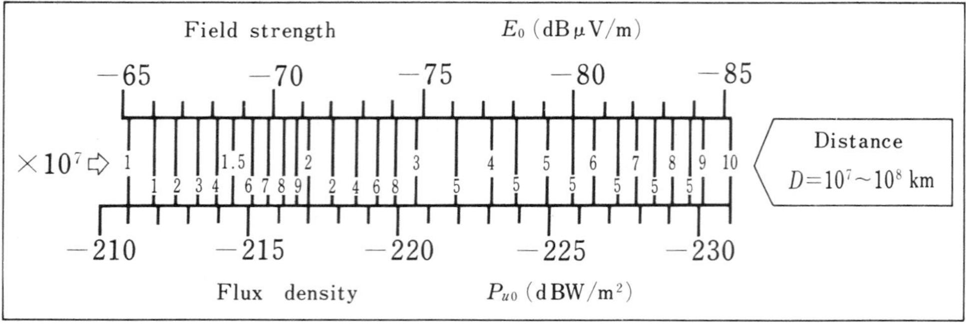

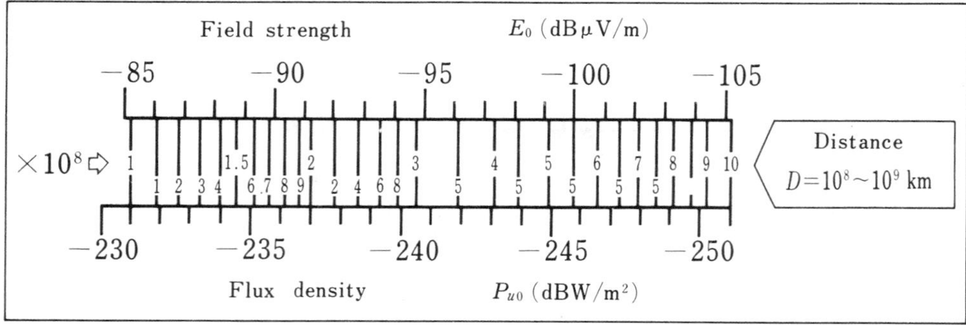

(1) Add +2.15 dB to the indicated value in case of a halfwave dipole (1 watt radiation).

(2) For ordinary radiation condition, add the correction factor found in Fig. 4.

Effective Isotropically Radiated Power (e.i.r.p.)

(1) Figure 3 is used to determine the free space field strength E_0 and the flux density P_{u0} for reference radiation (isotropic antenna, 1 watt radiated). Therefore add the e.i.r.p. ratio P_e for ordinary case.

(2) Correction factor

P_e (dB) = radiated power P (dBW) + transmit antenna gain G_t (dB)

(3) Use the table below for conversion between antilogarithm and dB value of P_e.

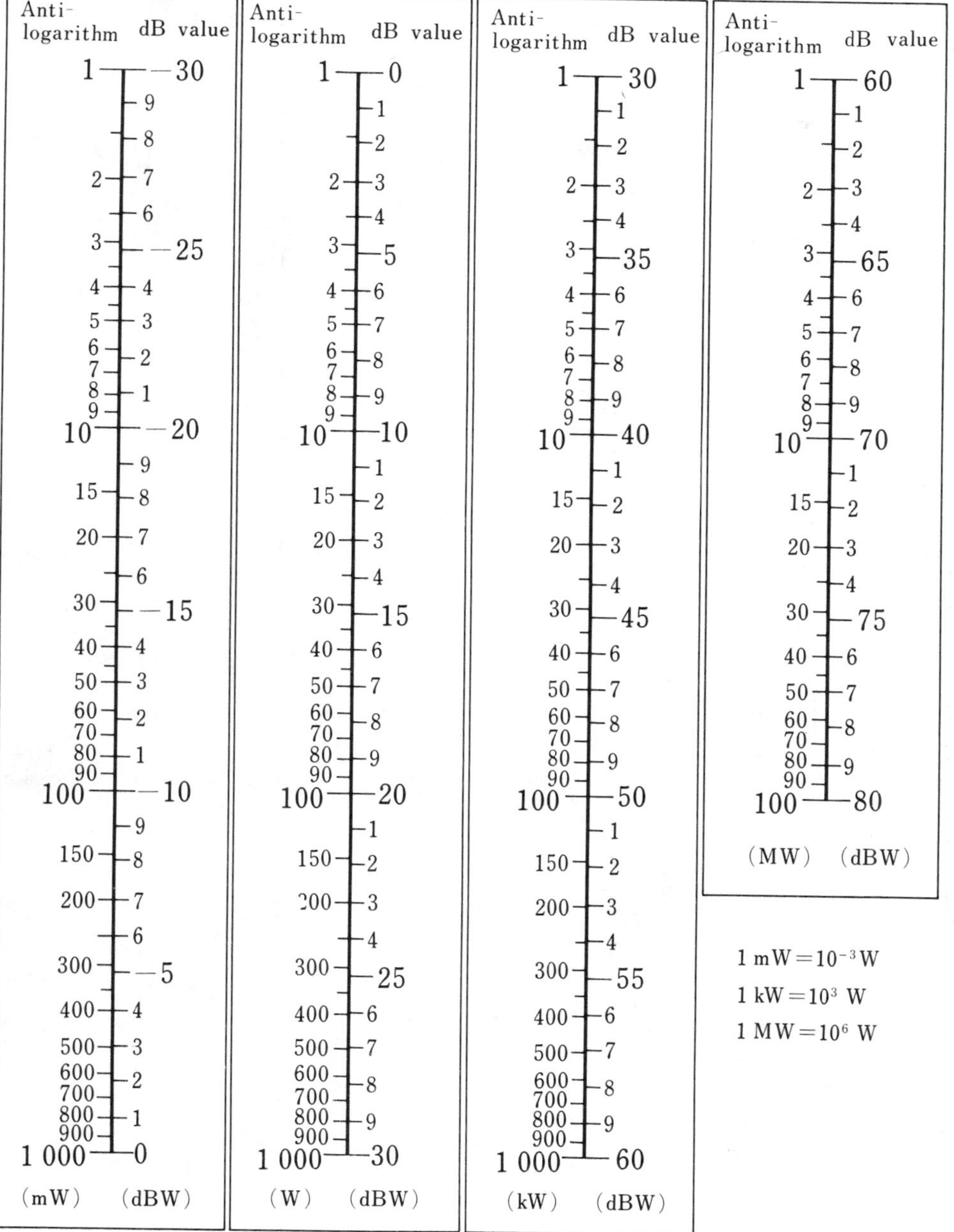

1 mW = 10^{-3} W

1 kW = 10^{3} W

1 MW = 10^{6} W

5 Field Strength Antilogarithm to dB Conversion Table

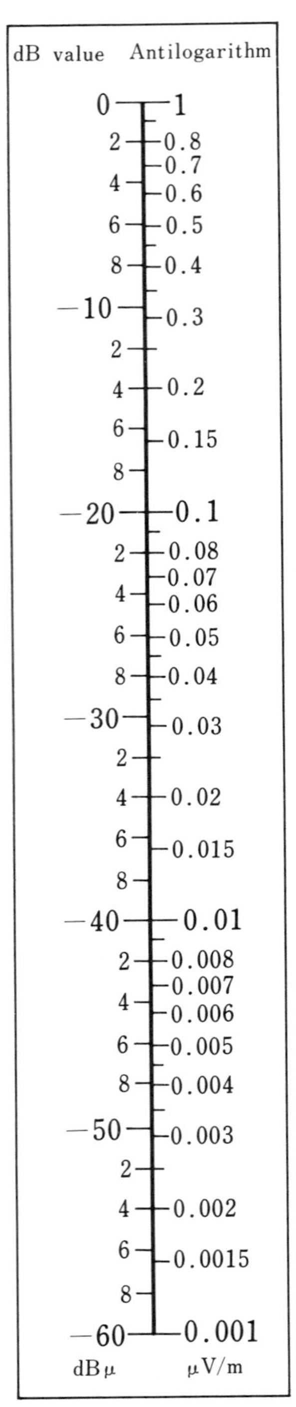

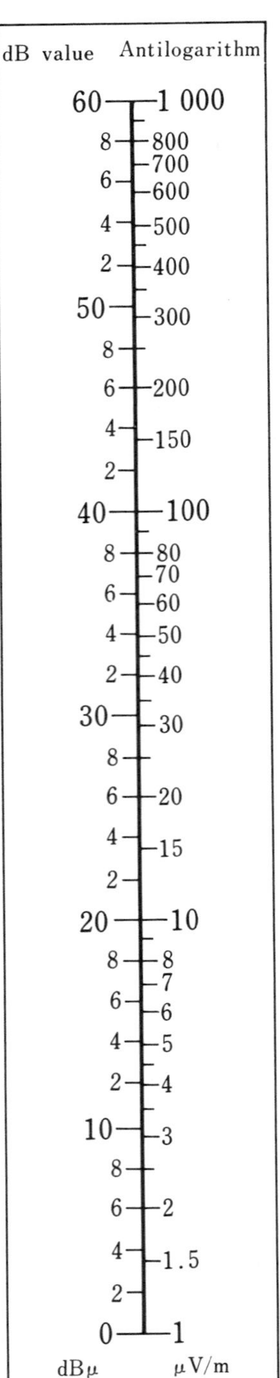

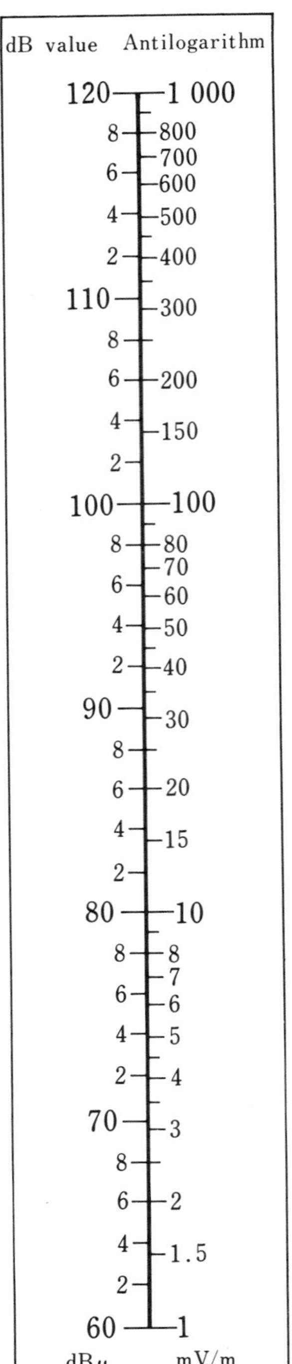

Effective Length of a Half-Wave (Length) Dipole

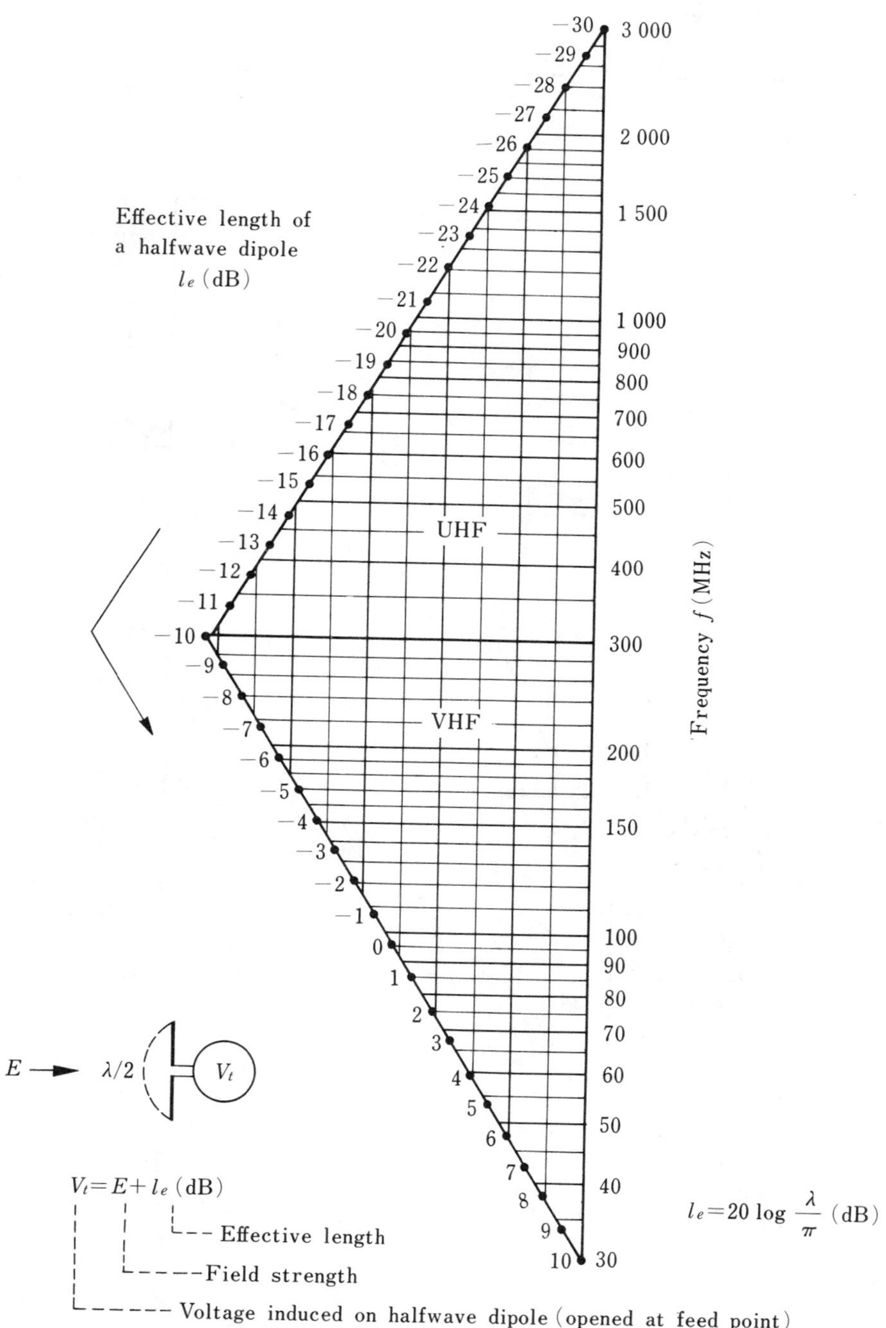

$V_t = E + l_e$ (dB)

- l_e --- Effective length
- E --- Field strength
- V_t --- Voltage induced on halfwave dipole (opened at feed point)

$$l_e = 20 \log \frac{\lambda}{\pi} \text{ (dB)}$$

Free-Space Propagation Loss, VHF Band

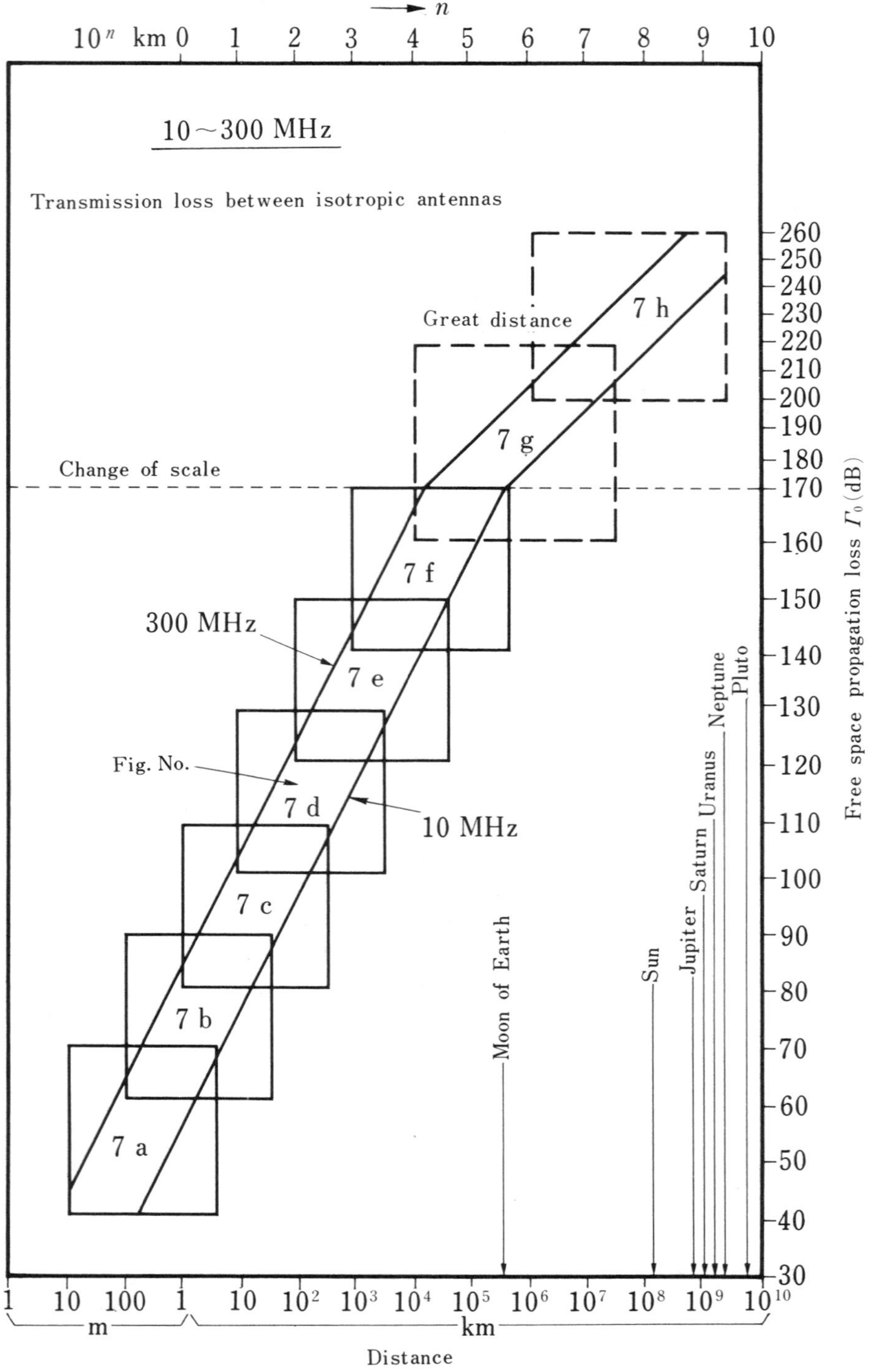

Free-Space Propagation Loss, UHF Band

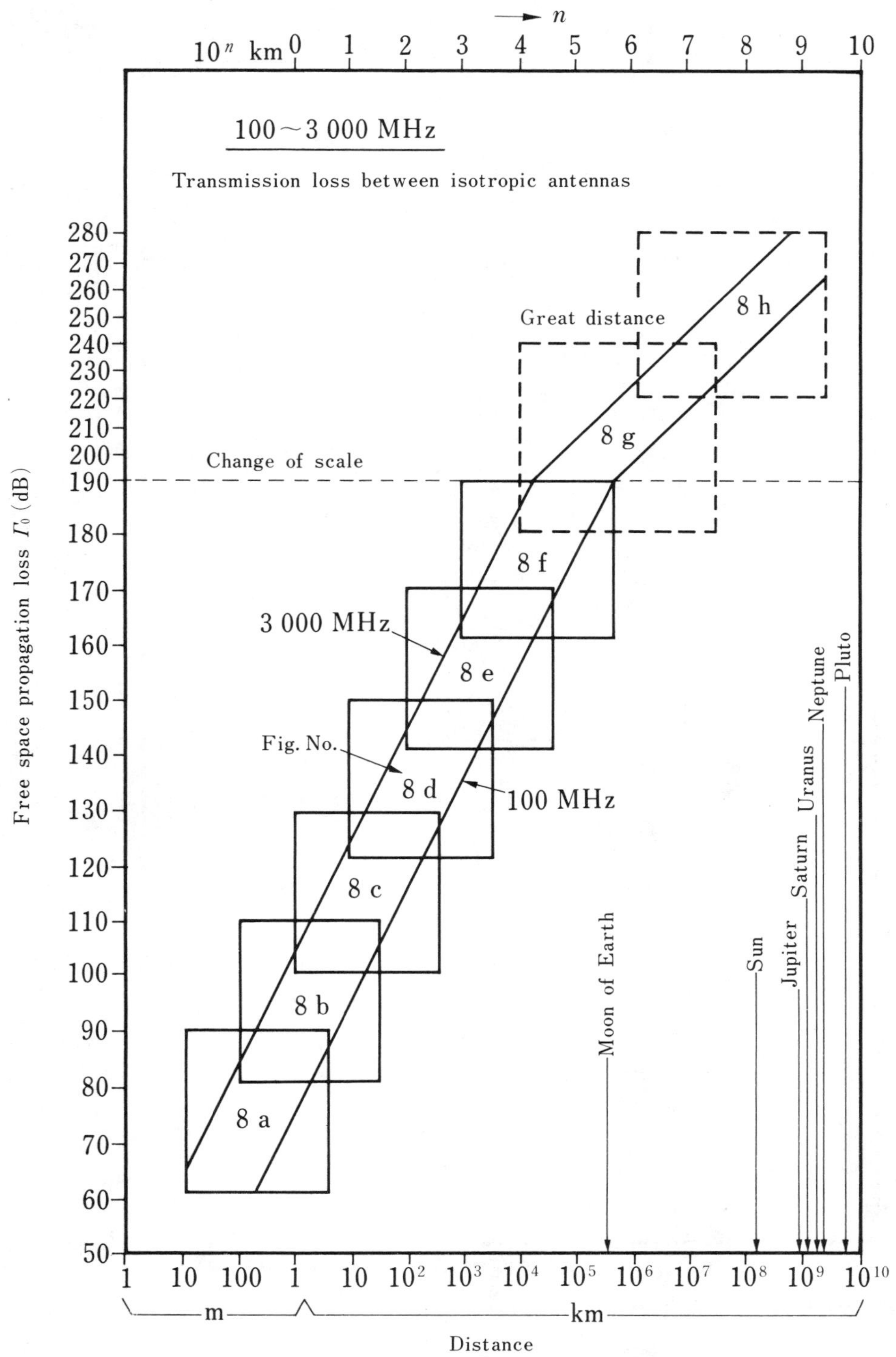

Free-Space Propagation Loss, SHF Band

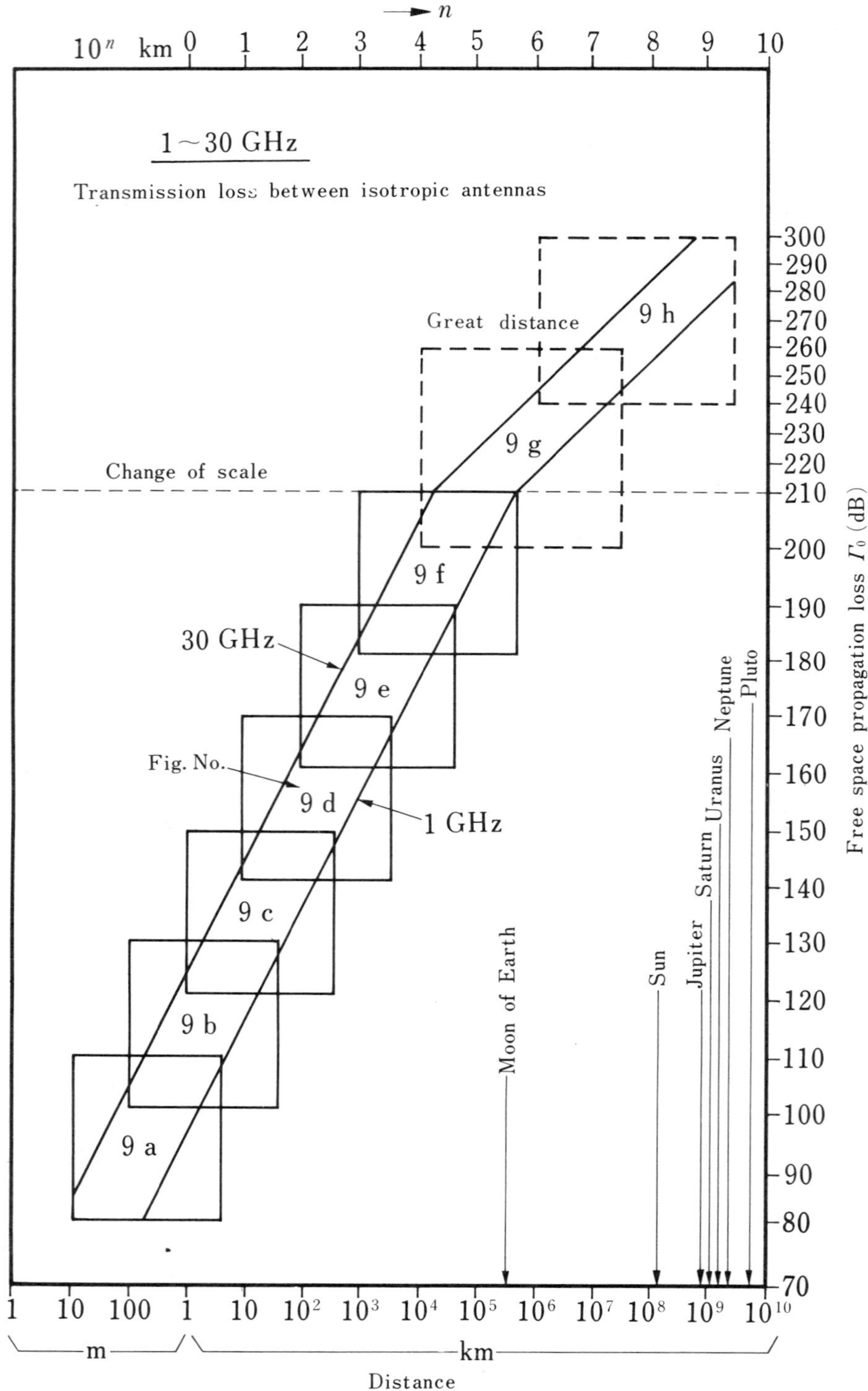

Free-Space Propagation Loss, EHF Band

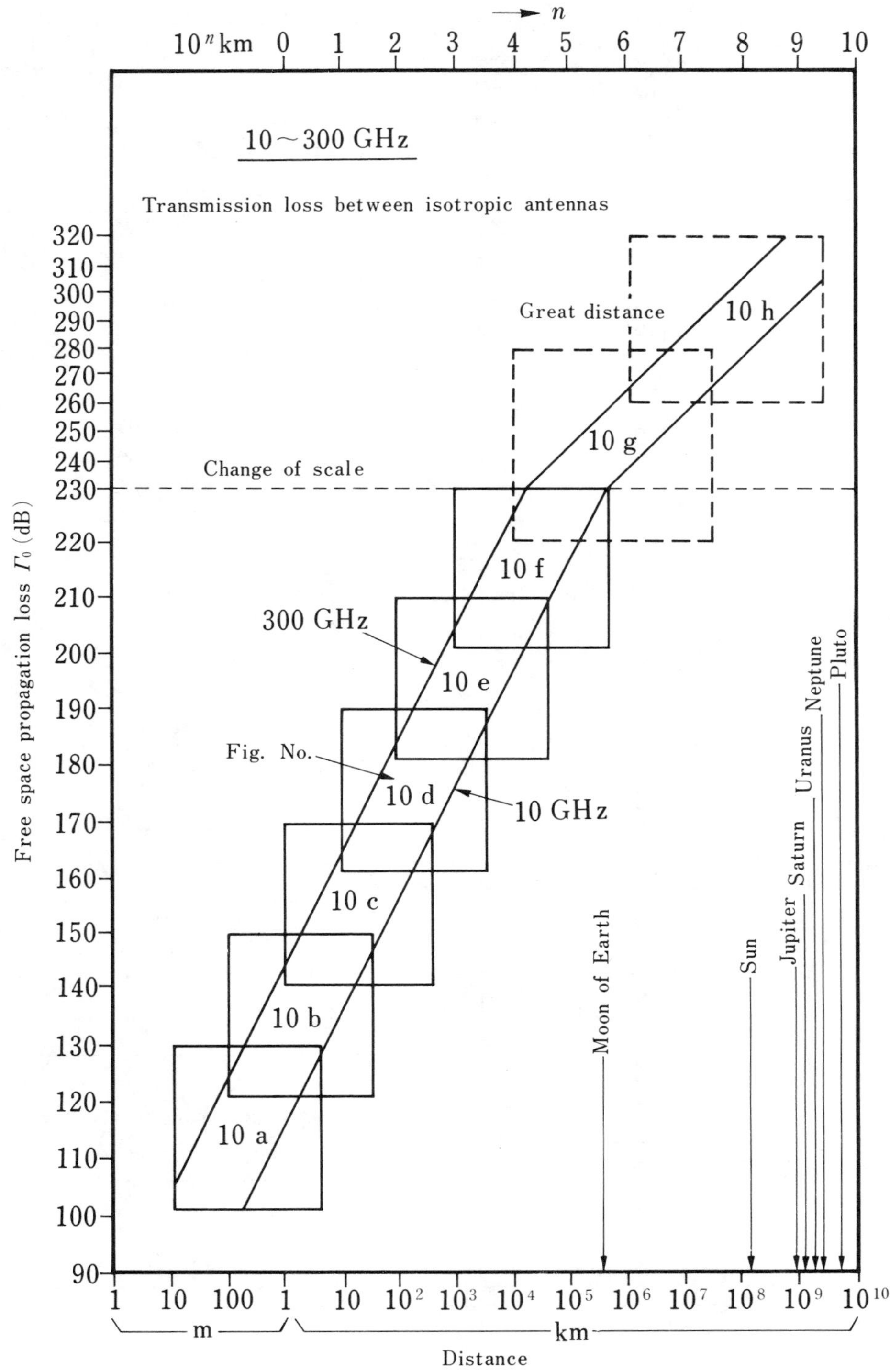

Free-Space Propagation Loss, VHF Band, 10–10³ m

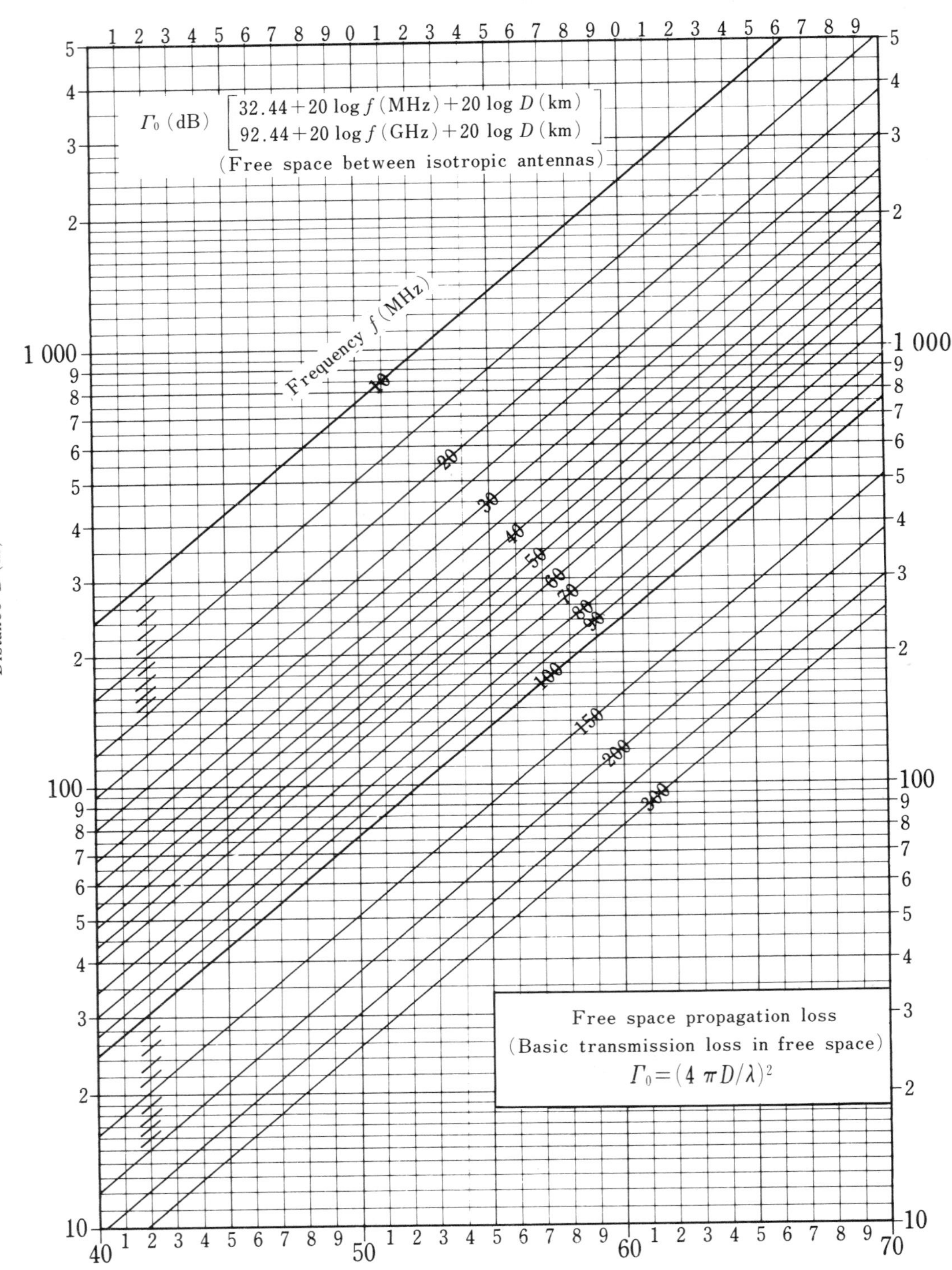

Free-Space Propagation Loss, VHF Band, 10^{-1}–10 km

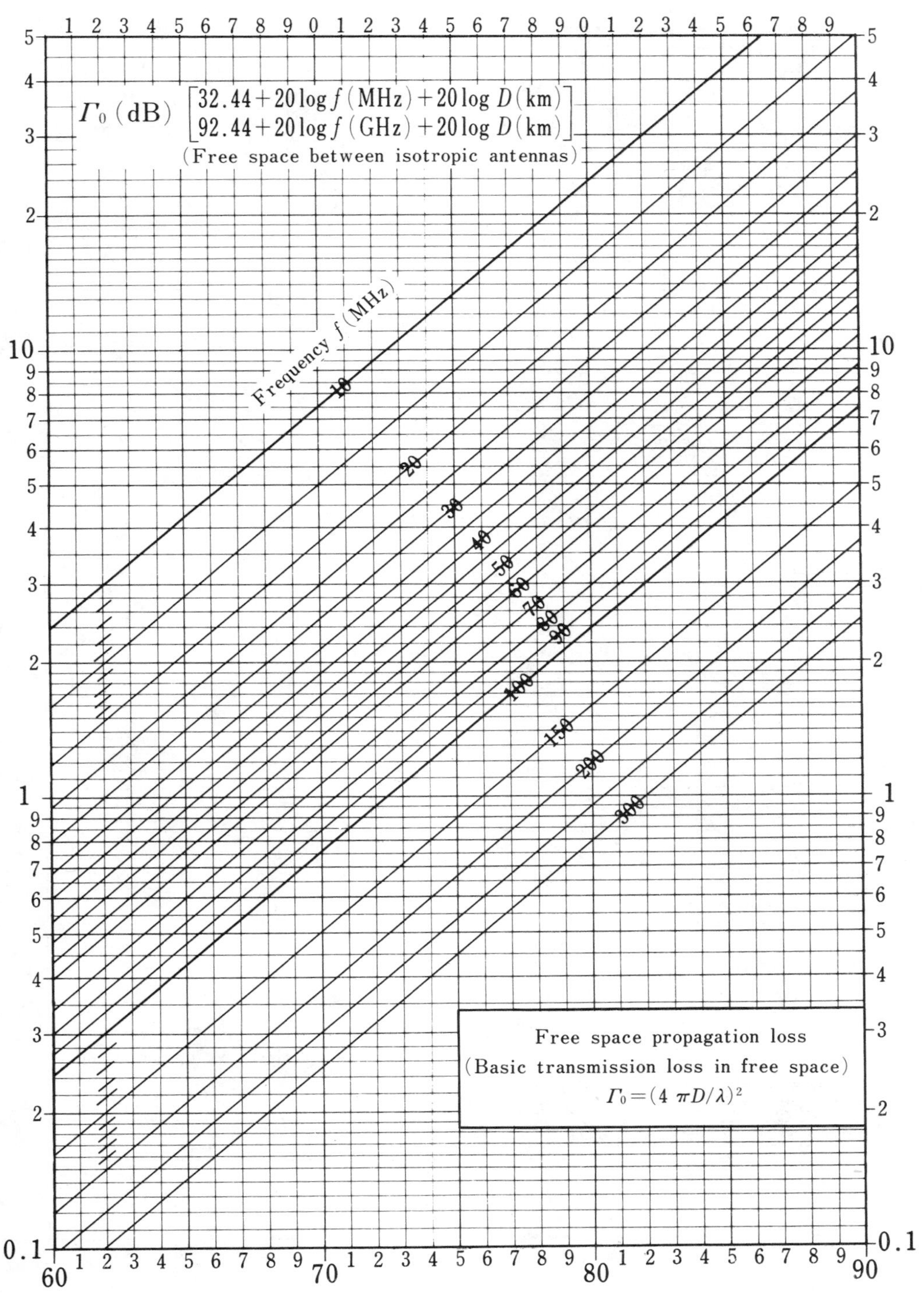

Free-Space Propagation Loss, VHF Band, 1–10^2 km

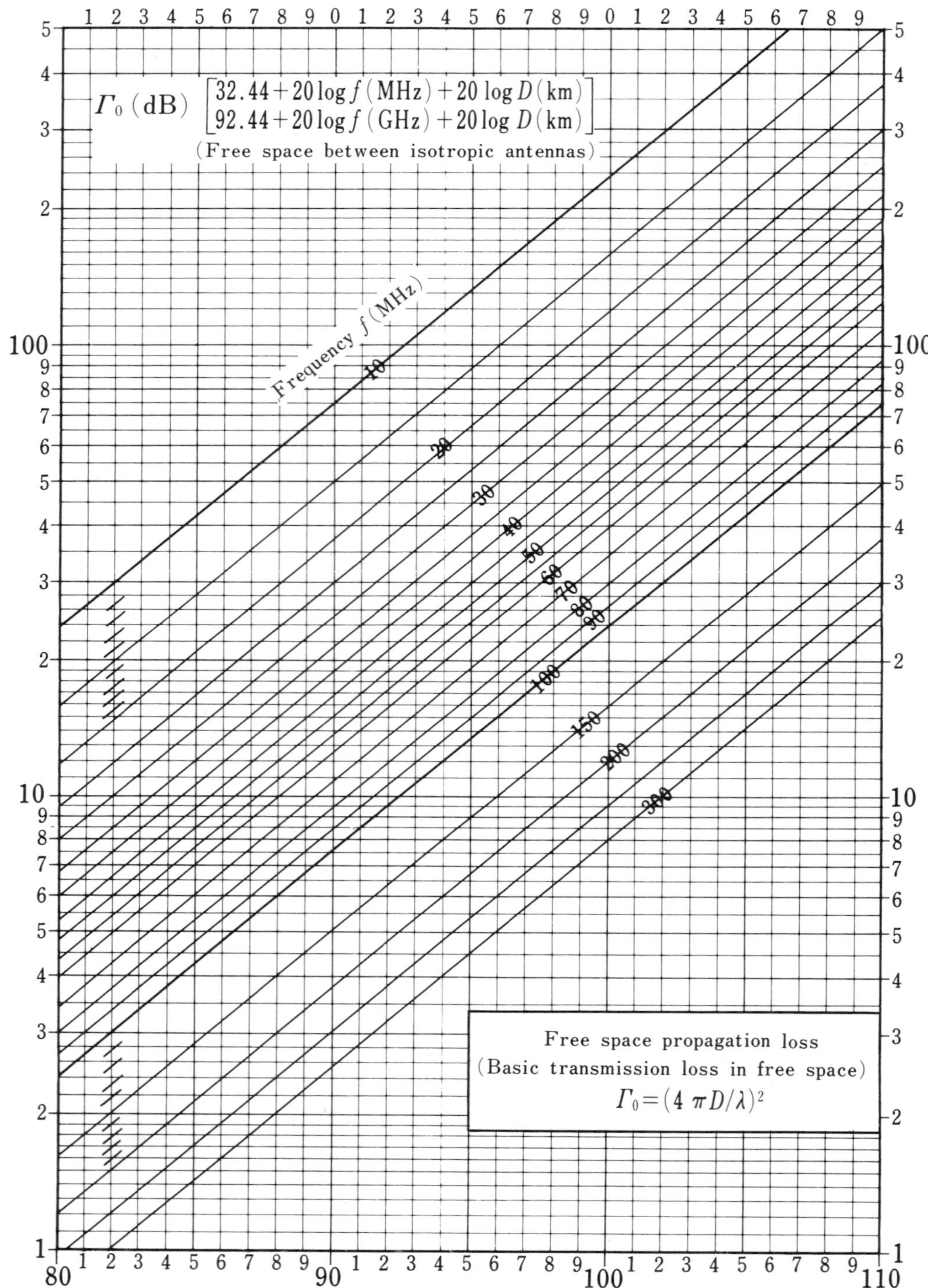

Free-Space Propagation Loss, VHF Band, 10–10³ km

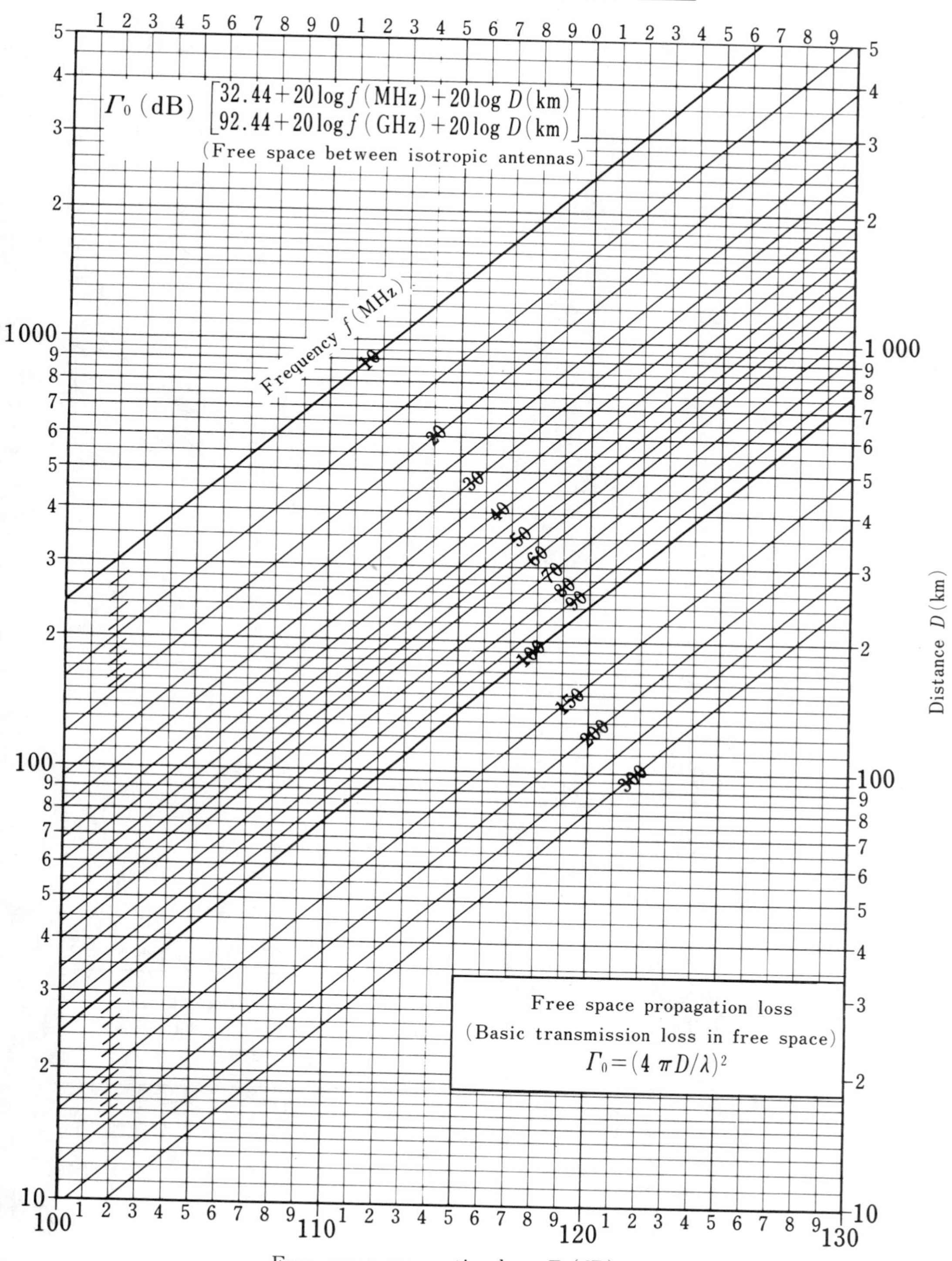

Free-Space Propagation Loss, VHF Band, 10^2–10^4 km

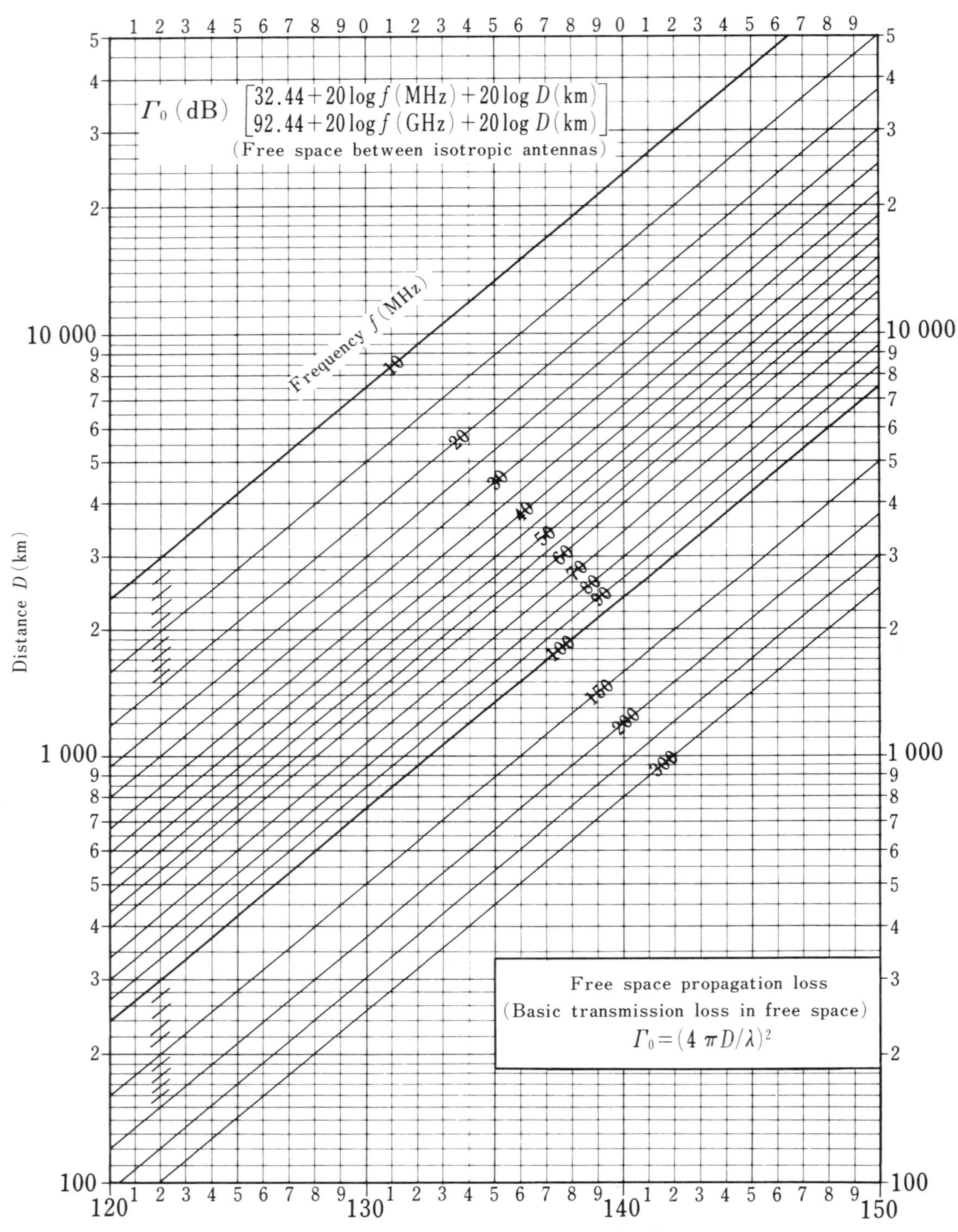

Free-Space Propagation Loss, VHF Band, 10^3–10^5 km

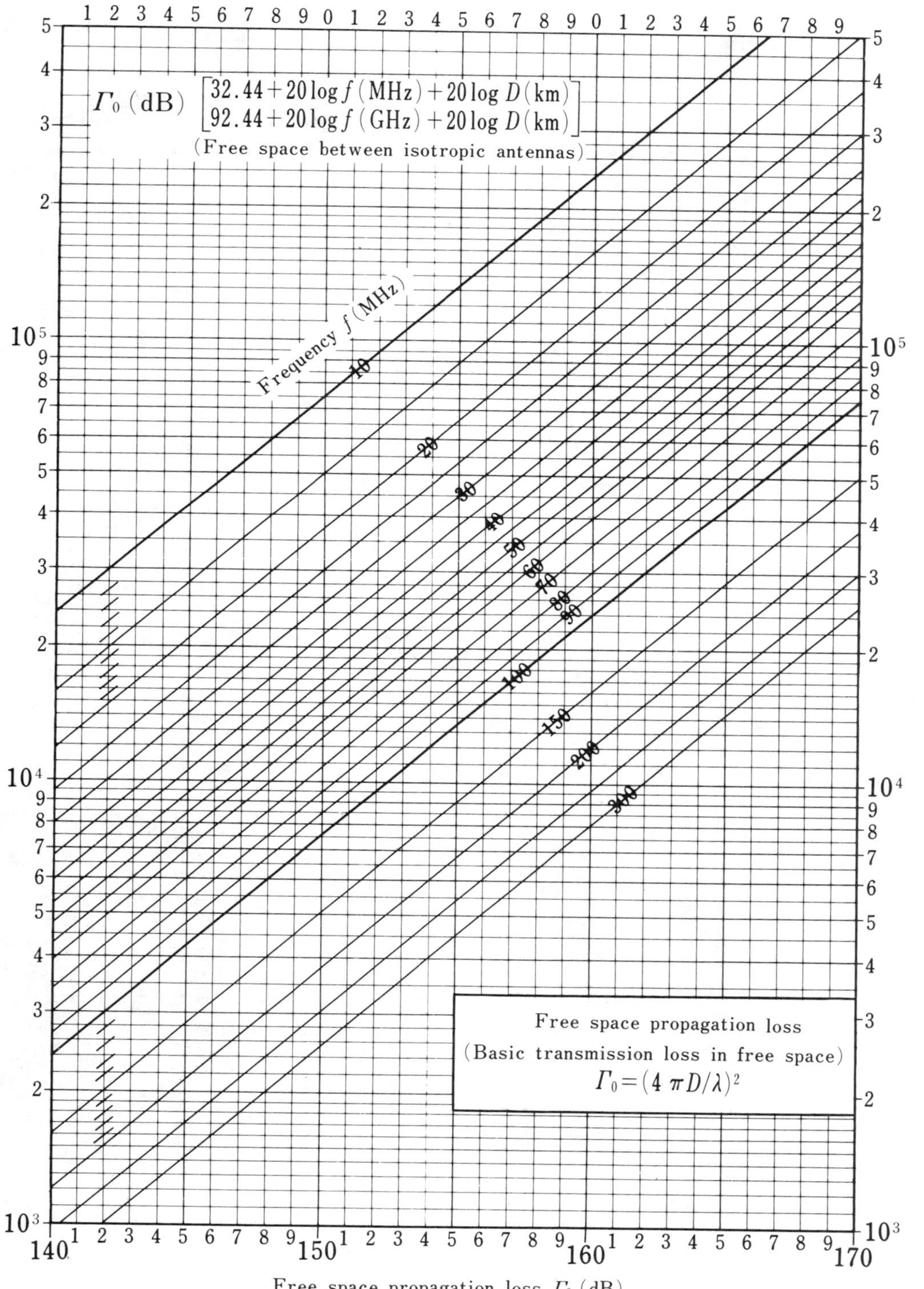

Free-Space Propagation Loss, VHF Band, 10^4–10^7 km

For Great Distance

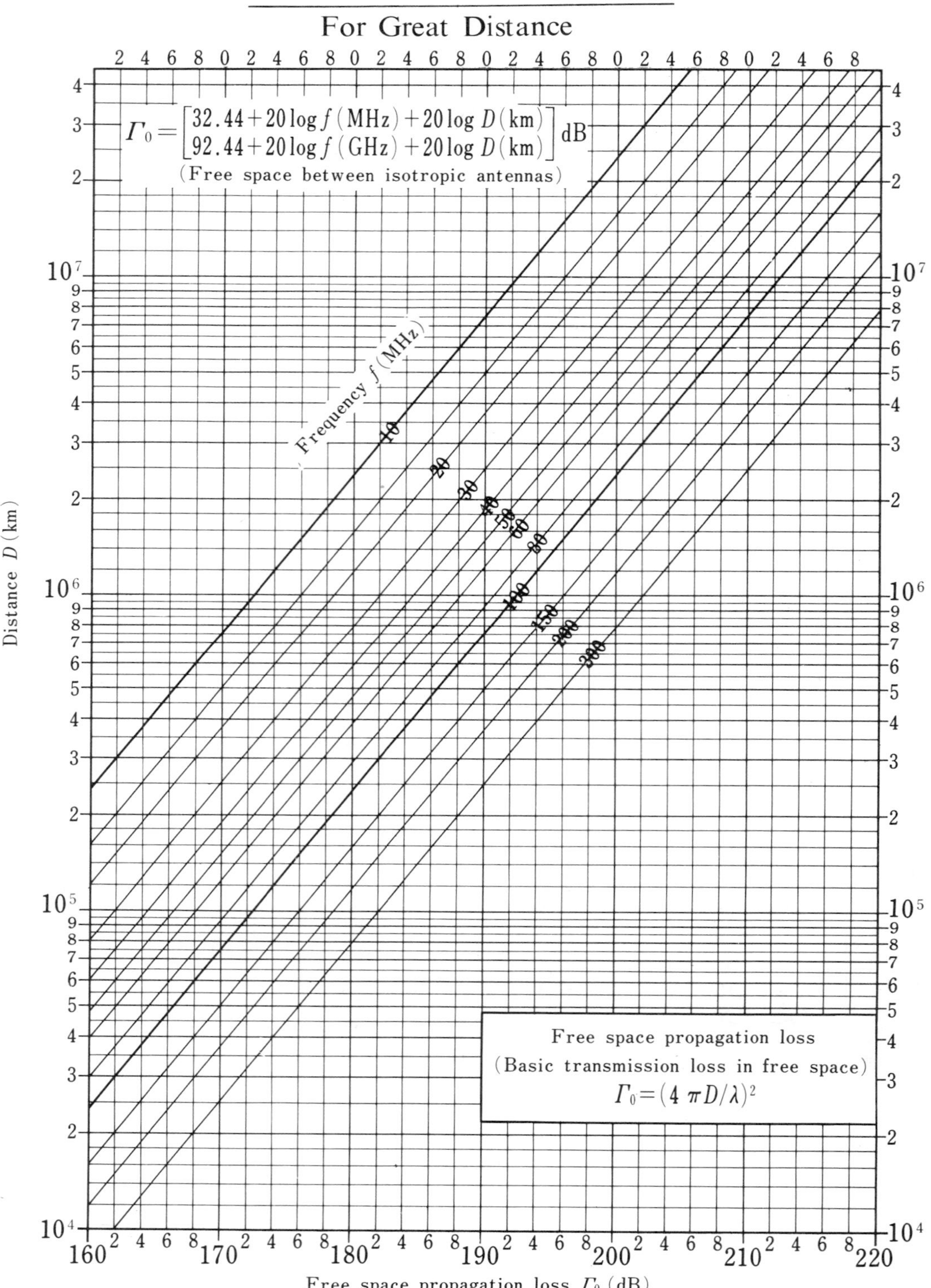

Free-Space Propagation Loss, VHF Band, 10^6–10^9 km

For Great Distance

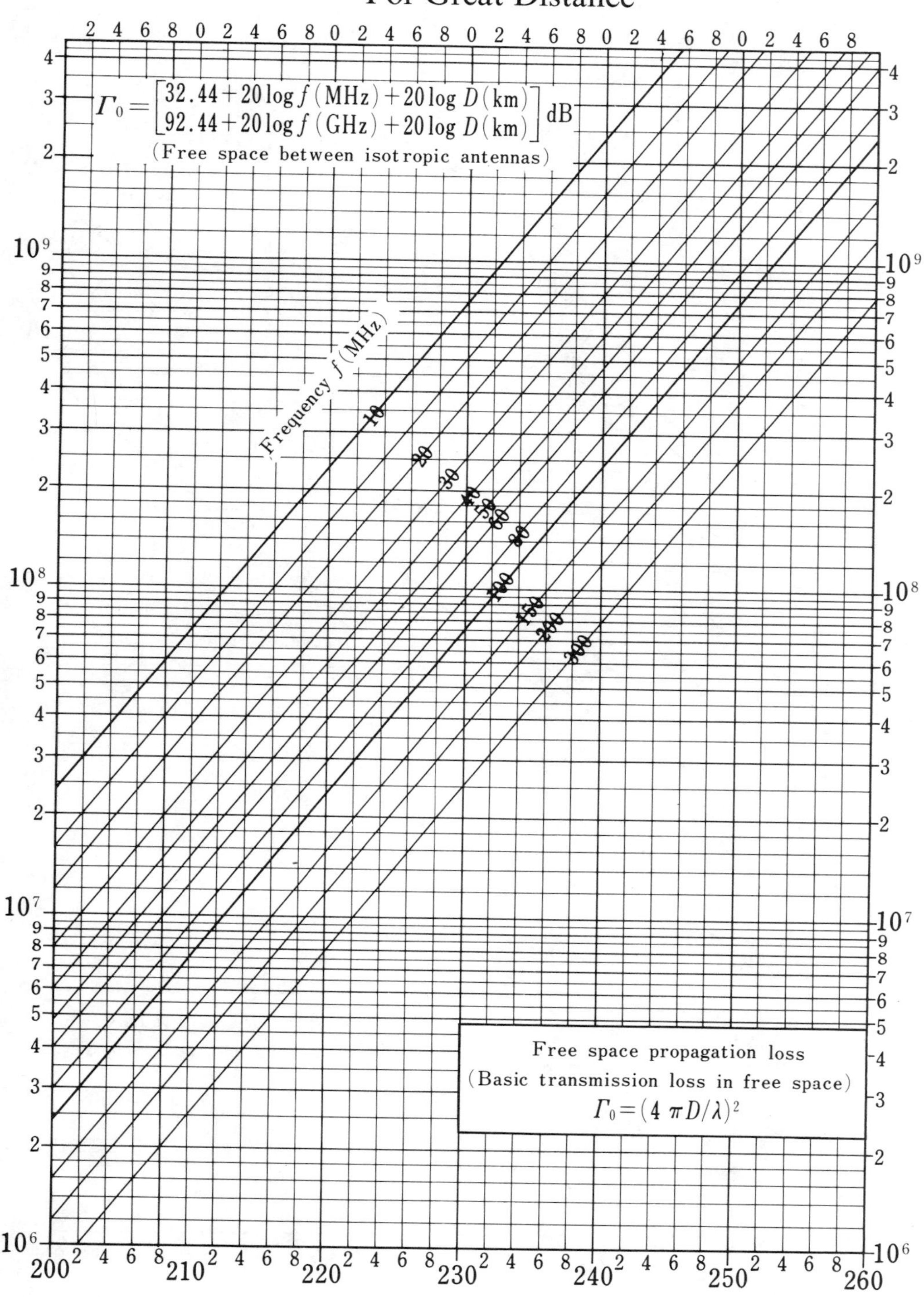

Free-Space Propagation Loss, UHF Band, 10–10³ m

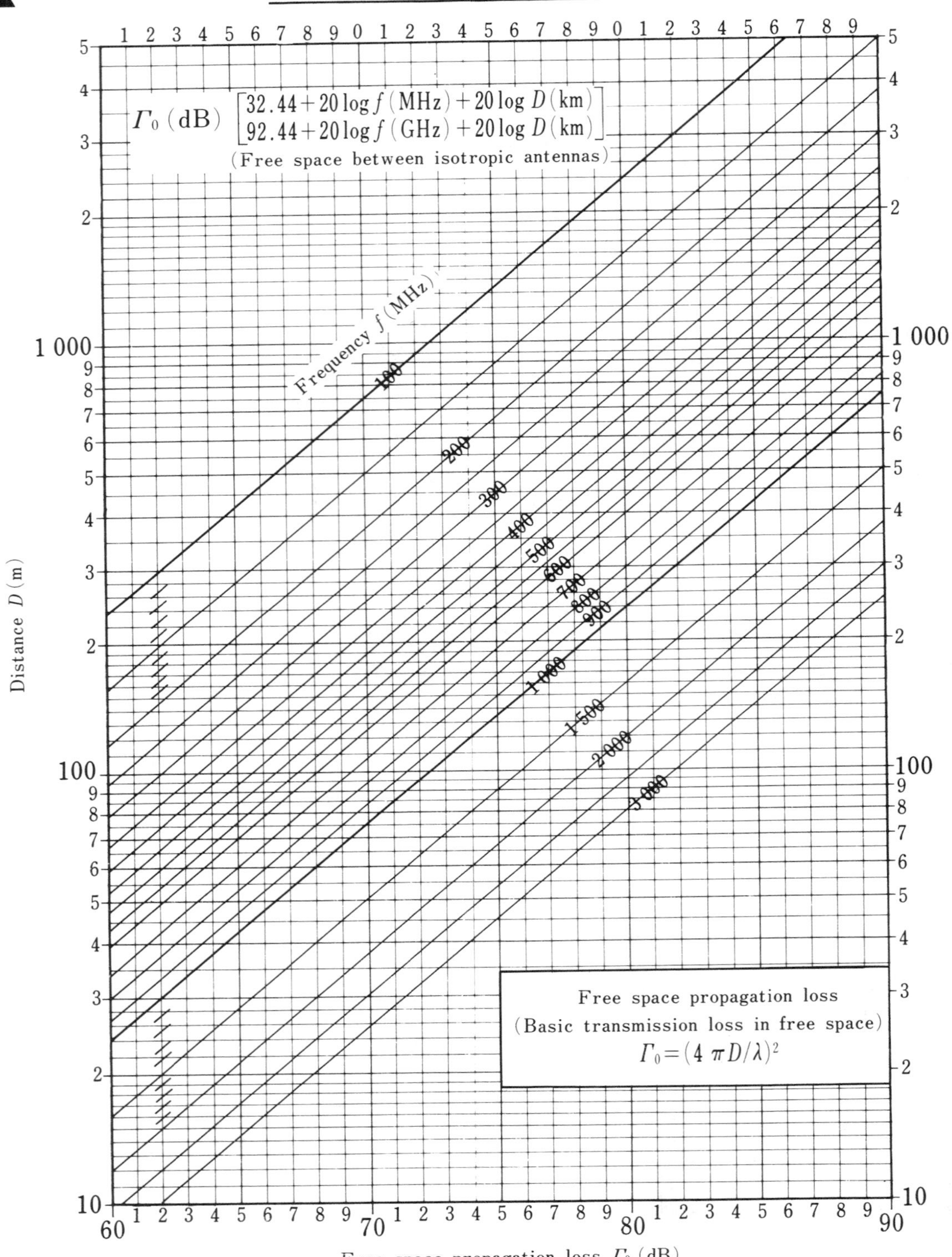

Free-Space Propagation Loss, UHF Band, 10^{-1}–10 km

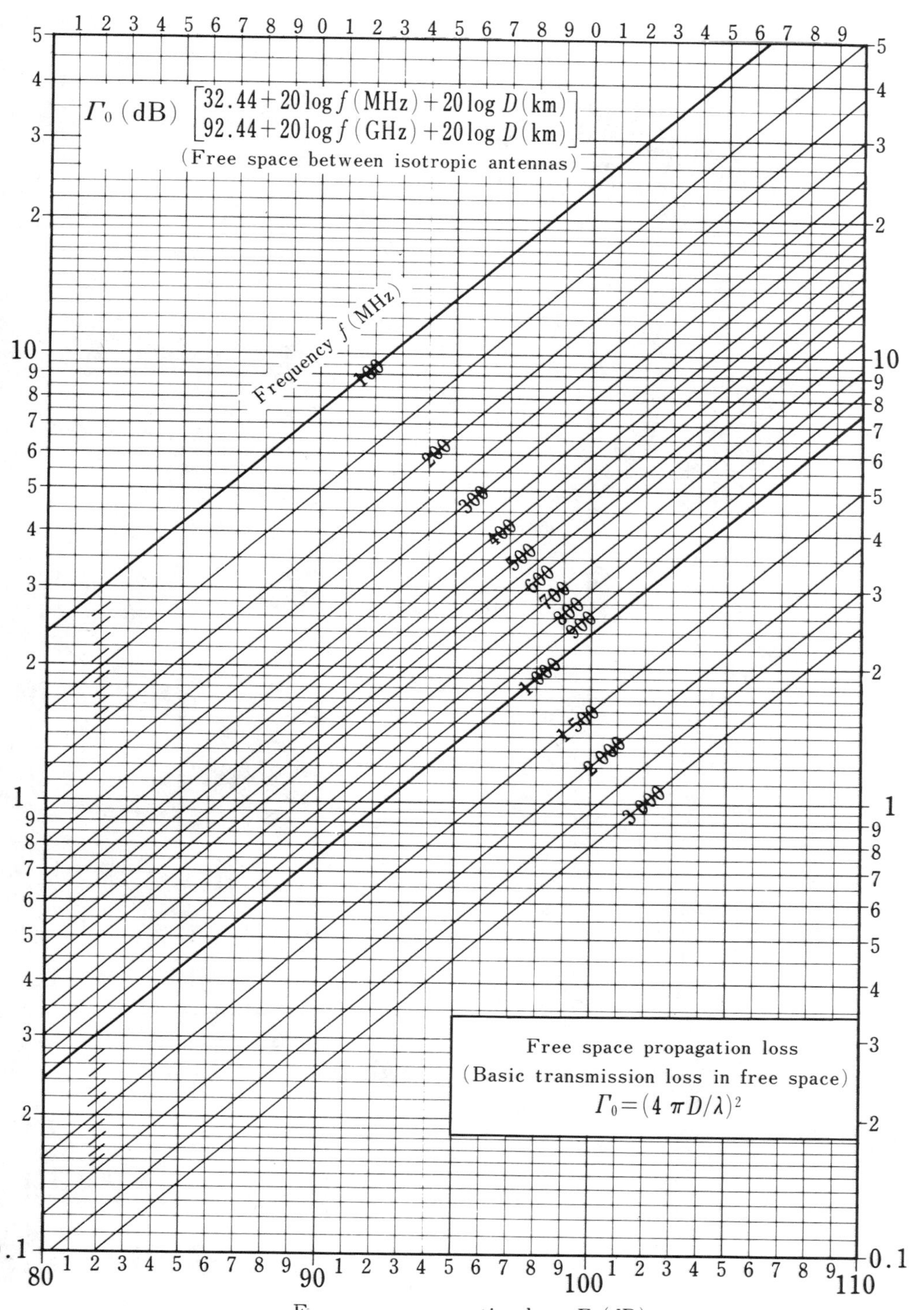

Free-Space Propagation Loss, UHF Band, 1–10² km

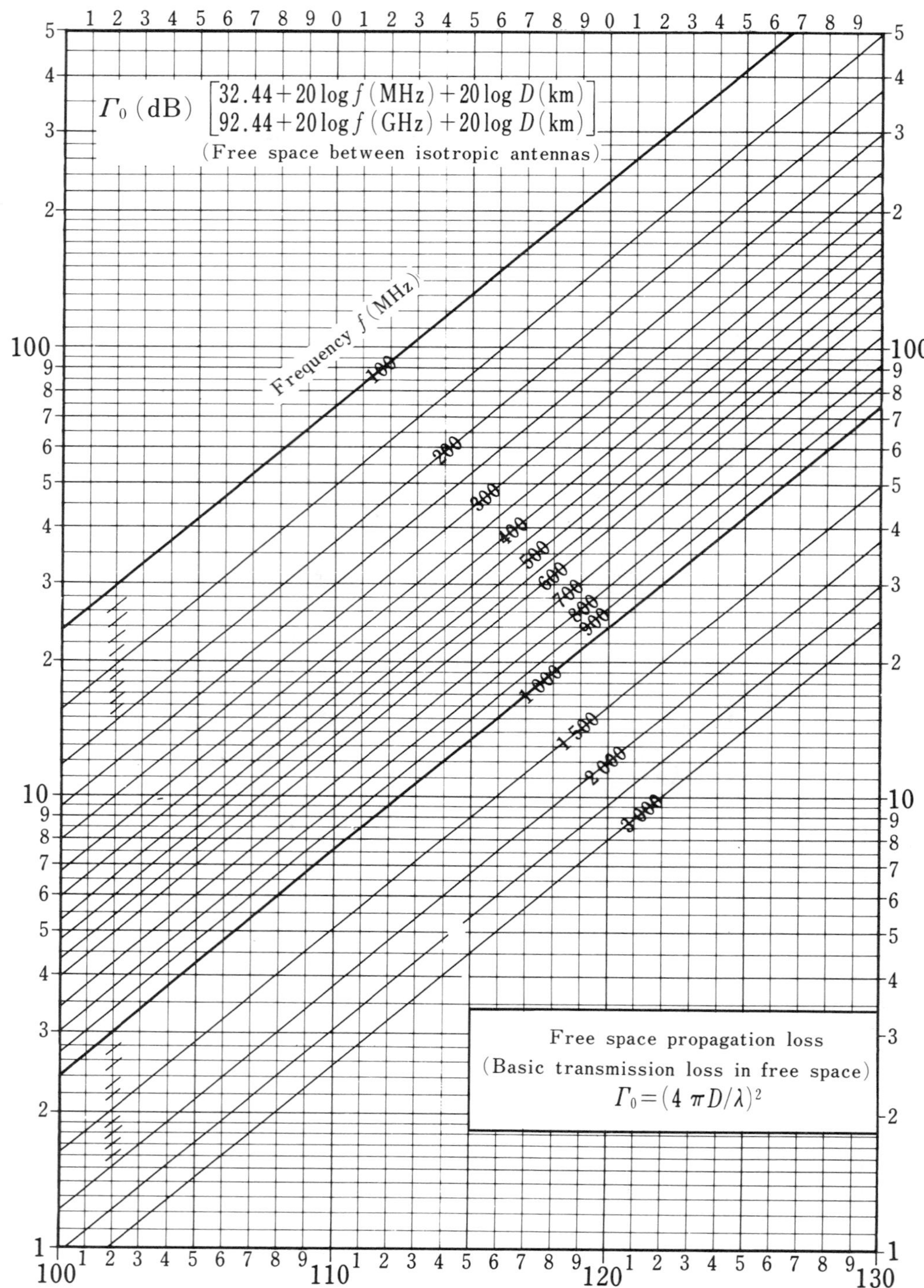

Free-Space Propagation Loss, UHF Band, 10–10³ km

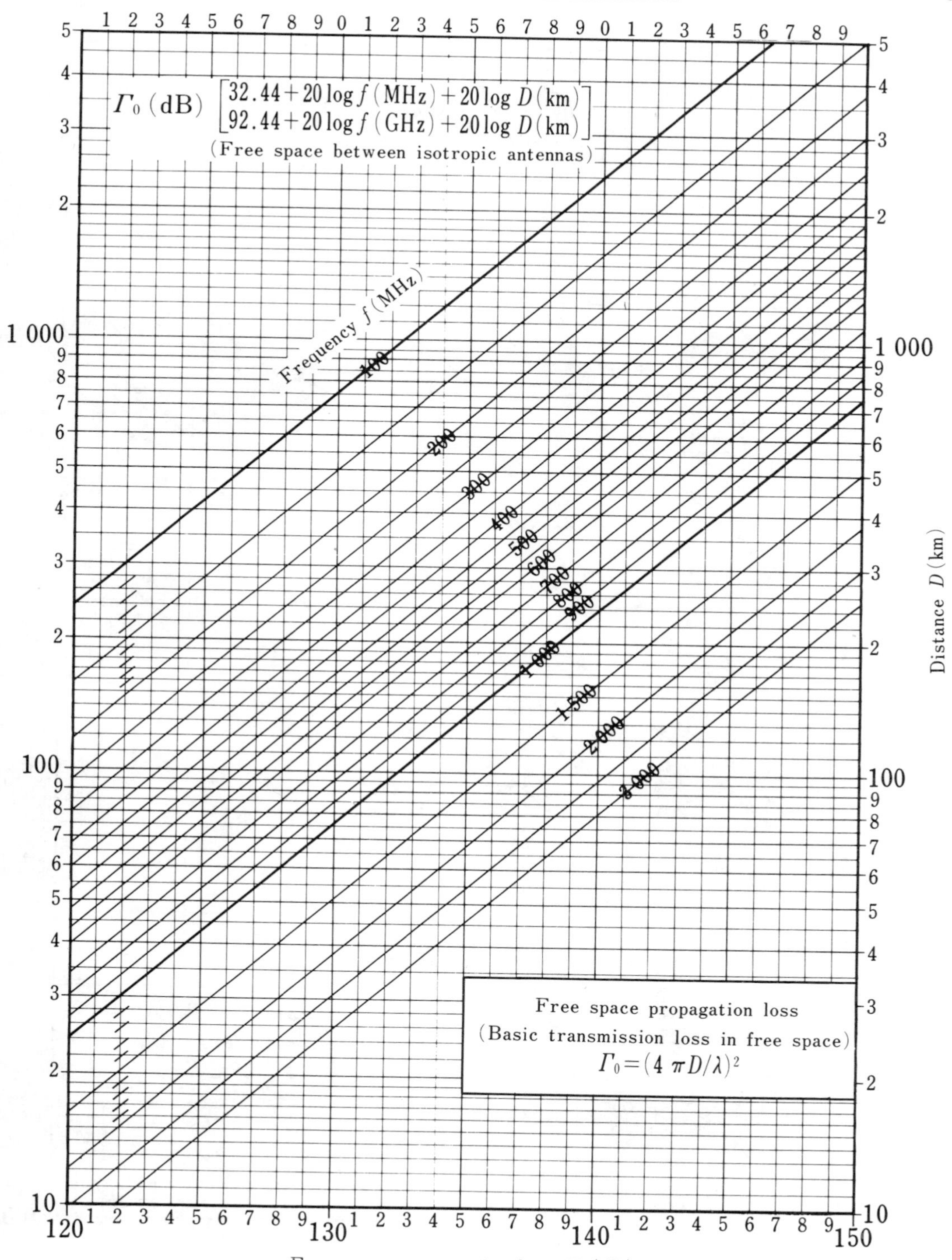

Free-Space Propagation Loss, UHF Band, 10^2–10^4 km

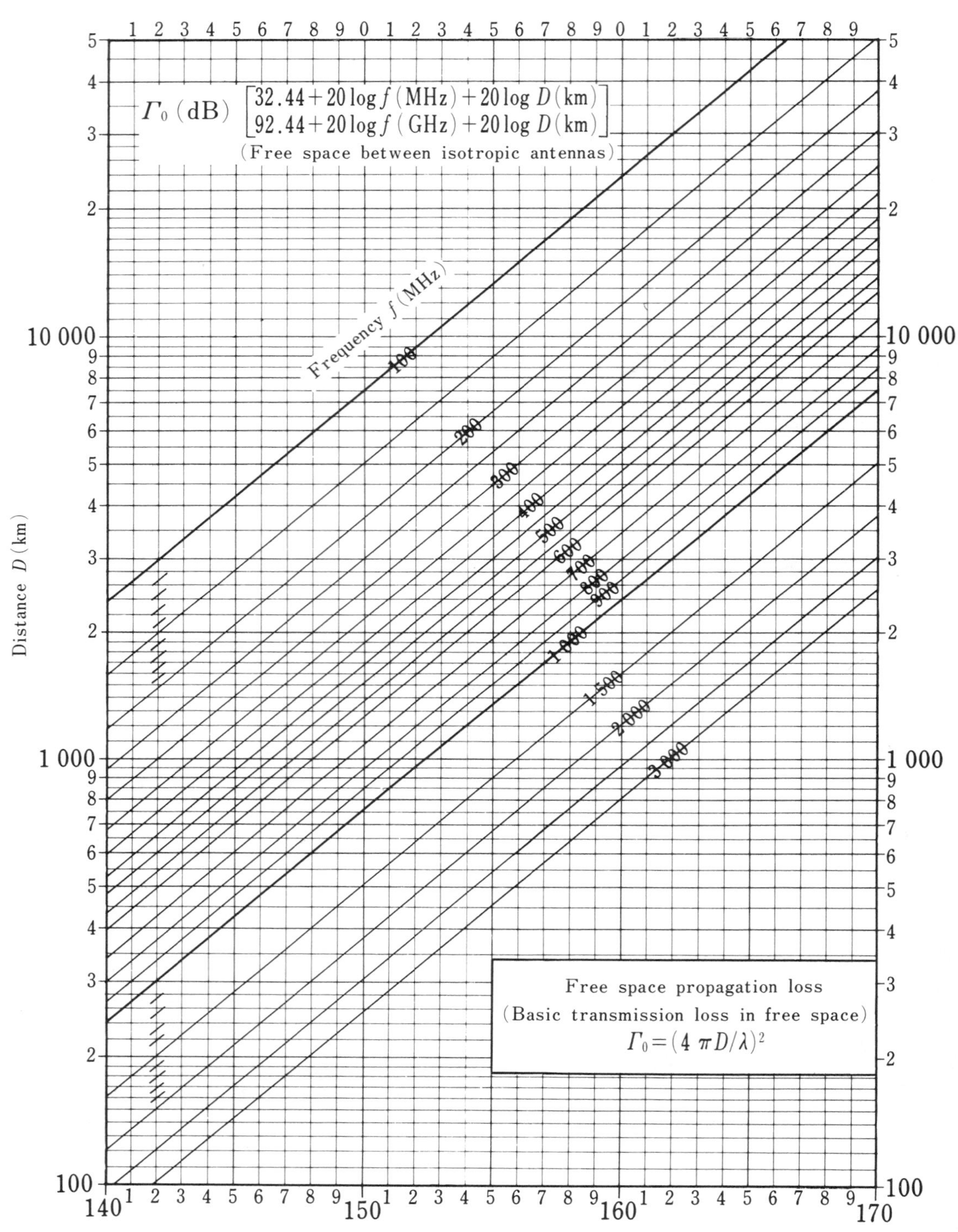

Free-Space Propagation Loss, UHF Band, 10^3–10^5 km

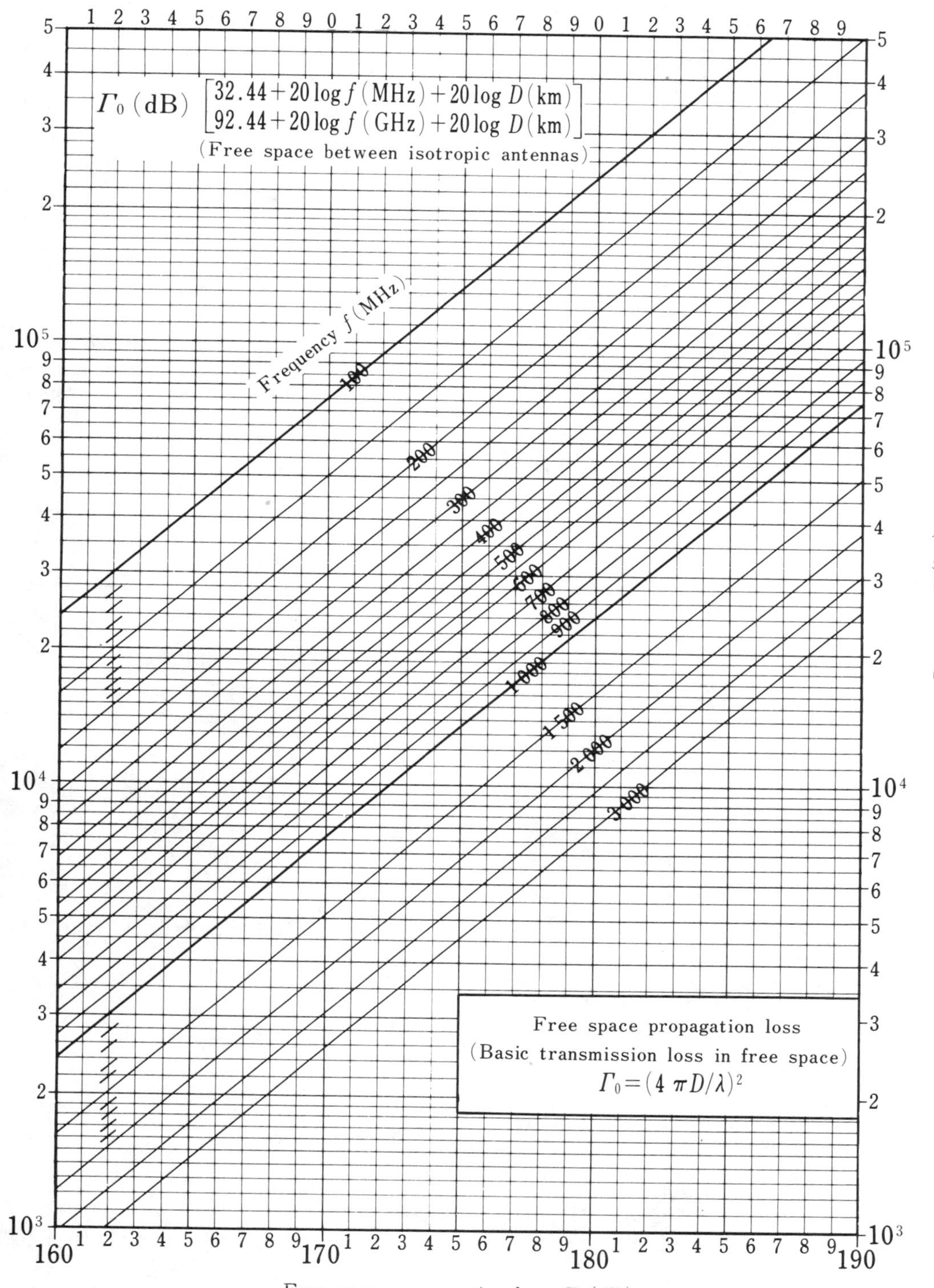

Free-Space Propagation Loss, UHF Band, 10^4–10^7 km

For Great Distance

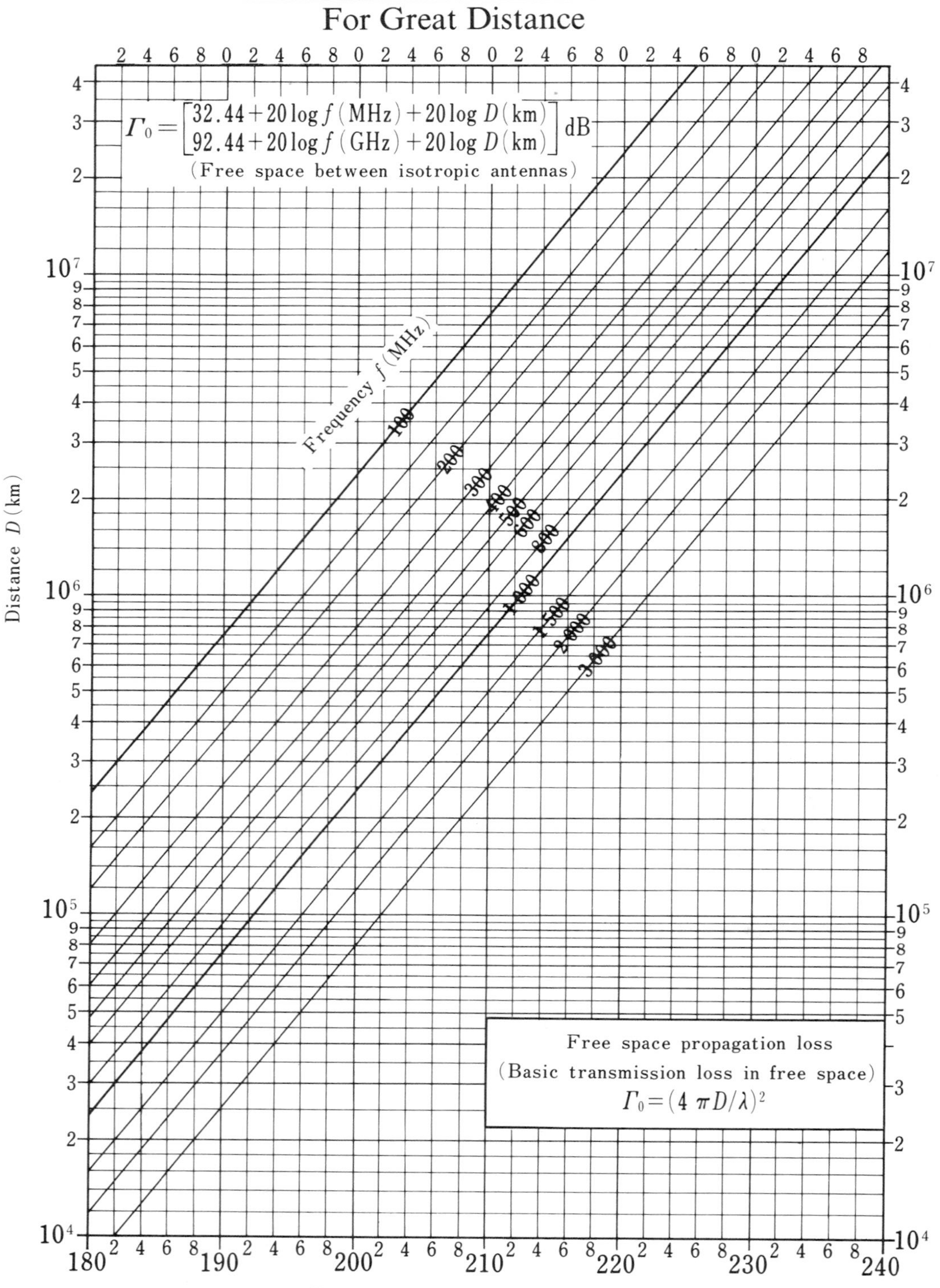

Free-Space Propagation Loss, UHF Band, 10^6–10^9 km

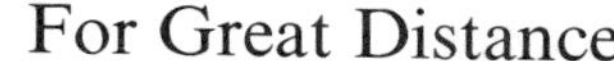

For Great Distance

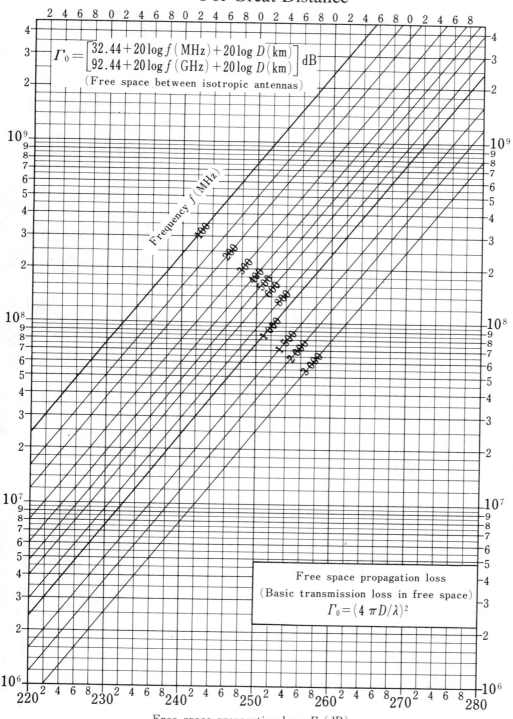

$$\Gamma_0 = \begin{bmatrix} 32.44 + 20\log f\,(\text{MHz}) + 20\log D\,(\text{km}) \\ 92.44 + 20\log f\,(\text{GHz}) + 20\log D\,(\text{km}) \end{bmatrix} \text{dB}$$

(Free space between isotropic antennas)

Frequency f (MHz)

100, 200, 300, 400, 500, 600, 800, 1 000, 1 500, 2 000, 3 000

Free space propagation loss
(Basic transmission loss in free space)

$$\Gamma_0 = (4\pi D/\lambda)^2$$

Distance D (km)

10^6, 10^7, 10^8, 10^9

220, 230, 240, 250, 260, 270, 280

Free space propagation loss Γ_0 (dB)

Free-Space Propagation Loss, SHF Band, 10–10^3 m

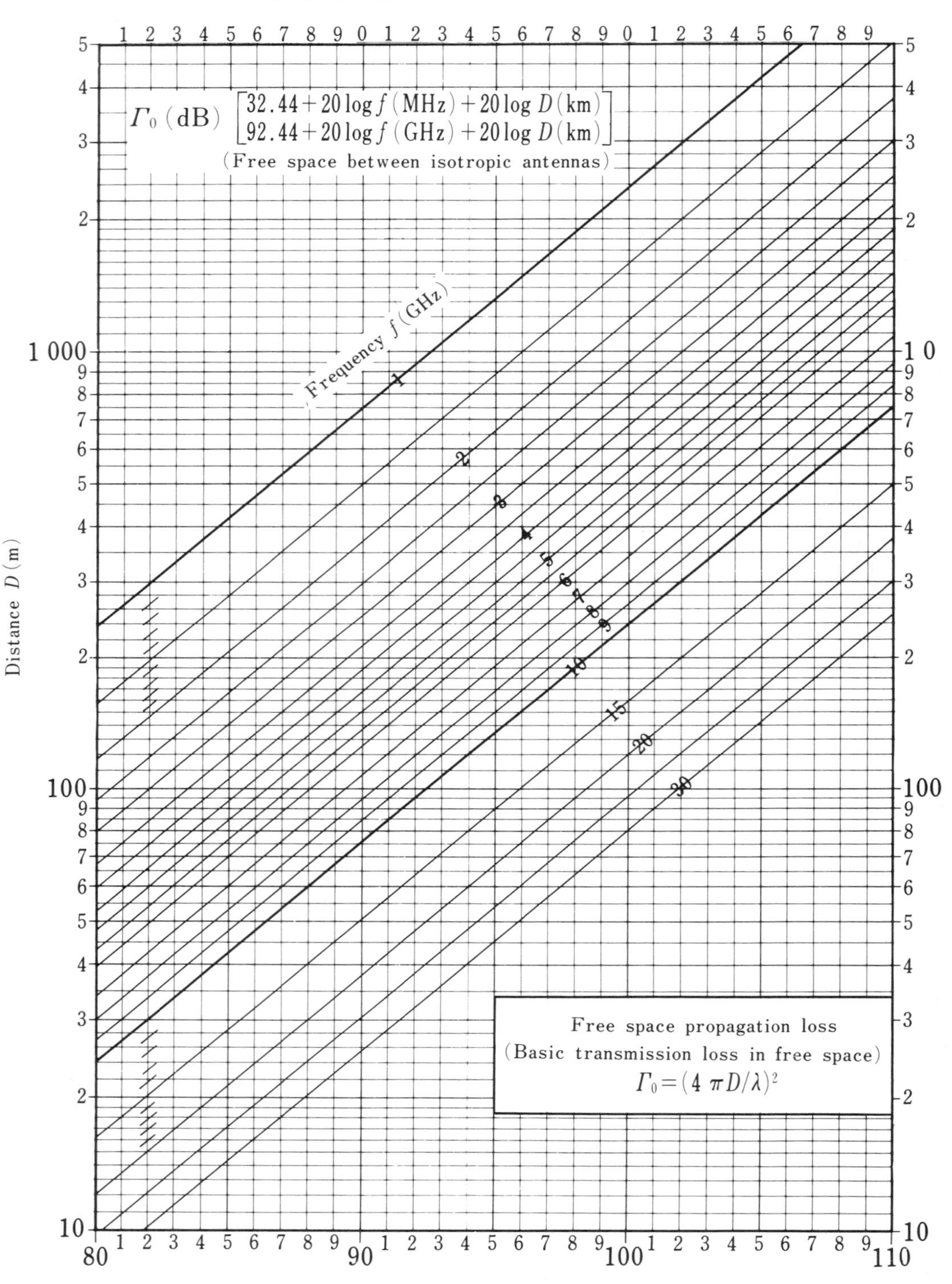

Free-Space Propagation Loss, SHF Band, 10^{-1}–10 km

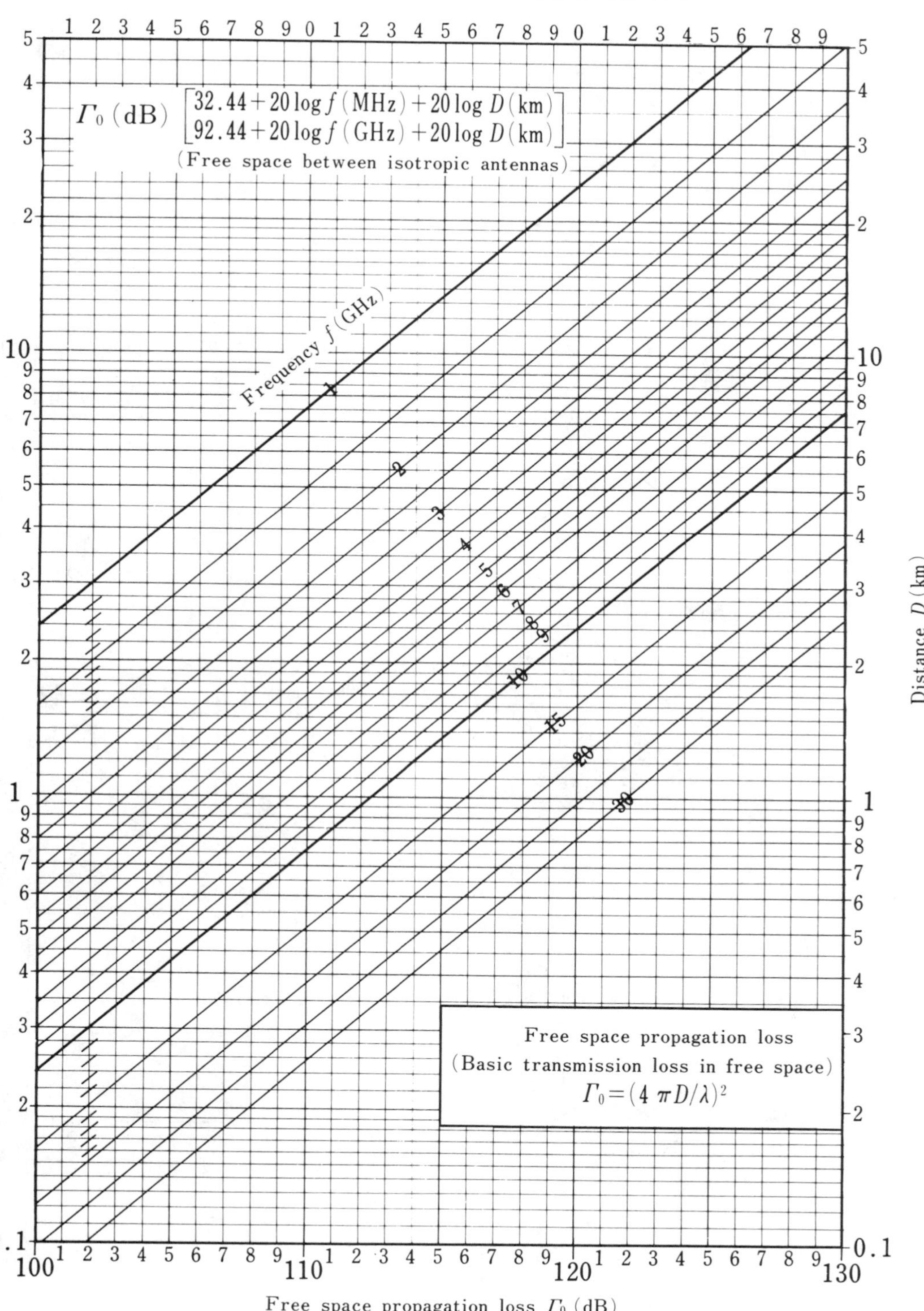

Free-Space Propagation Loss, SHF Band, 1–10² km

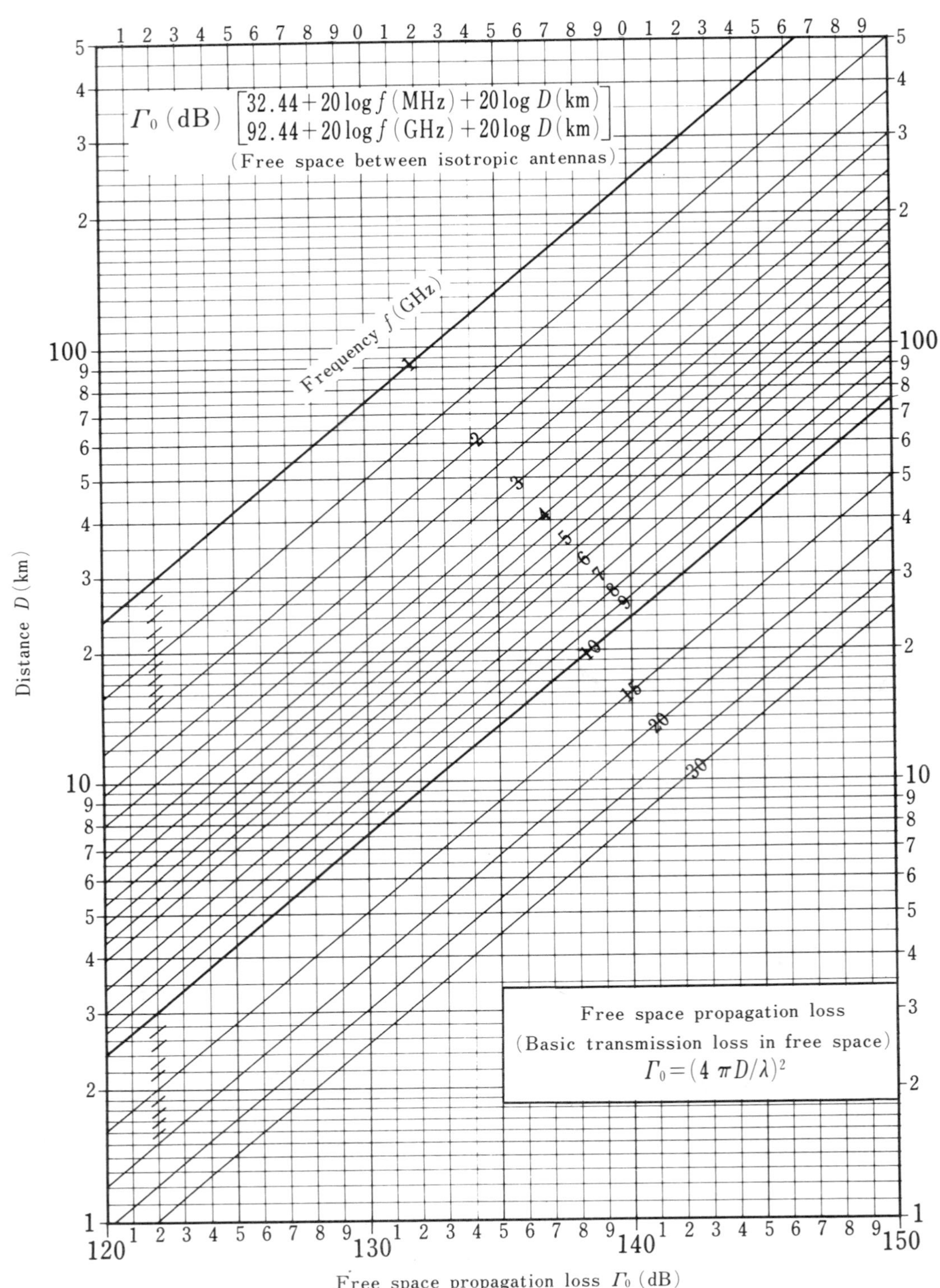

Free-Space Propagation Loss, SHF Band, 10–10³ km

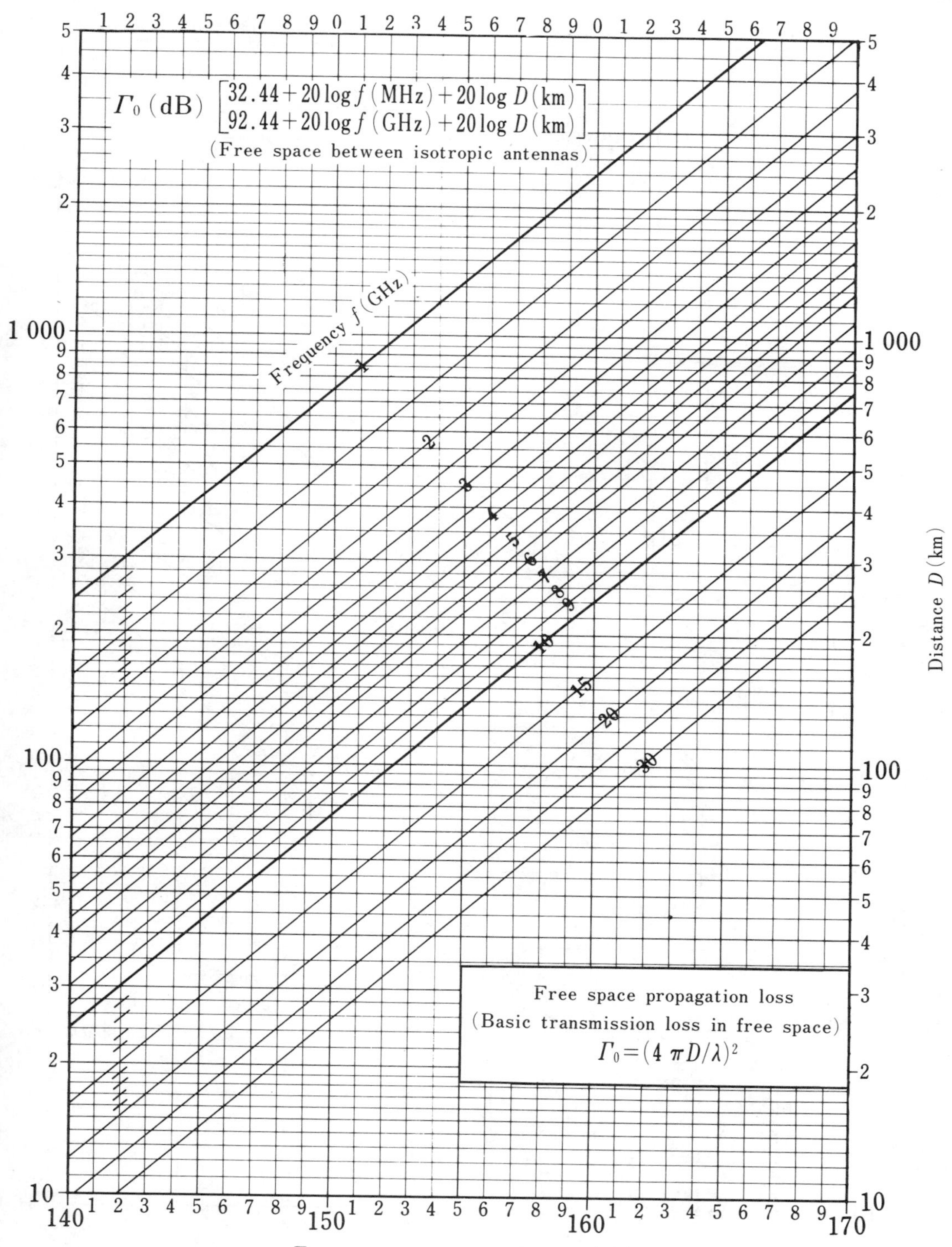

Free-Space Propagation Loss, SHF Band, 10^2–10^4 km

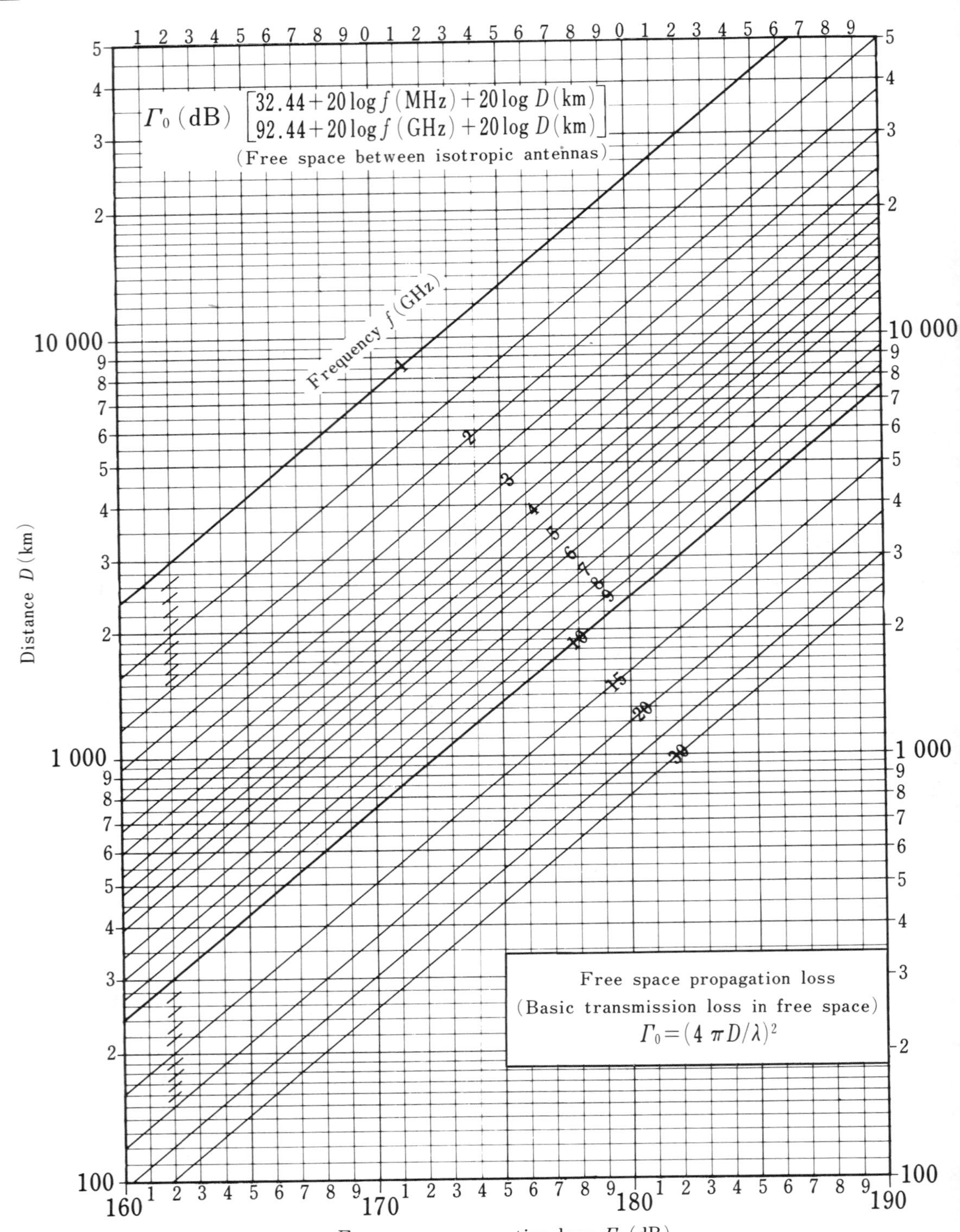

Free-Space Propagation Loss, SHF Band, 10^3–10^5 km

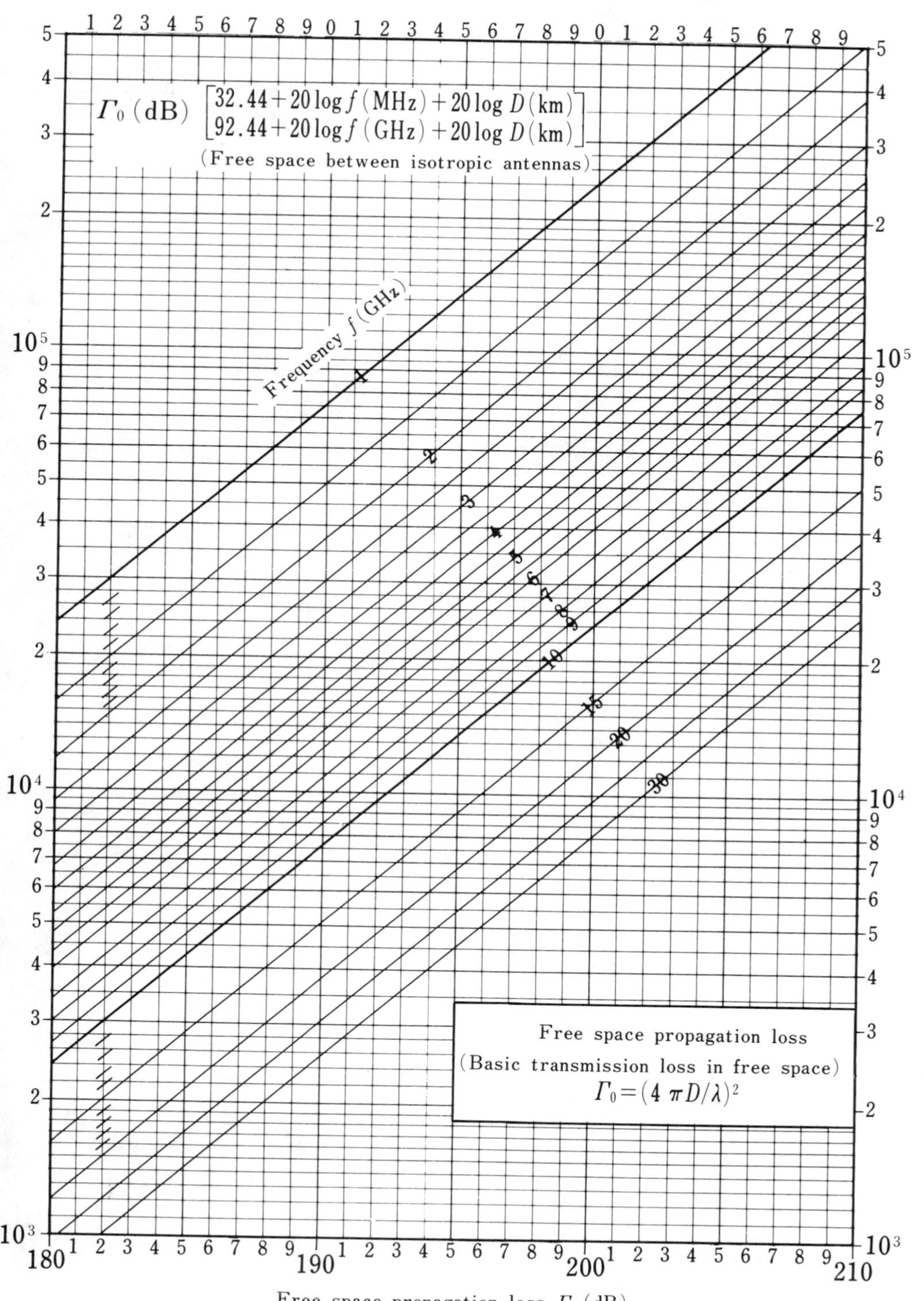

Free-Space Propagation Loss, SHF Band, 10^4–10^7 km

For Great Distance

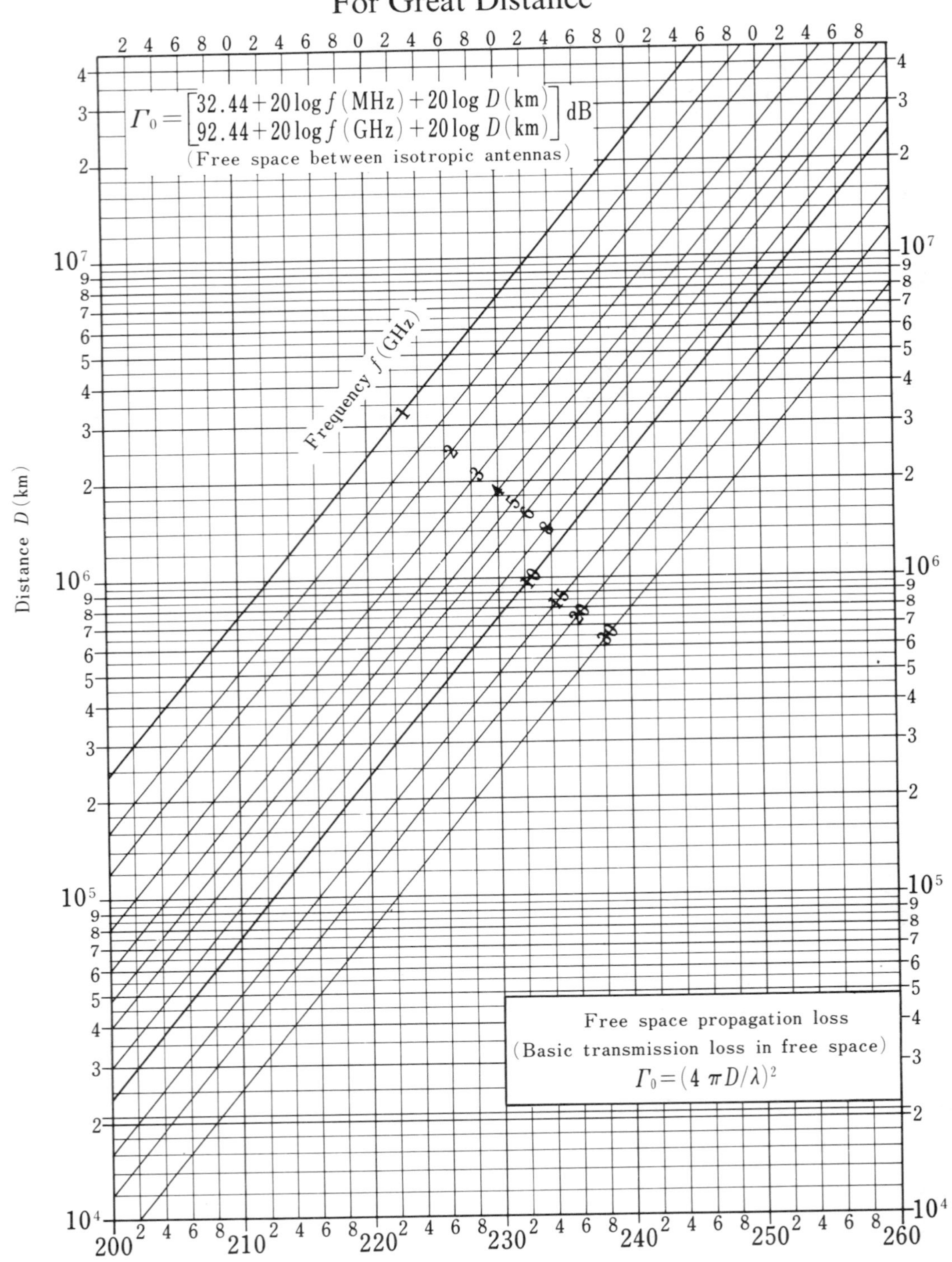

Free-Space Propagation Loss, SHF Band, 10^6–10^9 km

For Great Distance

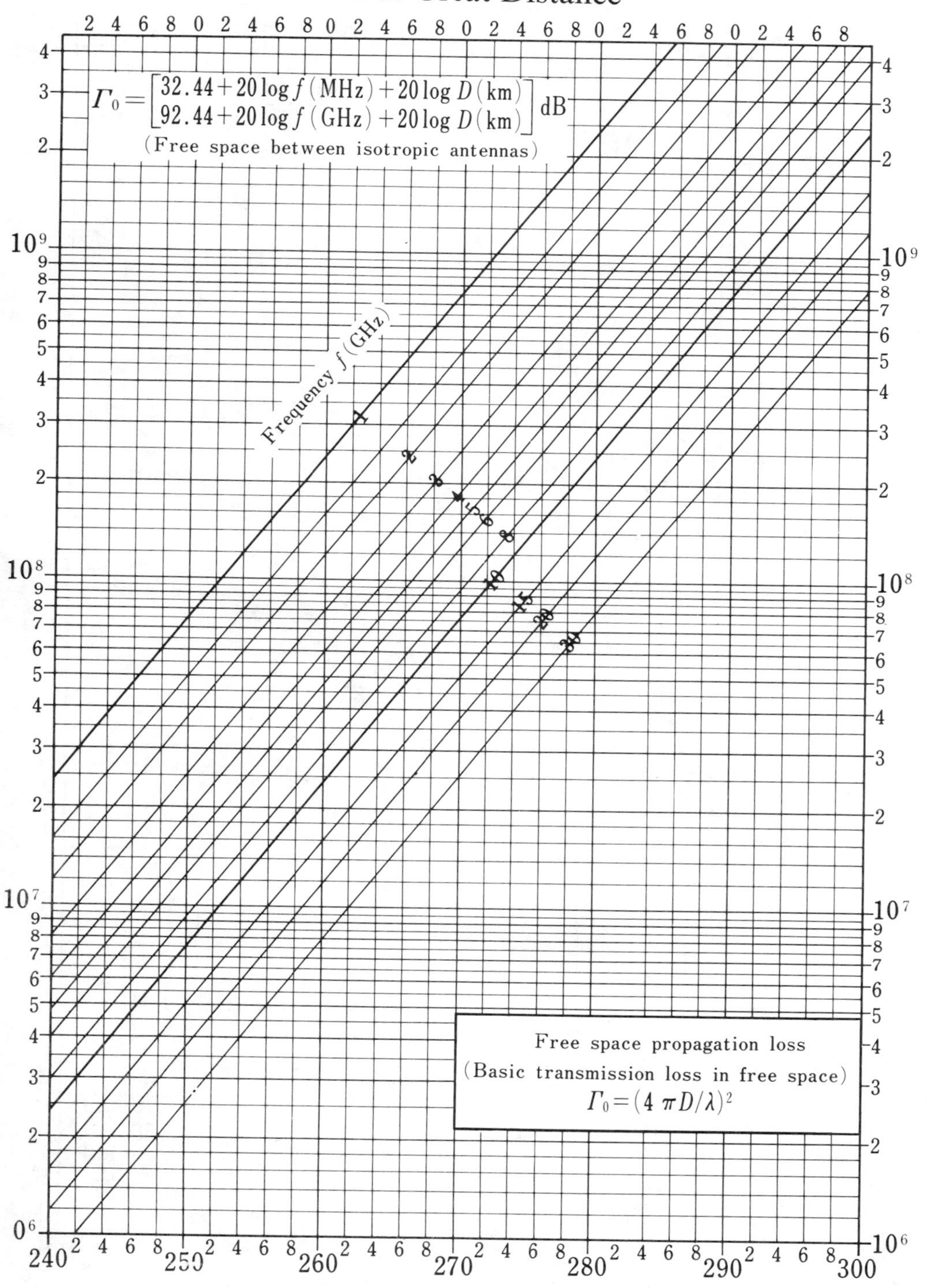

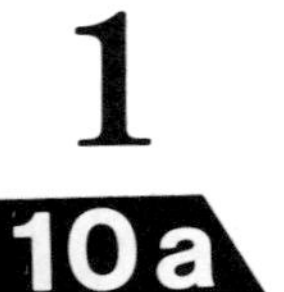

Free-Space Propagation Loss, EHF Band, 10–10³ m

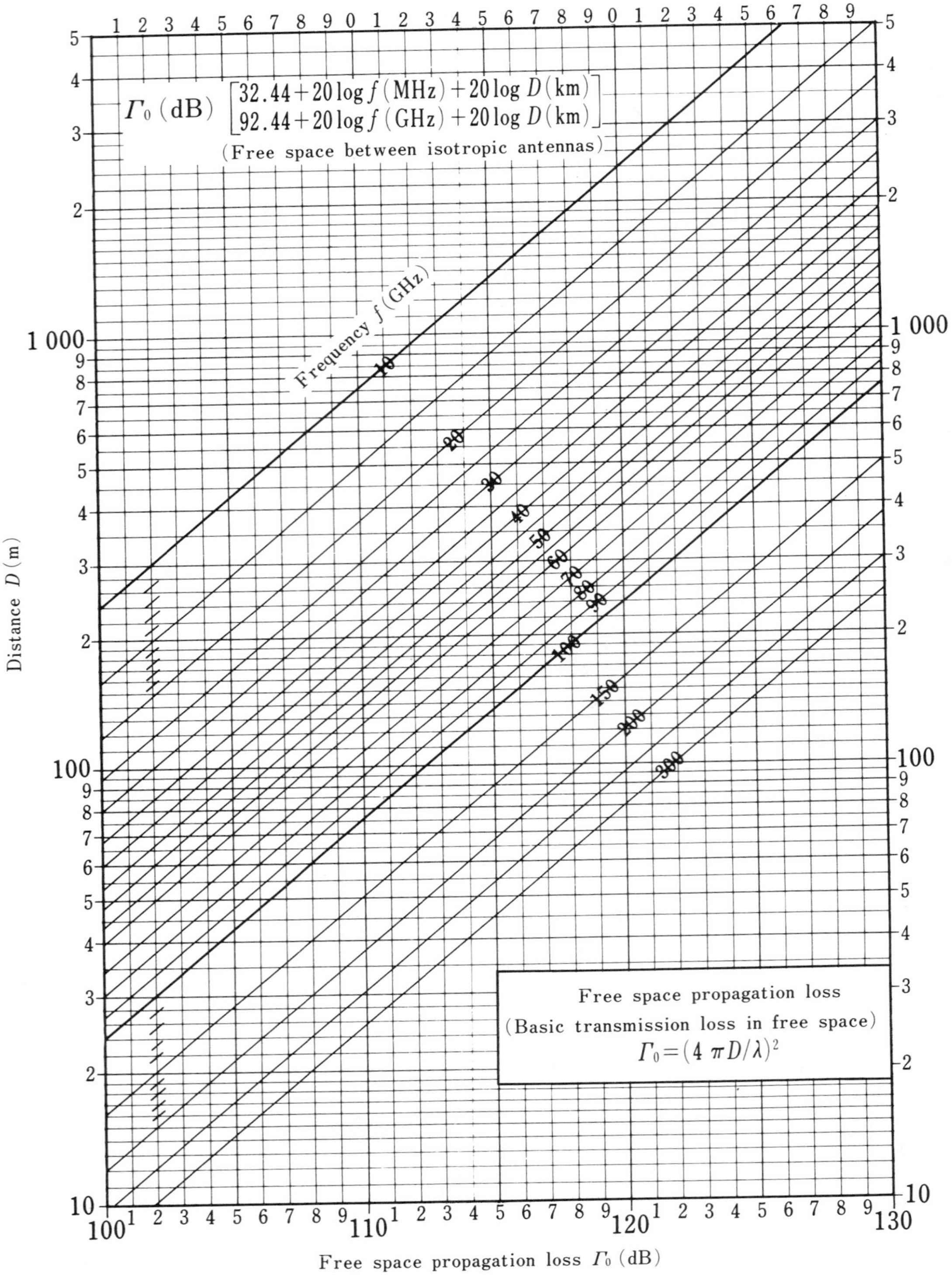

Free-Space Propagation Loss, EHF Band, 10^{-1}–10 km

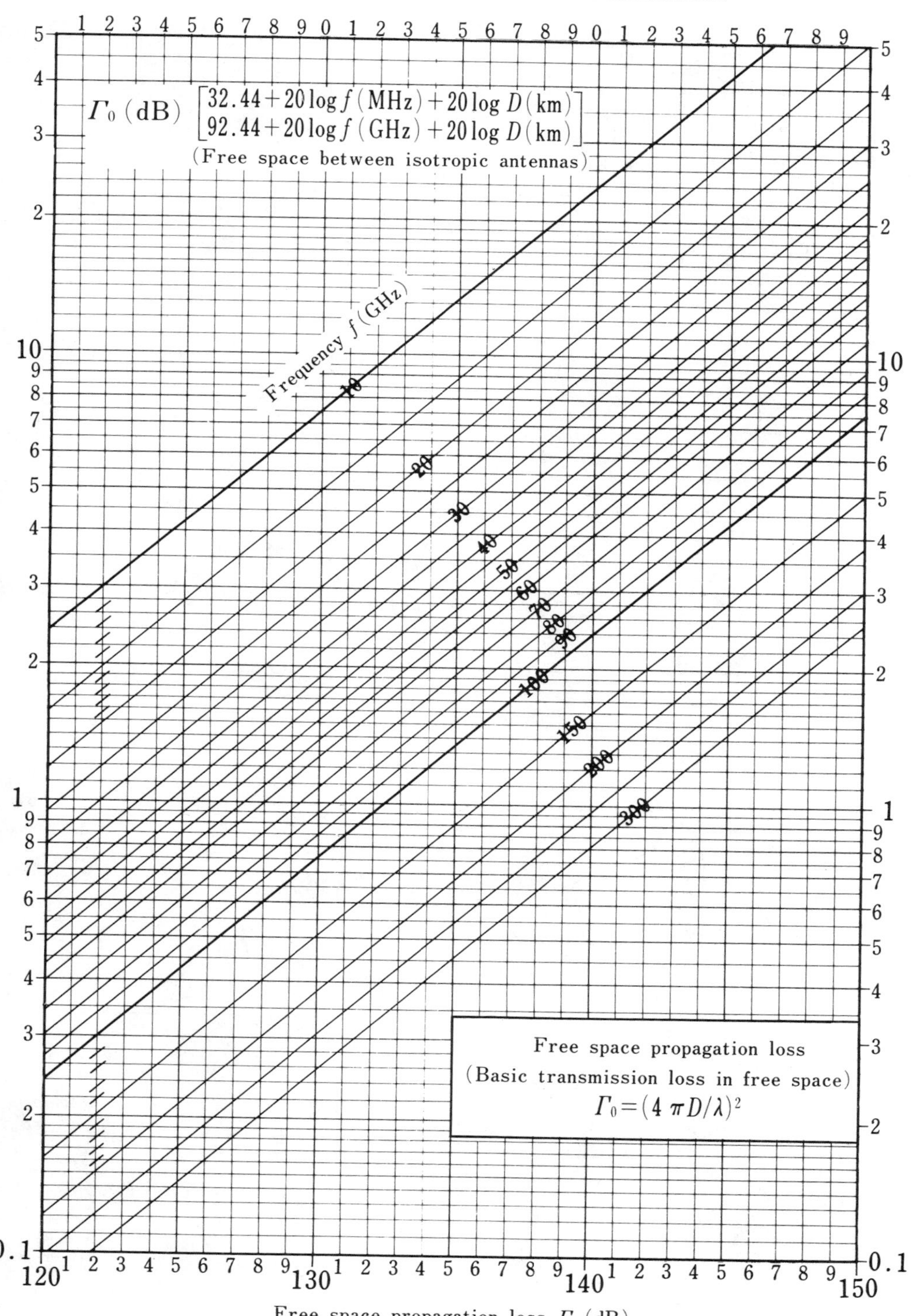

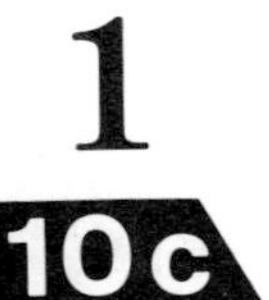

Free-Space Propagation Loss, EHF Band, 1–10² km

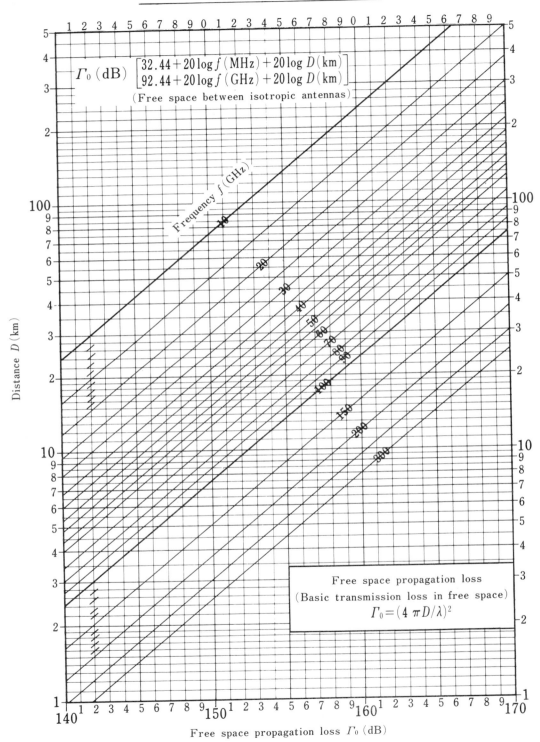

$$\Gamma_0\ (\text{dB}) \begin{bmatrix} 32.44 + 20\log f\,(\text{MHz}) + 20\log D\,(\text{km}) \\ 92.44 + 20\log f\,(\text{GHz}) + 20\log D\,(\text{km}) \end{bmatrix}$$

(Free space between isotropic antennas)

Frequency f (GHz)

10, 20, 30, 40, 50, 60, 70, 80, 90, 100, 150, 200, 300

Distance D (km)

Free space propagation loss
(Basic transmission loss in free space)
$\Gamma_0 = (4\pi D/\lambda)^2$

Free space propagation loss Γ_0 (dB)

140, 150, 160, 170

Free-Space Propagation Loss, EHF Band, 10–10³ km

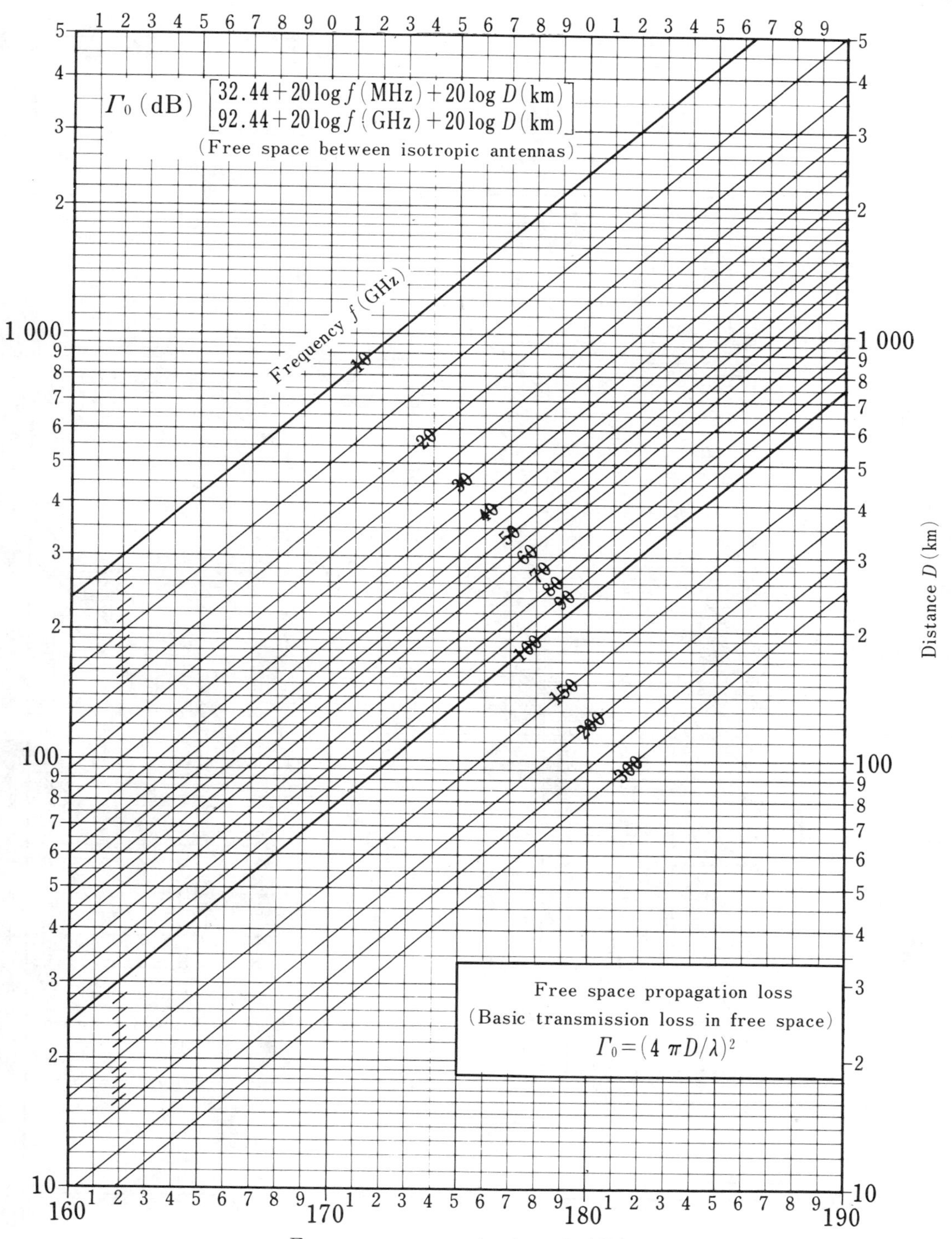

Free-Space Propagation Loss, EHF Band, 10^2–10^4 km

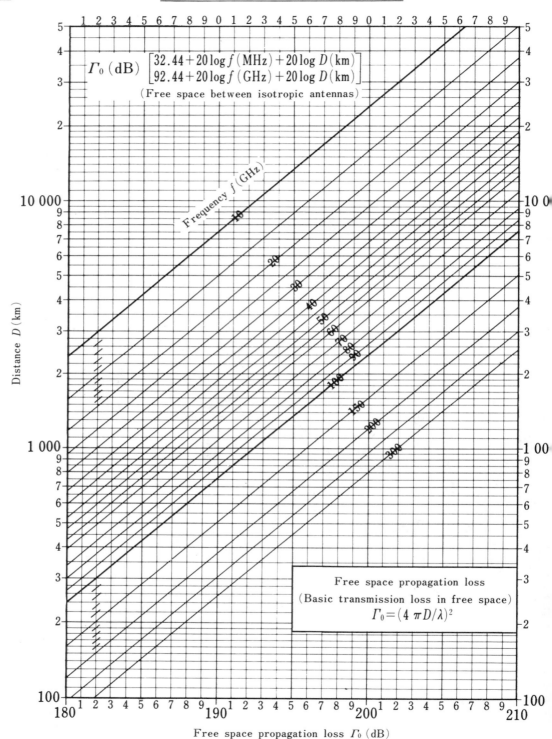

Γ_0 (dB) $\left[\begin{array}{l} 32.44 + 20 \log f\,(\text{MHz}) + 20 \log D\,(\text{km}) \\ 92.44 + 20 \log f\,(\text{GHz}) + 20 \log D\,(\text{km}) \end{array}\right]$

(Free space between isotropic antennas)

Frequency f (GHz): 10, 20, 30, 40, 50, 60, 70, 80, 90, 100, 150, 200, 300

Distance D (km): 100 – 10 000 – 50 000

Free space propagation loss Γ_0 (dB): 180 – 190 – 200 – 210

Free space propagation loss
(Basic transmission loss in free space)
$\Gamma_0 = (4\,\pi D/\lambda)^2$

Free-Space Propagation Loss, EHF Band, 10^3–10^5 km

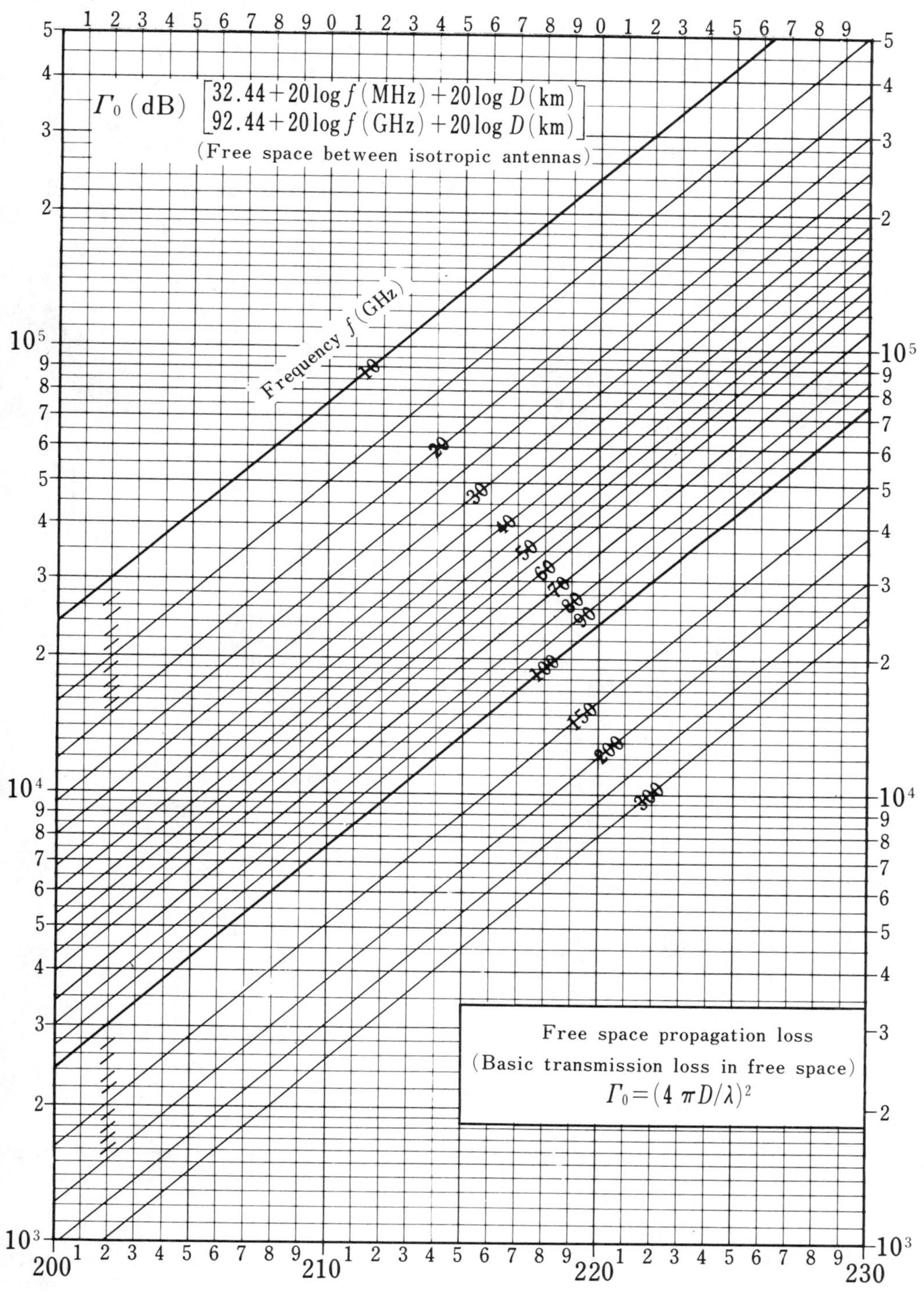

Free-Space Propagation Loss, EHF Band, 10^4–10^7 km

For Great Distance

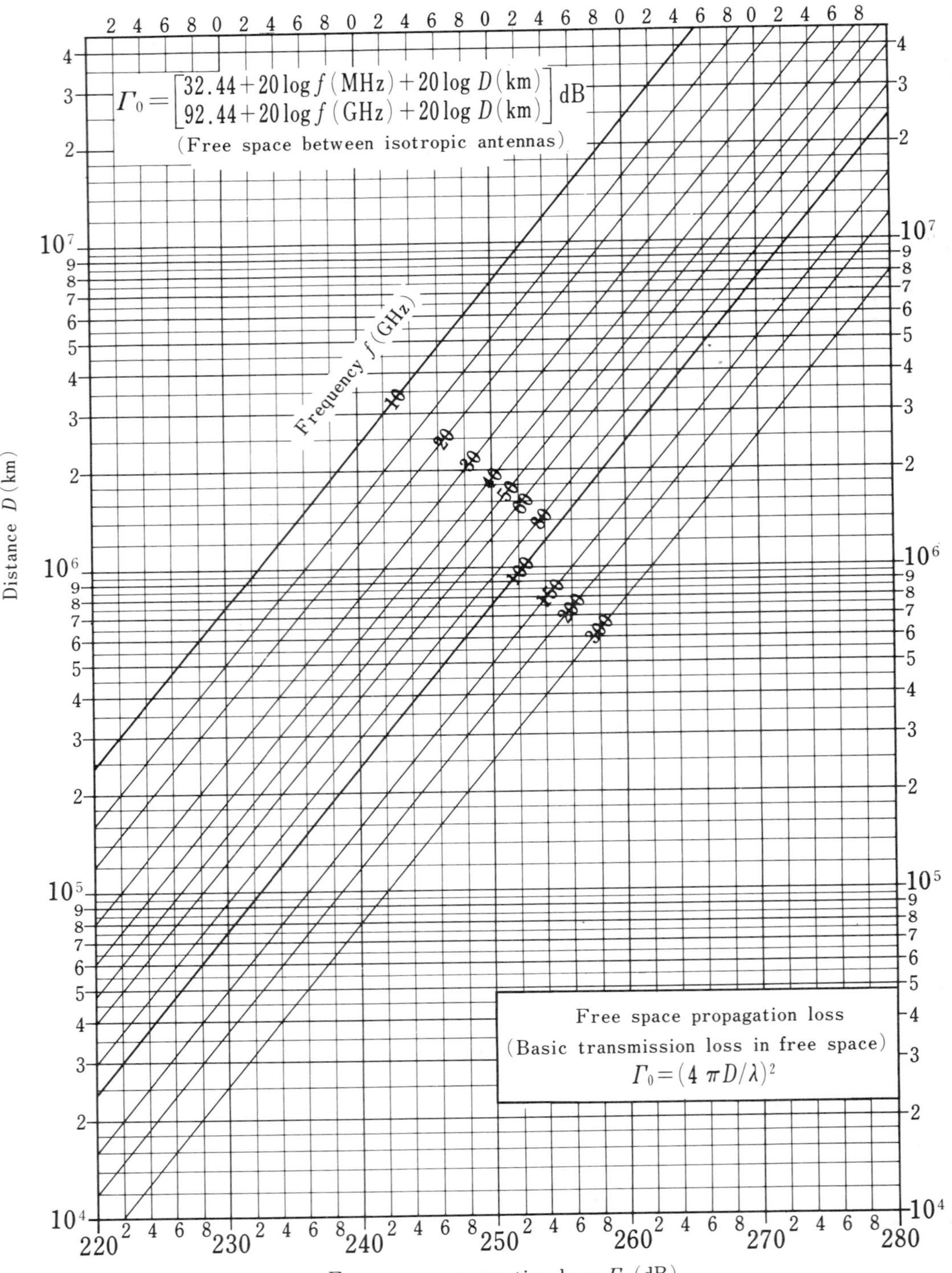

Free-Space Propagation Loss, EHF Band, 10^6–10^9 km

For Great Distance

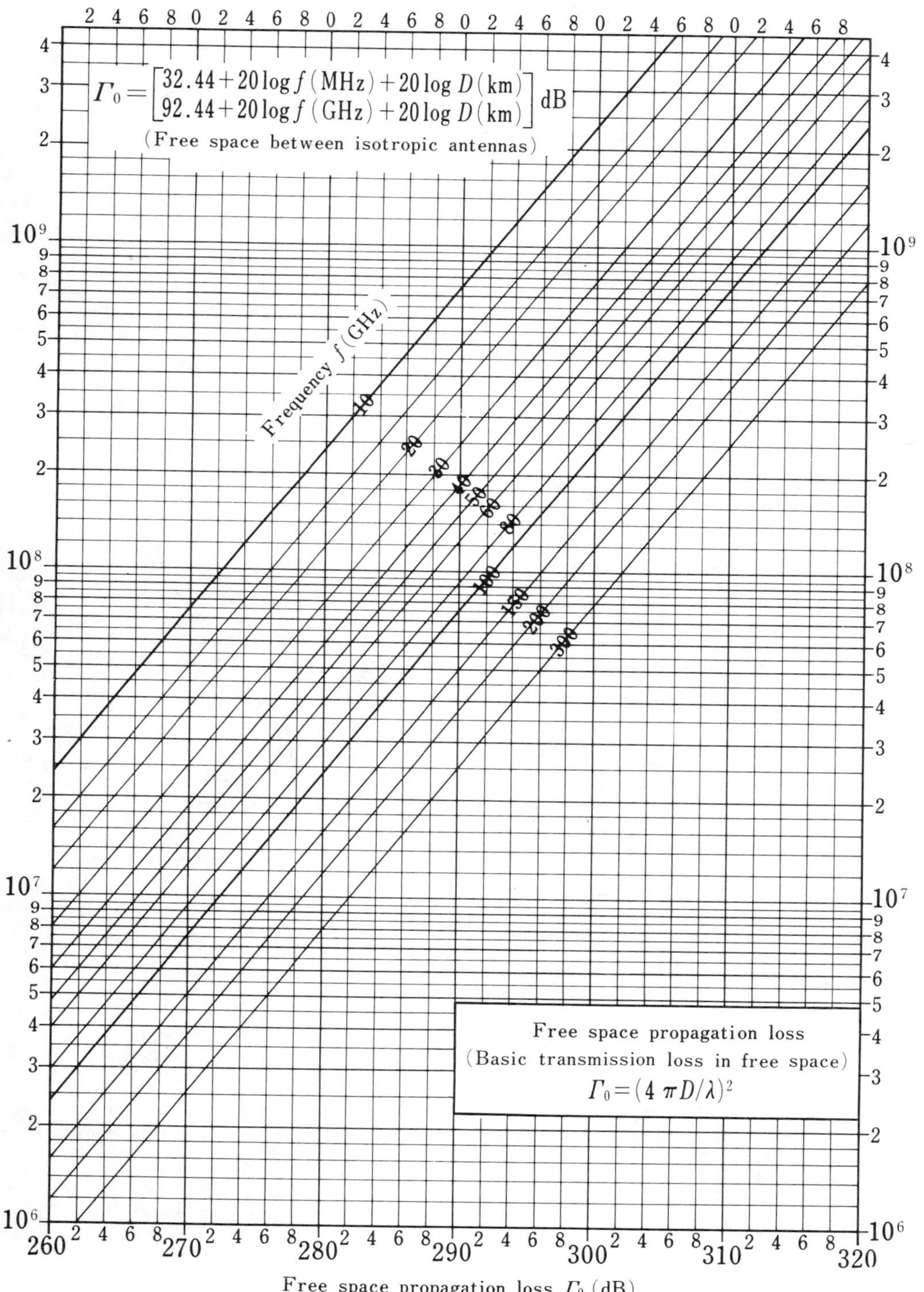

Application of Transmission Loss between Unit-Area Antennas

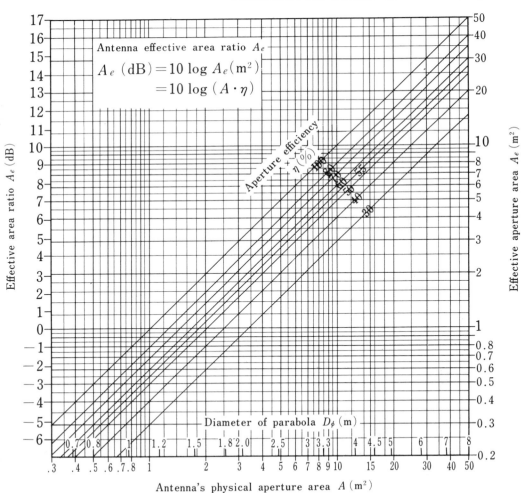

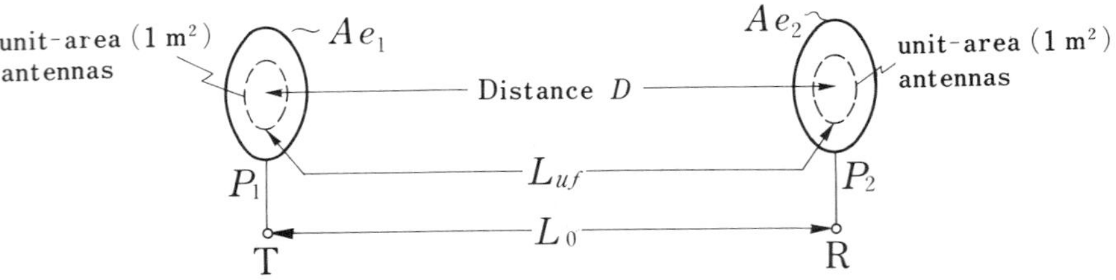

Transmission loss between transmit and receive antennas (Free space)

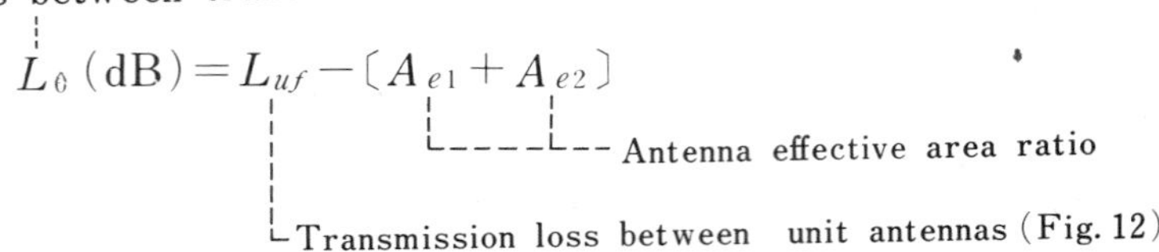

$$L_0\,(\mathrm{dB}) = L_{uf} - [A_{e1} + A_{e2}]$$

Antenna effective area ratio A_e

$$A_e\,(\mathrm{dB}) = 10 \log A_e(\mathrm{m}^2) = 10 \log (A \cdot \eta)$$

Aperture efficiency η (%): 100, 80, 70, 60, 55, 50, 40, 30

Diameter of parabola D_ϕ (m): 0.7, 0.8, 1, 1.2, 1.5, 1.8, 2.0, 2.5, 3, 3.3, 4, 4.5, 5, 6, 7, 8

Effective area ratio A_e (dB)

Effective aperture area A_e (m²)

Antenna's physical aperture area A (m²)

Transmission Loss between Unit-Area Antennas (1)

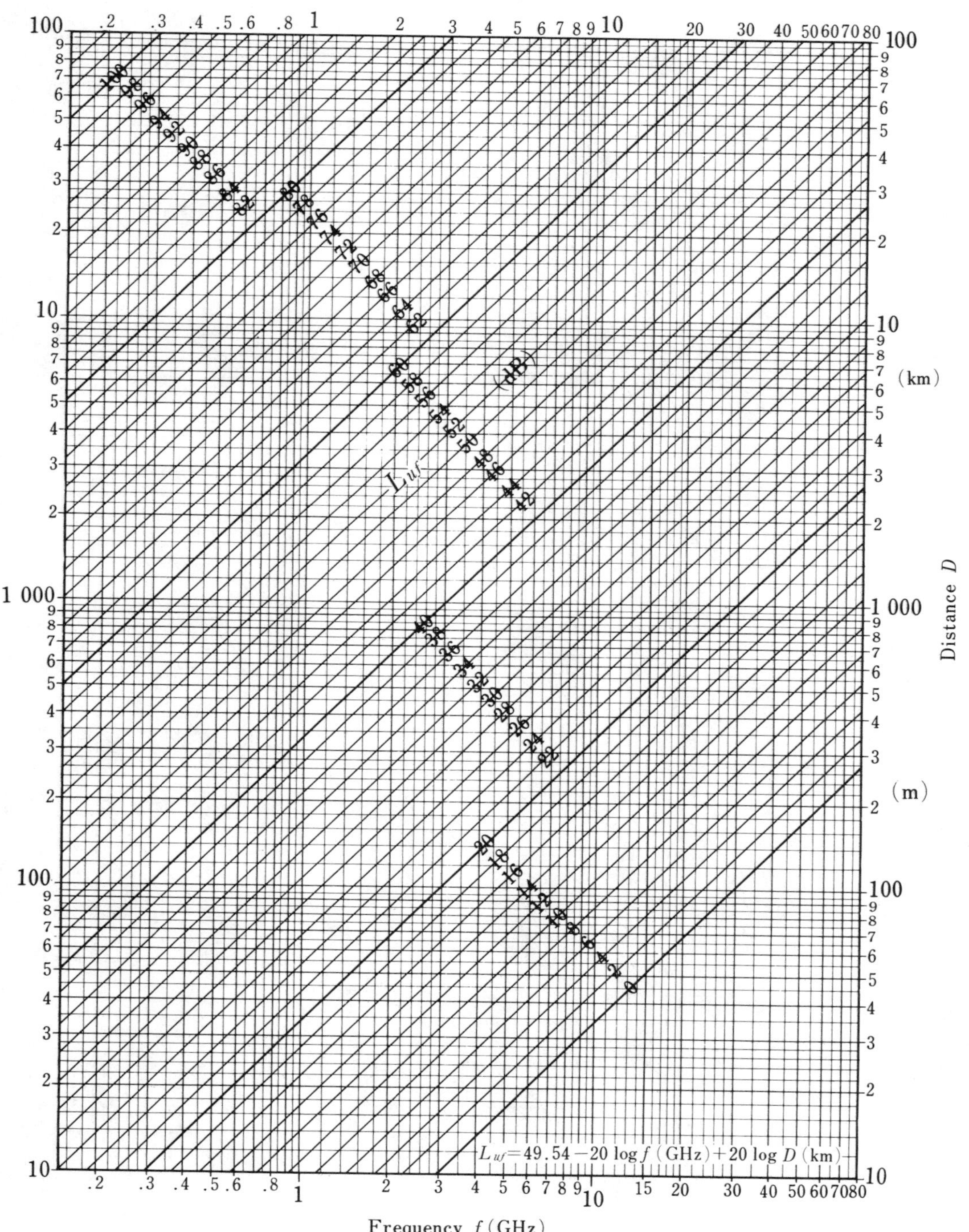

Transmission Loss between Unit-Area Antennas (2)

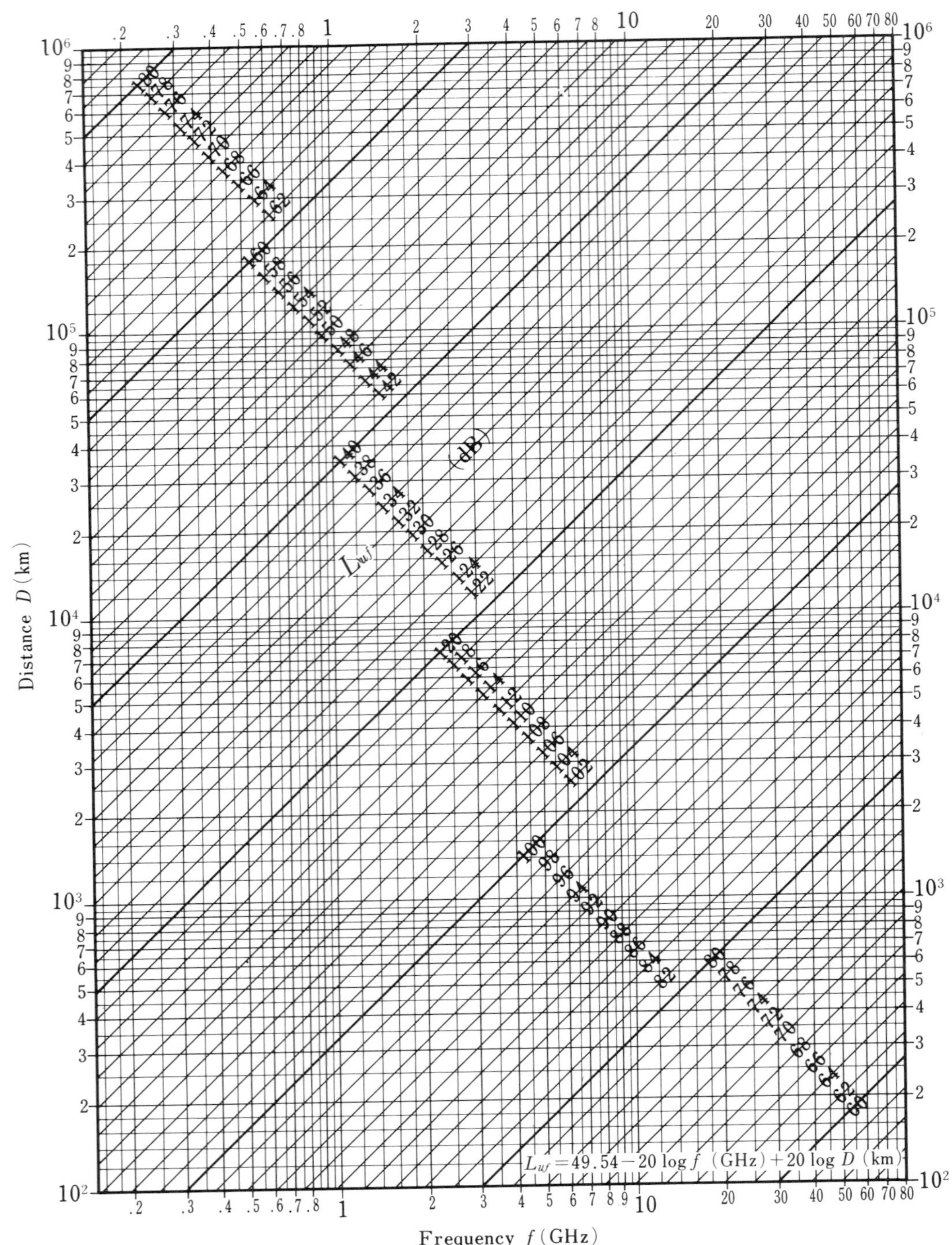

Transmission Loss between Unit-Area Antennas (3)

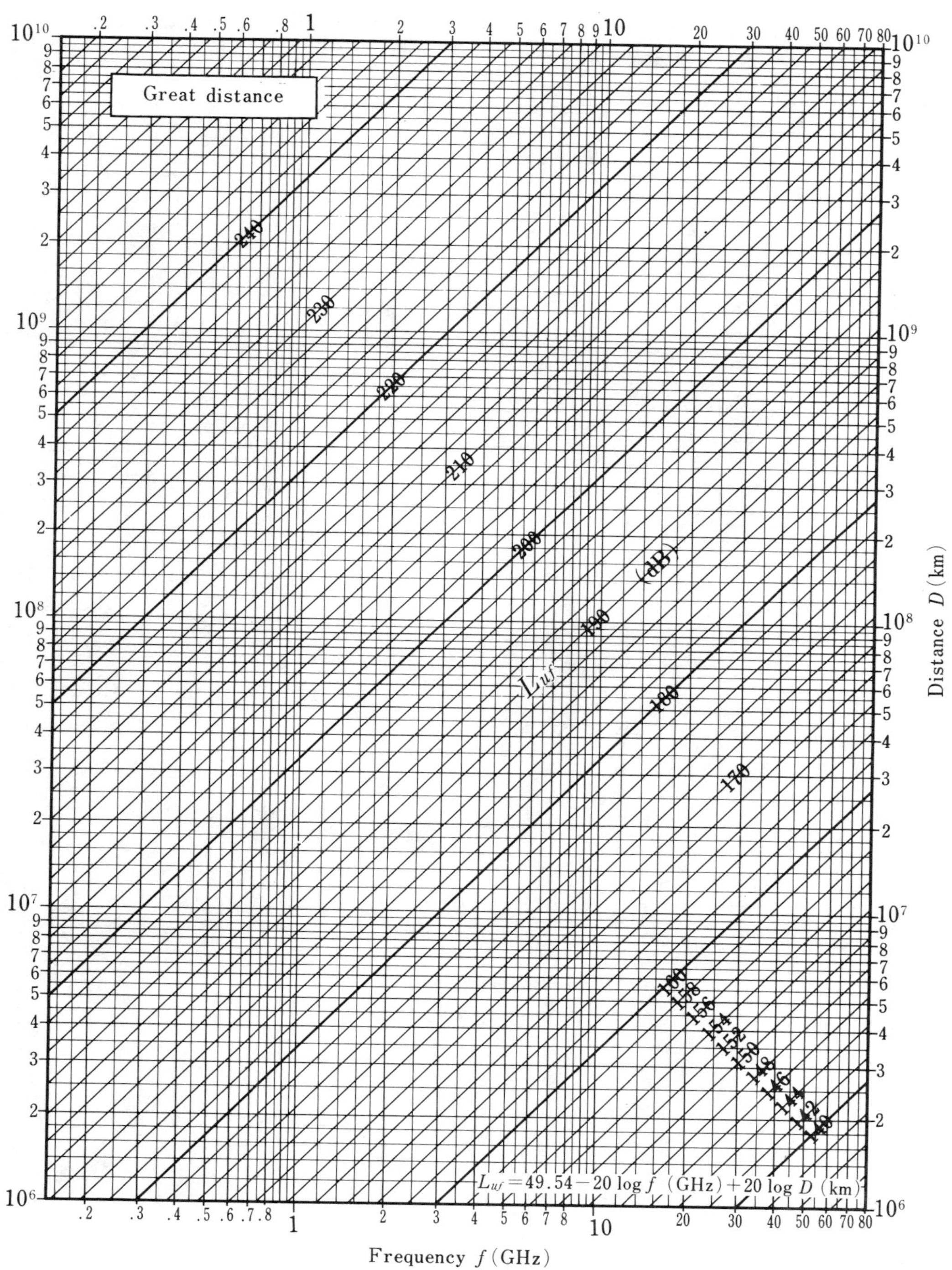

Antenna Gain (1)

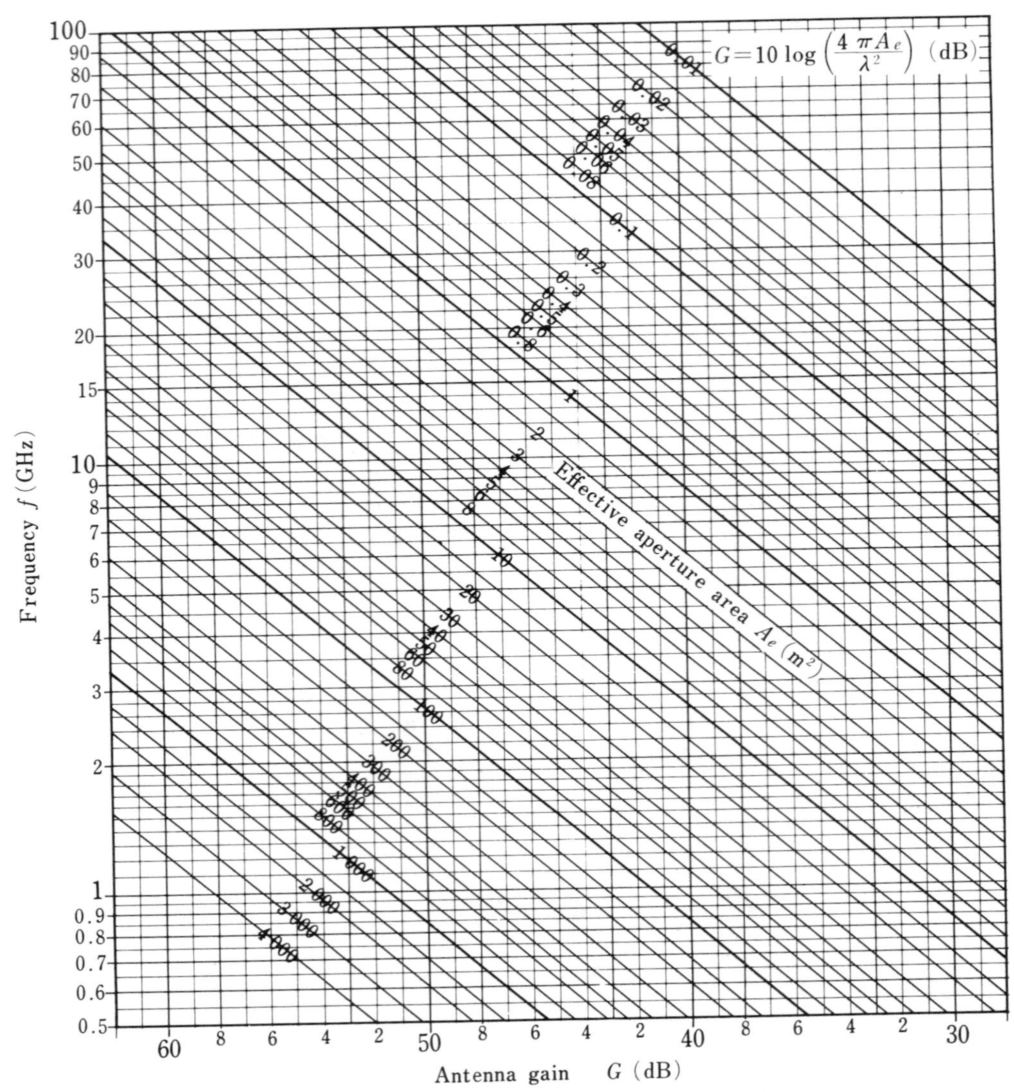

Antenna Gain (2)

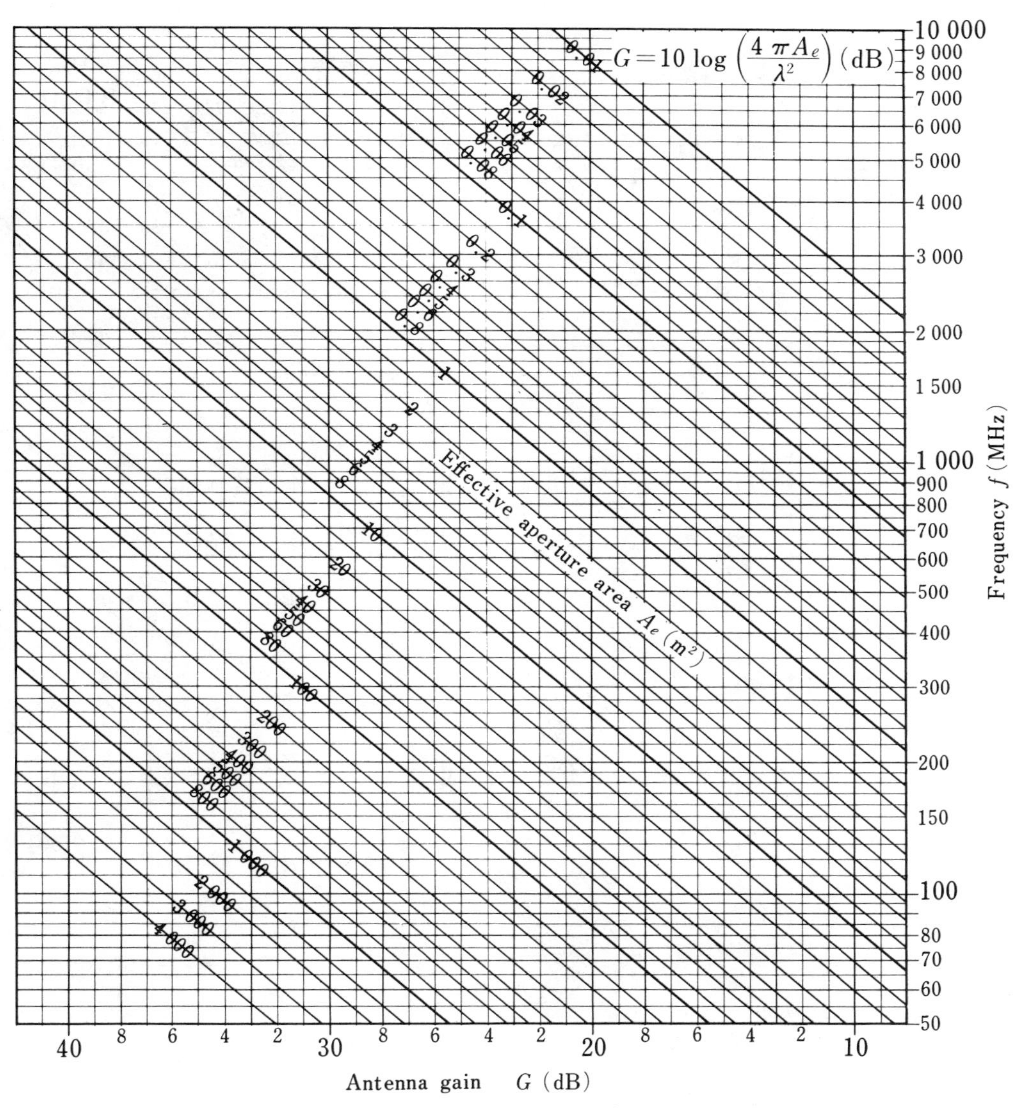

Effective Aperture Area of Parabola

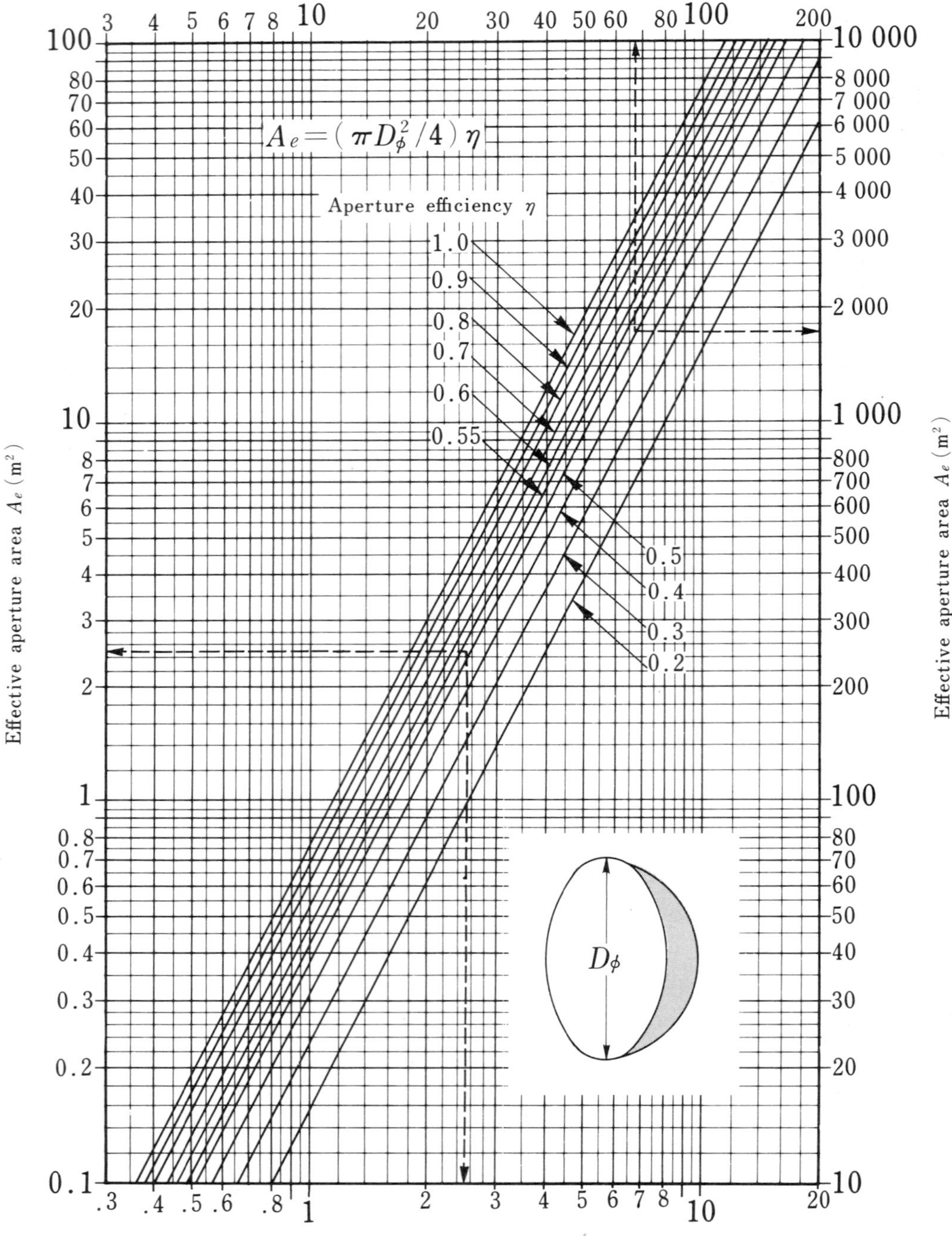

Nomograph for a Plane Wave, Simplified Calculations (1)

Free-Space Field Strength

$$E_0\ (\mu\mathrm{V/m}) = 10^3 \cdot \sqrt{30}\sqrt{P_W}/D_{\mathrm{km}}$$

$$E_0\ \mathrm{dB}(\mu\mathrm{V/m}) = 10\log P_W - 20\log D_{\mathrm{km}} + 74.77$$

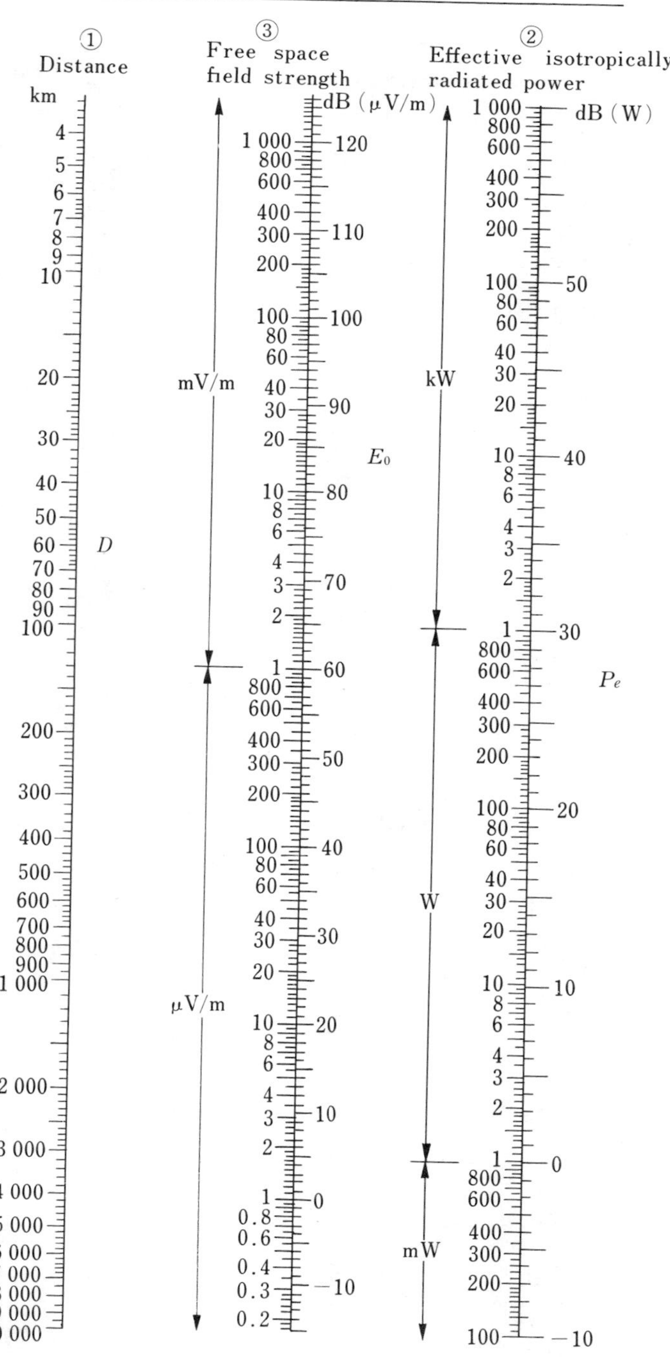

Nomograph for a Plane Wave, Simplified Calculations (2)

Free-Space Propagation Loss

$$\Gamma_0\,(\text{dB}) = 20 \log D_{km} - 20 \log \lambda_{cm} + 121.98$$

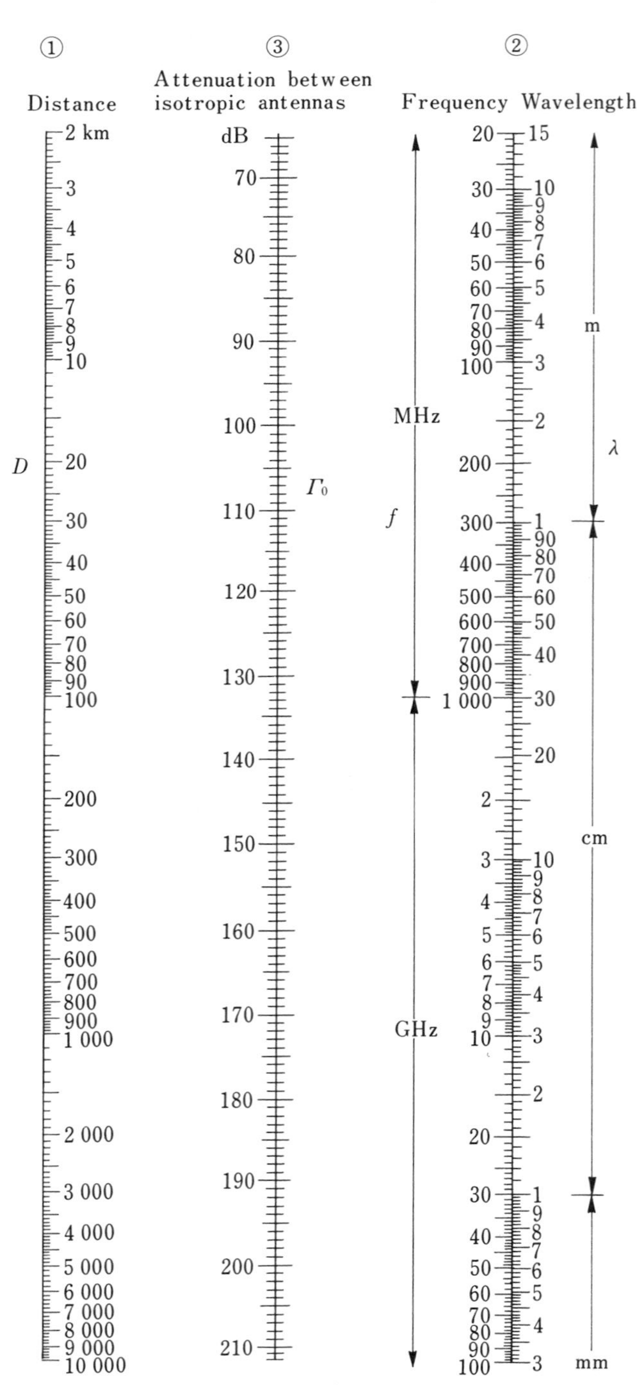

Nomograph for a Plane Wave, Simplified Calculations (3)

Free-Space Unit Aperture Area Antenna Transmission Loss

$$L_{uf}\,(\mathrm{dB}) = 20 \log D_{\mathrm{km}} + 20 \log \lambda_{\mathrm{cm}} + 20$$

(Unit-area antenna: $A_e = 1\ \mathrm{m}^2$, $G = 4\pi/\lambda^2 \mathrm{m}$)

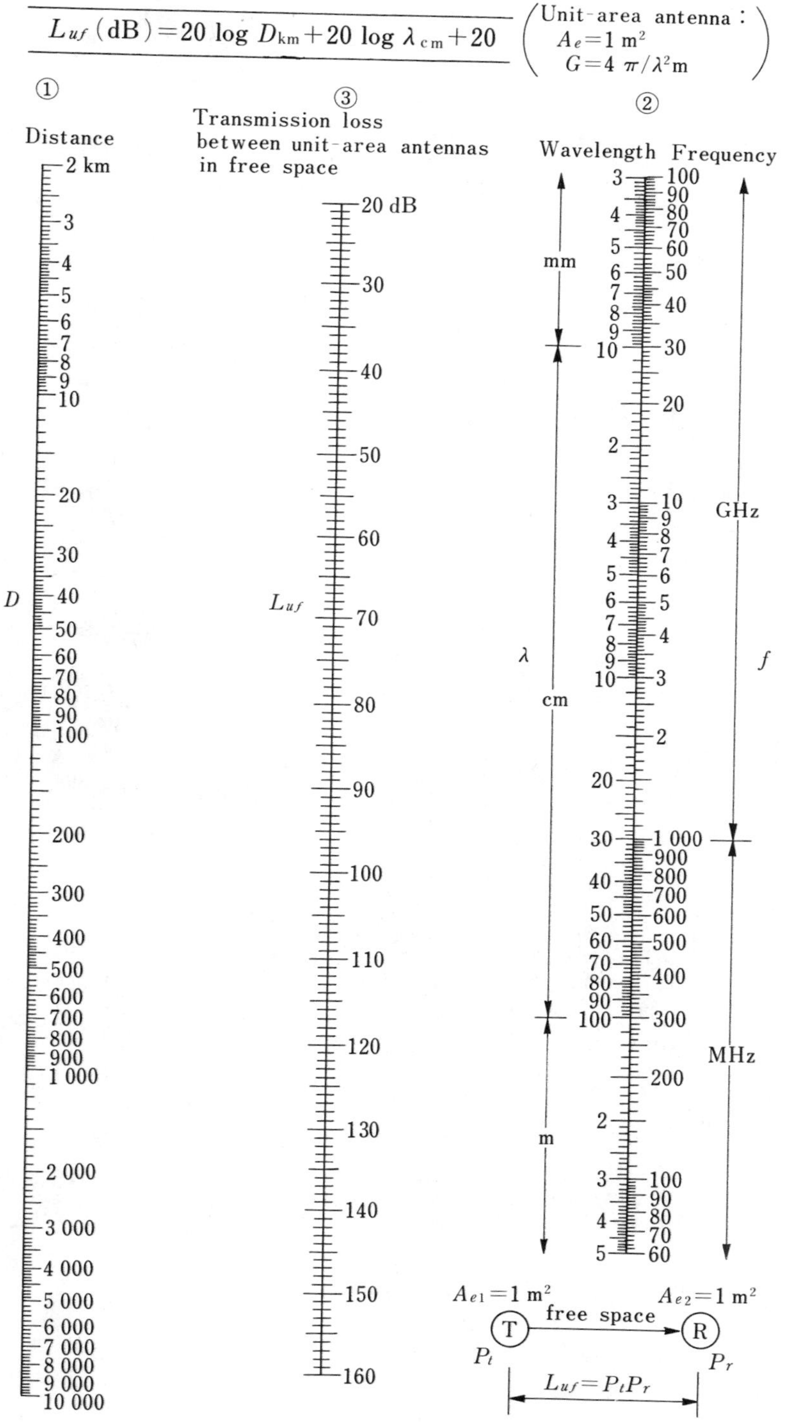

Nomograph for a Plane Wave, Simplified Calculations (4)

Receiving Characteristics

$$V_{td}\,\text{dB}(\mu\text{V}) = E\,\text{dB}(\mu\text{V/m}) + 20\log\lambda_{\text{m}} - 9.94$$
$$P_i\,\text{dB}(\text{mW}) = E\,\text{dB}(\mu\text{V/m}) + 20\log\lambda_{\text{m}} - 126.75$$

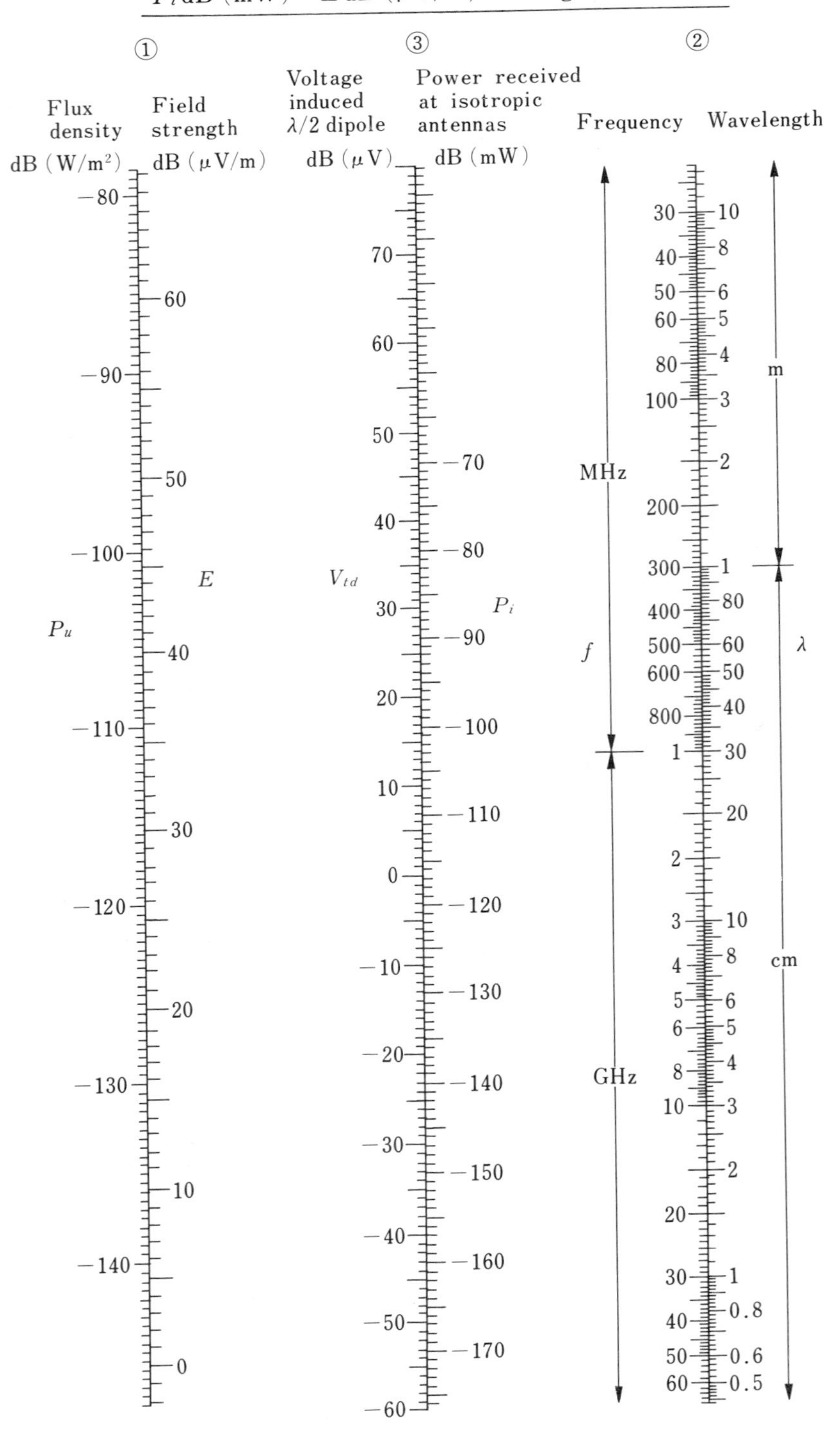

2

REFLECTION INTERFERENCE PROPAGATION PATH

2.1 TYPES OF INTERFERENCE PROPAGATION PATHS

The propagation path whose field strength at the receiving point is altered by the presence of reflected waves that combine vectorially with the direct wave is called a reflection interference path or simply, an interference path.

A. Factors Causing Variations in the Mean Received Signal Level

If the transmitting and the receiving points are in a line of sight as far as radio propagation is concerned, the strongest and most stable wave reaching the receiving point is the direct wave.

It is known empirically that if a good line of sight is obtained and atmospheric absorption is negligible, the propagation level of the direct wave equals approximately the theoretical level for free-space propagation. The following factors cause variations in the mean field strength of the direct wave, excluding variations due to short-term fading.

1. Primary factors.
   - (a) Reflection by ground surface, including sea, land, forests, and man-made structures.
   - (b) Atmospheric absorption (at approximately 10 GHz or higher).
2. Secondary factors.
   - (a) Ground waves.
   - (b) Diffraction by obstacles adjacent to the radio path.
   - (c) Abnormal reflection by man-made structures, hills, or mountains.
   - (d) Co-channel interference from another system.

B. Comparison of Effects

The discussion that follows excludes co-channel interference since it is not of interest in this atlas.

Atmospheric Absorption

The air layer surrounding the earth is called atmosphere. The water vapor and the oxygen contained in the atmosphere act continuously to cause absorption and attenuation of the radio wave. However, this may be ignored at frequencies below approximately 10 GHz. Absorption at frequencies above 10 GHz is discussed in Chapter 5.

Ground Waves

When the radio path for a vertically polarized wave whose frequency is below approximately 100 MHz passes close to the sea surface, which has high conductivity, the wave component traveling along the surface combines with the direct wave. In other words, unless all the conditions, that is, passing over seawater, vertical polarization, frequency below 100 MHz, and lower antenna mounting level occur simultaneously, there will be no serious problem. Consequently this atlas includes the necessary charts without detailed description.

Diffraction by Nearby Obstacles

Effects caused by such obstacles as mountain ridges and buildings are intensified as the radio path approaches the obstacle. However, the maximum loss to the direct wave is approximately 6 dB. This figure corresponds to the theoretical value except when a knife-edge obstacle that is wide transversally but narrow longitudinally comes close to the line of sight. A church spire, a chimney, or a steel tower for a power line, for example, will produce a loss of only 0 to 2 dB. When several high-rise buildings are located across the radio path or if there are hills close to and on the radio path, the loss will rarely reach 6 to 10 dB. However, normally it will not exceed 6 dB as long as the line of sight is secured.

Abnormal Reflection by Man-Made Structures, Mountains, Etc.

This type of reflection occurs alongside such structures as buildings, towers, and billboards. However, the effective reflection coefficient toward the receiving point of these man-made structures erected at random is normally below -30 dB and rarely in excess of -10 dB. This differs from the passive-relay reflector that intentionally causes reflection.

The reflection caused by structures refers to the forward (direction-to-destination) reflection and differs from front-to-back reflection frequently encountered in reflection interference by a nearby obstacle in an urban area and co-channel interference. The latter is treated separately under system design.

Normal Reflection by the Earth

The effective reflection coefficient ρ_e of smooth ground (the most typical is a calm water surface) will reach a value of 0.7 to 0.95 or higher, though it varies according to frequency, polarization, and grazing angle.

Thus it follows that the amplitude of the interfering reflected wave is more than 0.7 to 0.95 times (-3 to 0 dB) that of the direct wave. When combined in phase, the resultant amplitude reaches 1.7 to approximately 2.0 times ($+4.6$ to 6 dB) the amplitude of the direct wave. But when combined in the inverted phase, the direct and reflected waves cancel each other which results in an equilibrium of amplitude, that is, 0.3 to 0.05 times (-10 to -26 dB) the amplitude of the direct wave, thus seriously damaging transmission performance.

The ρ_e of a rolling terrain or forest remains low because of irregular reflections. However, in an urban area, where there are many buildings, or in mountainous terrain, ρ_e will decrease to an insignificant level. In propagation path design it is very important to consider how ρ_e can be reduced and to check if the anticipated ρ_e is tolerable.

It has been concluded that in line-of-sight propagation at frequencies above 30 MHz, the signal level is mostly influenced by the ground-reflected wave. Other factors will cause only a secondary effect, except for the atmospheric absorption at 10 GHz or higher. Figure 1 shows the geometric parameters of an interference propagation path drawn on a terrain profile.

C. Types of Propagation Paths with Interference Caused by Ground-Reflected Waves

In the theory of propagation, ground is the surface of the earth in general. The natural ground may be classified into sea and land. On the land there are grasses, trees, etc., and the undulating terrain includes such features as plains, hills, and mountains. The ground also includes man-made structures such as towers, buildings, roads, dams, villages, towns, and cities. Figures 2 and 3 show the types of propagation paths for the analysis of propagation characteristics.

Type 0—Smooth Flat Grounds

A short-haul over-seawater propagation path may be treated approximately as passing over flat ground, since the curvature of the earth is ignored. Over calm seawater, flat land, or a snow and ice field, mirror reflection will occur, and the reflection coefficient R is determined by the electrical characteristics (dielectric constant ε_r and conductivity σ) of the reflection surface. Estimated reflection coefficients R for seawater, damp ground, fertile land, and wasteland are shown in Fig. 8. A typical actually encountered propagation path is an over-water propagation path normally 5 to 10 km long, and can be approximately 20 km if the antenna is of sufficient height.

As shown in Fig. 2, the grazing angle ψ is equal for both sides A and B. Consequently the location d_1, d_2 of the reflection point is easily calculated if h_1 and h_2 are known.

Type I—Smooth Spherical Earth

This applies when the earth curvature cannot be ignored as in the case of type 0. However, it can be treated similarly to type 0 by establishing an imaginary tangential plane T at the reflection point and using antenna heights h_1' and h_2' (equivalent antenna heights) instead of h_1 and h_2. The reflection coefficient ρ_e is calculated by taking into account the spherical-ground divergence factor D_v (≤ 1), which gives the degree of energy divergence by a spherical surface, in addition to the reflection coefficient R for the flat ground.

The graph in Fig. 9 serves to determine D_v. Type I_1 applies to the case of seawater, type I_2 to the case of a lake surface, flat land, and a snow- and ice-covered surface (elevation above sea level > 0).

Type II—Rolling Spherical Earth

In this type of propagation path an irregular reflection by rippling water, rough ground, etc., cannot be ignored. Whether or not a radio path may be treated as a mirror reflection (type I) depends not only on the surface roughness, but also on the wavelength and the grazing angle. Less roughness, smaller grazing angle, and longer wavelength are likely to cause mirror reflection. Criteria for this are given in Fig. 6.

The characteristics of the reflection surface should be studied over the area where the first Fresnel zone of the reflected wave is intercepted. A half-major axis of this intercepted area is denoted by T_L and determined in Fig. 7.

The terrain outside the intercepted area will not contribute to the reflection. In other words, the presence of mountains or cities outside the reflection area will not alter the classification of the path type. (The same consideration applies to other path types.)

Type III—Rolling Terrain

In the case of hilly country, wooded regions, savannas, etc., the effective reflection coefficient is lower than that expected in the type II path, and the terrain of the reflection area is treated as rough ground. Approximate values of the effective reflection coefficient are given in Fig. 11.

Type IV—Urban Area

Here the reflection area is in a built-up residential area, an industrial area, etc. The reflected waves are weakened further through irregular reflection or a shielding effect. Approximate values of ρ_e are shown in Fig. 11.

Type V—Terrain with Reflected-Wave Shielding Ridges

This terrain has mountains, buildings, etc., which obstruct the path of the reflected wave. Even an over-seawater path can be categorized as type V if the locations of the transmitting and receiving points and the antenna heights are determined in such a way that the reflection occurs at a point near an island or a cape.

If such a reflected-wave shielding ridge is available at or near the center of the reflection area, the effective reflection coefficient ρ_e is approximately 0.1 (−20 dB) or less, and sufficient clearance is secured for the direct wave, the path is considered equivalent to a free-space propagation path. However, if the reflected-wave shielding ridge is not sufficiently high, ρ_e may approach 0.5 (−6 dB). Therefore the path should be treated as an interference path. The effect is evaluated using Fig. 12.

Type VI—Mountainous Propagation Path

Here the reflection point is in a mountainous region. The reflected wave is obstructed almost completely by the reflected-wave shielding ridge, and normally ρ_e is less than 0.01, or −40 dB. (It rarely exceeds 0.1.)

Reflection over a mountain peak is lower than −40 dB. Accordingly this type of path is excluded from the categories of interference propagation path and diffraction propagation path.

D. Classification and Treatment of the Reflected-Wave Interference Path

The interference path of a reflected wave changes according to the terrain profile of the propagation path, thereby causing a change in the receiving signal. Figure 4 summarizes the classification of interference paths and indicates how to treat the problems.

Although the actual terrain is more complicated, it will be useful to make a preliminary analysis to obtain a rough pattern of the path.

2.2 BASIC EQUATIONS OF INTERFERENCE PROPAGATION

In the presence of a ground-reflected wave over the line-of-sight propagation path, the field strength $\dot{E}$ at the receiving point is the vector sum of $\dot{E}_d$ and $\dot{E}_r$, the field strengths of the direct wave and the reflected wave, respectively,

$$\dot{E} = \dot{E}_d + \dot{E}_r \tag{2.1}$$

Assuming $\dot{E}_d$ to be the reference vector, Eq. (2.1) is rewritten as

$$\dot{E} = E_d + E_r e^{-j\Theta} \tag{2.2}$$

where E_d = amplitude of direct wave
E_r = amplitude of reflected wave
Θ = lagging angle of reflected wave behind direct wave

From Fig. 5 the amplitude E of $\dot{E}$ is calculated as

$$E = \sqrt{E_d^2 + E_r^2 - 2E_dE_r\cos(\Theta - \pi)} \tag{2.3}$$

or

$$E = \sqrt{E_d^2 + E_r^2 + 2E_d E_r \cos\Theta} \tag{2.4}$$

If the amplitude ratio of the reflected to the direct wave is expressed by ρ_e, the effective reflection coefficient,

$$E = E_d\sqrt{1 + \rho_e^2 - 2\rho_e \cos(\Theta - \pi)} \tag{2.5}$$

or

$$E = E_d\sqrt{1 + \rho_e^2 + 2\rho_e \cos\Theta} \tag{2.6}$$

then the ratio of the resultant received wave level to the received direct-wave level is

$$\boxed{A = E/E_d = \sqrt{1 + \rho_e^2 + 2\rho_e \cos\Theta}} \tag{2.7}$$

Figure 15 shows the relationship of Eq. (2.7).

REMARK. When E_d is the free-space value, $(E_d/E)^2$ or $1/A^2$ is expressed by Γ_p and is called additional propagation loss (loss in excess of the free-space loss). When expressed in decibels, Γ_p has the same magnitude as A, but the opposite sign.

2.3 SMOOTH FLAT GROUND

A. Equations of Propagation

The most typical smooth flat ground is an over-water propagation path over a calm water surface 5 to 10 km long, for which the earth's curvature may be neglected.

Equations of propagation over smooth flat ground are obtained substituting the free-space field strength E_0 for E_d in Eq. (2.5) and also the plane reflection coefficient R for ρ_e,

$$A = E/E_0 = \sqrt{1 + R^2 - 2R\cos(\Theta - \pi)} \tag{2.8}$$

In Eq. (2.8) it is assumed that the direct wave is not affected by diffraction, refraction, absorption, etc. The ratio of the path lengths of the direct wave to the reflected wave r_1/r_2 is approximately 1, and the ratio of the amplitudes of the reflected wave to the direct wave is determined only by the flat-ground reflection coefficient R.

Since the reflected wave undergoes a phase shift at reflection, the reflection coefficient is expressed as

$$\dot{R} = Re^{-j\phi_R} \tag{2.9}$$

where ϕ_R is the lagging angle at reflection.

Accordingly the total lagging angle Θ of the reflected wave is the sum of ϕ_R and the lagging angle δ, due to the path-length difference

$$\boxed{\Theta = \delta + \phi_R} \tag{2.10}$$

Substituting this into Eq. (2.8),

$$\boxed{A = E/E_0 = \sqrt{1 + R^2 - 2R\cos(\delta + \phi_R - \pi)}} \tag{2.11}$$

For flat ground the following relationship between the distance D, the transmitting and receiving antenna heights h_1 and h_2, the path-length difference Δ $(= r_2 - r_1)$, and the lagging angle δ at reflection is established, when $h_1, h_2 \ll D$,

$$\boxed{\begin{aligned} \Delta &= 2h_1h_2/D \\ \delta &= 2\pi\Delta/\lambda = 4\pi h_1 h_2/\lambda D \end{aligned}} \tag{2.12}$$

B. Equations of Propagation with a Small Grazing Angle ψ ($\dot{R} = -1$, Arbitrary δ)

Except for such specific cases as space communications, aeronautical communications, etc., the grazing angle ψ on an over-the-ground propagation path is usually below 3° and rarely exceeds 5°.

In such a range of angles, the smooth flat ground reflection coefficient $\dot{R}$, which is independent of polarization characteristics, frequency, and the electrical characteristics of the ground, is given by

$$\dot{R} \doteqdot -1 \tag{2.13}$$

Expressed in the form of Eq. (2.9),

$$\dot{R} = Re^{-j\phi_R} = 1 \cdot e^{-j\pi} \tag{2.14}$$

Applying this to Eq. (2.11),

$$A = E/E_0 = \sqrt{2(1 - \cos\delta)} = 2\sin(\delta/2) \tag{2.15}$$

Then from Eq. (2.12),

$$\boxed{A = E/E_0 = 2\sin(2\pi h_1 h_2/\lambda D)} \tag{2.16}$$

This relation is illustrated in Fig. 16*b*, where ε instead of δ is used; ε indicates the residual angle that is smaller than 360° (principal angle).

C. Case of $R = -1$, $\delta \leq 1$ rad

In this case

$$\sin(\delta/2) \doteqdot \delta/2$$

Accordingly,

$$\boxed{A = E/E_0 = 4\pi h_1 h_2/\lambda D = \delta < 1} \tag{2.17}$$

Figure 16*a* is drawn using Eq. (2.17) and ε instead of δ.

D. Effect of Ground Wave in Low Antenna Region (Over-Seawater Propagation, Vertical Polarization, in or below VHF Band)

Here the propagation path is between antennas at a very low level. Using vertical polarization and the lower region of the VHF band, it passes over seawater or a saltwater lake with high ground conductivity. Almost constant field intensity is obtained for a range of antenna heights up to a certain level. Such a limit value of the antenna height is called minimum effective antenna height h_0 and is shown in Fig. 17.

The equation of propagation is

$$\boxed{A = E/E_0 = 4\pi h_{1e} h_{2e}/\lambda D < 1} \tag{2.18}$$

where h_0 = minimum effective antenna height
h_e (h_{1e}, h_{2e}) = surface-wave effective antenna height

$$h_{1e} = \sqrt{h_1^2 + h_0^2}$$
$$h_{2e} = \sqrt{h_2^2 + h_0^2} \tag{2.19}$$

Figure 18 gives h_e.

E. Flat-Ground Reflection Coefficient

The flat-ground reflection coefficient $\dot{R}$ is expressed by the following equations for horizontal (HP) and vertical (VP) polarization:

$$\dot{R}_{\text{HP}} = \frac{\sin\psi - \sqrt{\varepsilon_c - \cos^2\psi}}{\sin\psi + \sqrt{\varepsilon_c - \cos^2\psi}} \tag{2.20}$$

$$\dot{R}_{\text{VP}} = \frac{\varepsilon_c \sin\psi - \sqrt{\varepsilon_c - \cos^2\psi}}{\varepsilon_c \sin\psi + \sqrt{\varepsilon_c - \cos^2\psi}} \tag{2.21}$$

where ψ = grazing angle (complementary angle of incidence)
ε_c = complex dielectric constant,

$$\varepsilon_c = \varepsilon_r - j60\sigma\lambda \tag{2.22}$$

ε_r = dielectric constant of reflection surface
σ = conductivity of reflection surface (℧/m)
λ = wavelength (m)

Figure 8 shows the absolute value R and the phase ϕ_R of the flat-plane reflection coefficient for seawater, damp ground, fertile land, and dry land.

The absolute value of the reflection coefficient for horizontal polarization is normally large, that is, $|R| \doteqdot 1$, $\phi_R = \pi$ for a small ψ. This means $\dot{R} \doteqdot -1$ for horizontal polarization. However, for vertical polarization, the reflection coefficient becomes minimum within the grazing angle range of 5 to 30°. At a certain grazing angle the lagging angle ϕ_R shows a rapid change. Such a grazing angle is called the Brewster angle.

2.4 SMOOTH SPHERICAL GROUND

A. Equation of Propagation

The tangential plane at a reflection point is denoted by T, as shown in Fig. 1. Using antenna heights above the T plane h_1' and h_2' (that is, equivalent antenna heights), an equation similar to that for a smooth flat ground is derived,

$$\boxed{A = E/E_0 = \sqrt{1 + \rho_e^2 - 2\rho_e \cos(\delta + \phi_R - \pi)}} \tag{2.23}$$

Here

$$\rho_e = RD_v \tag{2.24}$$

$$\delta = 4\pi h_1' h_2' / \lambda D \tag{2.25}$$

where D_v = spherical-ground divergence factor
R = absolute value of flat-plane reflection coefficient
ϕ_R = lagging angle of flat-plane reflection coefficient
ρ_e = effective reflection coefficient

$$D_v = 1/\sqrt{1 + 2h_1' h_2' / KaD \tan^3\psi} \tag{2.26}$$

or

$$D_v = 1/\sqrt{1 + 2d_1 d_2 / KaD \tan\psi}$$

$$D_v = 1/\sqrt{1 + 2d_1^2 d_2 / KaDh_1'} \tag{2.27}$$

$$D_v = 1/\sqrt{1 + 2d_1 d_2^2 / KaDh_2'}$$

where K = effective earth radius factor (standard value = 4/3)
a = earth radius (= 6370 km)
ψ = grazing angle
Ka = effective earth radius (standard value = 8500 km)

Modifying Eq. (2.26),

$$\boxed{D_v = \frac{1}{\sqrt{1 + 4p^2(1 - y)/(1 - p^2)}}} \tag{2.28}$$

where

$$p = d_1/d_{t1} = d_1/\sqrt{2Kah_1} \tag{2.29}$$

$$y = d_1/D \tag{2.30}$$

and d_{t1} is the line-of-sight distance from the antenna height h_1 (relative to spherical ground) above the spherical ground.

Figure 9 gives a chart to estimate the spherical-ground divergence factor from the parameters p and y.

B. Small Grazing Angle ψ

For a small grazing angle ψ, $\dot{R} = -1$, and if $D_v \doteqdot 1$ (the spherical divergence factor is small),

$$p_e \doteqdot 1, \qquad \phi_R \doteqdot \pi \tag{2.31}$$

Hence, as in Eq. (2.16),

$$\boxed{A = E/E_0 = 2\sin(2\pi h_1' h_2'/\lambda D)} \tag{2.32}$$

C. Small ψ and $\delta \leq 1$ rad

$$\boxed{A = E/E_0 = 4\pi h_1' h_2'/\lambda D} \tag{2.33}$$

Equations (2.32) and (2.33) indicate that Fig. 16 is applicable if h_1' and h_2' are substituted for h_1 and h_2, respectively.

D. Line-of-Sight Distance over Spherical-Ground d_t

The limit of a distance d_t visible electromagnetically from the level of height h above the spherical ground is given by

$$\boxed{d_t = \sqrt{2Kah}} \tag{2.34}$$

where d_t = line-of-sight distance
Ka = effective earth radius
h = antenna height (above the spherical ground)

As shown in Fig. 1, the heights u_1 and u_2 above the tangential flat plane at the point of reflection correspond to line-of-sight distances d_1 and d_2, respectively,

$$d_1 = \sqrt{2Kau_1}$$
$$d_2 = \sqrt{2Kau_2} \tag{2.35}$$

Hence

$$u_1 = d_1^2/2Ka$$
$$u_2 = d_2^2/2Ka \tag{2.36}$$

Figures 22 and 23 give the line-of-sight distances for various values of K and various antenna heights.

E. Equivalent Antenna Heights h_1' and h_2'

As shown clearly in Fig. 1,

$$h_1' = h_1 - u_1 = h_1 - d_1^2/2Ka$$
$$h_2' = h_2 - u_2 = h_2 - d_2^2/2Ka \tag{2.37}$$

F. Method to Determine the Reflection Point

Based on laws of the theories of electromagnetism, the grazing angles ψ for incidence and reflection are equal to each other, that is,

$$\tan\psi = h_1'/d_1 = h_2'/d_2 \tag{2.38}$$

Then using Eq. (2.37),

$$h_1/d_1 - d_1/2Ka = h_2/d_2 - d_2/2Ka \tag{2.39}$$

The reflection point is determined by calculating d_1 and d_2.

Now let M be the midpoint of D, and d' the distance between M and the reflection point, as shown in Fig. 24. Then

$$d_1 = D/2 + d' = (D/2)(1+b)$$
$$d_2 = D/2 - d' = (D/2)(1-b) \tag{2.40}$$

where

$$b = 2d'/D \tag{2.41}$$

Similar equations are obtained for h_1 and h_2,

$$\begin{aligned} h_1 &= [(h_1 + h_2)/2](1 + c) \\ h_2 &= [(h_1 + h_2)/2](1 - c) \end{aligned} \tag{2.42}$$

where

$$\boxed{c = (h_1 - h_2)/(h_1 + h_2)} \tag{2.43}$$

Substituting Eqs. (2.40) and (2.42) into Eq. (2.39), which defines the reflection point, and eliminating d_1 and d_2,

$$c = mb(1 - b^2) + b \tag{2.44}$$

where

$$\boxed{m = D^2/4Ka(h_1 + h_2)} \tag{2.45}$$

Since Eq. (2.44) is a linear equation, a group of parallel lines is obtained when drawn graphically for various values of b, taking m on the abscissa and C on the ordinate. The value of b is found at the coordinate determined by m and C, and then the location of the reflection point d_1 is obtained as

$$\begin{aligned} d_1 &= D(1 + b)/2 = Dy \\ d_2 &= D - d_1 \end{aligned} \tag{2.46}$$

where

$$y = (1 + b)/2 \tag{2.47}$$

Figures 24 and 25 show graphs prepared using parameters m, c, and y to determine the reflection point.

For convenience of calculation, $h_1 \geq h_2$, and accordingly, $d_1 \geq d_2$ is to be assumed in the application.

It is noted that h_1 and h_2 are the antenna heights measured from the mean elevation of the reflection surface above sea level h_r. Thus if the antenna heights above sea level are h_{s1} and h_{s2},

$$\begin{aligned} h_1 &= h_{s1} - h_r \\ h_2 &= h_{s2} - h_r \end{aligned} \tag{2.48}$$

No reflection point exists if the conditions $h_1, h_2 > 0$ do not hold.

2.5 ROUGH GROUND

A. Equation of Propagation

Even if the path profile drawn according to the topological map appears to be over smooth terrain, the actual ground of the reflection area will normally be somewhat rough so that the actual reflection coefficient may be low. Undulating seawater and rolling cultivated fields may represent such a terrain. However, we should remember that sound reflection may occur when the seawater becomes calm, the field is mowed, the ground is leveled, water fills the field, etc.

The equation of propagation for a rough reflection surface is obtained from Eq. (2.7),

$$\boxed{A = E/E_0 = \sqrt{1 + \rho_e^2 + 2\rho_e \cos\Theta}}$$

$$\rho_e = S \tag{2.49}$$

$$\Theta = \delta + \phi_s$$

where S = irregular reflection factor (effective reflection coefficient of rough-ground surface)

ϕ_s = irregular reflection phase shift (lagging angle)

δ = path-length difference

Θ = phase of effective reflection coefficient (lagging angle)

Generally in the case of rough ground it may be difficult to locate the reflection point since the reflected-wave components reaching the receiving point are considered to be a combination of a number of partially reflected waves. Thus it is only possible to make a rough estimate of S, ϕ_s, and δ.

B. Criteria for Mirror Reflection

No linearity exists between the degree of undulation and the reflection characteristics. Once the roughness reaches a certain level, the mirror reflection decreases rapidly and the reflection coefficient drops to a fraction of unity, as would be the case if steam were blown on a mirror surface.

As shown in Fig. 6, the phase difference between two reflected waves, that is, one from the bottom and the other one from the peak of the undulation in the reflection area, is produced corresponding to the path-length difference,

$$\varphi_\Delta = (2\pi/\lambda)\,\Delta_s$$

From Fig. 6,

$$\Delta_s = 2T_\Delta \sin\psi \tag{2.50}$$

Hence

$$\varphi_\Delta = (4\pi T_\Delta/\lambda)\sin\psi \tag{2.51}$$

where φ_Δ = phase difference between two reflected waves, one from the bottom and the other from the peak of the undulated surface

Δ_s = path-length difference corresponding to above phase difference

ψ = grazing angle

T_Δ = amplitude of undulation

The criterion of the phase difference of $\pi/4$ above which no mirror reflection occurs was proposed by Rayleigh. If the amplitude of the undulation producing the phase difference corresponding to the above criterion is expressed by

$$\pi/4 = (4\pi T_c/\lambda)\sin\psi$$

then (2.52)

$$T_c = \lambda/16\sin\psi$$

where T_c is the criterion for the critical amplitude of undulation for mirror reflection.

Equation (2.52) assumes that the convex and concave portions are regularly distributed so that equal areas are occupied by each portion, and that T_Δ is uniform over the entire reflection area. However, the actual terrain of the reflection area would be more irregular.

Generally, as is typical in practical systems, the convex portion of the ground is round and the concave portion is screened by the adjacent convex portion for $\psi \leq 5$. As a result, the reflection coefficient decreases further. Accordingly, if the approximate mean value of T_c at the reflection point and neighboring ground exceeds T_Δ, the ground may be judged to be rough,

$$\begin{array}{ll} T_\Delta < T_c, & \text{smooth ground} \\ T_\Delta \geq T_c, & \text{rough ground} \end{array} \quad \text{(criterion for mirror reflection)} \qquad (2.53)$$

C. Size of Effective Reflection Area

The dimension of the reflection area is considered to contribute effectively to the determination of the radio-wave energy of the reflected wave. In this area the first Fresnel zone of the reflected wave intersects with the ground.

Since the grazing angle ψ is generally within a few degrees, the intersection area becomes an ellipse prolonged along the propagation path. Thus

$$\boxed{T_L = r_R/\sin\psi} \qquad (2.54)$$

where T_L = half-major axis of effective reflection area

r_R = first Fresnel zone radius at reflection point

The reflection coefficient for possible mirror reflection of rough ground must be estimated for the effective reflection area as shown in Fig. 7.

D. Effective Reflection Coefficient of Ordinary Ground

On overland paths the reflection phenomenon is usually very complicated because of possible topographical changes, such as the presence of plants, trees, rocks, and man-made structures. Therefore reflected waves reaching the receiving point are composed of a number of irregularly reflected waves. Consequently a theoretical estimation is very difficult. As a result, tests can form the basis for estimation.

Figure 11 shows charts for estimating the rough-ground reflection coefficient on the basis of empirical data. For convenience, the ground with a reflection coefficient $S = 0.7$ or greater (-3 dB or higher) is defined as smooth ground, while rough ground is a ground with a lower reflection coefficient. The effective reflection coefficient ρ_e is calculated in Fig. 11 according to the ground type, using the following equations:

$$\begin{aligned} \rho_e &= RD_v \qquad \text{smooth ground} \\ \rho_e &= S, \qquad \text{rough ground} \end{aligned} \tag{2.55}$$

Since ground types IV through VI in Fig. 11 include improvements by suppressing the reflected waves using undulation of the terrain, $\rho_e = S$ is assumed. However, if there exists a separately located shielding ridge, as shown in Fig. 3 (type V), the diffraction coefficient Z can be calculated.

In the case where a shielding ridge is available there are two types of effective reflection coefficients,

$$\begin{aligned} \rho_e &= RD_vZ, \qquad \text{smooth spherical ground plus shielding ridge} \\ \rho_e &= SZ, \qquad \text{rough ground plus shielding ridge} \end{aligned} \tag{2.56}$$

Figure 12*b* gives the geometrical parameters necessary for estimating Z, while Fig. 12*a* presents evaluation procedures and graphs for estimation.

The path height is usually estimated graphically from the path profile (on a large scale) after the reflection point has been located using Figs. 24 and 25. This is because calculation of the path height h_p is time consuming. Then using the ridge height h_M found on a map, the clearance C and the ridge depth C_s are determined. The evaluation procedure is outlined as follows.

E. Effect of Shielding Ridge

Nearby Ridges

1. A space determined by one-half of the first Fresnel zone radius is called the protection zone, and any ridges that exist outside of this zone, as shown in Fig. 12*b* [1], that is, M_1 and M_2, can be neglected. Any ridge M that occupies the protection zone is called a nearby ridge and has to be taken into account in the calculation.
2. Read clearance C on the path profile.
3. Determine the first Fresnel zone radius r_s at a point of the ridge.

4. Calculate the clearance parameter $U_c = C/r_s$ and determine Z from Fig. 12*a*.
5. When the number of existing nearby ridges exceeds 1, estimate Z for each ridge, assuming that each ridge exists independently. Then multiply the results to obtain the total Z. However, if the total is less than 0.3 (corresponding to a loss of 10 dB), assume that it is 0.3 (10 dB).

Shielding Ridges

1. If there is a ridge that does not obstruct the direct wave but shields the reflected wave, determine the reflection point between points M and B and the reflected-wave path $A-M-R-B$ (Fig. 12*b* [2]).
2. Connect imaginary antenna point B' with point A and denote the ridge height above the path, or the depth of the ridge, by C_s.
3. Calculate the first Fresnel zone radius r_s at the ridge point.
4. Calculate the diffraction parameter $U = C_s/r_s$ and determine Z from Fig. 12*a*.
5. If there is a nearby ridge within the protection zone of the path $A-M-R-B$, add the resultant effects according to the procedure for nearby ridges.
6. When there is more than one shielding ridge, the path may be regarded as a reflectionless propagation path and no calculation of the reflected wave is needed.

Direct Wave Undergoes Ridge Loss

If a ridge shields not only the reflected wave but also part of the protection zone of the direct wave, calculate the loss by the following steps.

1. Determine the diffraction coefficient Z_0 of the direct wave, employing the method given under nearby ridges for the direct wave.
2. Calculate the diffraction coefficient Z_r of the reflected wave employing the method given under shielding ridges.
3. The diffraction coefficient Z is given by

$$Z = Z_r/Z_0 \tag{2.57}$$

2.6 INTERFERENCE PATTERN

A. Classification of Interference Patterns

In interference propagation paths the receiving field strength varies with the amplitude ratio of the direct to reflected waves P_e and the phase difference Θ between them.

The following factors cause variations of P_e.

1. Shift of reflection point with change of antenna height or path length.
2. Change in reflection surface characteristics, caused by tides, waves, changes in vegetation, snowfall, icing, etc.

Factors causing variations of Θ are given by the following equations:

$$\begin{aligned}\Theta &= \delta + \phi_c \\ \delta &= 4\pi h_1' h_2'/\lambda D \\ &= (4\pi/\lambda D)\left(h_1 - d_1^2/2Ka\right)\left(h_2 - d_2^2/2Ka\right)\end{aligned} \tag{2.58}$$

where ϕ_c is the lagging angle at ground reflection ($\phi_c = \phi_R \doteqdot \pi$ when the grazing angle ψ is small).

Types and causes of variation patterns of the field strength are classified, based on Eq. (2.58), as follows.

1. Space pattern.

Distance pattern, due to change in D.
Height pattern, due to change in heights h_1 and h_2.
Transverse pattern, due to topographical change by transverse change of reflection point.

2. K pattern, due to change in K.
3. f pattern, due to change in f, and accordingly in λ.

Figure 27 shows four typical patterns.

B. Path-Length Difference

The path-length difference can be calculated by

$$\Delta = 2h_1' h_2'/D$$

or

$$\boxed{\Delta = (2/D)\left(h_1 - d_1^2/2Ka\right)\left(h_2 - d_2^2/2Ka\right)} \tag{2.59}$$

where Δ = path-length difference between direct and reflected waves
h_1', h_2' = equivalent antenna height
h_1, h_2 = antenna heights
Ka = effective earth radius
D = distance

Figure 26 presents nomographs to determine the path-length difference.

C. Path-Length Difference between Maximum and Minimum Points

Regardless of the types of interference patterns, at the dip the field strength is low and unstable (at the minimum point where the reflected wave is inverted to the directed wave), while at the peak the field strength is high and stable (at the maximum point where the reflected and direct waves are in phase). Consequently, mounting the antenna at the level corresponding to the maximum point is often helpful in increasing the margin for fading.

However, predicting the maximum and minimum points with good accuracy is possible only for the VHF and lower bands, and only for paths over water (over seawater or lakes) or over land categorized as smooth flat ground.

If n is a natural number, the general equations for determining the maximum and minimum points are

$$\begin{aligned} \Theta &= \delta + \phi_R \to 2n\pi, && \text{maximum point} \\ \Theta &= \delta + \phi_R \to (2n-1)\pi, && \text{minimum point} \end{aligned} \tag{2.60}$$

When the grazing angle ψ is small, $\phi_R \doteqdot \pi$ (approximately constant). Then with respect to δ

$$\left.\begin{aligned} \delta &= 2\pi\Delta/\lambda \to (2n-1)\pi, && \text{maximum} \\ \delta &= 2\pi\Delta/\lambda \to 2n\pi, && \text{minimum} \end{aligned}\right\} \quad \text{if } \phi_R = \pi \tag{2.61}$$

where Θ = delay angle of reflected wave from direct wave
δ = delay angle due to path-length difference
ϕ_R = delay angle in reflection coefficient
Δ = path-level difference

As indicated in Eqs. (2.60) and (2.61), the maximum and minimum points are repeated alternately every π rad of phase change.

It follows that each time the path-length difference changes by $\Delta/2$ (half-wavelength) the maximum and the minimum are repeated alternately.

Then if Δ is normalized by $\lambda/2$,

$$\left.\begin{aligned} \Delta/(\lambda/2) &\to (2n-1) \ \ \text{(odd)}, && \text{maximum} \\ \Delta/(\lambda/2) &\to 2n \ \ \text{(even)}, && \text{minimum} \end{aligned}\right\} \quad \text{if } \phi_R = \pi \tag{2.62}$$

Thus when the path-length difference Δ and its ratio to a half-wavelength are obtained, the maximum and minimum points are determined. However, if the wavelength becomes shorter, the estimation error will become significant compared with the half-wavelength. As a result, this estimation is applicable only to VHF and UHF bands for over-water (over sea and lake) propagation paths and to the VHF band for overland propagation paths.

D. Range of Variation of Interference Pattern

The effect of interference on the quality of the system is determined by the relative strength of the reflected wave and also by the phase delay.

The following relations are established between the relative strength, that is, the effective reflection coefficient ρ_e, and the range of variation of the interference pattern, referred to the direct wave level (0 dB).

1. Depth to minimum point (dB),

$$D_p = 20\log(E_{\min}/E_0) = 20\log(1-\rho_e) \tag{2.63}$$

2. Rise to maximum point (dB),

$$R_p = 20\log(E_{max}/E_0) = 20\log(1 + \rho_e) \tag{2.64}$$

3. Ranges of variations of pattern (dB),

$$W_p = 20\log(E_{max}/E_{min})$$
$$= 20\log[(1 + \rho_e)/(1 - \rho_e)] = R_p + |D_p| \tag{2.65}$$

Conversely, if D_p, R_p, or W_p is known by a measurement of the interference pattern, ρ_e is estimated. Figure 29 shows this relation.

If the directivity of the antenna is taken into account, the propagation echo ratio ρ_θ is used instead of ρ_e.

When expressed in decibels,

$$\rho_\theta = \rho_e D_{\theta 1} D_{\theta 2}$$

$$\rho_\theta\ (\text{dB}) = \underbrace{20\log\rho_e}_{\substack{\text{effective}\\ \text{reflection}\\ \text{coefficient}}} + \underbrace{20\log(D_{\theta 1} D_{\theta 2})}_{\substack{\text{effect of}\\ \text{antenna}\\ \text{directivity}}} \tag{2.66}$$

where $D_{\theta 1}$, and $D_{\theta 2}$ are the directivities of the transmitting and receiving antennas in the direction of the reflected wave (the relative field strength referred to the direction of the direct wave). Figure 10 shows D_θ in decibels (see Section 2.7 for details).

E. Half-Pitch in Height Pattern

As mentioned in Section 2.6C, the antenna should preferably be mounted near the level at which the height pattern (h pattern) has its maximum value. In a microwave frequency band where the wavelength is short, it is difficult to determine the antenna mounting level because of a possible large estimation error.

This means that the phase shift resulting from changing the height is significant and the height pattern pitch, that is, the level difference between two consecutive peaks or two consecutive depths, is small.

The important characteristics of the height pattern pitch are as follows.

1. When the half-pitch (the difference in level between a maximum point and the adjacent minimum point, or one-half the pitch) is smaller than or close to the aperture diameter, suppression of the reflected wave by the antenna directivity can be achieved effectively.
2. By installing two antennas at a spacing equivalent to a half-pitch and combining the two received direct-wave signals in phase, the reflected wave can be canceled (so-called multiantenna combining reception).
3. If the half-pitch is large, an estimation of the peak (maximum point) is easy, and the required accuracy of calculation becomes less stringent.

Consequently, estimating the pitch is indispensable in interference propagation path design. Specifically, the height pattern pitch has to be estimated as a level difference between the maximum and minimum points on the height pattern. In this way the total field strength is calculated using various parameters, such as the reflection point (d_1, d_2), the amplitude R and phase ϕ_R of the reflection coefficient, the spherical-surface divergence factor D_v, the delay angle due to the path-length difference δ or the effective reflection coefficient ρ_e, and the phase Θ for each antenna. However, this is a time-consuming task without a computer.

Here a simplified method is given for approximating the pitch p or the half-pitch p_h in the height pattern using the included angles between direct and reflected waves θ_1 and θ_2

$$\begin{aligned} p_{h1} &\doteqdot \lambda/(2\sin\theta_1) \\ p_{h2} &\doteqdot \lambda/(2\sin\theta_2) \end{aligned} \quad \text{(half-pitches in height pattern)} \tag{2.67}$$

where p_{h1}, p_{h2} = half-pitches in height pattern at points A and B
θ_1, θ_2 = included angles between direct and reflected waves at points A and B

Figure 30 gives curves for estimating the pitches and the half-pitches.

2.7 EFFECT OF ANTENNA DIRECTIVITY

If the included angle θ_1 or θ_2 is large (the height pattern pitch is small), the effect caused by the reflected wave can be suppressed by the antenna directivity as mentioned in Section 2.6D and E.

Denoting the directivities in the direction of the reflected waves $D_{\theta 1}$ and $D_{\theta 2}$, where the included angles θ_1 and θ_2 are measured from the direction of the direct wave, the total effect of suppression $D_{\theta T}$ is expressed as

$$D_{\theta T} = D_{\theta 1} D_{\theta 2} \tag{2.68}$$

or

$$\underbrace{D_{\theta T}\ (\text{dB})}_{\substack{\text{total effect of}\\ \text{directivity}}} = \underbrace{20 \log D_{\theta 1}}_{\substack{\text{transmitting}\\ \text{antenna}\\ \text{directivity}}} + \underbrace{20 \log D_{\theta 2}}_{\substack{\text{receiving}\\ \text{antenna}\\ \text{directivity}}}$$

In general D_θ is obtained from the directivity curve in the vertical plane of the antenna to be used. However, if such a curve is not available, approximate figures can be estimated from the half-power width or aperture diameter.

Figure 10[1] gives the graph to estimate the approximate directivity of a parabola using the diameter D_ϕ and the angle θ as parameters, while Fig. 10[2] is a similar graph using the half-power width and the angle θ.

2.8 VERTICAL ANGLE

Figures 31 to 34 are graphs to estimate the vertical angle of the direct wave α relative to the horizon, the vertical angle of the reflected wave β relative to the horizon, the included angle between direct and reflected waves θ, and the grazing angle ψ.

First the required values x, y, and z are obtained from Figs. 32 to 34 using the parameters of propagation distance D, distances to reflection point d_1 and d_2, and antenna heights h_1 and h_2. Then the calculation chart in Fig. 31 gives the results.

It is noted that the sign of α is positive for angles above the horizon, the sign of β is always negative, and the signs of θ and ψ are always positive.

In regard to the grazing angle ψ, if the calculated values of ψ_1 at station A and of ψ_2 at station B differ by more than a certain limit, such as 10%, the reflection point may not exist, or the estimation may be erroneous. Therefore a reexamination will be necessary.

REMARK. Figures 19–21 give nomographs for simplified estimation of propagation loss (including diffraction loss due to a smooth ground surface) between antennas mounted within a low antenna region. For use of these nomographs, see the explanation given in the figures.

EXAMPLES

EXAMPLE 2.1

Estimate the receiving signal level relative to the free-space condition for propagation, and the smooth, flat ground, surface reflection coefficient $R = -1$. Also calculate the path-length difference Δ, the phase of the reflected wave Θ, and the propagation loss Γ.

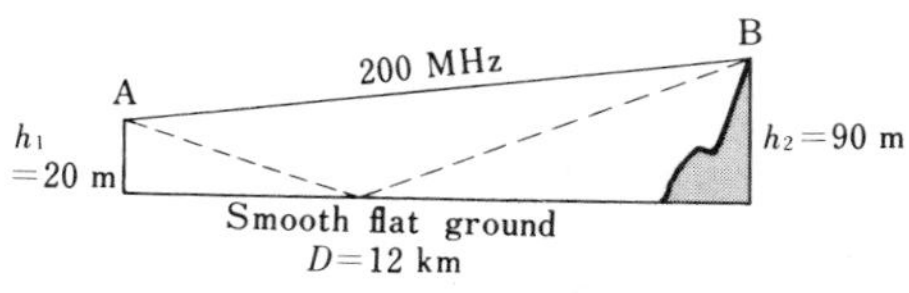

Solution

The receiving signal level relative to the free-space value, that is, the additional propagation loss Γ_p, is determined by the reflection coefficient $|R|$ and the lagging angle of the reflected wave from the direct wave Θ.

1. For $|R| = 1$ and 200 MHz, $\lambda = 1.5$ m.
2. Θ is the sum of the phase ϕ_R of the reflection coefficient and the phase corresponding to the path-length difference.
3. The path-length difference Δ is obtained using Fig. 26,

$$\Delta/\lambda = 0.3/1.5 = 0.2, \qquad \Delta = 0.3 \text{ m}$$

4. Since $R = -1$, the phase ϕ_R of the reflection coefficient is

$$\phi_R = 180°$$

5. ε and Θ are obtained using Fig. 14,

$$\varepsilon = \delta = 72°$$

$$\Theta = 252°$$

6. Since the reflection is the total reflection, using Fig. 16,

$$A = E/E_d = +1.4 \text{ dB}$$

7. The free-space propagation loss Γ_0 is obtained from Fig. 7c in Chapter 1,

$$\Gamma_0 = 100 \text{ dB}$$

Then

$$\Gamma = \Gamma_0 + \Gamma_p$$

$$= 98.6 \text{ dB}$$

where $\Gamma_p = -A \quad \text{(dB)}$

NOTE. If only an approximate value of Γ is needed, a simplified graph for a low antenna height region is used (Fig. 19), and $\Gamma \doteqdot 98$ dB.

EXAMPLE 2.2

Estimate the relative receiving level for 30 MHz, vertical polarization. For 30 MHz, $\lambda = 10$ m.

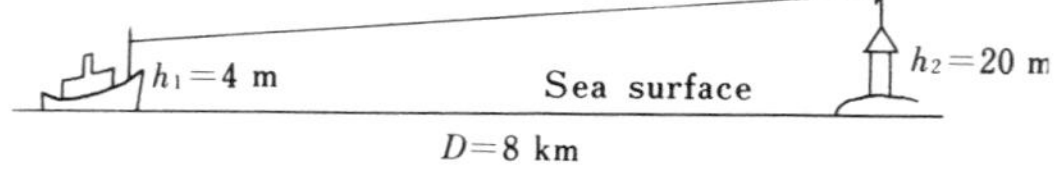

Solution

1. From Fig. 17, $h_0 > h_1, h_2$. From Fig. 18, $h_{1e} = h_{2e} = 60$ m (ground-wave region).
2. From Fig. 26, the path-length difference $\Delta = 0.9$ m,

$$\Delta/\lambda = 0.09, \qquad \delta = 360° \times 0.09 = 32.4°$$

3. From Fig. 16, the relative receiving field strength is obtained,

$$E/E_d = -5.2 \text{ dB}$$

NOTE. In case δ is smaller than approximately 60°, the equation for the linear region for h_{1e}, h_{2e} is applicable.

EXAMPLE 2.3

Determine the reflection point d_1, d_2 and calculate the path-length difference Δ using the given path profile. The distance $D = 74$ km is assumed.

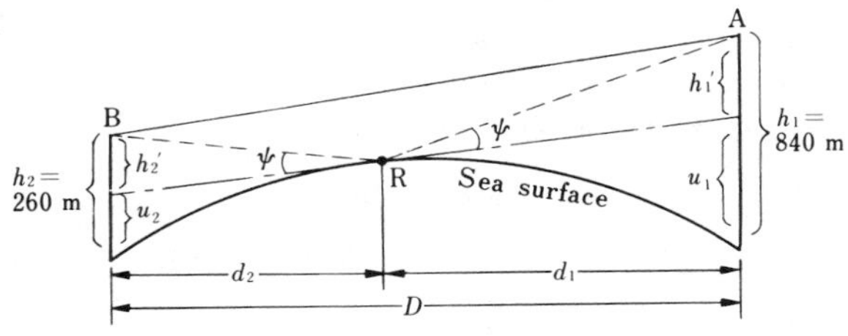

Solution

Proceed with the solution referring to the chart index and the figures.

1. Use Figs. 24 and 25 initially. The higher antenna height is h_1 and the lower is h_2. Accordingly, $d_1 > d_2$. $h_1 = 840$ m and $h_2 = 260$ m.

$$c = (h_1 - h_2)/(h_1 + h_2) = 0.53$$

$$m = 0.15 \qquad \text{(from Fig. 24)}$$

$$y = 0.737 \qquad \text{(from Fig. 25}b\text{)}$$

$$d_1 = Dy = 74 \times 0.737 = 54.54 \text{ km}$$

$$d_2 = D - d_1 = 74 - 54.54 = 19.46 \text{ km}$$

2. The effective antenna heights h_1 and h_2 are calculated to estimate the path-length difference,

$$h_1' = h_1 - u_1 = 840 - 175 = 665$$

$$h_2' = h_2 - u_2 = 260 - \underbrace{22.3}_{\text{Fig. 23}} \doteqdot 238$$

3. The path-length difference Δ is obtained using h_1' and h_2' (Fig. 26b),

$$\Delta = 4.3 \text{ m}$$

EXAMPLE 2.4

Estimate the effective reflection coefficient ρ_e assuming 1-GHz vertical polarization in addition to the conditions in Example 2.3. Also estimate the maximum wave amplitude up to which the seawater surface may be treated as a smooth spherical surface.

Solution

The effective reflection coefficient ρ_e equals the flat-plane reflection coefficient $|R|$ times the spherical-ground divergence factor D_v.

1. The grazing angle ψ is calculated to obtain $|R|$,

$$\psi = (h_1'/d_1)0.057 = (665/54.54)0.057 = 0.695°$$

$$|R| = 0.82 \qquad \text{(Fig. 8}a\text{)}$$

2. The parameters required for the graph of D_v are obtained. The line-of-sight distance d_{t1} for h_1 is found ($h_1 = 840$ m),

$$d_{t1} \doteqdot 120 \text{ km} \qquad \text{(Fig. 23)}$$

$$y = d_1/D = 54.54/74 = 0.74$$

$$p = d_1/d_{t1} = 54.54/120 = 0.45$$

$$D_v = 0.885 \qquad \text{(Fig. 9)}$$

3. $$\rho_e = |R|D_v = 0.82 \times 0.885 = 0.726$$

4. Judgment of whether it is a smooth sphere or a rough sphere is made using the graph of Fig. 6, the criterion for maximum height of undulation causing a mirror reflection. The maximum height for 1 GHz and a grazing angle $\psi = 0.695° = 42'$ is approximately 1.5 m. Therefore if the wave amplitude (from bottom to peak) is less than 1.5 m, the surface may be treated as a smooth sphere.

EXAMPLE 2.5

Using the same conditions as in Example 2.3, any radio frequency is to be avoided at which the reflected wave cancels the direct wave in the frequency range of 250 to 400 MHz in order to prevent the receiving signal level from dropping to the minimum and also from undergoing severe fading. Indicate the radio frequencies to be avoided and the possible level of the dip. Then select the frequencies that will guarantee a maximum receiving signal level and indicate the possible amount of rise in the receiving signal level. Horizontal polarization is to be used.

Solution

The flat-plane reflection coefficient in horizontal polarization is $R = -1$, that is, $|R| = 1$ and $\phi_R = \pi$. Accordingly, in Fig. 28 the maximum and minimum points of the receiving signal combined with interference can be found, and Fig. 29 is used to estimate the depth and the rise in the interference pattern.

1. Effective reflection coefficient ρ_e,

$$\rho_e = |\dot{R}|D_v \doteqdot 1 \times 0.885 = 0.885 = 0.88$$

2. Range of wavelength,

$$f = 250\text{–}400 \text{ MHz}, \qquad \lambda = 1.2\text{–}0.75 \text{ m}$$

3. Range of path-length difference normalized by a half-wavelength ($\Delta =$ 4.3 m),

$$\Delta/(\lambda/2) = 4.3/(1.2/2)\text{–}4.3/(0.75/2) = 7.17\text{–}11.47$$

4. From Fig. 28*a*,

$$\text{minimum point } n = 8, 10$$

$$\text{maximum point } n = 9, 11$$

5. $n = \Delta/(\lambda/2)$ then $\lambda = 2\Delta/n$. Hence for each n in step 4,

$$\text{minimum point } \lambda = 1.075 \text{ m}, 0.86 \text{ m}, \qquad f = 279 \text{ MHz}, 349 \text{ MHz}$$

$$\text{maximum point } \lambda = 0.956 \text{ m}, 0.782 \text{ m}, \qquad f = 314 \text{ MHz}, 384 \text{ MHz}$$

6. The depth D_p and rise R_p are estimated for $\rho = 0.88$ using

$$D_p = -18.4 \text{ dB}, \qquad R_p = +5.5 \text{ dB}$$

7. Tendency of frequency pattern f,

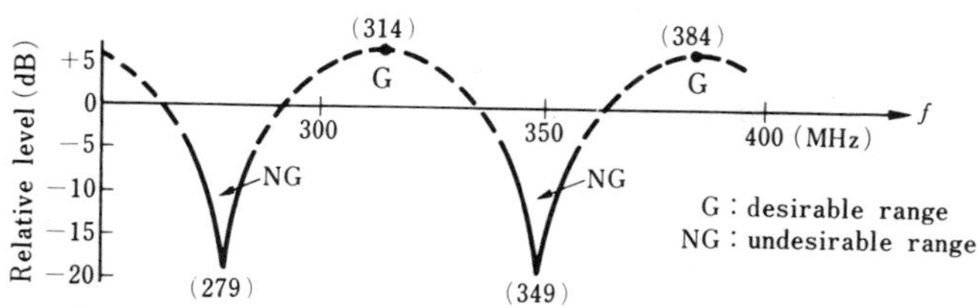

REMARKS

1. If the frequency has been allotted previously and the analysis shows the signal in the NG range, it is necessary to change the propagation path conditions such as antenna height and path length.

2. The f pattern given in step 7 exhibits a pitch of 70 MHz. Then the undesirable frequency range is approximately +10 to 15 MHz of the center frequency of the dip (279 MHz, 349 MHz).

Example 2.6

Calculate the propagation echo ratio ρ_θ, the half-pitches p_{h1} and p_{h2}, and the amount of depth for the propagation path shown in the figure, assuming that transmitting and receiving parabolic antennas of 3 m are used, the effective reflection coefficient $\rho_e = 0.8$, and the frequency is 4 GHz ($\lambda = 7.5$ cm).

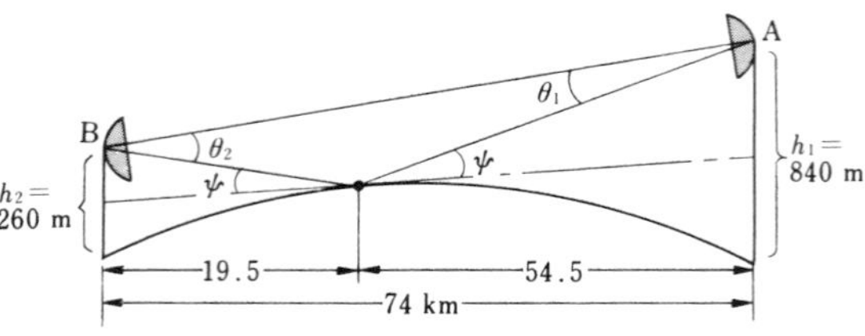

Solution

The included angles between direct and reflected waves θ_1 and θ_2 are calculated. Afterward the half-pitch, the directivity attenuation, and the depth are calculated.

1. From Figs. 31 to 34,

$$\alpha_1 = x_1 - z = -0.45 - 0.25 = -0.7^\circ$$

$$\alpha_2 = x_2 - z = +0.45 - 0.25 = +0.2^\circ$$

$$\beta_1 = -(y_1 + z_1) = -(0.88 + 0.18) = -1.06^\circ$$

$$\beta_2 = -(y_2 + z_2) = -(0.76 + 0.07) = -0.83^\circ$$

$$\theta_1 = \alpha_1 - \beta_1 = -0.7 - (-1.06) = 0.36^\circ$$

$$\theta_2 = \alpha_2 - \beta_2 = +0.2 - (-0.83) = 1.03^\circ$$

2. From Fig. 10 [2] the directivity attenuations $D_{\theta 1}$, and $D_{\theta 2}$ are calculated,

$$\theta_{1/2} = 57k\lambda/D_\phi = 57 \times 1.2 \times 7.5/300 = 1.74^\circ$$

$$\theta_1/\theta_{1/2} = 0.36/1.74 = 0.21$$

$$\theta_2/\theta_{1/2} = 1.03/1.74 = 0.59$$

From Fig. 10 [2],

$$D_{\theta 1} \doteqdot -0.3 \text{ dB}$$

$$D_{\theta 2} \doteqdot -4.6 \text{ dB}$$

$$D_{\theta 1} + D_{\theta 2} = -4.9 \text{ dB}$$

Antilogarithm is 0.57

3. Propagation echo ratio,

$$\rho_\theta = \rho_e D_{\theta 1} D_{\theta 2} = 0.8 \times 0.57 = 0.46 = 6.7 \text{ dB}$$

4. Depth of interference pattern (Fig. 29*a*),

$$D_p = -5.4 \text{ dB}$$

5. Half-pitches of interference pattern (Fig. 30),

$$p_{h1} \doteqdot 5.8 \text{ m}$$

$$p_{h2} \doteqdot 2.1 \text{ m}$$

EXAMPLE 2.7

There are two islands on the propgation path trasversing a seawater channel, as shown in the figure, and the study on the path profile shows the geometrical parameters indicated in the figure. Estimate the effective reflection coefficients for 80 MHz and 4 GHz, assuming the sea surface reflection coefficient $RD_v = 0.98$.

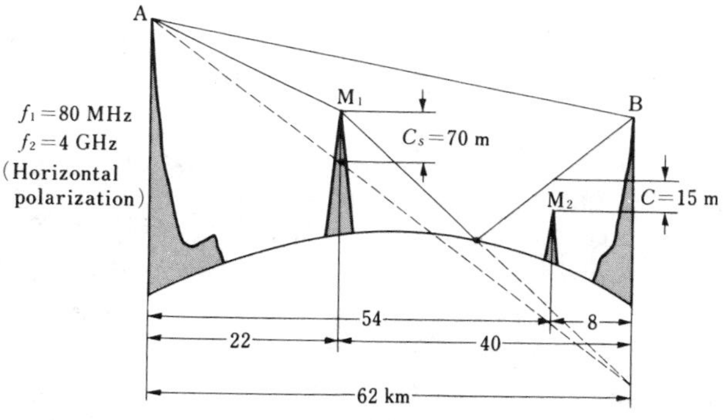

Solution

Since this is a problem on the shielding ridge, Fig. 12 is used.

1. Since the depth of ridge C_s for ridge M_1 and the clearance C for ridge M_2 are known, the first Fresnel zone radius r_s is calculated for the frequencies f_1 and

f_2. Using U and U_c and calculated Z_1 and Z_2, the effective reflection coefficient $\rho_e = (RD_v)Z$ is estimated. In the estimation $RD_v = 0.98$ and $Z = Z_1 Z_2$ are used.

2. First Fresnel zone radius r_s

λ (m)	Ridge	d_1 (km)	d_2 (km)	D (km)	$r_s = \sqrt{\lambda d_1 d_2 / D}$
3.75	M_1	22	40	62	231
3.75	M_2	8	54	62	162
0.075	M_1	22	40	62	32.6
0.075	M_2	8	54	62	22.9

Figures 12 and 13 in Chapter 3 may be used.

3. Diffraction coefficient,

f	Ridge	r_s	C_s (m)	C (m)	Trans-horizon $U = C_s/r_s$	Line of sight $U_c = C/r_s$	Diffraction Coefficient
80 MHz	M_1	231	70	—	0.3	—	$Z_1 = 0.35$
80 MHz	M_2	162	—	15	—	0.09	$Z_2 = 0.58$
4 GHz	M_1	32.6	70	—	2.15	—	$Z_1 = 0.075$
4 GHz	M_2	22.9	—	15	—	0.66	1 (no loss)

4. Ridge effect $Z = Z_1 Z_2$.
5. Effective reflection coefficient ρ_e,

$$\rho_e = 0.98 \times Z \begin{cases} = 0.2, & 80\text{ MHz} \\ = 0.07, & 4\text{ GHz} \end{cases}$$

Remarks

1. For 4 GHz, M_2 has no influence on the calculation, but due to the shielding ridge effect by M_1, the free-space propagation path ($\rho_e < 0.1$) is realized.
2. For 80 MHz, both M_1 and M_2 contribute to screening the reflected wave. However, the effect is not so great because of its larger Fresnel zone radius, and it is judged to be a quasi-free-space propagation path because ρ_e is approximately 0.2. For 80 MHz, if the ridge M_2 did not exist, $\rho_e = 0.34$ and the path would be categorized as interference propagation path in spite of the shielding effect by ridge M_1.

Chart Index (1),
Reflected-Wave Interference Propagation Path

Chart Index (2), Reflected-Wave Interference Propagation Path

Geometrical Parameters on Interference Propagation Path

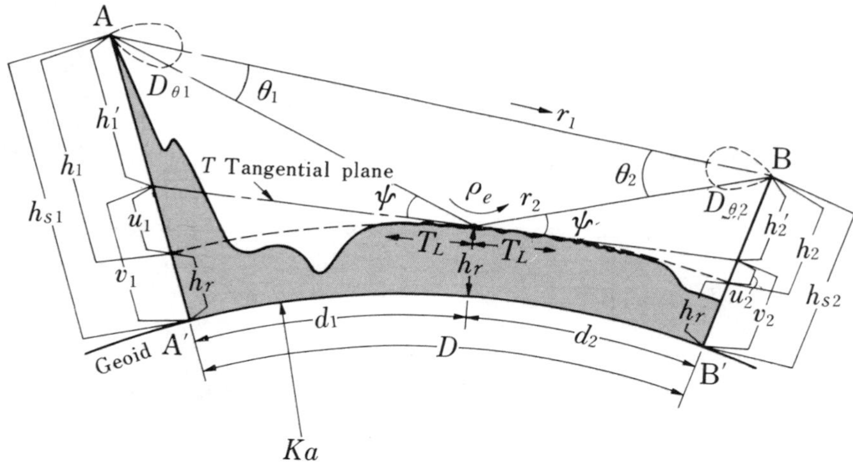

D : Great circle distance

d_1, d_2 : Great circle distance to the reflection point from point A or B

r_1 : Path length of direct wave

r_2 : Path length of reflected wave

h_{s1}, h_{s2} : Antenna height above sea level at point A or B

h_r : Mean elevation above sea level of the reflection point

h_1, h_2 : Antenna height relative to the elevation of the reflection point

h_1', h_2' : Antenna height above the tangential plane at the reflection point (Equivalent antenna height over the flat ground)

ψ : Grazing angle of reflected wave

T_L : A half-major axis of the area projected by the first Fresnel zone on the reflection surface (Major radius of the effective reflection area)

θ_1, θ_2 : Included angle between direct and reflected waves at point A or B

u_1, u_2 : Height corresponding to the line-of-sight distance of d_1 or d_2

v_1, v_2 : Height of the reflection point above sea level where the tangential plane T contacts

---

Δ : Path length difference

$$\Delta = r_2 - r_1 = 2\, h_1' h_2' / D$$

K : Effective earth radius factor (Standard $K=4/3$)

a : True radius of earth (6.37×10^3 km)

Ka : Effective earth radius (8.5×10^3 km for $K=4/3$)

ρ_e : Effective reflection eoefficient ($\rho_e \leqq 1$)

ρ_θ : Propagation echo ratio ($\rho_\theta = \rho_e \cdot D_{\theta 1} \cdot D_{\theta 2}$)

$D_{\theta 1}, D_{\theta 2}$: Directivity coefficient in the direction of the reflected wave of antenna at point A or B

$$Ka \gg D \gg h_{s1}, h_{s2}$$

Smooth-Ground Model

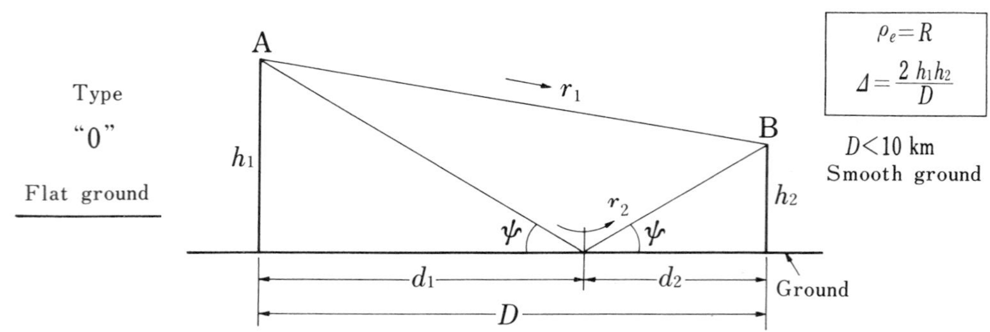

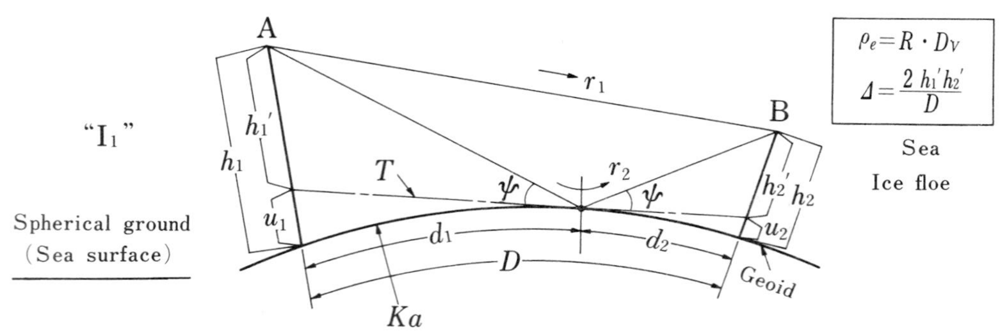

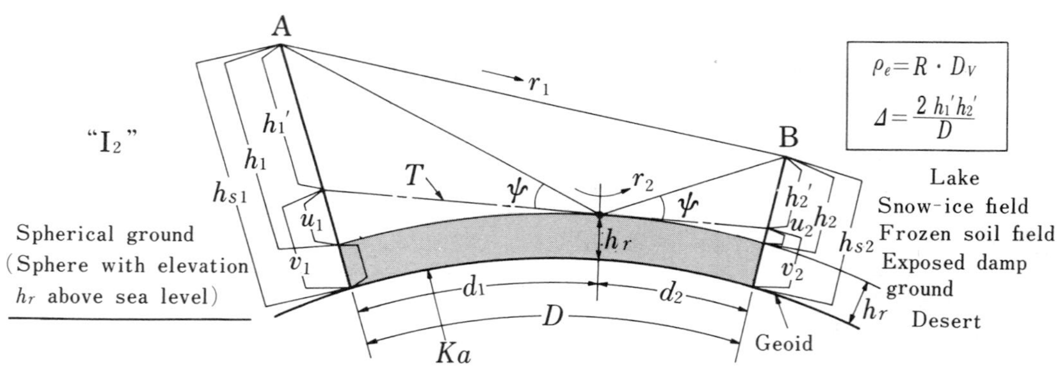

ρ_e : Effective reflection coefficient
R : Flat plane reflection coefficient
D_V : Spherical ground divergence factor
Δ : Path length difference between direct and reflected waves

Rough-Ground Model

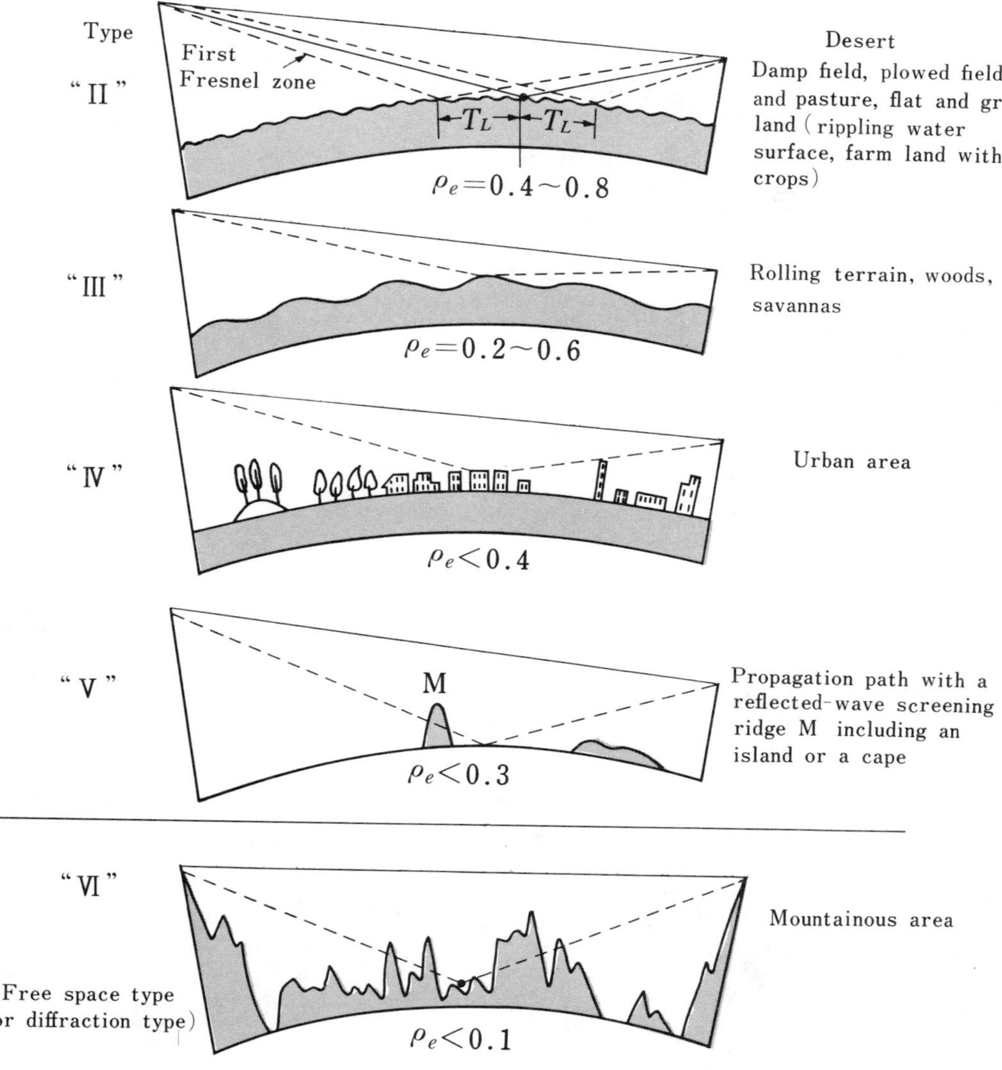

T_L : A half-major axis of projected area by 1st Fresnel zone of the reflected wave (See Fig. 7 for the half-major axis of the effective reflection area). Characteristics of the reflection area should be examined over the range of T_L. The same should apply for the other type of path.

Types IV through VI will become "free space type," if ρ_e is small and there is no diffraction loss.

2-4 Categories of Reflected-Wave Interference Paths and Their Treatment

Two waves (direct and reflected waves) interference with singular reflection

Normal reflection

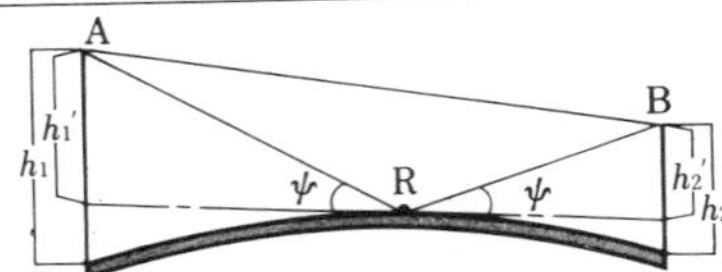

This is a reflection occurring at the sea surface, lake surface or the plain with an almost uniform elevation above sea level, and the reflection point R is determined from the effective earth radius Ka, antenna heights h_1, h_2 as shown in the figure and the distance D. The path length difference Δ is proportional to the equivalent antenna heights h_1', h_2'.

Quasinormal reflection

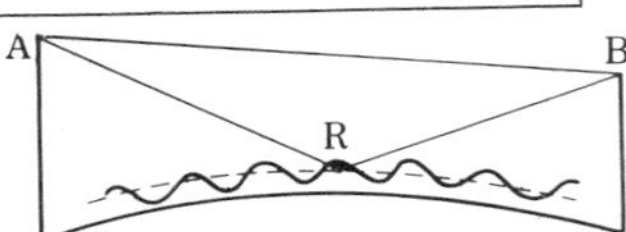

For an undulating terrain, the mean elevation is assumed to form a spherical surface which produces normal reflection. However, the estimation of the effective reflection coefficient ρ_e should take into account the effect of undulation.

Reflection on sloped surface

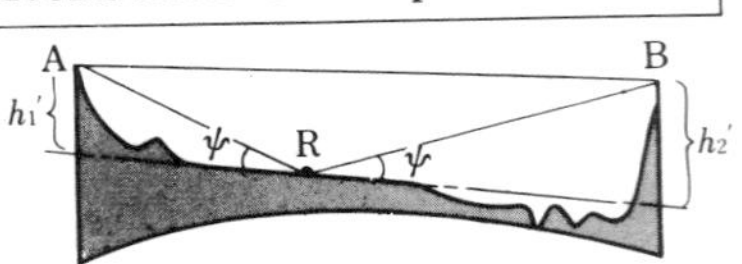

The reflection point is at point R where the grazing angle ψ is equal in both sides A and B on the sloped surface. The heights h_1' and h_2' above the tangential plane at R are used as the effective antenna heights and from which the path length difference Δ is calculated.

Half-reflection

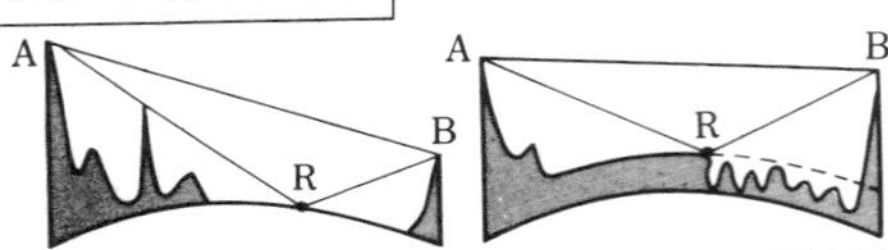

One-half of the first Fresnel zone is shielded or an irreqular reflection occurs, reducing the effect of the reflection in either case in the figures. Effective reflection coefficient $\rho_e \leqq -6$ dB.

Interference screening type

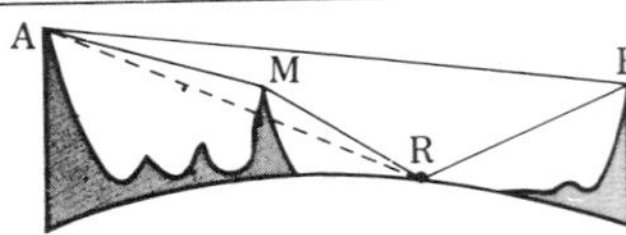

In case the shielding ridge M is deep, the propagation path is of the free space type or the diffraction type.
If M shields fully the first Fresnel zone of the reflected wave, the effective reflection coefficient $\rho_e \leqq -16$ dB. Hence, examination is necessary only when M is shallower than that. The reflection point is between M and B.

Multireflections and multiwave interference

Specific types

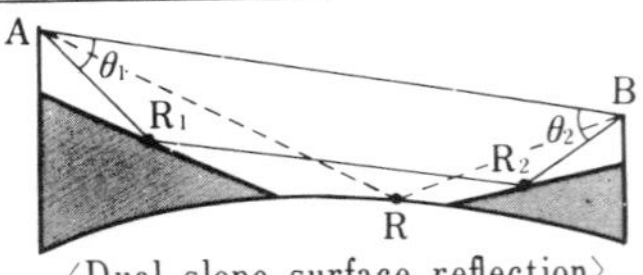

〈Dual slope surface reflection〉

This type of reflection rarely becomes serious because the dual reflection reduces the effective reflection coefficient appreciably ($\rho_e = \rho_{e1} \cdot \rho_{e2}$) and it is very rare to provide a large effective reflection area at these two points. Generally the reflected-wave is attenuated by an antenna directivity effect in the microwave frequency band.

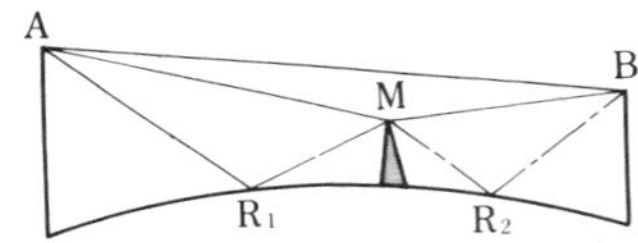

〈Single ridge over sea water〉

Reflected wave taking the path A R_1 M R_2B undergoes reflection twice. However, the diffraction loss by M is so large that the reflected wave can be neglected. Then it is normally sufficient to check the paths AR_1MB and AMR_2B with a single reflection, and generally only the more influential one.

Remarks

Non-interference type

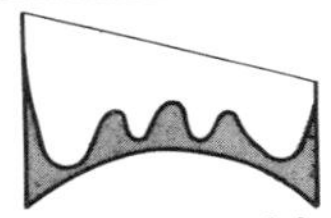

Terrain with an undulation greater than the first Fresnel zone.

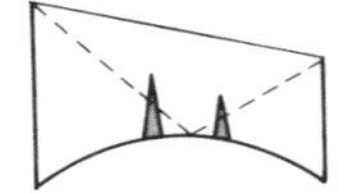

Presence of two or more shielding ridges

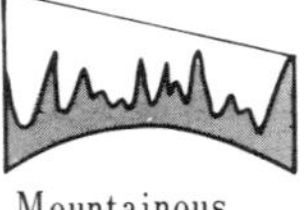

Mountainous propagation path

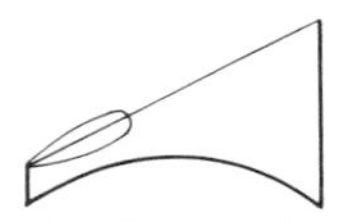

Suppression due to antenna directivity pattern is expected (in satellite communication, etc.)

Vector Diagram of Two-Wave Interference

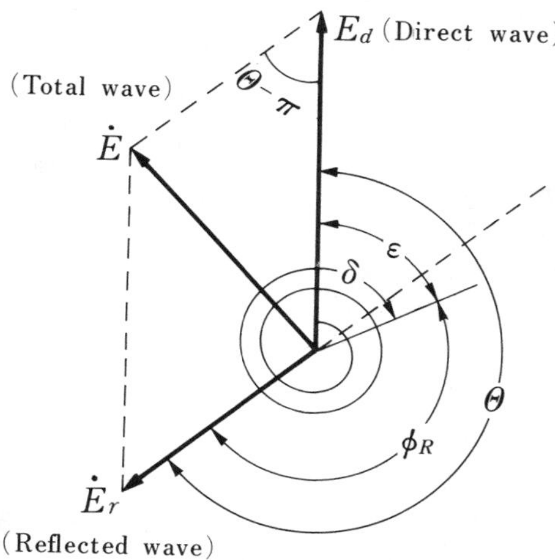

E_d : Direct-wave field strength vector (reference), (Normally $E_d = E_0$)

$\dot{E}_r$: Reflected-wave field strength vector

δ : Delay angle corresponding to path length difference

$$\delta = 2\pi\Delta/\lambda$$

Δ : Path length difference between direct and reflected waves

ε : Effective angle (a residual angle when δ is divided by 2π radian)

ϕ_R : Delay angle (phase of the reflection coefficient) at ground reflection

Θ : Delay angle of the reflected wave to the direct wave (relative phase of reflected wave)

$$\Theta = \varepsilon + \phi_R$$

ρ_e : Amplitude ratio of the direct to reflected waves

$$\rho_e = |E_r/E_d| \qquad E_r = \rho_e \cdot E_d$$

E : Received field strength vector (sum of E_d and $\dot{E}_R$)

$$E = E_d\sqrt{1+\rho_e^2 - 2\rho_e\cos(\Theta-\pi)}$$

Criterion for Mirror-Face Reflection

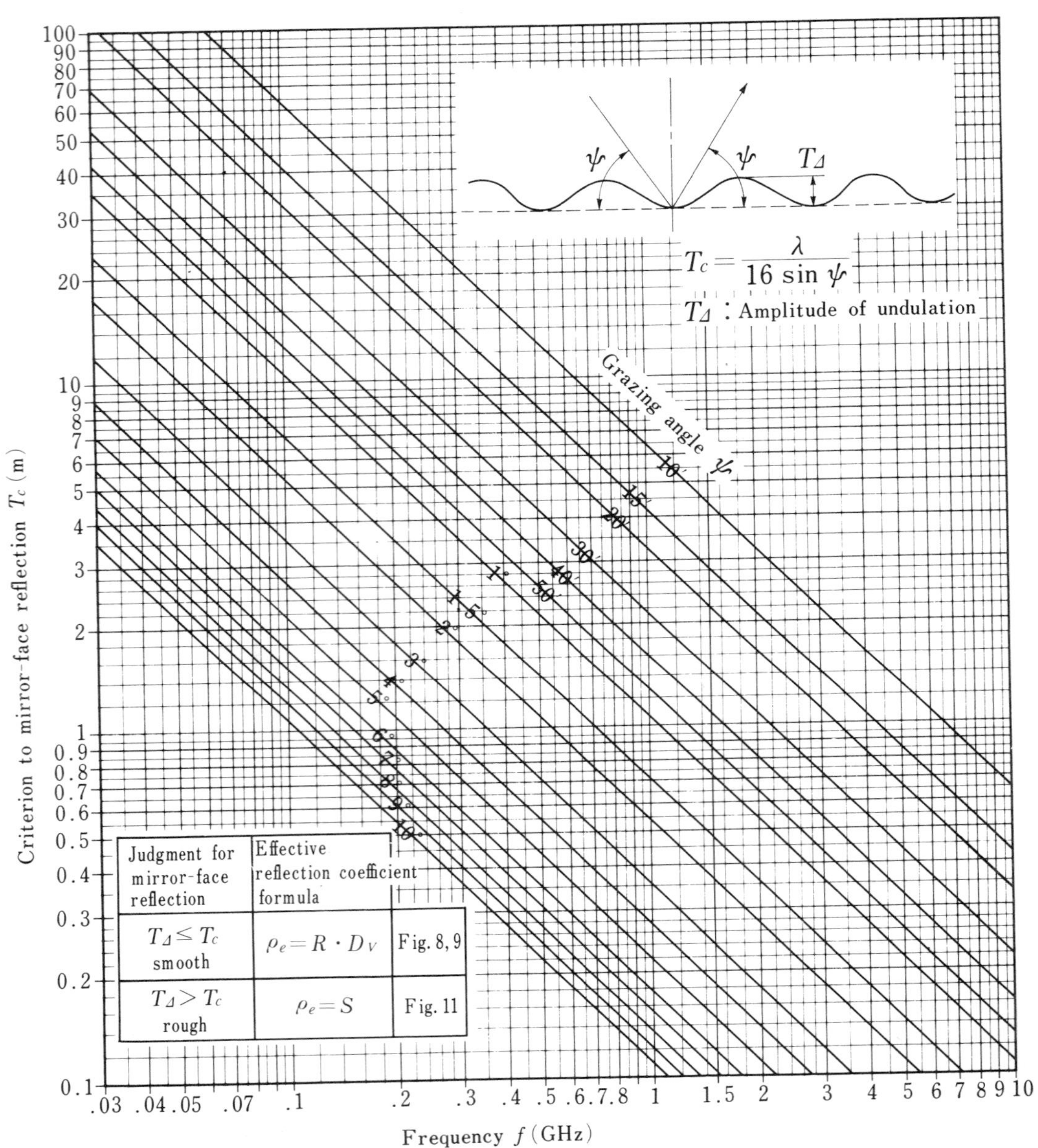

Judgment for mirror-face reflection	Effective reflection coefficient formula	
$T_\Delta \leq T_c$ smooth	$\rho_e = R \cdot D_V$	Fig. 8, 9
$T_\Delta > T_c$ rough	$\rho_e = S$	Fig. 11

A Half-Major Axis of Effective Reflection Area

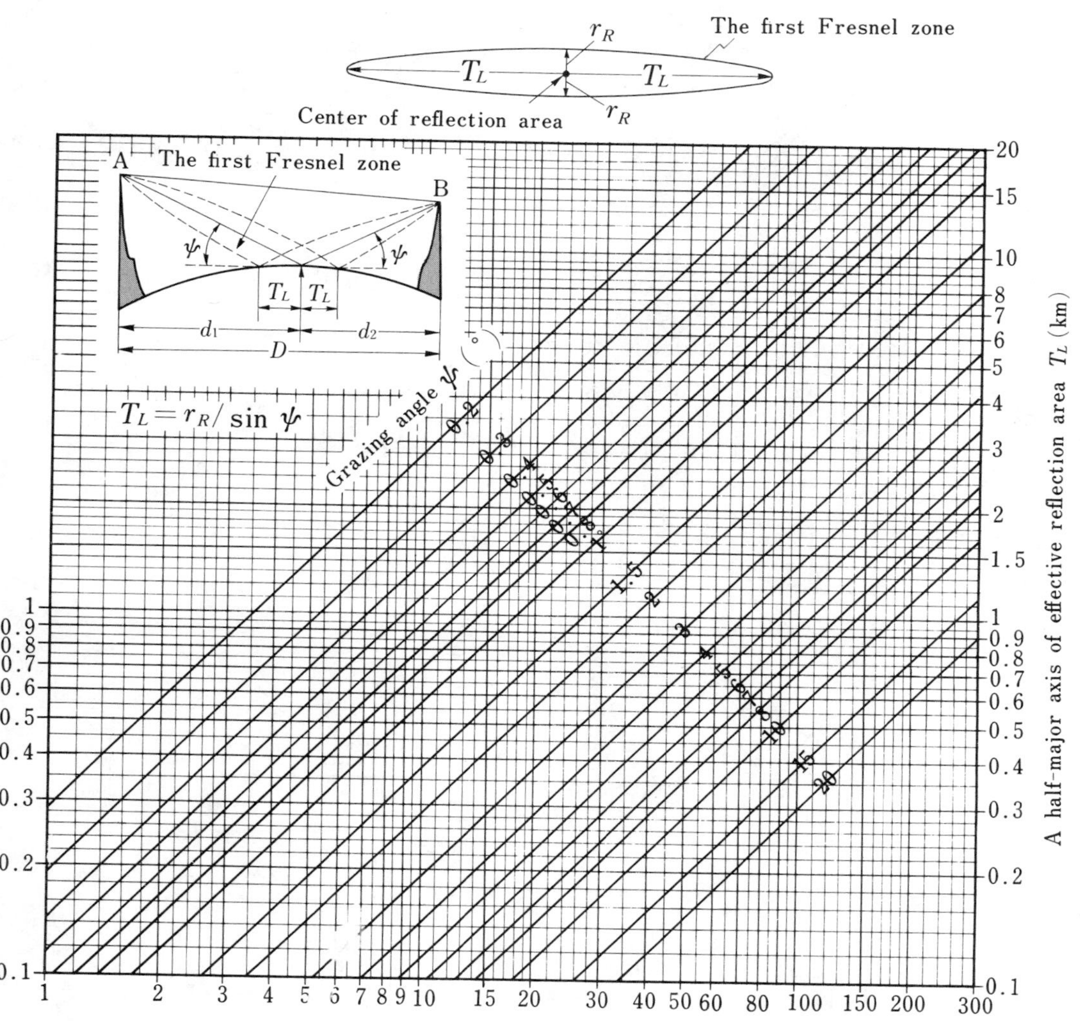

Flat-Ground Reflection Coefficient (1)

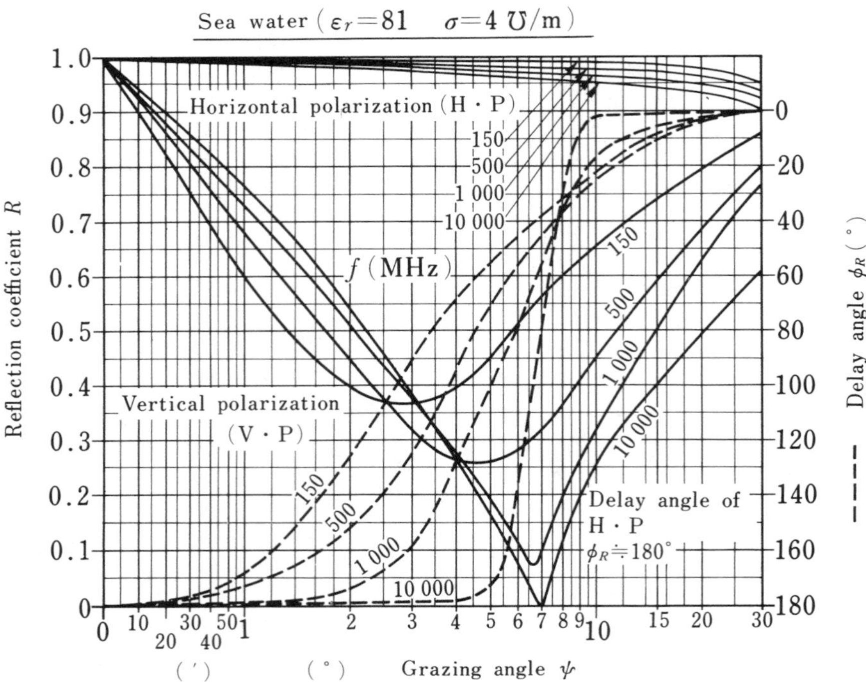

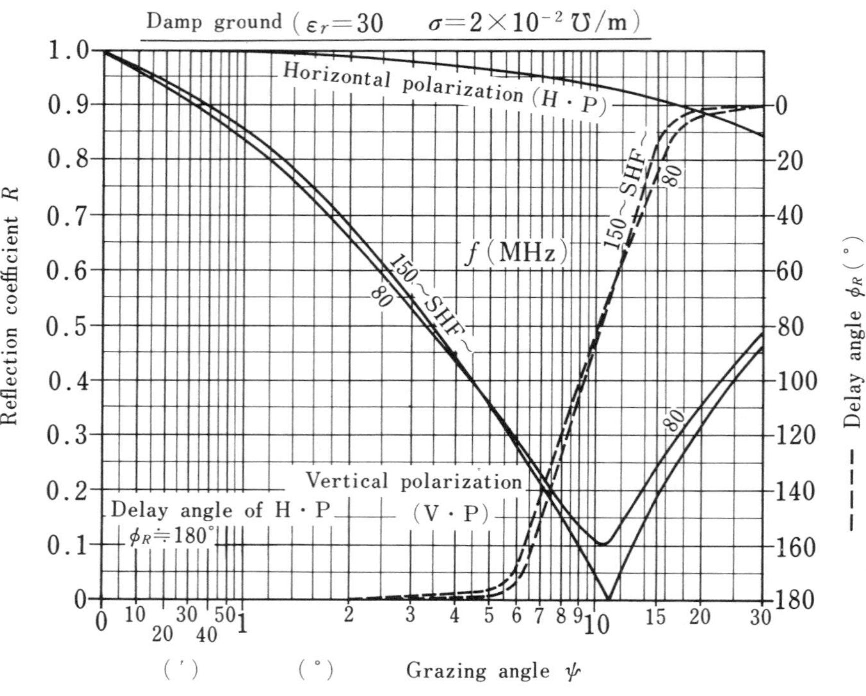

Flat-Ground Reflection Coefficient (2)

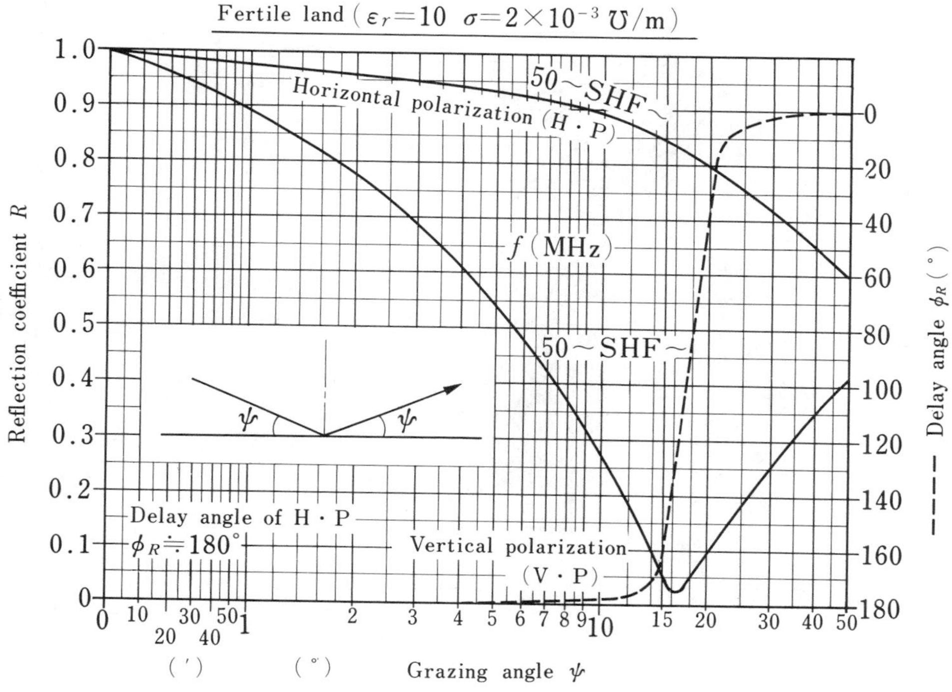

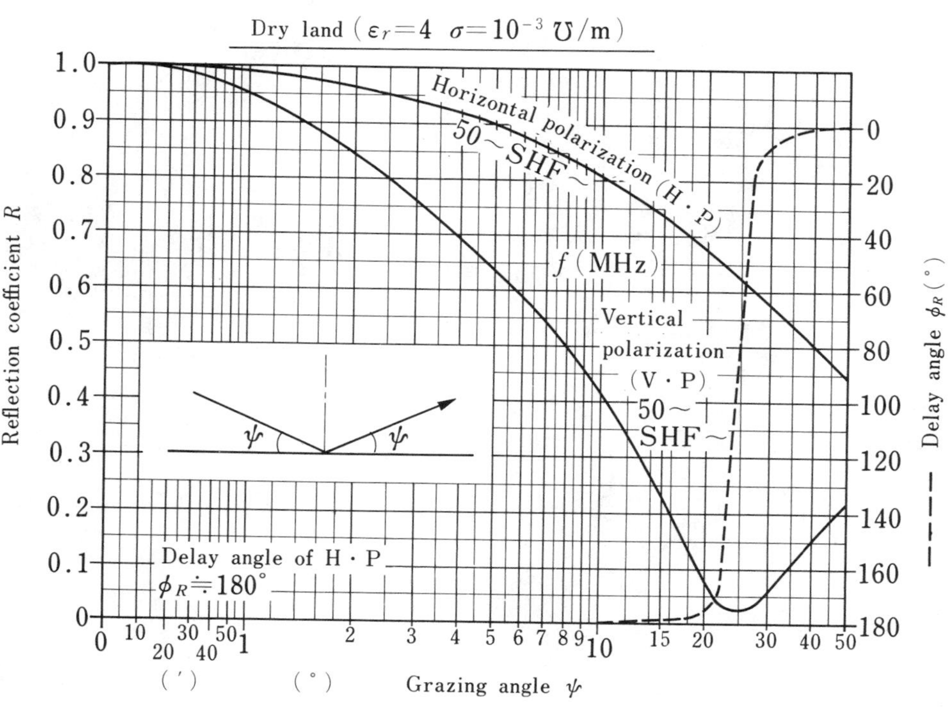

Spherical-Ground Divergence Factor

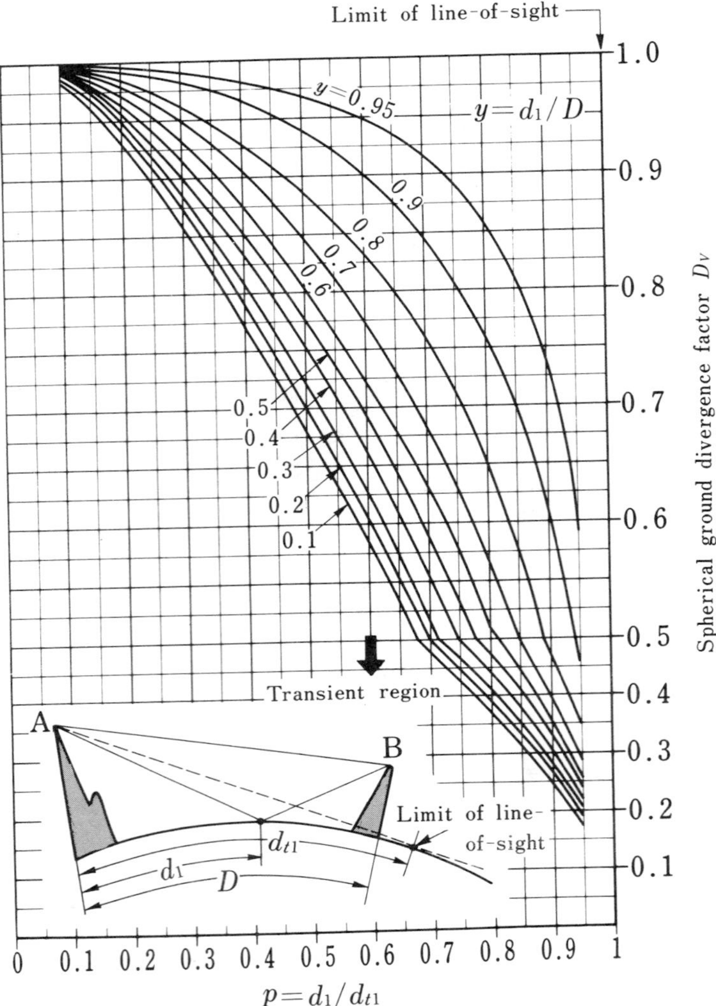

Antenna Directivity Attenuation

[1] Reference directivity of parabolic antenna

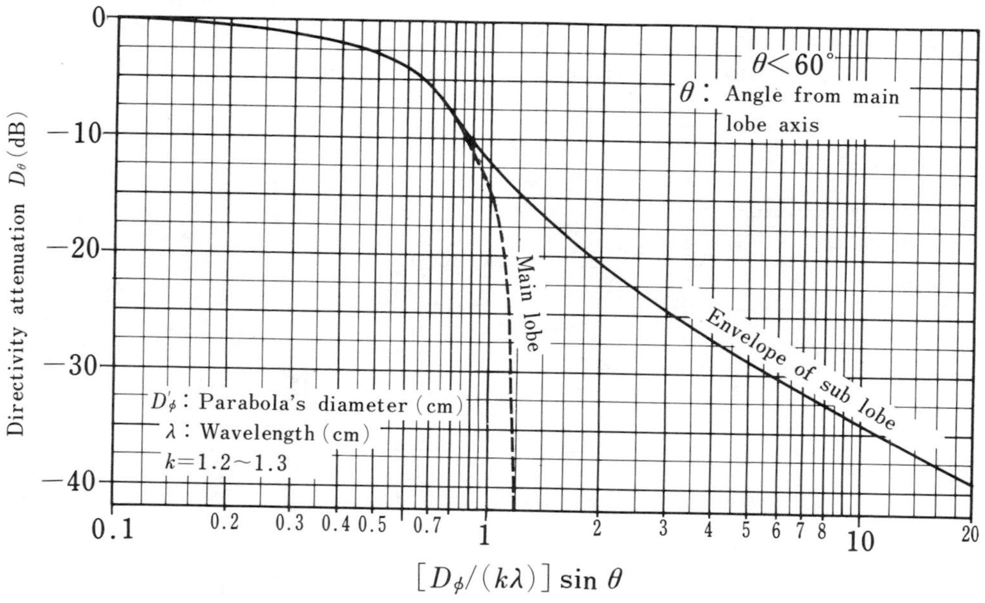

[2] Normalized directivity of antennas (including parabola)

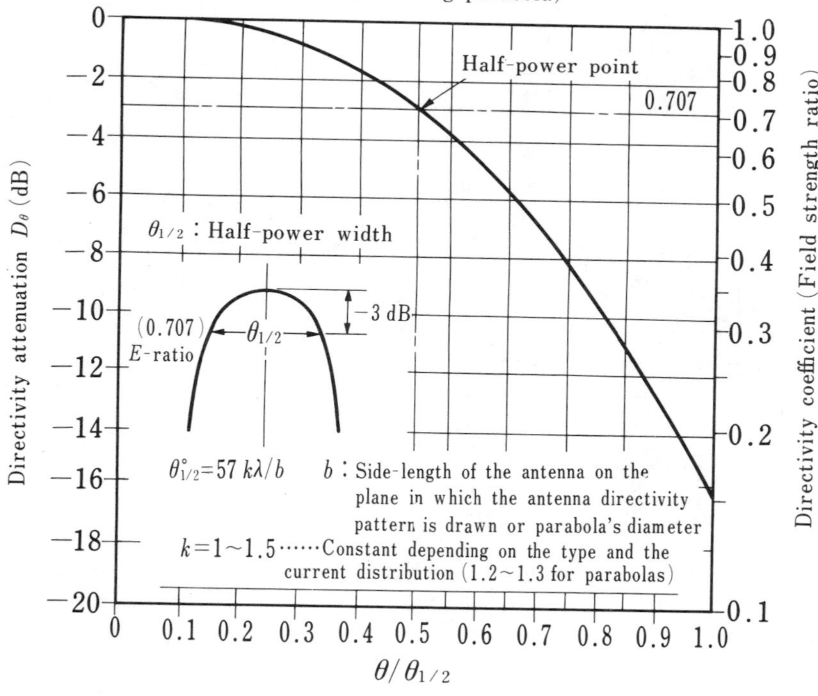

Effective Reflection Coefficient of Ordinary Ground (1)

1. Approximate ranges of the coefficients for topography and conditions of the earth surface.
2. Grazing angle, $0 < \psi < 5°$.
3. Figures apply for the characteristics of the effective reflection area (projection of the first Fresnel zone) of the reflected wave.

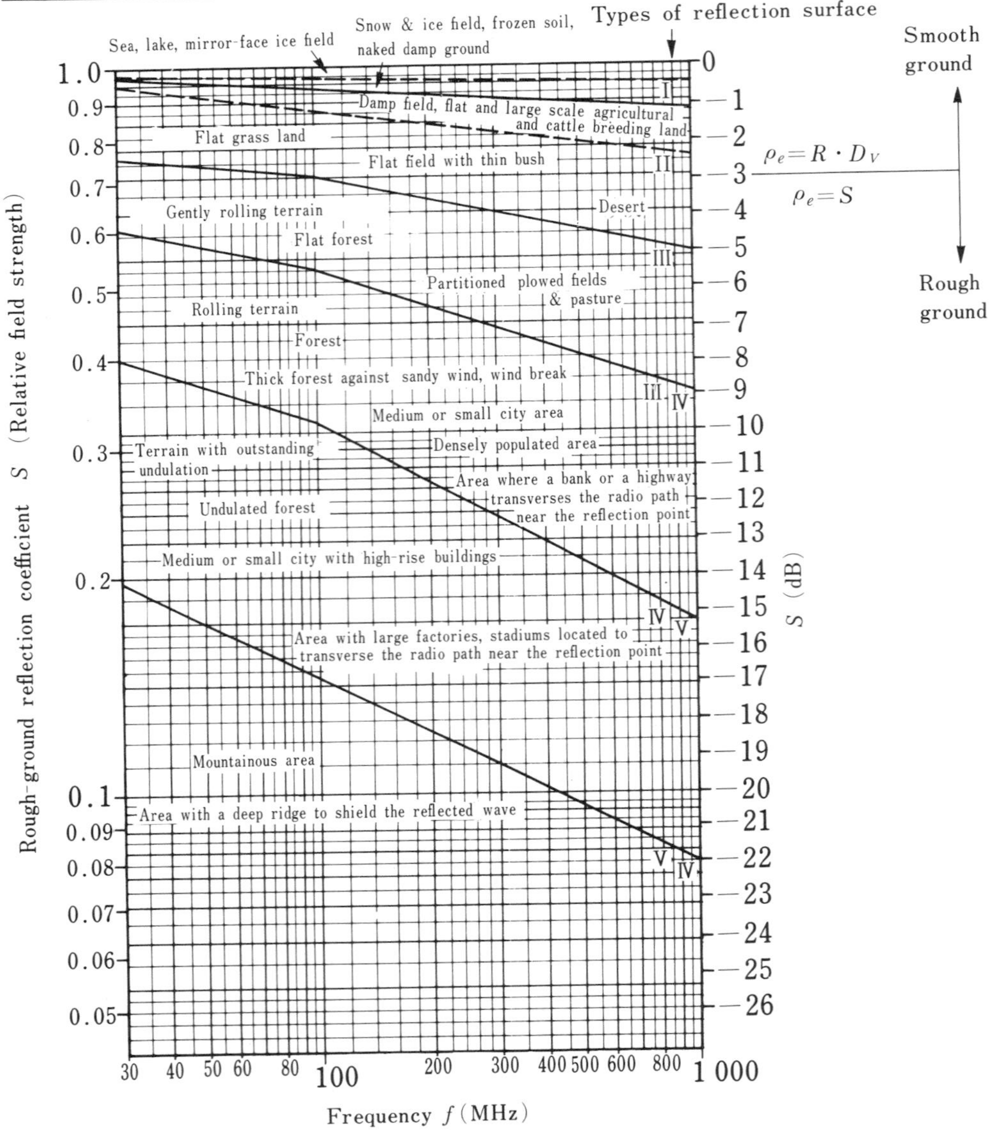

Effective Reflection Coefficient of Ordinary Ground (2)

4. Use Fig. 8 for types I or II in vertical polarization and for $\psi > 3°$ in vertical polarization.
5. Two equations for ρ_e are used, one for smooth terrain the other one for rough terrain.
   ρ_e : Effective reflection coefficient R : Flat plane reflection coefficient
   D_V : Spherical ground divergence factor
6. Curves in the figure include the effect by the reflected-wave shielding ridge (Use Fig. 12 to obtain the theoretical value to single reflected-wave shielding ridge).

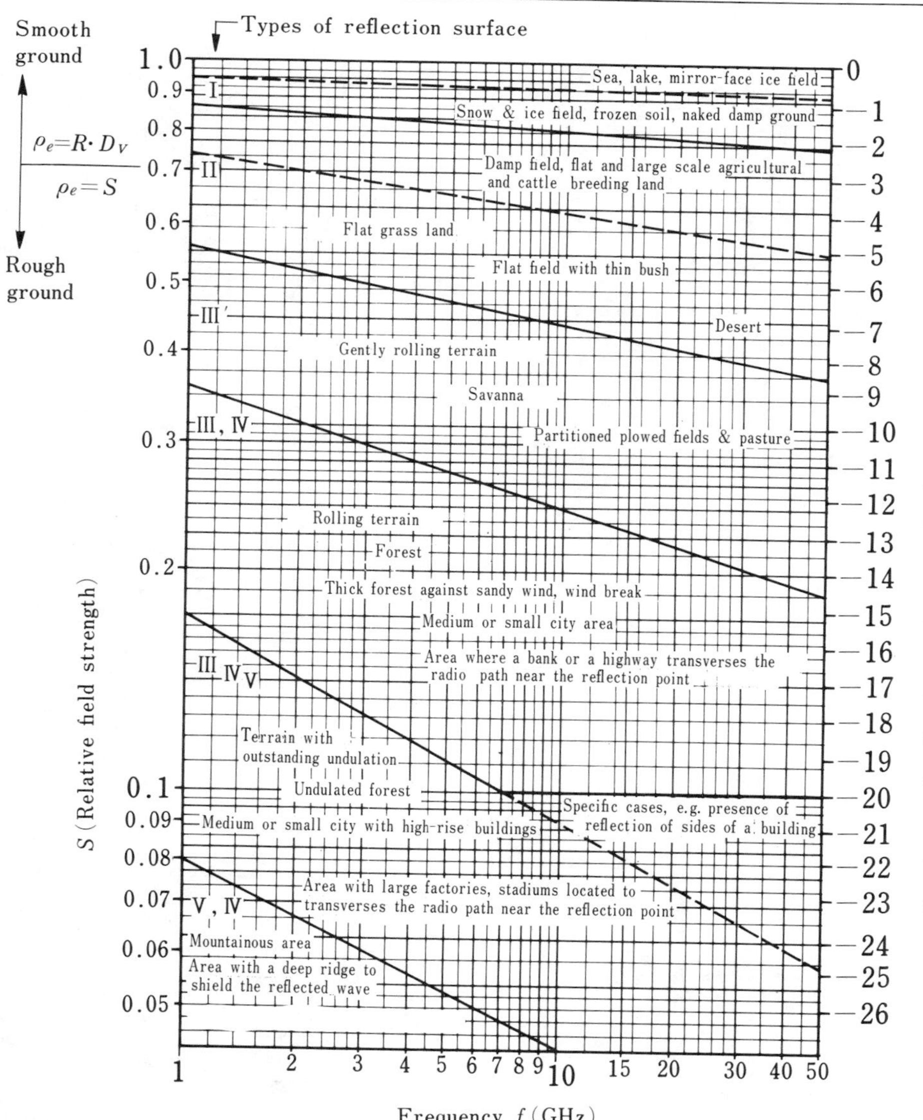

Reflected-Wave Shielding Ridge (Evaluation of Effect)

[1] Effective reflection coefficient ρ_e

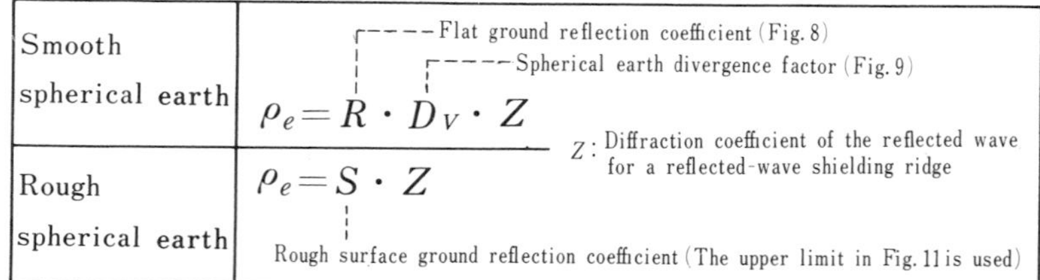

Smooth spherical earth	$\rho_e = R \cdot D_V \cdot Z$ R: Flat ground reflection coefficient (Fig. 8) D_V: Spherical earth divergence factor (Fig. 9)
Rough spherical earth	$\rho_e = S \cdot Z$ S: Rough surface ground reflection coefficient (The upper limit in Fig. 11 is used)

Z: Diffraction coefficient of the reflected wave for a reflected-wave shielding ridge

[2] Diffraction coefficient Z

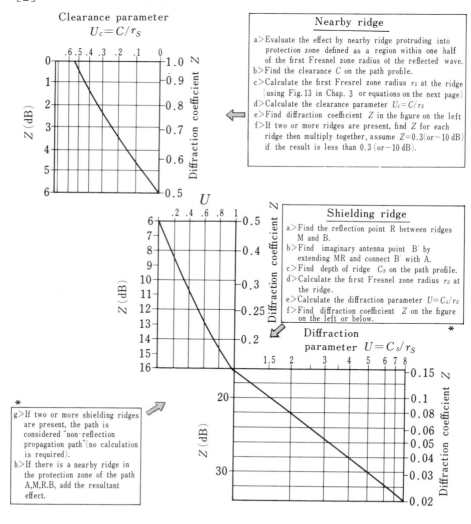

Clearance parameter $U_c = C/r_S$

Nearby ridge

a> Evaluate the effect by nearby ridge protruding into protection zone defined as a region within one half of the first Fresnel zone radius of the reflected wave.
b> Find the clearance C on the path profile.
c> Calculate the first Fresrel zone radius r_S at the ridge (using Fig. 13 in Chap. 3 or equations on the next page)
d> Calculate the clearance parameter $U_c = C/r_S$
e> Find diffraction coefficient Z in the figure on the left
f> If two or more ridges are present, find Z for each ridge then multiply together, assume $Z=0.3$ (or -10 dB) if the result is less than 0.3 (or -10 dB).

Shielding ridge

a> Find the reflection point R between ridges M and B.
b> Find imaginary antenna point B′ by extending MR and connect B′ with A.
c> Find depth of ridge C_S on the path profile.
d> Calculate the first Fresnel zone radius r_S at the ridge.
e> Calculate the diffraction parameter $U = C_S/r_S$
f> Find diffraction coefficient Z on the figure on the left or below.

Diffraction parameter $U = C_S/r_S$

*

*
g> If two or more shielding ridges are present, the path is considered "non-reflection propagation path" (no calculation is required).
h> If there is a nearby ridge in the protection zone of the path A,M,R.B, add the resultant effect.

Reflected-Wave Shielding Ridge (Path Geometry)

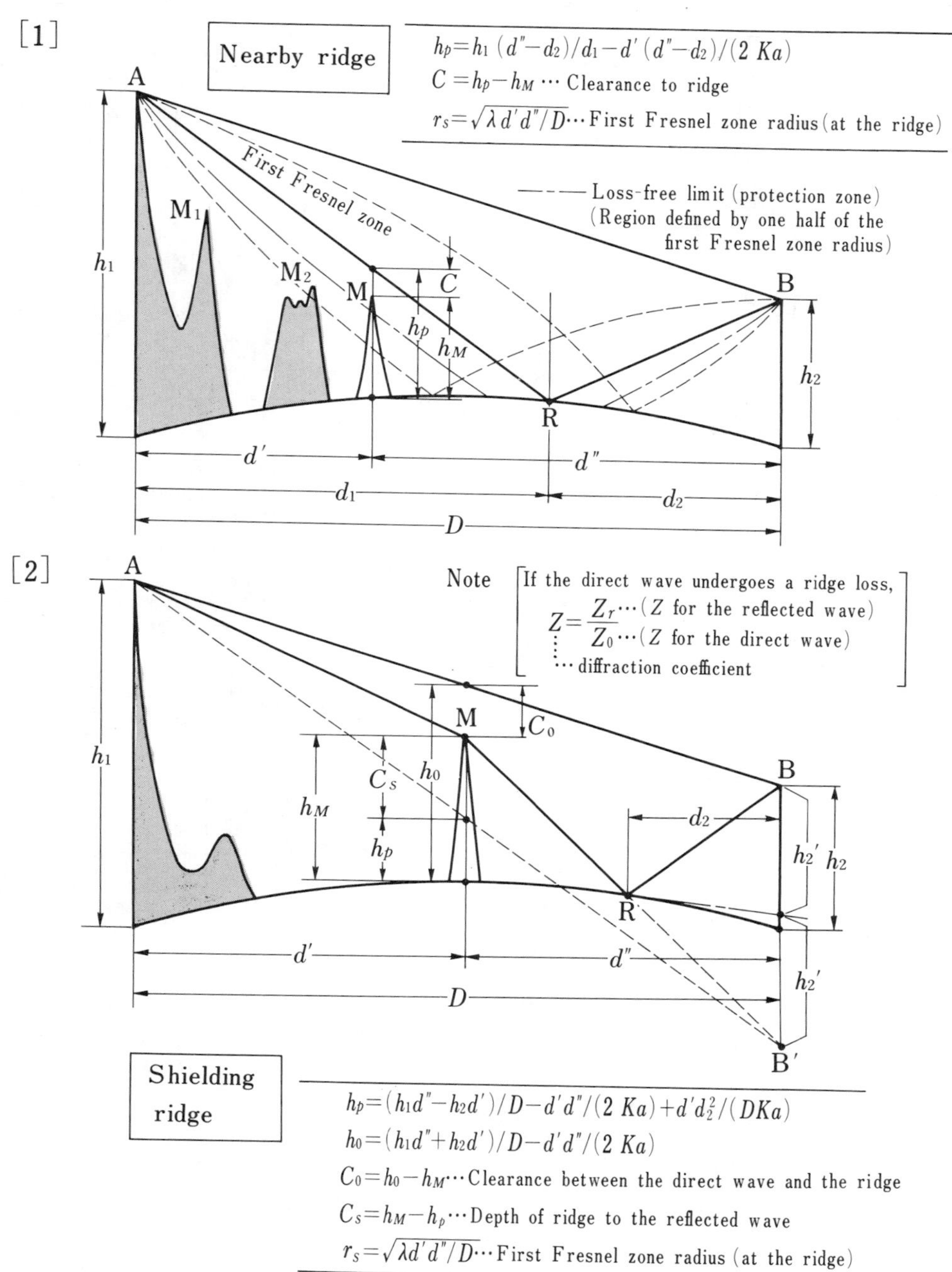
[1]
Nearby ridge
$h_p = h_1(d'' - d_2)/d_1 - d'(d'' - d_2)/(2Ka)$
$C = h_p - h_M$ ··· Clearance to ridge
$r_s = \sqrt{\lambda d' d''/D}$ ··· First Fresnel zone radius (at the ridge)
Loss-free limit (protection zone)
(Region defined by one half of the first Fresnel zone radius)
First Fresnel zone
A
B
R
M
M1
M2
C
hp
hM
h1
h2
d'
d''
d1
d2
D
[2]
Note [If the direct wave undergoes a ridge loss,
$Z = \frac{Z_r \cdots (Z \text{ for the reflected wave})}{Z_0 \cdots (Z \text{ for the direct wave})}$
··· diffraction coefficient]
A
M
B
R
B'
C0
Cs
h0
hM
hp
h1
h2'
h2
d2
d'
d''
D
Shielding ridge
$h_p = (h_1 d'' - h_2 d')/D - d'd''/(2Ka) + d'd_2^2/(DKa)$
$h_0 = (h_1 d'' + h_2 d')/D - d'd''/(2Ka)$
$C_0 = h_0 - h_M$ ··· Clearance between the direct wave and the ridge
$C_s = h_M - h_p$ ··· Depth of ridge to the reflected wave
$r_s = \sqrt{\lambda d' d''/D}$ ··· First Fresnel zone radius (at the ridge)

Effective Reflection Coefficient and Propagation Echo Ratio, Antilogarithm-to-dB Conversion Chart

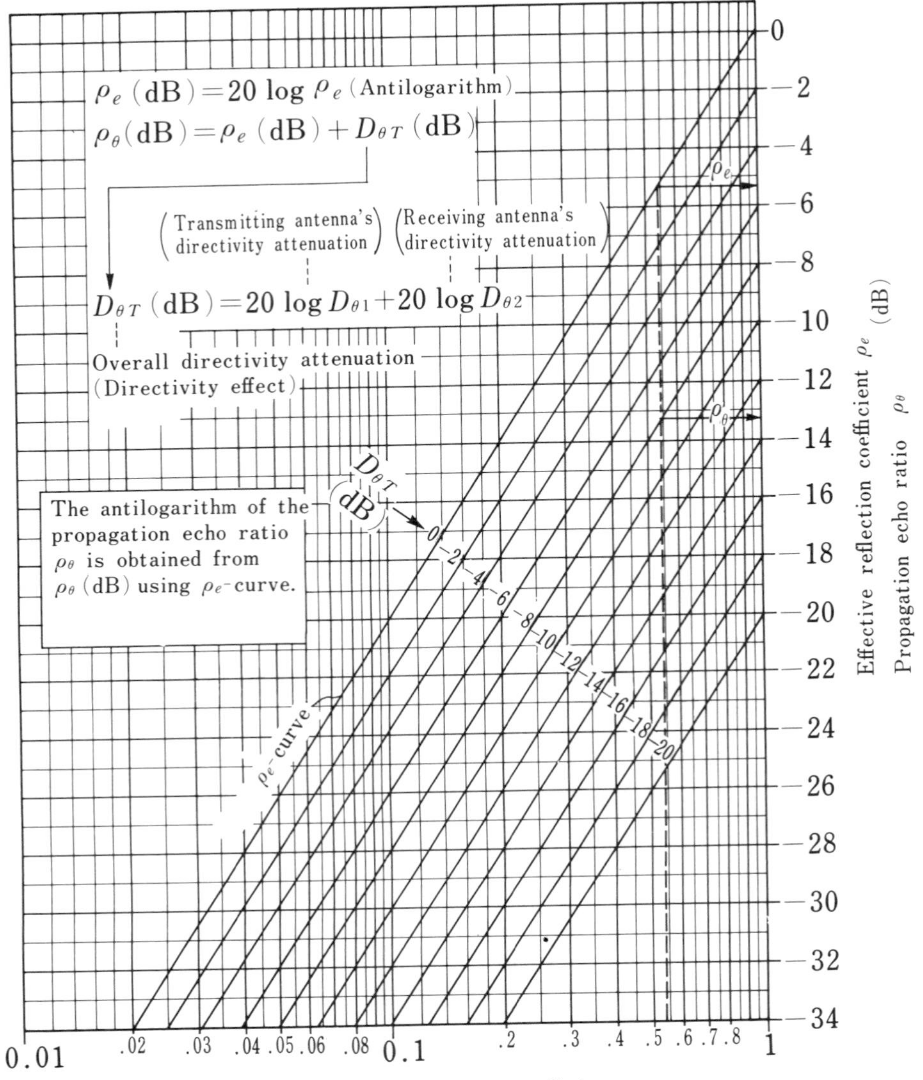

Phase Angle of Reflected Wave (Delay from Direct Wave)

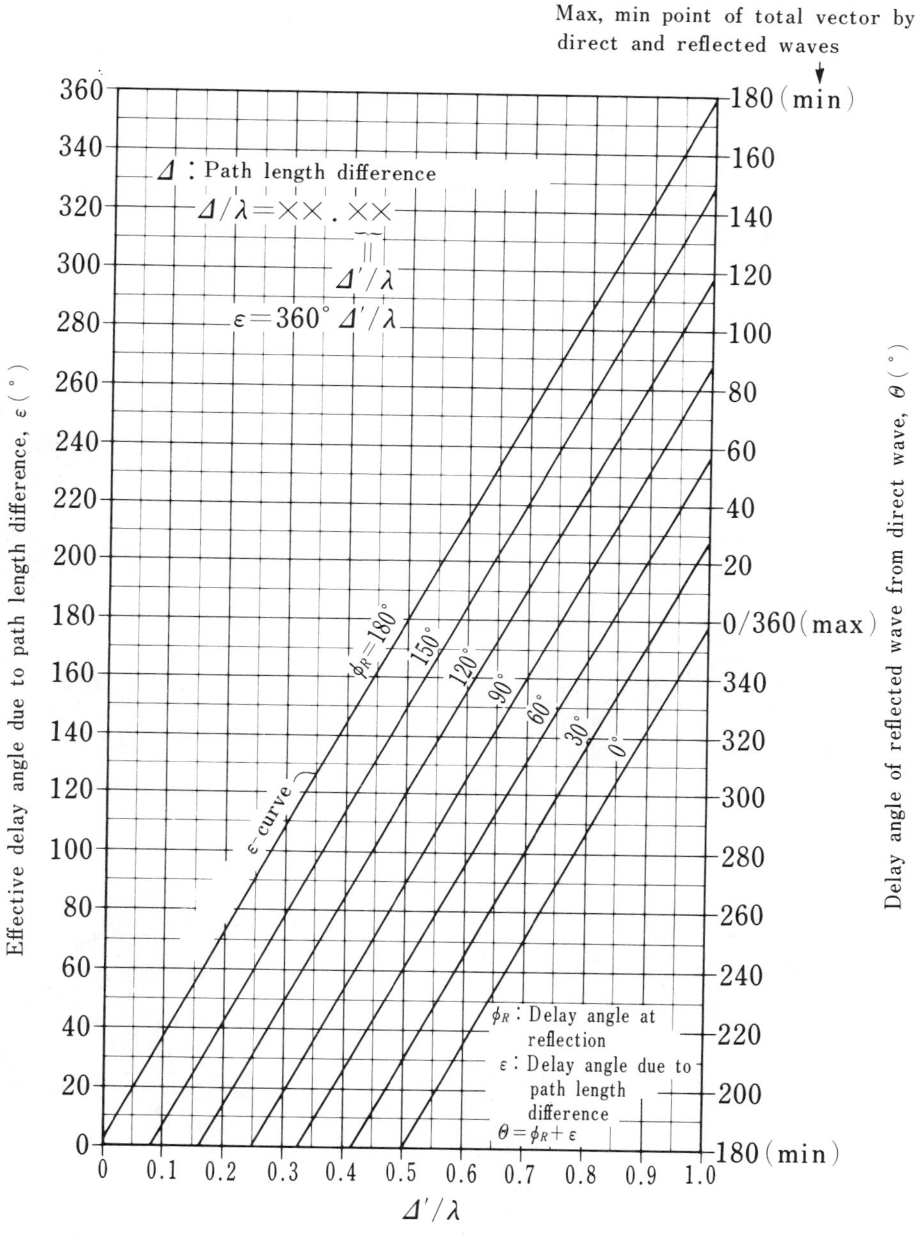

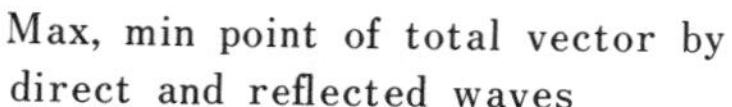

Vector Sum of Field Strength

For Universal Use

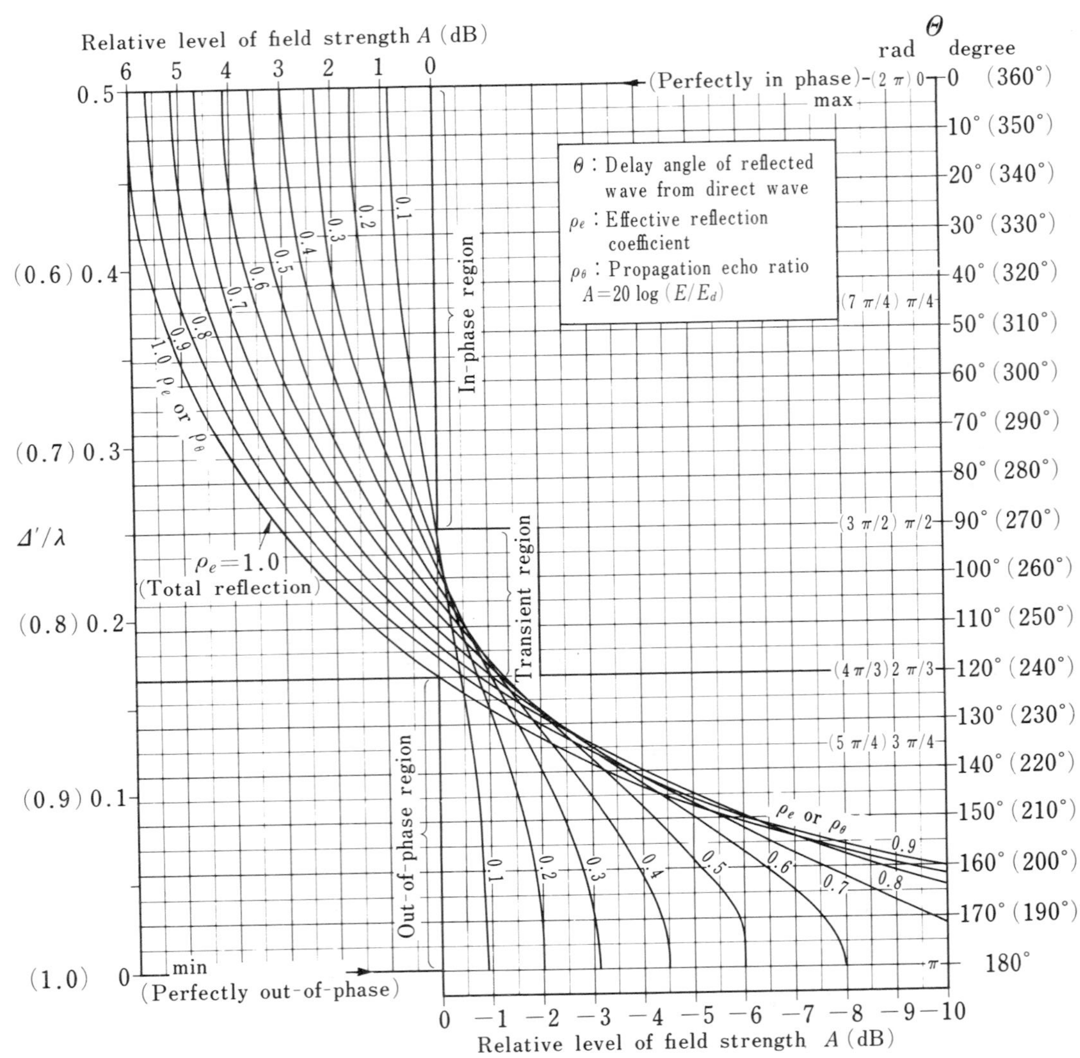

Vector Sum of Field Strength

Details of Out-of-Phase Region For Universal Use

1. This is a graph to estimate a ratio of the levels in dB of the total wave E to the direct wave E_d from the effective reflection coefficient ρ_e (or propagation echo ratio ρ_θ) and the phase angle Θ. If $\theta > 2\pi$ (360°), reduce it to an angle below 2π (360°).

2. Vector diagram

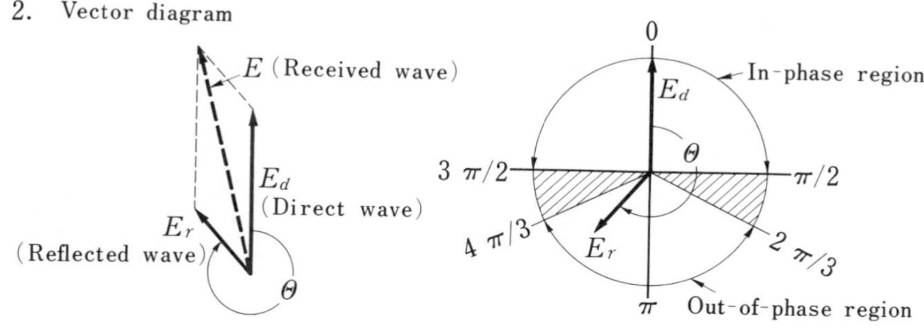

3. Scale for Δ'/λ is used when the phase shift at reflection (phase of the reflection coefficient) is 2π radian (180°).

4. If the path length difference between direct and reflected waves is Δ and the wave length λ,

$$\Delta/\lambda = \times\times.\underbrace{\times\times}_{\Delta'/\lambda}$$

Δ'/λ ··· Figure below a decimal point

5. Effective angle for path length difference $\varepsilon = 2\pi \cdot \Delta'/\lambda$ (to be used in Fig. 16)

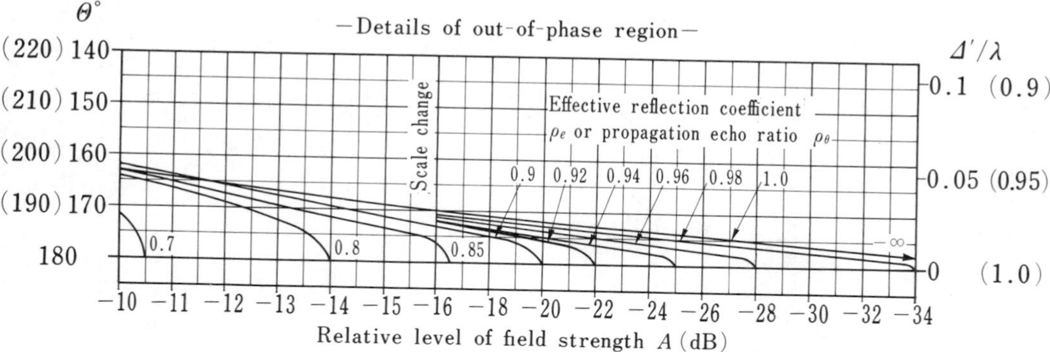

2

16a

Total Reflection (Linear Region)

$$E/E_d \doteqdot \frac{4\pi h_1 h_2}{\lambda D} = \frac{2\pi}{\lambda}\Delta = \varepsilon$$

Effective reflection coefficient $\rho_e = -1$
(Absolute value $\rho_e = 1$
Delay angle $\phi_R = 180°$)

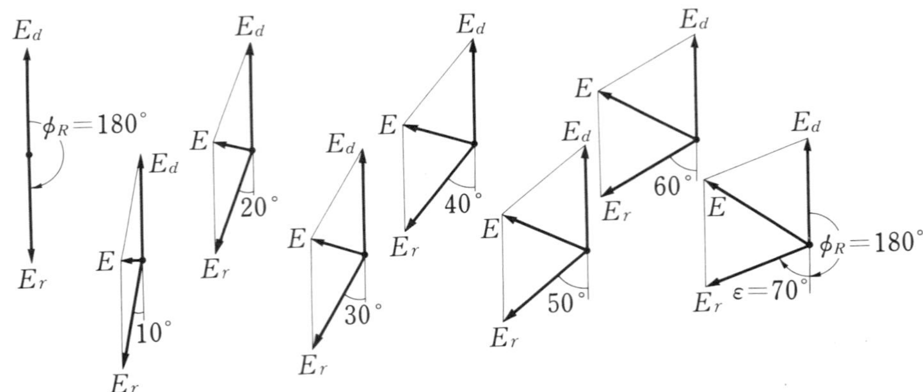

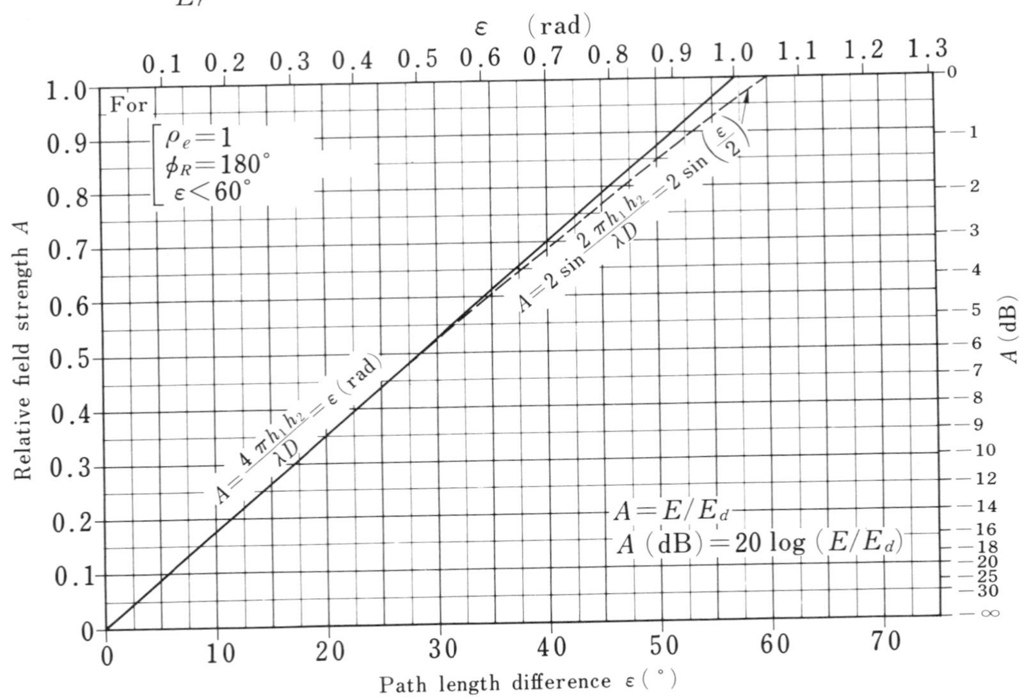

Total Reflection (General)

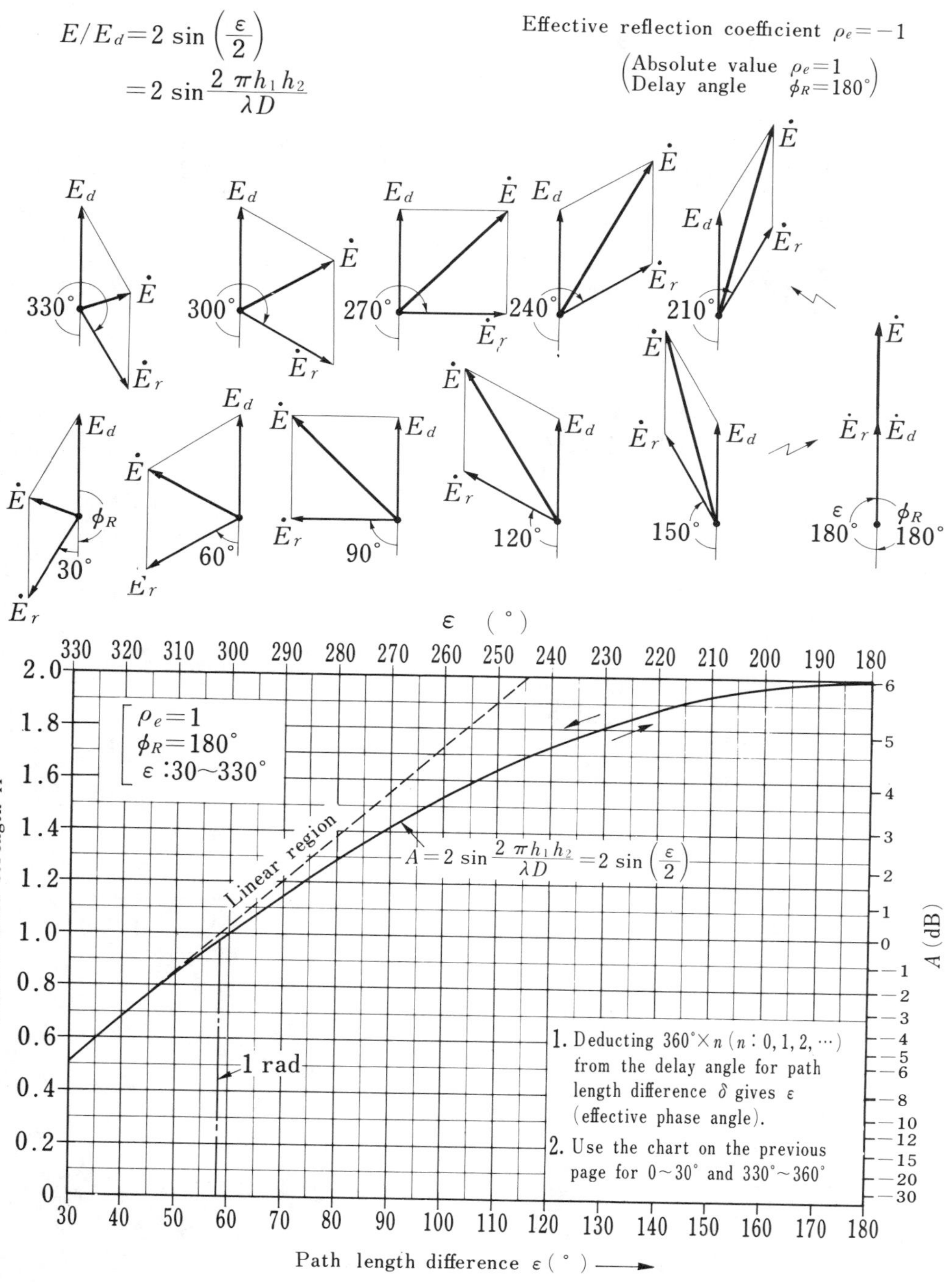

Minimum Effective Antenna Height (Ground Wave)

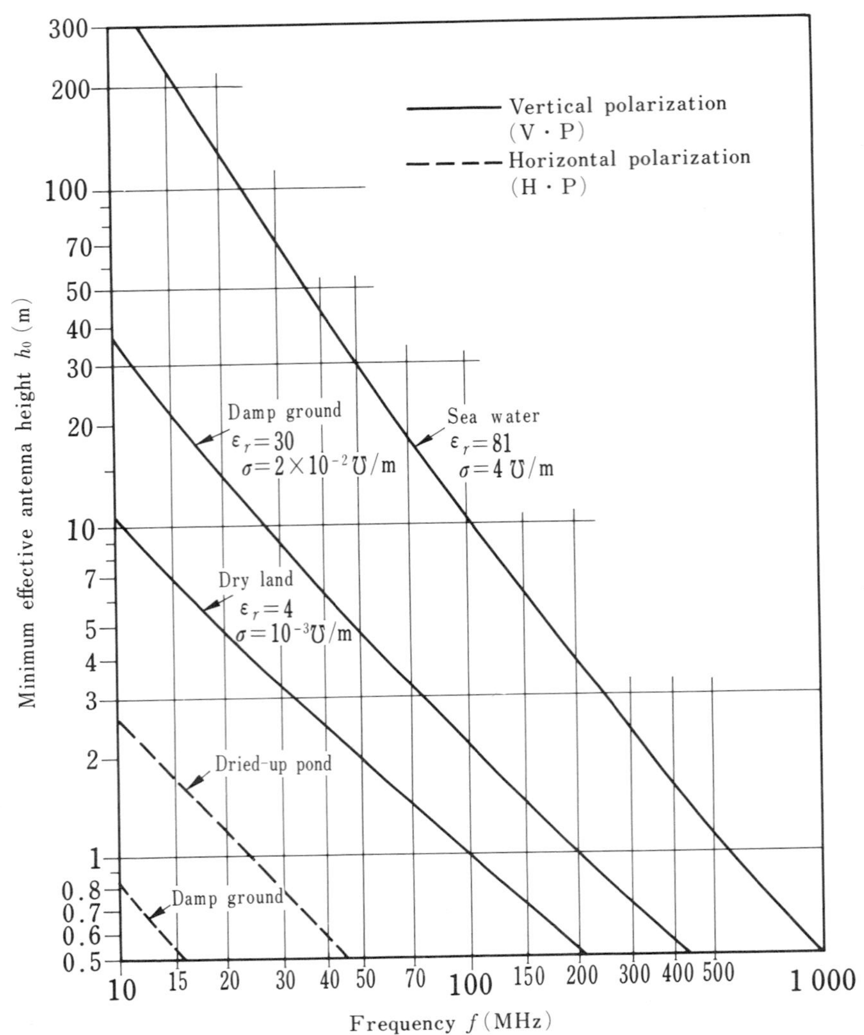

Effective Antenna Height in Ground-Wave Region

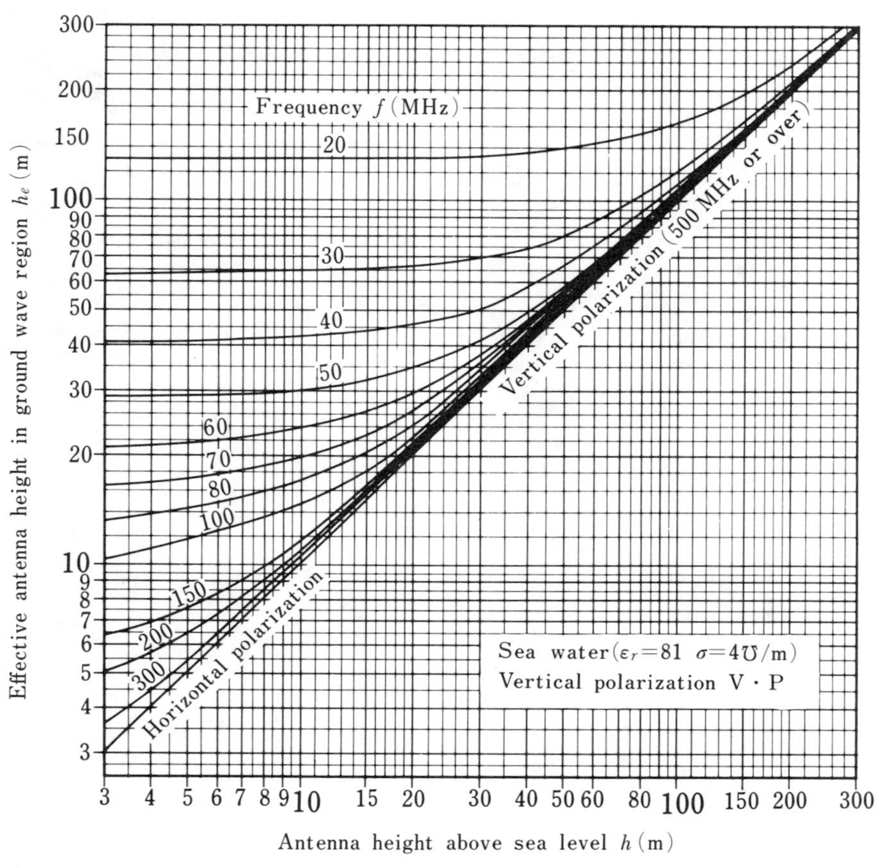

Simplified Nomograph for Low Antenna Height Region

Flat-Ground Propagation Loss

(Spherical earth propagation loss)

$$\Gamma_s = \Gamma_{pl} + A_s$$

Spherical earth additional loss (next figure) (A_s)
Flat ground propagation loss (Γ_{pl})

1. This nomograph is applicable only when flat ground propagation loss Γ_{pl} (dB) estimated here is greater than free space propagation loss Γ_0 (dB).

2. For antenna height, the actual antenna height or the effective antenna height in ground wave region, whichever is greater, shall apply.

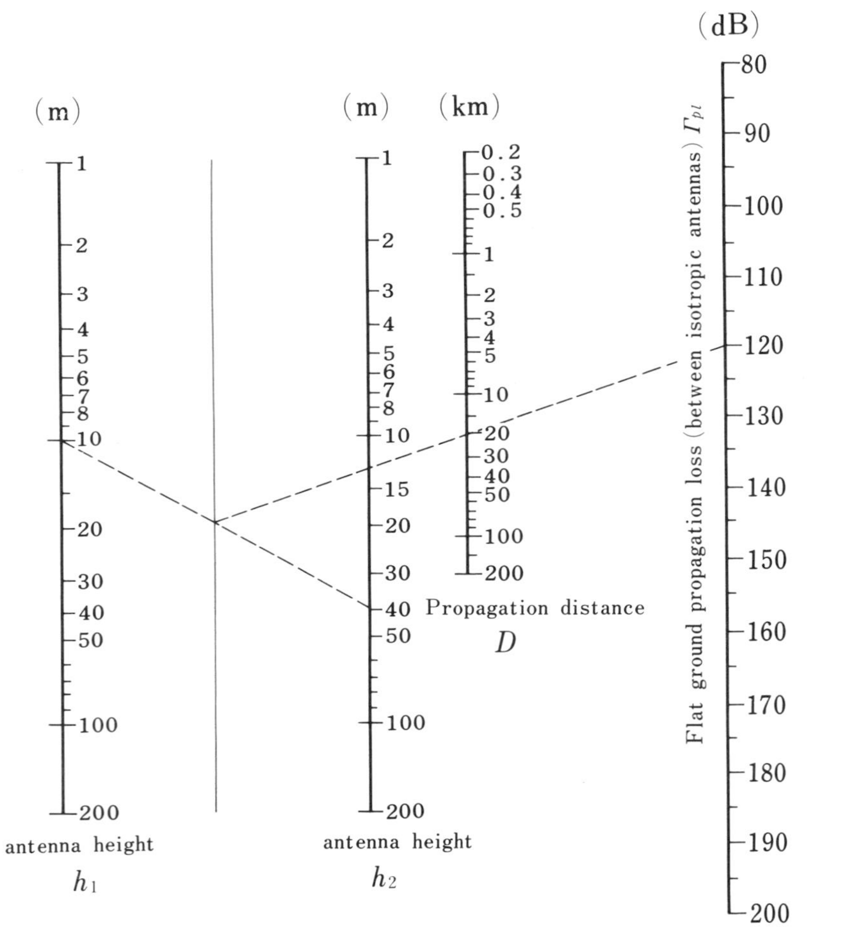

Simplified Nomograph for Low Antenna Height Region

Spherical-Ground Additional Loss

(Spherical earth propagation loss)

$$\Gamma_s = \Gamma_{pl} + A_s \text{ (dB)}$$

└----From the preceding figure

Applicable antenna height limit (m)

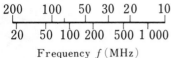

200 100 50 30 20 10

20 50 100 200 500 1 000

Frequency f (MHz)

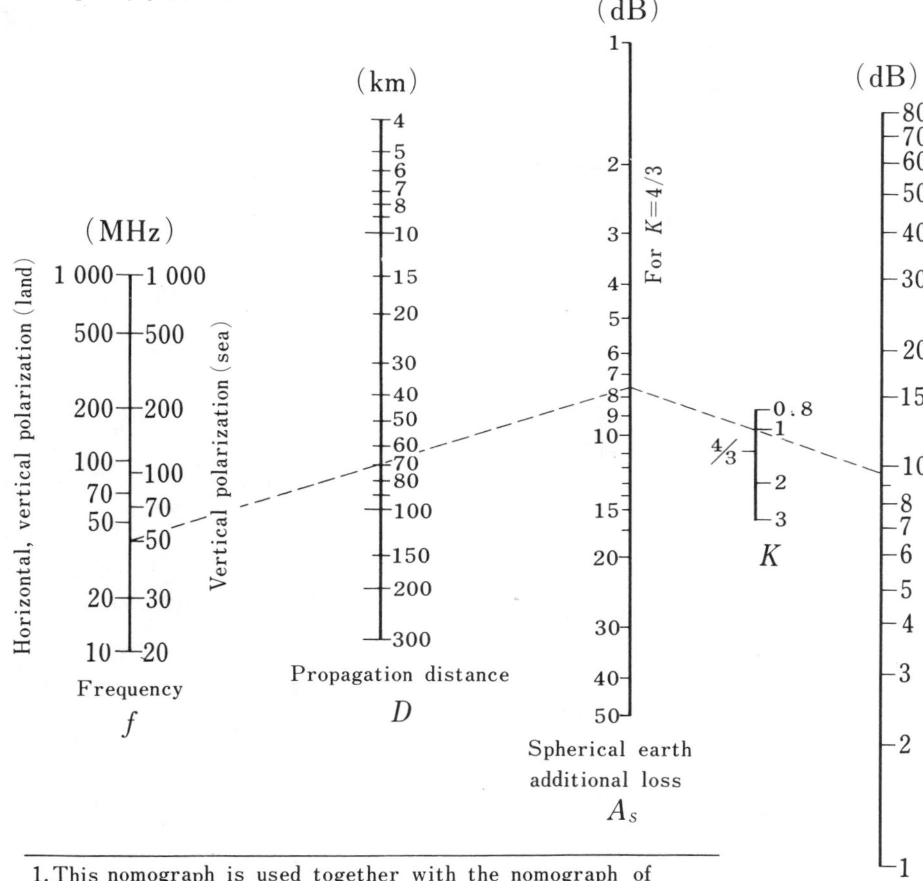

Spherical earth additional loss A_s

1. This nomograph is used together with the nomograph of Fig. 19.
2. This nomograph is applicable only when the antenna height is less than the above limit.
3. This is a nomograph to estimate the additional loss required when the spherical earth is assumed as a flat ground.
4. This is an approximate estimation when the phase angle δ corresponding to the actual path length difference is approximately 1 radian or less.

Simplified Nomograph for Low Antenna Height Region

Free-Space Parameters

1. Isotropic antenna (both transmitting and receiving)
2. Radiated power 1 watt

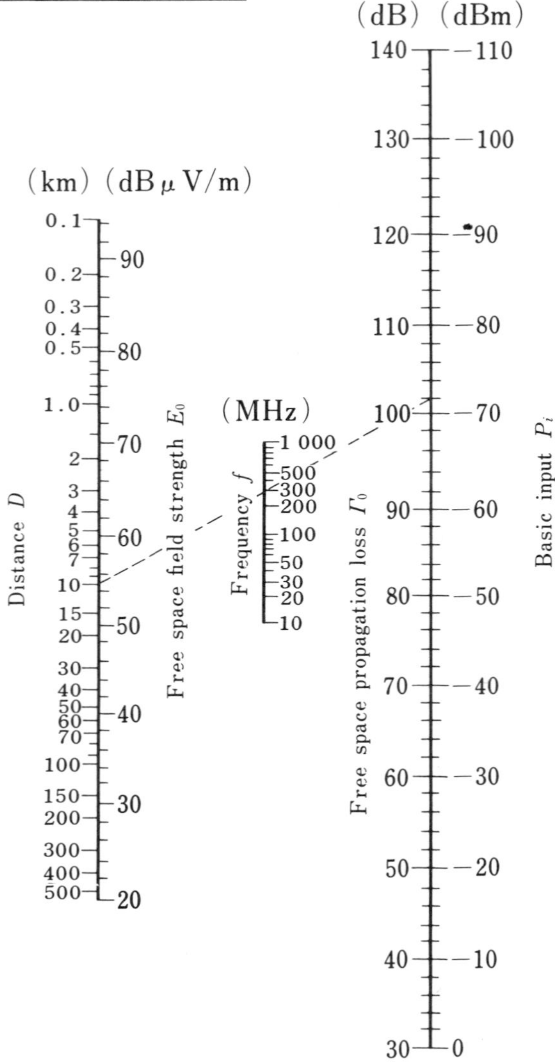

Spherical-Ground Line-of-Sight Distance

For Macroscopic Study

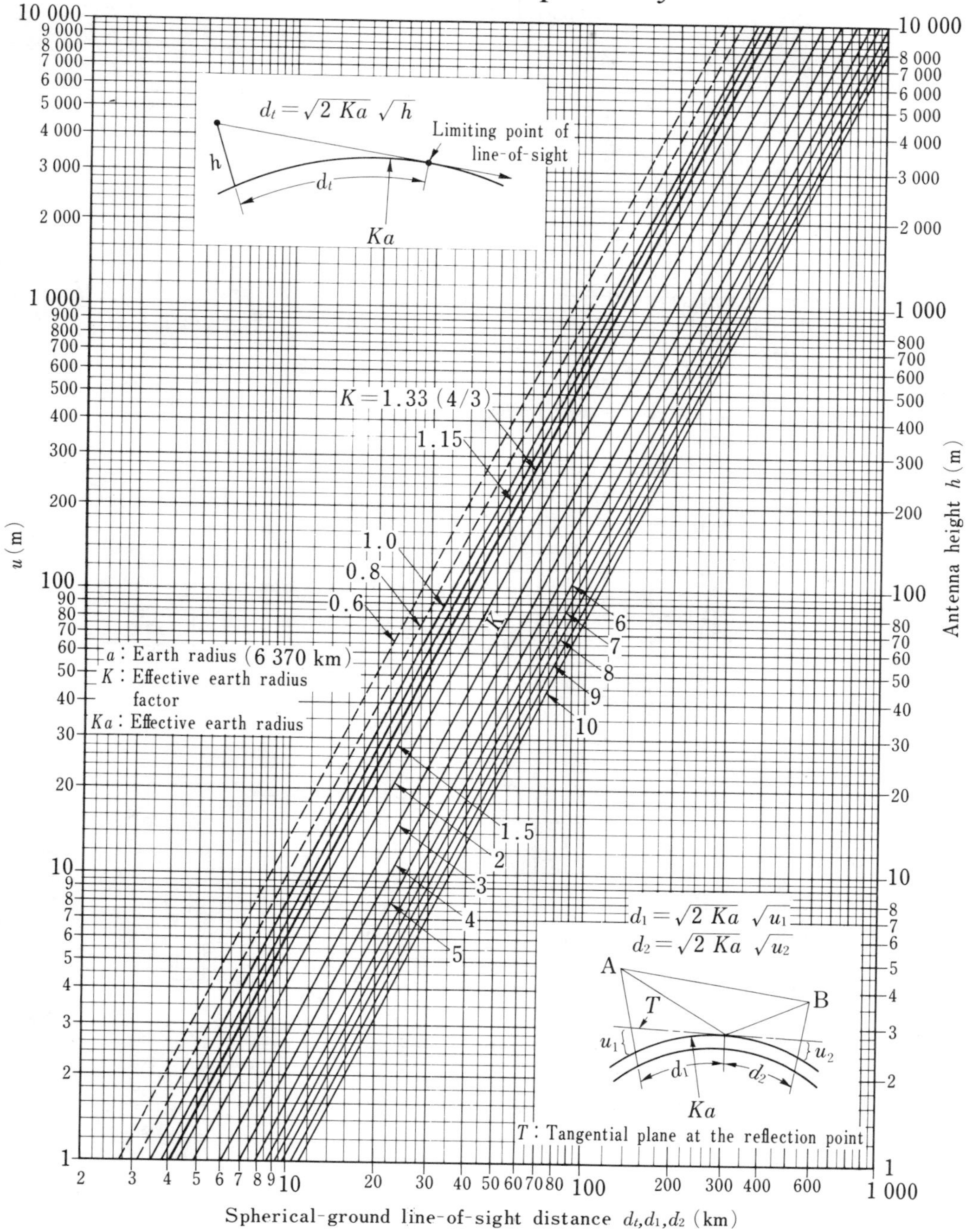

Spherical-Ground Line-of-Sight Distance

Detailed Design (1)

$$d_t = \sqrt{2\ Ka \cdot h}$$
$$d_{1,2} = \sqrt{2\ Ka \cdot u}$$

K	Condition
1.0	Vacuum, uniform atmosphere
1.15	Mean value in optics
1.33	Standard propagation

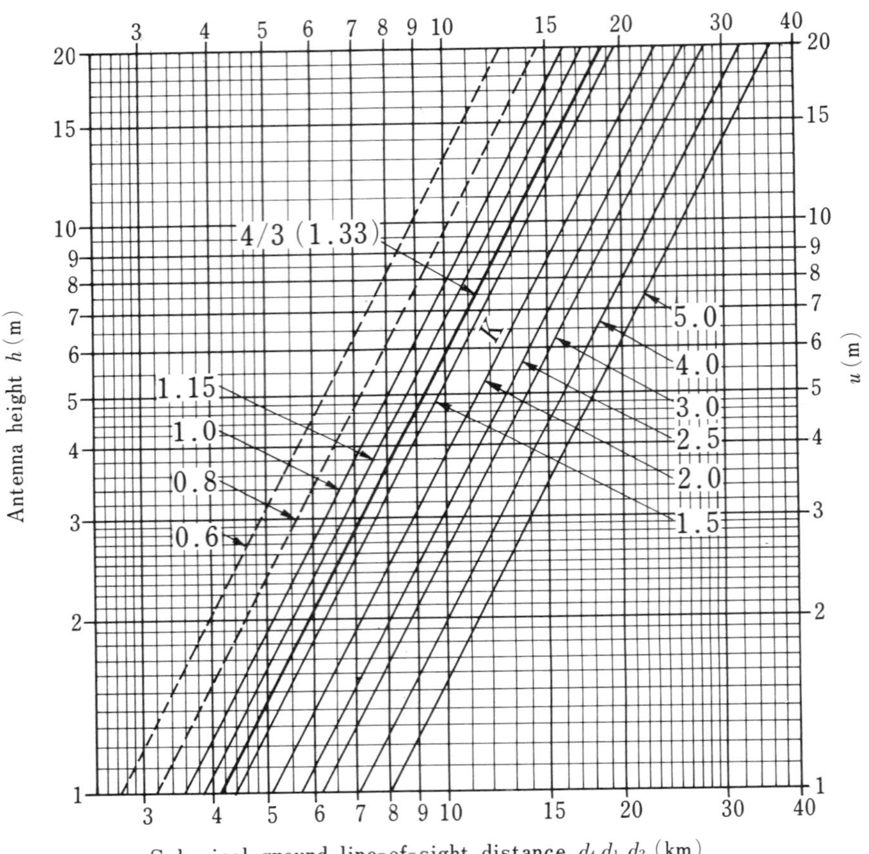

Spherical-Ground Line-of-Sight Distance

Detailed Design (2)

$$d_t = \sqrt{2\,Ka \cdot h}$$

$$d_{1,2} = \sqrt{2\,Ka \cdot u}$$

K	Condition
1.0	Vacuum, uniform atmosphere
1.15	Mean value in optics
1.33	Standard propagation

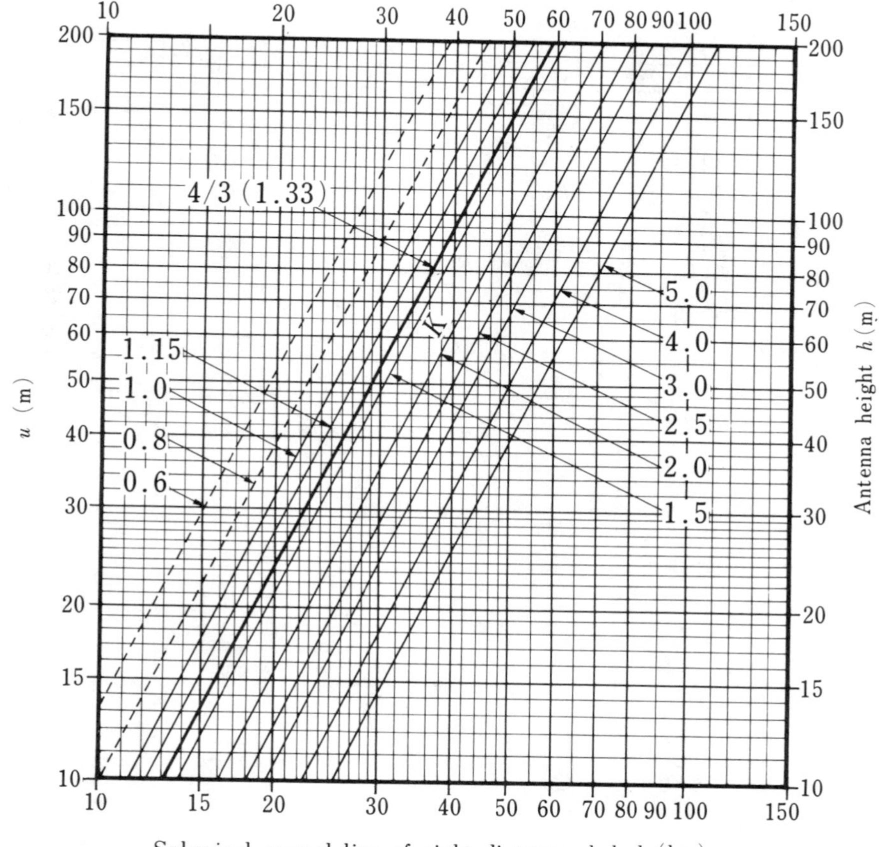

Spherical-Ground Line-of-Sight Distance

Detailed Design (3)

$$d_t = \sqrt{2\ Ka \cdot h}$$
$$d_{1,2} = \sqrt{2\ Ka \cdot u}$$

K	Condition
1.0	Vacuum, uniform atmosphere
1.15	Mean value in optics
1.33	Standard propagation

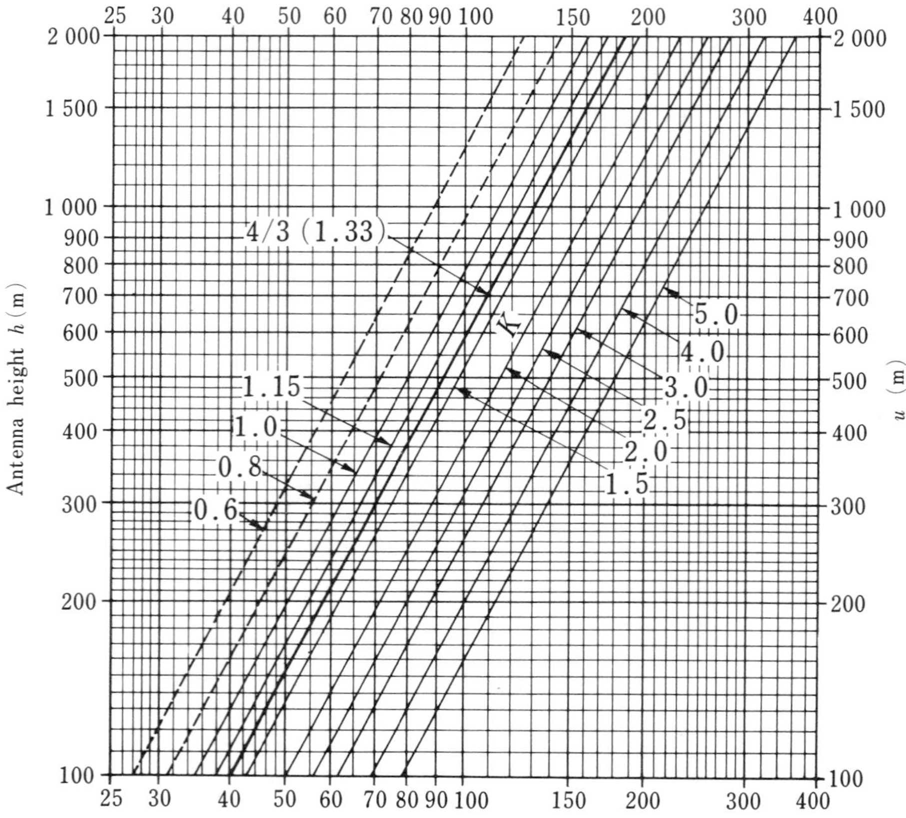

Spherical-Ground Line-of-Sight Distance

Detailed Design (4)

$$d_t = \sqrt{2\,Ka \cdot h}$$
$$d_{1,2} = \sqrt{2\,Ka \cdot u}$$

K	Condition
1.0	Vacuum, uniform atmosphere
1.15	Mean value in optics
1.33	Standard propagation

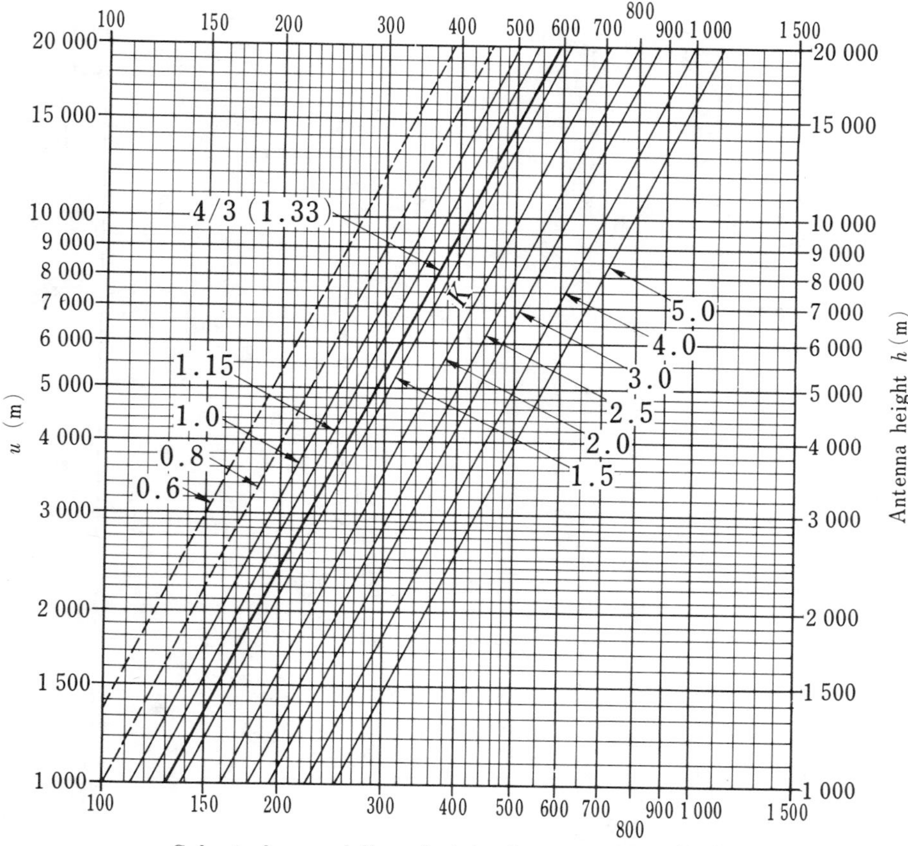

2

24

Nomograph for *m* (For Determination of Reflection Point)

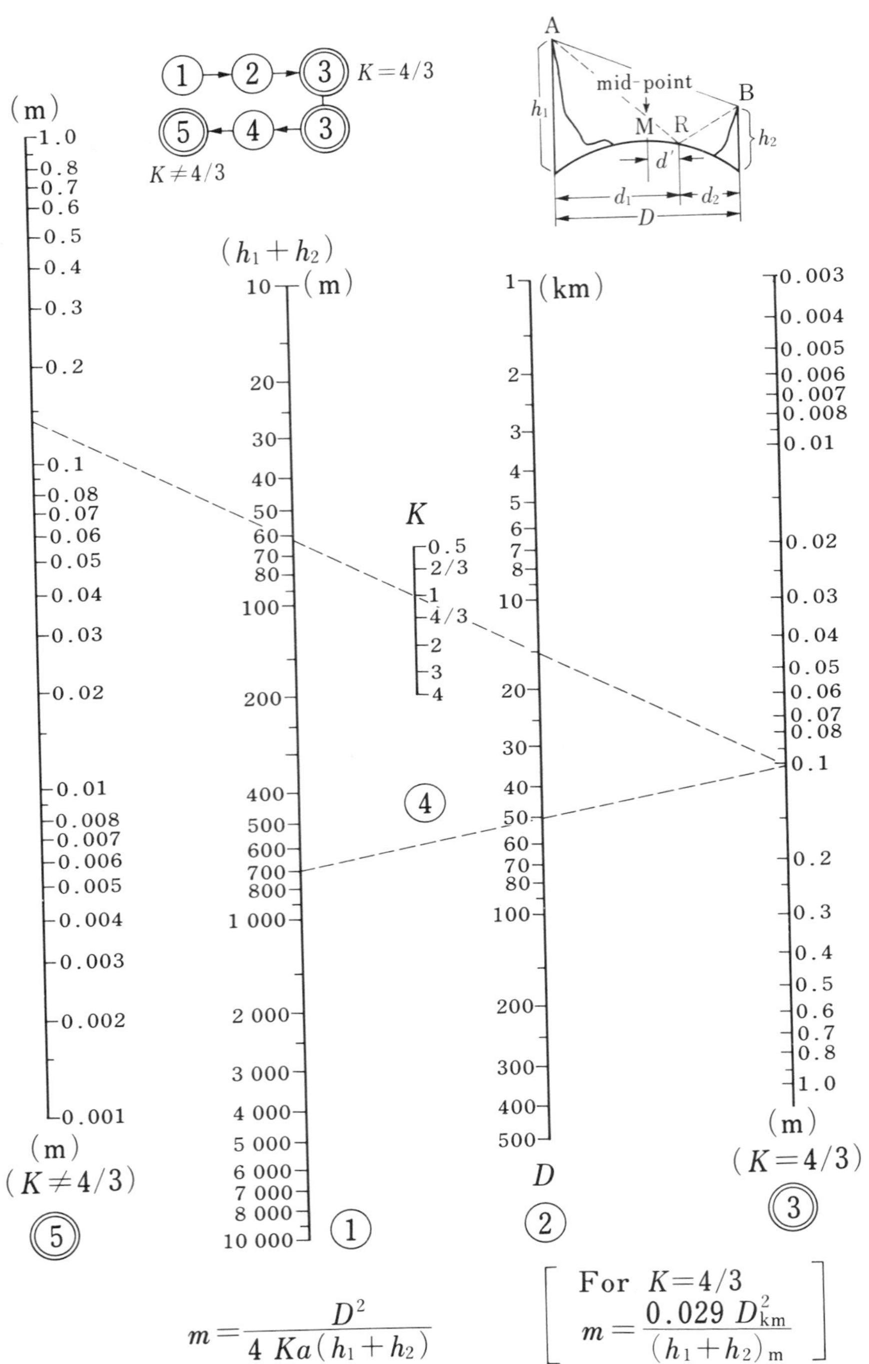

$$m = \frac{D^2}{4\,Ka(h_1+h_2)}$$

$$\left[\text{For } K=4/3 \quad m = \frac{0.029\,D_{\text{km}}^2}{(h_1+h_2)_{\text{m}}}\right]$$

Graph for *y* (For Determination of Reflection Point) (1)

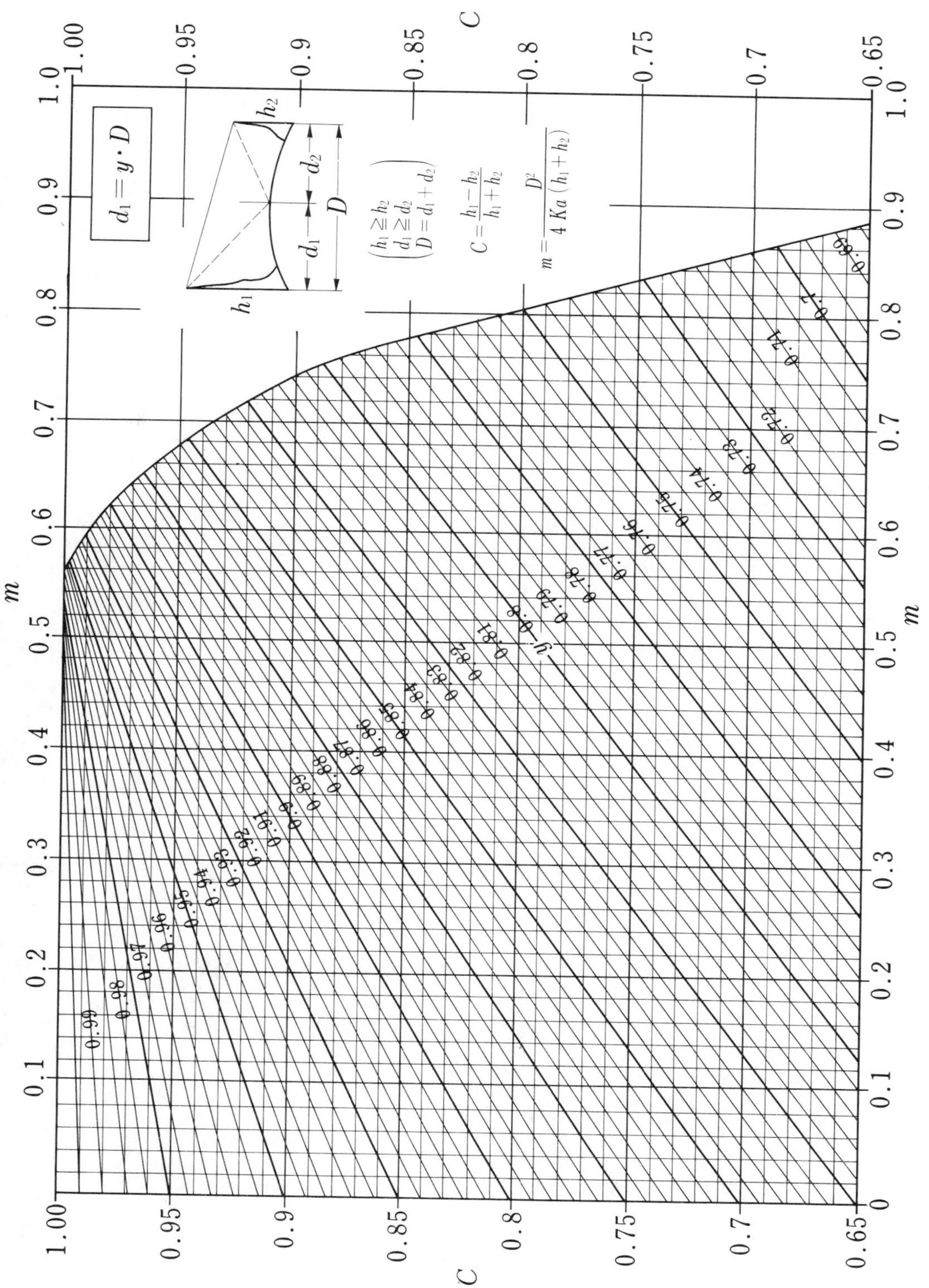

Graph for *y* (For Determination of Reflection Point) (2)

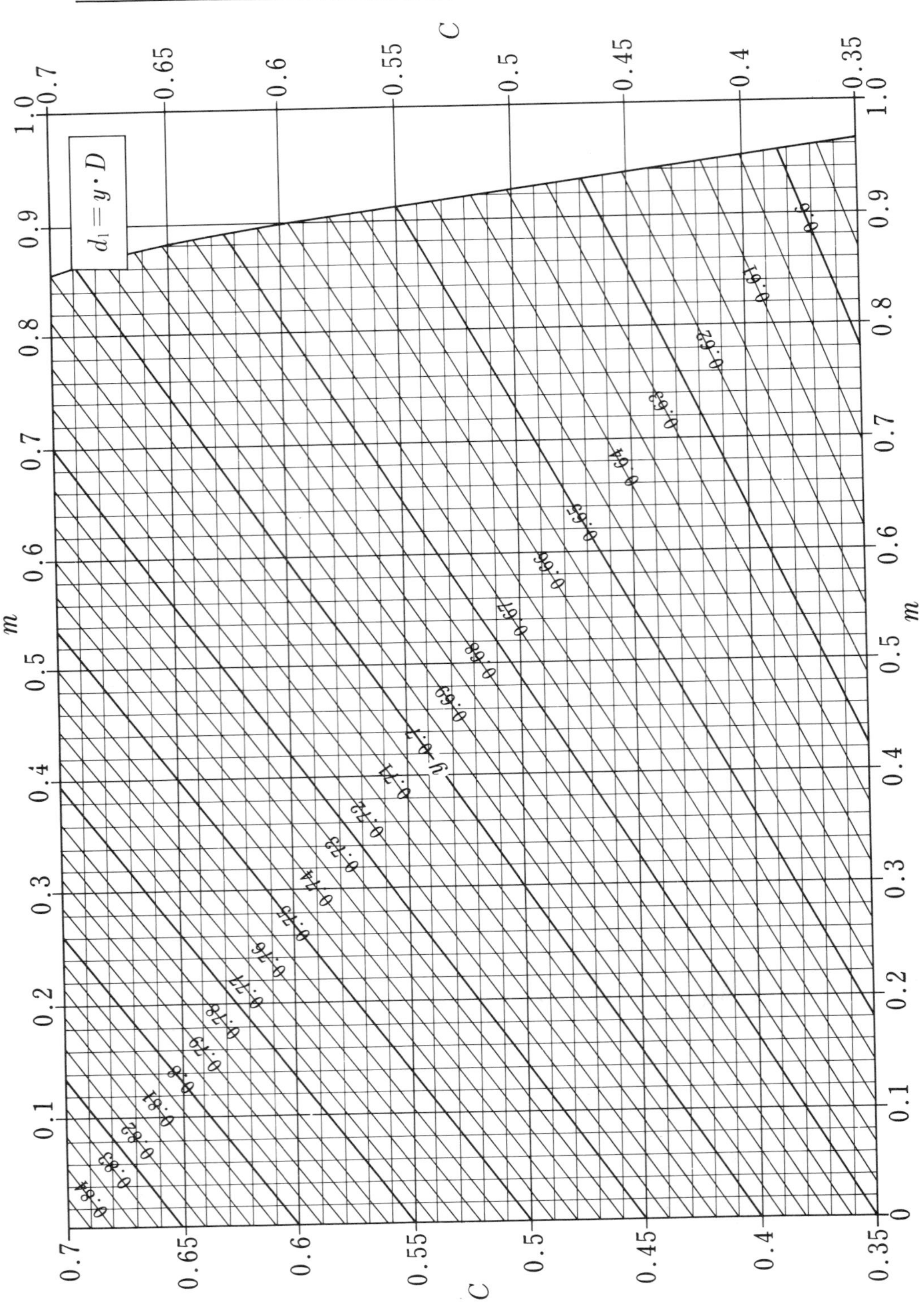

Graph for y (For Determination of Reflection Point) (3)

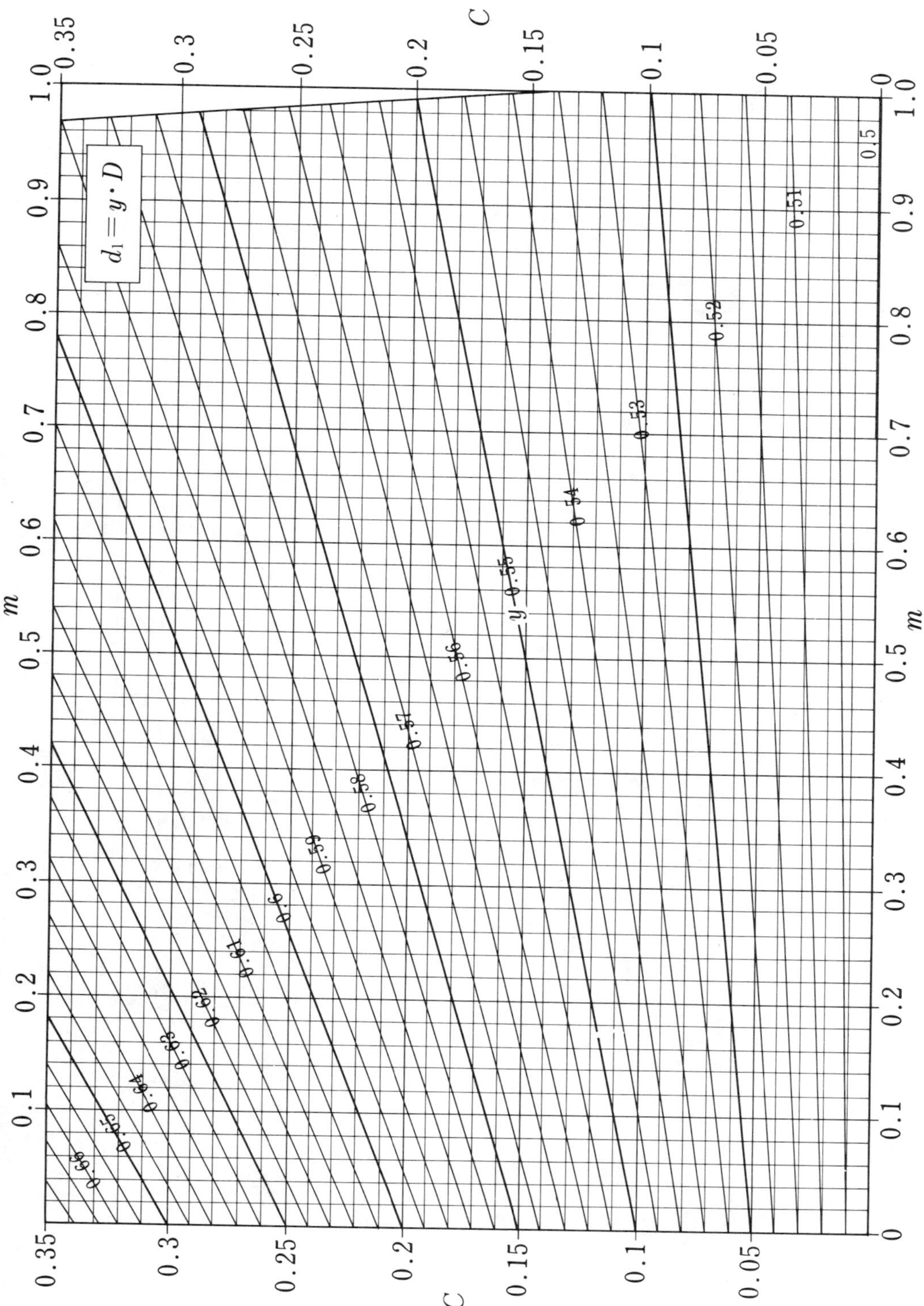

Nomograph for Path-Length Difference

For Small Δ

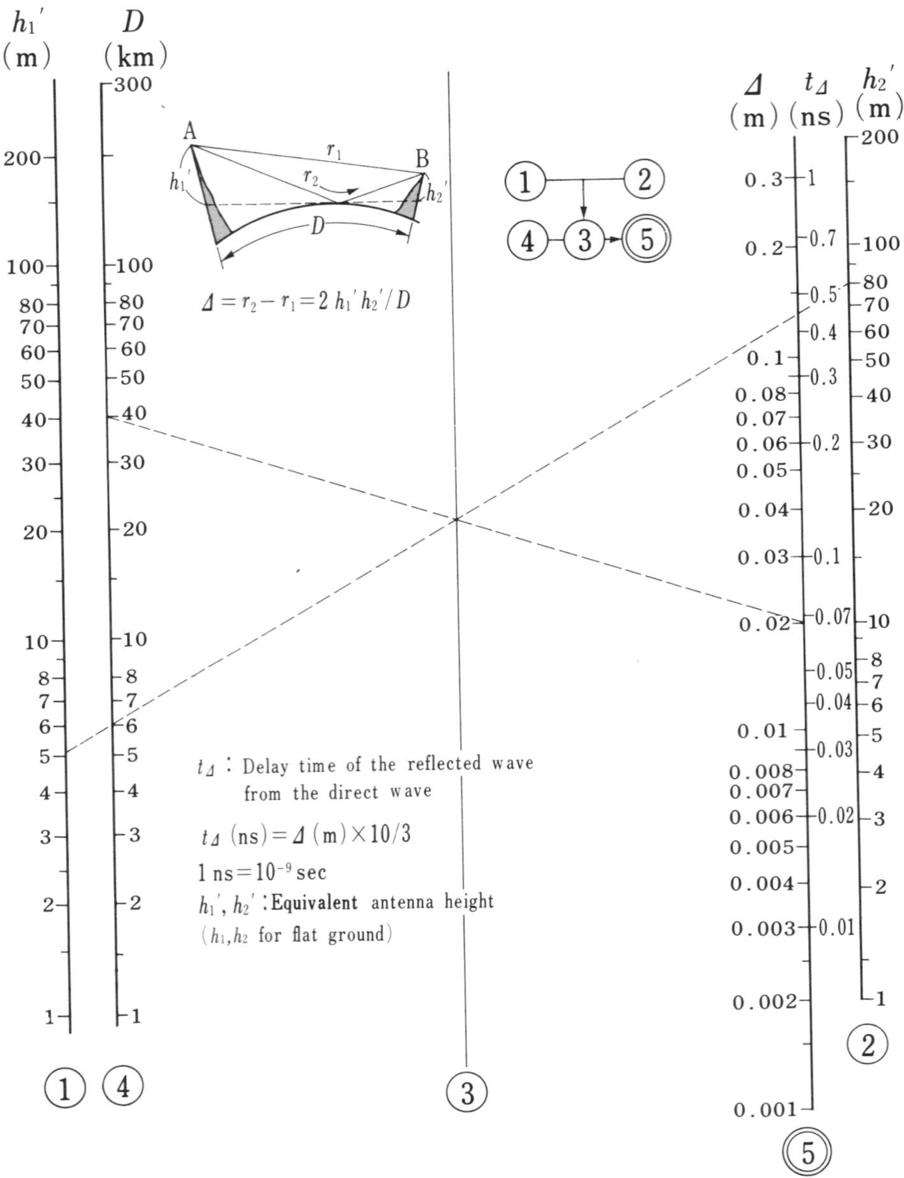

Nomograph for Path-Length Difference

For Large Δ

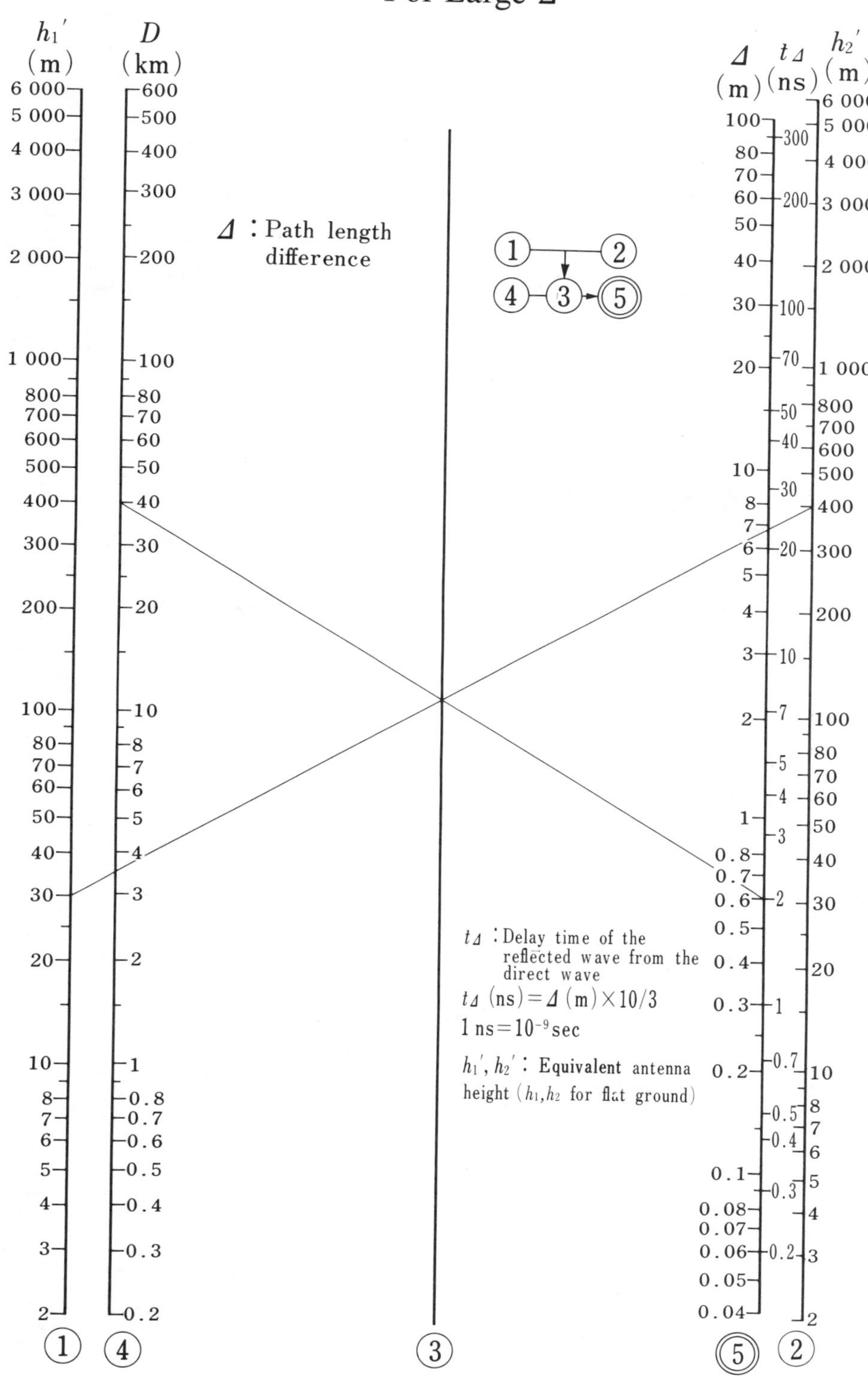

Types of Interference Patterns

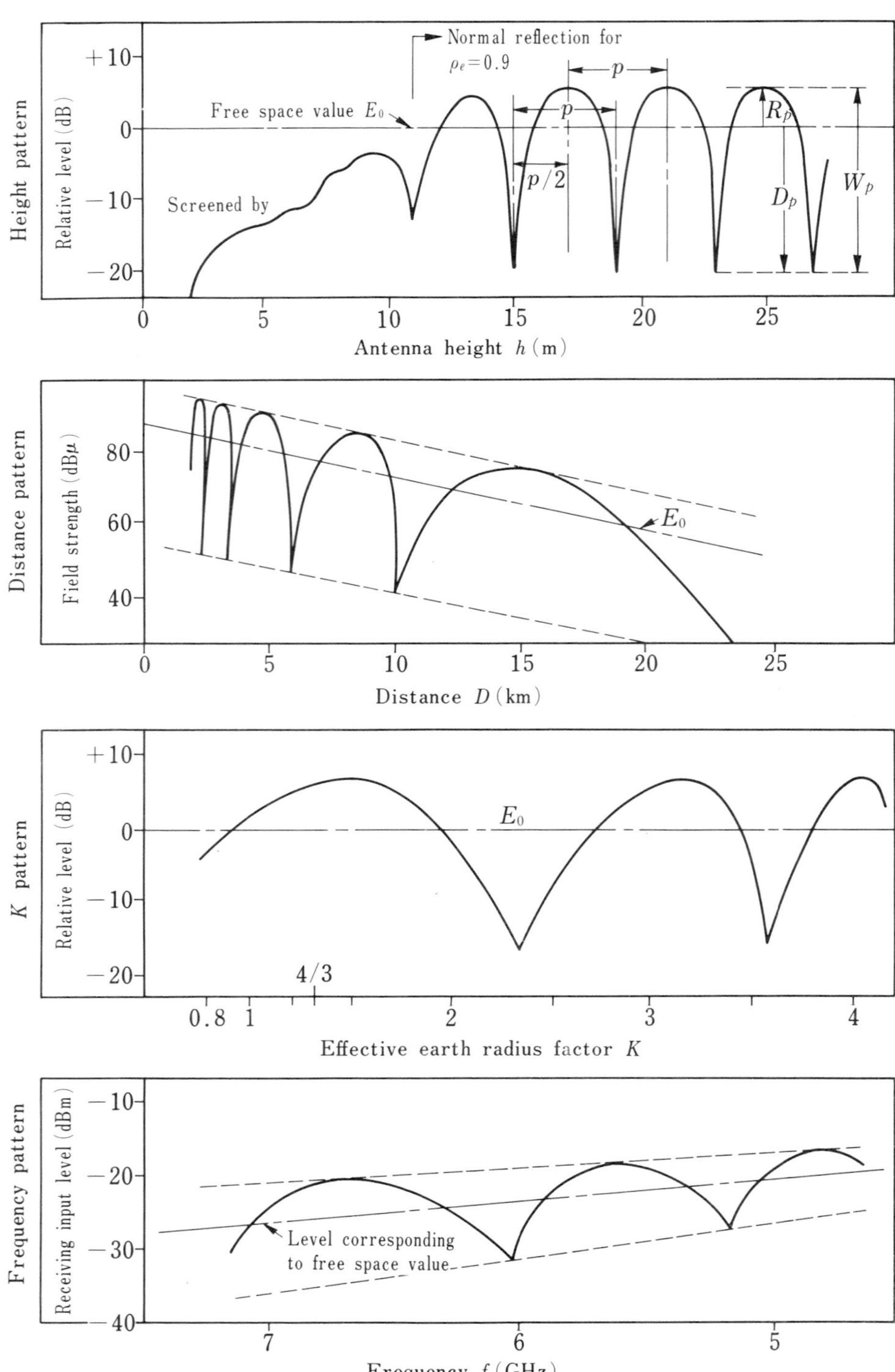

Maximum and Minimum Points of Direct Wave Combined with Interference (1)

1. Phase delay of reflection coefficient is $\phi_R = \pi$ (=180°)
2. For vertical polarization over sea water reflection and vertical polarization with a grazing angle $\psi > 5°$, use Fig. 8 to find ϕ_R for correction.

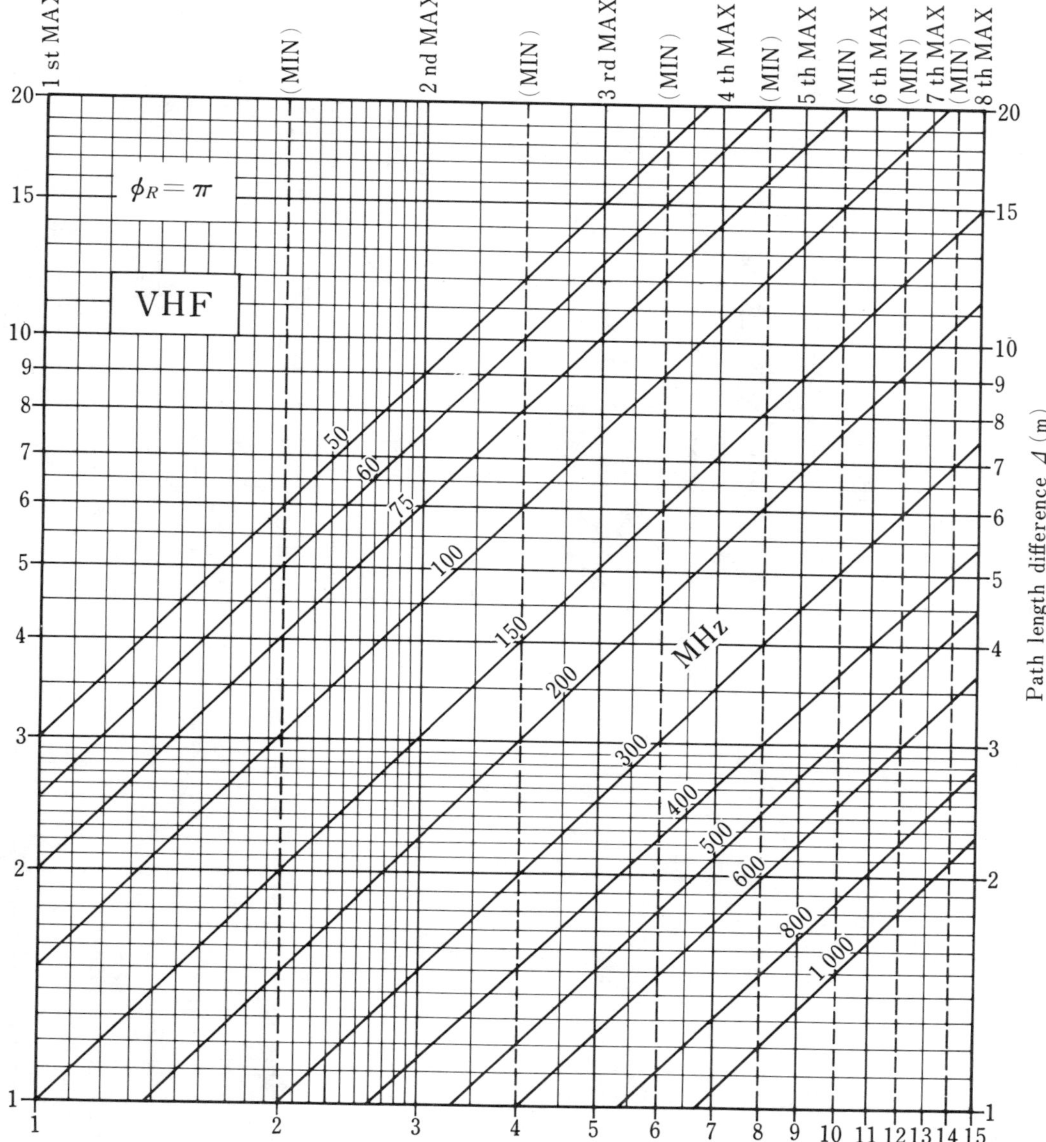

Maximum and Minimum Points of Direct Wave Combined with Interference (2)

1. In and above SHF band and for $n>5$, the error for estimation of Max, Min points becomes significant.
2. In this case the major aim of this study is to know the number of repetitions and pitches of the pattern.

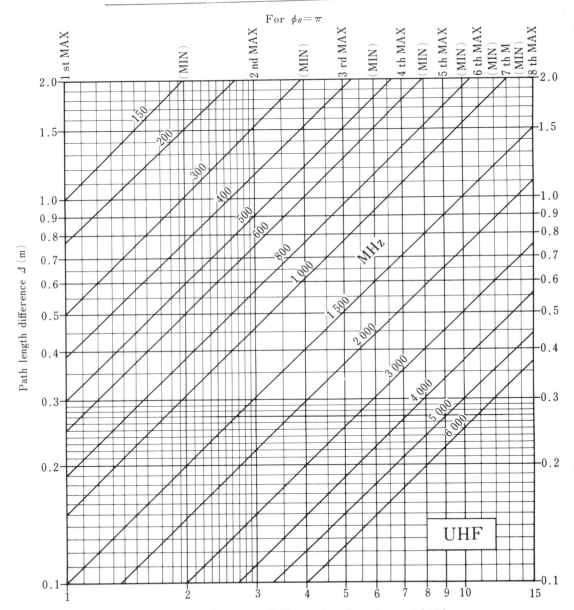

For $\phi_R=\pi$

1 st MAX (MIN) 2 nd MAX (MIN) 3 rd MAX (MIN) 4 th MAX (MIN) 5 th MAX (MIN) 6 th MAX (MIN) 7 th M (MIN) 8 th MAX

MHz: 150, 200, 300, 400, 500, 600, 800, 1 000, 1 500, 2 000, 3 000, 4 000, 5 000, 6 000

UHF

Path length difference $\varDelta$ (m): 0.1, 0.2, 0.3, 0.4, 0.5, 0.6, 0.7, 0.8, 0.9, 1.0, 1.5, 2.0

Path length difference to half-wavelength ratio $n=\varDelta\,(\lambda/2)$: 1, 2, 3, 4, 5, 6, 7, 8, 9, 10, 15

Maximum and Minimum Points of Direct Wave Combined with Interference (3)

1. In and above SHF band the error for estimation of Max, Min points becomes significant.
2. In this case the major aim of this study is to know the number of repetitions and pitches of the pattern.

For $\phi_R = \pi$

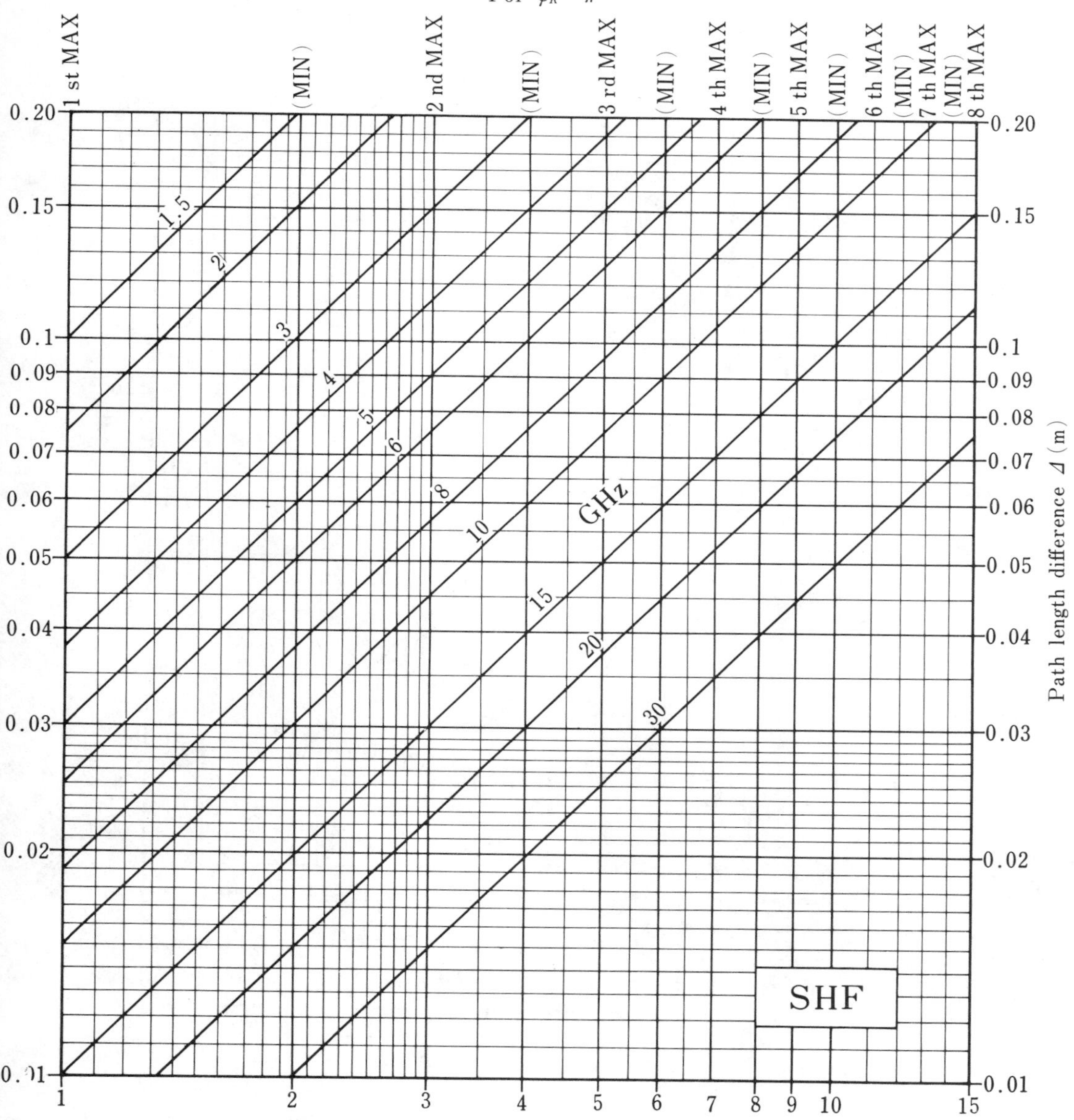

Path length difference to a half-wavelength ratio $n = \Delta/(\lambda/2)$

Range of Variations of Interference Pattern (1)

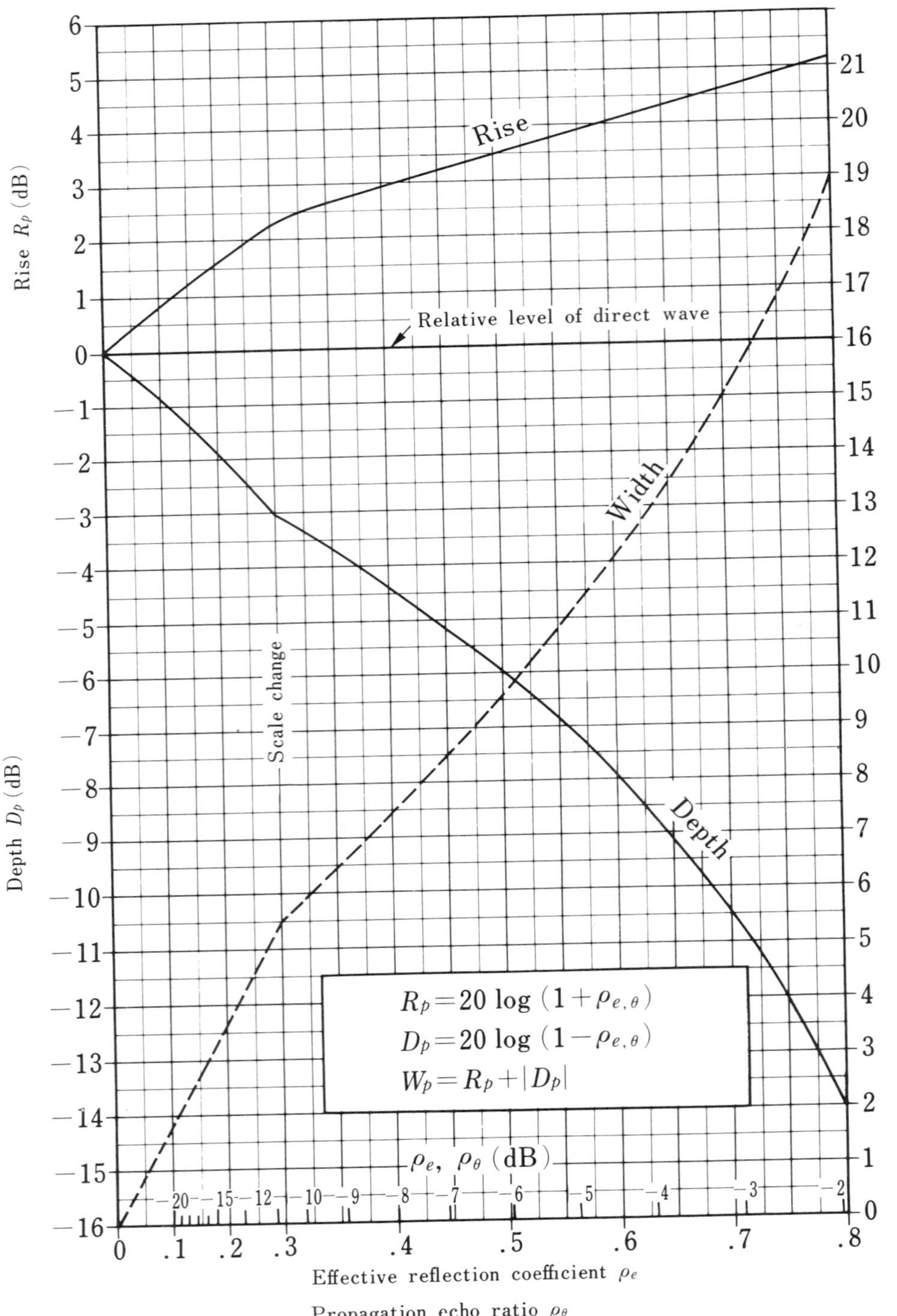

Range of Variations of Interference Pattern (2)

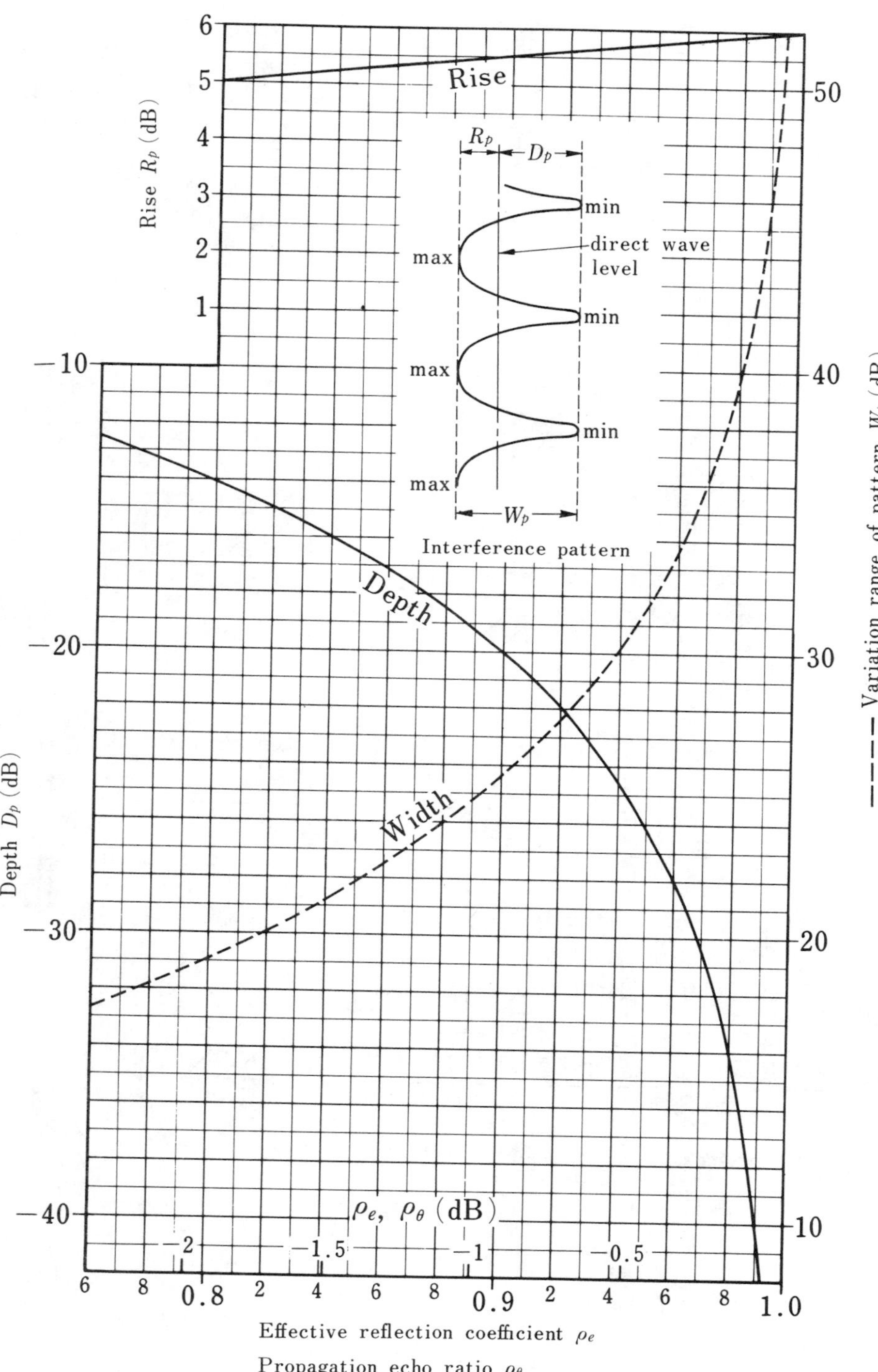

Pitches in Height Pattern (1)

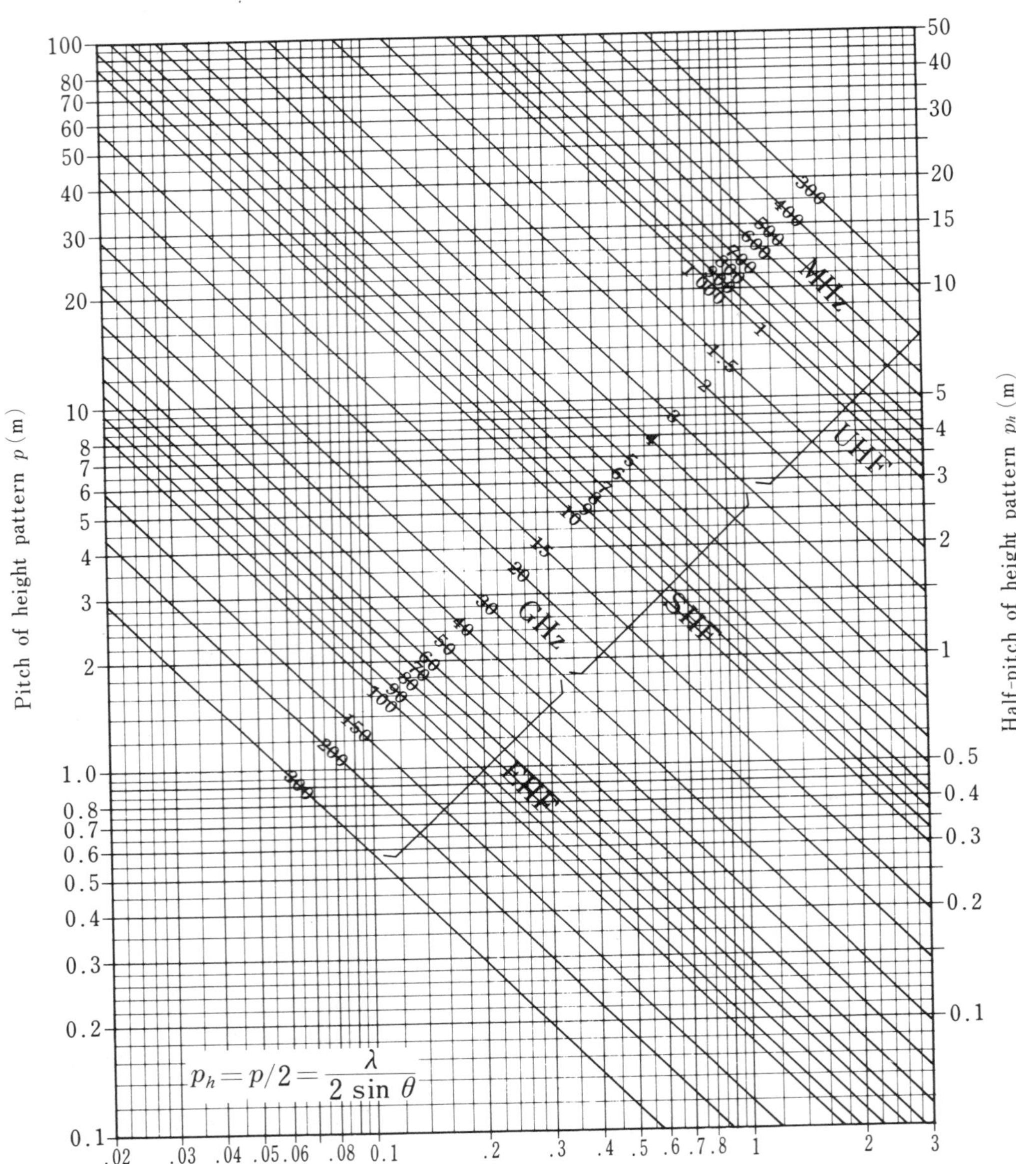

Pitches in Height Pattern (2)

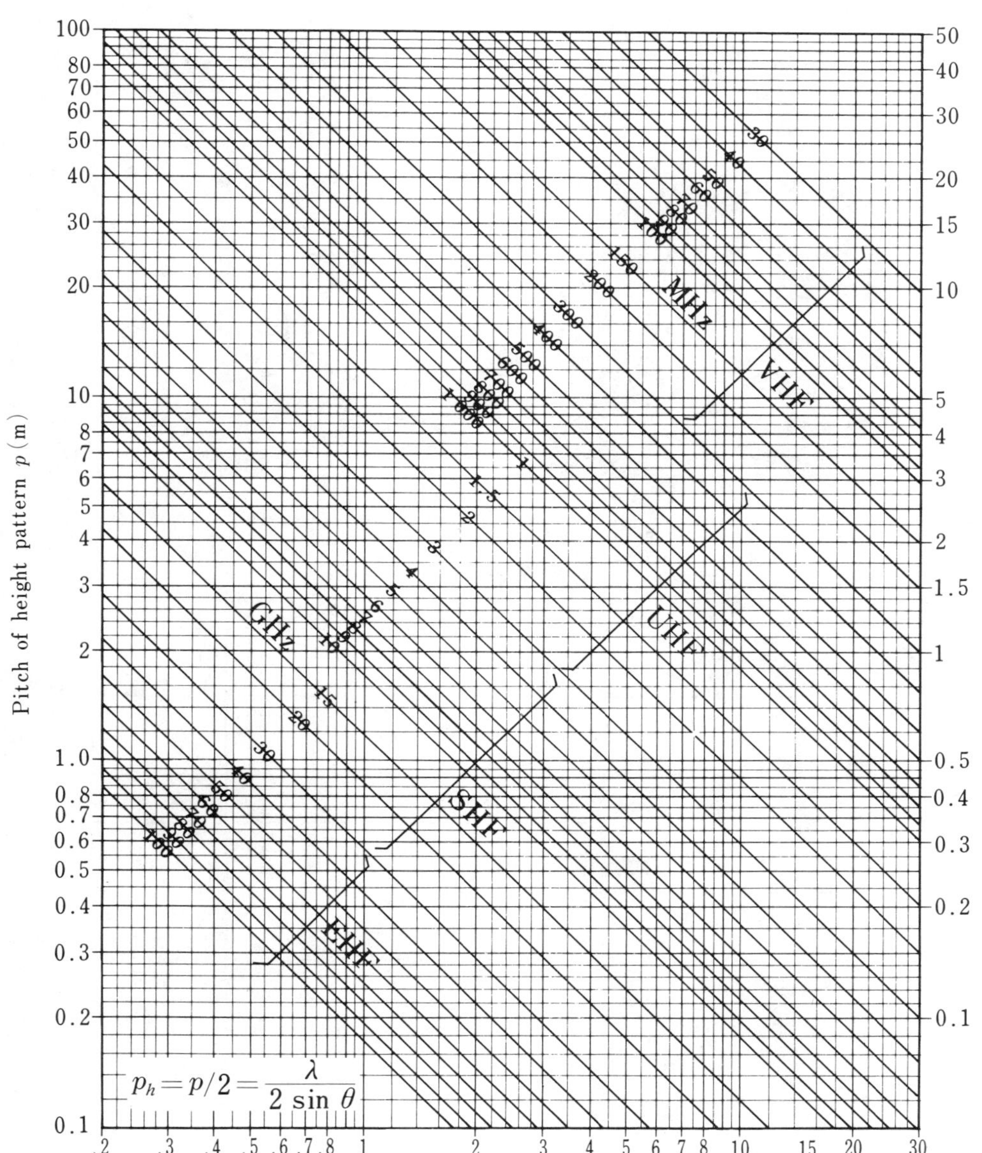

2

31

Radiation Vertical Angle

Basic Formulas and Calculation Charts

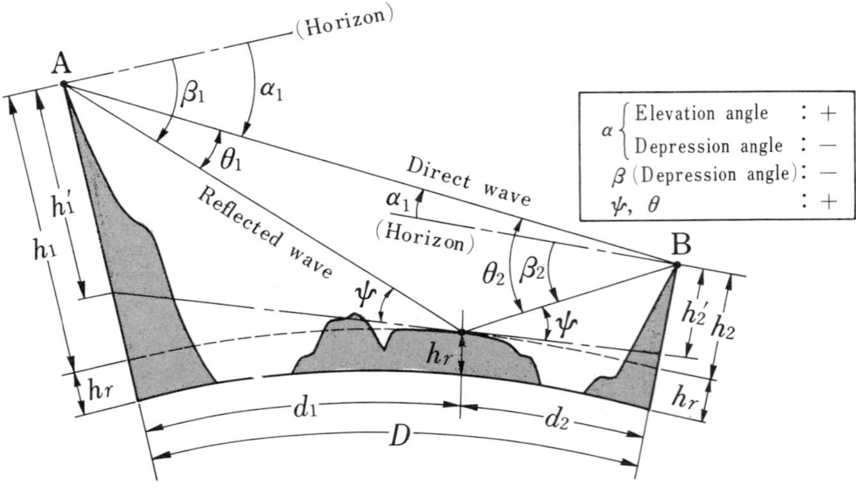

Vertical angle basic formula (Approximate equation)	Calculation chart (Figs. 32~34)	
Vertical angle of direct wave	[$x-Z$]	$\alpha°$
$\alpha_1 \doteqdot [(h_2-h_1)/D - D/(2\,Ka)]$	x_1 Z: ☐ − ☐ =	$\pm$. α_1
$\alpha_2 \doteqdot [(h_1-h_2)/D - D/(2\,Ka)]$	x_2 Z: ☐ − ☐ =	$\pm$. α_2
Vertical angle of reflected wave	$-$ [$y+Z$]	$\beta°$
$\beta_1 \doteqdot -[h_1/d_1 + d_1/(2\,Ka)]$	y_1 Z_1: − ☐ + ☐ =	− . β_1
$\beta_2 \doteqdot -[h_2/d_2 + d_2/(2\,Ka)]$	y_2 Z_2: − ☐ + ☐ =	− . β_2
Grazing angle	[$y-Z$]	$\psi°$
$\psi \doteqdot [h_{1,2}/d_{1,2} - d_{1,2}/(2\,Ka)]$	y_1 Z_1: ☐ − ☐ =	. ψ_1
	y_2 Z_2: ☐ − ☐ =	. ψ_2
	$(\psi_1+\psi_2)/2=$	. ψ

Approximate equation of $\alpha, \beta, \psi < 10°$

Included angle between direct and reflected waves θ

$\theta_1 =$ [$\pm$.] (α_1) − [− .] (β_1) = [.]

$\theta_2 =$ [$\pm$.] (α_2) − [− .] (β_2) = [.]

[Remarks]

If $\psi_1, \psi_2, \theta_1, \theta_2 \leqq 0$, the path is of transhorizon, while if $|(\psi_1-\psi_2)/\psi_1| \geqq 0.1$, reexamine the reflection points, (d_1, d_2), etc.

Radiation Vertical Angle (1)

Graph of x

$$x = [|h_1 - h_2| / D] \times 0.0573^\circ$$

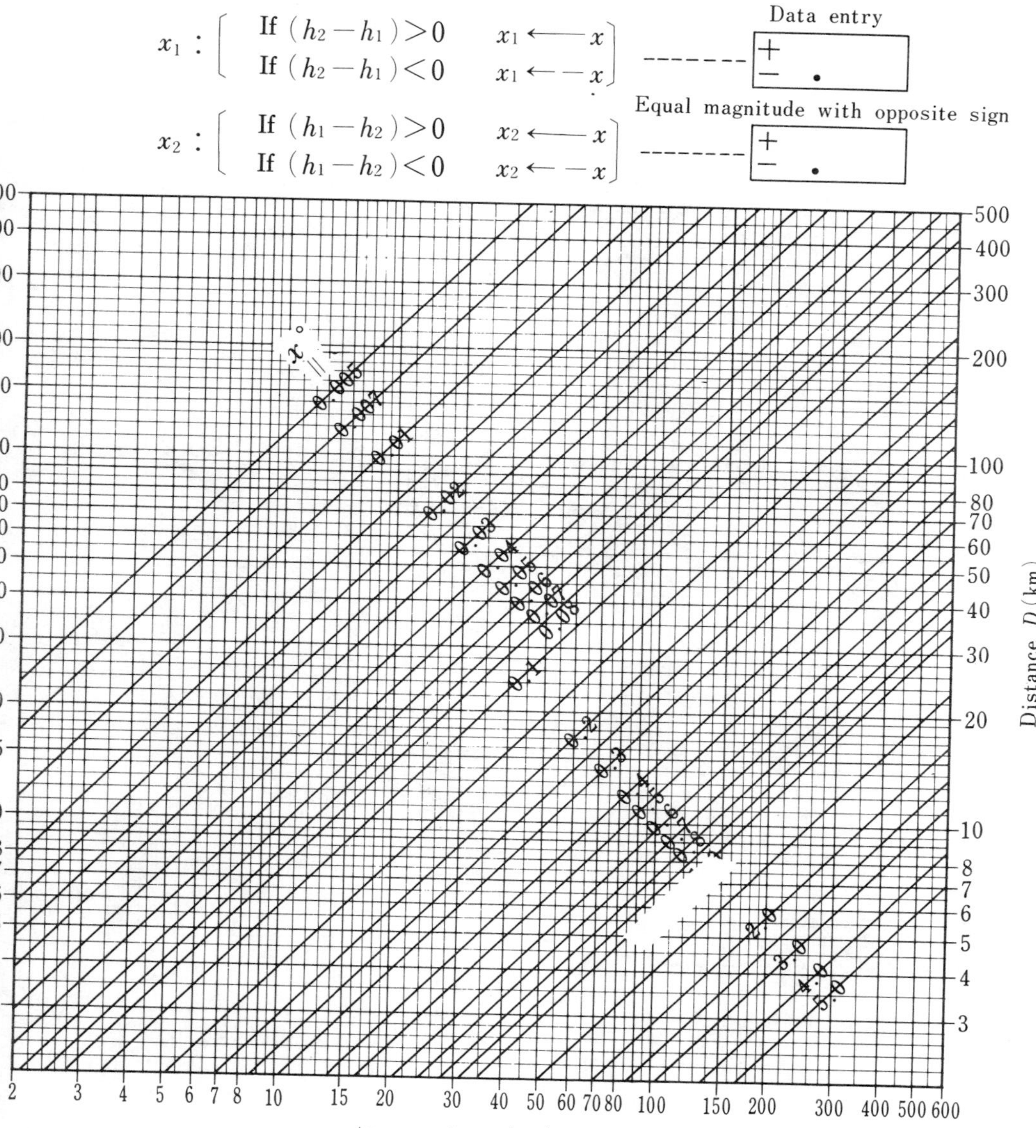

Radiation Vertical Angle (2)

Graph of x

$$x = [|h_1 - h_2|/D] \times 0.0573^\circ$$

Data entry

x_1 : [If $(h_2 - h_1) > 0$ $x_1 \leftarrow x$; If $(h_2 - h_1) < 0$ $x_1 \leftarrow -x$] ----- + − •

Equal magnitude with opposite sign

x_2 : [If $(h_1 - h_2) > 0$ $x_2 \leftarrow x$; If $(h_1 - h_2) < 0$ $x_2 \leftarrow -x$] ----- + − •

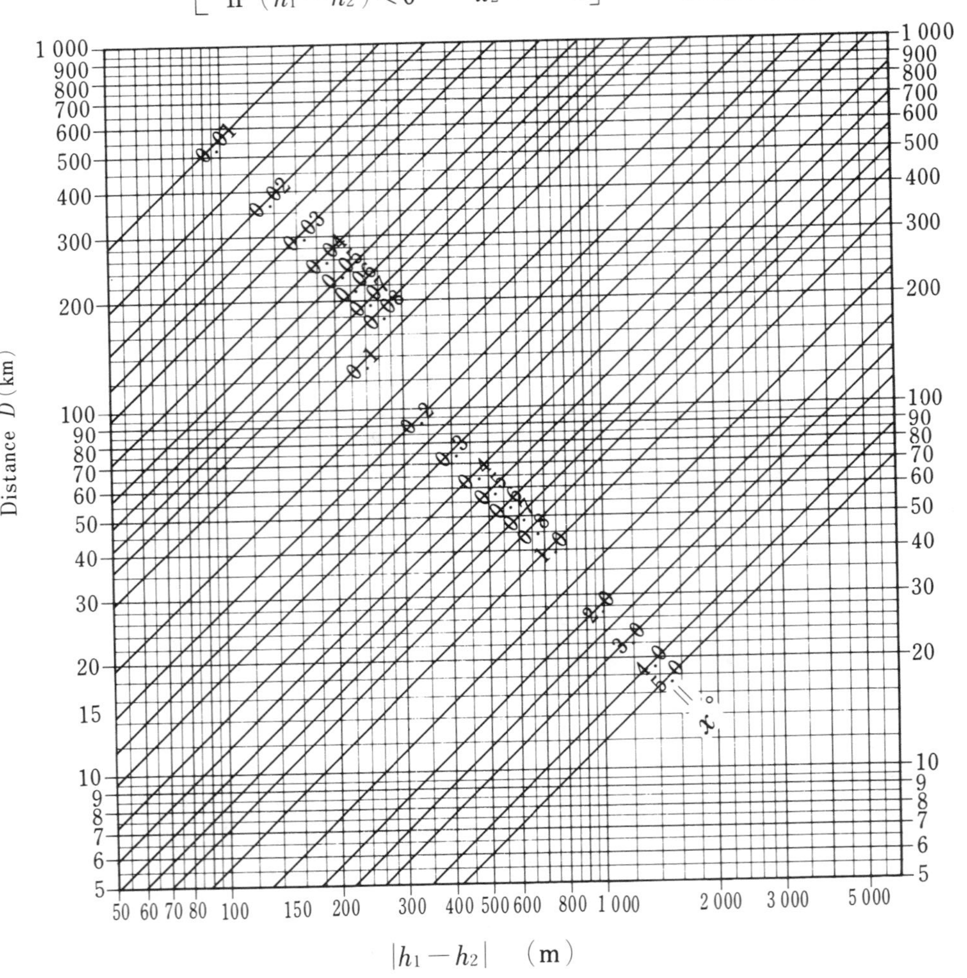

Radiation Vertical Angle

Graph of y

Data entry

$y_1 = (h_1/d_1) \times 0.0573^\circ$ ----- [.]

$y_2 = (h_2/d_2) \times 0.0573^\circ$ ----- [.]

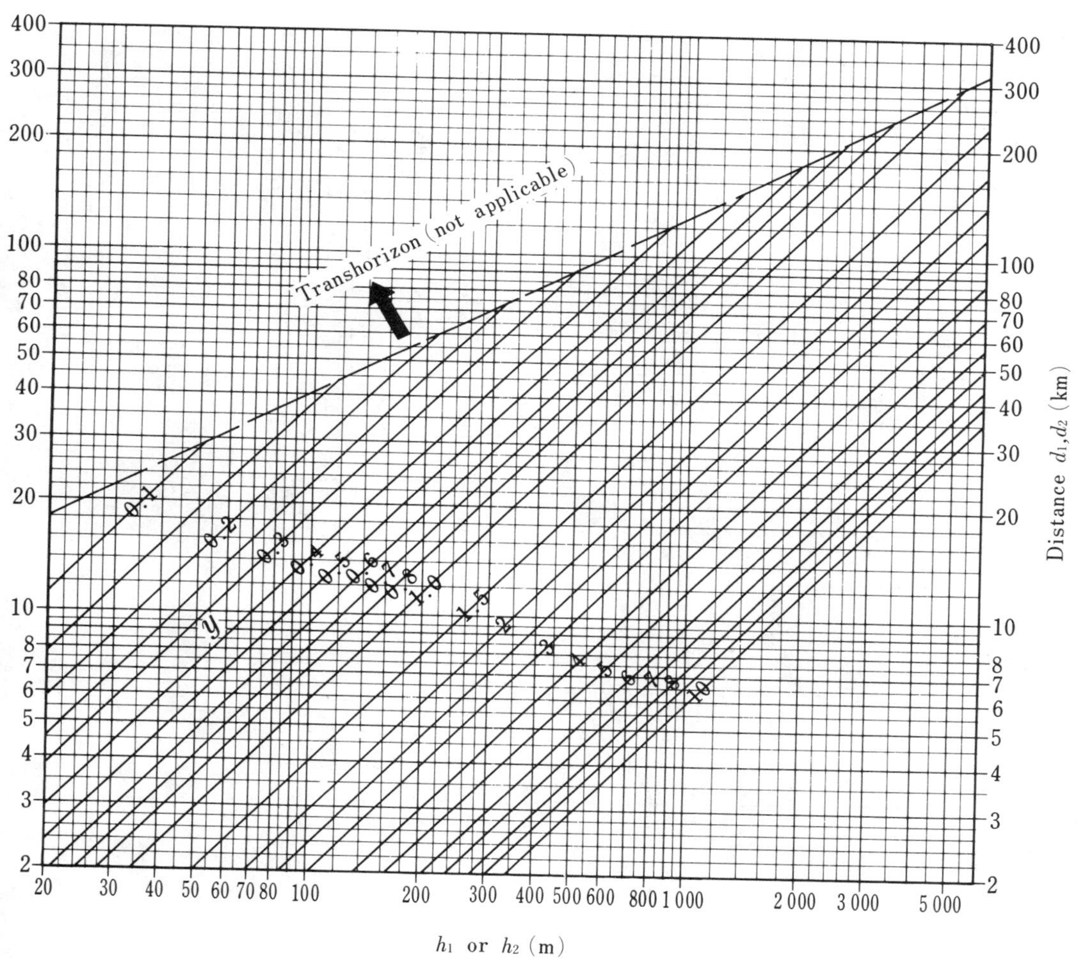

Radiation Vertical Angle

Graph of Z

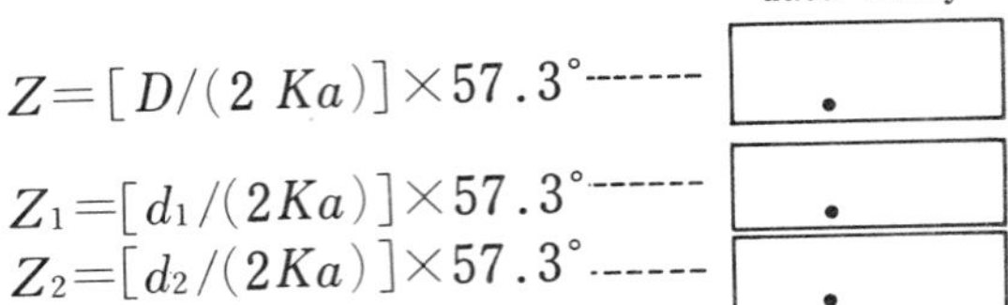

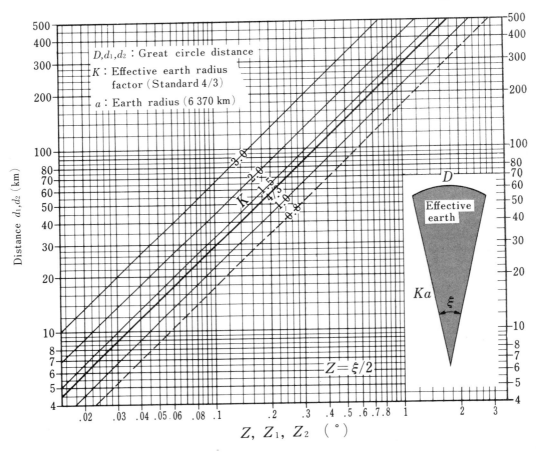

D, d_1, d_2 : Great circle distance

K : Effective earth radius factor (Standard 4/3)

a : Earth radius (6 370 km)

K: 3.0, 2.0, 1.5, 4/3, 1.0, 0.8

D

Effective earth

Ka

ξ

$Z = \xi/2$

Distance d_1, d_2 (km)

Z, Z_1, Z_2 (°)

3

DIFFRACTION PROPAGATION PATH

3.1 BASIC TYPES OF DIFFRACTION PROPAGATION PATHS

When obstacles such as mountains and buildings obstruct direct-wave transmission between the transmitting and the receiving points, the field strength produced at the receiving point is considered to have been formed by those waves coming over the top or side of the obstacles. This type of radio-wave transmission is called diffraction propagation.

A. Smooth-Spherical-Ground Diffraction Model

Since the earth is spherical in shape, the line of sight between the transmitting and the receiving points is obstructed by the horizon at great distances.

Type I in Fig. 1*a* is similar to a transhorizon path over seas or lakes, where the effect of the waves can be ignored. This is called smooth-spherical-ground diffraction model. Included in this type are extensive land masses such as savannas and ice fields.

The electromagnetic line-of-sight distances d_{t1} and d_{t2} from the antennas with heights h_1 and h_2, respectively, over spherical ground (sea surface in the figure) are from Fig. 22 in Chapter 2,

$$d_{t1} = \sqrt{2Kah_1}, \qquad d_{t2} = \sqrt{2Kah_2} \tag{3.1}$$

Let $K = 4/3$ (standard value); the distance is expressed in kilometers and the height in meters,

$$d_{t1} = 4.12\sqrt{h_1}, \qquad d_{t2} = 4.12\sqrt{h_2} \tag{3.2}$$

If station A is on a ship where $h_1 = 20$ m, and station B is at a maritime base station where $h_2 = 400$ m, the resultant line-of-sight distance d_t is given by

$$d_t = d_{t1} + d_{t2} \doteqdot 18 + 82 = 100 \text{ km}$$

Hence if the ship travels more than 100 km from the base station, the propagation path will become the smooth-spherical-ground diffraction model type.

The radio-wave field strength in a diffraction region of the diffraction model can be determined when the electrical constants of the ground (dielectric constant, conductivity), the frequency, the transmitting and receiving antenna heights, and the propagation distance are known, and the proposed calculation methods (including nomographs and propagation charts) are employed.

However, recent technological advances have shown that there is a large difference (up to several tens of decibels) between experimental data and published theoretical propagation curves.

Now it is believed that the propagation-wave components (that is, abnormally refracted waves and scattered waves) predominate over the spherical-ground diffraction wave because of irregularities of the atmosphere near the ground surface.

Consequently various theories and propagation charts applicable only to spherical-ground diffraction have lost their credibility. In other words, empirical propagation curves, including statistical parameters such as time and local climatic conditions, have become more and more important.

Since it is proposed here that the classical diffraction theory, which has been used widely for a long time, has already lost its value, it will be discussed briefly at the end of this chapter.

B. Rough-Spherical-Ground Diffraction Model

The earth surface is generally not level because of trees, rocks, and man-made structures such as buildings. It is often categorized as a rough-spherical-ground diffraction model.

There is no general solution for this model, and usually empirical corrections are made using smooth-spherical-ground diffraction as the basis for the calculations.

In most cases the troposcatter-wave components are stronger than the diffraction-wave components. Therefore analysis for the diffraction propagation model is considered meaningless.

C. Thick-Ridge Diffraction Model

When the transmission path crosses a transverse mountain ridge and is short, and the curvature effect of the spherical ground is negligible, the dominant waves reaching the receiving point are diffracted waves since the troposcatter-wave component is not sufficiently strong.

An example is given in type III of Fig. 1*b*.

A thick ridge is a ridge such that there is no single point on it from which the transmitting and the receiving points are visible simultaneously.

One type of thick ridge, called a smoothly rounded ridge, can be analyzed theoretically if the following three conditions are satisfied.

1. The ridge surface is smooth.
2. The electrical characteristics (dielectric constant ε and conductivity σ) are uniform.
3. The ridge cross section is semicircular with a uniform curvature radius.

The electromagnetic field can be analyzed mathematically using the available calculation charts. However, no matter how accurate the theory is, if the actual conditions and assumptions do not match, use of the charts cannot be justified.

Most of the ridges encountered have a nonuniform rough surface due to the presence of rocks, plants, trees, etc. Therefore the electrical characteristics are rarely uniform.

Although, for example, Mt. Fuji is approximately cone shaped, the surfaces of its sides undulate from the viewpoint of microwave propagation.

Therefore instead of approximating the rounded ridge, a simplified multiknife-edge approximation method is adopted.

In Fig. 1*b* a tangent is drawn from the transmitting and the receiving points to the ridge, and it is assumed that there are two separate imaginary knife-edges M_1 and M_2.

Consequently the thick but irregularly shaped ridge usually is replaced by a single knife-edge M_e, as shown in the figure, and the nomograph for estimating has been prepared.

However, it should be noted that this substitution is possible only when M_1 and M_2 are very close to one another and the propagation loss is apt to be underestimated.

D. Knife-Edge Diffraction Model

A screen-shaped radio-wave barrier which theoretically is infinitely thin and extends infinitely downward and transversally is called a knife-edge.

If there is only one knife-edge, as shown by type IV of Fig. 1*b*, it generates single knife-edge diffraction. This is applicable when the width of the ridge is longer than the width of the first Fresnel zone, and both the transmitting and the receiving points are almost visible from the top of the ridge.

Under such conditions, the theoretical diffraction loss for the substituted knife-edge is able to give an approximate value for the actual loss.

E. Multiridge Diffraction Model

There may be more than one ridge between the transmitting and the receiving points. When point A is connected to point B by lines passing successively through intermediate ridge tips which form maximum elevation angles with other ridge tips, a convex-shaped path of the diffracted radio ray is obtained, as shown by type V in Fig. 1*b*. The figure is an example of triple-ridge diffraction.

However, the diffraction loss increases rapidly as the number of ridges increases, and as a result, the received signal level becomes increasingly lower. Consequently in practical applications, the number of ridges must first be determined.

3.2 SINGLE-RIDGE DIFFRACTION LOSS CALCULATION METHOD

A. Parameters for Approximate Solution of Single-Ridge Knife-Edge (Fig. 2)

1. Ridge depth (diffraction height) C_s. Find the height above sea level h_M of a ridge and let the distances from the ridge to the transmitting and the receiving points be d_1 and d_2, respectively. Then the imaginary radio path height h_p of the radio wave, which may occur in the absence of the ridge, is obtained as

$$h_p = (h_1 d_2 + h_2 d_1)/D - d_1 d_2/2Ka \tag{3.3}$$

where K = effective earth radius factor (standard value 4/3)
a = real earth radius (= 6370 km)
h_1, h_2 = transmitting and receiving antenna heights

The ridge depth C_s is then

$$C_s = h_M - h_p \tag{3.4}$$

Figure 4 shows a nomograph for the imaginary radio path height h_p.

2. The first Fresnel zone radius at the ridge point r_s is given by

$$r_s = \sqrt{\lambda d_1 d_2/D} \tag{3.5}$$

3. The diffraction parameter U is equal to the ridge depth C_s normalized by the first Fresnel zone radius r_s at that point. It is used as a variable to determine diffraction loss

$$U = C_s/r_s \qquad \text{(diffraction parameter)} \tag{3.6}$$

Remark. The negative diffraction parameter U is called clearance parameter U_e to indicate the degree of visibility.

B. Diffraction Loss Z of Single Ridge (Approximated by a Knife-Edge)

Diffraction loss Z (dB) is estimated directly using Fig. 7 or Fig. 8. The curve on the left-hand side of Fig. 8*a* shows the state immediately before the radio line of sight is obstructed by a knife-edge, and a small ripple (about 0-dB loss) converging at 0 dB with improving visibility.

If the knife-edge is only on the line of sight, the diffraction loss at $U = 0$ is only 6 dB.

When the path is transhorizon, that is, $U > 0$, the loss increases as the ridge depth increases, and if $U > 1$, the ridge diffraction loss Z is expressed by

$$Z \doteqdot 16 + 20 \log U \quad (\text{dB}) \tag{3.7}$$

as shown in Fig. 8*b*.

If $U \leq -0.5$ or the path clearance between the line of sight and the tip of the knife-edge is one-half or larger than the first Fresnel zone radius, almost no obstruction propagation (free-space propagation if there are no other factors producing the loss) can be assumed, as shown in the curves.

Figure 8*b* shows the diffraction factor that is the field strength ratio or the antilogarithm of $-Z$ (dB).

C. Nomograph for Direct Estimation of Diffraction Loss Due to a Single Ridge

The nomograph used to estimate the diffraction loss Z from distances d_1, d_2, D $(= d_1 + d_2)$, ridge depth C_s, and frequency f is shown in Fig. 7. Using the data in this figure, the first Fresnel zone radius r_s is also obtained.

This nomograph will be particularly useful for analyzing multiridge diffraction, which requires complex calculation procedures.

For the effect of a nearby ridge, J_s, or of a reflected wave, J_R, refer to Section 3.4.

3.3 CALCULATION OF DIFFRACTION LOSS CAUSED BY MULTIPLE RIDGES

Although the calculation procedures are slightly more complicated for multiple ridges, the solution can be obtained by repeating the same procedure for a path with more than one ridge. Figure 3 shows an example for a triple ridge.

First determine the direction of analysis, that is, $A \rightarrow B$ or $A \leftarrow B$. In the figure, $A \rightarrow B$ is assumed.

1. Estimate the single-ridge diffraction loss Z_1 by the first ridge M_1 for propagation path A to M_2.
2. Then, as in step 1, estimate the diffraction loss Z_2 by the second ridge M_2, which will be a diffraction ridge to the radio wave coming from M_1 and traveling toward M_3. It is assumed that the imaginary transmitting point is located at A_1 in the direction toward M_1 and is separated by the same distance as that between A and M_2. It is noted that in calculating the parameters using Eqs. (3.3) to (3.5), $d_1 + d_2$ is used instead of d_1, and d_3 instead of d_2.
3. The third ridge M_3 is assumed to be a diffraction ridge lying on the radio path M_2 to B. As in step 2, the imaginary transmitting point is located at A_2 and the single-ridge model between A_2, M_3, and B. Substitute $d_1 + d_2 + d_3$ for d_1 and also d_4 for d_2 in Eqs. (3.3) to (3.5) to obtain diffraction loss Z_3.

4. The total additional propagation loss (loss in addition to free-space loss) between points A and B is given by

$$Z_t = (Z_1 + Z_2 + Z_3) + [J_s + J_R] \quad \text{(dB)} \tag{3.8}$$

For J_s and J_R refer to Section 3.4.

3.4 EFFECTS CAUSED BY A NEARBY RIDGE AND A REFLECTED WAVE

Even when there is a line of sight, if one or more ridges are close to the radio line of sight, there will be approximately a 6 to 10-dB diffraction loss.

Also, the effect should be examined that will be caused by a reflected wave when a radio section between a ridge and the transmitting or the receiving point, or between ridges, gives an interference propagation path.

A. Additional Loss J_s caused by Nearby Ridge

It is difficult to determine additional loss when there are other nearby ridges close to the radio line of sight between an obstruction ridge and the transmitting or the receiving point, or between consecutive obstructing ridges.

The following is a simplified calculation method developed for determining this loss (see Fig. 5*a*).

1. Draw the first Fresnel zone for each side of a polygon formed by the individual path of a diffracted radio ray.
2. Then draw the boundary determined by half the first Fresnel zone radius. The region within this boundary is called protection zone.
3. For section i, in which no ridge is in the protection zone among those diffraction sections, it is assumed that the additional loss $J_{st} = 0$ (dB).
4. A ridge in the protection zone is called nearby ridge.
5. If there is more than one nearby ridge in a diffraction section, determine the effect assuming that these nearby ridges exist independently. Let C denote the clearance between the main transmission path and the top of the nearby ridge and r the first Fresnel zone radius at the ridge point. Then the additional loss t is obtained from Fig. 6 [1] using the parameter $U_c = C/r$.
6. All the additional losses as a result of nearby ridges located within the same diffraction section are added. If the sum exceeds 10 dB, assume 10 dB as the additional loss for the diffraction section,

$$J_{st} = t_1 + t_2 + \cdots \leq 10 \text{ dB} \tag{3.9}$$

7. Add the additional losses for all the diffraction sections to obtain J_s,

$$J_s = J_{s1} + J_{s2} + J_{s3} + \cdots \tag{3.10}$$

Note that it is not always necessary to draw the protection zone. Check only unclarified sections.

B. Additional Loss J_R Due to Reflected-Wave Interference

In the multiple-ridge diffraction propagation path, ground reflection in the interridge diffraction section (excluding the diffraction sections that include the transmitting or the receiving point) has a negligible effect on the total propagation loss.

This was determined through trial calculations carried out on various kinds of models, even when there existed a sound reflection whose effective reflection coefficient was nearly unity.

Accordingly, when examining the effects of reflected waves, practically any intermediate interridge diffraction section can be ignored.

Only the sections that include the transmitting or the receiving point at an end of the section need to be examined.

1. When effective reflection coefficients ρ_e for the first section (#1) and the last section (#$i + 1$) in Fig. 5*b* are below 0.2, the effect caused by these sections can be ignored, that is,

$$J_R = 0 \quad \text{(dB)} \tag{3.11}$$

   This is determined considering that diffraction loss by the adjacent ridge is greater for the reflected wave than for the main-path wave.
2. For a section where the effective reflection coefficient ρ_e exceeds 0.2, estimate the level ratio of the main-path wave to the reflected wave, then calculate the additional loss J_R.
   (a) Let X_1 be the relative level (field strength ratio) of the single-reflection wave for section #1 over the main wave.
   (b) Similarly, let X_2 be the same for section #$i + 1$ over the main wave.
   (c) Let X_3 be the relative level of the wave undergoing a double reflection, that is, in sections #1 and #$i + 1$.
   (d) Although the resultant receiving level is the vectorial sum of the main-path wave and the three different waves taking paths (a), (b), and (c) as mentioned in the foregoing, the worst value occurs when X_1, X_2, and X_3 are in phase with each other but in opposite phase with the main-path wave. In this case the additional loss J_R (dB) is

$$J_R = -20\log(1 - X_t) \quad \text{(dB)} \tag{3.12}$$

   where

$$X_t = X_1 + X_2 + X_3$$

$$X_t = 1, \qquad \text{for } X_t > 1 \tag{3.13}$$

   The X not taken into consideration is assumed to be zero.
   (c) Methods for estimating X_1, X_2, and X_3 from the known parameters for a propagation path are given in Fig. 5*c* and *d*.

3. X_t may be unity or greater, according to the circumstances, because there are three kinds of reflected-wave paths. When they are combined with the main-path wave, the resultant level is

$$0 \text{ minimum to } (1 + X_t) \text{ maximum}$$

in field strength ratio, depending upon the phases, and it is unstable in general. J_R, which represents additional loss in the worst case, reaches infinite decibels, theoretically.
4. To determine the total level under the conditions given in step 3, first the relative phases of the main-wave X_1, X_2, and X_3 and the resultant vector quantities $\dot{X}_1$, $\dot{X}_2$, and $\dot{X}_3$ need to be determined. Then the vector summation is carried out sequentially, as shown in Fig. 15 of Chapter 2, and the results are calculated.
5. There are two calculation methods for X_3, depending on whether there are one or more ridges. The methods are shown in Fig. 5*c* and *d*.

3.5 FRESNEL ZONE

A set of points, determined in such a manner that the sum of the distances from two fixed points, namely, the transmitting and the receiving points, has a constant path-length difference from the shortest path length between these two points, forms a prolonged ellipsoid with an axis passing through the two points. The region within the ellipsoid surface which gives the path-length difference $(\lambda/2)$ is called the Fresnel zone.

Similarly, the region enclosed by the two ellipsoids, giving path-length differences $n(\lambda/2)$ and $(n-1)(\lambda/2)$ respectively, is called the *n*th Fresnel zone.

If a spatial region corresponding to the first Fresnel zone is secured in a radio-wave transmission, an energy almost equivalent to that expected in free-space propagation can be transmitted to the receiving point.

The size of the Fresnel zone usually is expressed by a radius of circle r as a transverse cross section at a point of interest, such as a ridge point or a reflection point.

A. First Fresnel Zone Radius r_0 for Transmission between Points Separated by an Extremely Long Distance

The most typical case is where a solar or lunar light is received on a global or intraroom scale. However, the Fresnel zone actually encountered in practical radio engineering is found in transmission between earth and a radio star or an artificial planet.

Such a case can be approximated even when the total distance is from a few kilometers to tens of kilometers, as shown in Fig. 10, if the distance D_0 to the point under consideration from the transmitting or the receiving point is extremely short compared with the total distance D_f, and the wavelength λ is sufficiently short compared with D_0.

In this case, the first Fresnel zone radius r_0 is given by

$$r_0 \doteqdot \sqrt{\lambda D_0} \tag{3.14}$$

where $D_f \gg D_0 \gg \lambda$.

The nomograph for a simplified calculation of r_0 when D_0 is 500 km or shorter is given in Fig. 10.

Figure 11 gives calculation graphs covering great distances into space.

B. First Fresnel Zone Radius r_1

Denoting the distances to a point of interest from the transmitting and the receiving points by d_1 and d_2, respectively, the first Fresnel zone radius r_1 at that point is

$$r_1 = \sqrt{\lambda d_1 d_2 / D} \tag{3.15}$$

where D is the distance between the transmitting and the receiving points,

$$D = d_1 + d_2, \qquad d_1, d_2 \gg \lambda$$

Figure 12 gives a nomograph convenient for estimating r_1 using the parameters d_1, d_2, and D and the frequency f.

The following procedure is useful for determining r_1 more accurately for frequencies covering a wide range, which is also shown in Figs. 13 and 14.

1. Find the first Fresnel zone radius at midpoint r_{max}. (It takes the maximum value at this point.)

$$r_{max} = \tfrac{1}{2}\sqrt{\lambda D} \qquad \text{(Fig. 13)} \tag{3.16}$$

2. Estimate coefficient p using d_1/d or d_2/D,

$$p = 2\sqrt{d_1 d_2}/D \qquad \text{(Fig. 14)} \tag{3.17}$$

3. Calculate the first Fresnel zone radius r_1 using r_{max} and p,

$$r_1 = p r_{max} \tag{3.18}$$

NOTE. The notation r_1, r_0, or r for the first Fresnel zone radius at an arbitrary point is used taking specific points into account to indicate its position, namely, r_s for the ridge point and r_R for the reflection point.

3.6 *n*th FRESNEL ZONE RADIUS

A set of points which give the path-length difference of $n\lambda/2$ (when connected to the transmitting and the receiving points) from the direct path length can form the surface of a prolonged ellipsoid about the axis that coincides with the line of sight.

The cross section of the plane that is perpendicular to the major axis of the prolonged ellipsoid shows a circle with radius r_n,

$$r_n = \sqrt{n\lambda d_1 d_2/D} \qquad \text{(general equation)} \tag{3.19}$$

where r_n = nth Fresnel zone radius
d_1, d_2 = distances from transmitting and receiving points to point under consideration
D = total distance, $= d_1 + d_2$

If either the transmitting or the receiving point is located at a great distance,

$$r_{0n} = \sqrt{n\lambda D_0} \qquad \text{(for extremely far points)} \tag{3.20}$$

where D_0 is the distance between the point under consideration and the transmitting or the receiving point, whichever is nearer.

Applying Eqs. (3.14) and (3.15), we obtain

$$\begin{aligned} r_n &= r_1\sqrt{n} \\ r_{0n} &= r_0\sqrt{n} \end{aligned} \tag{3.21}$$

When the first Fresnel zone radius is known, the nth Fresnel zone radius is easily obtained using the multiplier

$$k_n = \sqrt{n} \tag{3.22}$$

Figure 9 gives a table of k_n values for $n = 1$ to 99.

3.7 DIFFRACTION LOSS CALCULATION CHART FOR A MULTIRIDGE ($i \leq 5$)

Since calculating the multiridge diffraction loss is time consuming and difficult, simplified calculating procedures are given in the charts.

To proceed with the calculation, Fig. 15*b* is used, referring to Fig. 15*a*. If the number of ridges is 4 or less, the unnecessary columns are deleted.

For a multiridge path, the scattering difference between calculated and measured values (estimation error) will become large. Thus it is pointless to pursue accuracy excessively.

Accordingly, most troublesome calculations may be omitted by measuring graphically the ridge depth C_s ($C_{s1}, C_{s2}, \ldots$) on a path profile chart of a greater scale, estimating the diffraction parameter U that is the ratio between ridge depth C_s and the first Fresnel zone radius r_s (refer to the description in Fig. 3) at the ridge point and by utilizing Figs. 7 and 8.

3.8 DIFFRACTION LOSS OVER SMOOTH SPHERICAL GROUND (SOLUTION ACCORDING TO CCIR RECOMMENDATION 715)

A. Propagation Loss

The propagation loss-between A and B over smooth spherical ground (basic transmission loss, transmission loss between isotropic antennas of transmission and reception) is given by

$$\Gamma = \Gamma_0 + A_P \quad \text{(dB)} \tag{3.23}$$

where Γ_0 = free-space propagation loss
A_P = additional loss over smooth spherical ground (additional propagation loss)

B. Estimation of Additional Loss A_P over Smooth Spherical Ground

1. Method using nomographs of Fig. 16,

$$A_P = F(D) - \{G(h_1) + G(h_2)\} \quad \text{(dB)} \tag{3.24}$$

where $F(D)$ = attenuation due to distance (dB)
$G(h_1), G(h_2)$ = height gain factor (dB)
D = distance (km)
h_1, h_2 = height of antenna above ground (m)

2. An alternate solution uses Fig. 17,

$$A_P = G(\chi_0) - \{F(\chi_1) + F(\chi_2)\} - 20.5 \quad \text{(dB)} \tag{3.25}$$

with

$$\chi_0 = DB_0, \qquad \chi_1 = d_{t1}B_0, \qquad \chi_2 = d_{t2}B_0, \qquad B_0 = 670\left(f/K^2a^2\right)^{1/3}$$

where f = frequency (MHz)
K = effective earth radius factor
a = mean radius of earth (= 6370 km)
d_{t1}, d_{t2} = line-of-sight distance (km) over spherical ground for h_1 and h_2 (Chapter 2, Figs. 22 and 23)

C. Application of Graphs

The nomographs of Figs. 16 and 17 are used as follows.

Figure 16*a*	$F(D)$	HP land and sea and VP land
Figure 16*b*	$G(h_1), G(h_2)$	HP land and sea and VP land
Figure 16*c*	$F(D)$	VP sea
Figure 16*d*	$G(h_1), G(h_2)$	VP sea
Figure 17	$G(\chi_0), F(\chi_1), F(\chi_2)$	HP land and sea

where HP = horizontal polarization
VP = vertical polarization

The difference between VP sea and the others is greater when h_1 and h_2 are lower or when the frequency is lower. However, it can be ignored at frequencies of approximately 100 MHz or higher. Therefore the practical range of application by this method is restricted to an extremely narrow diffraction region (lying between the line-of-sight region and the scattering region).

D. Notes on Application Method

The forementioned solution is based on the traditional methods for smooth-spherical-ground diffraction. It is considered to give a better estimate (with an error of 1 to 2 dB at most) for the deep diffraction (the greater distance) region.

This is true only when the atmosphere is assumed to be homogeneous. However, in the actual path over a great distance, the error may frequently reach up to several tens of decibels if abnormal diffraction or tropospheric scatter occurs.

Accordingly, the distance range considered practical by these methods is limited to a very narrow diffraction region, where the diffraction-wave field strength is approximately equal to or less than the median field strength of the troposcatter waves.

EXAMPLES

Example 3.1

Estimate the propagation loss at $f = 1.5$ GHz assuming that the effects of geographical objects other than ridge M can be ignored.

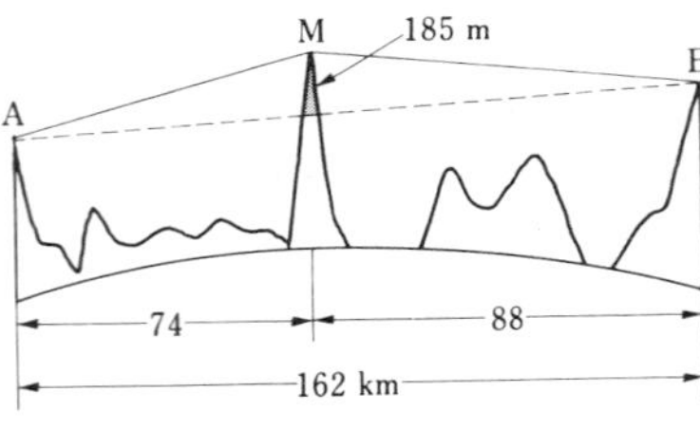

Solution

The diffraction loss Z is estimated and added to the free-space propagation loss Γ_0 to obtain the propagation loss Γ assuming that there is a single ridge M in a free space,

$$\Gamma = \Gamma_0 + Z \quad \text{(dB)}$$

1. First Fresnel zone r_s (Figs. 13 and 14),

$$D = 162 \text{ km}$$

$$d_1 = 74 \text{ km}, \quad d_2 = 88 \text{ km}, \quad d_1/D = 0.46$$

$$r_{\max} = 90 \text{ m}, \quad p \doteqdot 0.99$$

$$r_s \doteqdot 89 \text{ m}$$

2. Diffraction parameter U;

$$U = C_s/r_s = 185/89 = 2.08$$

3. Diffraction loss Z (Fig. 8b),

$$Z = 22.2 \text{ dB}$$

4. Free-space propagation loss Γ_0 (Fig. 8d in Chapter 1),

$$\Gamma_0 = 140.2 \text{ dB}$$

5. Propagation loss Γ,

$$\Gamma = 140.2 + 22.2 = 162.4 \text{ dB}$$

Example 3.2

Elevations above sea level of this point and the distant point are 2400 m and 1850 m, respectively, and the distance between them is 85 km. However, there is a mountain 1830 m high at a point 66 km from this point. Is it possible for this path to be line of sight? If it is possible, examine quantitatively whether or not there is any obstruction at 750 MHz.

Solution

$$h_1 = 2400 \text{ m}, \quad h_2 = 1850 \text{ m}, \; h_M = 1880, \quad \lambda = 0.4 \text{ m},$$

$$d_1 = 66 \text{ km}, \quad d_2 = 19 \text{ km, and } D = 85 \text{ km}.$$

1. Imaginary path height h_p. Through macroscopic study using the nomograph in Fig. 4, y_1, y_2, and x are obtained,

$$\begin{pmatrix} h_1 \to h' & h_2 \to h'' \\ d_1 \to d' & d_2 \to d'' \end{pmatrix}$$

Thus, $y_1 = 520$, $y_2 = 1450$, and $x = 75$. Hence

$$h_p = (y_1 + y_2) - 75 = 1895 \text{ m}$$

2. Clearance C,

$$C = h_p - h_M = 1895 - 1880 = 15 \text{ m} > 0 \qquad \text{(line-of-sight)}$$

3. First Fresnel zone radius at ridge point r_s. From Fig. 13, $r_{max} = 94$ m and $p = 0.82$. Hence

$$r_s = r_{max}\, p = 77 \text{ m}$$

4. Clearance parameter U_c,

$$U_c = c/r_s = 15/77 = 0.19 < 0.5 \qquad \text{(nearby ridge)}$$

5. In step 4 the path is barely in the light of sight. A more accurate computation is made with a calculator,

$$h_p = \frac{h'd'' + h''d'}{D} - \frac{d'd''}{2Ka} \qquad \text{(Figs 2, 4, and 6)}$$

$$= \frac{(2400 \times 19 + 1850 \times 66)}{85} - \frac{19 \times 66 \times 10^3}{2 \times 1.33 \times 6370} = 1973 - 74 = 1899 \text{ m}$$

$$C = 1899 - 1880 = 19 \text{ m}$$

$$r_s = \sqrt{\lambda d'd''/D} = \sqrt{0.4 \times 66 \times 19 \times 10^3/85} = 76.8 \text{ m}$$

$$U_c = 19/76.8 = 0.25$$

The nearby-ridge additional loss J_s is obtained from Fig. 6 [1]. From this figure, $t = 3$ dB. Since there is only one nearby ridge, $t = J_s = 3$ dB.

EXAMPLE 3.3

Estimate the additional propagation loss (double-ridge diffraction), where the frequency $f_1 = 5$ GHz ($\lambda_1 = 0.06$ m) and $f_2 = 200$ MHz ($\lambda_2 = 1.5$ m). $D = 46$ km, $d_1 = 14$ km, $d_2 = 20$ km, $d_3 = 12$ km; $h_1 = 220$ m, $h_2 = 350$ m, $h_{M1} = 370$ m, $h_{M2} = 405$ m; $C_{s1} = 90$ m, and $C_{s2} = 75$ m.

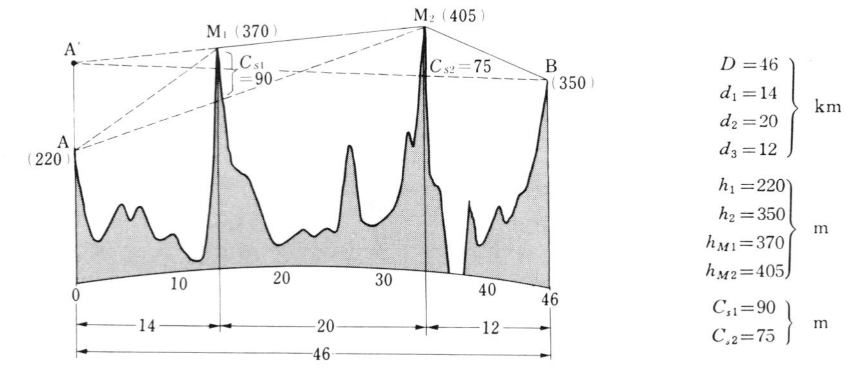

Solution

Figure 3 is used.

1. First Fresnel zone radius at ridge point r_s. For $\lambda_1 = 0.06$ m, at point M_1 on path $\overline{AM_2}$,

$$r_{s1} = \sqrt{\lambda_1 d_1 d_2/(d_1 + d_2)} = \sqrt{0.06 \times 14 \times 20 \times 10^3/34} = 22.2 \text{ m}$$

and at point M_2 on path $\overline{A'B}$,

$$r_{s2} = \sqrt{\lambda_1(d_1 + d_2)d_3/D} = \sqrt{0.06(14 + 20)12/46} = 23.1 \text{ m}$$

Since the first Fresnel zone radius is proportional to the square root of the wavelength, the radius at λ_2 is calculated as follows:

$$\sqrt{\lambda_2/\lambda_1} = \sqrt{1.5/0.06} = 5$$

Hence

$$r'_{s1} = r_{s1} \times 5 = 22.2 \times 5 = 111 \text{ m}$$

$$r'_{s2} = r_{s2} \times 5 = 23.1 \times 5 = 115.5 \text{ m}$$

2. Then diffraction parameter U and diffraction loss Z and Z_t, from Fig. 8, are as follows:

f	Ridge	r_s	$U = C_s/r_s$	Z (dB)	Z_t (dB)
5 GHz	M_1	22.2	4.05	28.1	54.3
	M_2	23.1	3.25	26.2	
200 MHz	M_1	111.0	0.81	15.2	28.5
	M_2	115.5	0.65	13.3	

Additional propagation loss between A and B is shown in the last column.

EXAMPLE 3.4

Estimate the additional loss due to the reflected wave in the case where the first diffraction section has a lake water surface reflection ($\rho_e \doteqdot 1.0$), assuming that the frequency is 200 MHz and the other parameters are as in Example 3.3.

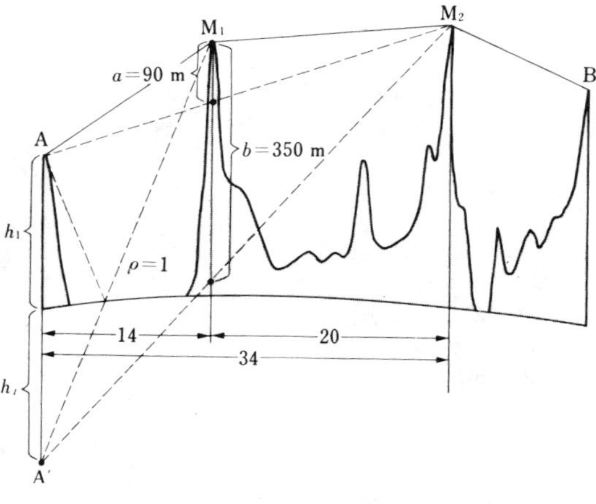

Solution

Figure 5*d* is used.

1. The reflection point is located between A and M_1. The distance is 14 km, and h_1 and the height of M_1 are very large. Thus the earth's curvature may be ignored and the reflection point is easily obtained graphically under the assumption that $h_1 \doteqdot h_2$.
2. Ridge depth b at point M_1 and $\overline{A'M_2}$ section,

$$b = 350 \text{ m}$$

3. Since $\overline{A'M_1} \doteqdot \overline{AM_1}$ and $\overline{A'M_2} \doteqdot \overline{AM_2}$, the first Fresnel zone radius r_{s1} is obtained from the result of Example 3.3. Accordingly, $r_{s1} = 111$ m (200 MHz).
4. Diffraction parameters and diffraction coefficients are obtained,

$$\begin{aligned} U_a &= 90/111 = 0.81, \quad & Z_a &= 0.18 \\ U_b &= 350/111 = 3.15, \quad & Z_b &= 0.05 \end{aligned} \qquad \text{(Fig. 8}a\text{ and }b\text{)}$$

5.

$$X_t = \rho_e Z_b / Z_a = 1 \times 0.05/0.18 = 0.278$$

6. Additional loss due to reflected wave,

$$J_R = -20 \log A = -20 \log(1 - X_t) = 2.8 \text{ dB}$$

Example 3.5

Estimate the propagation loss and the receiving input (dBm) for a smooth-spherical-earth diffraction propagation path where the influence from tropo-scatter waves may be ignored. $f = 800$ MHz, $K = 4/3$, $h_1 = 120$ m, $h_2 = 90$ m. Antenna gain $G_t = G_r = 16$ dB, transmitting power $P_t = 1$ kW $= 60$ dBm, and horizontal polarization is assumed.

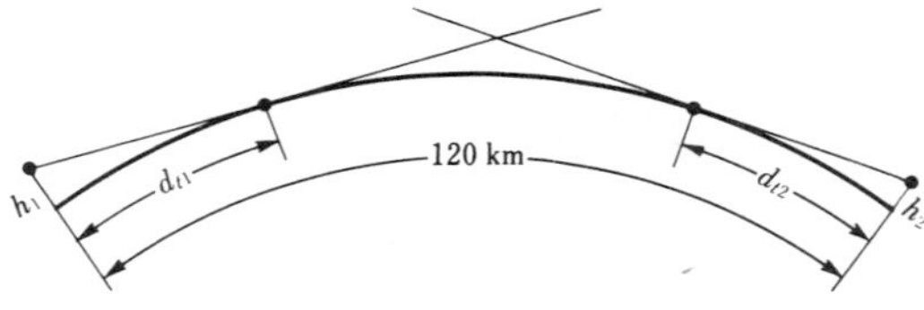

Solution

Although it is easier to use the nomograph of Fig. 16, another solution employing Fig. 17 is given in the following.

1. Parameter B_0,

$$B_0 = 15 \qquad \text{for } f = 800 \text{ MHz, } K = 4/3$$

2. Line-of-sight distances d_{t1} and d_{t2} for h_1 and h_2, respectively, are computed.

$$d_{t1} = 4.12\sqrt{h_1} = 4.12\sqrt{120} = 45 \text{ km}$$
$$d_{t2} = 4.12\sqrt{h_2} = 4.12\sqrt{90} = 39 \text{ km}$$

(Fig. 23 in Chapter 2 may be used)

3. Parameters χ_0, χ_1, and χ_2 are computed.

$$\chi_0 = 1800, \qquad \chi_1 = 675, \qquad \chi_2 = 585$$

4. Find parameters $G(\chi_0)$, $F(\chi_1)$, and $F(\chi_2)$ in Fig. 17,

$$G(\chi_0) = 70, \qquad F(\chi_1) = 6, \qquad F(\chi_2) = 2$$

5. Additional loss over smooth spherical earth A_p,

$$A_p = G(\chi_0) - \{F(\chi_1) + F(\chi_2)\} - 20.5$$
$$= 70 - (6 + 2) - 20.5 = 41.5 \text{ dB}$$

6. Free-space propagation loss Γ_0 ($f = 800$ MHz, $D = 120$ km),

$$\Gamma_0 = 132.1 \text{ dB (Fig. } 8d \text{ in Chapter 1)}$$

7. Propagation loss Γ,

$$\Gamma = \Gamma_0 = A_p = 132.1 + 41.5 = 173.6 \text{ dB}$$

8. Received power P_r,

$$P_r(\text{dBm}) = P_t(\text{dBm}) + G_t + G_r - \Gamma$$
$$= 60 + 16 + 16 - 173.6 = -81.6 \text{ dBm}$$

Chart Index (1), Diffraction Propagation Path

Chart Index (2), Diffraction Propagation Path

Auxiliary Index, Fresnel Zone Radius

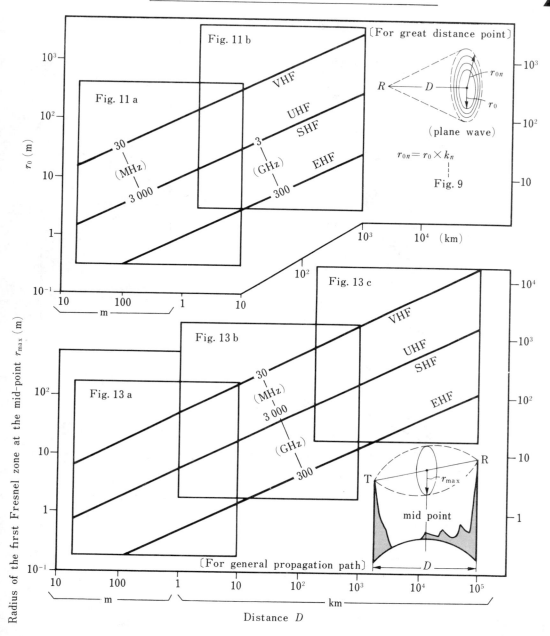

〔For great distance point〕

Fig. 11 b

Fig. 11 a

VHF UHF SHF EHF

30 (MHz) 3 000   3 (GHz) 300

r_0 (m)

R — D — r_{0n}, r_0

(plane wave)

$r_{0n} = r_0 \times k_n$ — Fig. 9

10^{-1}, 1, 10, 10^2, 10^3

10, 100 (m); 1, 10, 10^2, 10^3, 10^4 (km)

Fig. 13 c

Fig. 13 b

Fig. 13 a

VHF UHF SHF EHF

30 (MHz) 3 000 (GHz) 300

T R r_{max} mid point D

〔For general propagation path〕

Radius of the first Fresnel zone at the mid-point r_{max} (m)

10^{-1}, 1, 10, 10^2, 10^3, 10^4

10, 100 (m); 1, 10, 10^2, 10^3, 10^4, 10^5 (km)

Distance D

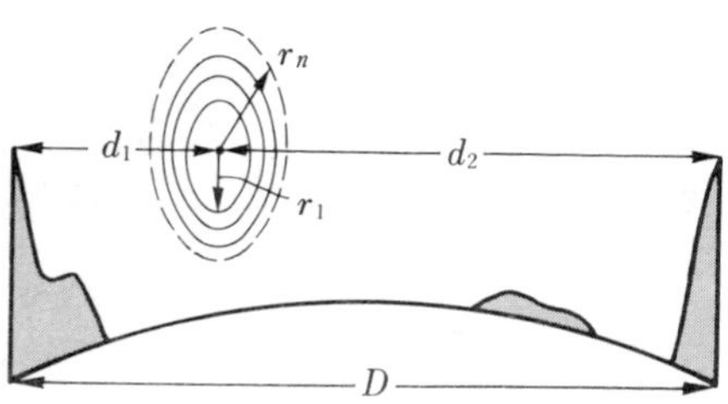

First Fresnel zone radius — Fig. 14

$r_1 = r_{max} \times p$

n-th Fresnel zone radius --- Fig. 9

$r_n = r_1 \times k_n$

3

1a

Basic Model of Diffraction Propagation Path

Spherical-Ground Type

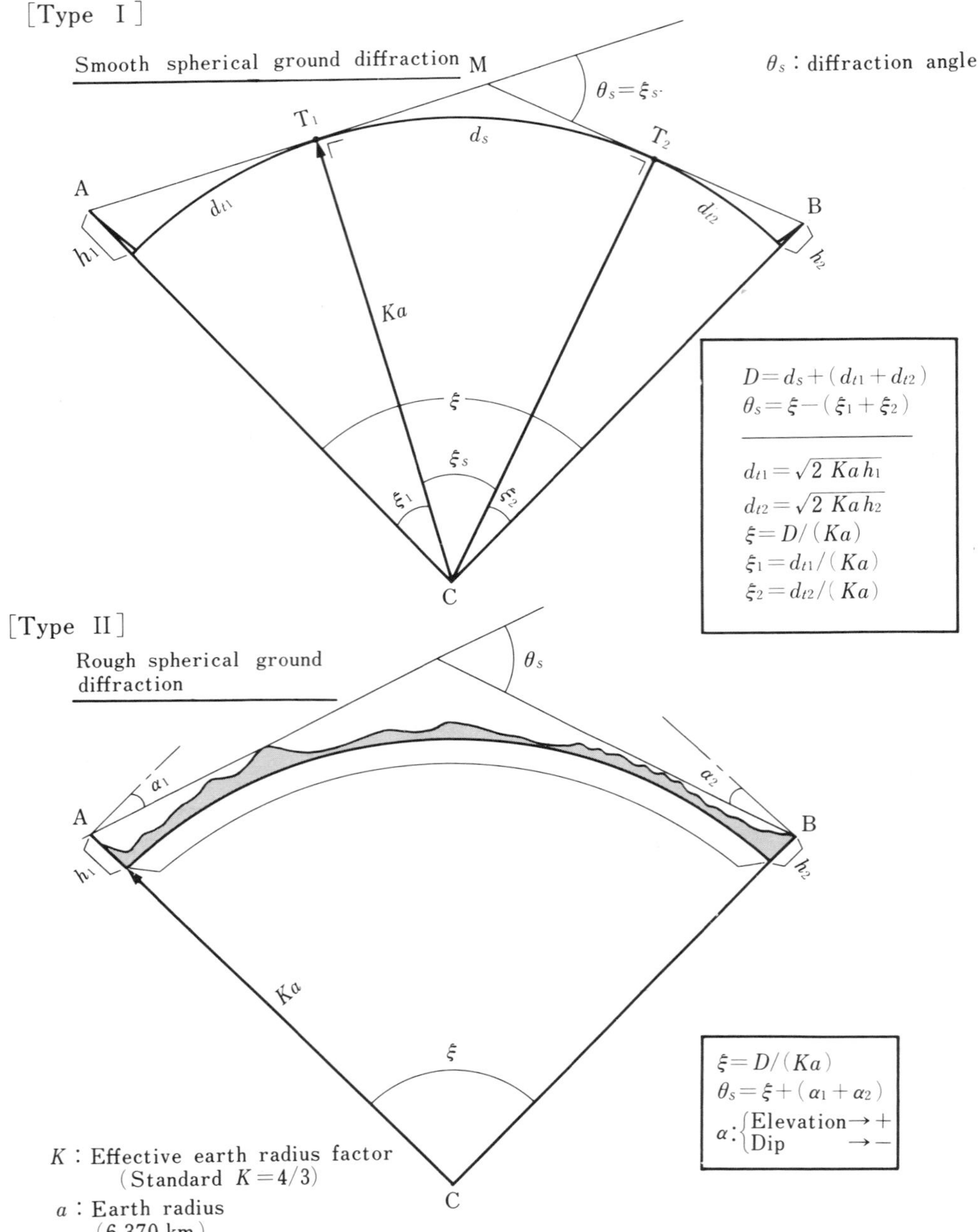

Basic Model of Diffraction Propagation Path

Ridge Type

[Type Ⅲ]

Thick-ridge diffraction

• Round type (smooth)
Indefinite type (rough)

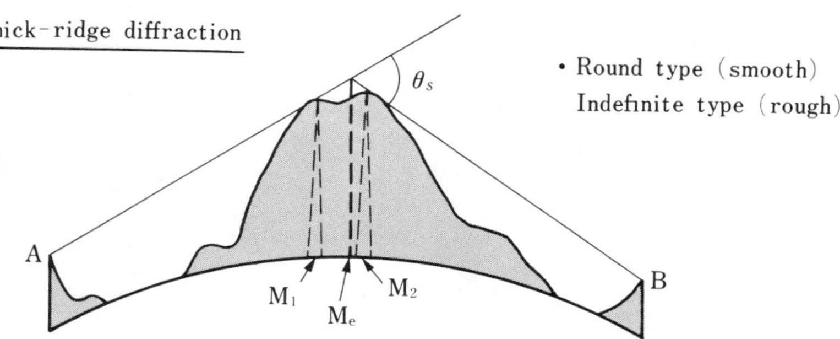

[Type Ⅳ]

Knife-edge diffraction
(thin-ridge)

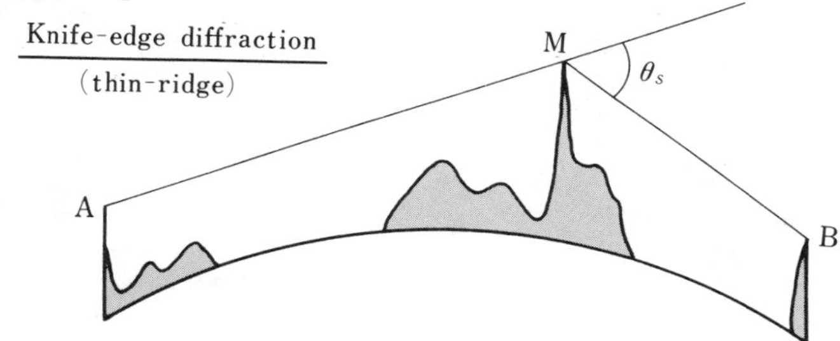

[Type Ⅴ]

Multiridge

$\theta_{si} > 0$

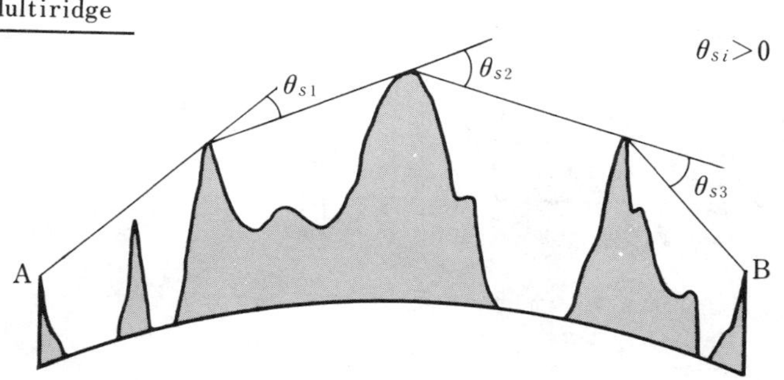

Ridge : Obstruction transversal to the radio path,
the width of the obstruction $\omega \gg 2\,r_S$ is assumed.
r_S : First Fresnel zone radius at the ridge point

Analysis of Single-Ridge Diffraction Propagation Path

[1] Profile

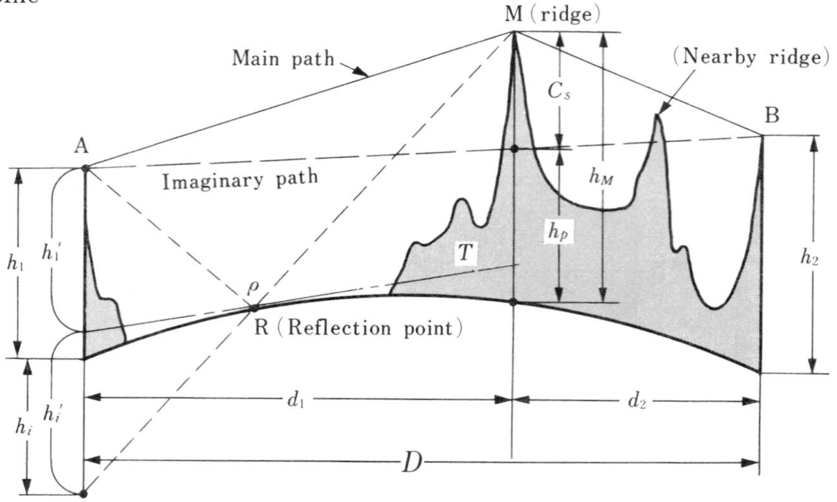

[2] Parameters required for analysis of a single ridge (treated as a knife-edge).

- Depth of ridge (diffraction height) $C_s = h_M - h_p$ where h_M ridge height above sea level
- Imaginary path height $h_p = \dfrac{h_1 d_2 + h_2 d_1}{D} - \dfrac{d_1 d_2}{2\,Ka}$ (Fig. 4)
- First Fresnel zone radius at the ridge point $r_S = \sqrt{d_1 d_2 / D}$ (Figs. 13 & 14)
- Diffraction parameter $U = C_s / r_S$ then obtain Z using U (Fig. 8)

[3] Propagation loss $\Gamma = \Gamma_0$ (Free space loss) $+ Z_t$ (Additional loss)

$$Z_t = Z + [J_s + J_R] \quad \text{(dB)}$$

- Z: Ridge diffraction loss (Figs. 7 & 8)
- J_R: Additional loss due to reflected wave interference (Figs. 5 & 6)
- J_s: Additional loss due to nearby ridge (Figs. 5 & 6)

---

1. C_s may be obtained directly from the path profile drawn on a large scale.
2. Effects of nearby ridge and reflected wave may be ignored depending on actual conditions (refer to Fig. 5).
3. The Clearance parameter $-U = U_c$ shows the degree of visibility.
4. The h_M is to be read from the map or to be measured in the field survey.

Analysis of Multiridge Diffraction Propagation Path

[1] Profile

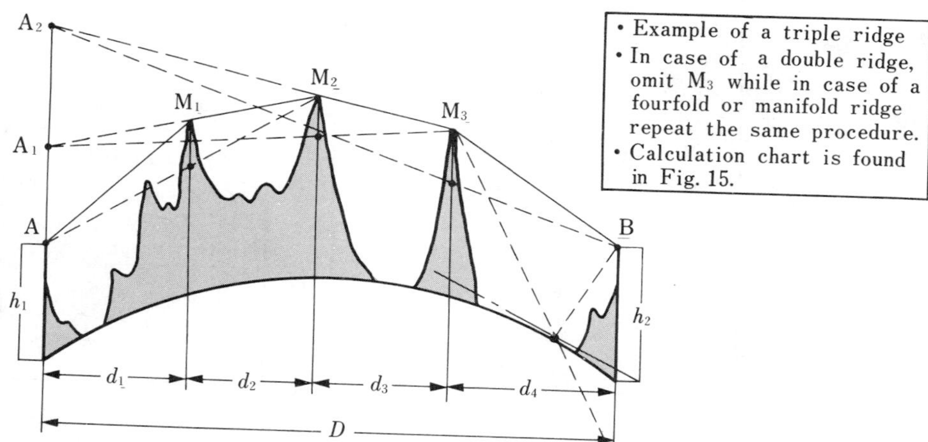

[2] Method of analysis

(a)

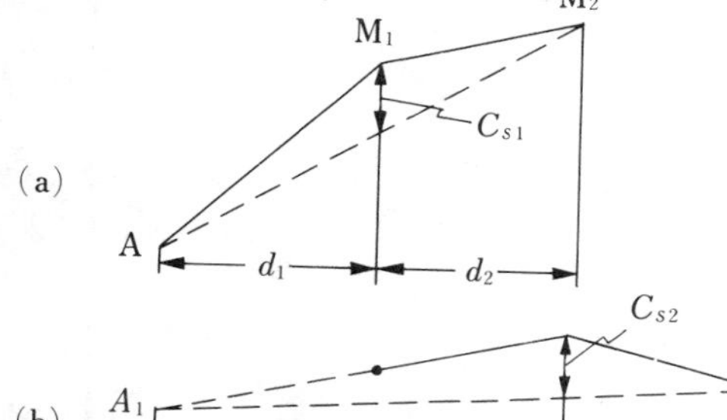

- Ridge height C_{s1} at M_1 above the imaginary path $\overline{AM_2}$ is obtained.
- Diffraction parameter $U_1 = C_{s1}/r_{s1}$
  $r_{s1} = \sqrt{\lambda d_1 d_2/(d_1+d_2)}$
- Diffraction loss Z_1……(Fig. 8)

(b)

C_{s2}, A_1, (d_1+d_2), d_3

- Imaginary antenna point A_1 is determined by extending line $\overline{M_2\ M_1}$.
- Ridge height C_{s2} at M_2 above the imaginary path $\overline{A_1\ M_3}$ is obtained.
- Diffraction parameter $U_2 = C_{s2}/r_{s2}$
  $r_{s2} = \sqrt{\lambda (d_1+d_2) d_3/(d_1+d_2+d_3)}$
- Diffraction loss Z_2……(Fig. 8)

(c)

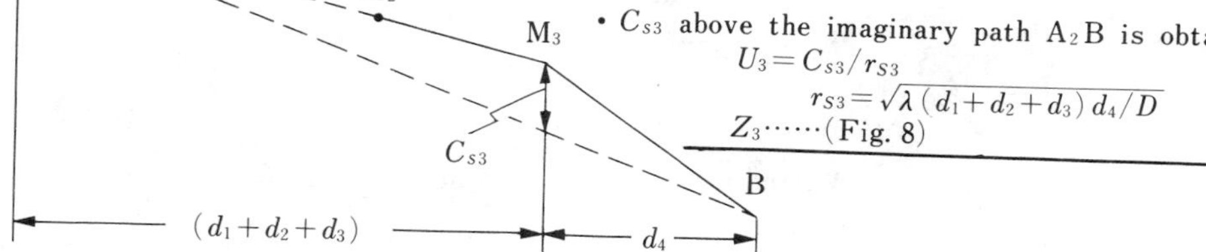

- Point A_2 is determined by extending the line $\overline{M_3 M_2}$.
- C_{s3} above the imaginary path $A_2 B$ is obtained.
  $U_3 = C_{s3}/r_{s3}$
  $r_{s3} = \sqrt{\lambda (d_1+d_2+d_3) d_4/D}$
  Z_3……(Fig. 8)

[3] Propagation loss $\Gamma = \Gamma_0$ (Free space loss) $+ Z_t$ (Additional loss)

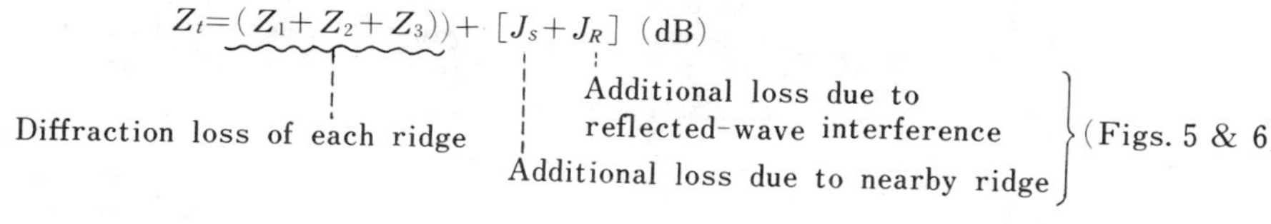

$$Z_t = (Z_1 + Z_2 + Z_3)) + [J_s + J_R] \text{ (dB)}$$

Nomograph for Imaginary Path Height (1)

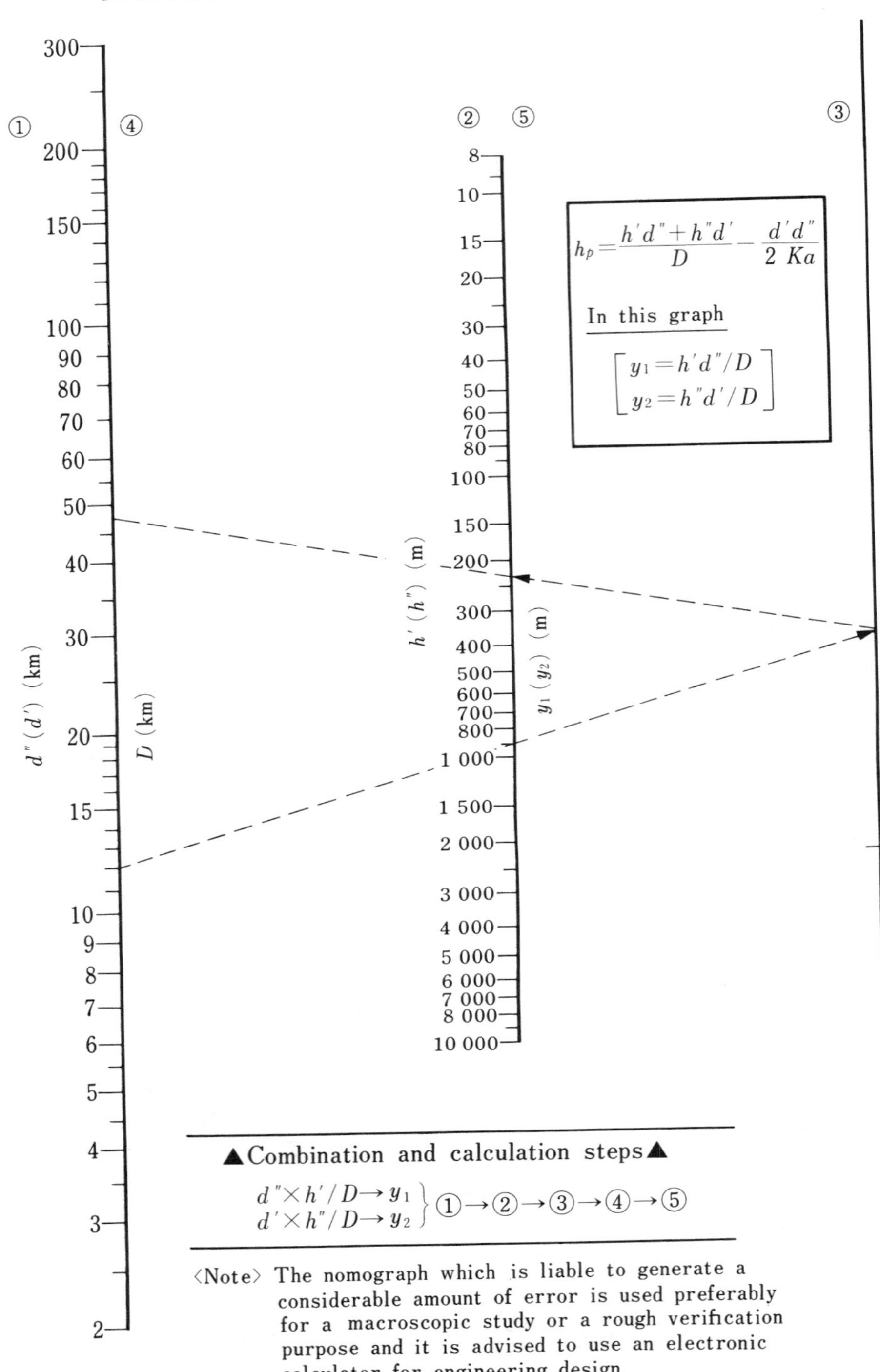

▲Combination and calculation steps▲

$d'' \times h'/D \rightarrow y_1$
$d' \times h''/D \rightarrow y_2$
} ①→②→③→④→⑤

〈Note〉 The nomograph which is liable to generate a considerable amount of error is used preferably for a macroscopic study or a rough verification purpose and it is advised to use an electronic calculator for engineering design.

Nomograph for Imaginary Path Height (2)

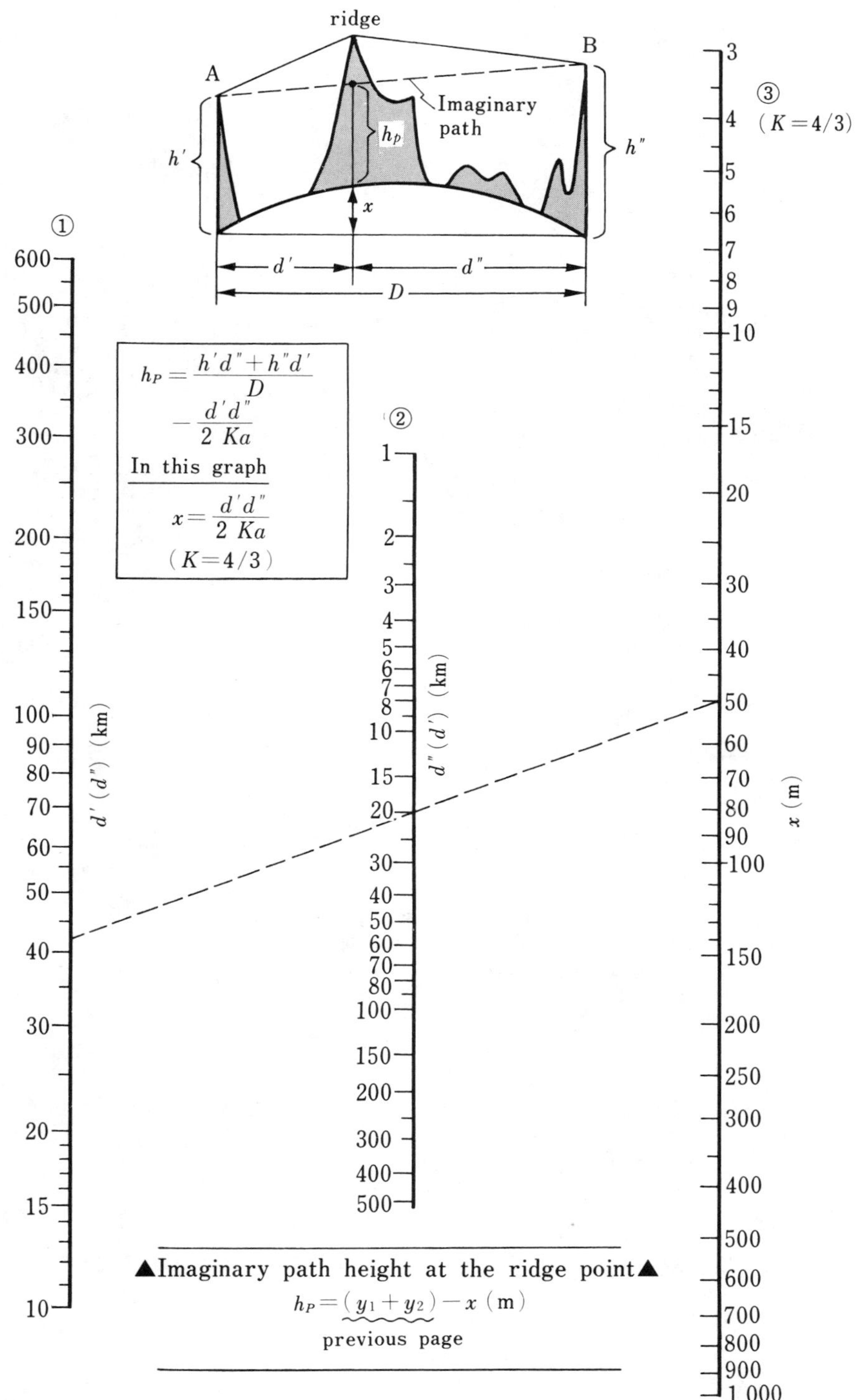

Estimation of Additional Loss Due to Nearby Ridge J_s

Estimation method

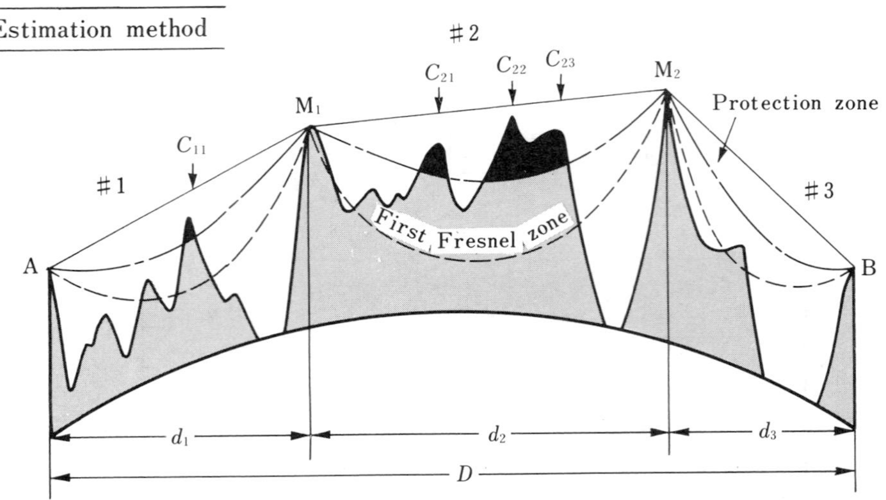

(1) Determining the protection zone
The first Fresnel zone is drawn for each diffraction section as shown by the dotted line in the figure. The region defined by half the radius of the first Fresnel zone is drawn and the inner region is called the protection zone as shown by the line —·— in the figure.

(2) Definition of nearby ridge
A ridge that penetrates into the protection zone is called a nearby ridge.

(3) Nearby ridge loss for each diffraction section J_i (dB)
a>Non-loss section (Ex. #3 section in the above figure)
The section where no obstruction exists in the protection zone

$$J_i = 0 \text{ dB}$$

b>Section with a single nearby ridge (Ex. #1 section)
The first Fresnel zone radius at the nearby-ridge point is denoted by r_i and the clearance between the path and the peak of nearby ridge C_i is estimated. Then, clearance parameter $U_{ci} = C_i / r_i$ is calculated and diffraction loss t_i is estimated by Fig. 6.

$$J_i = t_i \text{ (dB)}$$

c>Section with two or more nearby ridges (Ex. #2 section)
Assuming that each nearby ridge exists independently, each t_{ik} (k : 1, 2, ···) is obtained in the same manner as b>above.

$$J_i = t_{i1} + t_{i2} + \cdots\cdots \leqq 10 \text{ dB}$$

In case J_i exceeds 10 dB, the 10 dB is to be used.

(4) Total additional loss due to nearby ridges for A-B section, J_s

$$J_s = J_1 + J_2 + J_3 + \cdots\cdots \text{(dB)}$$

Estimation of Additional Loss Due to Reflected-Wave Interference J_R

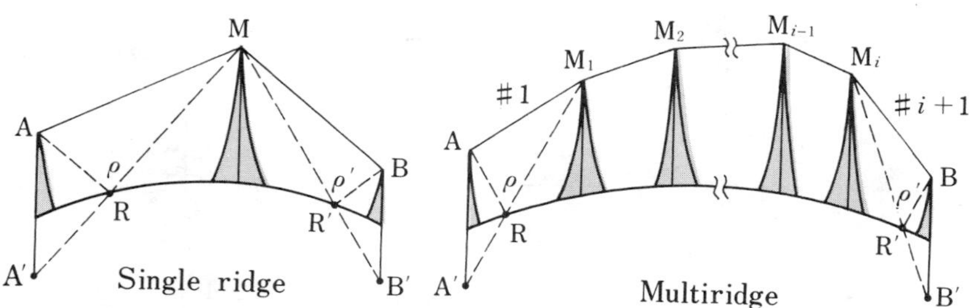

(1) Determination of the sections to be examined for reflected-wave interference.

a>Only the reflections occurring in both end sections are to be taken into account. Reflections caused in the intermediate sections are to be ignored.

b>Only the end section with an effective reflection coefficient (ρ, ρ') of 0.2 (=−14 dB) or more is to be examined.

(2) Paths to be examined (excluding those requiring no consideration)

Type of path	Single ridge	Multiridge
Main path	$\overline{AMB}$	$\overline{AM_1M_2\cdots M_{i-1}M_iB}$
Single reflection path A	$\overline{ARMB} \rightarrow \overline{A'MB}$	$\overline{ARM_1\cdots M_iB} \rightarrow \overline{A'M_1\cdots M_iB}$
Single reflection path B	$\overline{AMR'B} \rightarrow \overline{AMB'}$	$\overline{AM_1\cdots M_iR'B} \rightarrow \overline{AM_1\cdots M_iB'}$
Double reflection path C	$\overline{ARMR'B} \rightarrow \overline{A'MB'}$	$\overline{ARM_1\cdots M_iR'B} \rightarrow \overline{A'M_1\cdots M_iB}$

(3) Steps to estimate additional loss J_R

a>Relative levels (in the field strength) obtained for the paths A, B and C to that of the main path are denoted by X_1, X_2 and X_3 respectively.

b>X_1, X_2 and X_3, are estimated by Figs. 5 c and 5 d.

c>The additional loss generated when these three types of interfering reflected waves are combined in phase with each other but out of phase with the field strength obtained for the main path (the worst condition) is denoted by J_R (dB).

d>Total relative level of interference waves

$X_t = X_1 + X_2 + X_3$

Note : $X_t = 1$ is assumed if $X_1 + X_2 + X_3 > 1$

Additional loss $J_R = -20 \log (1 - X_t)$ (dB) (Fig. 6)

[Remark]

If $X_1 + X_2 + X_3 \fallingdotseq 1$ or $X_1 + X_2 + X_3 > 1$, the wavelengths used are considerably long and calculation of the relative phases against the main path is possible. Thus it is possible to estimate $\dot{X}_1$, $\dot{X}_2$ and $\dot{X}_3$ in order to calculate $\dot{X}_t = \dot{X}_1 + \dot{X}_2 + \dot{X}_3$, and to find the total level ratio using Fig. 15 of Chapter 2 by putting $|\dot{X}_t| = \rho_e$ and the phase of $\dot{X}_t = \theta$. In this case, it should be noted that the additional loss thus obtained differs from the J_R (corresponding to the worst condition).

3

5c

Reflected-Wave Interference over Single-Ridge Propagation Path

[1] Profile

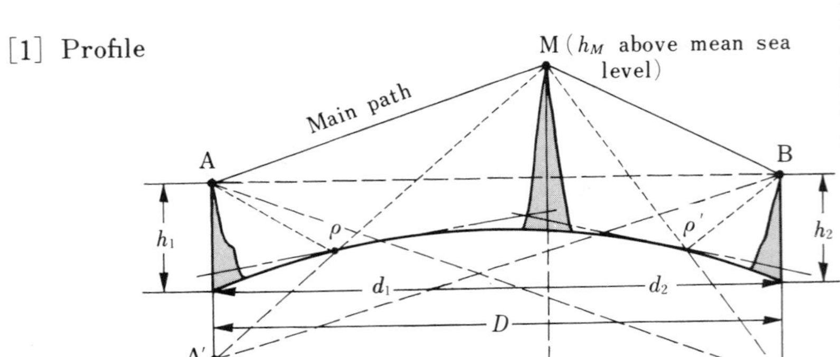

- The first Fresnel zone at M
  $r_S = \sqrt{\lambda d_1 d_2 / D}$
- Diffraction parameter (main path)
  $U_a = a / r_S$
  ↓
- Diffraction coefficient
  Z_a (Fig. 8)

[2] Analysis

(a) Single reflection path A

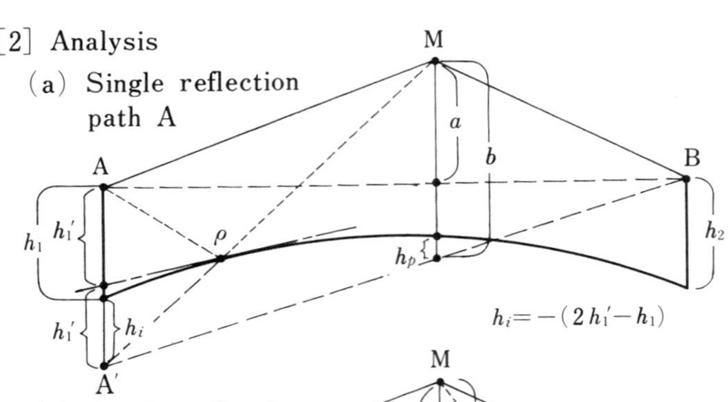

- Diffraction parameter (path A′MB)
  $U_b = b / r_S$
  ↓
- Diffraction coefficient
  Z_b (Fig. 8)
- Relative level of reflected wave
  $X_1 - \rho Z_b / Z_a$

(b) Single reflection path B

M
A
B
a
b'
h_1
h_p
ρ'
h_2'
h_2
h_i'
h_2'
B′

- Diffraction parameter (path AMB′)
  $U_b' = b' / r_S$
  ↓
- Diffraction coefficient
  Z_b' (Fig. 8)
- Relative level of reflected wave
  $X_2 = \rho' Z_b' / Z_a$

(c) Double reflection path C

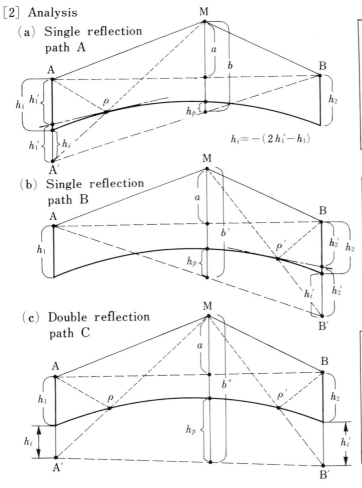

M
A
B
a
b''
h_1
ρ
ρ'
h_2
h_p
h_i
h_i'
A′
B′

- Diffraction parameter (path A′MB′)
  $U_b'' = b'' / r_S$
  ↓
- Diffraction coefficient
  Z_b'' (Fig. 8)
- Relative level of reflected wave
  $X_3 = \rho\rho' Z_b'' / Z_a$

Reflected-Wave Interference over Multiridge Propagation Path

[1] Profile

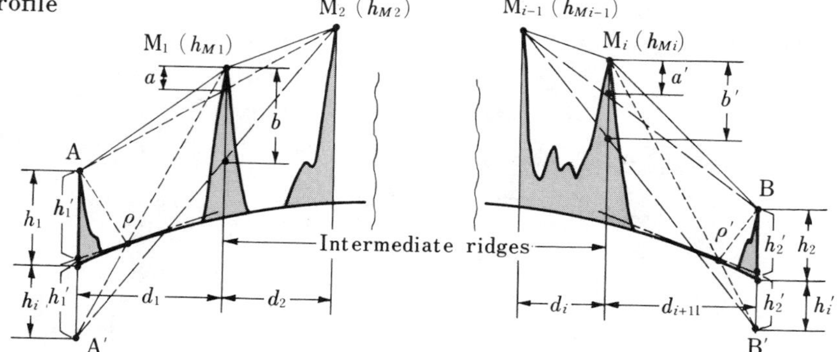

h_M	Ridge height
h_p	Imaginary path height
ρ	Effective reflection coefficient

[2] Calculation formula

- First Fresnel zone radius at a ridge point
  - Point M_1 $r_S=\sqrt{\lambda d_1 d_2/(d_1+d_2)}$
  - Point M_2 $r_S=\sqrt{\lambda d_i d_{i+1}/(d_i+d_{i+1})}$

  } according to Figs. 13 & 14
- Ridge depth

$$a=h_{M1}-h_{pa}=h_{M1}-\left[\frac{h_1 d_2+h_{M2} d_1}{d_1+d_2}-\frac{d_1 d_2}{2\,Ka}\right]$$

$$b=h_{M1}-h_{pb}=h_{M1}-\left[\frac{h_i d_2+h_{M2} d_1}{d_1+d_2}-\frac{d_1 d_2}{2\,Ka}\right]$$

$$a'=h_{Mi}-h_{pa}'=h_{Mi}-\left[\frac{h_2 d_i+h_{Mi-1} d_{i+1}}{d_i+d_{i+1}}-\frac{d_i d_{i+1}}{2\,Ka}\right]$$

$$b'=h_{Mi}-h_{pb}'=h_{Mi}-\left[\frac{h_i' d_i+h_{Mi-1} d_{i+1}}{d_i+d_{i+1}}-\frac{d_i d_{i+1}}{2\,Ka}\right]$$

} The reflection point is determined by Figs. 24 and 25 in Chapter 2 but estimation is readily carried out if a large scale profile is used.

- Estimation of h_i and h_i'
  - $h_i=-(2h_1'-h_1)\cdots h_1'$: Effective antenna height
  - $h_i'=-(2h_2'-h_2)\cdots h_2'$: Effective antenna height
- Diffraction parameter U, diffraction parameter Z (Field strength ratio)
  - $U_a=a/r_S \longrightarrow Z_a$
  - $U_b=b/r_S \longrightarrow Z_b$
  - $U_a'=a'/r_S' \longrightarrow Z_a'$
  - $U_b'=b'/r_S' \longrightarrow Z_b'$

} Fig. 1 and Fig. 23 in Chapter 2. $h'=h-d_r^2/(2\,Ka)$ ⋮ (Distance from A or B to the reflection point) Fig. 8 a or Fig. 8 b

[3] Relative level X of reflected-wave path for the main path

Single reflection A (Path $A'M_1M_2\cdots B$)	$X_1=\rho Z_b/Z_a$
Single reflection B (Path $A\cdots M_{i-1}M_iB'$)	$X_2=\rho' Z_b'/Z_a'$
Double reflection C (Path $A'M_1M_2\cdots M_{i-1}M_iB'$)	$X_3=X_1\cdot X_2$

Auxiliary Graph for Additional Loss J_s, J_R

[1] Auxiliary graph for additional loss caused by nearby ridge

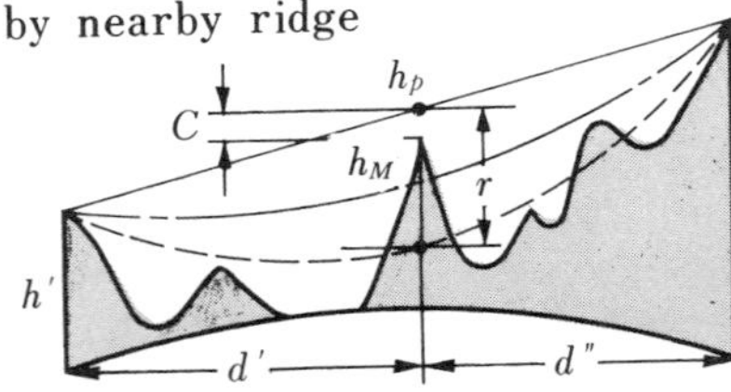

Height above sea level of nearby ridge

$C = h_p - h_M$

$h_p = \dfrac{(h'd'' + h''d')}{d' + d''} - \dfrac{d'd''}{2\,Ka}$

$r = \sqrt{\lambda d'd''/(d'+d'')}$

(Clearance parameter) $U_c = C/r$

0.6
0.5
0.4
0.3
0.2
0.1
0

6 5 4 3 2 1 0

t (dB)

(Additional loss caused by nearby ridge)

C : In practice it may be measured directly on the path profile

r : Fig. 12 or Fig. 13 & 14

$U_c = -U$

Diffraction parameter

Clearance parameter

[2] Auxiliary graph for additional loss caused by reflected-wave interference

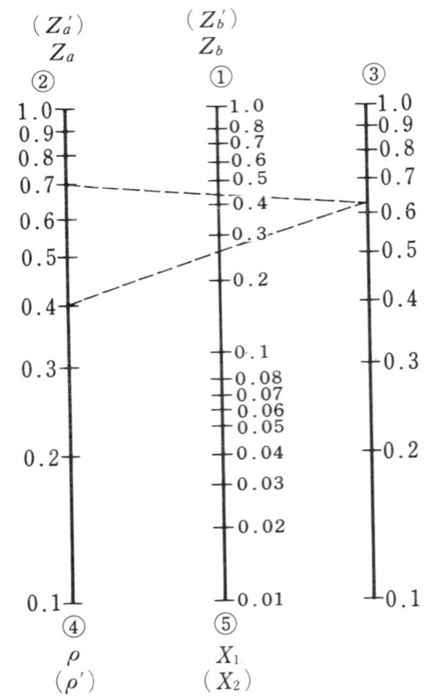

(a) X for a single ridge

$(Z_b/Z_a) \cdot \rho = X_1$	0.
$(Z_b'/Z_a) \cdot \rho' = X_2$	0.
①→②→③→④→⑤	
$(Z_b''/Z_a) \cdot \rho \cdot \rho' = X_3$	0.

(b) X for multiridge

$(Z_b/Z_a) \cdot \rho = X_1$	0.
$(Z_b'/Z_a') \cdot \rho' = X_2$	0.
①→②→③→④→⑤	
$X_1 \cdot X_2 = X_3$	0.

(c) Calculation of J_R

$(X_1 + X_2 + X_3) = X_t$	.
if $X_t > 1$ $X_t = 1$	
$J_R = -20 \log (1 - X_t)$	dB

Nomograph for Single-Ridge (Treated as Knife-Edge) Diffraction

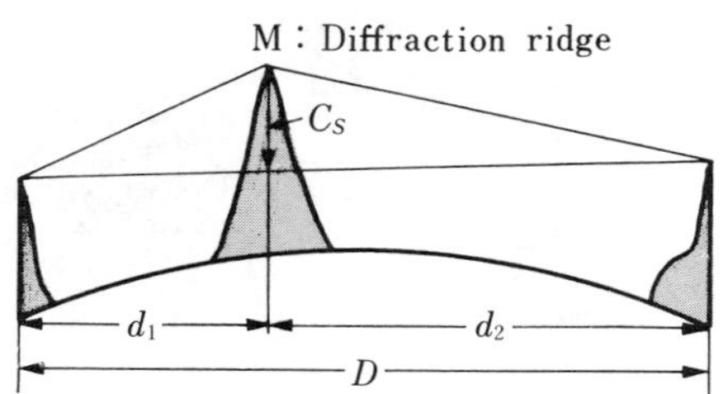

r_S : First Fresnel zone radius (at ridge point)
$r_S = \sqrt{\lambda \cdot d_1 d_2 / D}$
C_s : Depth of ridge

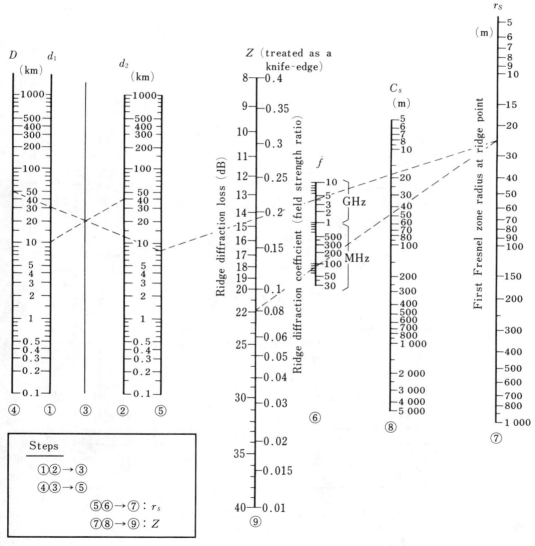

① d_1 (km): 1000, 500, 400, 300, 200, 100, 50, 40, 30, 20, 10, 5, 4, 3, 2, 1, 0.5, 0.4, 0.3, 0.2, 0.1

② d_2 (km): 1000, 500, 400, 300, 200, 100, 50, 40, 30, 20, 10, 5, 4, 3, 2, 1, 0.5, 0.4, 0.3, 0.2, 0.1

③

④ D (km): 1000, 500, 400, 300, 200, 100, 50, 40, 30, 20, 10, 5, 4, 3, 2, 1, 0.5, 0.4, 0.3, 0.2, 0.1

⑤

⑥ f: 10, 5, 3, 2, 1 GHz; 500, 300, 200, 100, 50, 30 MHz

⑦ r_S (m) First Fresnel zone radius at ridge point: 5, 6, 7, 8, 9, 10, 15, 20, 30, 40, 50, 60, 70, 80, 90, 100, 150, 200, 300, 400, 500, 600, 700, 800, 1 000

⑧ C_s (m): 5, 6, 7, 8, 10, 20, 30, 40, 50, 60, 70, 80, 100, 200, 300, 400, 500, 600, 700, 800, 1 000, 2 000, 3 000, 4 000, 5 000

⑨ Z (treated as a knife-edge)
Ridge diffraction loss (dB): 8, 9, 10, 11, 12, 13, 14, 15, 16, 17, 18, 19, 20, 22, 25, 30, 35, 40
Ridge diffraction coefficient (field strength ratio): 0.4, 0.35, 0.3, 0.25, 0.2, 0.15, 0.1, 0.08, 0.06, 0.05, 0.04, 0.03, 0.02, 0.015, 0.01

Steps

①②→③
④③→⑤
⑤⑥→⑦ : r_s
⑦⑧→⑨ : Z

Knife-Edge Diffraction Loss (Close to Line of Sight)

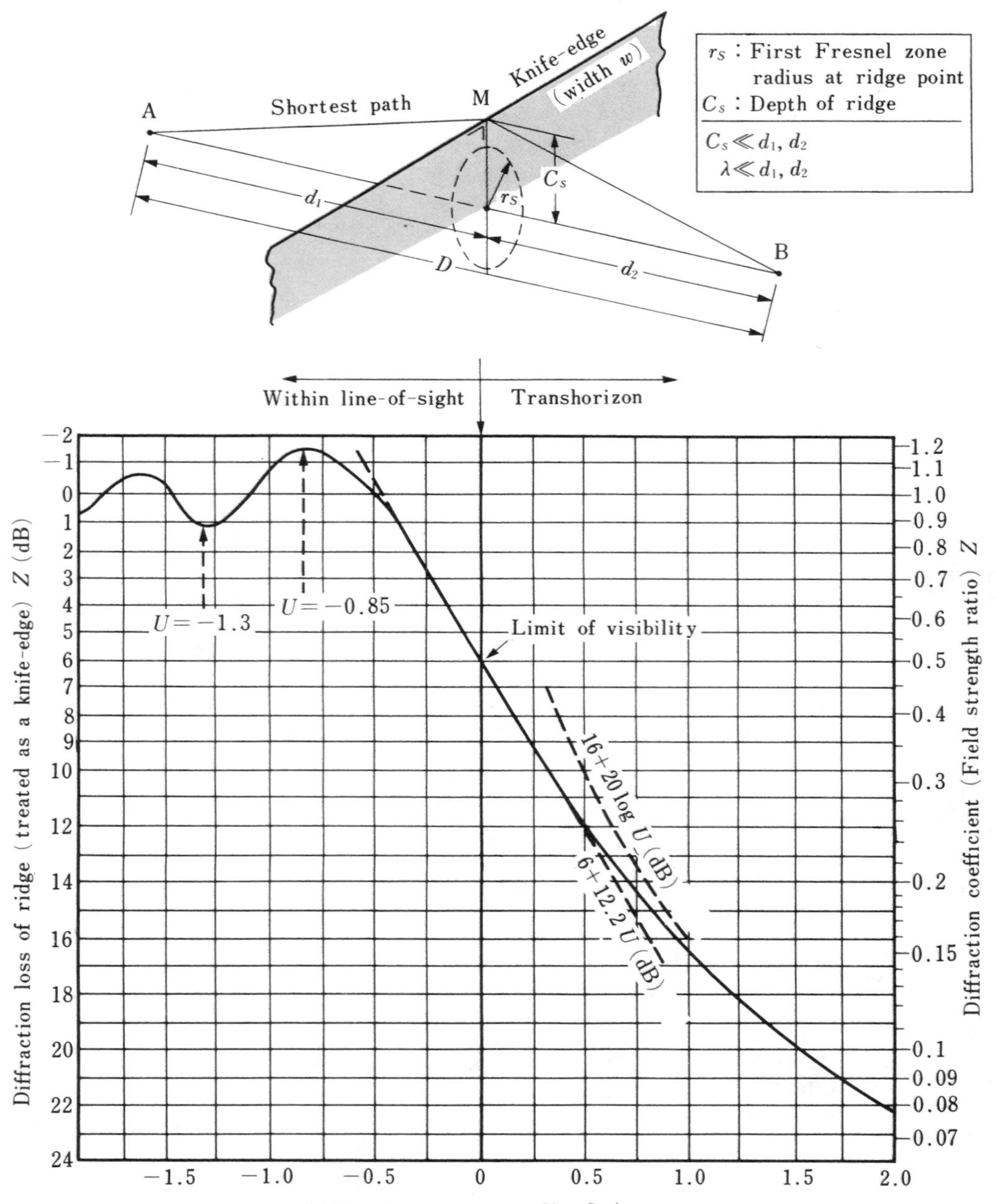

Knife-Edge Diffraction Loss (Deep Diffraction Region)

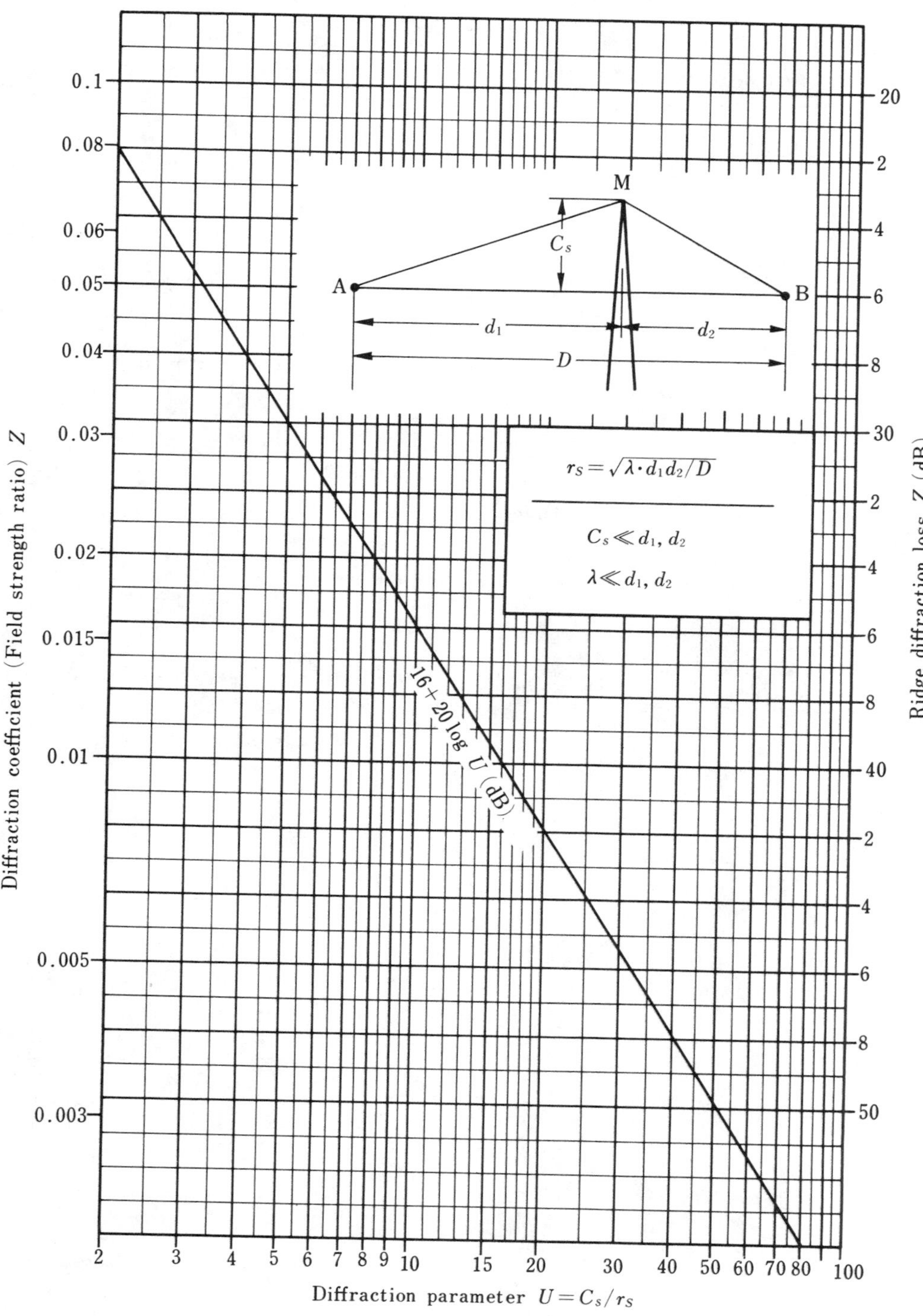

3-9 nth Fresnel Zone Radius and Conversion Factor k_n

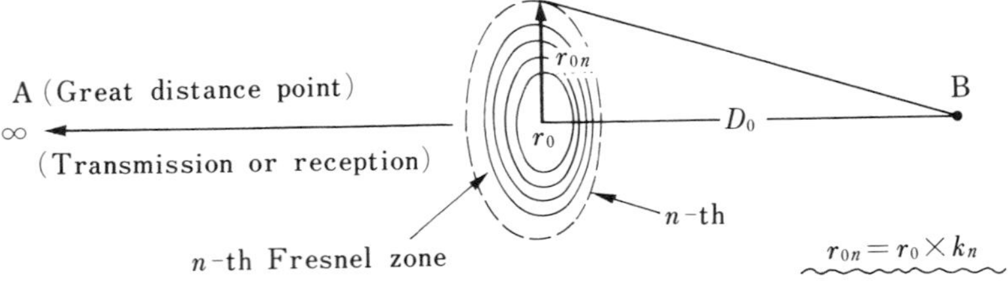

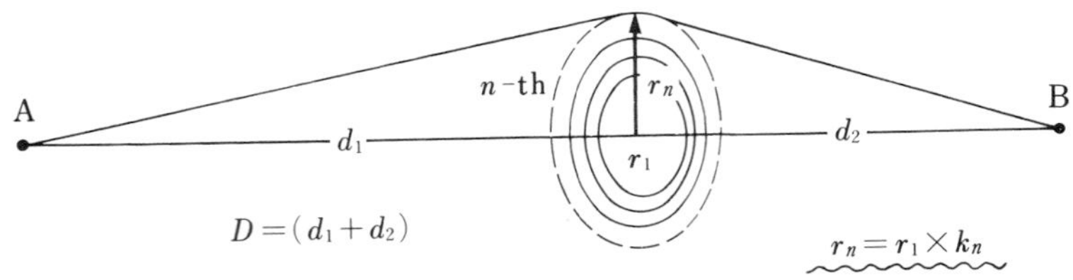

Table of conversion factor k_n

n	0	1	2	3	4	5	6	7	8	9
0	—	1.00	1.41	1.73	2.00	2.24	2.45	2.65	2.83	3.00
10	3.16	3.32	3.46	3.61	3.74	3.87	4.00	4.12	4.24	4.36
20	4.47	4.58	4.69	4.80	4.90	5.00	5.10	5.20	5.29	5.39
30	5.48	5.57	5.66	5.74	5.83	5.92	6.00	6.08	6.16	6.24
40	6.32	6.40	6.48	6.56	6.63	6.71	6.78	6.86	6.93	7.00
50	7.07	7.14	7.21	7.28	7.35	7.42	7.48	7.55	7.62	7.68
60	7.75	7.81	7.87	7.94	8.00	8.06	8.12	8.19	8.25	8.31
70	8.37	8.43	8.49	8.54	8.60	8.66	8.72	8.78	8.83	8.89
80	8.94	9.00	9.06	9.11	9.17	9.22	9.27	9.33	9.38	9.43
90	9.49	9.54	9.59	9.64	9.70	9.75	9.80	9.85	9.90	9.95

First Fresnel Zone Radius (for Great Distance Point)

Nomograph

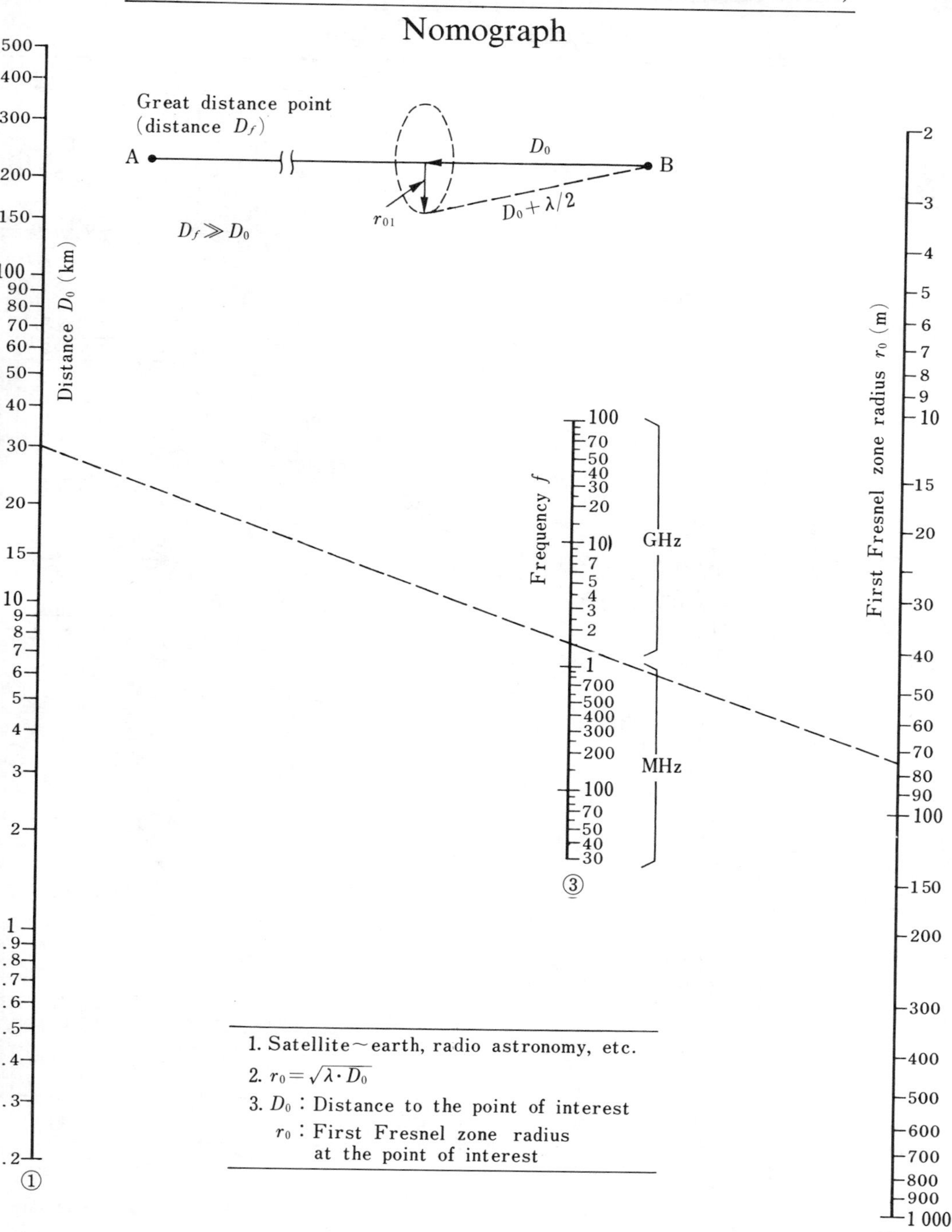

1. Satellite～earth, radio astronomy, etc.
2. $r_0=\sqrt{\lambda \cdot D_0}$
3. D_0 : Distance to the point of interest
   r_0 : First Fresnel zone radius at the point of interest

First Fresnel Zone Radius (for Great Distance Point) (1)

Direct Reading Graph

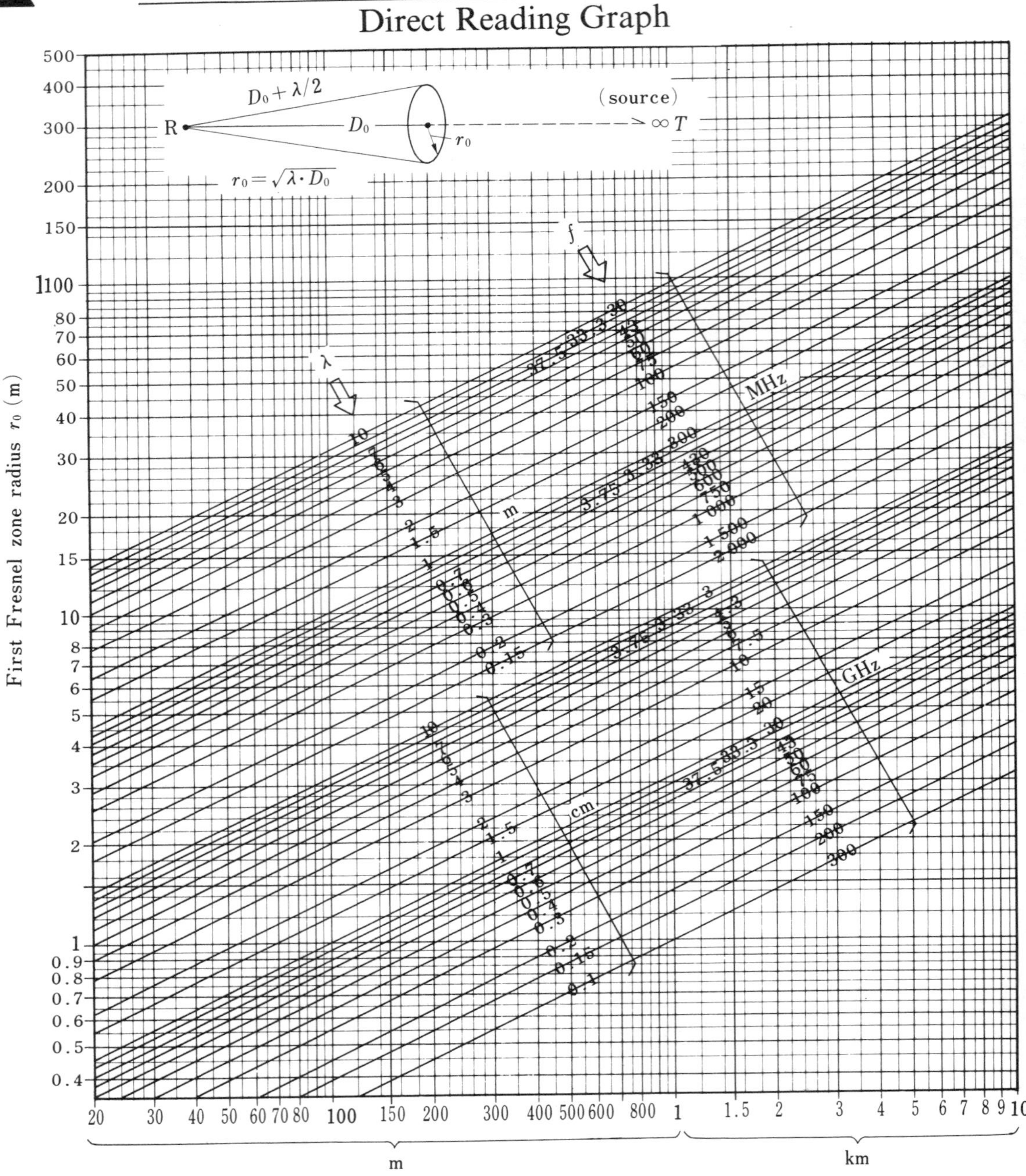

First Fresnel Zone Radius (for Great Distance Point) (2)

Direct Reading Graph

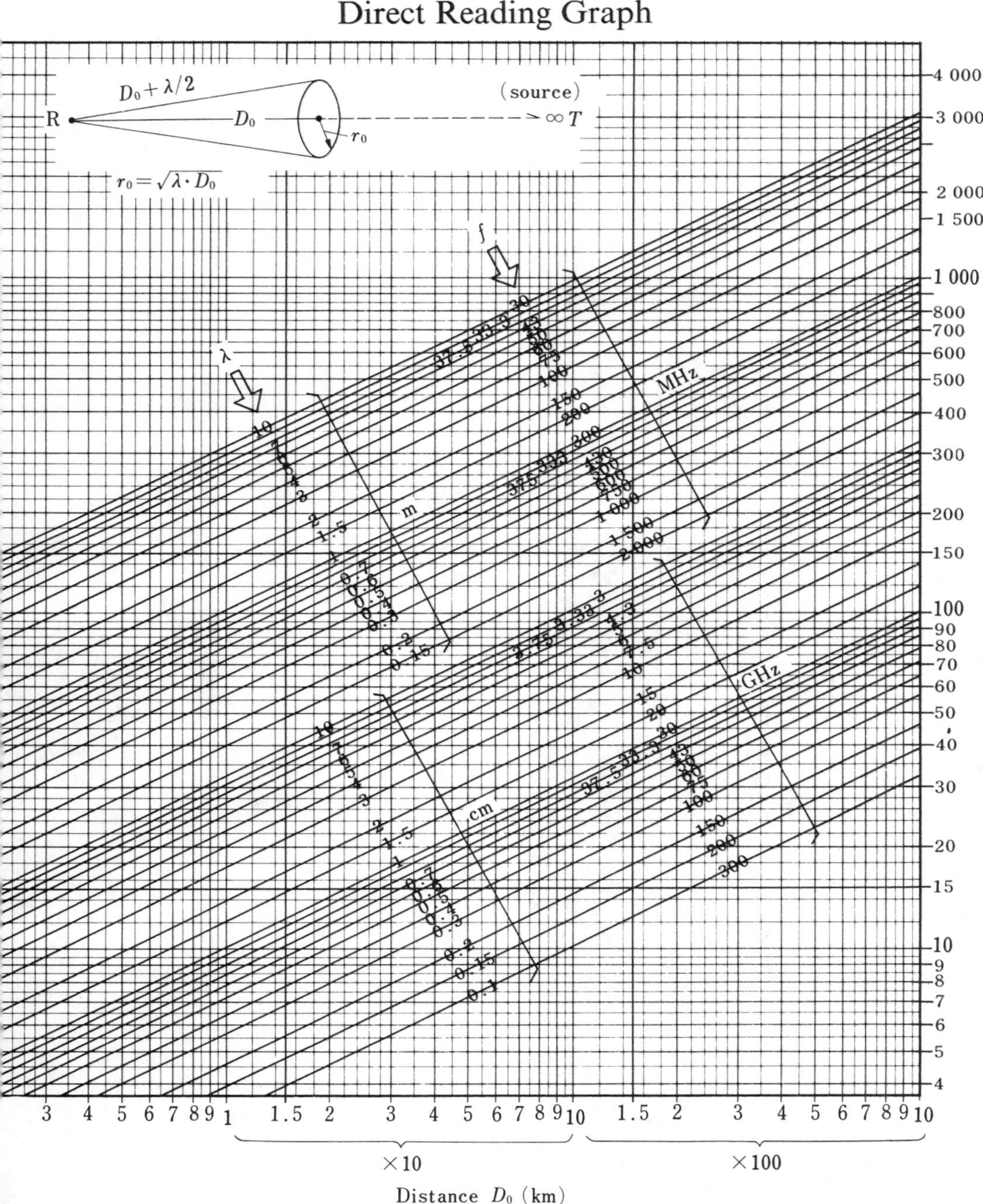

First Fresnel Zone Radius (for General Use)

Nomograph

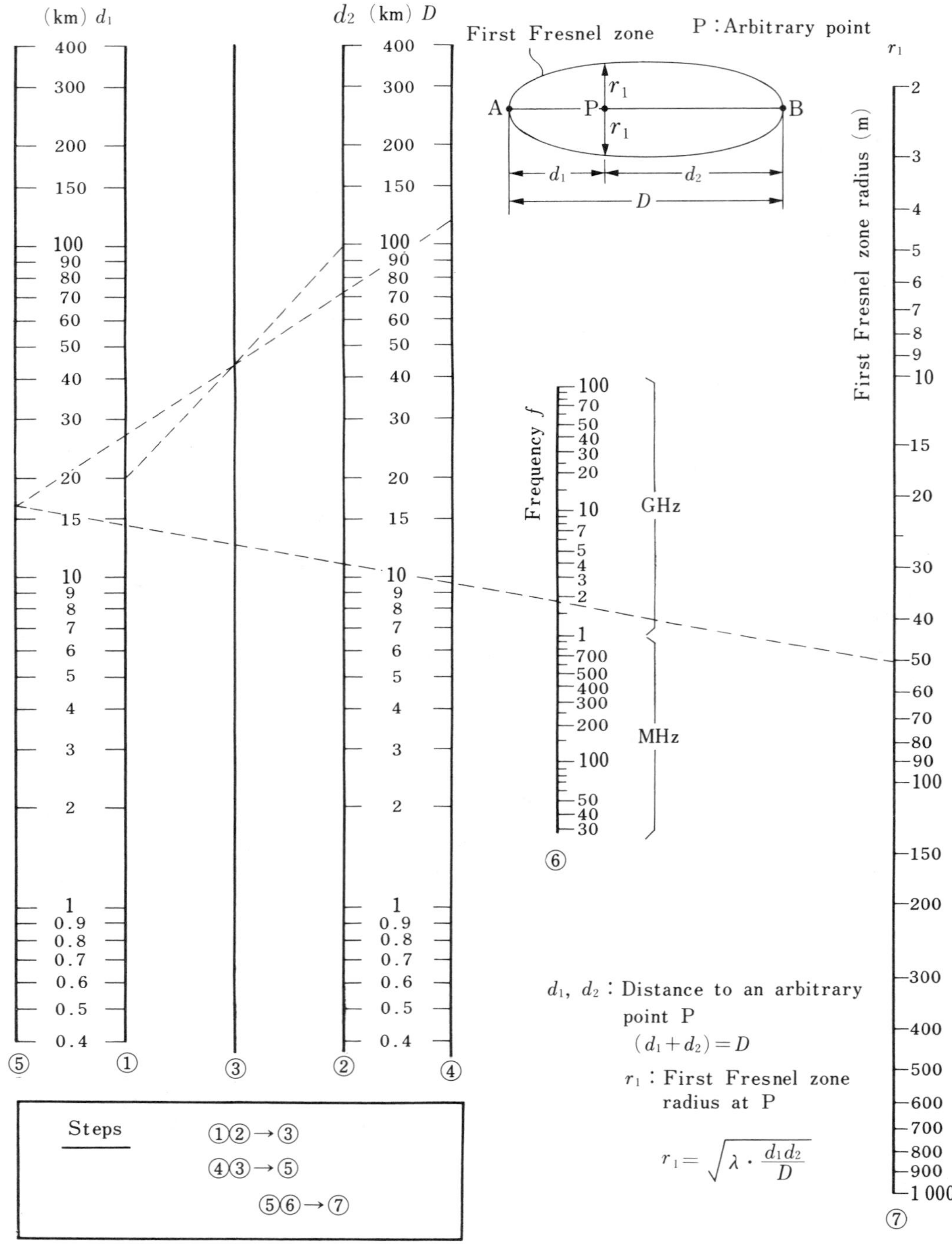

First Fresnel Zone Radius (1)

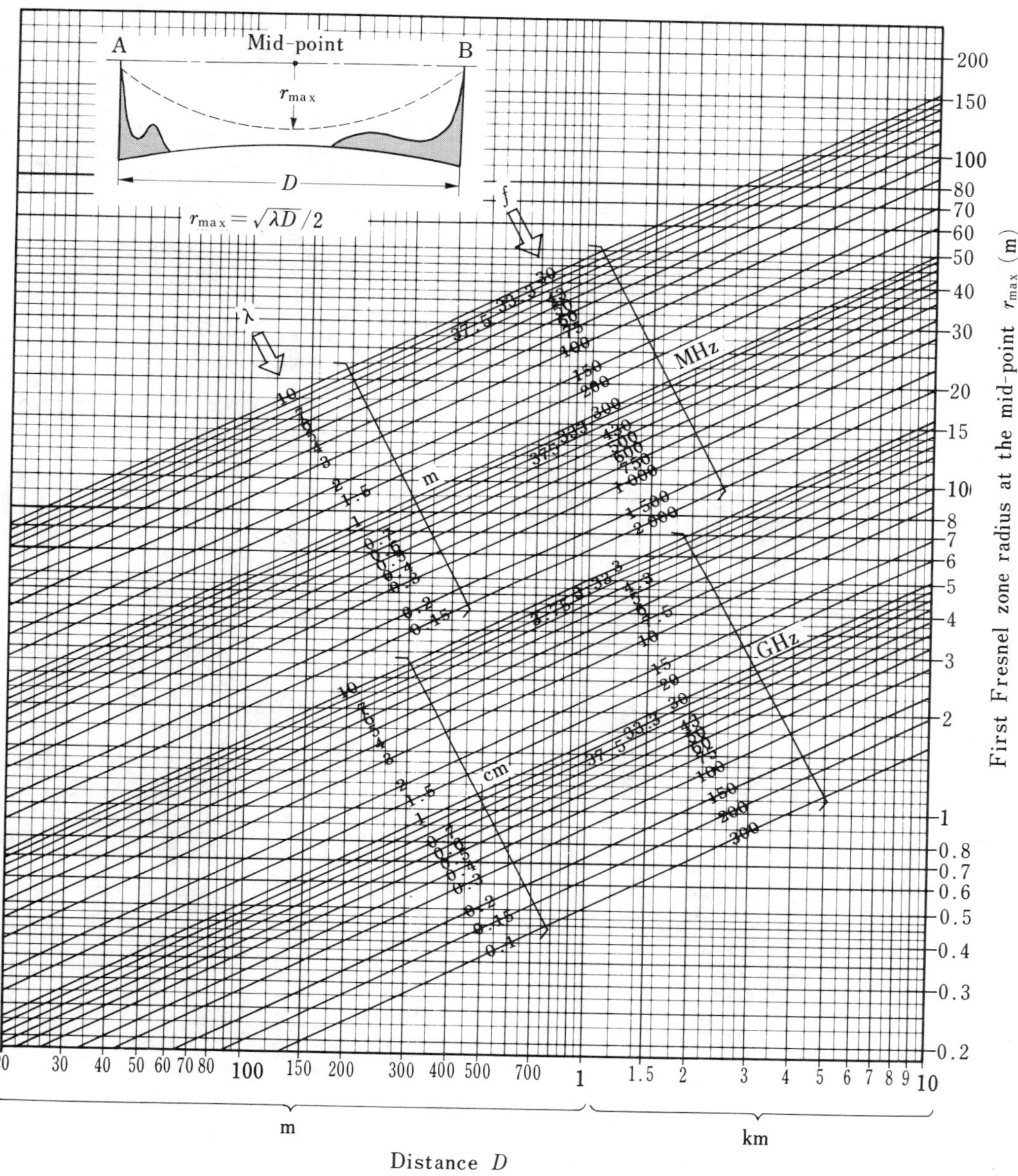

First Fresnel Zone Radius (2)

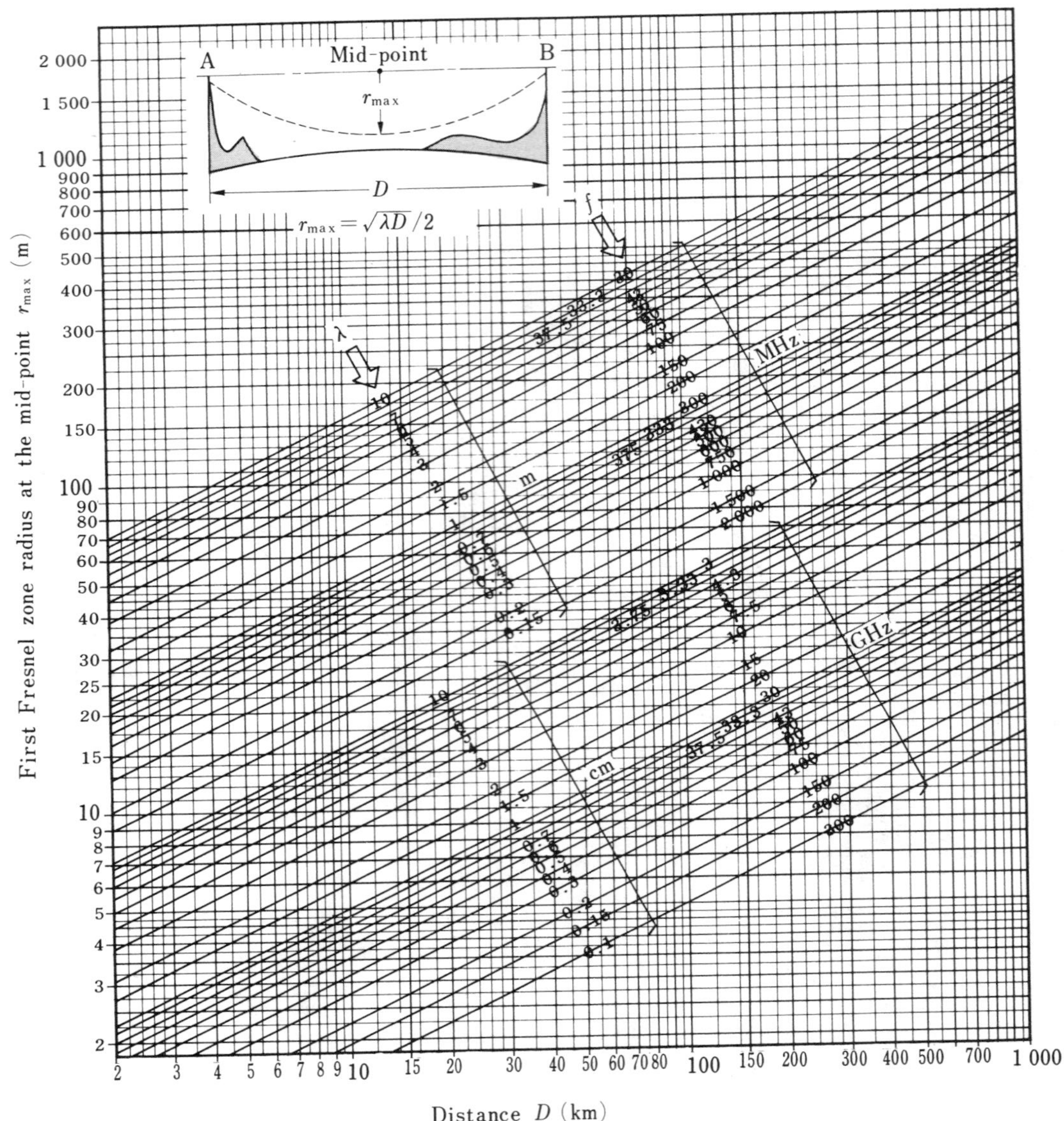

First Fresnel Zone Radius (3)

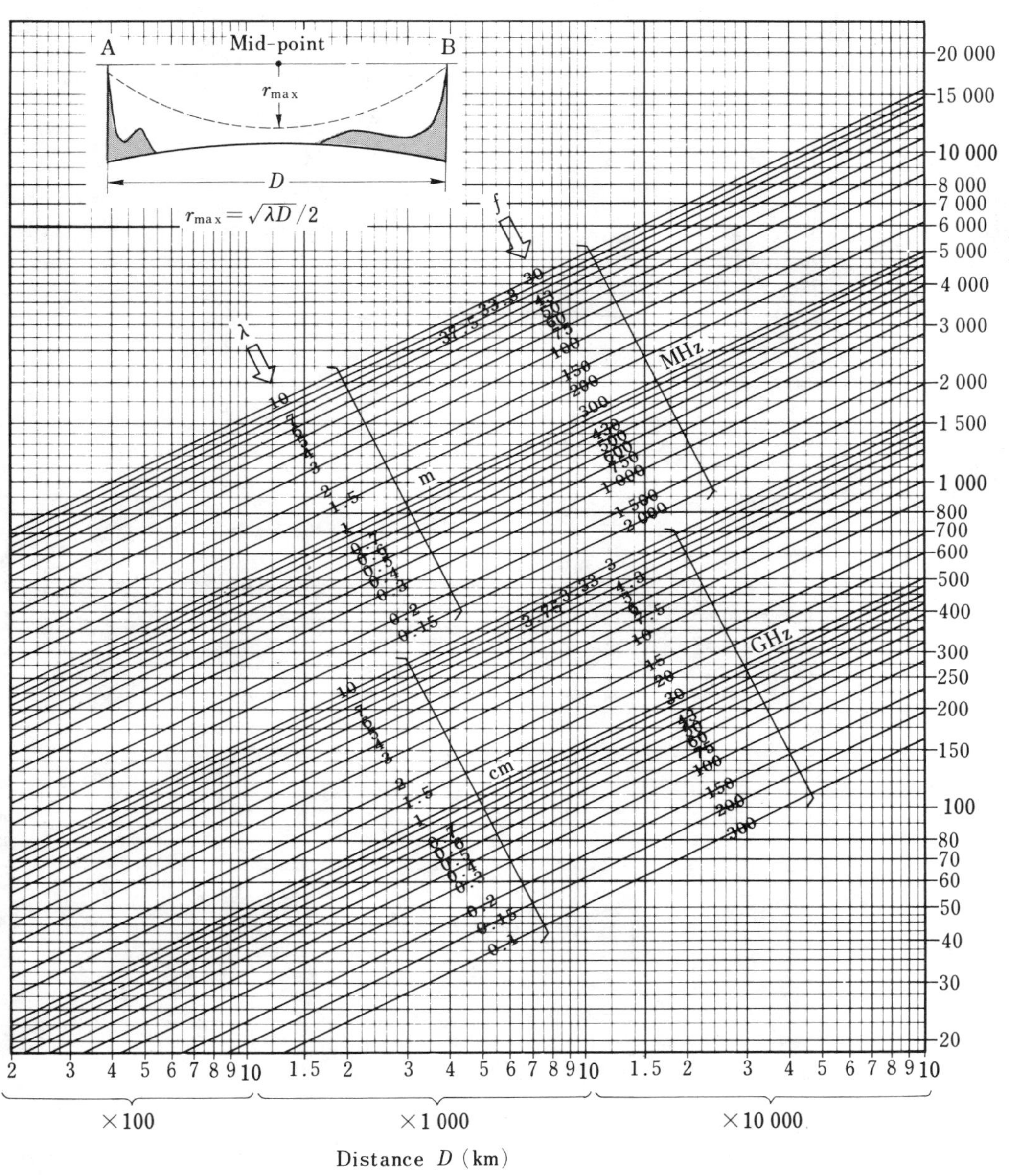

First Fresnel Zone Radius (1)

Graph of p

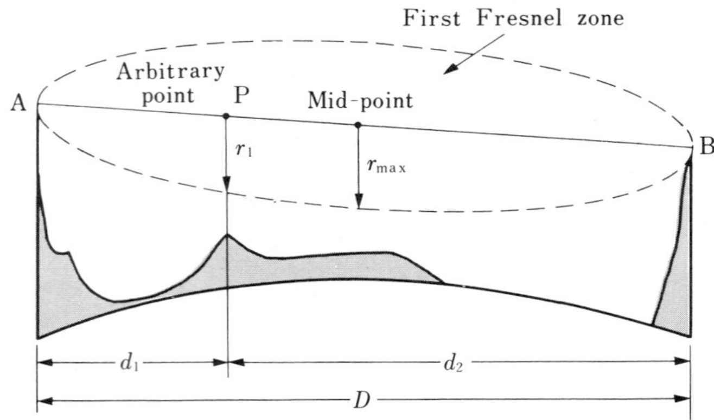

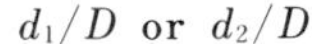

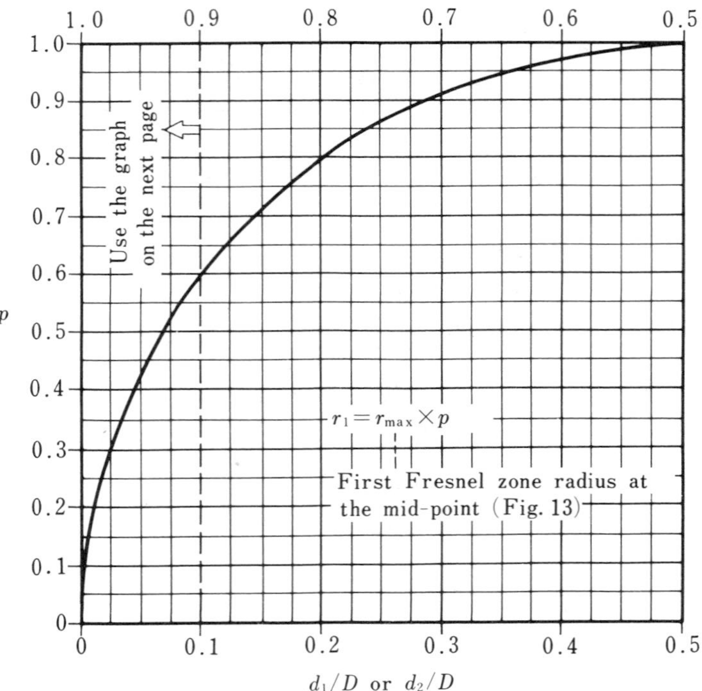

First Fresnel Zone Radius (2)

Graph of p

For nearby point

$$\left[\begin{array}{l} d_1/D \text{ or } d_2/D \geqq 0.9 \\ d_1/D \text{ or } d_2/D \leqq 0.1 \end{array}\right]$$

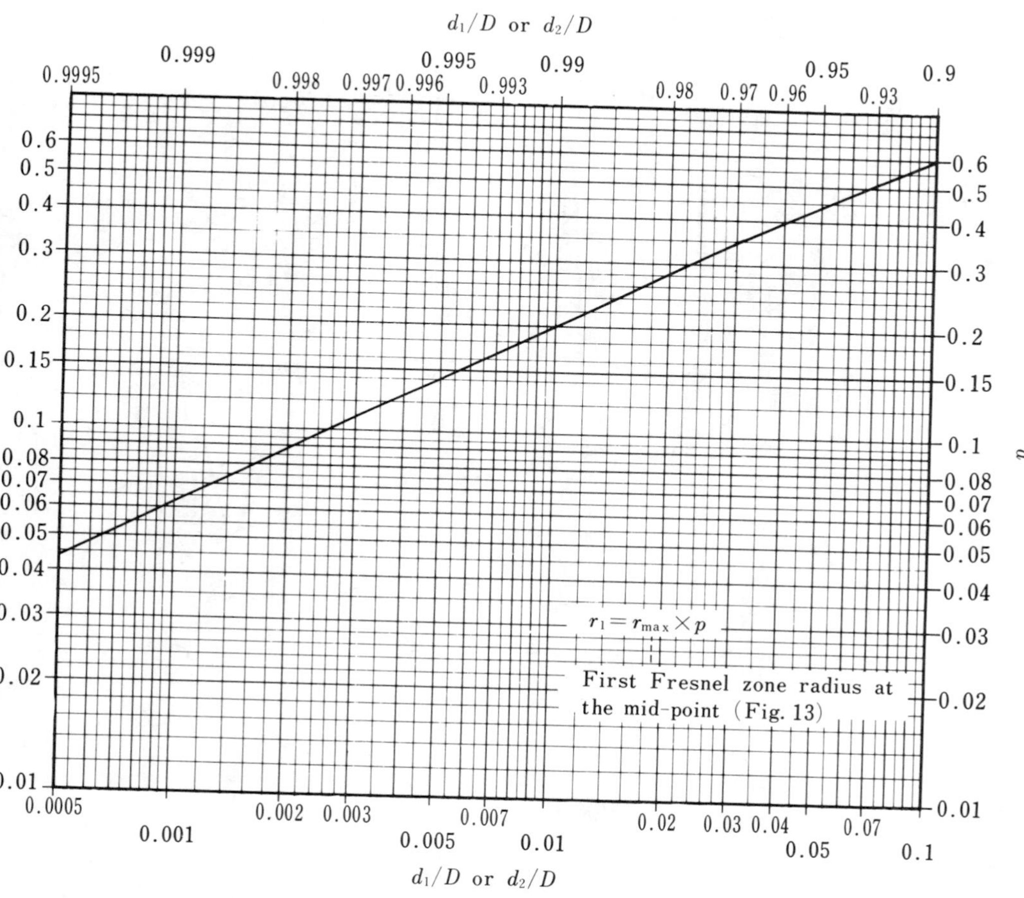

Calculation Chart for Multiridge ($i \leq 5$) Path (1)

Parameters

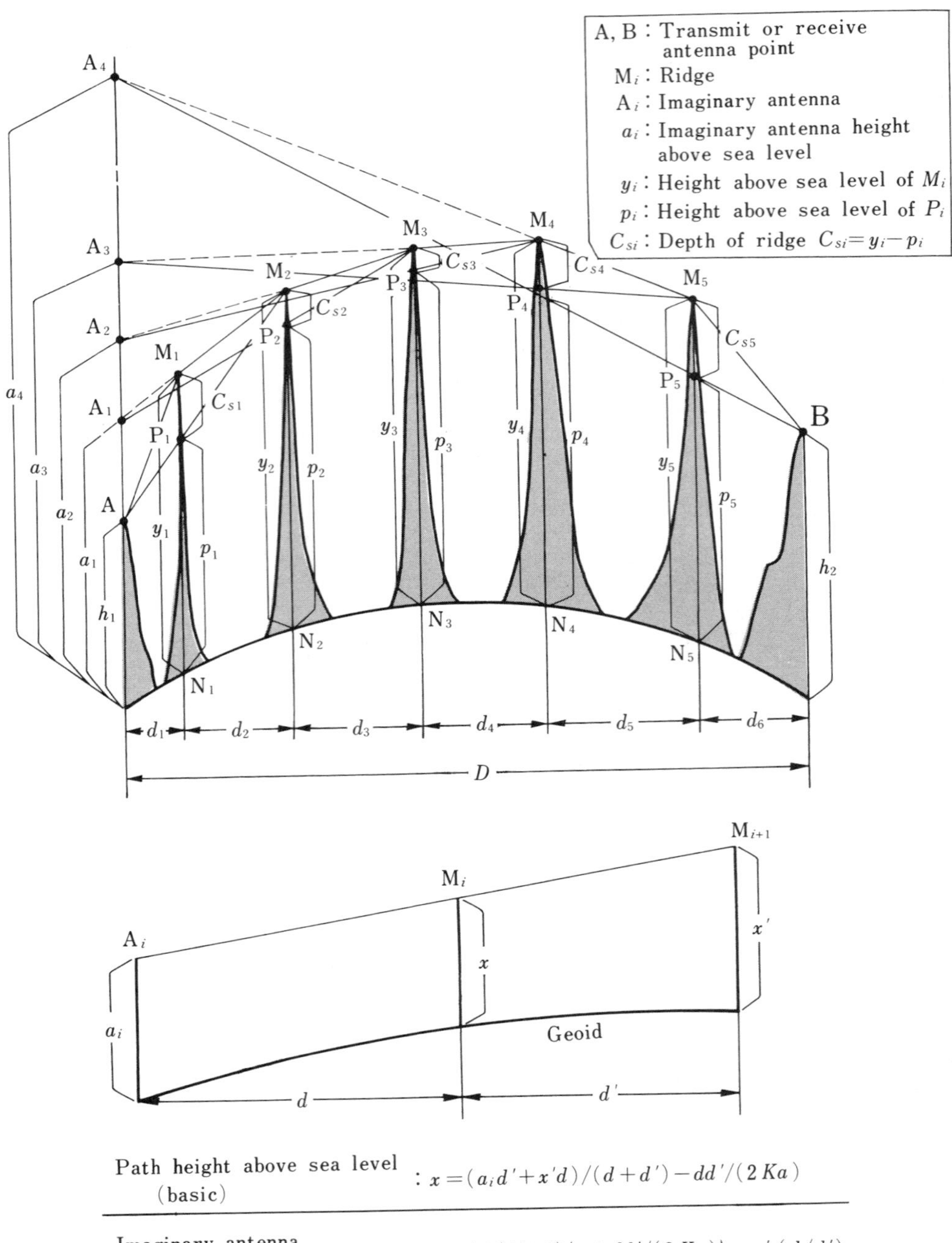

Path height above sea level (basic)	: $x=(a_i d'+x'd)/(d+d')-dd'/(2Ka)$
Imaginary antenna height above sea level	: $a_i=(d/d'+1)\{x+dd'/(2Ka)\}-x'(d/d')$

Calculation Chart for Multiridge ($i \leq 5$) Path (2)

Calculation Form

[Steps]

Since it is time consuming to find the path height and/or antenna height, it is more practical to find them graphically on the path profile of a considerably large scale.

Depth of ridge C_{si} and distance d_i are required to determine the heights. Then, in the order of first Fresnel zone radius r_{Si} for individual ridge → diffraction parameter U_i → diffraction loss Z, the calculations are carried out.

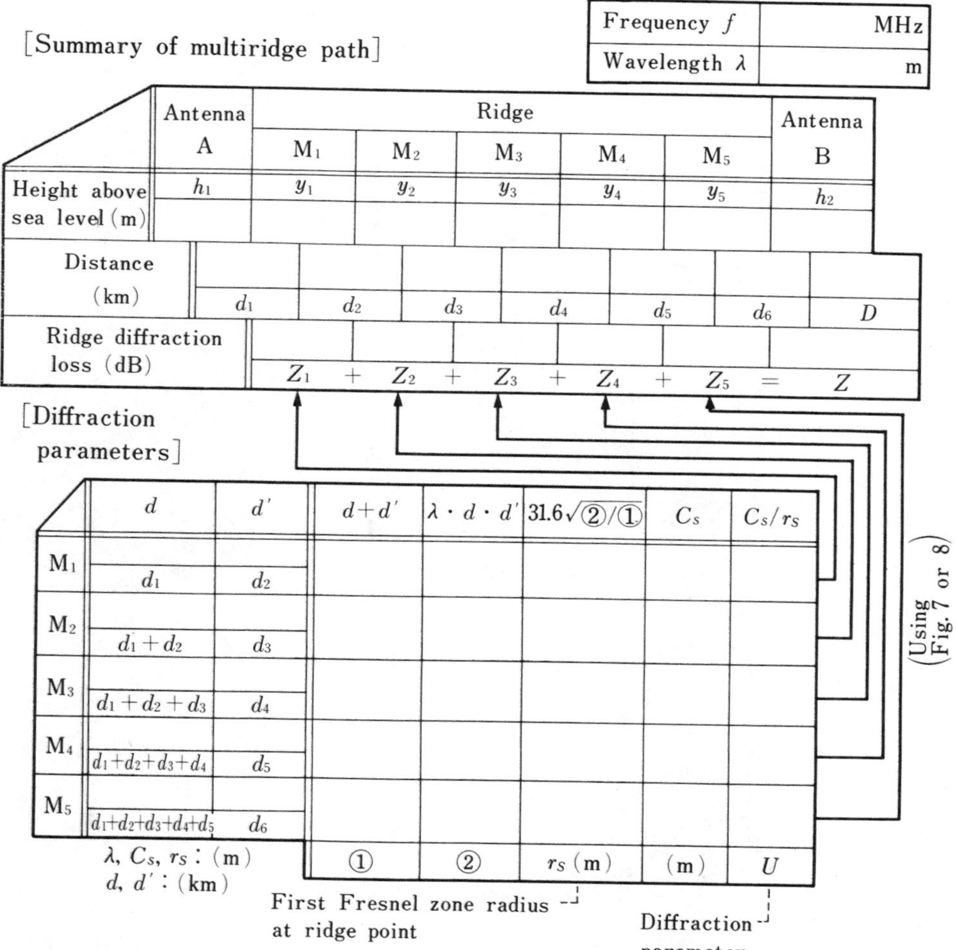

[Summary of multiridge path]

Frequency f	MHz
Wavelength λ	m

	Antenna A	Ridge M_1	Ridge M_2	Ridge M_3	Ridge M_4	Ridge M_5	Antenna B
Height above sea level (m)	h_1	y_1	y_2	y_3	y_4	y_5	h_2
Distance (km)	d_1	d_2	d_3	d_4	d_5	d_6	D
Ridge diffraction loss (dB)		Z_1 +	Z_2 +	Z_3 +	Z_4 +	Z_5 =	Z

[Diffraction parameters]

	d	d'	$d+d'$	$\lambda \cdot d \cdot d'$	$31.6\sqrt{②/①}$	C_s	C_s/r_s
M_1	d_1	d_2					
M_2	d_1+d_2	d_3					
M_3	$d_1+d_2+d_3$	d_4					
M_4	$d_1+d_2+d_3+d_4$	d_5					
M_5	$d_1+d_2+d_3+d_4+d_5$	d_6					
			①	②	r_s (m)	(m)	U

1) When the number of ridges is 4 or less, inapplicable M's must be deleted.
2) A ridge depth is calculated using the equations given at the bottom of the previous page (an additional calculation chart is necessary).
3) Total additional loss for the section between A and B, Z_t is given by

$$Z_t = Z + (J_s + J_R) \text{ (dB)}$$

where J_s and J_R are additional terms representing effects caused by the presence of a nearby ridge and reflected wave between ridges, respectively (refer also to Fig. 5).

Smooth-Spherical-Ground Diffraction Loss (1)

H · P Land and Sea, V · P Land

$$A_P = F(D) - \{G(h_1) + G(h_2)\} \quad (\text{dB})$$ ------ Additional loss for spherical earth

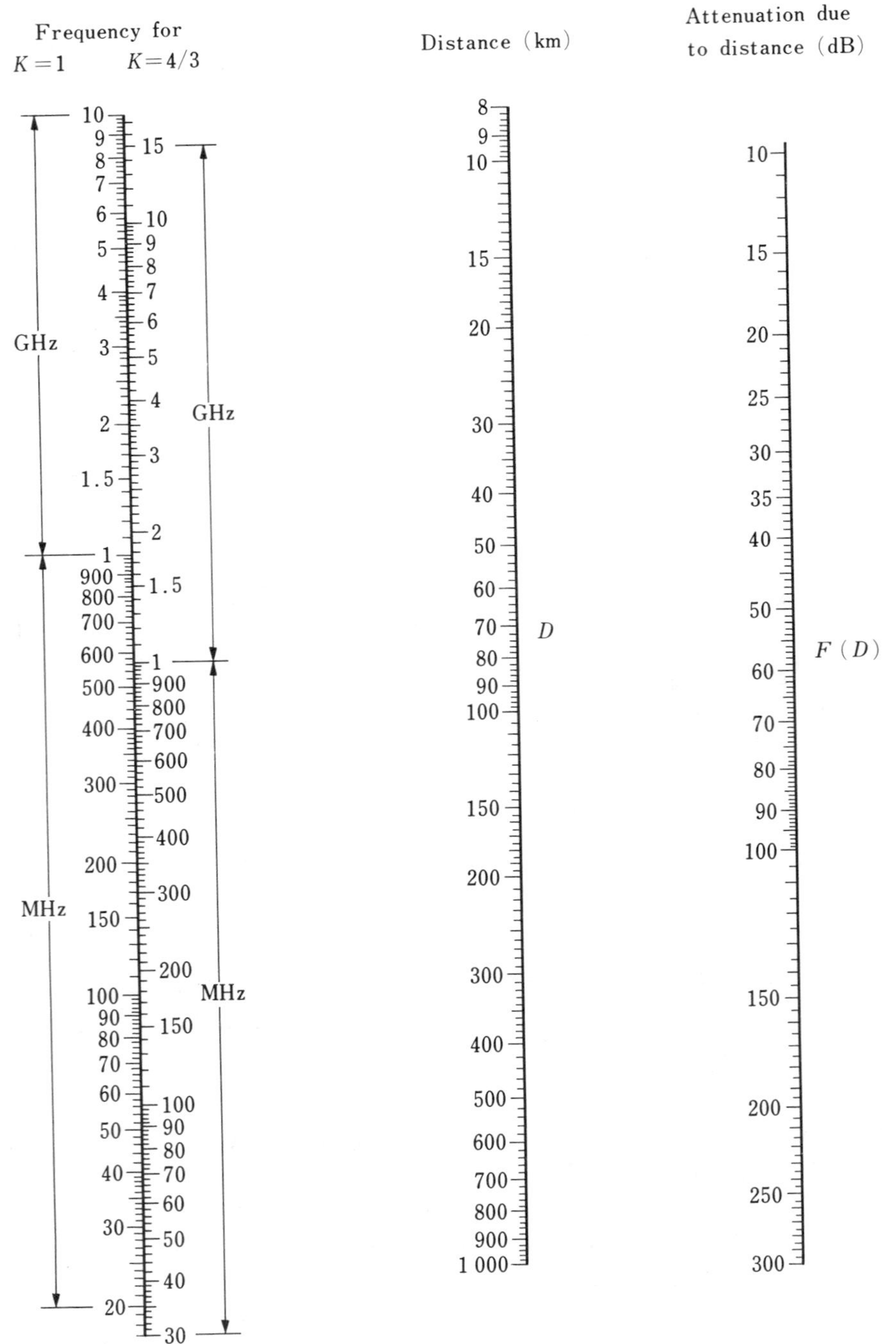

Smooth-Spherical-Ground Diffraction Loss (2)

H · P Land and Sea, V · P Land

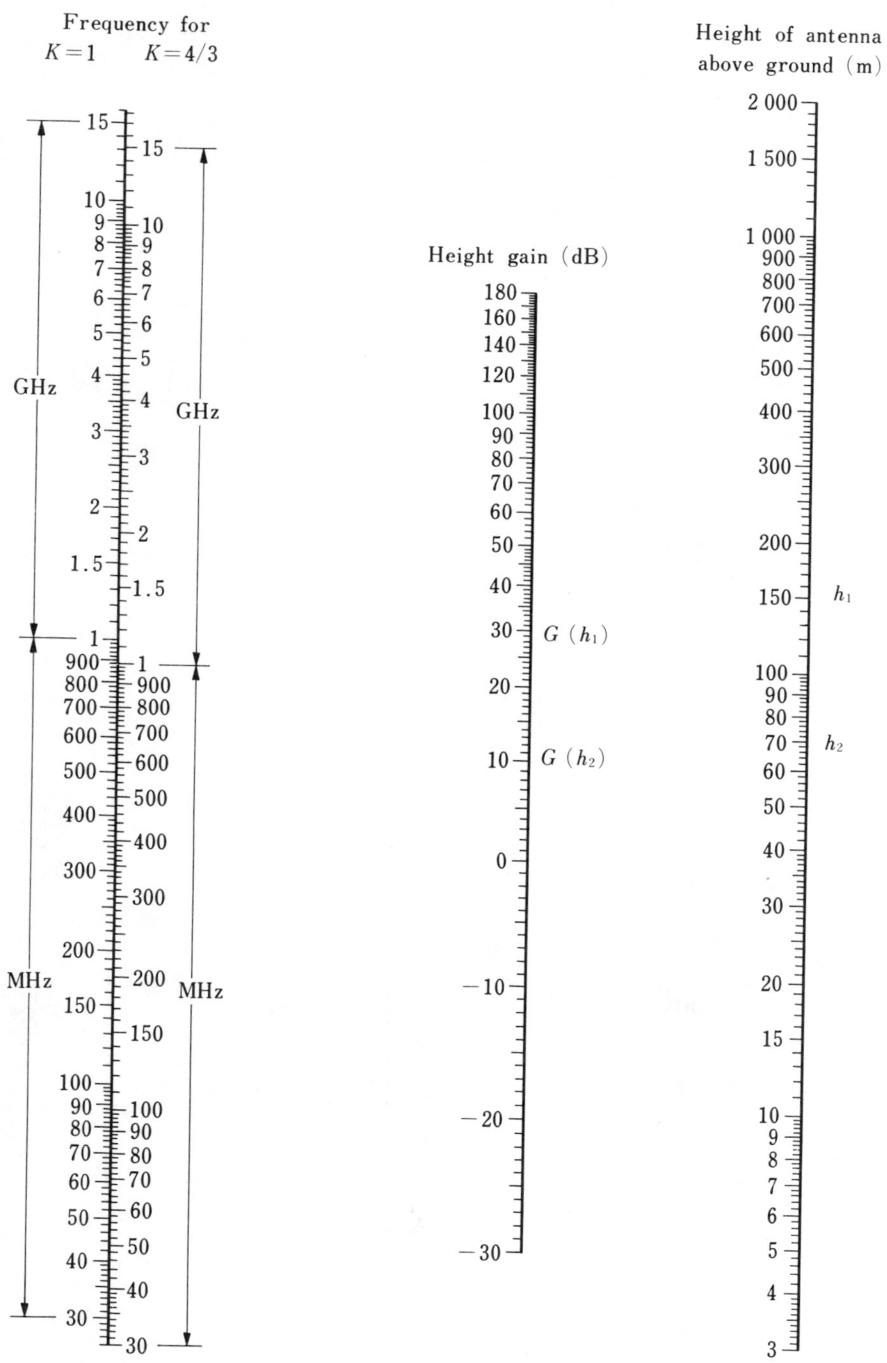

Smooth-Spherical-Ground Diffraction Loss (3)

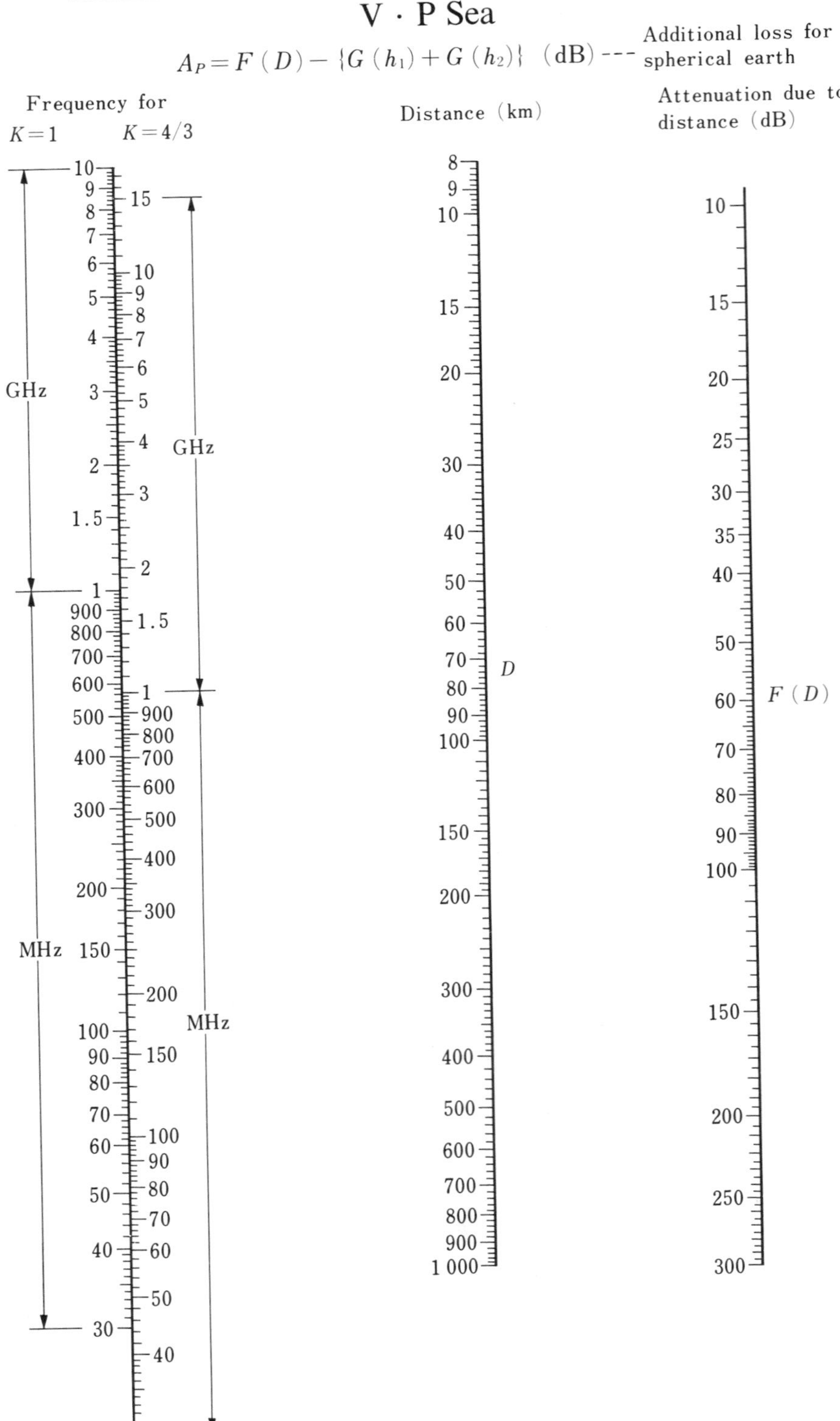

Smooth-Spherical-Ground Diffraction Loss (4)

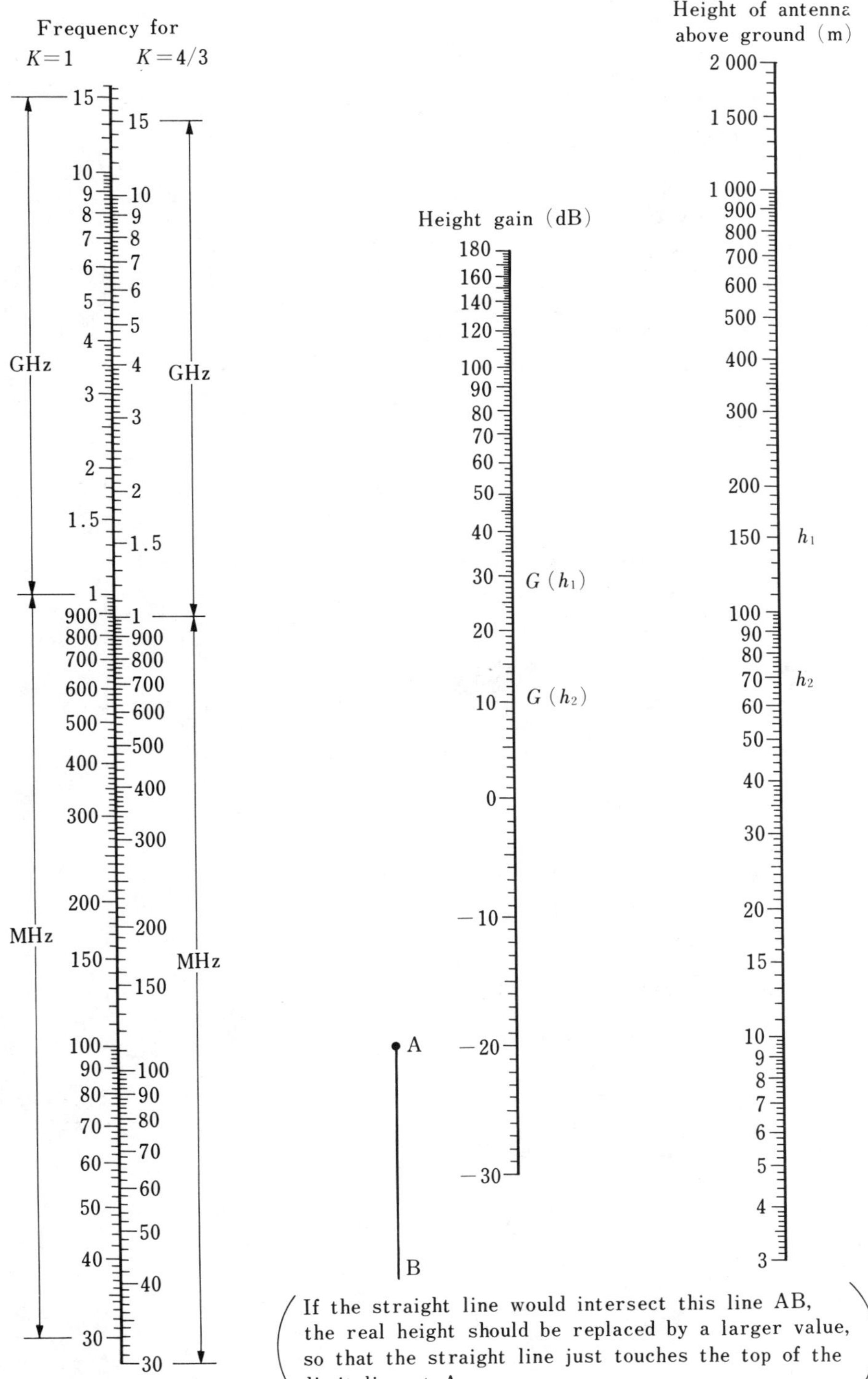

Spherical-Ground Diffraction Loss (Alternative Solution)

$A_P = G(\chi_0) - \{F(\chi_1) + F(\chi_2)\} - 20.5$ (dB)

└-- Additional loss caused by the spherical earth

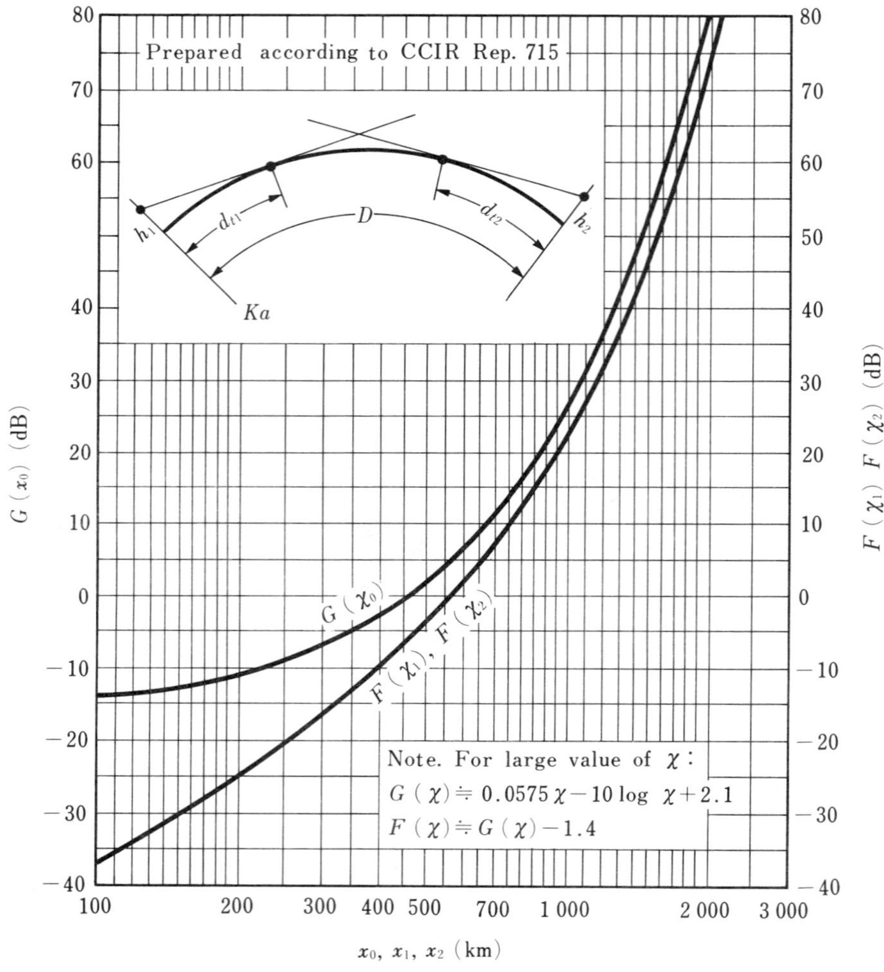

$\chi_0 = B \cdot D$ (km)

$\chi_1 = B \cdot d_{t1} = B \cdot \sqrt{2Kah_1} = B\,(3.57\sqrt{K}\sqrt{h_1(\mathrm{m})})$

$\chi_2 = B \cdot d_{t2} = B \cdot \sqrt{2Kah_2} = B\,(3.57\sqrt{K}\sqrt{h_2(\mathrm{m})})$

$B = 670\,\{f/(Ka)^2\}^{1/3} = 1.95\,(f\,(\mathrm{MHz})/K^2)^{1/3}$

4

TROPOSCATTER PROPAGATION PATH

4.1 APPROACH TO TROPOSCATTER PROPAGATION PATH

The radio-wave field strength at points a great distance above the horizon, at frequencies above 100 MHz, attains higher levels than can be supported by the diffraction theory.

In this frequency region the ionospheric wave or the ground wave undergoes a higher attenuation, and hence, except for a special case, it can be ignored. Therefore there should be some other type of propagation mechanism to explain such a high field strength.

As a result of studies on radio-wave propagation in the VHF to SHF bands and on high-altitude radio meteorology during the past 30 years, it has almost been made clear that such a high field strength is caused by scattering through abnormal refraction, partial refraction, partial reflection, etc., because of the inhomogeneity of the tropospheric atmosphere, which surrounds the earth up to a height of 12 km.

Radio meteorological phenomena in the troposphere vary greatly, depending on the latitude, on the character of the earth surface, such as sea, coast, inland, swamp, desert, or icy fields, and also on the seasons and on time (day or night). Consequently the field strength, or the propagation loss, should be treated statistically as geographical or temporal variables.

Because of the complexity of its propagation mechanism, no reliable analytic method has been established for the troposcatter propagation path.

Especially in two extremes of a rate of time or space, such as 99 to 99.9% on one end and 1.0 to 0.1% on the other, the difference between measured and estimated values may frequently reach 20 to 40 dB. Hence it may not be very meaningful to employ a sophisticated equation or to follow a troublesome analytic procedure. This chapter emphasizes practicability and introduces an approach in accordance mainly with the method given in CCIR Report 238-3. Many useful charts are also provided.

4.2 METHOD OF PREDICTING THE MEDIAN TRANSMISSION LOSS Γ(50)

A. Prediction Formula

The long-term yearly median (50% value) obtained from samples of the hourly median transmission loss is expressed approximately by

$$\Gamma(50) = F_1(f, D) + F_2(\theta D) - V(d_e) \tag{4.1}$$

$$F_1(f, D) = 30 \log f - 20 \log D \tag{4.2}$$

where $F_1(f, D)$ = function of frequency f and propagation distance D (see Fig. 6)
$F_2(\theta D)$ = function of the product of angular distance θ and propagation distance d (see Fig. 7)
$V(d_e)$ = function of effective distance d_e (see Fig. 8[1])

The parameter d_e is determined according to regional radio meteorological conditions as indicated by type numbers. The type number for each region is given in Fig. 8[2].

B. Angular Distance (Scatter Angle)

The angle θ indicated in Figs. 1 and 2 is called angular distance or scatter angle. For a smooth spherical ground (the most typical is a calm water surface), the central angle, which includes the transhorizon portion of the effective earth ξ is equal to the angle θ, that is, the angular distance θ is proportional to the transhorizon distance d_s (the distance between points of the radio horizon).

This formula explains clearly the geometrical relation of the angular distance. In the actual ground with a very complicated profile, θ is different from and the propagational behavior is not always equivalent to that expected in a smooth spherical ground.

However, for convenience of experimental and theoretical approximation, the angular distance θ in the actual ground is assumed to cause an effect similar to that in a smooth spherical ground.

θ of Smooth Spherical Ground

On smooth spherical ground with a uniform elevation h_0 ($h_0 = 0$ for the sea) above sea level, when the antenna heights above ground at points A and B are h_1 and h_2, the line-of-sight distances (radio horizon) are d_{t1} and d_{t2}, as was given in Fig. 22 of Chapter 2, that is,

$$d_{t1} = \sqrt{2Kah_1}$$
$$d_{t2} = \sqrt{2Kah_2} \tag{4.3}$$

where K = effective earth radius factor (4/3 as standard)
a = average earth radius (= 6370 km)

Hence the transhorizon distance d_s is given by

$$d_s = D - (d_{t1} + d_{t2}) \tag{4.4}$$

From Fig. 1a,

$$\theta = \xi = d_s/Ka \tag{4.5}$$

Figure 2 is a chart to estimate θ directly from h_1, h_2, and D by the following steps.

1. In the lower half of the figure, the intersection of h_2 (or h_1) and D is obtained.
2. From the intersection, the line is drawn vertically upward, and the second intersection with a corresponding h_1 (or h_2) curve is obtained.
3. The value of θ on the ordinate is read in milliradians (10^{-3} rad).

θ of Actual Ground

In Fig. 1b the distances from A and B to obstacles such as a mountain obstructing the line of sight are denoted by d_{L1} and d_{L2}, the heights above sea level of the obstacles are denoted by g_1 and g_2, the vertical angle of the direction from A to the obstacle is denoted by α_1, that from B by α_2, and the central angle determined by A and B of the effective earth (with the radius of Ka) is denoted by ξ. Then

$$\theta = \xi + (\alpha_1 + \alpha_2) \tag{4.6}$$

where the polarities of α_1 and α_2 are positive for the elevation above the horizon and negative for the dip below the horizon. (Both α_1 and α_2 are negative in the figure.)

If ξ, α_1, and α_2 are expressed in geometrical parameters,

$$\xi = D/Ka \tag{4.7}$$

$$\begin{aligned} \alpha_1 &= -\left[(h_{s1} - g_1)/d_{L1} + d_{L1}/2Ka\right] \\ \alpha_2 &= -\left[(h_{s2} - g_2)/d_{L2} + d_{L2}/2Ka\right] \end{aligned} \tag{4.8}$$

Substituting the above relations into Eq. (4.6) and rearranging,

$$\theta = \begin{matrix} -(h_{s1} - g_1)/d_{L1} \to X_{a1} \\ -(h_{s2} - g_2)/d_{L2} \to X_{a2} \end{matrix} \quad \text{(Fig. 3}a\text{)} \tag{4.9}$$

$$\begin{matrix} -d_{L1}/2Ka \to X_{b1} \\ -d_{L2}/2Ka \to X_{b2} \end{matrix} \quad \text{(Fig. 3}b\text{, left-hand side)} \tag{4.10}$$

$$+D/Ka \to X_c \quad \text{(Fig. 3}b\text{, right-hand side)} \tag{4.11}$$

Thus we obtain

$$\theta\ (\text{mrad}) = (X_{a1} + X_{a2}) + (X_{b1} + X_{b2}) + X_c \tag{4.12}$$

It is recommended that the result column in the figure be used to avoid errors. Since $\theta > 0$, if the calculated θ shows a negative value, there must be an error in the computation, or the path is line of sight. Therefore it is necessary to reexamine the path profile.

C. $F_2(\theta D)$

The parameter $F_2(\theta D)$ is shown in Fig. 7, and Fig. 4 gives a nomograph to obtain the variable θD, that is, the product of angular distance (scatter angle) θ in radians and propagation distance D in kilometers.

If the refractive index on the ground surface N_s is not available, use the curve for $N_s = 300$. If N_s is not available, find the value of K in Fig. 3*b*, referring to the top and the bottom abscissa scales. Then apply Fig. 7.

D. $V(d_e)$

The parameter $V(d_e)$ as a function of the effective distance d is shown in Fig. 8. The effective distance d_e is obtained as follows.

1. The line-of-sight distances d_{t1} and d_{t2} are obtained from Fig. 22 in Chapter 2.
2. The parameter d_c as a function of f is obtained from Fig. 5*a*. d_c is a portion of the transhorizon at which the levels of the diffracted wave and the troposcatter wave are almost equal, that is,

$$d_c = 65(100/f)^{1/3} \tag{4.13}$$

3. Then

$$G = (d_{t1} + d_{t2}) + d_c \tag{4.14}$$

is calculated.

4. The effective distance d_e for $G \geq D$ is

$$d_e = 130D/G \tag{4.15}$$

5. $G < D$ corresponds to a sufficiently deep transhorizon region where the diffracted waves attenuate rapidly and the troposcatter waves are predominant. Then the following equations can hold:

$$d_e = (D - G) + 130 \quad \text{(km)} \tag{4.16}$$

or

$$d_e = (130 + D) - (d_{t1} + d_{t2} + d_c) \tag{4.17}$$

4.3 LONG-TERM VARIABILITY OF HOURLY MEDIAN TRANSMISSION LOSS $\Gamma(q)$

For q (%) of the time in a year, let the upper limit of the transmission loss not exceed the hourly median denoted by $\Gamma(q)$. Then

$$\Gamma(q) = \Gamma(50) - Y(q) \quad \text{(dB)} \tag{4.18}$$

$$Y(q) = g(f)Y_0(q) \quad \text{(dB)} \tag{4.19}$$

A. $g(f)$

The parameter $g(f)$ is determined by the frequency f and the regional meteorological conditions, as shown in Fig. 9.

B. $Y_0(q)$

The q variation of the field strength relative to the hourly medians estimated for various time rates, that is, 0.01, 0.1, 1, 10, 50, 90, 99.9, or 99.99% for each regional meteorological condition, is shown in Fig. 10. It is noted that these curves always represent regional characteristics because of the limited amount of collected data, the period of measurement, the classification of topographical conditions, etc. Moreover, there has not yet been an acceptable explanation for the great variability of the data. However, at present there is no other way than to use these curves. Therefore care must be taken in using these curves to avoid error.

Values of $\Gamma(q)$ and $E(q)$ for smaller time rates, such as $q = 0.01, 0.1, 1, 10\%$, give the corresponding lower limit of transmission loss, or the higher limit of field strength, in the higher field strength region expected q percent of the time (or location) under consideration. Consequently the estimated field strength values for these time rates are always greater than the median ($q = 50\%$).

Accordingly, an estimate in this region for the smaller time rates is useful for an analysis of possible interference to other systems, while estimates for larger time rates, that is, $q = 90, 99, 99.9, 99.99\%$, etc., fall in the region of lower field strength and are used to evaluate the performances of the system proper, such as signal-to-noise ratio and circuit outage rate.

The transmission loss for a time rate not provided in the figure is obtained by graphic interpolation or by computation using an approximated mathematical distribution model.

Whether the field strength level is to be indicated for a smaller or a larger percentage depends upon the method of data analysis, and there is no essential difference. However, it is necessary that this be defined clearly to prevent confusion. Here it is defined as in CCIR Recommendation 311-1 (refer to Fig. 5, Chapter 8).

4.4 SIMPLIFIED METHOD OF ESTIMATING TRANSMISSION LOSS (ALTERNATIVE METHOD)

This method was prepared for extensive use to reduce the calculation process as much as possible and still use ample empirical data.

Since the rate of time (or the location factor) normally referred to is either a large or a small percentage figure, graphs are provided showing 99% and 1% values for the data at 1 GHz.

In addition, supplemental graphs are given for correcting frequencies (200 to 4000 MHz) and for estimation at an arbitrary percentage.

A. 99% Value of Transmission Loss for the Worst Month in a Year

Since the transmission of troposcatter waves presents large seasonal variations, there is a great difference in reliability between two cases: for a 1-year period and for a worst 1-month period. It is safer to design the system for the worst month in a year.

Figure 11*a*[1] gives the 99% value of distribution of the hourly median transmission losses in the worst month for the effective distance and for each meteorological region defined in Fig. 8.

B. 1% Value of Transmission Loss in a Year

A study of the interference problem requires knowing the tendency of the field strength in a small-q region corresponding to a region of high field strength. Figure 11*a*[2] gives 1% values and indicates that even in a transhorizon distance, the field strength may almost reach the free-space value.

C. Frequency Correction Coefficient

Figure 11*b*[3] gives the coefficient over the frequency band from 200 MHz to 4 GHz relative to that at 1 GHz. The approximate tendency is expressed as

$$C_f = 30 \log f \quad (f \text{ in GHz}) \tag{4.20}$$

D. Standard Deviation

Figure 11*b*[4] presents standard deviations of the variability for effective distances and classified regional meteorological conditions.

If the standard deviation is known and the original variability is assumed to follow the log-normal distribution model, the transmission loss for an arbitrary percentage of time can be estimated (refer also to Fig. 5, Chapter 8).

The described method requiring no calculation process is convenient. However, it is apt to lower the accuracy of estimation due to an excessive approximation and/or reduction.

4.5 TRANSMISSION LOSS

In the troposcatter propagation path with a very large scatter volume contributing to the generation of a receiving field strength, an antenna having too sharp a directivity leads to a reduction of the actual antenna gain by narrowing the scatter volume.

Such a reduction of antenna gain is called aperture-to-medium coupling loss L_c, and the effective gain, which includes the above effects for the transmitting and receiving antennas, is called path antenna gain G_c. Using them, the transmission loss L between transmitting and receiving antennas is expressed by

$$L = \Gamma + L_c - (G_1 + G_2) \tag{4.21}$$

$$L = \Gamma - G_c \tag{4.22}$$

or

$$G_c = G_1 + G_2 - L_c \tag{4.23}$$

where G_1, G_2 = gains of transmitting and receiving antennas in free space
L_c = aperture-to-medium coupling loss
G_c = path antenna gain

Figure 12 shows L_c for $G_1 + G_2$ in decibels.

4.6 METHOD USING VERTICAL DISTRIBUTION OF ATMOSPHERIC REFRACTIVITY

There is a method to estimate the transmission loss from various parameters in radio meteorology with emphasis on a troposcatter wave generating mechanism. However, this method cannot be used unless the high-altitude radio meteorological data (vertical distributions of the atmospheric refractive index measured by radiosonde) and their seasonal and diurnal variations are available.

It is rather difficult at present to determine whether or not this method gives better estimates and matches the actual condition better than the methods for direct analysis of the propagation data as mentioned in Sections 4.2 to 4.4.

REMARKS. Equation (1) proposed in France and Eq. (2) modified in the United States for estimating the transmission loss, using the vertical distribution of the atmospheric refractive index, are summarized below for reference (CCIR Reports 244-2 and 233-2):

$$\Gamma(q) = 110.5 + 30\log f + 30\log D - T(q) \quad \text{(dB)} \tag{1}$$

$$\Gamma(q) = 20\log(fD) + 10\log(fd_s) + K_1(d_s) - K_2(d_s\theta) - T(q) \quad \text{(dB)} \tag{2}$$

$$T(q) = -\tfrac{3}{8}G_e - \tfrac{5}{4}G_c$$

where $\Gamma(q)$ = q (%) value of propagation loss (basic transmission loss)
D = propagation distance (km)
d_s = transhorizon distance (km) at $K = 4/3$
f = frequency (MHz)
θ = angular distance (rad) at $K = 4/3$
$T(q)$ = q (%) value of atmospheric refractive index (according to statistical values of G_e and G_c)
G_e = effective slope of N between earth surface and common volume (or difference in N between earth surface and 1 km above ground)
G_c = difference in N between 1 km above the bottom of the common volume and the bottom
N = refractivity [$N = (n-1) \times 10^6$]
n = refractive coefficient of atmosphere

When G_e and G_c are not available, the following formulas are used:

$$G_e \doteqdot G_0$$

where G_0 is the difference in N between the surface and 100 m above the surface, multiplied by 10, and

$$G_c = \begin{cases} \frac{1}{3}\left(200 - N_s - \frac{2}{3}G_0\right), & 1 \text{ km} < d_s\theta < 2.7 \text{ km} \\ \frac{3}{7}\left(200 - N_s - \frac{2}{3}G_0\right), & d_s\theta \qquad > 2.7 \text{ km} \end{cases}$$

where N_s is the N value of the surface. When $d_s < 50$ km or $d_s\theta < 1$ km, the propagation will take the smooth spherical-ground diffraction model as given in Chapter 3. $K_1(d_s)$ and $K_2(d_s\theta)$ are determined by the graph given below.

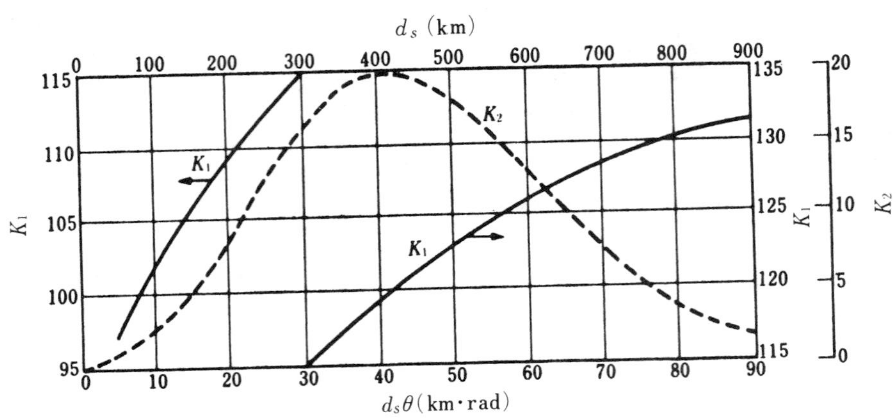

4.7 REMARKS ON THE APPLICATION OF THE CALCULATION CHARTS

1. It should be noted that since this type of propagation depends on the spatial and temporal irregularities of the tropospheric atmosphere, the following applies.
   (a) Regional differences may be significant and the given charts prepared based on a limited amount of data may not always be applicable to a specific region.
   (b) The seasonal variations of monthly medians may reach 2 to 10 dB, according to location.
   (c) There are large diurnal variations, and the diurnal variation of hourly medians may reach 5 to 15 dB.
2. The prepared charts have been drawn from samples consisting of hourly medians. The variations caused by rapid fading are superposed on the actual variations of the field intensity. The short-term variations in rapid fading within about 5 min can be approximated by a Rayleigh distribution, and the variations over a period of more than 1 hour, by log-normal distribution.
3. The system design of the actual link needs to take into account both items 1 and 2.

Normally, space diversity and frequency diversity are used for performance improvement. However, the resulting variability or the improvement factor sometimes falls short of expectation. Consequently it is desirable to make the design margin as wide as possible.

EXAMPLES

EXAMPLE 4.1

Determine the median propagation loss $\Gamma(50)$ for the troposcatter propagation path shown below.

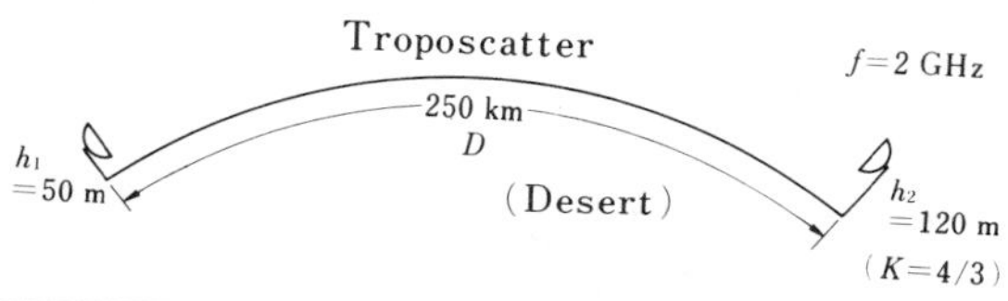

Solution

In Fig. 1 the troposcatter path over smooth spherical ground is referenced. The median propagation loss $\Gamma(50)$ is determined using related charts.

1. Line-of-sight distance d_t and beyond-the-horizon distance d_s,

$$d_{t1} = 4.12\sqrt{h_1} = 4.12\sqrt{50} = 29.1 \text{ km}$$
$$d_{t2} = 4.12\sqrt{h_2} = 4.12\sqrt{120} = 45.1 \text{ km} \qquad \text{(Fig. 23, Chapter 2)}$$

$$d_s = D - (d_{t1} + d_{t2}) = 250 - (29.1 + 45.1) = 175.8 \text{ km}$$

2. Angular distance θ,

$$\theta = 20.5 \text{ mrad} \qquad \text{(Fig. } 2a\text{)}$$

3. Parameter θD,

$$\theta D = 5.1 \qquad \text{(Fig. 4)}$$

4. Parameter d_c,

$$d_c = 24 \text{ km} \qquad \text{(Fig. } 5a\text{)}$$

5. Effective distance d_e,

$$(d_{t1} + d_{t2}) + d_c = 29.1 + 45.1 + 24 = 98.2 < D$$

Therefore the nomograph in Fig. 5*b* is not applicable and the equation given in the note is to be used,

$$d_e = (D + 130) - (d_{t1} + d_t + d_c) = (250 + 130) - 98.2 = 281.8 \text{ km}$$

6.

$$F_1(f, D) = 51 \text{ dB} \qquad \text{(Fig. 6)}$$

7.

$$F_2(\theta D) = 158 \text{ dB} \qquad \text{(Fig. 7)}$$

8.

$$V(d_e) = -7 \text{ dB} \qquad \text{(Fig. 8, type 4)}$$

9. Median propagation loss,

$$F_1(f, D) + F_2(\theta D) - V(d_e) = \Gamma(50) = 216 \text{ dB}$$

EXAMPLE 4.2

Determine $L(99)$, the transmission loss between A and B, for 99% reliability over the yearly distribution of the hourly medians.

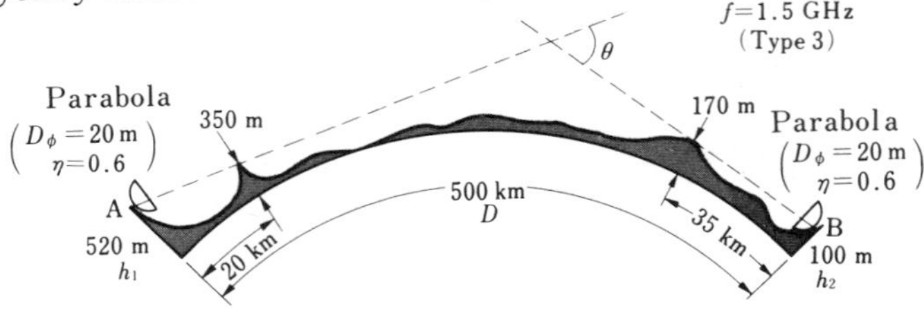

Solution

Using Fig. 1b and the associated charts, the 99% value of the propagation loss $\Gamma(99)$, and Figs. 13 and 14 in Chapter 1, the parabolic antenna gains are estimated. Thereafter the transmission loss $L(99)$ is determined using Fig. 11.

1. Line-of-sight distance d_t and beyond-the-horizon distance d_s,

$$d_{t1} = 4.12\sqrt{h_1} = 4.12\sqrt{520} = 94 \text{ km} \qquad \text{(Fig. 23, Chapter 2)}$$

$$d_{t2} = 4.12\sqrt{h_2} = 4.12\sqrt{100} = 41.2 \text{ km}$$

$$d_s = D - (d_{t1} + d_{t2}) = 500 - (94 + 41.2) = 364.8 \text{ km}$$

2. Angular distance θ. The basic parameters to be used are $h_1 = 520$, $h_2 = 100$, $g_1 = 350$, $g_2 = 170$, $d_{L1} = 20$, $d_{L2} = 35$, and $D = 500$,

$$(h_1 - g_1) = 170, \qquad d_{L1} = 20$$

$$(h_2 - g_2) = -70, \qquad d_{L2} = 35$$

$$X_{a1} = -8.5, \qquad X_{b1} = -1.2, \qquad X_c \doteqdot 59 \qquad \text{(Fig. 3)}$$

$$X_{a2} = +2, \qquad X_{b2} = -2,$$

$$\theta = (X_{a1} + X_{a2}) \pm (X_{b1} + X_{b2}) + X_c = 49.3 \text{ mrad}$$

3. Parameter θD,

$$\theta D = 24.7 \qquad \text{(Fig. 4)}$$

4. Parameter d_c,

$$d_c = 27 \qquad \text{(Fig. 5}a\text{)}$$

5. Effective distance d_e. The equation in the note in Fig. 5b is used,

$$d_e = (D + 130) - (d_{t1} + d_{t2} + d_c) = (500 + 130) - (94 + 41.2 + 27) \doteqdot 468 \text{ km}$$

6.

$$F_1(f, D) \fallingdotseq 42 \text{ dB} \qquad \text{(Fig. 6)}$$

$$F_2(\theta D) \fallingdotseq 186 \text{ dB} \qquad \text{(Fig. 7)}$$

$$V(d_e) \fallingdotseq 2 \text{ dB} \qquad \text{(Fig. 8, type 3)}$$

7. Median propagation loss $\Gamma(50)$,

$$\Gamma(50) = F_1(f, D) + F_2(\theta D) - V(d_e) = 226 \text{ dB}$$

8.

$$g(f) = 0.98 \qquad \text{(Fig. 9)}$$

$$Y_0(q) \doteqdot -17.5 \text{ dB} \qquad \text{(Fig. 10}b\text{, 99\% type 3)}$$

9. $Y(q)$, correction to the time rate 99%,

$$Y(q) = g(f) Y_0(q) \doteqdot -17 \text{ dB}$$

10. 99% value of propagation loss $\Gamma(99)$,

$$\Gamma(99) = \Gamma(50) - Y(99) = 226 - (-17) \doteqdot 243 \text{ dB}$$

11. Parabolic antenna, diameter $D_\phi = 20$ m, $\eta = 0.6$, free-space gain G_t, G_r,

$$A_e \doteqdot 185 \text{ m}^2 \qquad \text{(Fig. 14, Chapter 1)}$$

$$G_t = G_r = 47.6 \text{ dB} \qquad \text{(Fig. 13}a\text{ Chapter 1)}$$

12. Path antenna gain G_p,

$$G_p = (G_t + G_r - L_c) = 95.2 - 13 \doteqdot 82 \text{ dB} \qquad \text{(Fig. 12)}$$

13. Transmission loss between A and B (99% of time) $L(99)$,

$$\Gamma(99) - G_p = 243 - 82 = L(99) = 161 \text{ dB}$$

EXAMPLE 4.3

Solve Example 4.2 with the simplified method of estimation (Section 4.4 and Fig. 11) and compare the results.

Solution

Figure 11 is used.

1. Effective distance d_e. Referring to steps 1, 2, and 5 of the previous solution, $d_e = 468$ km.
2. 99% value of propagation loss (1 GHz) $\Gamma_1(99)$,

$$\Gamma_1(99) = 242 \text{ dB} \qquad \text{(Fig. 11}a\text{[1], type 3)}$$

3. Frequency correction coefficient (1.5 GHz) C_f,

$$C_f = 6 \text{ dB} \qquad \text{(Fig. 11}b\text{[3])}$$

4. 99% value of propagation loss (1.5 GHz) $\Gamma(99)$,

$$\Gamma(99) = \Gamma_1(99) + C_f = 242 + 6 = 248 \text{ dB}$$

5. 99% value of transmission loss $L(99)$ between antennas at A and B. Using the result from steps 11 and 12 in the previous solution, $G_p = 82$ dB,

$$\Gamma(99) - G_p = 248 - 82 = L(99) = 166 \text{ dB}$$

6. Comparison. There is a 5-dB difference between this result and the previous one. Since the simplified method is applied for the worst month in a year, some error is inevitable. Use of this method may be justified, noting that in general the estimate of the troposcatter propagation path loss produces a difference of 5 to 10 dB or more, depending on the solution and the basic conditions.

Example 4.4

The $G_e = -38N$ and $G_c = -41N$ median values have been obtained from data on the vertical distribution of the atmospheric refractive index. Determine the median propagation loss, assuming that the path profile and the frequency are the same as in Example 4.2 and using the methods given under Remarks in Section 4.6.

Solution

1. From the solution to Example 4.2, the propagation distance $D = 500$ km, the beyond-the-horizon distance $d_s = 375$ km, the angular distance $\theta = 0.05$ rad, and the frequency $f = 1500$ MHz.
2. Calculation.

   (a) Method 1,

$$\begin{array}{rr} 30\log(1500) \doteqdot & 95.5 \\ 30\log(1500) = & 81 \\ & 110.5 \\ -3(3 \times 38/8 + 5 \times 41/4) = & -65.5 \\ \hline \text{Median propagation loss} & 221.5 \text{ dB} \end{array}$$

   (b) Method 2,

$$\begin{array}{rr} 20\log(1500 \times 500) = & 117.5 \\ 10\log(1500 \times 375) = & 57.5 \\ K_1(375) \doteqdot & 118 \\ K_2(375 \times 0.05) \doteqdot & -8 \\ T = & -65.5 \\ \hline \text{Median propagation loss} & 219.5 \text{ dB} \end{array}$$

Chart Index, Troposcatter Propagation Path

Troposcatter Propagation Path over Smooth Spherical Surface

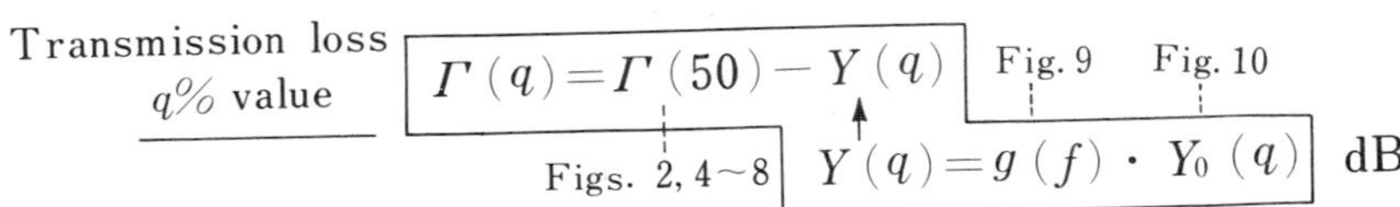

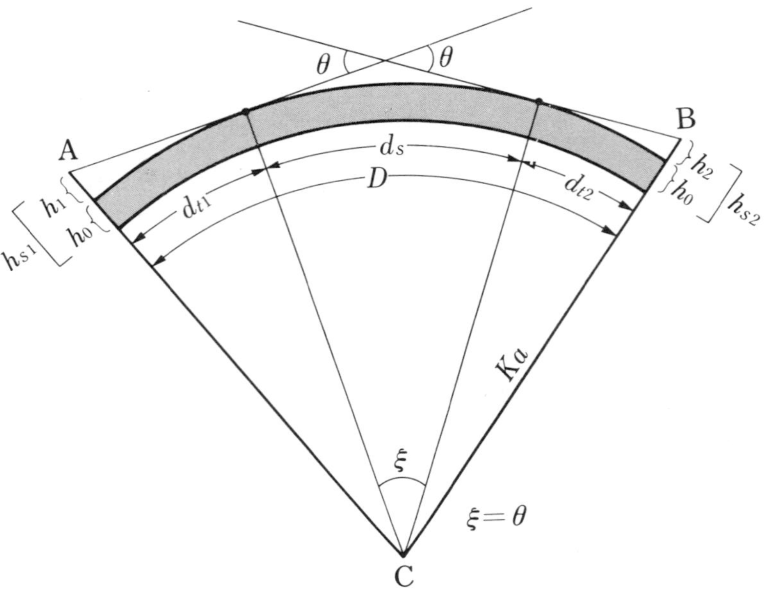

Angular distance $\theta=d_s/(Ka)=(D-d_{t1}-d_{t2})/(Ka)$

Line-of-sight distance over spherical earth
$$d_{t1}=\sqrt{2\,Kah_1}\cdots 4.12\sqrt{h_1\,(\mathrm{m})}\ (\mathrm{km})$$
$$d_{t2}=\sqrt{2\,Kah_2}\cdots 4.12\sqrt{h_1\,(\mathrm{m})}\ (\mathrm{km})$$
$(K=4/3)$

Transmission loss (median) $\Gamma(50)=F_1(f,\ D)+F_2(\theta D)-V(d_e)$

〔Note〕 The θ may be determined as follows:

1) $\theta=D/(Ka)-\{\sqrt{2\,h_1/(Ka)}+\sqrt{2\,h_2/(Ka)}\}$

2) If $\theta\rightarrow$mrad $D\rightarrow$km $h_1, h_2\rightarrow$m

- for $K=4/3$

  $\theta=0.118\,D-0.485\,(\sqrt{h_1}+\sqrt{h_2})$

- for $K=1$

  $\theta=0.157\,D-0.560\,(\sqrt{h_1}+\sqrt{h_2})$

Troposcatter Propagation Path over Ordinary Land

Propagation loss $q\%$ value $\Gamma(q) = \Gamma(50) - Y(q)$ (Figs. 3~8) ; $Y(q) = g(f) \cdot Y_0(q)$ dB (Fig. 9, Fig. 10)

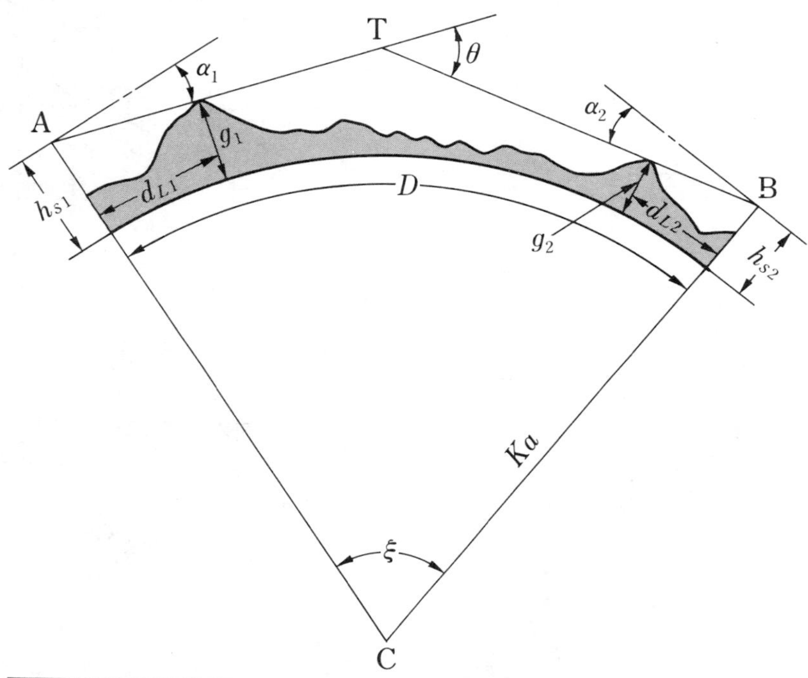

Angular distance $\theta = \xi + (\alpha_1 + \alpha_2)$ Sign of α { ⊕ above the horizon; ⊖ below the horizon }

Central angle of equivalent earth $\xi = D/(Ka)$

Vertical angle to main-beam direction

$$\alpha_1 = -[(h_{s1} - g_1)/d_{L1} + d_{L1}/(2\,Ka)]$$
$$\alpha_2 = -[(h_{s2} - g_2)/d_{L2} + d_{L2}/(2\,Ka)]$$

Propagation loss (median) $\Gamma(50) = F_1(f, D) + F_2(\theta D) - V(d_e)$

〔Note〕 If $\theta \to$ mrad $D, d_{L1}, d_{L2} \to$ km $h_{s1}, h_{s2}, g_1, g_2 \to$ m

- for $K = 4/3$

$$\theta = 0.118\,D - [(h_{s1} - g_1)/d_{L1} + (h_{s2} - g_2)/d_{L2} + 0.059\,(d_{L1} + d_{L2})]$$

- for $K = 1$

$$\theta = 0.157\,D - [(h_{s1} - g_1)\,d_{L1} + (h_{s2} - g_2)/d_{L2} + 0.078\,(d_{L1} + d_{L2})]$$

4

2a

Angular Distance θ (Smooth Spherical Ground)

100–500 km

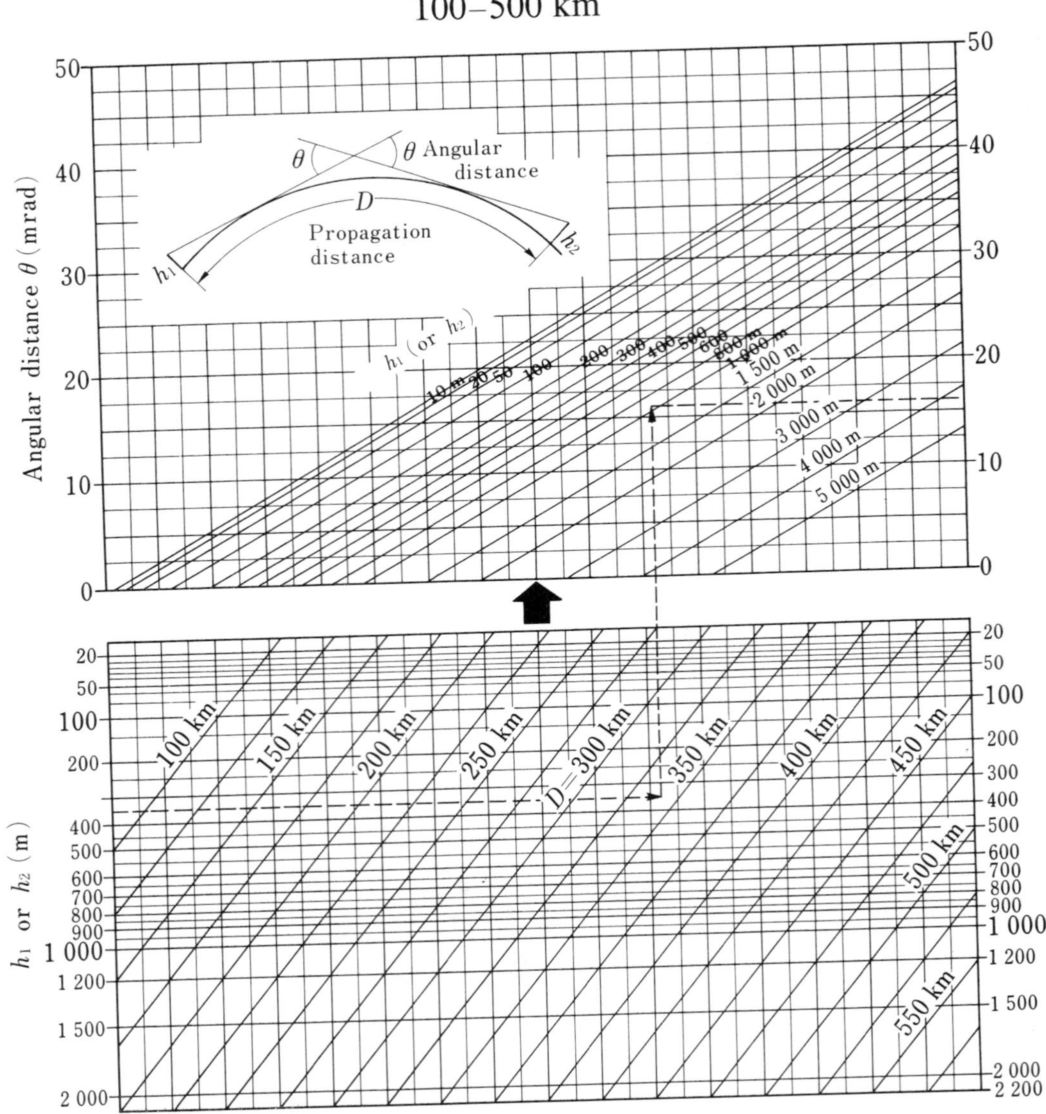

Angular Distance θ (Smooth Spherical Ground)

500–1000 km

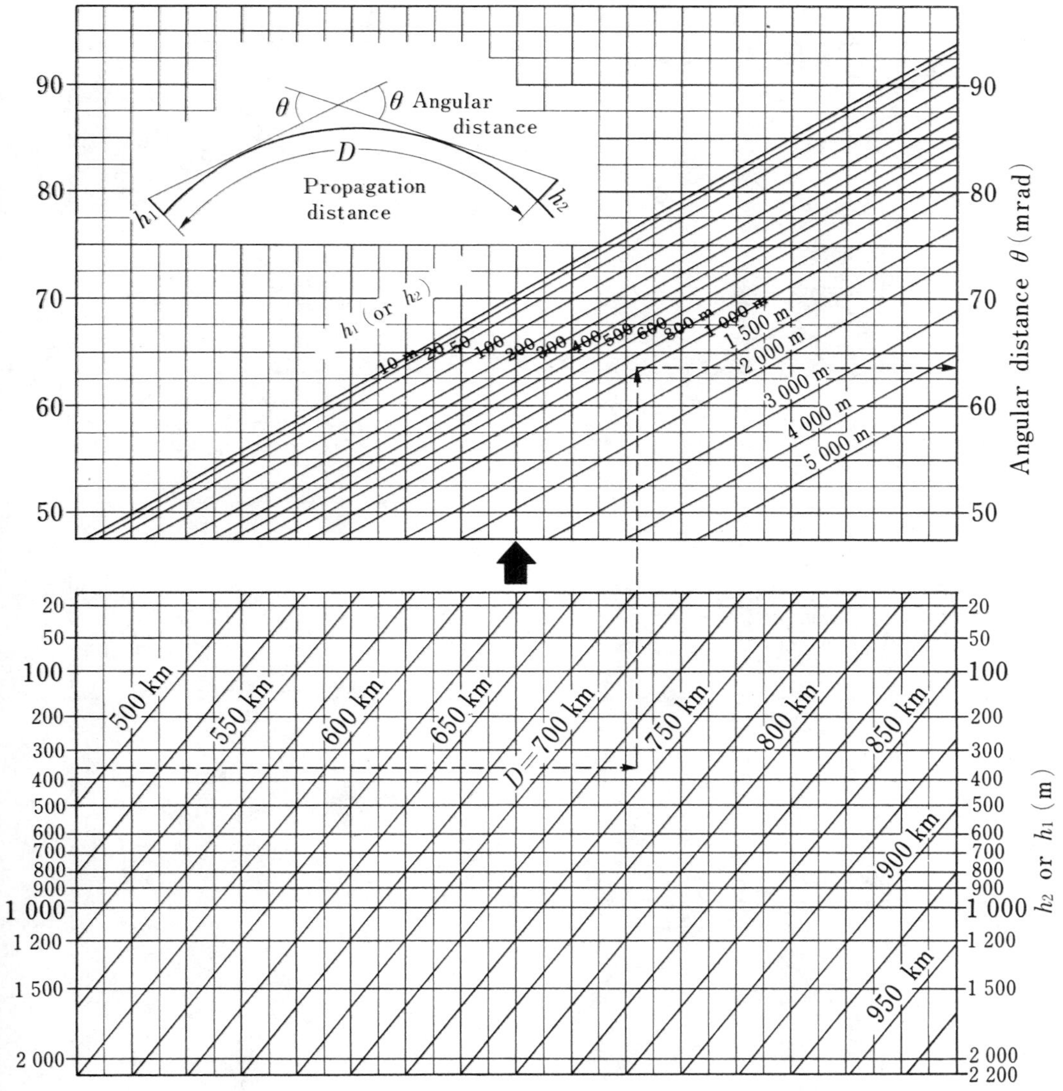

θ Angular distance
D Propagation distance
h_1 h_2

h_1 (or h_2)
10 m 20 50 100 200 300 400 500 600 800 m 1 000 m 1 500 m 2 000 m 3 000 m 4 000 m 5 000 m

Angular distance θ (mrad): 50 60 70 80 90

500 km 550 km 600 km 650 km D=700 km 750 km 800 km 850 km 900 km 950 km

h_2 or h_1 (m): 20 50 100 200 300 400 500 600 700 800 900 1 000 1 200 1 500 2 000 2 200

Angular Distance for Ordinary Earth

Parameter X_a

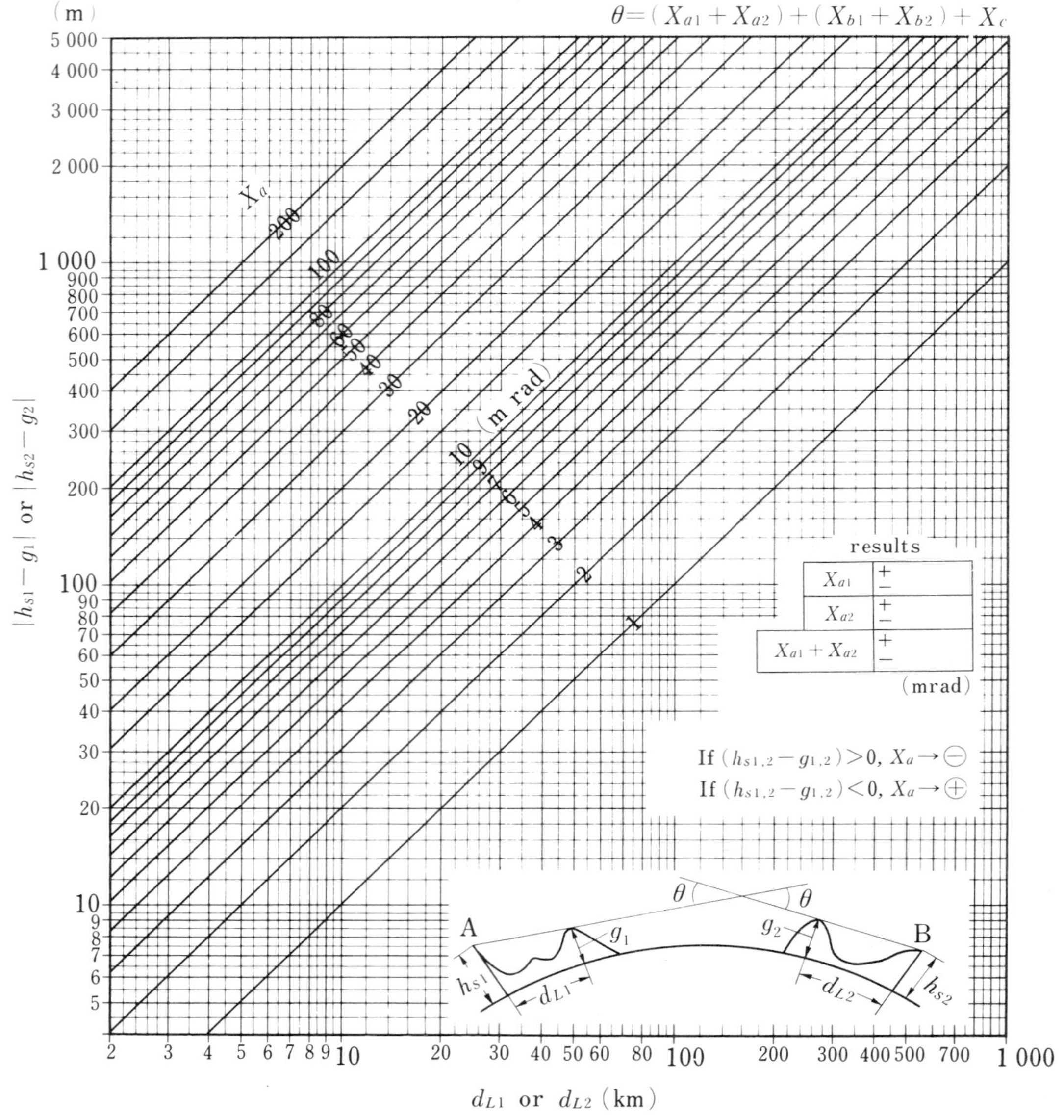

〔Note〕 • The following formulas are used when $g_1 = 0$ or $g_2 = 0$.

$$\left.\begin{array}{l} X_{a1} = -0.56\sqrt{h_{s1}/K} \\ X_{b1} = 0 \end{array}\right\} \text{ or } \left\{\begin{array}{l} X_{a2} = -0.56\sqrt{h_{s2}/K} \\ X_{b2} = 0 \end{array}\right.$$

• When $K = 4/3$,

$X_{a1} = -0.485\sqrt{h_{s1}}$ or $X_{a2} = -0.485\sqrt{h_{s2}}$

Angular Distance for Ordinary Earth

Parameters X_b and X_c

$$\theta = [X_{a1} + X_{a2}] + [X_{b1} + X_{b2}] + X_c$$

From previous page

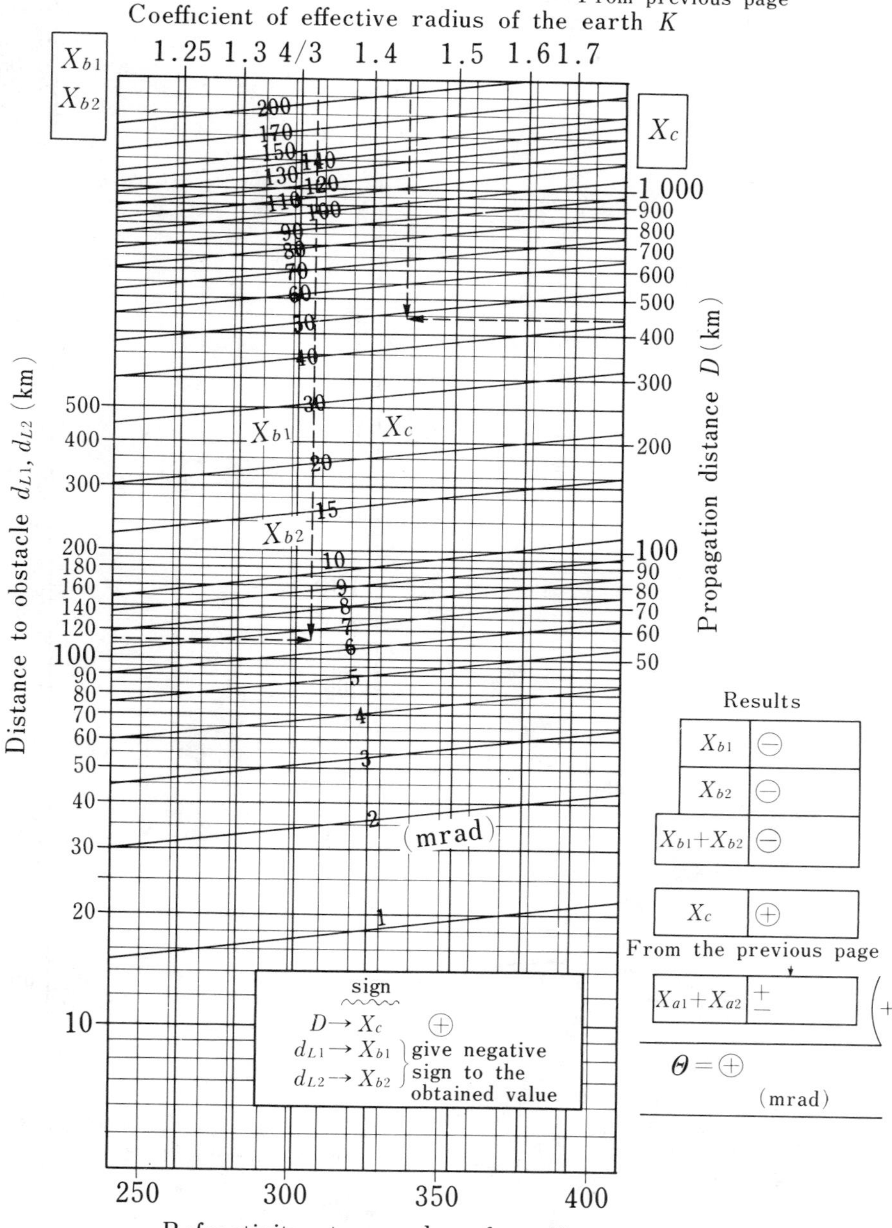

Nomograph for Parameter θD

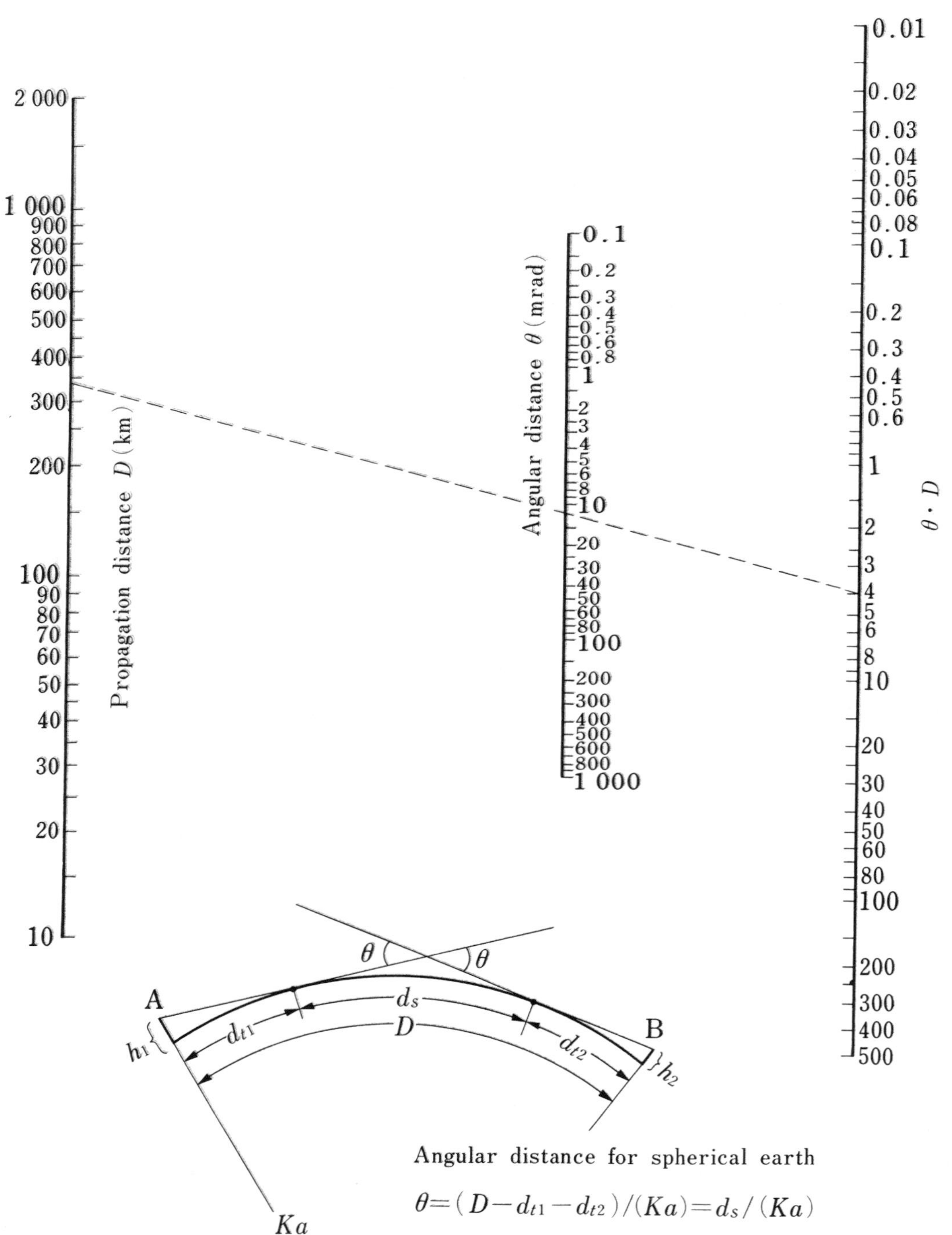

Angular distance for spherical earth

$\theta = (D - d_{t1} - d_{t2})/(Ka) = d_s/(Ka)$

Parameter for Effective Distance d_c

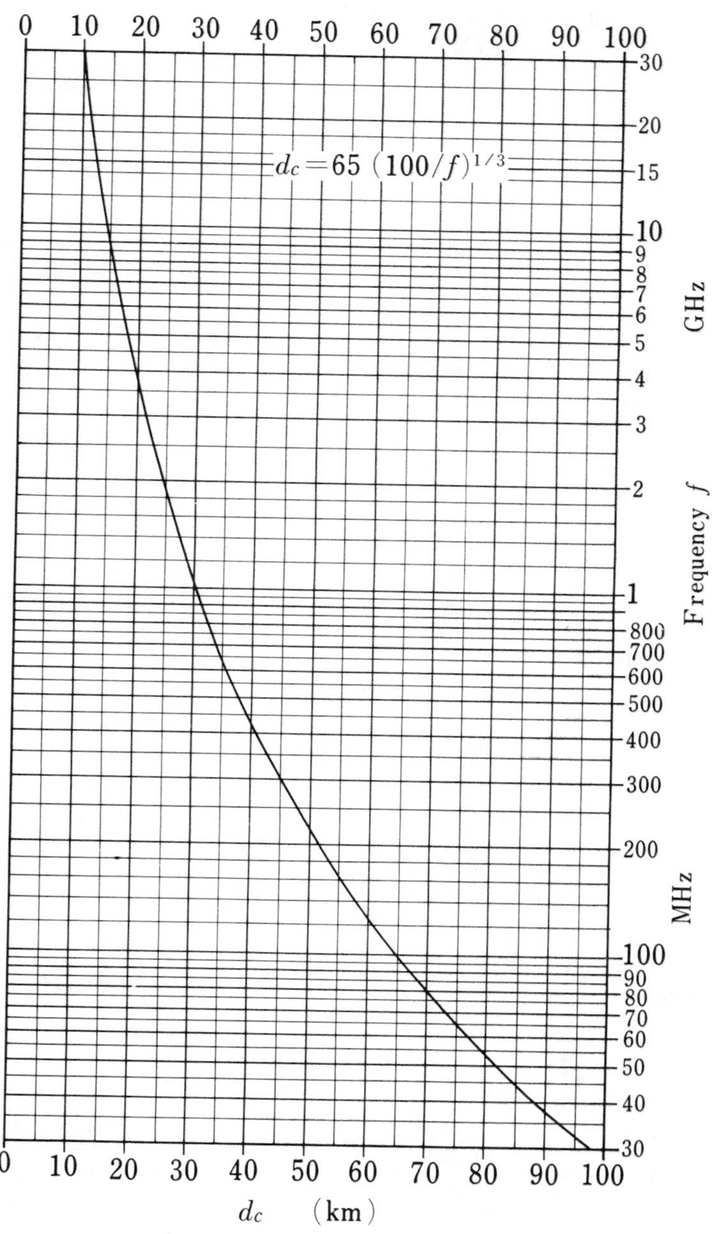

Effective Distance d_e

Nomograph for $D \leqq (d_{t1} + d_{t2}) + d_c$

Note : If $D > (dt_1 + dt_2) + d_c$ the following formula is used.
Effective distance $d_e = (D + 130) - (d_{t1} + d_{t2} + d_e)$

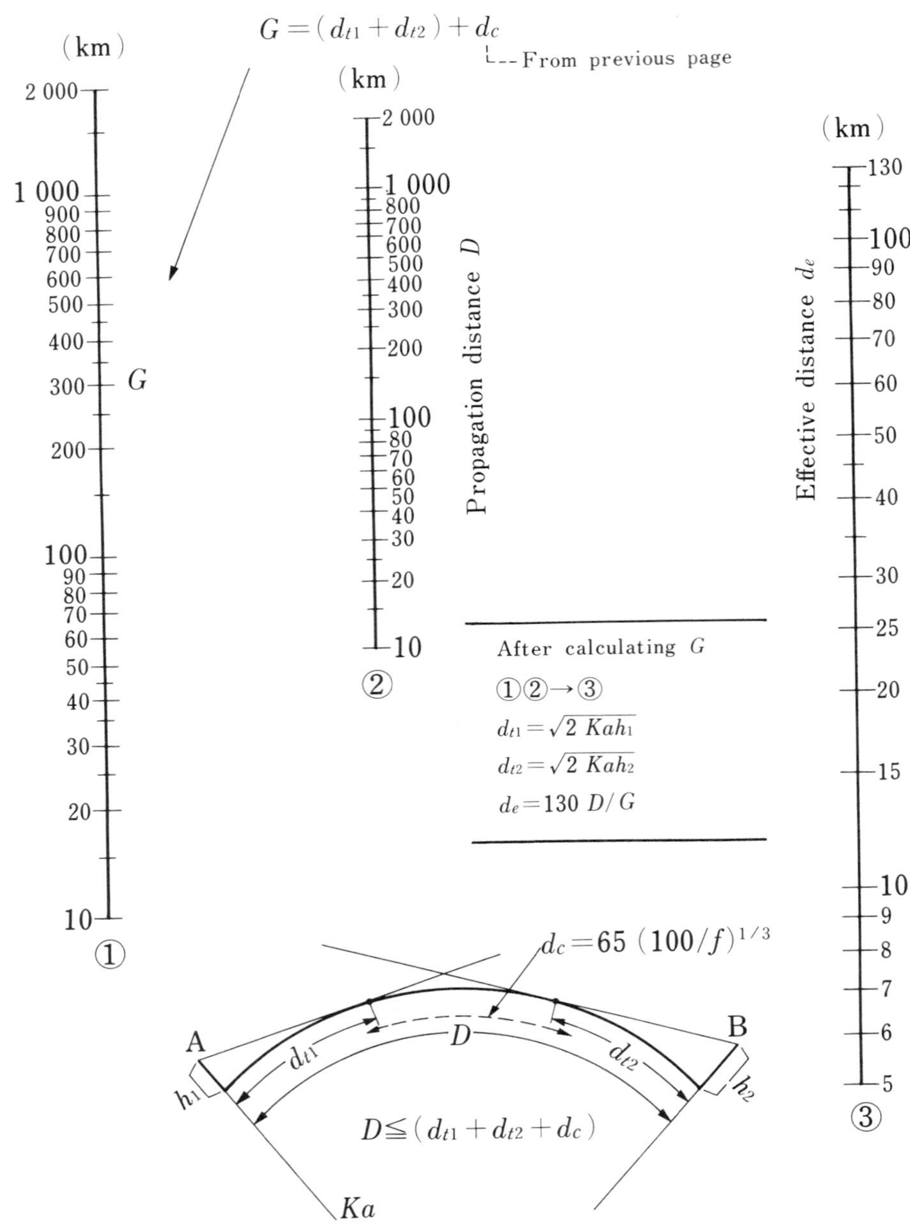

Graph for $F_1(f, D)$

$$\Gamma(50) = F_1(f, D) + F_2(\theta D) - V(d_e)$$

Fig. 7   Fig. 8

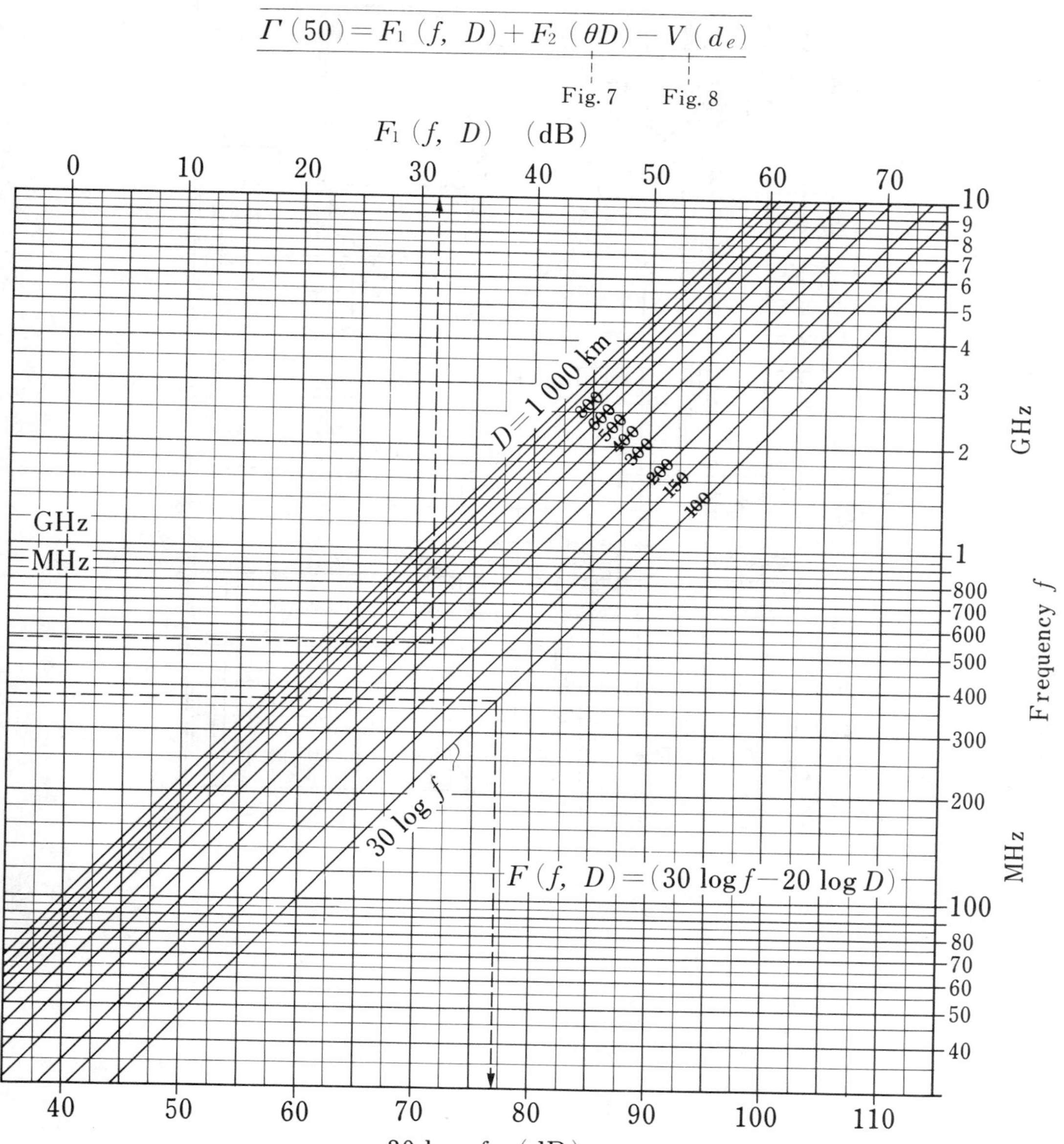

Graph for $F_2(\theta D)$

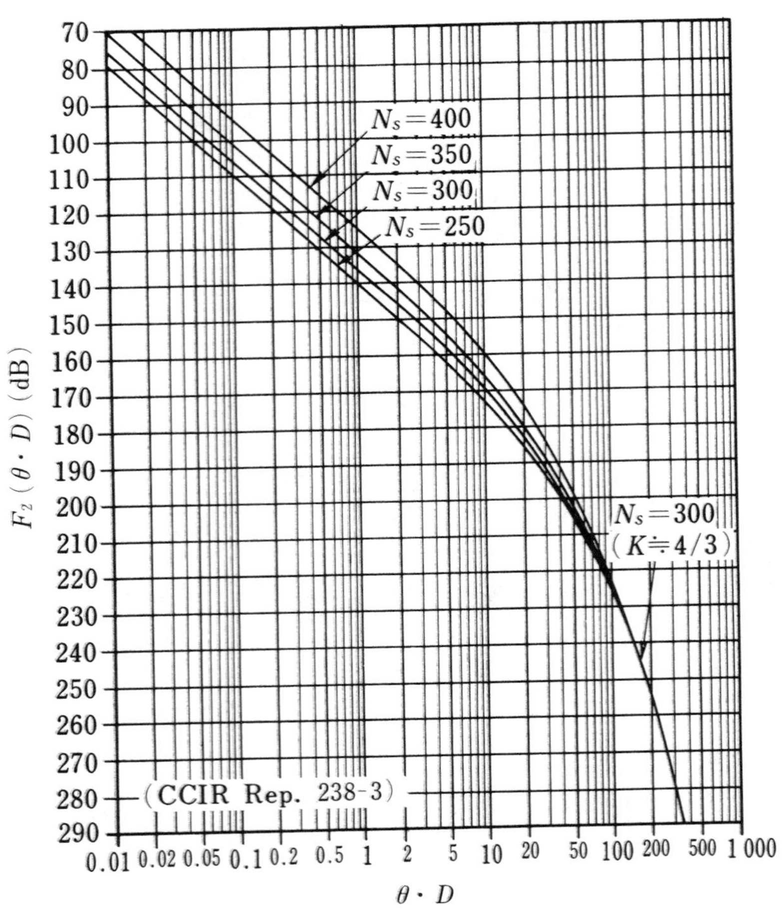

$V(d_e)$ and Radio Climatic Regions

[1]

Graph for $V(d_e)$

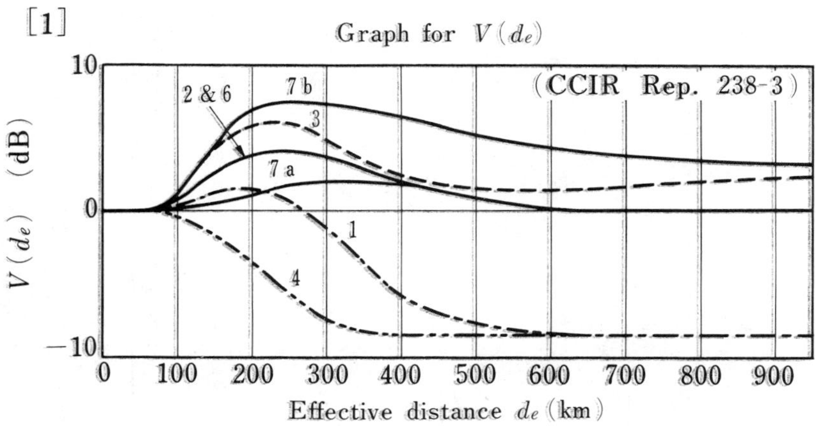

[2]

Radio climatic regions

Type	Climate	Latitude sphere ±	Note
1	equatorial	0 ∼ 10°	data from Congo and Ivory Coast
2	continental subtropical	10° ∼ 20°	data from Sudan
3	maritime subtropical		data from west coast of Africa
4	desert	20° ∼ 30°	data from Sahara
5	mediterranean	30° ∼ 40°	no curves available
6	continental temperate	30° ∼ 60°	data from France, FR. Germany, USA
7_a	maritime temperate (land)		data from United Kingdom
7_b	maritime temperate (sea)		data from United Kingdom
8	polar	60° ∼ 90°	no curves available

Note: Fig. 11 to be used as an alternative includes curves for Type 5.

Correction Factor for Variation $g(f)$

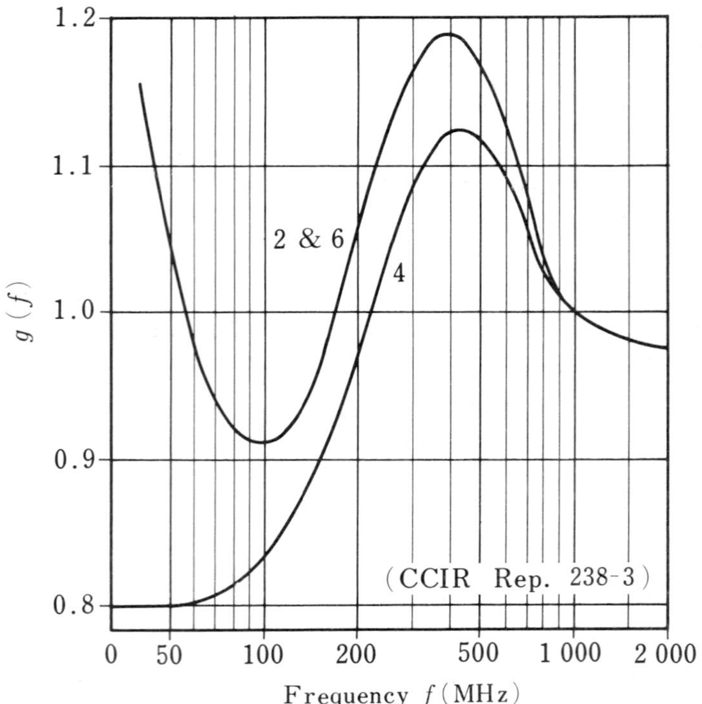

〔Remark〕
1. For any other types, select either of the curves shown, both of which are similar in nature, or make interpolation.
2. Since the variation range of the factor is not large, the result and error have little significance.
3. Regardless of types, $g(f)=1.0$ may be assumed for frequencies of 1 GHz or higher.

Graph for Variation $Y_0(q)$ (1)

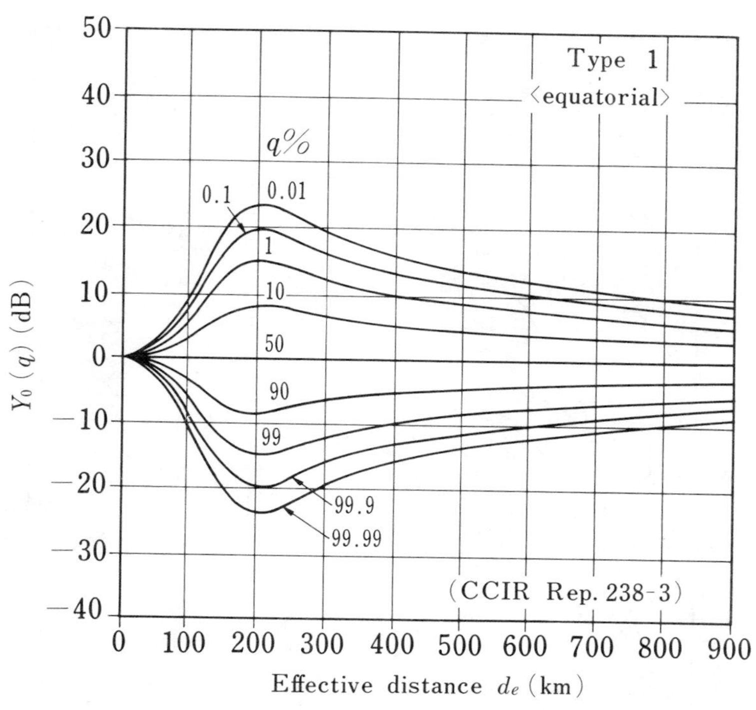

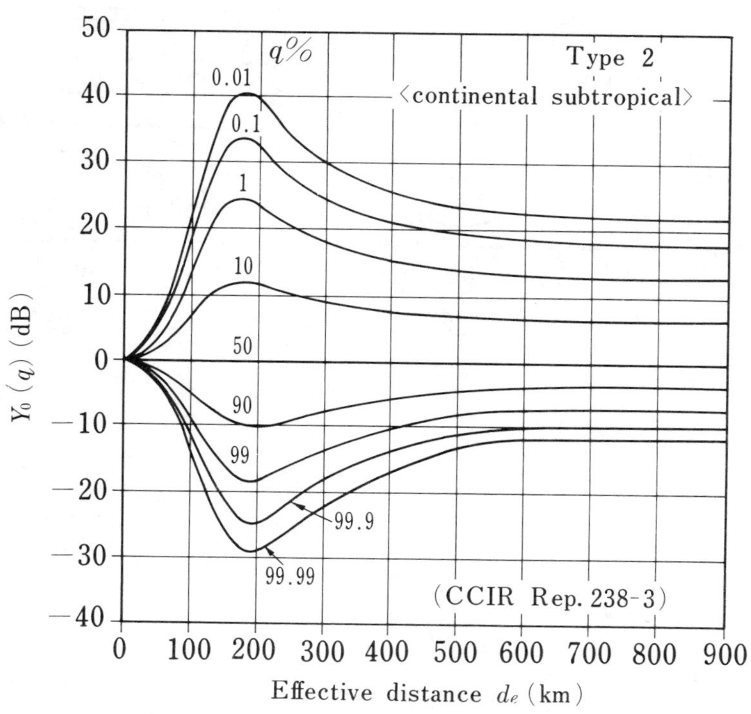

Graph for Variation $Y_0(q)$ (2)

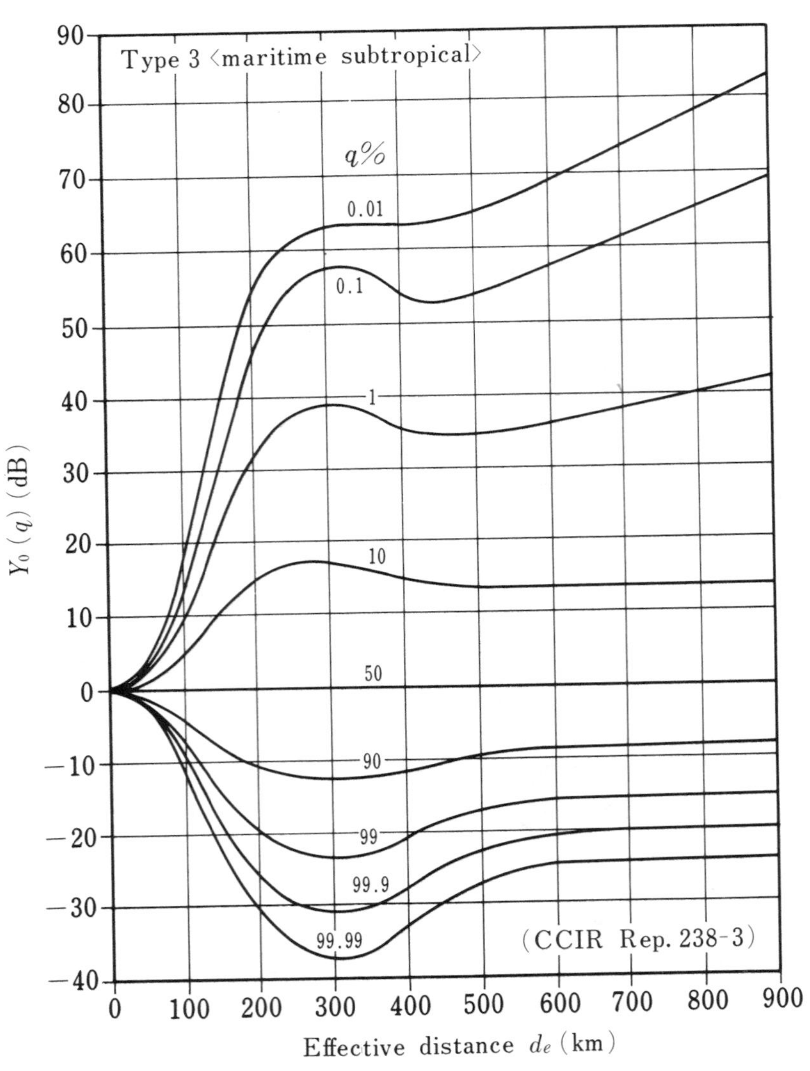

Graph for Variation $Y_0(q)$ (3)

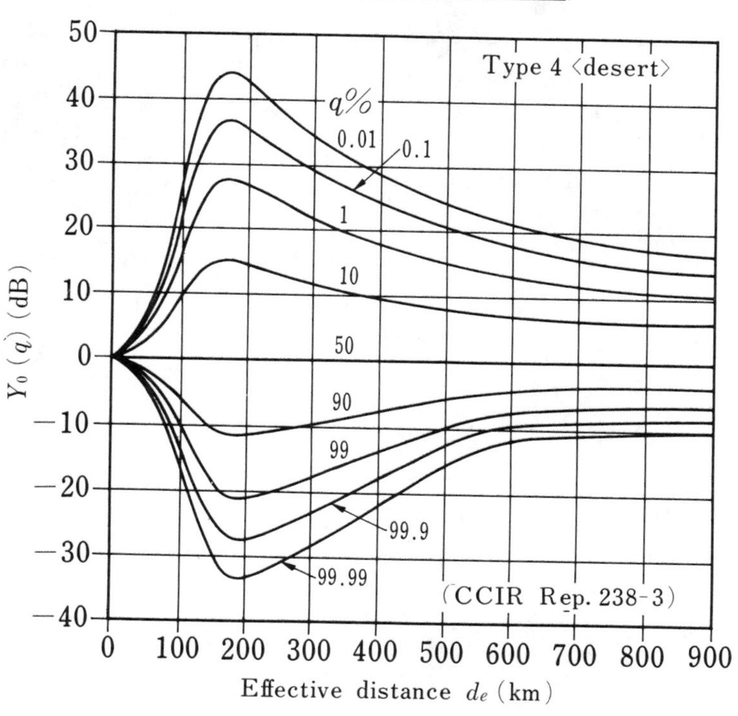

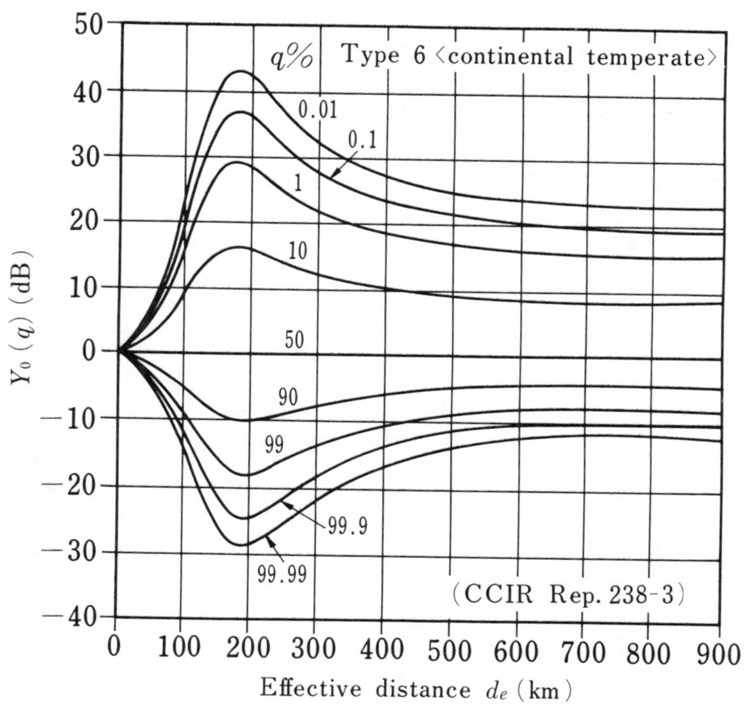

Graph for Variation $Y_0(q)$ (4)

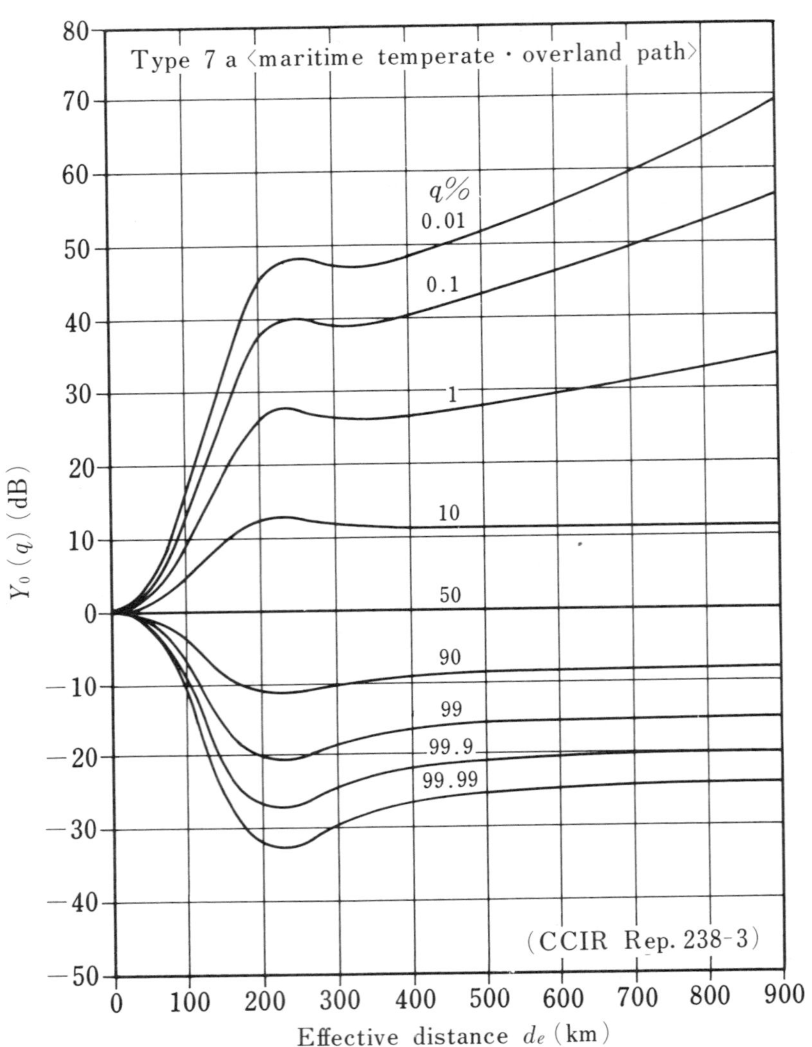

Graph for Variation $Y_0(q)$ (5)

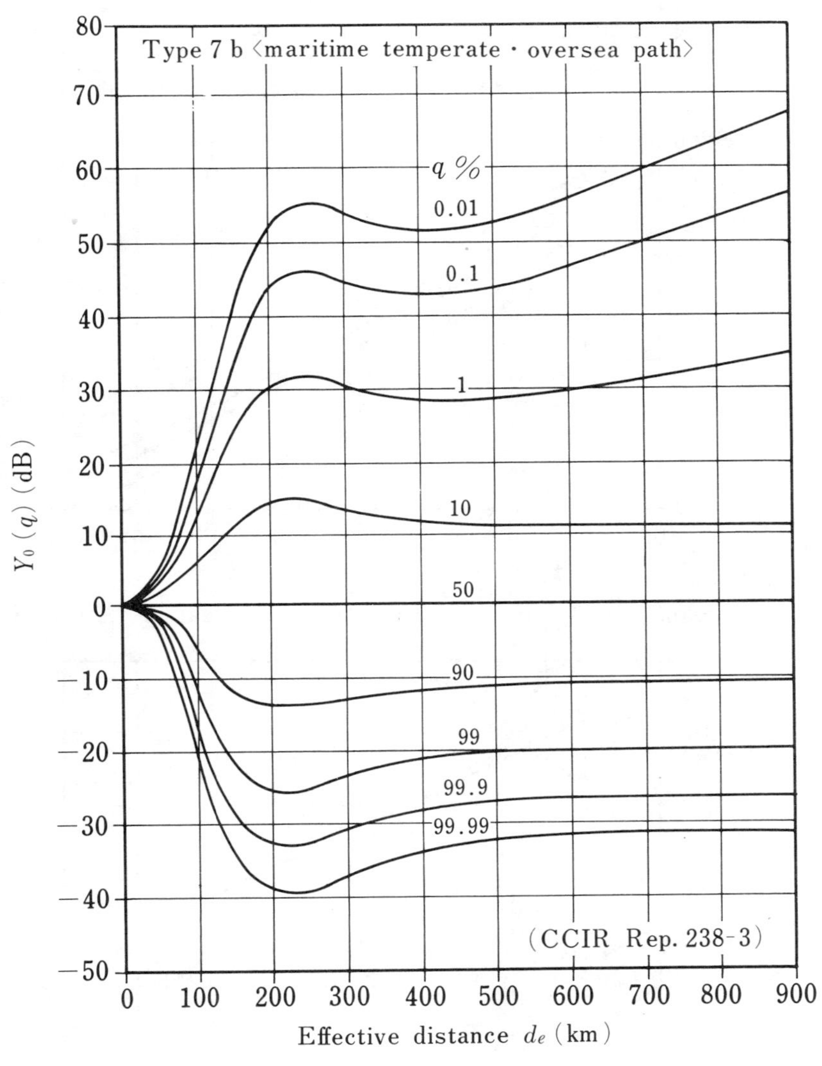

Simplified Method for Estimation (1)

[1] 99% value of propagation loss (worst month in a year, 1 GHz) F_1 (99)

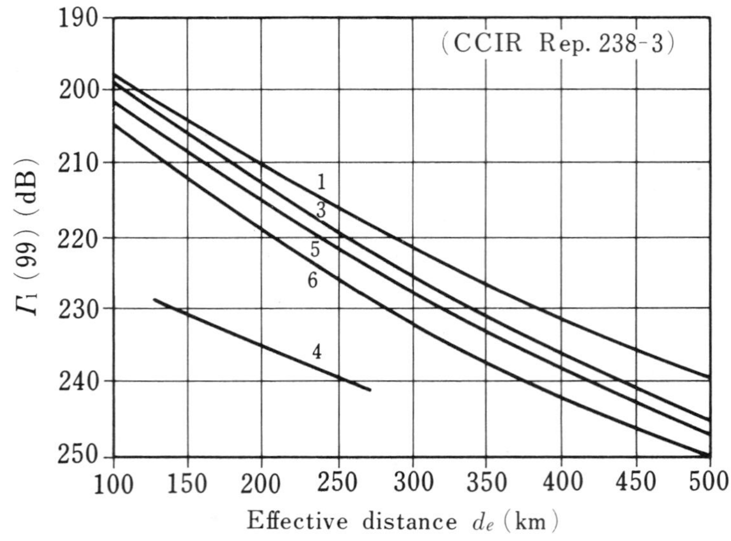

[2] 1% value of propagation loss (1 GHz in a year) F_1 (1)

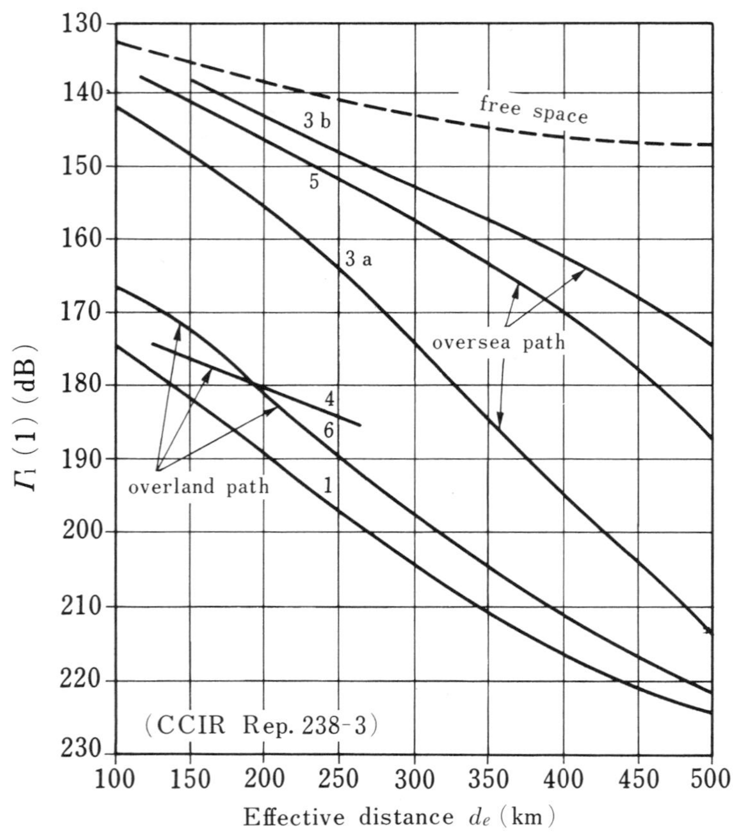

Simplified Method for Estimation (2)

[3] Frequency correction coefficient C_f

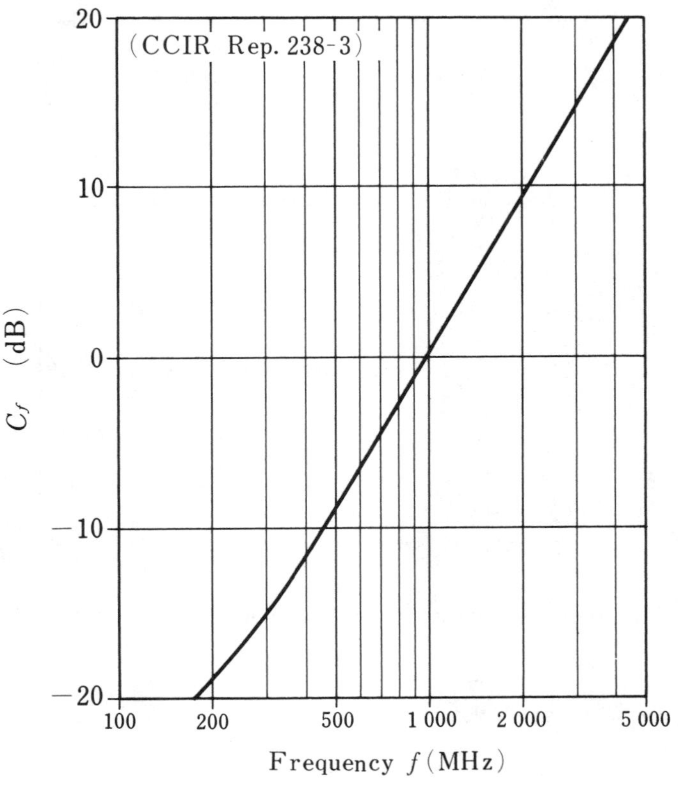

[4] Standard deviation σ

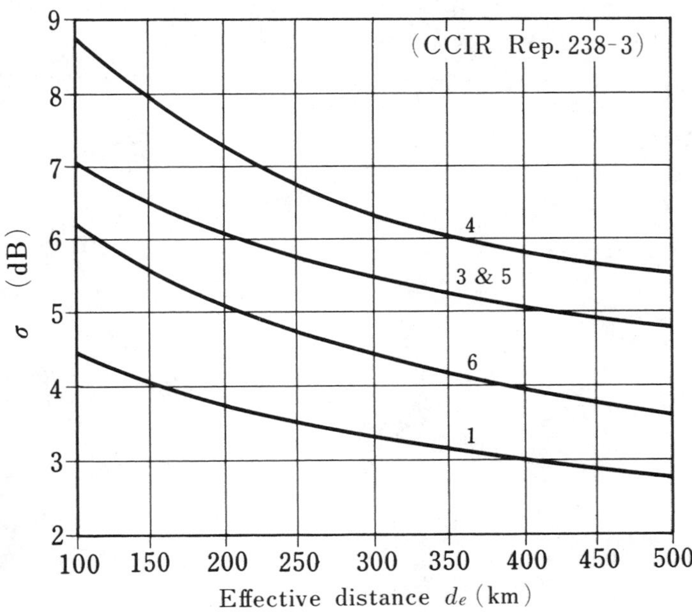

4

12

Path Antenna Gain and Transmission Loss

[1] Path antenna gain G_c

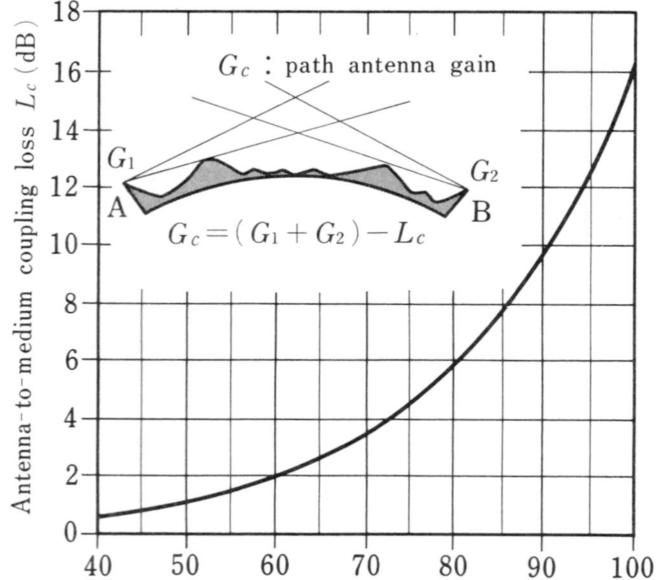

[2] Transmission loss L

$L = \Gamma - G_C = \Gamma - (G_1 + G_2 - L_C)$ (dB)

where L : Transmission loss between A and B

Γ : Propagation loss (basic transmission loss)

G_1, G_2 : Free space antenna gain

G_C : Path-antenna gain (troposcatter path antenna gain)

L_C : Antenna-to-medium coupling loss

$L_C = 0.07 \exp[0.055(G_1 + G_2)]$ (dB)

$\left\{ \begin{array}{ll} G_1 < 50 \text{ dB} & G_2 < 50 \text{ dB} \\ \text{distance } D = 150 \sim 500 \text{ km} \end{array} \right\}$

5

ABSORPTION PROPAGATION PATH

5.1 ABSORPTION ATTENUATION IN THE TROPOSPHERE

A boundless vacuum is the ideal medium for radio-wave propagation. The energy-density reduction due to unrestricted diffusion is the sole factor causing radio-wave attenuation. This can be explained by the fact that stars more than several million light-years away can be observed through light as one form of electromagnetic wave.

On the other hand, in close proximity to the earth surface, the presence of the atmosphere containing water vapor, water drops, and ice particles may cause absorption or scattering. Therefore it is necessary to determine whether or not such effects can be ignored in radio propagation design.

In general such effects can be ignored in or below the UHF band, regardless of the service mode of the radio wave. In and above the SHF band, the absorption more or less affects terrestrial point-to-point communication, interaircraft communication, ground-to-ground communication, radar, satellite-to-ground communication, radio astronomical observation, and so on. For radio waves above 10 GHz in particular, the absorption is a determining factor for system performance.

This chapter presents a quantitative analysis of these effects. However, large parts of this study lack sufficient data. The type of attenuation mechanism discussed here can be classified, in terms of causes, into absorption type and scattering type which includes partial reflection, refraction, diffraction, etc. However, both these types (absorption and scattering) are also called absorption type because they do not give a spatial interference pattern. Instead, they exhibit absorption attenuation characteristics with less frequency selectivity within normally employed radio channel bandwidths.

5.2 COMPOSITION OF THE ATMOSPHERE

The type of atmosphere is described before the degree of absorption is discussed. As is known empirically, the atmosphere contains oxygen, carbon dioxide, water vapor, and also clouds, which may generate rain and snow.

First, the composition of dry air, that is, the atmosphere without water vapor, is examined. Interestingly enough, the composition of the dry air exhibits little vertical variation between the surface near the ground and the upper portion of the stratosphere (approximately 54 km above ground), and little locational variation over the earth.

As shown in the table in Fig. 1*a*, nitrogen accounts for approximately 78% and oxygen for approximately 21%. Specifically, these two components account for 99%, while the remaining 1% is shared by carbon dioxide and a number of rare elements such as argon, neon, helium, krypton, and xenon.

These proportions are considered to have resulted from sufficient mixing in the atmosphere. However, the pressure decreases with altitude, and hence the absolute amount of gases becomes less (thinner gas) and the absorption and the resulting attenuation of radio waves decrease with altitude.

Water in the atmosphere may take the form of gas (water vapor), liquid (clouds, fog, or rain drops), or solid (snow or hail). Thus the effects caused by water are very complicated, depending on altitude, season, and geographic conditions.

5.3 CAUSES AND CHARACTERISTICS OF ATMOSPHERIC ATTENUATION

The frequency range under consideration is limited to within 100 GHz. In the case of dry air, only oxygen causes a substantial attenuation due to radio-wave absorption.

For humid air, the effects of the liquid components (rain, fog) are most significant, while gaseous components such as water vapor, exhibit a secondary effect at a specific frequency. The solid components such as snow and hail show lesser effects.

A. Oxygen

Since the density of oxygen in the atmosphere is proportional to the density of the atmosphere, its effect may be determined broadly for a given radio-wave path.

Frequencies that may cause a problem in this regard range from approximately 50 to 70 GHz, with peak absorption near 60 GHz.

Figure 2 shows attenuation per kilometer due to absorption as well as other causes. It is found that the water vapor as well as the fog or rain formed as a result of condensation of the water vapor generally have a greater effect on atmospheric attenuation than the oxygen gas.

In the desert, where there is little rain, attenuation due to absorption by oxygen predominates. However, this attenuation may be ignored since it is small except in the specific frequency band 50 to 70 GHz.

B. Water Vapor

The amount of water vapor in the atmosphere varies according to whether the radio path is over land or over seawater, and also according to the altitude and the season. Therefore estimation of the absorption attenuation is not as simple as for oxygen.

The absorption attenuation has a frequency characteristic with a peak at 22.2 GHz. However, the absorption attenuation at this frequency is only 0.2 dB/km for an average water vapor content (760 mm Hg with a mixing ratio of 7.6 g/kg) and causes only a secondary effect compared with other types of fading and attenuation by rain.

C. Rain

The dielectric constant of the atmosphere containing water vapor is much closer to unity (e.g., 1.00064), while the dielectric constant of the liquid H_2O is approximately 80. Consequently when a large amount of fog, raindrops, etc., are present over the propagation path, the radio waves traveling there undergo attenuation due to scattering or absorption.

The raindrop sizes are distributed widely between 0.1 mm in diameter, as in fog, and more than several millimeters (possibly 5 mm) in a torrential rain. Hence compared with a wavelength ranging from 30 to 3 mm or radio frequencies from 10 to 100 GHz, they cannot be ignored.

On the other hand, since the type and behavior of rainfall differ greatly according to region and time, and its behavior so far is only partially understood in spite of meteorological observations, no effective means to relate radio-wave attenuation directly to precipitation has yet been established.

Figure 2 shows the relation between radio-wave attenuation (dB/km) and hourly precipitation (depth of the rain water accumulated in a bucket). These data are comparatively easier to obtain from a number of precipitation parameters.

The precipitation in an hour may be called intensity of precipitation in 1 h. For example, a precipitation of an intensity of 50 mm/h will produce approximately 0.3 dB/km of radio-wave attenuation at 6 GHz according to Fig. 2. Thus it follows that if the propagation distance is 50 km, the attenuation reaches $0.3 \times 50 = 15$ dB.

However, a rainfall as heavy as 50 mm/h occurs very rarely, maybe once in a few years or in a 20- to 30-year period, even in a region of heavy precipitation.

Since the ordinary 50-km system on the 6-GHz band is generally designed to provide a margin of 15 dB or more (20 to 40 dB or more in certain cases), even a rainfall as heavy as the one mentioned will hardly produce system outage. Also, on a rainy day it is not likely to generate the fading caused by abnormal refractivity.

According to experience, even in a region of high humidity and heavy rainfall, the propagation path length will seldom be restricted by precipitation attenuation if the frequency is 10 GHz or less.

By increasing the radio frequency to more than 10 GHz, the precipitation, even when its intensity is low, will increase the attenuation, and the probability of a system outage rapidly worsens. Therefore rainfall is an essential item to be studied in system design, the details of which are discussed in later sections.

D. Snow, Hail, and Hailstones

The dielectric constant of ice is rather small, 2 to 3, while snow may be likened to floss silk made of ice and inflated with air. Therefore the effect on radio-wave attenuation will be only several tenths that caused by raindrops.

The opaque and ball-shaped soft hail with characteristics similar to snow and small semitransparent hail in the solid state produce more attenuation than snow. However, no substantial effect is produced because the amount of precipitation produced at a time is not very great.

Hailstones usually are approximately 5 mm in diameter and rarely reach 100 to 150 mm. Since hailstones are formed inside cumulonimbuses, they are observed more frequently in a warm period and occasionally in tropical regions such as Africa and Central and South America. However, they will hardly become serious for propagation design because of their trivial amount and frequency.

Such H_2O as rain, snow, and hail falling from the sky is designated generically as precipitation. Of these, only rain may become a problem for propagation.

E. Clouds and Fog

Since clouds and fog may remain over the propagation path for a long time, they cannot be ignored if they are of very high density and frequency. Figure 3 shows the attenuation Z_0 per kilometer of distance, caused by clouds or fog with an atmospheric water content of 1 g/m^3, versus frequency.

If the water drops are 0.01 mm or less in diameter, they usually drift in the atmosphere and form clouds or fog. Such humid atmosphere prevailing near the ground surface is called fog, while that drifting at medium or high altitude is called a cloud, but there is no substantial difference between them.

F. Smoke, Sandstorms, Etc.

As for the possible effects caused by sandstorms, smoke exhausted from large factories, and so on, there are no definite data available at present to support the correlation. Nevertheless, this is not harmful if the density, volume, and frequency of occurrence are sufficiently small.

Carbonic acid gas can be ignored in view of its density. Pulverized carbon on the other hand may be harmful only in the form of a smoke screen immediately in front of an antenna aperture; otherwise it can be neglected.

Since the dielectric constants of dried sand are rather low and the size of sand grains in a sandstorm is quite small, their effect may usually be ignored. However, because of insufficient experimental data, satisfactory system performances cannot be guaranteed if a strong sandstorm covers the whole propagation path.

5.4 ESTIMATION OF ATTENUATION

As a general theory, the absorption attenuation for the entire path length may be calculated through integration of the absorption attenuation (including the effect due to scattering) by infinitesimal portions over the entire path length. However, the actual and approximate estimation is carried out on a per kilometer basis, using past meteorological data and the results of physical and chemical model experiments, propagation tests, etc.

A. Attenuation Caused by Oxygen and Water Vapor

Terrestrial Microwave System

$$A_a = D(\gamma_{xo} p_{ho} + \gamma_{xw} p_{hw}) \quad \text{(dB)} \tag{5.1}$$

where A_a = attenuation by oxygen and water vapor
γ_{xo} = attenuation by oxygen (dB/km)
γ_{xw} = attenuation by water vapor (dB/km) (see Fig. 2)
D = propagation distance (km)
p_{ho}, p_{hw} = correction factors for change of contents of oxygen and water vapor, resulting from change of average altitude above mean sea level of the propagation path as compared to Fig. 2 curves prepared for zero-mean sea level. (The factor should include an effect due to temperature change in the strict sense.)

When $p_{ho} = p_{hw} = 1$, Eq. (5.1) reduces to

$$A_a = D(\gamma_{xo} + \gamma_{xw}) \quad \text{(dB)} \tag{5.2}$$

The expression suffices for estimating the path at an average altitude of 500 m or less above mean sea level under normal meteorological conditions. The equation may be applied safely for a path under other conditions.

Satellite Communication System (Earth Station to Satellite Station)

As shown in Fig. 1*b*, most of the propagation path is in a near vacuum with little atmosphere, and in the troposphere, the higher the altitude, the thinner are both oxygen and water vapor densities,

$$A_a = \gamma_{xo} d_{ea} + \gamma_{xw} d_{ew} \quad \text{(dB)} \tag{5.3}$$

where d_{eo} = equivalent distance for oxygen
d_{ew} = equivalent distance for water vapor

Particular attention should be paid to the peaks at 22.2 and 60 GHz.

Assuming that the water vapor whose density is indicated in Fig. 2 is distributed uniformly, the distances corresponding to the actual losses can be obtained; the path length of the wave traveling within the atmosphere layer varies according to the elevation angle from the earth station toward the satellite station, as shown in Fig. 1*b*. Then d_{eo} and d_{ew} for $\alpha = 90°$ will be approximately 4 and 2 km, respectively.

Figure 4 shows the atmospheric absorption attenuation for $\alpha = 0°$ (horizontal) and $\alpha = 90°$ (vertical).

B. Attenuation Caused by Clouds and Fog

Employing the attenuation factor Z_c given in Section 5.3E, the attenuation A_c for the distance D (km), where the water content is M (g/m^3), is

$$A_c = Z_c MD \quad \text{(dB)} \tag{5.4}$$

C. Attenuation Due to Rainfall

Since, among various atmospheric phenomena, rainfall has a dominant effect on radio-wave attenuation, careful studies are needed for both terrestrial microwave and satellite communication systems when they are operated at about 10 GHz or higher. However, such studies may be omitted for many areas in the world, such as regions with scant rainfall (the Sahara and Gobi deserts) and those with annual rainfall of less than 100 to 300 mm. (There are 1000 to 3000 mm per year in Japan.)

The attenuation due to rainfall within a short period is determined by the total amount of water, its density distribution, and the sizes of the raindrops contained in the first Fresnel zone of the radio path at the same time.

However, the rainfall phenomenon itself is unstable, both spatially and temporally, and it is very hard to extract the influential factor quantitatively. In practice, the attenuation is estimated on the basis of statistical figures on average rainfall intensity for a certain period, such as 1 h, 10 min, or 1 min.

Figure 5 shows estimating graphs prepared as follows.

1. Intensity of 10-min Rainfall. Indicated as R on the abscissa in Fig. 5*a*.
2. Attenuation per Kilometer Z_r. The attenuation per kilometer of a radio wave at each frequency has been obtained experimentally and is indicated on the ordinate against the rainfall intensity on the abscissa in Fig. 5*a*.
3. Rainfall Equivalent Distance *Dr*. It is unlikely that rainfall occurs simultaneously and uniformly throughout the propagation path. The heavier the rainfall, the narrower, in general, is the rainfall area. In other words, the assumption that heavy rain occurs only 0.1 to 0.01% of the time under consideration, and over the entire radio path, could overestimate the attenuation, thereby requiring appropriate adjustment. The equivalent distance on the basis of uniform rainfall is shown in Fig. 5*b*.
4. Estimation of attenuation,

$$A_r = Z_r D_r \tag{5.5}$$

where A_r = rainfall attenuation in decibels exceeded for P % of the time under consideration

Z_r = attenuation per kilometer (dB/km) for 10-min rainfall intensity

D_r = rainfall equivalent distance (km)

D. System Time Reliability and Statistical Value of Rainfall Intensity

For 10-min rainfall intensity R, the corresponding figure from data collected at the observatory that is nearest to the path and located in an area with a similar rainfall condition should be used. Moreover, the value should also be a statistical value, corresponding to the specified time rate, to maintain the performance required for the system. For example, when the time rate required for maintaining the performance is q (%), the complement

$$p = 100 - q \quad (\%) \tag{5.6}$$

indicates the time rate for which the performance may deteriorate to below the requirement.

The rainfall tends to be concentrated within several months in a year. In the absorption radio propagation path, emphasis is placed on maintaining performance during this period, and it is more realistic to express the reliability as 99.9% of the time during the rainiest 2 to 4 months, 99.99% of the same, or S/N > y (dB) during the rainiest 2 to 4 months, except x min.

The 99.99% of the rainiest 3 months may be converted to a period expressed in minutes for exclusion,

$$p\ (\%) = (100 - 99.99) = 0.01\%$$

$$\begin{aligned} p\ (\text{mins}) &= \{3\ (\text{months}) \times 30\ (\text{days}) \times 24\ (\text{hs}) \times 60\ (\text{mins})\}\, p/100 \\ &= 13 \text{ mins} \end{aligned} \tag{5.7}$$

Conversely, the 10-min exclusion corresponds to $p \doteqdot 0.0075\%$.

Therefore the system design that will ensure the required circuit performance is carried out in such a manner that excluding the heavier rainfall accounting for $p\%$ of the total rainfall data in the rainiest season, the value of the maximum rainfall intensity in the remaining portion is applied to Fig. 5.

Method to Determine the x-min Rainfall Intensity for p Percent Cumulative Probability

The periods of observation are 5 to 10 years. Each year consists of 1 to 3 rainier months, and the precipitation is determined for every 10 min (10-min rainfall intensity) for the entire period to be observed.

The rainfall intensity for any 10-min period with no rain or snow is indicated as 0 (zero). The total number of data is 6 per hour, 144 per day, 4320 for 30 days, and 129,600 for 3 months times 10 years.

Rearrange the total number of these data in the order of decreasing intensity and extract the one corresponding to or nearest the ten-thousandth of the total number of data. The resulting data present the 0.01% 10-min rainfall intensity.

The reference data for 0.01% 10-min rainfall intensity shown in Fig. 5b[2] were prepared from data collected from all meteorological observatories in Japan.

5.5 DESIGN CHARTS FOR 8- TO 15-GHZ BANDS

Figure 6a, b shows the charts for free-space propagation loss Γ_0 covering the frequencies of 8 to 15 GHz and the distances 2.5 to 50 km; Fig. 6c shows free-space propagation loss of Γ_0 (for satellite communication) covering 10 to 50 GHz and 10 to 10^5 km.

Figure 7 gives the charts for rainfall attenuation Γ_p (additional propagation loss) covering 0 to 50 km in distance and 4 to 16 mm per 10-min rainfall intensities.

Thus the propagation loss Γ is given by

$$\Gamma = \Gamma_0 + \Gamma_p \quad \text{(dB)} \qquad \text{(with } q \text{ percent reliability)} \tag{5.8}$$

When the statistical value for 10-min rainfall intensity R is given for $p\%$ rainfall intensity, Γ_p is obtained as a value corresponding to $q = (100 - p)\%$. Accordingly the thus obtained Γ gives the value for the $q\%$ reliability.

The following supplemental remarks refer to Fig. 7.

1. If the rainfall intensity is not provided as a statistical value for a 10-min unit period, it shall be converted to the value for the 10-min unit period.
2. Large errors occur when p is not in the range of 0.05 to 0.005%. In general, the propagation loss may be underestimated if $p > 0.05$ and overestimated if $p < 0.005$.
3. Estimated results differ appreciably, depending on the methods of estimation, and it is difficult to provide an objective evaluation of these methods. Accordingly, it is not wise to request excessively high accuracy.
4. Figure 7 should be employed effectively for the purpose of system design, taking into account the points mentioned above.

5.6 WORLD RAINFALL

Figure 9 summarizes the precipitation reported during a 30-year period. The data were obtained from about 450 observatories throughout the world for high-rainfall-rate areas with a monthly mean precipitation of 300 mm or more and for low-rainfall-rate areas with a yearly mean precipitation of less than 100 mm.

The time reliability for designing a microwave propagation path using a frequency of 10 GHz or higher is between 99.9 and 99.99% in general, depending on the purpose.

As mentioned in Section 5.4D, the necessary rainfall data should be between 0.1 and 0.01%, and the data-measuring time unit should normally be several minutes to 1 hr.

As shown in Fig. 9, there is no clear correlation between yearly precipitation in high-rainfall-rate months. Similarly, there is poor correlation between monthly precipitation and precipitation on high-rainfall-rate days as well as between daily precipitation and short-term precipitation (rainfall intensity to be applied in the system design).

Although Fig. 9 cannot serve directly in system design, it is nevertheless useful for grasping local weather conditions and establishing a method of approach, including data collection. Moreover, Fig. 9 may indicate that planning a system in a high-rainfall-rate area requires a well-organized preliminary survey. In a low-rainfall-rate area, collection of simple data suffices, while a detailed survey is required when a long-span radio path is to be designed.

5.7 COUNTERMEASURES IN RADIO SYSTEM DESIGN AGAINST OUTAGE DUE TO HEAVY RAIN

Since frequency diversity for 10-GHz or higher frequency bands or space diversity is not effective in reducing the system outage due to rainfall, suitable countermeasures must be applied.

A. Route Diversity

The high-rainfall-rate area tends to move and concentrate in a limited local area. Therefore the probability that two separate radio routes suffer simultaneously from a system outage is extremely low. The effect is enhanced further if the two routes are separated by intermediate mountains and designed to form a loop.

In the case of a public communication system composed of a meshlike network, route diversity is easily attained via equipment to distribute the traffic and also to switch over the routes. However, if the system is composed of only one route, the required investment for the route diversity system will be doubled compared with that for the single-route system.

B. Auxiliary System Using Lower Frequency Band

In this case an auxiliary system is installed along the main route, using an 8-GHz or lower frequency band that is almost free from rainfall attenuation. The system can be established economically by a hop, deleting one or several intermediate repeaters operating in a 10-GHz or higher frequency band. Since no duct-type fading occurs during rainfall, the hop length may be extended to 60 to 80 km if good line of sight is obtained.

C. Lower Frequency Auxiliary System with Antenna Sharing

The main beamwidth of the antenna is inversely proportional to frequency. If station sites are determined so as to avoid overreach interference at frequencies of 10 GHz or higher for the main route, and if the auxiliary system using the lower frequency band is designed in such a way that one or more intermediate repeater stations are deleted and in addition any antenna is directed to share use with the main system within the half-power width, the low-frequency auxiliary system can be provided economically (refer also to Example 5.5).

Although there will be topographical restrictions, such design consideration is useful not only as a remedy for rainfall attenuation, but also for installing two different systems in parallel and for establishing an emergency radio communication system.

5.8 ABSORPTION ATTENUATION IN MILLIMETER WAVES

Since the electromagnetic wave with millimeter wavelength (corresponding to the EHF band) or shorter causes very extensive attenuation due to the atmosphere and the rain in the troposphere, this wave is practical only for short-distance systems in the case of a terrestrial radio system. The development of millimeter and shorter waves has recently progressed rapidly for satellite–earth radio communication systems because the radio path takes a higher elevation angle to the horizontal. Accordingly, the radio wave travels through a smaller portion of the atmospheric layer, resulting in smaller atmospheric attenuation; the wave is suitable for broadband transmission systems. Figure 10 shows the attenuation per kilometer for centimeter, millimeter, and submillimeter waves near the ground surface.

EXAMPLES

Example 5.1

An 8-GHz, 75-km propagation path is designed to provide a 25-dB margin against the fading in the receiver input. Determine the maximum rainfall intensity expressed in millimeters per hour allowable for the system.

Solution

1. The 25-dB margin of receiver input may be allocated fully to absorption fading because no refraction fading caused by duct, superrefraction, etc., occurs on a rainy day.
2. Allowable attenuation per kilometer M_0,
$$M_0 = 25 \text{ dB}/75 \text{ km} = 0.333 \text{ dB/km}$$
3. Referring to Fig. 2 (8 GHz on the abscissa and 0.333 dB/km on the ordinate), the obtained rainfall intensity is 25 mm/h.
4. Accordingly, the system can be operated normally at least up to 25 mm/h. (It is very rare to have a uniformly distributed heavy rainfall of 25 mm/h over the 75-km path.)

Remarks. No reference is made to time probability in Fig. 2. Therefore it is necessary to estimate the time probability that the hourly rainfall intensity exceeds 25 mm from the rainfall data collected in the area under consideration. The greater the amount of data, and the longer the period over which the data were collected, the higher the reliability of the estimate. The data are collected normally during a period of two to four of the rainiest months over 10 or more years. When the time probability is denoted by $p\%$, the corresponding circuit performance is ensured for $100 - p\%$, the complementary period.

Example 5.2

A meteorological satellite transmits information on 3, 10, and 20 GHz. Estimate the absorption attenuation in decibels to be taken into account in engineering the receiving facilities of an earth station excluding the effect of precipitation. The satellite is assumed to be in random orbit.

Solution

1. Figure 4 is used.
2. Since a random-orbit satellite is employed, a 0° elevation angle toward the horizon has to be assumed.
3. A one-way attenuation is determined by Fig. 4 as

$$\approx 2 \text{ dB} \quad \text{at } 3 \text{ GHz}$$

$$\approx 4 \text{ dB} \quad \text{at } 10 \text{ GHz}$$

$$\approx 20 \text{ dB} \quad \text{at } 20 \text{ GHz}$$

Example 5.3

Determine the maximum length in kilometers of the propagation path where $f = 12$ GHz, transmitter output $P_t = 200$ mW (23 dBm), antenna gains $G_t = G_r = 38$ dB, feeder system losses $L_{ft} = 2$ dB and $L_{fr} = 4.5$ dB, squelch threshold level $P_{\text{cut}} = -72$ dBm. Required system reliability $q = 99.99\%$, and 0.01% 10-min rainfall intensity $R = 10$ mm must be assumed.

Solution

1. Receiver input power P_r,

$$P_r = P_t - (\Gamma_0 + \Gamma_p) + (G_t + G_r - L_{ft} - L_{fr})$$

$$P_r = 23 - (\Gamma_0 + \Gamma_p) + (38 \times 2 - 2 - 4.5) = 92.5 - (\Gamma_0 + \Gamma_p)$$

2. To maintain the required circuit performance, P_r shall be greater than the squelch operation threshold level P_{cut},

$$P_{\text{cut}} \leq P_r$$

$$-72 \leq 97.5 - (\Gamma_0 + \Gamma_p)$$

$$\Gamma_0 + \Gamma_p \leq 92.5 + 72$$

Hence

$$\Gamma_{\text{req}} \leq 164.5 \qquad \text{(allowable propagation loss)}$$

3. To check the possibility, an estimate is first made assuming $D_1 = 10$ km and $D_2 = 20$ km. For $D_1 = 10$ km,

$$\Gamma_0 = 134.0 \qquad \text{(Fig. 6}a\text{)}$$

$$\Gamma_p = 21.4 \qquad \text{(Fig. 7}g\text{)}$$

$$\Gamma_0 + \Gamma_p = 155.4 < 164.5$$

For $D_2 = 20$ km,

$$\Gamma_0 = 140.0 \qquad \text{(Fig. } 6a\text{)}$$

$$\Gamma_p = 33.5 \qquad \text{(Fig. } 7g\text{)}$$

$$\Gamma_0 + \Gamma_p = 173.5 > 164.5$$

Therefore the applicable distance is between 10 and 20 km.

4. The more detailed relation is drawn below as a curve.

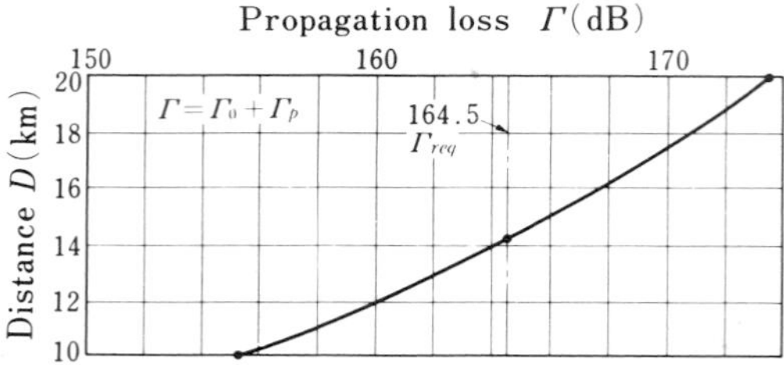

5. From this curve, maximum applicable distance = 14.3 km

Example 5.4

Determine the transmitter output power required to expand the applicable range in Example 5.3 to 18 km.

Solution

1. The propagation loss Γ for 18 km is obtained from the figure in Example 5.3,

$$\Gamma = 170.7$$

2. This is compared with the allowable propagation loss of 164.5 dB in Example 5.3. Then

$$\text{excessive loss} = 170.7 - 164.5 = 6.2 \text{ dB}$$

Accordingly the increase in required power P_t is 6.2 dB.

3. Calculation of P_t,

$$P_t \text{ (dBm)} = 23 + 6.2 = 29.2$$

Then

$$P_t \text{ (mW)} = 832 = 0.85 \text{ watt}$$

EXAMPLE 5.5

Plans are made to install a 12-GHz 1800-channel trunk transmission system between telephone exchanges at city *A* and city *G*. Between these two cities are a mountain ridge and an isolated mountain. On the ridge and on the mountain, there are an existing radio relay station *C* and a meteorological observatory *E*, respectively. Prepare a station site selection plan making maximum use of the existing facilities. It is assumed that the applicable distance to the 12-GHz system is 18 km at maximum and the 4-GHz auxiliary system is to be installed in parallel with the main system to protect it. Intermediate terrains consist of a flat field and farmland. The station site may be chosen at any location. Assumed distances are $AC = 35$ km, $CE = 28$ km, and $EG = 35$ km.

Solution

1. Key points for the station site selection plan:
   (a) Utilization of existing facilities.
   (b) Minimization of the number of hops not exceeding the feasible hop length.
   (c) Prevention of overreach interference in two-frequency plan in principle.
   (d) Provision of a lower frequency auxiliary system as well as shared use of the antenna.
2. The elevated point *C* or *E* is chosen for installation of the auxiliary station using the lower frequency band. An examination of the contours on the map indicate that the line of sight is not available for the section *EA*, while it is available for sections *CA* and *CG*. Consequently point *C* is given first priority.
3. Site location plan.

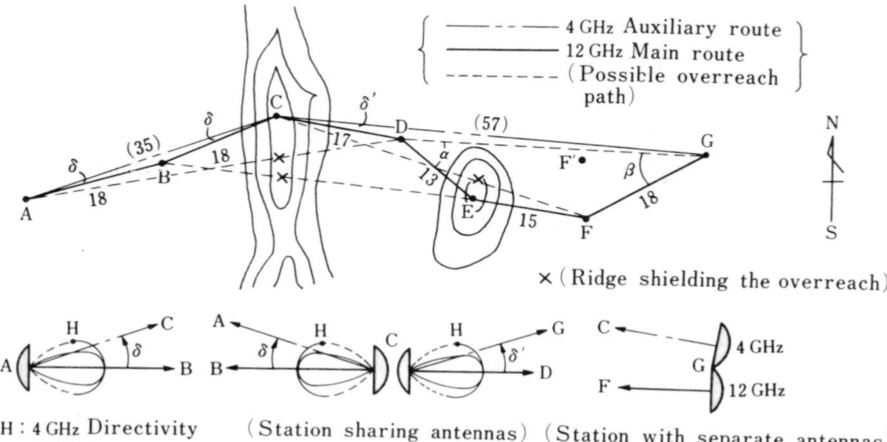

4. Comments.
   (a) A minimum of 6 hops are needed, with each hop length within the allowable limit.
   (b) The auxiliary system is realizable with a single repeater (including two hops).
   (c) Sharing use of one antenna for 12- and 4-GHz bands is possible at stations *A* and *C* ($\delta, \delta' \leq$ half power beamwidth/2 of directivity at 4 GHz).
   (d) Overreach interference is effectively prevented for path *DG* due to large angles α and β, and for path *AD* due to the presence of a shielding ridge.
   (e) δ and δ' may be $0°$ at maximum. The antenna heights at stations *A* and *B* are to be determined so as to avoid the dip point of the antenna height pattern at 4 GHz.
   (f) Alternate plan. To share use of the antenna at station *G*, the station site *F* is to be moved to site *F'*, but in this case the overreach interference paths *GD* and *F'C* will become a problem. This alternate plan is the most economical if the overreach interference can be suppressed using the cross-polarization technique, for example.

Chart Index, Absorption Propagation Path

Causes of Absorption and Attenuation in Terrestrial Microwave Propagation

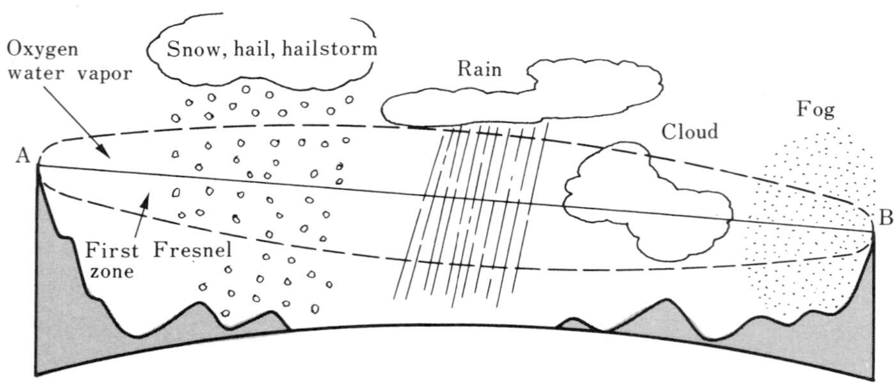

Causes of atmospheric absorption				
Gas	Oxygen	Peak attenuation is at 60 GHz. Excluding 60±10 GHz band, the attenuation caused by rainfall is predominant.		
	Water vapor	The peak attenuation is at 22 GHz.		
Liquid	Fog, cloud	Only the secondary effects are observed and the higher the frequency is, the greater the attenuation is.		
	Rain	Attenuation is higher at greater rainfall intensity or at higher frequency.	Precipitation	
Solid	Snow, freezing rain, hail, hailstorm	Attenuation is so small that the effect is secondary except in the case of moist snow (mixed with rain) even when the precipitation is great.		
	Smoke, sandy dust	Attenuation is negligible because of its small density.		

Composition of dry air										
Name of gas	N_2 Nitrogen	O_2 Oxygen	Ar Argon	CO_2 Carbon dioxide	Ne Neon	He Helium	Kr Krypton	Xe Xenon	H_2 Hydrogen
Volume %	78.08	20.95	0.93	0.03	0.0018	5×10^{-4}	1.1×10^{-4}	0.9×10^{-5}	Less than 10^{-3}

Horizontal and Vertical Paths

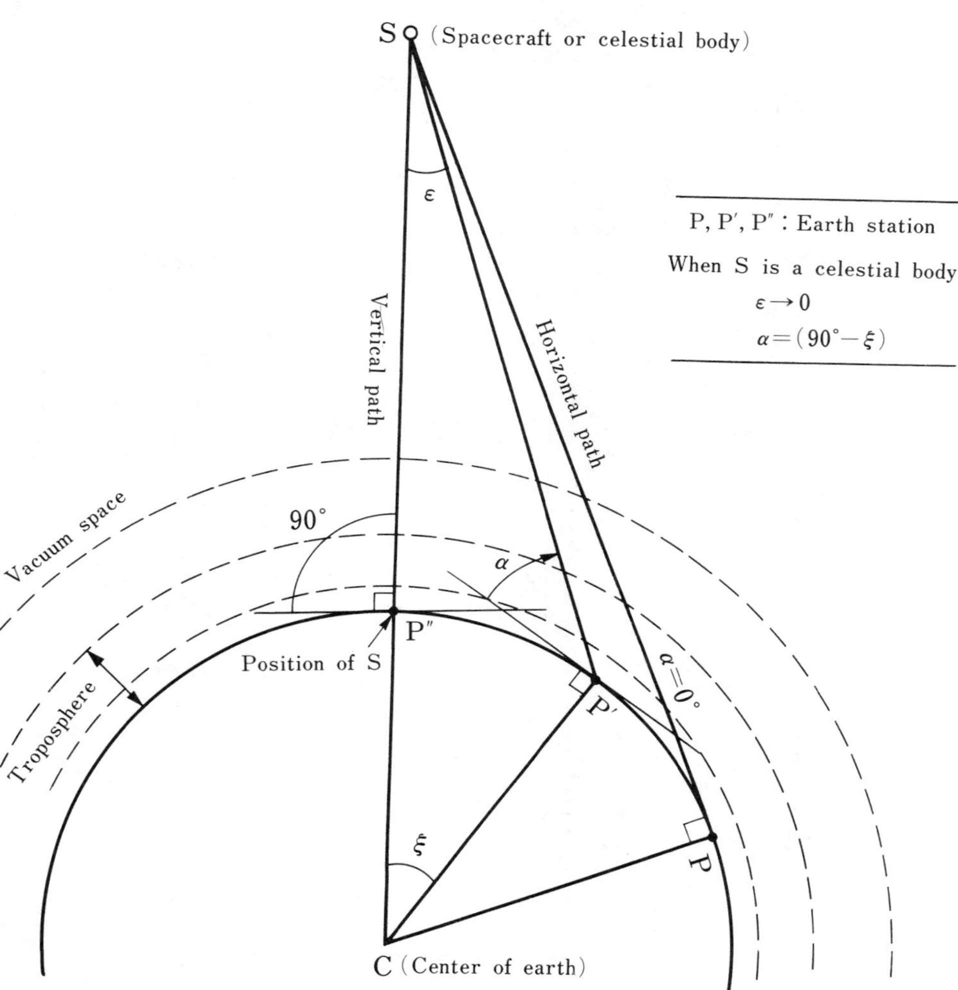

Attenuation Due to Oxygen, Water Vapor, and Rain

Extracted from CCIR Rep. 234-2

Attenuation per km γ_x (dB/km)

Frequency f (GHz)

Precipitation per hour R (mm/h): 100, 50, 25, 10, 1

Water vapor

Oxygen

*1

*2

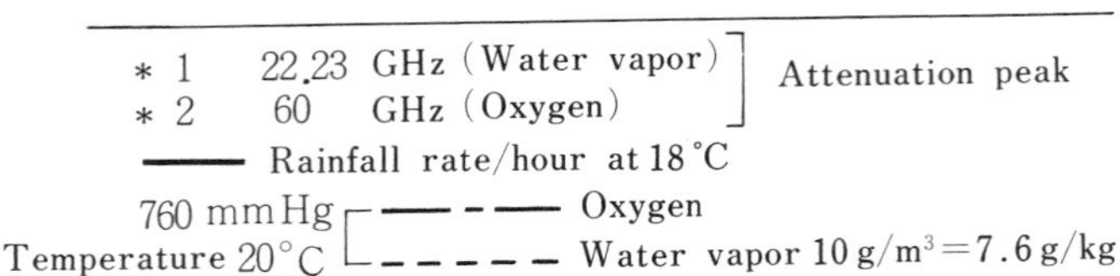

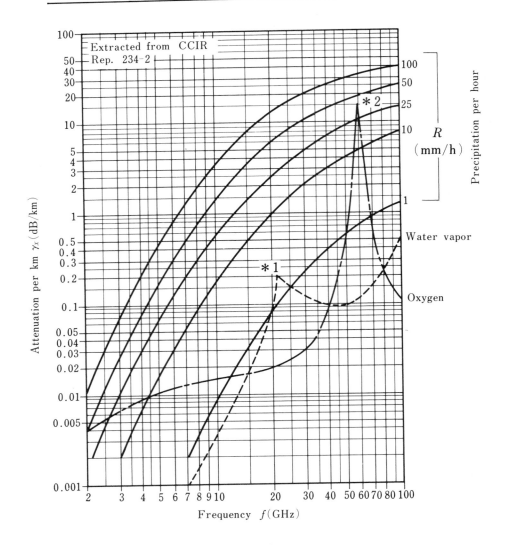

Attenuation Due to Clouds and Fog

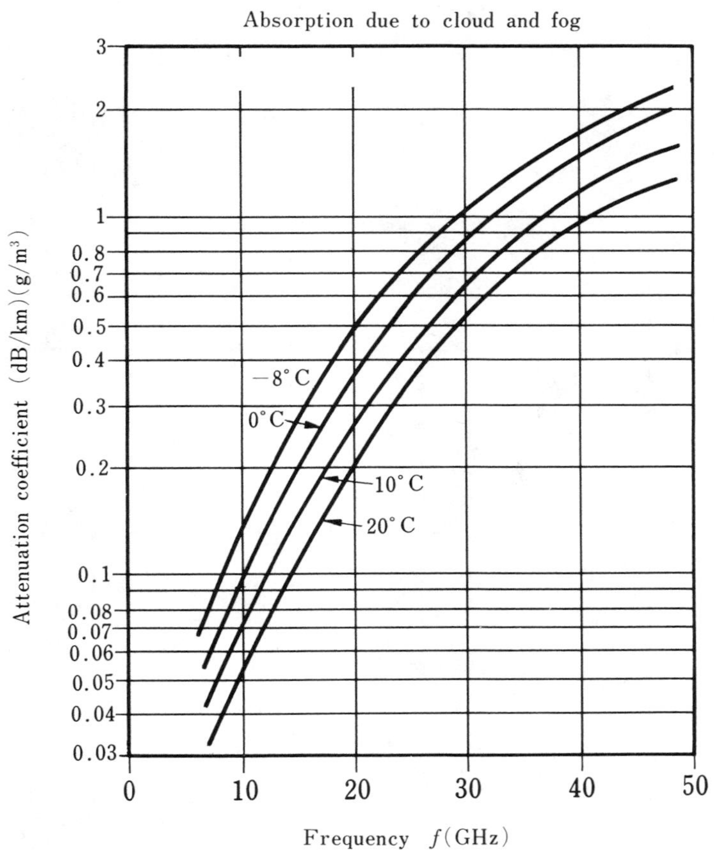

Atmospheric Absorption Loss in Satellite Communication

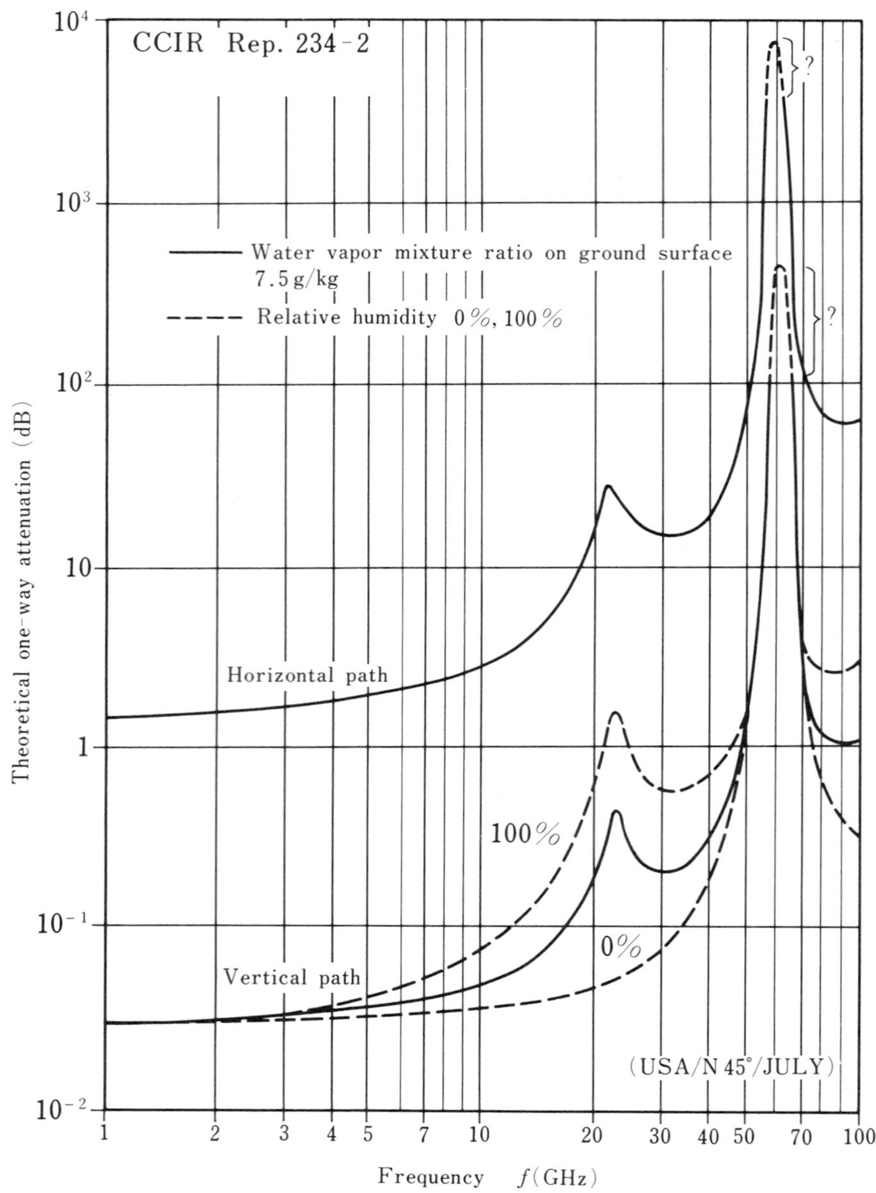

Rainfall Attenuation Estimation Graph (1)

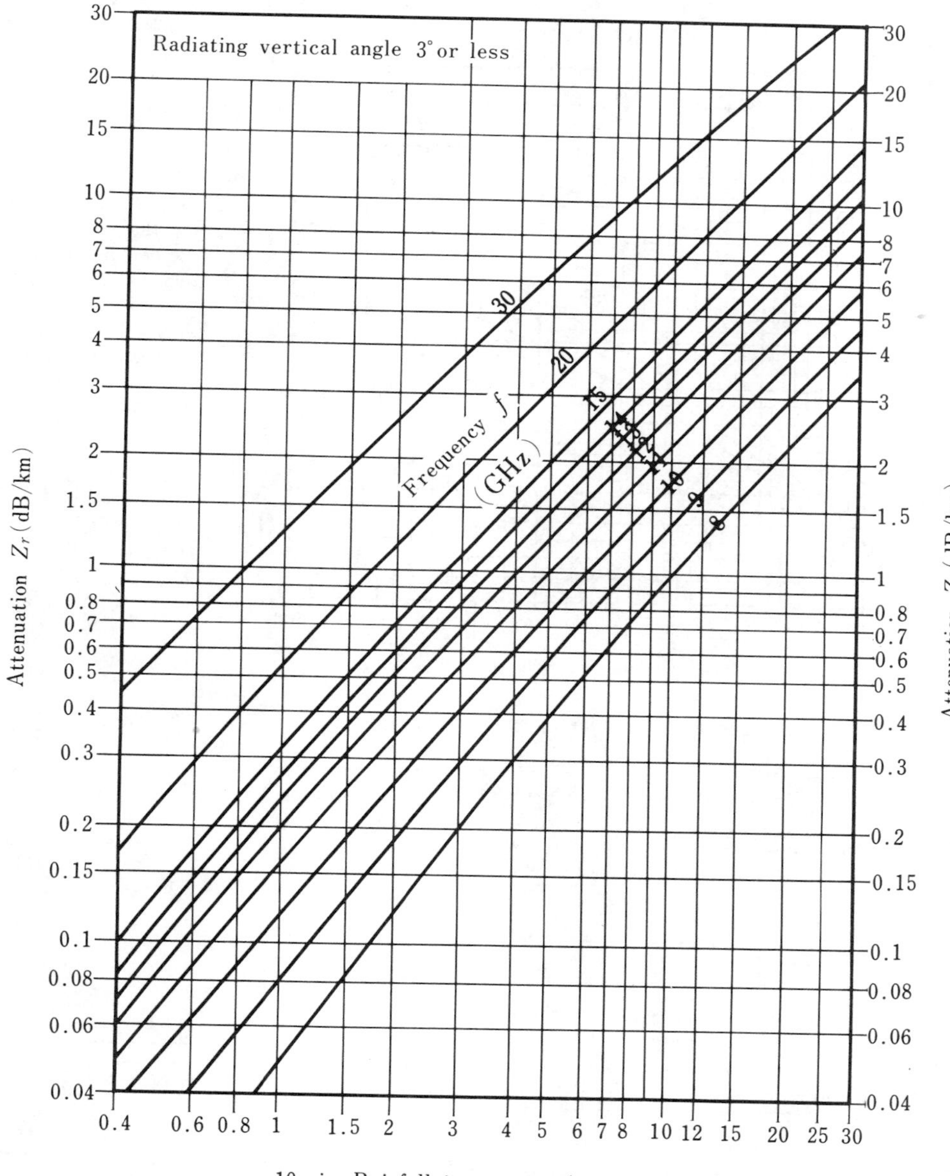

5

5b Rainfall Attenuation Estimation Graph (2)

[1] Rainfall equivalent distance D_r

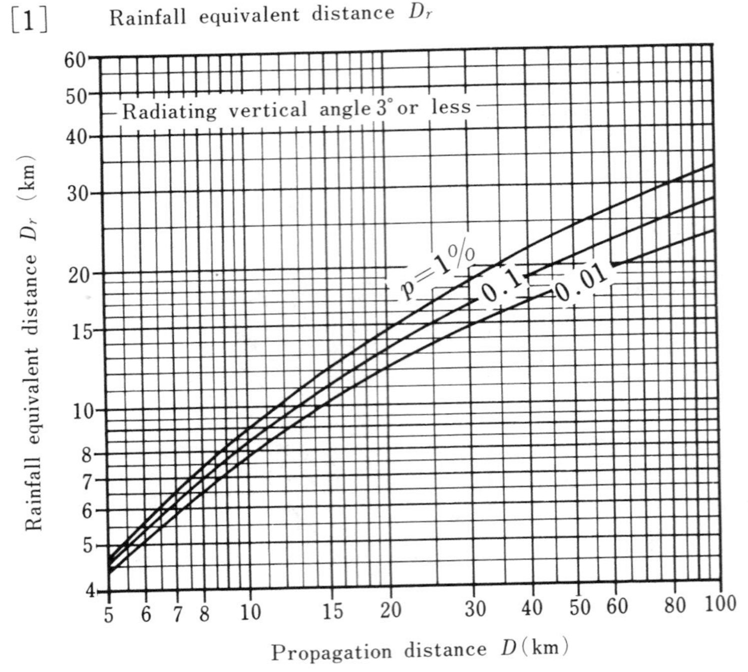

1. This graph is not applicable when the vertical angle is large (as in ground-satellite, ground-aircraft paths, etc.).
2. For $D > 30$ km, the diffraction fading is increased and this will restrict the applicable operating distance on a fine day.

[2] Example of 10-min rainfall intensity in Japan

mm/10 min	Location (meteorological observatory)	121
7	Kushiro, Haboro, Asahikawa, Obihiro, Urakawa, Iwa-mizawa, Otaru, Sapporo, Kucchan, Esashi, Tanabu	11
8	Mori, Aomori, Shinjo, Wakamatsu, Iida	5
9	Rumoi, Suttsu, Takada, Choshi, Kobe, Toyooka, Matsunaga, Hamada, Kure	9
10	Muroran, Hakodate, Yamagata, Fukushima, Onahama, Morioka, Yokohama, Kofu, Hikone, Tokushima, Matsuyama, Hashihama, Sakai	13
11	Akita, Ishinomaki, Niigata, Tomisaki, Nagano, Matsumoto, Mishima, Fushiki, Shizuoka, Hamamatsu, Takayama, Ibukiyama, Tsuruga, Osaka, Himeji, Tadotsu, Uwajima, Hiroshima, Shimonoseki	19
12	Tomakomai, Hachinohe, Tsukubasan, Ajiro, Toyama, Tsu, Wakayama, Sumoto, Tsurugisan, Tottori, Yonago, Tsuyama, Oita, Onsendake	14
13	Sakata, Sendai, Chugushi, Mito, Maebashi, Karuizawa, Funatsu, Nagatsuro, Kanazawa, Fukui, Nagoya, Gifu, Kameyama, Kyoto, Maizuru, Takamatsu, Shimizu, Sukumo, Hofu, Kumamoto, Asosan, Aburatsu	22
14	Shirakawa, Tokyo, Chichibu, Omaezaki, Kashihara, Hagi, Fukuoka, Makurazaki	8
15	Nagasaki, Tomie, Yakushima, Tanegashima, Akune, Hitoyoshi, Matsue, Owase, Kumagaya, Iizuka	10
16	Miyakonojo, Saga, Hita, Hirato	4
17	Kagoshima, Saseho, Odaigawara, Kochi	4
18	Shionomisaki, Utsunomiya	2

Note: The lower figures show the total number of weather observatories.

Free-Space Propagation Loss (1)

2.5–25 km (8–15 GHz)

Free-space propagation loss Γ_0 (dB)

f (GHz)

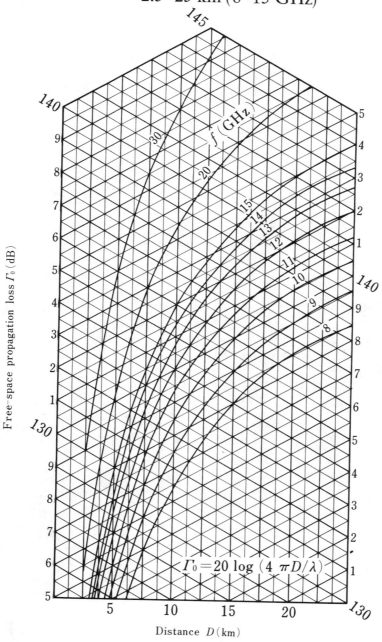

$\Gamma_0 = 20 \log (4\pi D/\lambda)$

Distance D (km)

Free-Space Propagation Loss (2)

25–50 km (8–15 GHz)

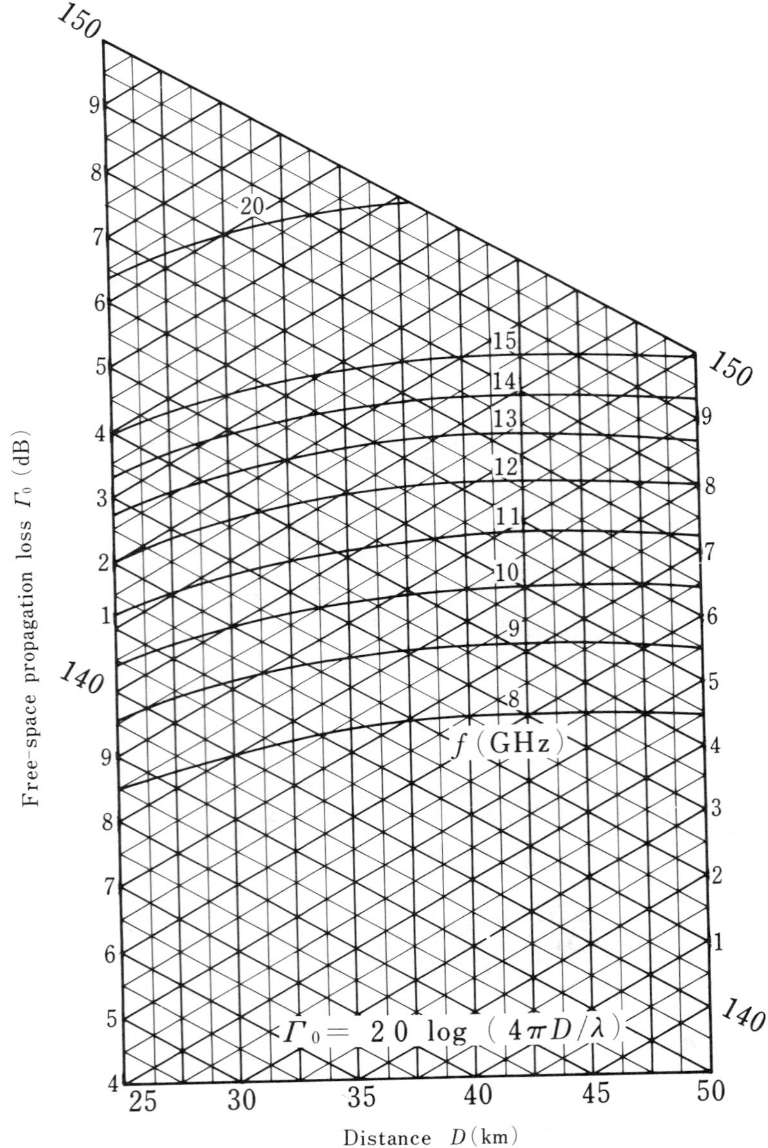

Free-Space Propagation Loss (3)

10–100,000 km (10–50 GHz)

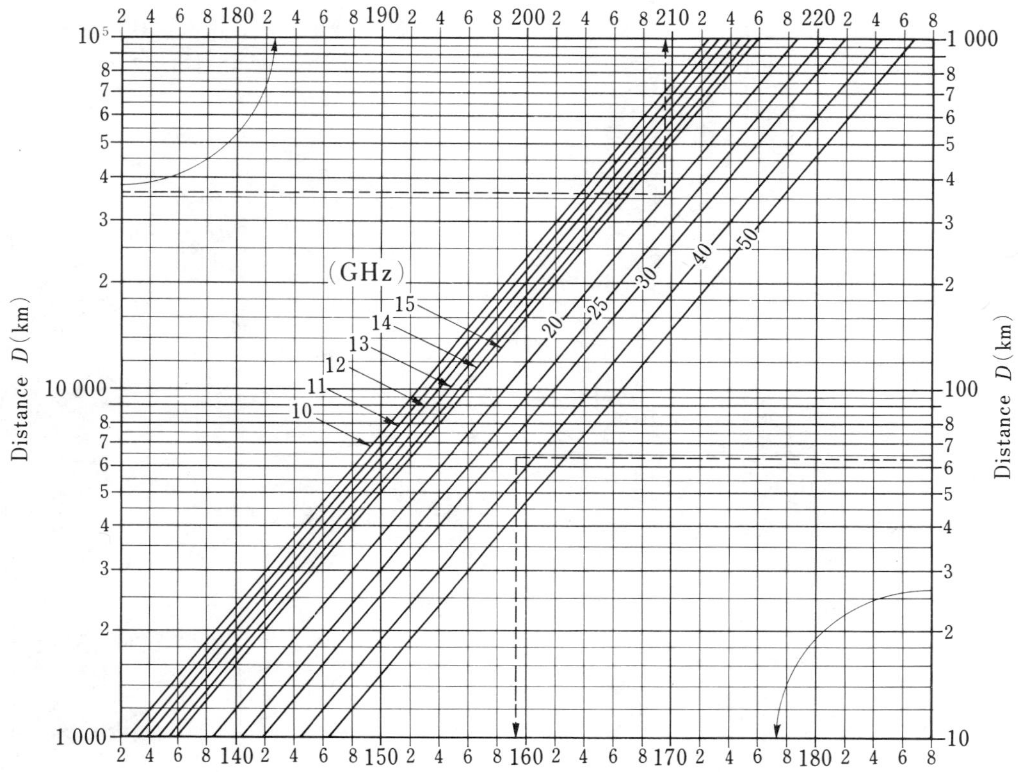

Design Chart for Rainfall Attenuation (1)

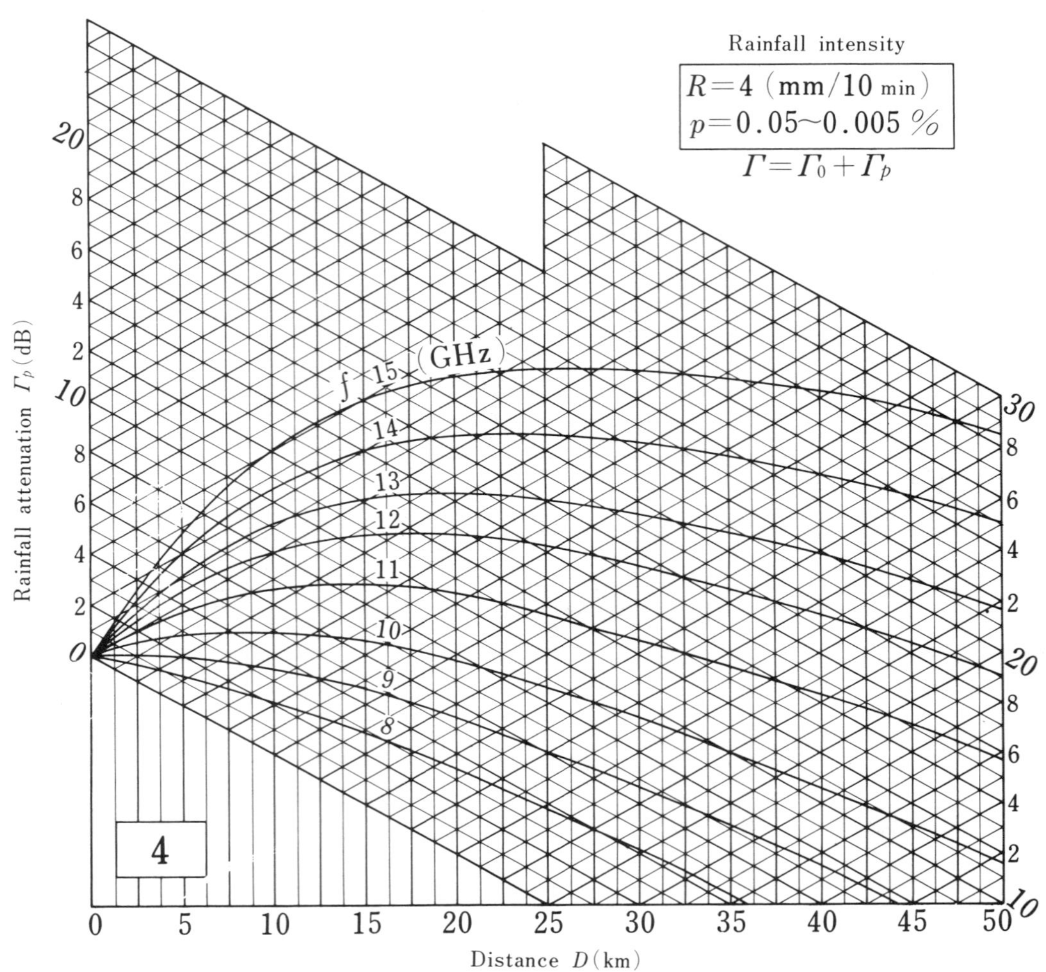

Design Chart for Rainfall Attenuation (2)

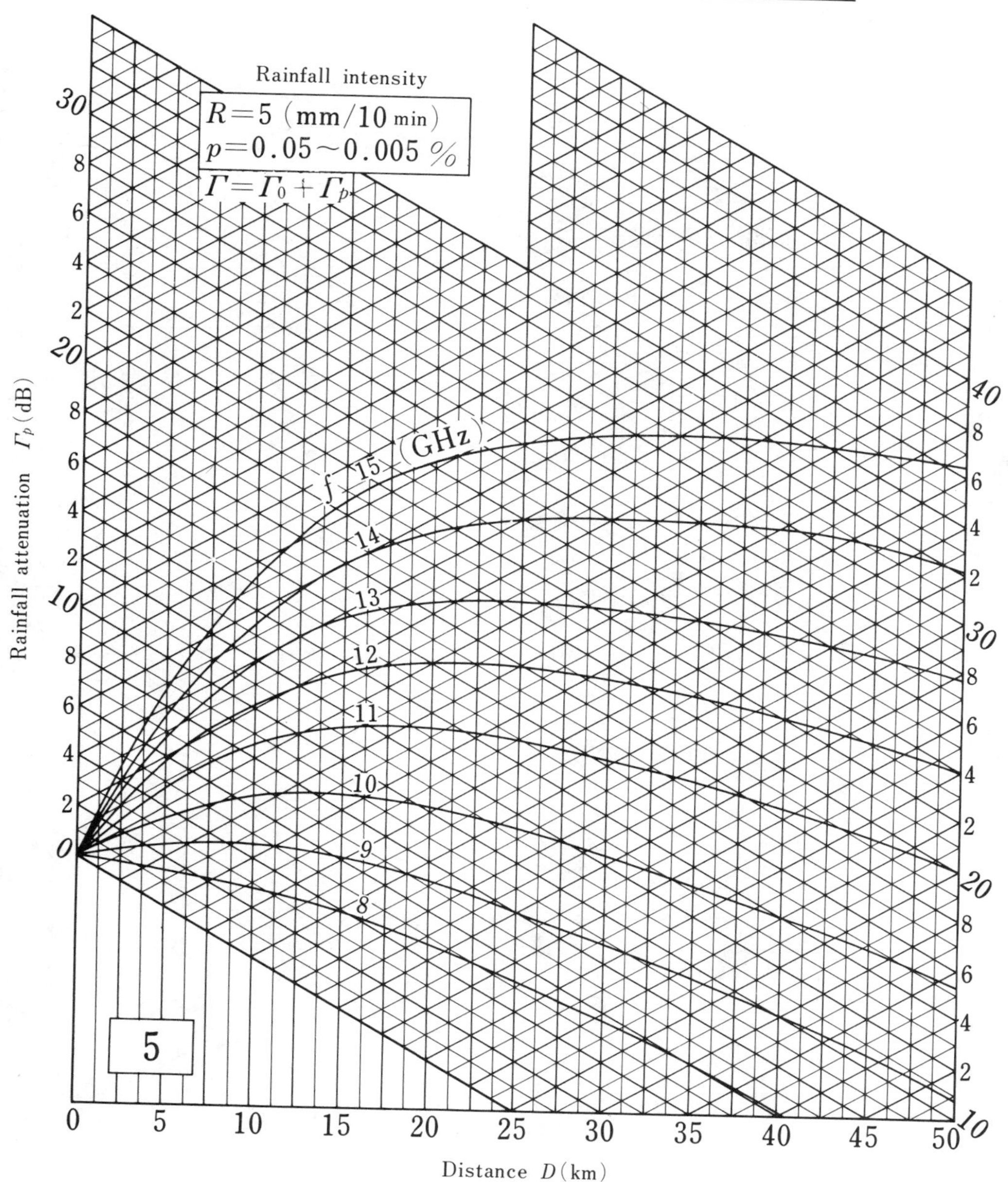

Design Chart for Rainfall Attenuation (3)

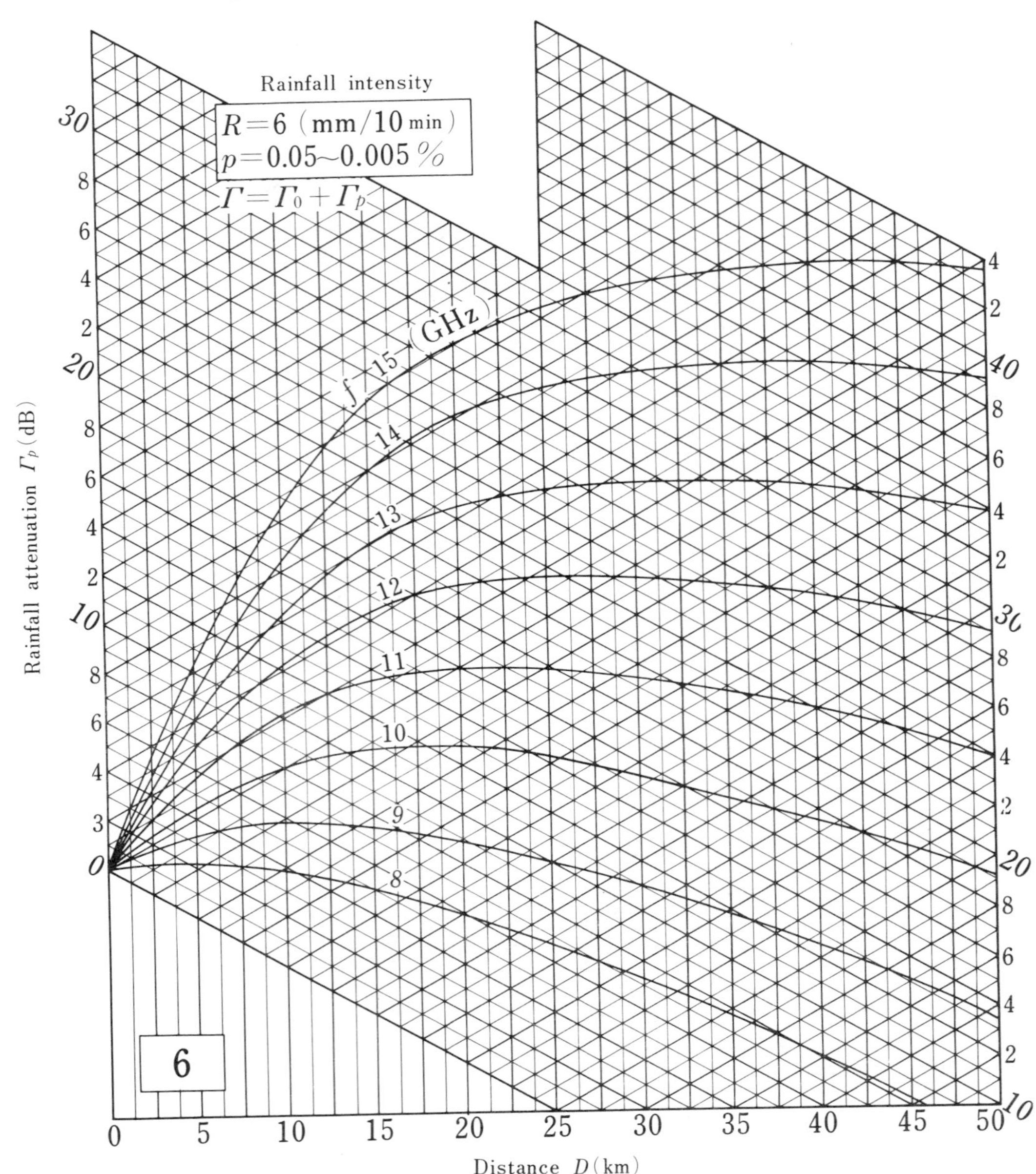

Design Chart for Rainfall Attenuation (4)

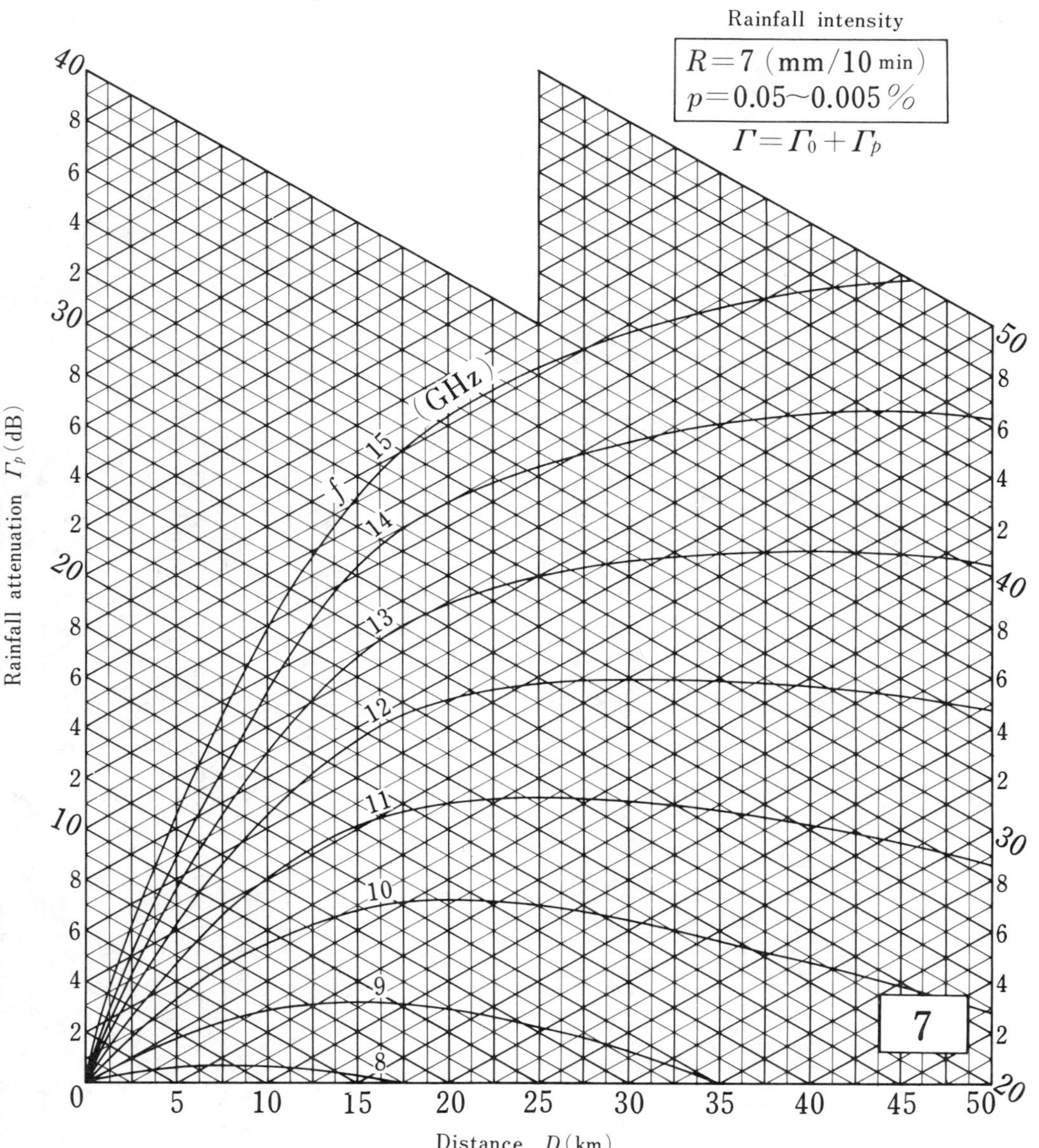

Design Chart for Rainfall Attenuation (5)

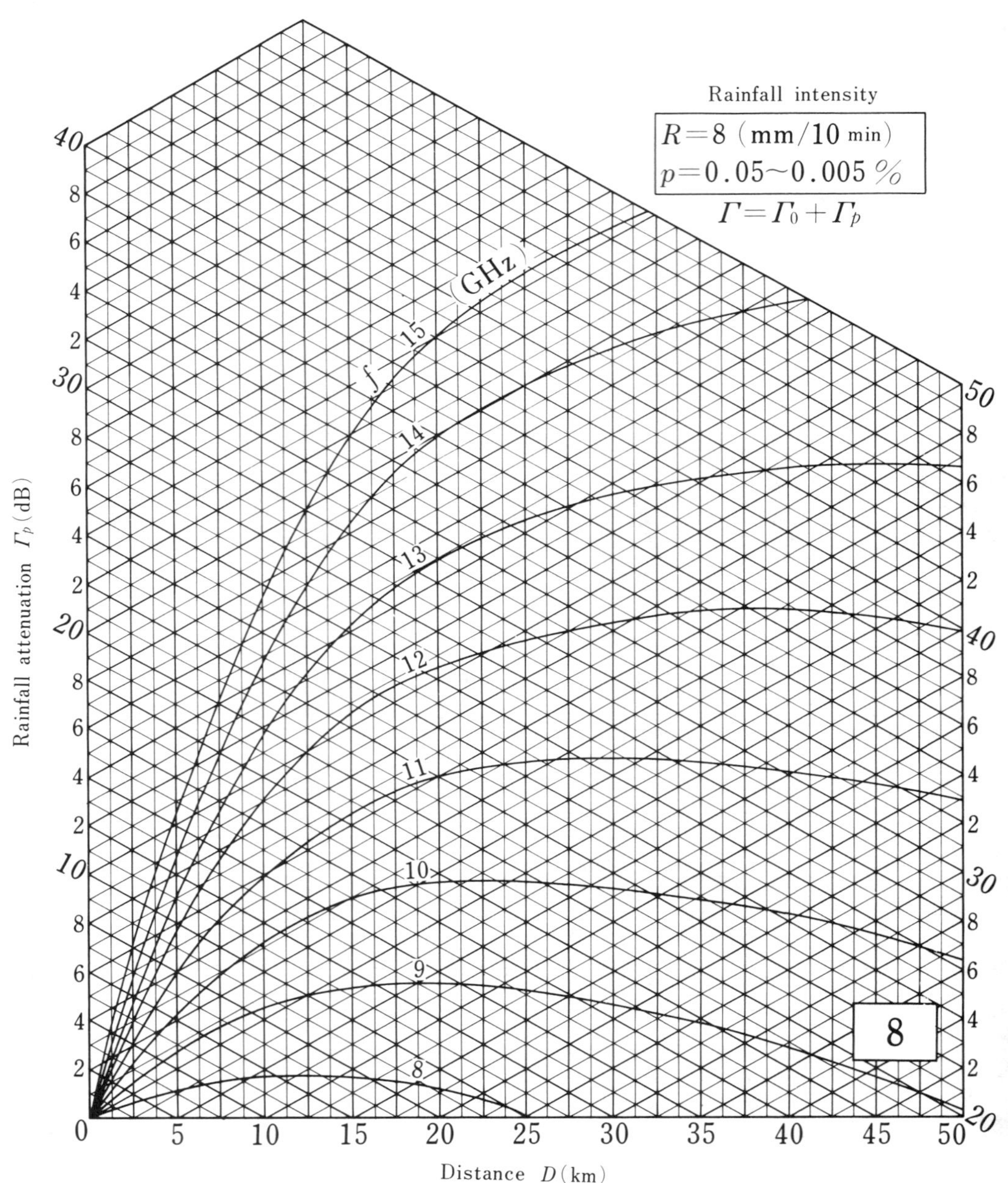

Design Chart for Rainfall Attenuation (6)

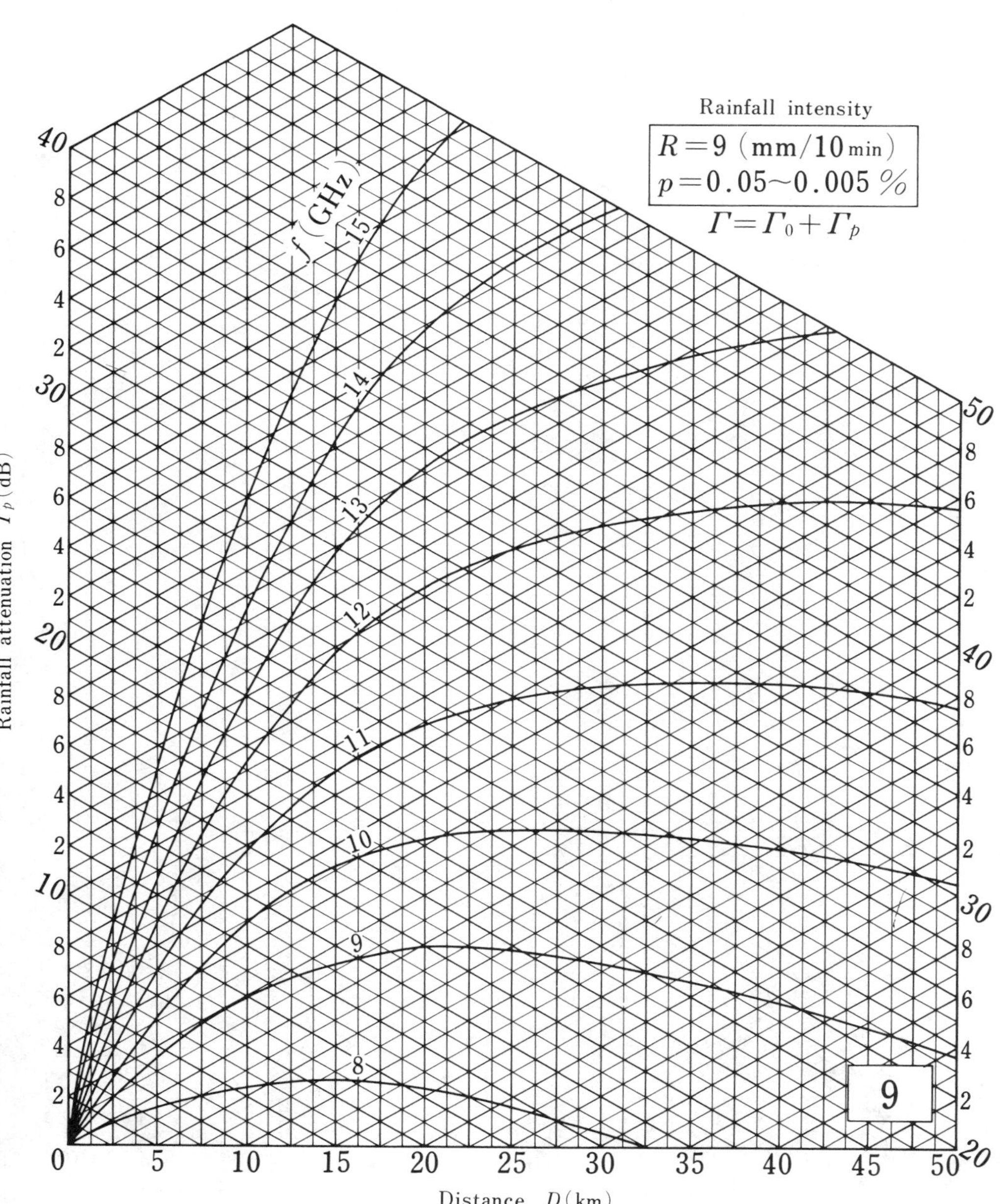

Design Chart for Rainfall Attenuation (7)

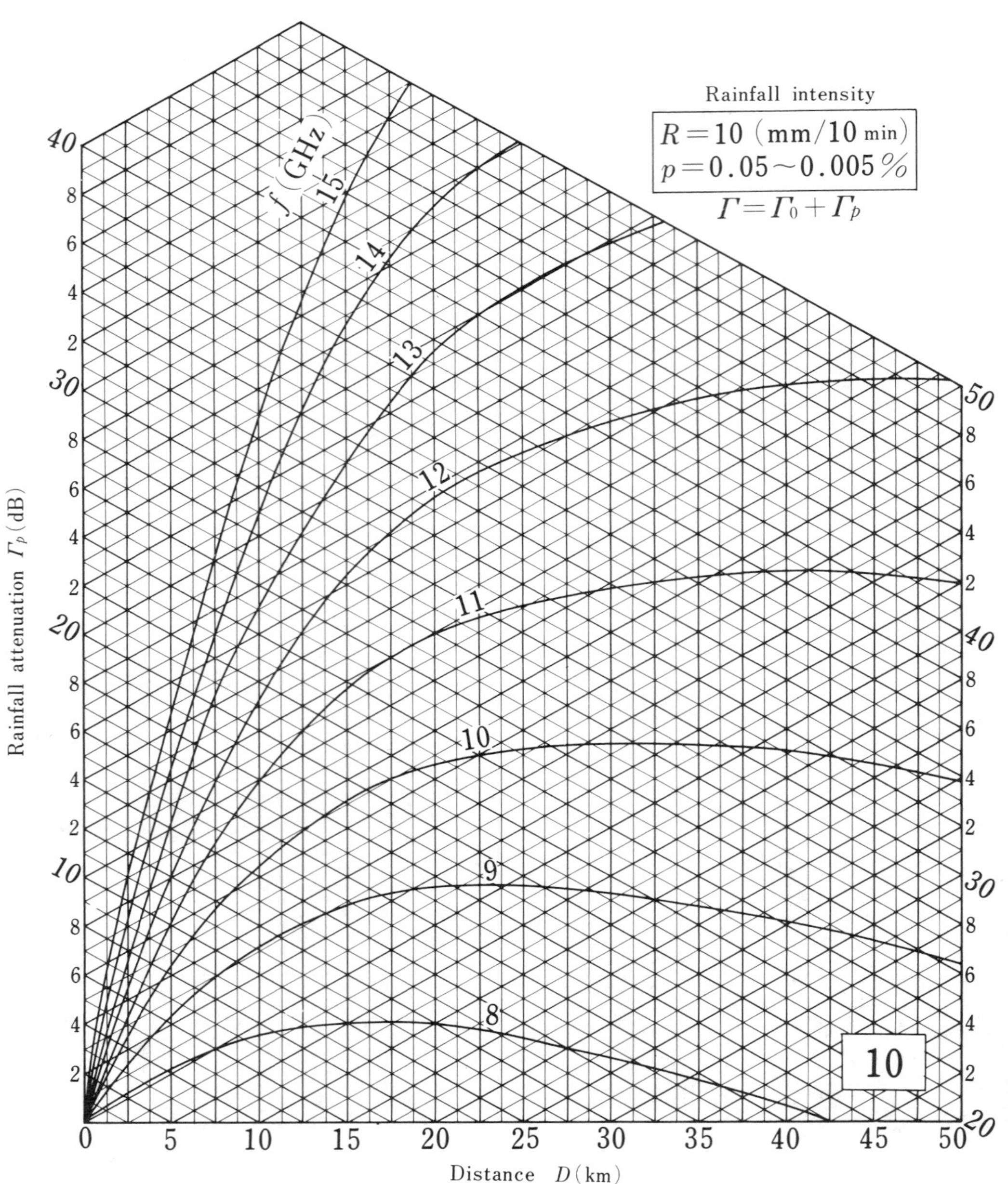

Design Chart for Rainfall Attenuation (8)

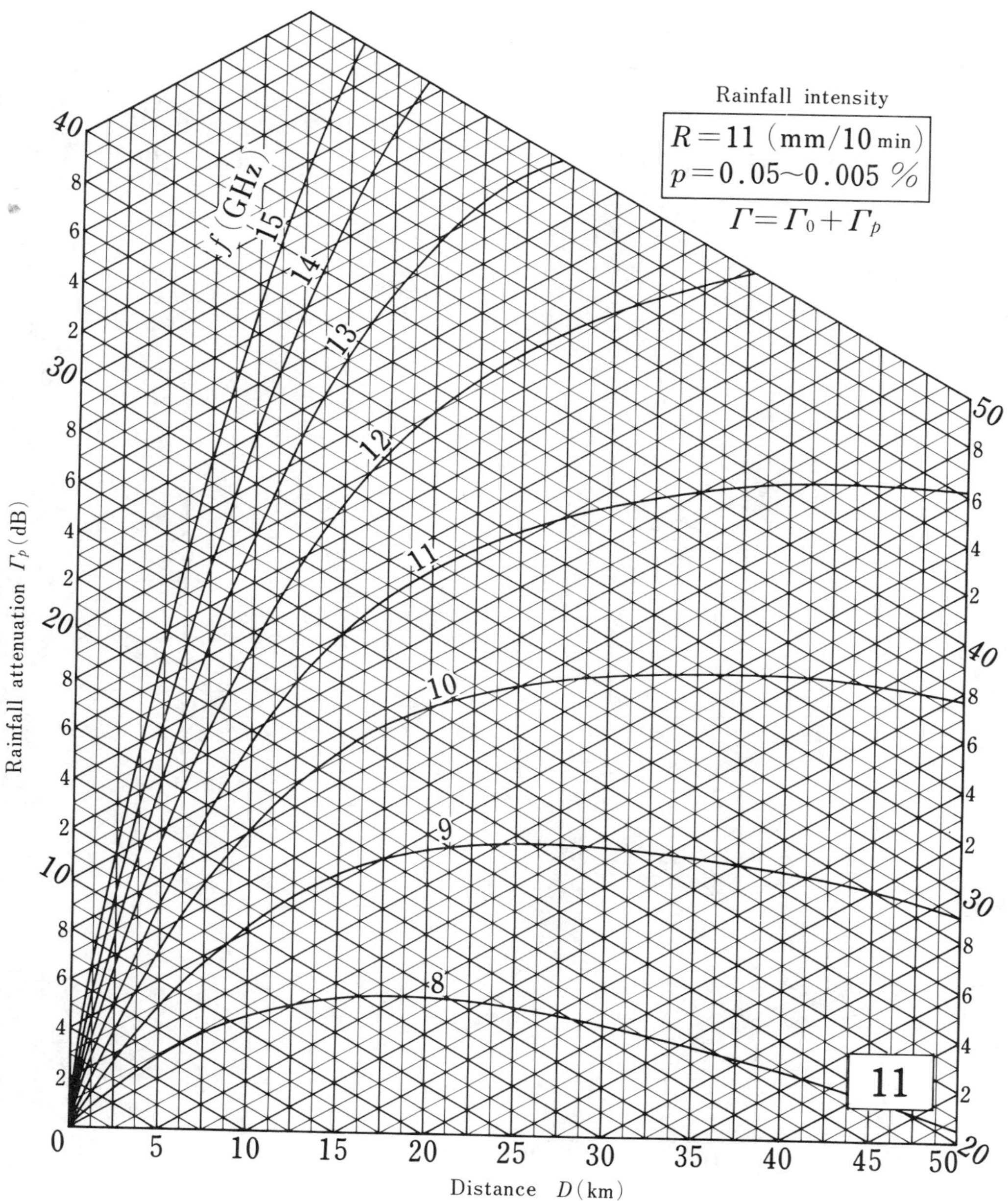

Design Chart for Rainfall Attenuation (9)

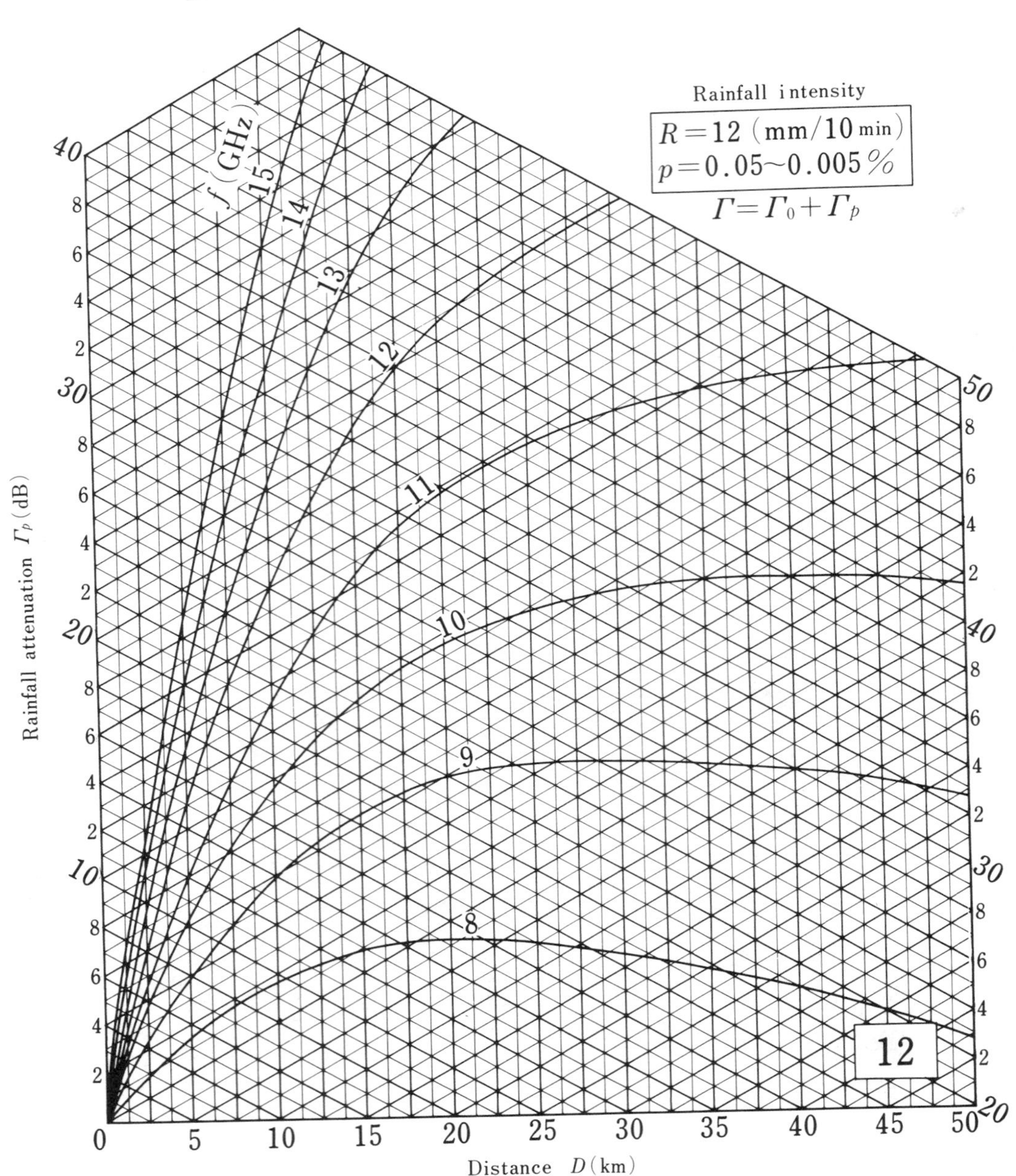

Design Chart for Rainfall Attenuation (10)

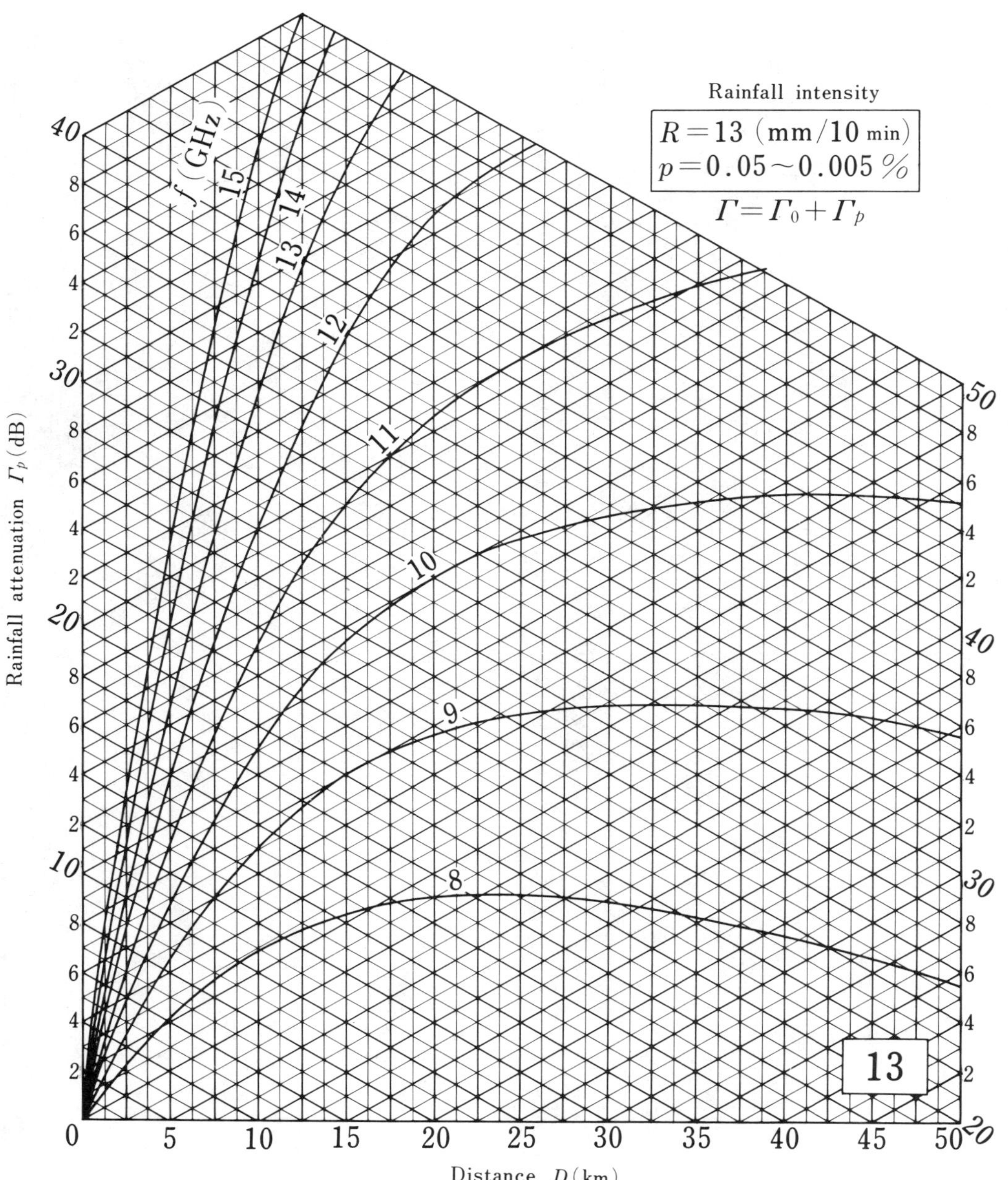

Design Chart for Rainfall Attenuation (11)

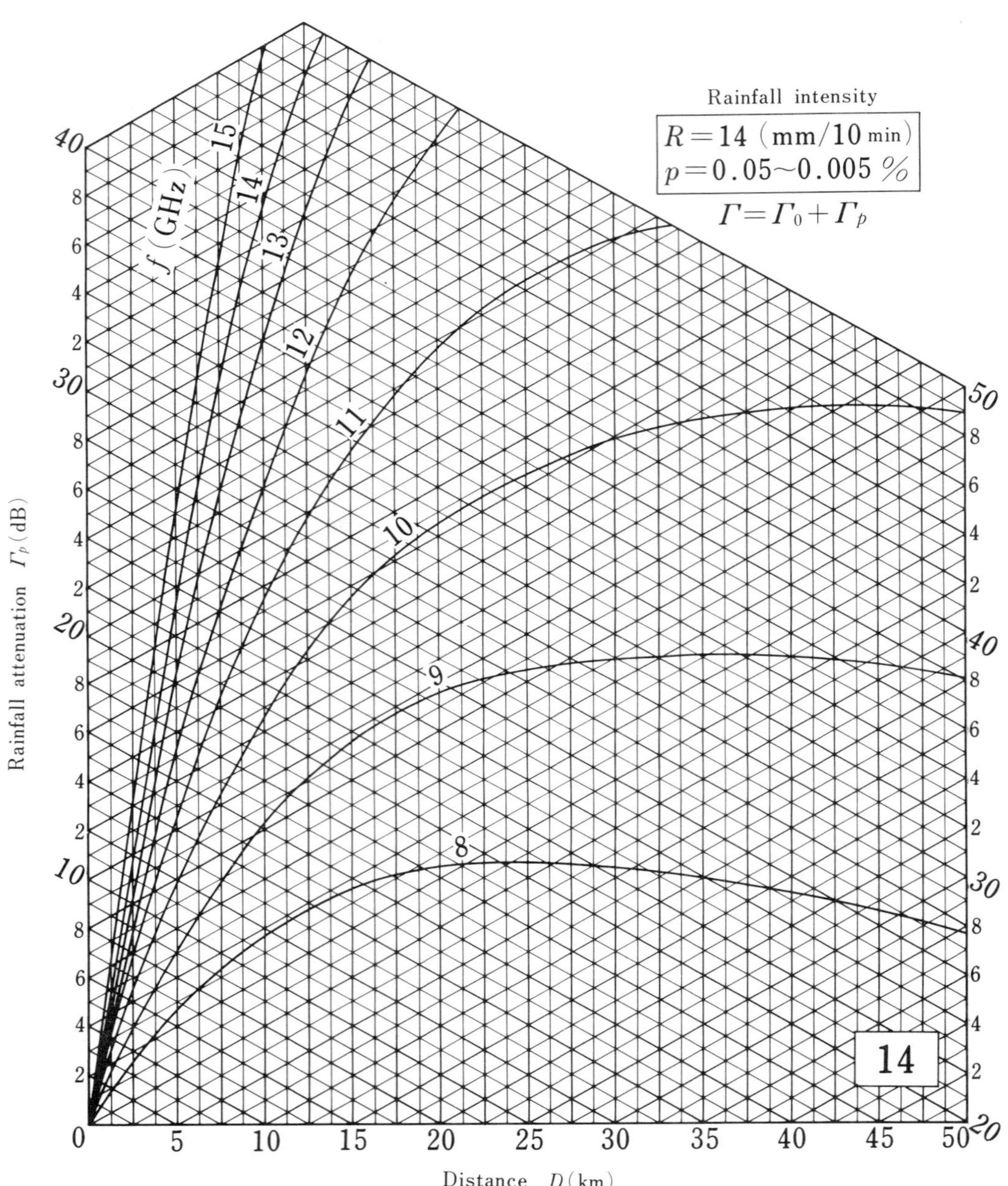

Design Chart for Rainfall Attenuation (12)

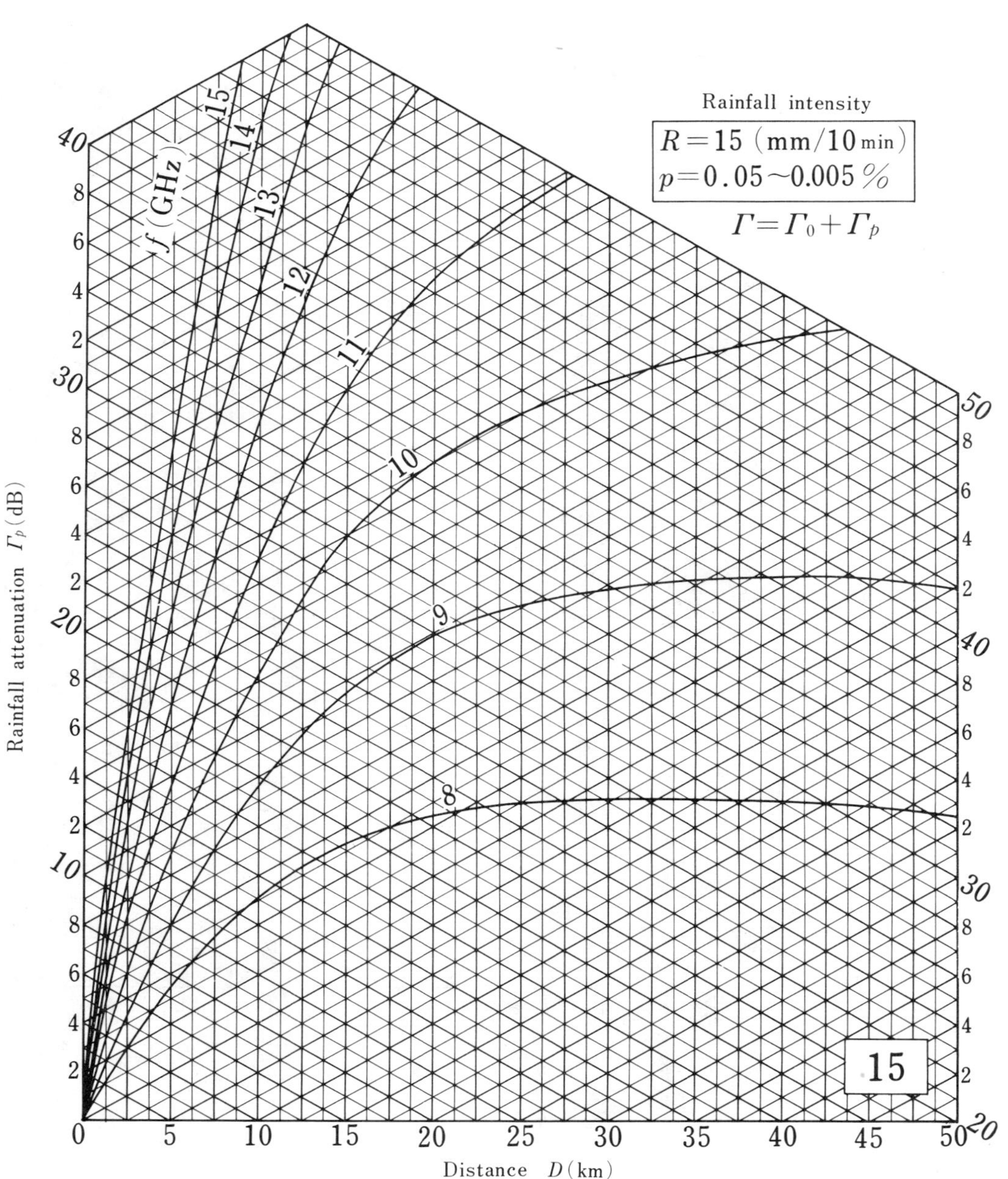

Design Chart for Rainfall Attenuation (13)

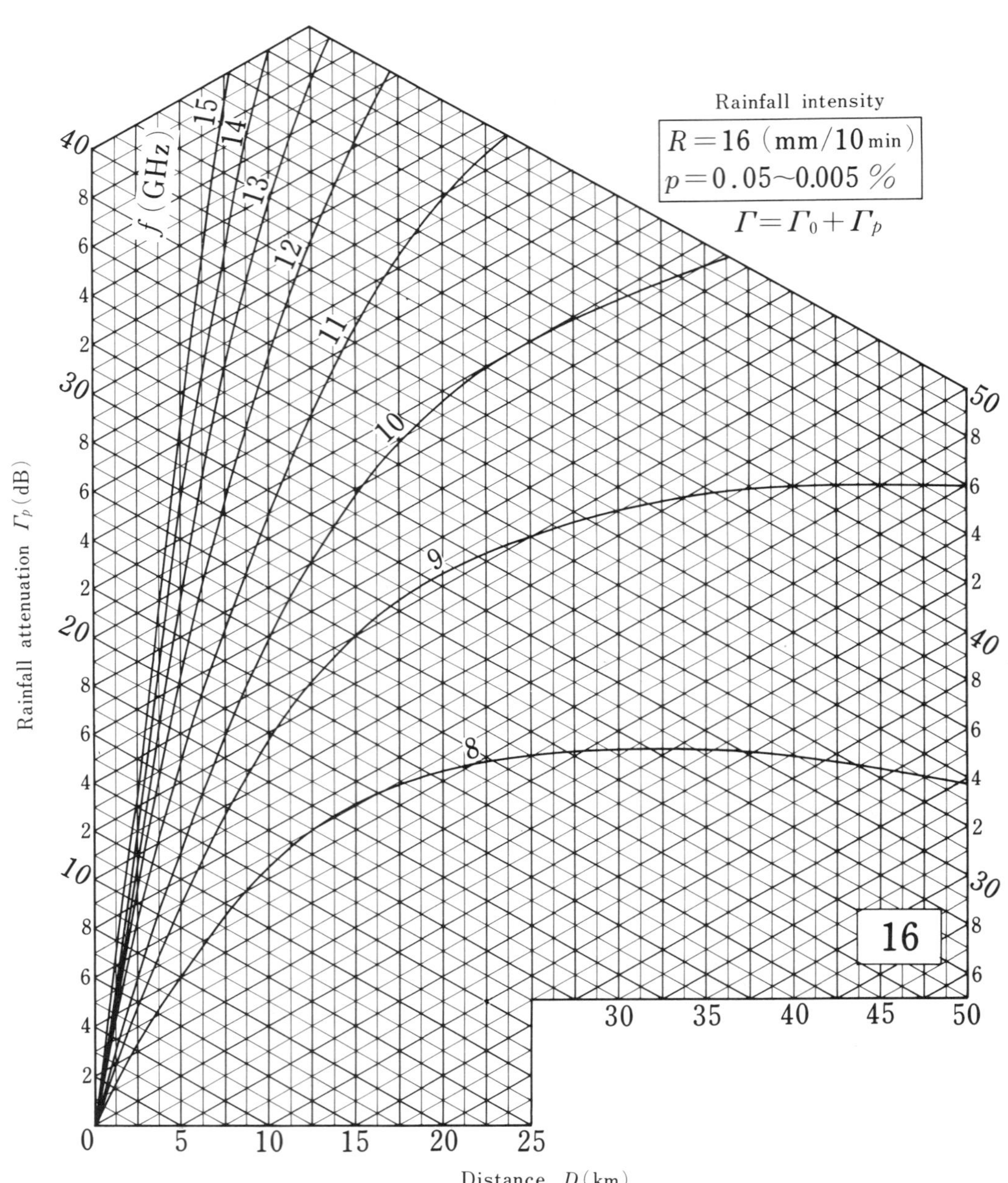

World Map for Heavy and Light Rain Observatories

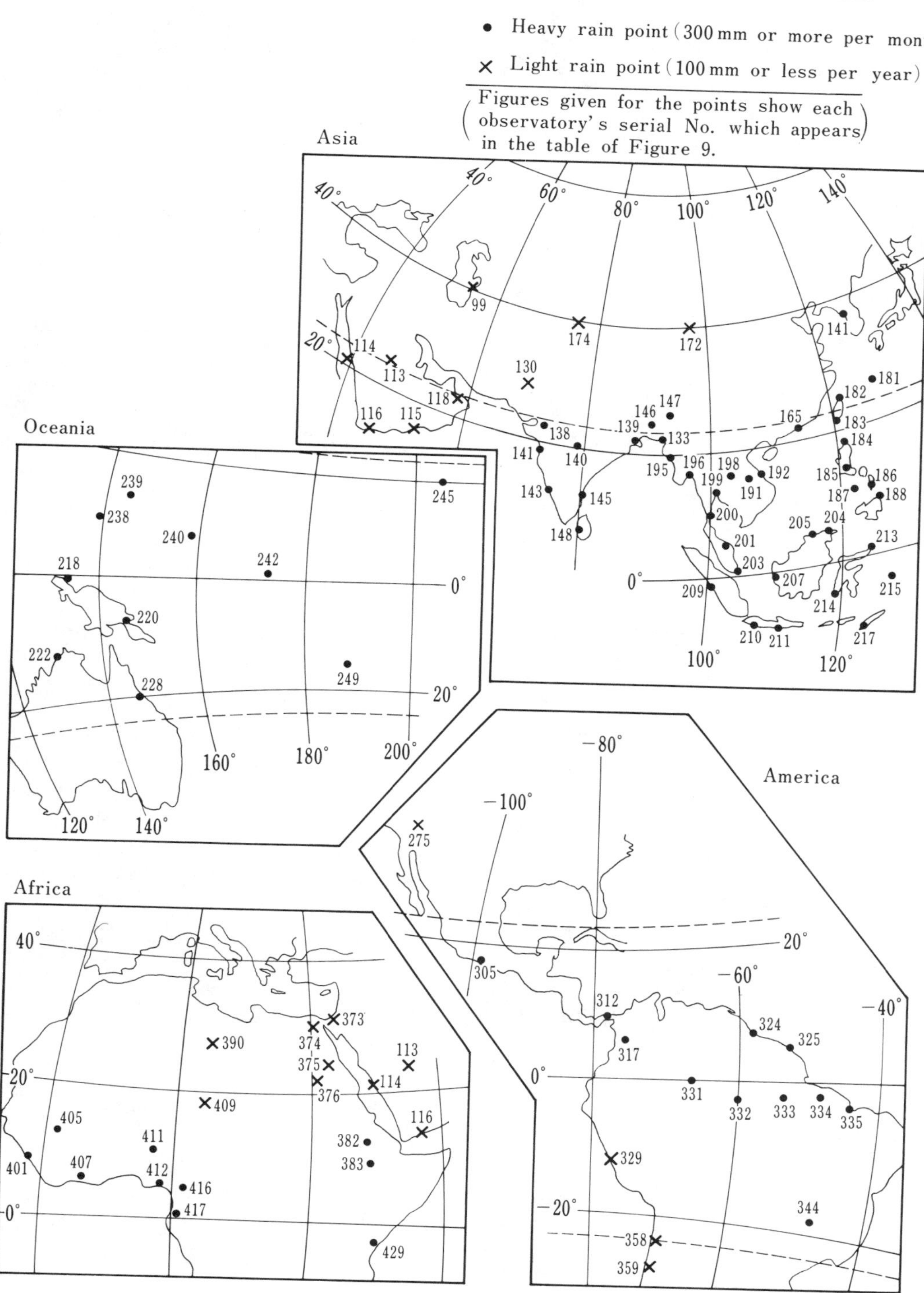

Precipitation in Heavy and Light Rain in the World (1)

[1] Precipitation in heavy rain month at heavy rain points in the world (300 mm or more/month, 70 points) (averaged over 30 years from 1931 to 1960)

No.	Name of location	Elevation above sea level (m)	Precipitation (mm)										□ 600 mm or more/month			Note.1
			1	2	3	4	5	6	7	8	9	10	11	12	Yearly total	
133	Chittagong	6						507	642	572	344				2858	
138	Ahmedabad	55							316						804	
139	Calcutta	6							301	306					1582	
140	Nagpur	310							407						1251	
141	Bombay	11						520	709	419					2078	
143	Mangalore	22						980	1059	577					3479	
145	Madras	16											308		1233	
146	Gauhati	55							301						1679	
147	Dibrugarh	106					356	507	523	419	359				2775	
148	Columbo	6					353					354	324		2307	
165	Hong Kong	129					332	479		415	364				2265	
179	Inch'ŏn	33							304						1089	
181	Naha	36						329							2222	
182	Taibei	9						322							2100	
183	Hêngchun	24						466	584	538	357				2462	
184	Aparri	4									307	390	386		2312	
185	Manila	15								480					1791	
186	Tacloban	21	300												2322	
187	Iloilo	14							302	360					2122	
188	Surigao	22	580	405	308							308	415	653	3863	
191	Pakse	93								415					1564	
192	Tourane	7										533	417		1970	
195	Akyab	5					362	966	1110	1162	655	398			4778	3
196	Rangoon	23						524	492	574	308				2530	
198	Udon Thani	178								313	310				1539	
199	Bangkok	16									306				1492	
200	Chumphon	3										318	327		2033	
201	Kota Bharu	9										326	617	546	2755	
203	Singapore	10												306	2282	
204	Sandakan	14	483										368	479	3150	
205	Labuan	31					345	351	318		417	465	419		3571	
207	Pontianak	4										366	389	323	3175	
209	Padang	3	351		307	363	315	307		348		495	518	480	4172	5
210	Djakarta	8	335												1755	
211	Djokja	107	307												1849	

Note 1. Decreasing order of the yearly total

Precipitation in Heavy and Light Rain in the World (2)

Precipitation (mm) — ☐ 600 mm or more/month (boxed values shown in bold)

No.	Name of location	Elevation above sea level (m)	1	2	3	4	5	6	7	8	9	10	11	12	Yearly total	Note.1
213	Menado	80	465	358	305									371	2662	
214	Makassar	14	**686**	536										**610**	2850	
215	Ambon	12					516	**638**	**602**	401					3459	
217	Kupang	108	389	366											1440	
218	Manokwari	2				334									2597	
220	Lae	8			330	420	387	414	538	542	415	320	326	351	4538	4
222	Darwin	27	341	338											1562	
228	Townsville	4	332	364											1333	
238	Yap I.	17							350	373	356	335			3086	
239	Guam I.	162								326	339	333			2249	
240	Ponape I.	4			370	509	516	424	412	415	402	406	428	466	4875	2
242	Tarawa I.	2												318	1996	
245	Hilo	11	300	329	373	303							340	386	3470	
249	Apia	2	424	364	352									385	2928	
305	Acapulco	3						325			353				1401	
312	Colon	11					318	353	396	389	323	401	566		3310	
317	Qnibdo	73	**635**	544	495	**663**	**648**	**655**	592	**643**	**625**	577	569	495	7140	1
324	Georgetown	2						302							2253	
325	Cayenne	9	431	423	432	480	590	457							3744	
331	Uaupes	86												305	2921	
332	Manaus	83			300										2095	
333	Santarem	72			358	362									2102	
334	Belem	14	317	413	436	382									2770	
335	San Luiz	53			450	406									2088	
344	Catalao	855	300											378	1739	
382	Bahordar	1 835							422						1316	
383	Addis Ababa	2 360								300					1237	
401	Freetown	27						363	**742**	**927**	566	361			3495	
405	Bamako	332								334					1099	
407	Abidjan	16					366	**608**							2144	
411	Kaduna	646								302					1273	
412	Port Harcourt	18							332		442				2421	
416	Yaounde	760										300			1547	
417	Libreville	10	331	305	410	363						384	506	389	3120	
429	Mombasa	55					319								1163	

Yearly precipitation and the precipitation in the greatest precipitation month (100 mm or less/year, 20 points)

No.	Location	Elevation (m)	Precipitation (mm) Monthly maximum (month)	Precipitation (mm) Year
99	Krasnovodsk	−20	15 (3)	92
113	Riyadh	594	25 (4)	81
114	Jidda	11	30 (12)	64
115	Salala	17	28 (7)	81
116	Aden	3	7 (1,9)	39
118	Muscut	6	28 (1)	99
130	Jacobadad	56	37 (7)	99
172	Kiuchuan	1 543	28 (8)	76
174	Kashgar	1 309	15 (1)	86
275	Las Vegas	664	13 (1,7)	99

No.	Location	Elevation (m)	Precipitation (mm) Monthly maximum (month)	Precipitation (mm) Year
329	Lima	34	7 (8)	31
358	Antofagasta	122	5 (7)	13
359	Potrerillos	2 850	18 (5)	56
373	Port Said	7	18 (12)	66
374	Cairo	139	7 (12)	25
375	Aswan	194	1 (5)	2
376	Wadi Halfa	155	1 (7.10)	3
390	Sebba	444	3 (5.6)	8
398	Villacisneros	6	36 (9)	76
409	Bilma	357	11 (8)	21

5

10

Absorption Attenuation Due in Millimeter Wave Band

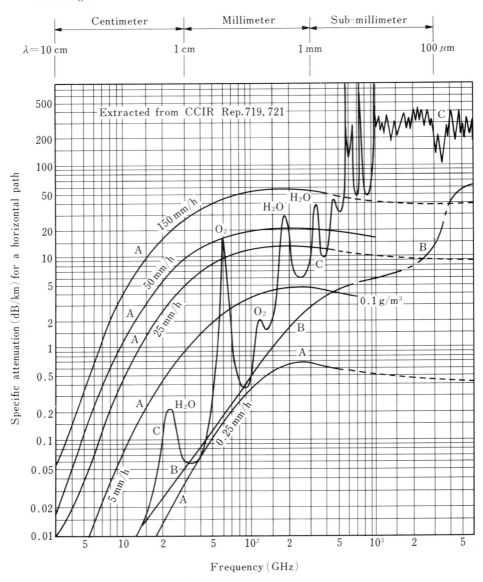

Attenuation due to gaseous constituents and precipitation for transmission through the atmosphere

Pressure : Sea level : 1 atm (1 013.6 mbar)	A : Rain
Temperature : 20°C	B : Fog
Water vapor : 7.5 g/m³	C : Gaseous

6

PASSIVE-RELAY PROPAGATION PATH

6.1 PASSIVE-RELAY SYSTEM

A radio relay system using a plane reflector or back-to-back antennas to change the direction of the radio wave is called a passive relay system. It contains no active elements such as transistors and vacuum tubes.

Passive relay systems can be classified broadly into three types: the plane reflector system, the back-to-back antenna system, and the radio prism system.

A single-plane reflector or a double-plane reflector (parallel or intersectional arrangement) is used extensively taking into account the transmission efficiency, the installation cost, the transfer angle, and the adaptability to the geographical and meteorological conditions.

The passive-relay propagation path is divided generally into two parts—the longer hop and the shorter hop. The longer hop is called the main section and the shorter one, the added passive section or, for short, the passive section.

When the passive section is sufficiently long compared with the size of the antenna or passive reflector, the radio wave can be treated as a plane wave. In this case the reflector is called a far-field type and, conversely, a near-field type.

Figure 5 shows the curves that determine the boundary between the far field and the near field. Figure 2 illustrates three methods to analyze the passive-relay propagation paths, methods I, II, and III.

This work uses method III, which may be applied to any type of propagation path. However, most of the published literature uses method II, which is applicable to free-space propagation paths. Nomographs applicable to method III and method II are given in Fig. 9.

A passive relay system having more than two hops produces a very large additional loss. It is used only in particular cases because of restrictions on the reflector size.

At present the use of passive relay systems is required more and more frequently because of the development of the upper region of the microwave frequency band. A summary of the solutions is presented in Fig. 4, where formulas are given to estimate the equivalent propagation loss (transmission loss between the transmitting and the receiving isotropic antennas when the section is replaced by a single propagation path).

The multistage relay system used most extensively is of the double-stage type. The double-plane reflector system (parallel type or intersection type) composed of two reflectors located close to one another is treated as a single-stage passive relay system.

Most fundamental to the analysis is an estimate of the additional loss of each passive section contained in the multistage passive relay system.

Figure 3 gives a definition of the additional passive-relay loss, a list of abbreviations, as well as the calculation formulas for a single-stage passive relay system. Figures 6 to 9 give the graphs for the actual design to be applied to various cases; Figs. 10 through 15 show related graphs that will facilitate the design of the geometrically difficult double-plane reflector system.

For the passive relay using ordinary antennas, replacement of the following parameters is required:

Effective aperture area = effective area of reflector B_e.
Gain (for a single direction) = single-direction gain of reflector G_p.
Bidirectional gain = bidirectional gain of reflector G_{REF} $(= G_p^2)$.

6.2 PASSIVE-RELAY ADDITIONAL LOSS

The receiving power P_r at point B in the top drawing of Fig. 3 is given by

$$P_r = P_t G_t G_r / \Gamma_D \tag{6.1}$$

From this, the span loss L_s between points A and B is given by

$$L_s = P_t / P_r = \Gamma_D / G_t G_r \tag{6.2}$$

where P_t = transmitting power (radiating power)
Γ_D = transmission loss between A and B (D is distance)
G_t, G_r = transmitting, receiving antenna gain

If point B is not suitable for installation of the radio station, a passive reflector is provided and the radio equipment is removed to point C, located at a distance R from point B (bottom drawing in Fig. 3). Then the receiving power P_r is calculated as follows.

First consider the plane reflector as a receiving antenna with gain G_p. The receiving power P_p at point B for the section $\overline{AB}$ is given by

$$P_p = P_t G_t G_p / \Gamma_D \tag{6.3}$$

Similarly for section $\overline{BC}$, it is considered that the transmitting antenna with gain G_p reradiates the output power P_p. Then

$$P_{r'} = P_p G_p G_r / \Gamma_R \tag{6.4}$$

where G_p = equivalent reflector gain per direction
Γ_R = transmission loss for section $\overline{BC}$ (R is distance)

By substituting Eq. (6.3) into Eq. (6.4),

$$P_{r'} = P_t G_t G_r G_p^2 / \Gamma_D \Gamma_R \tag{6.5}$$

The overall span loss, including the loss L_t for the added passive section $\overline{BC}$, is given by

$$L_t = P_t / P_{r'} = \Gamma_D \Gamma_R / G_t G_r G_p^2 = \left(\Gamma_D \Gamma_R / G_p^2 \right) / G_t G_r \tag{6.6}$$

and the equivalent transmission loss is

$$\Gamma_e = \Gamma_D L_p \tag{6.7}$$

In comparison with Eq. (6.2), it is found that Γ_e represents the equivalent transmission loss of the section between the transmitter and receiver, including the passive relay.

Here L_p is called the additional passive-relay loss because it is generated by providing the passive section. L_p is given by

$$L_p = \Gamma_R / G_p^2 \tag{6.8}$$

or

$$L_p = \Gamma_R / G_{\text{REF}} \tag{6.9}$$

Here

$$G_{\text{REF}} = G_p^2 = \left[\frac{4\pi}{\lambda^2} B_r \cos\phi \cdot \eta \right]^2 \tag{6.10}$$

where G_{REF} = bidirectional gain of reflector
G_p = unidirectional equivalent antenna gain of reflector
B_r = actual area of reflector
ϕ = angle of incidence
η = efficiency of area (for single direction)

When expressed in decibels,

$$G_{\text{REF}}\ (\text{dB}) = \underbrace{20\log(4\pi B_r/\lambda^2)}_{\substack{\text{bidirectional} \\ \text{actual area gain } G_B}} + \underbrace{20\log(\cos\phi\cdot\eta)}_{\substack{\text{bidirectional} \\ \text{performance} \\ \text{efficiency } \varepsilon_R}} \tag{6.11}$$

The design graph is shown in Fig. 6. However, it should be noted that Eqs. (6.10) and (6.11) should not be used when the discrimination chart (Fig. 5) indicates that the path is that of a near-field passive relay.

6.3 PASSIVE SECTION AS FREE-SPACE PROPAGATION PATH

Most of the passive sections may be treated as free space, since they are chosen on paths whose lengths are very short and where a good line of sight is obtained. Also, the interference caused by ground reflection may be suppressed effectively by means of the directivity patterns of the reflector and the antenna. Thus Eq. (6.9) is rewritten as

$$\begin{aligned} L_p &= \Gamma_{R0}/G_{\text{REF}} \\ &= \left(\frac{4\pi R}{\lambda}\right)^2 \Big/ \left[\frac{4\pi}{\lambda^2} B_r \cos\phi\cdot\eta\right]^2 \\ &= \left(\frac{\lambda R}{B_r\cos\phi\cdot\eta}\right)^2 \end{aligned} \tag{6.12}$$

where

$$\Gamma_{R0} = \left(\frac{4\pi R}{\lambda}\right)^2 \quad \text{(free-space propagation loss of passive section)}$$

and R is length. When the loss is expressed in decibels,

$$L_p\ (\text{dB}) = \underbrace{20\log(\lambda R)}_{\substack{\text{passive-relay} \\ \text{additional loss of} \\ \text{single-plane} \\ \text{reflector } (L_{pu})}} - \underbrace{20\log B_r\cos\phi\cdot\eta}_{\substack{\text{dual-area} \\ \text{parameter } [B_f]}} \tag{6.13}$$

The design graph is given in Fig. 7. The near-field-type passive relay is discussed in Section 6.4.

6.4 NEAR-FIELD-TYPE PASSIVE RELAY (BEAM FEEDING)

Sections 6.2 and 6.3 considered the region where the sizes of the flat reflector at the passive-relay site and both antennas facing the reflector are very small compared with the distance R ($D > R$ is assumed), to the extent that the wave front at the reflector can be treated as a flat plane. For the far-field-type passive relay, such a region is called a Fraunhofer region.

On the other hand, for the near-field-type passive relay, the region where the size of the reflector or the facing antenna is so large that the phase difference is significant for the given distance R and wavelength λ, is called a Fresnel region.

The following method is used to determine approximate values of various parameters on a widely employed common model. The method defines the relationship between the sizes of the antenna and the plane reflector and the beam-feeding loss (near-field-type additional passive-relay loss).

1. Parameter u (Fig. 8*a*),

$$u = 1/(\lambda R) \tag{6.14}$$

2. Parameter p (Fig. 8*b*),

$$p = uA_r \tag{6.15}$$

where A_r is the antenna aperture area,

$$A_r = \pi D_\phi^2/4 \qquad \text{(for a parabola with diameter } D_\phi\text{)}$$

3. Parameter q (Fig. 8*c*),

$$q = B_j/A_r \tag{6.16}$$

where B_j = projected area of reflector viewed from antenna,

$$B_j = B_r \cos\phi \tag{6.17}$$

B_r = actual area of reflector

ϕ = angle of incidence

4. Additional passive-relay loss L_p. Obtain L_p from Fig. 8*d* using p, q, and η. Normally L_p is first given as an allowable limit value (or an allocated value) for the total system, including the passive-relay section, to compute q ($= B_j/A_r$). The actual area of the reflector B_r is estimated from q using Fig. 8*c*.

6.5 METHOD OF DEFINITION OF FAR-FIELD AND NEAR-FIELD TYPES

Since the terms far-field type (Fraunhofer region) and near-field type (Fresnel region) are rather vague definitions, it is preferable to define the boundary taking into account the purpose of the reflector. Whether it is in the near field or in the far field may be determined by the maximum path-length difference between any point on the projected reflector area and the antenna feeding point. (The criterion measured by the wavelength may be $\lambda/6$, $\lambda/8$, or $\lambda/12$.)

Since the approximate calculation as mentioned above involves a number of assumptions and requires a rather complicated method of determination, the following method is used.

Condition of Determination

If C represents the projected area (limiting area) of an imaginary reflector whose far-field-type additional passive-relay loss L_p is 6 dB ($\eta = 1$ is assumed), the near-field type is defined as

$$A_r \geqq C \qquad \text{or} \qquad B_j \geqq C$$

and the far-field type is defined as Fig. 5 (6.18)

$$A_r < C \qquad \text{or} \qquad B_j < C$$

where A_r = actual aperture area of antenna
B_j = projected area of flat reflector
C = limiting area,

$$C = R/2 \tag{6.19}$$

R = passive-relay additional section length ($D \gg R \gg \lambda$)
λ = wavelength
D = main section length

6.6 EQUIVALENT PROPAGATION LOSS OF RADIO PATH WITH TOTALLY FREE-SPACE MAIN AND PASSIVE RELAY SECTIONS

In the radio section where the effective reflection coefficient of the earth surface is sufficiently small, the reflected wave is suppressed effectively through the directivity of the parabola or the passive reflector. Also, no obstacle exists within the zone formed by half of the first Fresnel zone radius, and the radio path may be treated practically as a free-space propagation path.

In microwave system design it is essential to secure as complete a free-space propagation loss as possible when both the main section and the passive-relay section are considered as free space.

A. General Equation

$$\Gamma_e = \underbrace{\Gamma_D}_{\text{main section}} \cdot \underbrace{\Gamma_{R1}\Gamma_{R2} \cdots \Gamma_{Rn}/(G_{p1}G_{p2} \cdots G_{pn})^2}_{\text{passive relay section, } n \text{ stages}} \quad \text{(general equation, Fig. 4)} \tag{6.20}$$

$$\Gamma_D = (4\pi D/\lambda)^2,\ \Gamma_{R1} = (4\pi R_1/\lambda)^2,$$

$$\Gamma_{R2} = (4\pi R_2/\lambda)^2, \ldots \Gamma_{Rn} = (4\pi R_n/\lambda)^2 \quad \text{(free-space propagation loss)} \tag{6.21}$$

$$G_{p1}^2 = (4\pi B_{e1}/\lambda^2)^2,$$

$$G_{p2}^2 = (4\pi B_{e2}/\lambda^2)^2, \ldots G_{pn}^2 = (4\pi B_{en}/\lambda^2)^2 \quad \text{(relation between reflector bidirectional gain } G_p^2 \text{ and effective area } B_e) \tag{6.22}$$

From Eqs. (6.20), (6.21), and (6.22),

$$\Gamma_e = (4\pi D R_1 R_2 \cdots R_n \lambda^{n-1}/B_{e1}B_{e2} \cdots B_{en})^2 \tag{6.23}$$

where D = main section length
$R_1, R_2, \ldots, R_n$ = individual passive-relay section length
$B_{e1}, B_{e2}, \ldots, B_{en}$ = effective area of individual reflectors ($B_e = B_r \eta \cos\phi$)

B. Double-Stage Passive Relay

The following result is obtained assuming $n = 2$ in Eq. (6.23),

$$\Gamma_e = (4\pi D R_1 R_2 \lambda/B_{e1}B_{e2})^2 \tag{6.24}$$

C. Single-Stage Passive Relay (Including Double-Stage Passive Relays Whose Reflectors Are Located Close to Each Other)

When $n = 1$, Γ_e is expressed simply by

$$\Gamma_e = (4\pi D R_1/B_{e1})^2 \tag{6.25}$$

where the frequency term vanishes. When it is expressed in decibels by putting $R_1 = R$ and $B_{e1} = B_e$, we obtain

$$\Gamma_e = 20\log(4\pi DR/B_e) \quad \text{(dB)} \tag{6.26}$$

or

$$\Gamma_e \doteqdot 20\log D\ \text{(km)} + 20\log R\ \text{(km)} - 20\log B_e\ (\text{m}^2) + 142 \quad \text{(dB)} \tag{6.27}$$

Figure 9 is the nomograph for Eq. (6.27).

6.7 APPLICATION OF REFLECTORS OF VARIOUS TYPES ACCORDING TO TRANSFER-SUPPLEMENTARY ANGLE

The effective area of the plane reflector B_e is given by

$$B_e = B_r \cos\phi \cdot \eta = B_j \eta \tag{6.28}$$

where B_r = actual area of reflector
ϕ = angle of incidence
η = aperture area efficiency
B_j = projected area of reflector ($B_j = B_r \cos\phi$)

The relationship between the angle of incidence ϕ and the transfer-supplementary angle θ is found by

$$\theta = 2\phi$$
$$\phi = \theta/2 \tag{6.29}$$

When $\phi \leq 30°$ ($\theta \leq 60°$), the decrease of the effective area from the actual area is below 14%. However, when $\phi > 60°$ ($\theta > 120°$), the decreasing rate is more intensified with increasing ϕ (θ), and if the decreasing rate exceeds 50%, the use of an extremely large passive reflector becomes inevitable. To remedy this a double-mirror arrangement is employed.

There are two types of arrangements for the double passive reflector—the intersection type, where radio paths intersect one another, and the parallel type, where they do not intersect. In either case two reflectors are erected with a certain space between them so that the radio paths will not be blocked. Also, it is preferable to choose the most economical type and size, knowing the relationship between the angles of incidence ϕ_1 and ϕ_2 at the reflector, the transfer-supplementary angle θ, and the minimum spacing. The relationship between transfer angles and reflector types is summarized as follows.

Transfer-Supplementary Angle θ	Angle of Incidence ϕ or (ϕ_1), (ϕ_2)	Reflector Type
0–120°	0°–60°	Single plane
100–150°	(30–10°)–(20–5°)	Intersection (double)
120–180°	(45–10°)–(15–10°)	Parallel (double)

NOTE. Angle ranges shown are subject to change according to site conditions and also to reflector size.

The intersection-type reflector, if $\phi_1 = \phi_2$, is called a symmetrical intersection type when two passive plane reflectors are equal to each other in both size and type, thereby facilitating the engineering design and the installation adjustment.

In the case of the parallel-type reflector, the condition $\phi_1 = \phi_2$ or $\theta = 180°$ is rarely encountered. If $\phi_1 \neq \phi_2$, the sizes of the two reflectors are not the same. However, the difference in horizontal width is kept within 10%, provided that the reflectors are used within the ranges shown above.

Figure 1 summarizes the reflector types and the applicable conditions.

6.8 GEOMETRICAL PARAMETERS OF A DOUBLE PASSIVE PLANE REFLECTOR SYSTEM

A. Relationship between the Angle of Incidence ϕ and the Transfer-Supplementary Angle θ

The single passive reflector presents a simple relationship $\theta = 2\phi$, as mentioned previously. However, the double passive reflector yields the following relationship provided that $\phi_1 \geq \phi_2$.

1. Intersection-type reflector,

$$\theta = \pi - 2(\phi_1 + \phi_2) \tag{6.30}$$

$$\begin{aligned} \phi_1 &= (\pi - \theta)/2 - \phi_2 \\ \phi_2 &= (\pi - \theta)/2 - \phi_1 \end{aligned} \tag{6.31}$$

If it is symmetrical, that is, $\phi_1 = \phi_2 = \phi$,

$$\begin{aligned} \theta &= \pi - 4\phi \\ \phi &= (\pi - \theta)/4 \end{aligned} \tag{6.32}$$

2. Parallel-type reflector,

$$\theta = \pi - 2(\phi_1 - \phi_2) \tag{6.33}$$

$$\begin{aligned} \phi_1 &= \phi_2 + (\pi - \theta)/2 \\ \phi_2 &= \phi_1 - (\pi - \theta)/2 \end{aligned} \tag{6.34}$$

If $\phi_1 = \phi_2 = \phi$,

$$\theta = \pi, \qquad \text{constant} \tag{6.35}$$

The preceding relation is shown in Fig. 10.

B. Horizontal Width W of Reflector

When the width of the reflector with a larger angle of incidence ϕ_1 is represented by W_1 and that of a reflector with a smaller angle of incidence ϕ_2 by W_2, the following equation is established,

$$g = W_2/W_1 = \cos\phi_1/\cos\phi_2 \leq 1 \tag{6.36}$$

This equation is applicable for both intersection- and parallel-type reflectors. In the case of the symmetrical intersection type or the parallel type, where $\theta = 180°$, $g = 1$ or $W_1 = W_2$; g rarely falls below 0.9 since the reflector is arranged so that ϕ_1 will not differ appreciably from ϕ_2.

The horizontal width of the reflector should be determined correctly. Then the radio-wave energy reflected from one of the plane reflectors can be most effectively projected onto the smaller reflector to change its direction.

C. Spacing between Two Reflectors

Two reflectors are suitably spaced when the radio path of one does not obstruct the radio path of the other. However, it is preferable to provide additional spacing on the order of 1 or several meters. This provides some clearance for the radio-wave path to avoid loss due to an alignment error of the reflectors or to radio-wave diffraction by the reflector edges. Hence

$$S = S_{\min} + S_c \tag{6.37}$$

where $S_{\min}$ = minimum spacing
S_c = additional spacing to give a margin to the radio-wave path

1. Equation to find $S_{\min}$,

$$S_{\min} = W_2(p_s + \alpha) \tag{6.38}$$

$$p_s = (\operatorname{cosec} \phi_2)/2 \tag{6.39}$$

where p_s = constant applicable to both intersection-type and parallel-type reflectors (see Fig. 11*a*)
W_2 = horizontal width of smaller reflector
ϕ_1, ϕ_2 = angles of incidence ($\phi_1 > \phi_2$)
α = correction required; given by

$$\alpha = -\cos\phi_2(\tan\phi_1 + \tan\phi_2)/2 \tag{6.40}$$

for intersection-type reflector, and by

$$\alpha = +\cos\phi_2(\tan\phi_1 - \tan\phi_2)/2 \tag{6.41}$$

for parallel-type reflector (see Fig. 11*b* to *e*)

2. Equation to find additional spacing S_c,

$$S_c = c_1\beta \tag{6.42}$$

where

$$\beta = 1/\sin 2\phi_2 \tag{6.43}$$

(applicable to both intersection-type and parallel-type reflectors; see Fig. 13), and c_1 is the transverse clearance between the radio path and the edge of the reflector. There exist two parameters c_1 and c_2 for two plane reflectors, but it is assumed that $c_1 < c_2$ since $\phi_1 > \phi_2$.

3. When $\phi_1 = \phi_2$,

$$S = Wp_0 + S_c \tag{6.44}$$

(a) For the symmetrical intersection type,

$$p_0 = \cos 2\phi/2 \sin \phi \tag{6.45}$$

or

$$p_0 = [\sin(\theta/2)/\sin(\pi/4 - \theta/4)]/2 \tag{6.46}$$

where p_0 is the coefficient to determine the minimum spacing; Eq. (6.46) can be derived using the relationship

$$\begin{aligned} \theta &= \pi - 4\phi \\ 2\phi &= \pi/2 - \theta/2 \\ \phi &= \pi/4 - \theta/4 \end{aligned} \tag{6.47}$$

$$S_c = c/\sin 2\phi, \qquad \text{when } c_1 = c_2 = c \tag{6.48}$$

(b) For the parallel type, $\theta = 180°$, no equation is given here, since this type has no practical application.

D. Geometrical Analysis of Double-Plane Reflector System

The analysis of the parameters given in the foregoing is summarized in Fig. 15 for reflectors of the intersection, symmetrical intersection, and parallel types. This analysis is made on the plane that includes the transmitting point, the receiving point, and the mean center of the reflector system. Therefore some correction will be required when it is applied to the actual propagation path which differs appreciably from the above assumptions.

A slight deviation may be corrected through a fine adjustment of the direction during installation work. However, if there is a considerable difference in levels of two plane reflectors, the calculation becomes difficult.

6.9 LOSS OF DOUBLE-REFLECTOR SYSTEM

A double-reflector system, where two reflectors are arranged closely, forms a passive antenna system and can be treated as follows in calculating the passive-relay additional loss.

A. Effective Area B_e and Bidirectional Gain G_{REF}

When properly designed, the effective areas of two plane reflectors are approximately equal to each other. Accordingly, in most cases it is sufficient to study any one of these two reflectors. Studies are also made for the following, including the cases where the effective areas of the two reflectors differ appreciably from one another.

1. The reflector located on the side of the passive relay section, to estimate the additional passive-relay loss, the equivalent propagation loss.
2. The reflector located on the side of the major section, to estimate the loss between these two reflectors.
3. For a double-reflector system,

$$B_e = B_1 \eta_1 \cos \phi_1 \tag{6.49a}$$

or

$$B_e = B_2 \eta_2 \cos \phi_2 \tag{6.49b}$$

4. For a double-reflector system,

$$G_{\text{REF}} = \left(4\pi B_e/\lambda^2\right)^2$$

B. Plane-Area Efficiency η

Approximate values of η are

$$\begin{aligned} &0.8\text{–}0.95, && \text{single-plane reflector system} \\ &0.6\text{–}0.9 && \text{double plane reflector system} \end{aligned} \tag{6.50}$$

To convert these figures into decibels, use Fig. 6*c*.

These figures are considered to include consideration for conductivity, surface roughness, alignment error, and orientation error provided that proper adjustments have been made in engineering design, manufacturing, and installation.

The calculation for the double-reflector system uses the value of η for any one of the reflectors as given in Eq. (6.49) to reduce the problem to that of a single-reflector system.

If the interreflector loss is ignored or the value of X_R (see Section 6.9C) is equal to zero, use η for a double-reflector system in calculating the additional passive-relay loss to compensate for the underestimated loss.

C. Additional Loss X_R between Two Plane Reflectors

In general this loss can be ignored unless the spacing of the two reflectors is very large. Sometimes a margin of 1 to 3 dB is applied. The calculation graphs given in Fig. 7*e* and *f* were prepared to avoid the errors caused by subjective judgment.

Assume $X_R = 0$ when the results from the graphs show a negative value,

$$X_R = X_u - B_f$$
$$X_R = 0 \text{ for } X_R < 0 \tag{6.51}$$

For the plane reflector area, the angle of incidence, the plane area efficiency, etc., use the data of the reflector on the side of the major section.

6.10 TWO-ANTENNA METHOD

This method is used only for specific cases because of the high cost and the low aperture efficiency of the antennas.

The equation to estimate the additional passive-relay loss L_p is given by

$$L_p = \Gamma_R / G_{p1} G_{p2} \tag{6.52}$$

where Γ_R = transmission loss in passive-relay additional section
G_{p1}, G_{p2} = antenna gain for each direction

An analysis of the near-field type is not very simple since it involves the aperture current distribution, the phase characteristics, and the shape of the aperture.

Graphs contained in this chapter dealing with plane reflectors can also be used for the two-antenna method by employing the effective aperture area and the mono- or bidirectional gain.

EXAMPLES

EXAMPLE 6.1

Calculate the overall transmission loss (equivalent propagation loss) Γ_e at 6 GHz of a radio section 60 km in length. Here the 5 by 6-m single-plane reflector is installed 2.5 km from a station of this section, assuming 60° of the transfer-supplementary angle θ and 3 m of the diameter of both the transmitting and receiving antennas.

Solution

To solve this problem the general expression is used since it is not clear whether or not free-space propagation can be assumed.

1. Judgment on whether it is far-field or near-field type. Area of plane reflector,

$$B_r = 5 \times 6 = 30 \text{ m}^2$$

Projected area,

$$B_j = 30 \times \cos 30° = 26 \text{ m}^2$$

Aperture area of parabola,

$$A_r \simeq 7 \text{ m}^2 \qquad \text{(Fig. } 8b\text{)}$$

$$C \simeq 63 \text{ m}^2 \qquad \text{(6 GHz, 2.5 km,} \qquad \text{Fig. 5)}$$

$$B_r < C \qquad \text{and } A_r < C$$

Judgment: far field.

2. Equation of equivalent propagation loss,

$$\Gamma_e = \Gamma_D \Gamma_R / G_{\text{REF}} \qquad \text{(Fig. 3)}$$

3. Bidirectional gain of reflector,

$$G_{\text{REF}} = G_B + \varepsilon_R \quad \text{(dB)} \qquad \text{(Fig. 6)}$$

$$G_B = 103.6 \text{ dB} \qquad \text{(Fig. } 6b\text{)}$$

$$\times \frac{\varepsilon_R = -3.2 \text{ dB} \qquad (\eta = 0.8 \text{ is assumed, Fig. } 6c)}{G_{\text{REF}} = 100.4 \text{ dB}}$$

4. Equivalent propagation loss,

$$\Gamma_e = 20 \log(\Gamma_D \Gamma_R) - 100.4 \text{ dB}$$

> **EXAMPLE 6.2**
>
> Calculate the equivalent propagation loss Γ_e for Example 6.1, assuming that the passive-relay additional section is in free space.

Solution

Related parameters are obtained from the formulas given in Auxiliary Index B_1.

1. L_p,

$$L_p = L_{pu} - B_f \quad \text{(dB)} \qquad \text{(additional passive-relay loss, Fig. 7)}$$

$$L_{pu} = 42 \text{ dB} \qquad \text{(Fig. } 7b\text{)}$$

$$\times \frac{B_f = 26.4 \text{ dB} \qquad (\eta^2 = -2 \text{ dB}, \eta = 0.8, \text{Fig. } 7d)}{L_p = 15.6 \text{ dB}}$$

2. Equivalent propagation loss,

$$\Gamma_e = \Gamma_D L_p$$

$$\Gamma_e = 20 \log \Gamma_0 + 15.6 \text{ dB}$$

EXAMPLE 6.3

Calculate the equivalent propagation loss Γ_e when both the major and the passive relay sections are of a free-space propagation path.

Solution A. The results of Example 6.2 are used.

1. Obtain the free-space propagation loss Γ_D at 6 GHz for the major section length $D = 60 - 2.5 = 57.5$ km as in Chapter 1,

$$\Gamma_D = 143.2 \text{ dB}$$

2. Using the results of Example 6.2,

$$\Gamma_e = 143.2 + 15.6 = 158.8 \text{ dB}$$

Solution B. The graph for the equivalent propagation loss in totally free space is used.

$$B_e = B_r \cos\phi \cdot \eta = 30 \times \cos 30° \times 0.8 = 20.8 \text{ m}^2$$

$$\Gamma_e = 158.8 \text{ dB} \qquad (D = 57.5, \quad R = 2.5, \quad B_e = 21, \text{Fig. } 9a)$$

EXAMPLE 6.4

Calculate the additional passive-relay loss L_p at 8 GHz assuming the length of passive relay section is 120 m, the plane reflector area $B_r = 6$ m^2, the aperture area of a 3-m parabola $A_r = 7$ m^2, and the transfer-supplementary angle $\theta = 90°$.

Solution

1. Judgment on whether it is near-field or far-field type.

$$C = 2.25 \text{ m}^2 \qquad (\text{Fig. } 5)$$

$$C < B_j \qquad \text{and} \qquad C < A_r$$

Judgment: near field.

2. Using the graphs for near field,

$$u = 0.23 \qquad (\text{Fig. } 8a)$$

$$p = 1.6 \qquad (\text{Fig. } 8b)$$

$$q = 0.63 \qquad (\text{Fig. } 8c)$$

$$L_p \doteqdot 5.2 \text{ dB} \qquad (\eta^2 = -3 \text{ dB is assumed, Fig. } 8d)$$

EXAMPLE 6.5

A preliminary study of the passive relay system has indicated that use of the double-plane reflector system with the transfer-supplementary angle $\theta = 180°$ and the horizontal width of the plane reflector = 5 m is appropriate. Determine the spacing S between the two plane reflectors.

Solution

1. Using Fig. 1,

$$\theta = 138° \qquad \text{(intersection or parallel type)}$$

2. Symmetrical intersection type where $\phi_1 = \phi_2 = \phi$ is chosen.
3. Angle of incidence,

$$\phi = 10.5° \qquad (\theta = 138°,\ \phi_1 = \phi_2, \qquad \text{Fig. } 10b)$$

4. Horizontal width,

$$w_1 = w_2 = w = 5 \text{ m}$$

5. Minimum spacing $S_{\min}$,

$$p_0 \doteqdot 2.55$$

$$S_{\min} = wp_0 = 5 \times 2.55 = 12.75 \text{ m} \qquad \text{(Fig. 12)}$$

6. Margin S_c,

$$\beta = 2.8 \qquad \text{(Fig. 13)}$$

Assuming clearance $C = 1$ m,

$$S_c = C\beta = 2.8 \text{ m}$$

7. Necessary spacing S,

$$S = S_{\min} + S_c = 12.75 + 2.8 \doteqdot 15.6 \text{ m}$$

EXAMPLE 6.6

Estimate the equivalent propagation loss Γ_e and the receiving antenna output power P_r in the parallel-type passive-relay propagation path whose parameters are shown in the figure. The transmitting power at the input of the transmitting antenna P_t is assumed to be 27 dBm (0.5 W), reflector 1 is designed to match reflector 2, and the propagation mode is of free space.

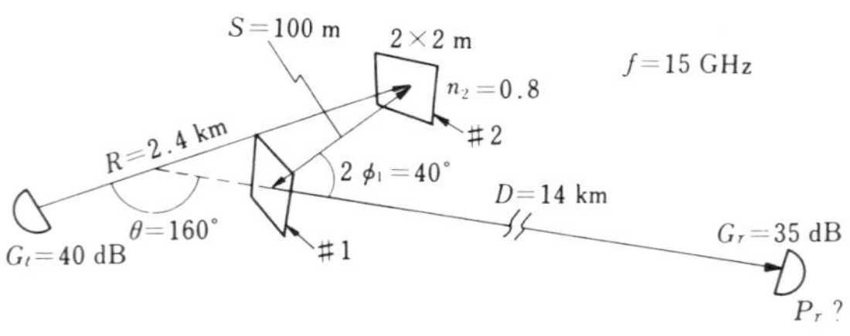

Solution

1. Calculate the angle of incidence for reflector 2,

$$\phi_2 = 10° \qquad (\theta = 160°,\ \phi_1 = 20°,\ \text{Fig. } 10a)$$

2. Interreflector loss (Figs. 7*e* and *f*),

$$\begin{array}{r} X_u = 6\ \text{dB} \\ -\ B_f = 10\ \text{dB} \\ \hline X_R = -4 < 0 \end{array}$$

Then $X_R = 0$ dB.

3. Equivalent propagation loss Γ_e. Since the interreflector loss has been ignored, use the plane area efficiency of the double-reflector system in the estimation of Γ_e. When

$$B_e = B_{r2}\eta \cos\phi_2 = (2 \times 2) \times 0.65 \times \cos 10° = 2.56\ \text{m}^2$$

$$D = 14\ \text{km}, \qquad R \doteqdot 2.4\ \text{km}$$

then from Fig. 9*a*,

$$\Gamma_e = 164.3\ \text{dB}$$

4. Receiving antenna output power P_r (dBm),

$$P_r = P_t + G_t + G_r - \Gamma_e$$

$$= 27 + 40 + 35 - 164.3 = -62.3\ \text{dBm}$$

Chart Index,
Passive-Relay Propagation Path

Parameters of Passive-Relay Propagation

Auxiliary Index

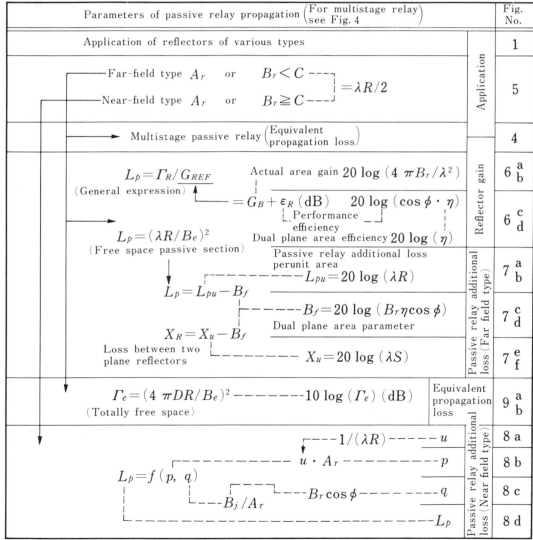

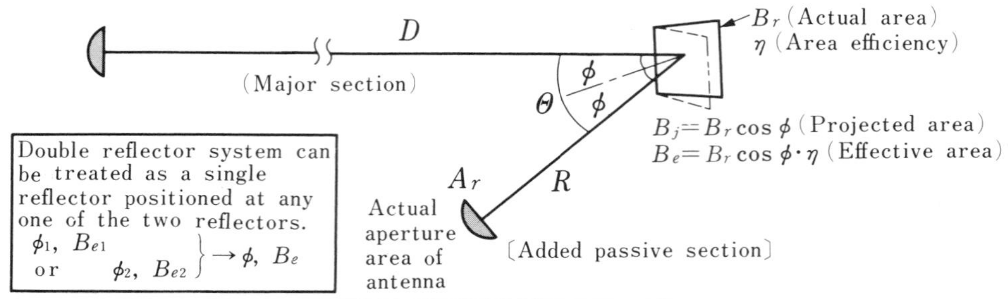

Parameters of passive relay propagation (For multistage relay see Fig. 4)		Fig. No.
Application of reflectors of various types	Application	1
Far-field type A_r or $B_r < C$; Near-field type A_r or $B_r \geqq C$; $= \lambda R/2$	Application	5
→ Multistage passive relay (Equivalent propagation loss)		4
$L_p = \Gamma_R / G_{REF}$ (General expression) ← $= G_B + \varepsilon_R$ (dB); Actual area gain $20 \log (4\pi B_r/\lambda^2)$	Reflector gain	6 a b
Performance efficiency: $20 \log (\cos\phi \cdot \eta)$; Dual plane area efficiency $20 \log (\eta)$; → $L_p = (\lambda R/B_e)^2$ (Free space passive section)	Reflector gain	6 c d
Passive relay additional loss perunit area: $L_{pu} = 20 \log (\lambda R)$; $L_p = L_{pu} - B_f$	Passive relay additional loss (Far field type)	7 a b
$B_f = 20 \log (B_r \eta \cos\phi)$; Dual plane area parameter	Passive relay additional loss (Far field type)	7 c d
$X_R = X_u - B_f$ Loss between two plane reflectors; $X_u = 20 \log (\lambda S)$	Passive relay additional loss (Far field type)	7 e f
$\Gamma_e = (4\pi DR/B_e)^2$ (Totally free space) — $10 \log (\Gamma_e)$ (dB)	Equivalent propagation loss	9 a b
$1/(\lambda R)$ — u	Passive relay additional loss (Near field type)	8 a
$u \cdot A_r$ — p	Passive relay additional loss (Near field type)	8 b
$L_p = f(p, q)$; B_j / A_r; $B_r \cos\phi$ — q	Passive relay additional loss (Near field type)	8 c
— L_p	Passive relay additional loss (Near field type)	8 d

Geometrical Parameters of Double Reflector System

Auxiliary Index

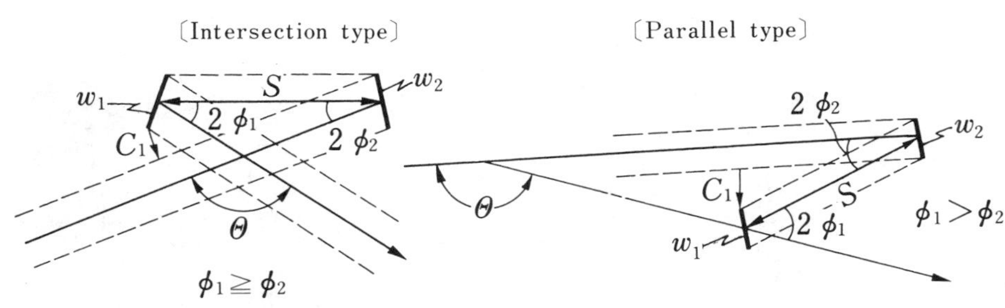

Geometrical parameters of double reflector system	Fig. No.	
Application of reflectors of various type		1
Transfer-supplementary angle Θ / Angle of incidence ϕ — Intersection type: $\phi_1=(180°-\Theta)/2-\phi_2$, $\phi_2=(180°-\Theta)/2-\phi_1$	Angle of incidence	10 a, b
Parallel type: $\phi_1=\phi_2+(180°-\Theta)/2$, $\phi_2=\phi_1-(180°-\Theta)/2$	Angle of incidence	10 c, d
Ratio of horizontal width $g=w_2/w_1 \leqq 1 \cdots\cdots\cdots \cos\phi_1/\cos\phi_2$	Ratio of horizontal width	14 a ~ d
$S=S_c+S_{min}$ (Minimum spacing) — $(p_s+\alpha)\,w_2$ — $(\mathrm{cosec}\,\phi_2)/2$	Spacing	11 a
Intersection type $-\cos\phi_2\,(\tan\phi_1+\tan\phi_2)/2$	Spacing	11 b, c
Parallel type $+\cos\phi_2\,(\tan\phi_1-\tan\phi_2)/2$	Spacing	11 d, e
(Additional spacing to give a margin) $C_1\cdot\beta$ — $\beta=1/(\sin 2\phi_2)$	Spacing	13
$S=S_c+S_{min}$ Symmetrical intersection type $\left\{\begin{matrix}\phi_1=\phi_2=\phi\\ w_1=w_2=w\\ C_1=C_2=C\end{matrix}\right\}$; S_c: $C/\sin 2\phi$; S_{min}: $w\cdot p_0$ — $(\cos 2\phi/\sin\phi)/2$ $\Vert$ $\sin\frac{\Theta}{2}\Big/\sin\left(45°-\frac{\Theta}{4}\right)/2$	Symmetrical intersection type	12
Geometrical analysis of intersection type	Analysis	15 a
Geometrical analysis of parallel type	Analysis	15 b

Application of Reflectors of Various Types

Type	Application range	Composition of propagation path
Single reflector type	$\theta \fallingdotseq 0 \sim 120°$ $\phi \fallingdotseq 0 \sim 60°$	θ: Transfer-supplementary angle ϕ: Angle of incidence 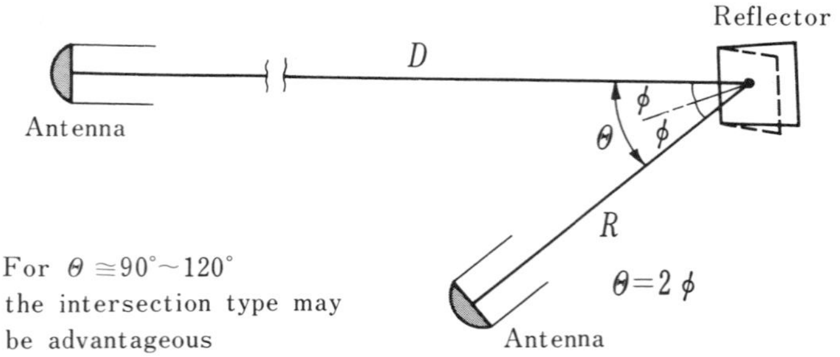 Reflector, Antenna, D, R, ϕ, θ, Antenna $\theta = 2\phi$ For $\theta \cong 90° \sim 120°$ the intersection type may be advantageous
Intersection type	$\theta \fallingdotseq 90° \sim 120°$	To be used dependent on the site conditions
	$\theta \fallingdotseq 100° \sim 150°$ $\phi_1 \fallingdotseq 30° \sim 10°$ $\phi_2 \fallingdotseq 20° \sim 5°$	Symmetrical type when $\phi_1 = \phi_2$ 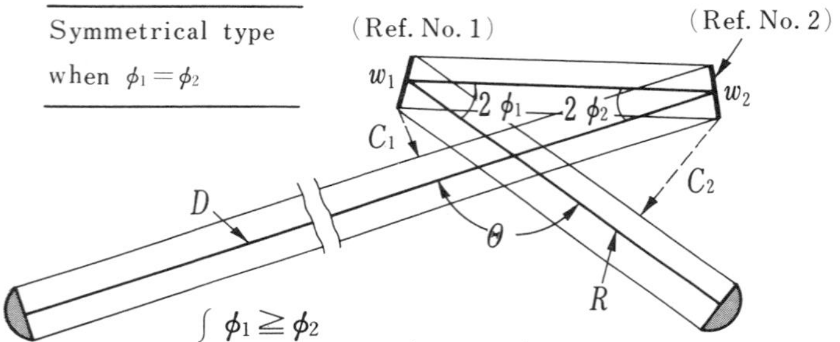 (Ref. No. 1) (Ref. No. 2) w_1 w_2 $2\phi_1$ $2\phi_2$ C_1 C_2 D θ R $\phi_1 \geqq \phi_2$ $\theta = 180° - 2(\phi_1 + \phi_2)$
Parallel type	$\theta \fallingdotseq 100° \sim 150°$	To be used dependent on the site conditions
	$\theta \fallingdotseq 120° \sim 180°$ $\phi_1 \fallingdotseq 45° \sim 10°$ $\phi_2 \fallingdotseq 15° \sim 10°$	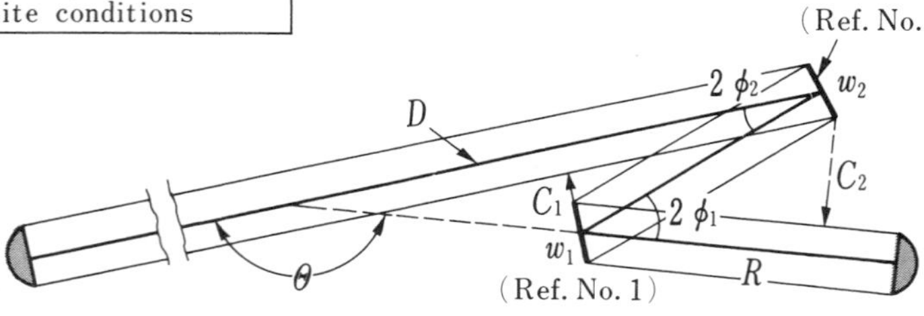 (Ref. No. 2) $2\phi_2$ w_2 D C_2 C_1 $2\phi_1$ w_1 (Ref. No. 1) R θ $\theta = 180°$ when $\phi_1 = \phi_2$ $\phi_1 \geqq \phi_2$ $\theta = 180° - 2(\phi_1 - \phi_2)$
Remark		$D > R$ D: Length of major section w: Width of reflector R: Length of added passive section For use of the calculation chart in principle; Assume $\phi_1 \geqq \phi_2$ $w_1 \geqq w_2$ $C_1 \leqq C_2$

Solution of Passive-Relay Propagation Path

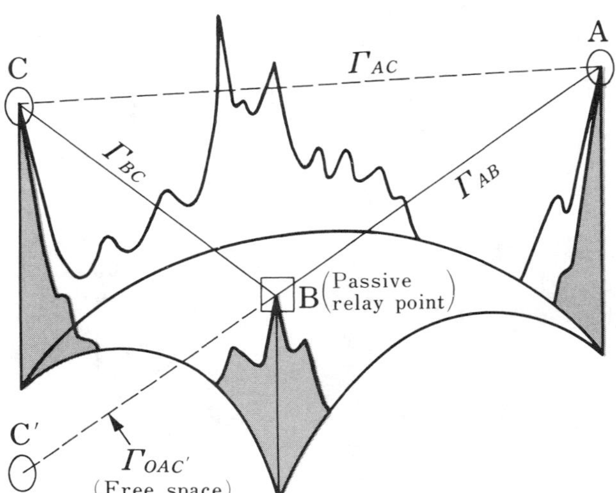

Method		Propagation equations
I	Calculation of improvement in reference to the propagation loss of the section AC of interest.	$\Gamma_{ABC} = \Gamma_{AC} - I_P$ (I_P: Passive improvement; Γ_{AC}: Propagation loss of the section AC)
II	Calculation of the additional loss in reference to the free space propagation loss of the imaginary section AC′ where AC′=AB+BC.	$\Gamma_{ABC} = \Gamma_{OAC'} + L_B$ (L_B: Additional loss; $\Gamma_{OAC'}$: Free space propagation loss of the imaginary section)
III	Calculation of the additional loss in which the BC section is treated as the extended section (added passive section) for the AB section.	Equivalent propagation loss of the section ABC. $\Gamma_e = \Gamma_{AB} + L_P$ (L_P: Passive relay additional loss; Γ_{AB}: Propagation loss of major section)

1. Method I is not used in practice since it may cause difficulty or inaccuracy in the calculation of the propagation loss of AC section.
2. Method II may be applied to the case where both AB and BC sections represent free space propagation. Accordingly, if these sections are not considered as free space, this method should not be used.
3. Method III presents a variety of applications to the ordinary propagation path.
4. This atlas uses Method III. Attention is drawn to this fact, since most published literature uses Method II.

Concept of Passive-Relay Additional Loss

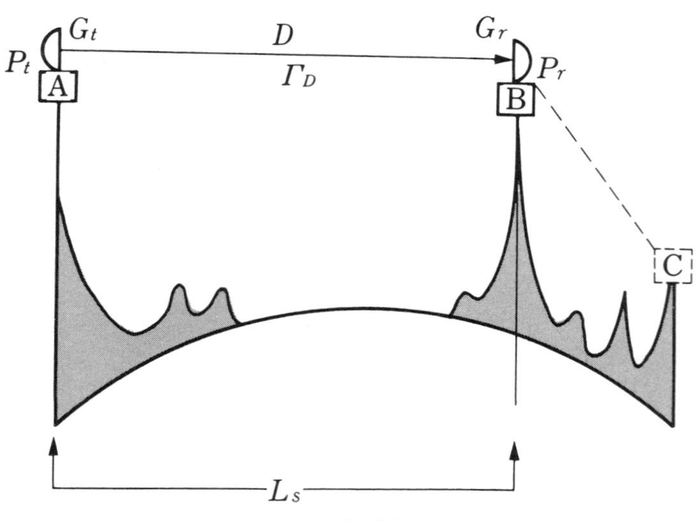

P_t : Transmitting power
P_r : Receiving power
G_t : Transmitting antenna gain
G_r : Receiving antenna gain
D : Section length
Γ_D : Propagation loss
L_s : P_t/P_r ; span loss

(Passive relay)

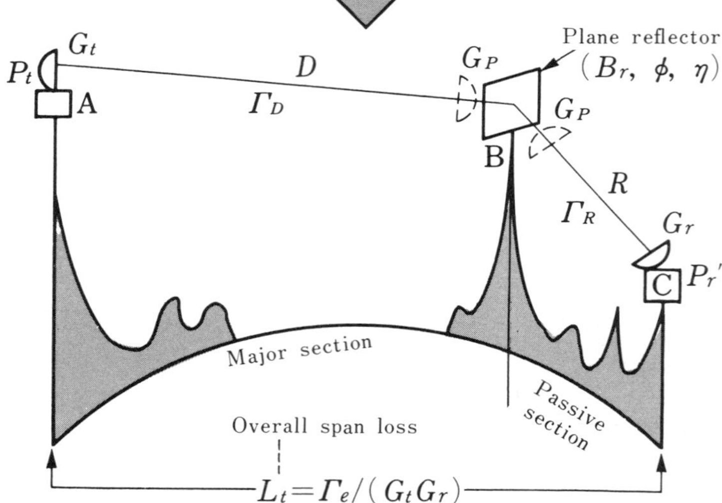

B_r : Actual area of reflector
ϕ : Angle of incidence
η : Plane area efficiency
B_j : Projected area ($B_r \cos \phi$
B_e : Effective area ($= B_j \cdot \eta$)
$G_P = (4\pi\ B_r \cos\phi \cdot \eta/\lambda^2)$
unidirectional gain of re
$G_{REF} = (G_P)^2$: Reflector gain
L_t : Overall span loss
L_s : Span loss of major sect
L_p : Passive relay additional
Γ_D : Propagation loss of maj
Γ_R : Propagation loss of pass
F_e : Equivalent propagation l

$$\Gamma_e = \Gamma_D \cdot L_p$$

Equivalent propagation loss where both the major and the passive sections are treated as free space.
$\Gamma_e = (4\pi D \cdot R/B_e)^2$

		Passive relay additional loss L_p
Far-field type	General expression	$\Gamma_R/G_{REF} = \Gamma_R/(4\pi B_r \cos\phi \cdot \eta/\lambda^2)$
	Free space	$(\lambda \cdot R/B_e)^2 = \{\lambda R/B_r \cos\phi \cdot \eta)\}^2$
Near-field type	$f(p, q)$ ···	$p = A_r/(\lambda R)$ $q = B_r \cos\phi/A_r$

Multistage Passive Relay

[1] Path profile

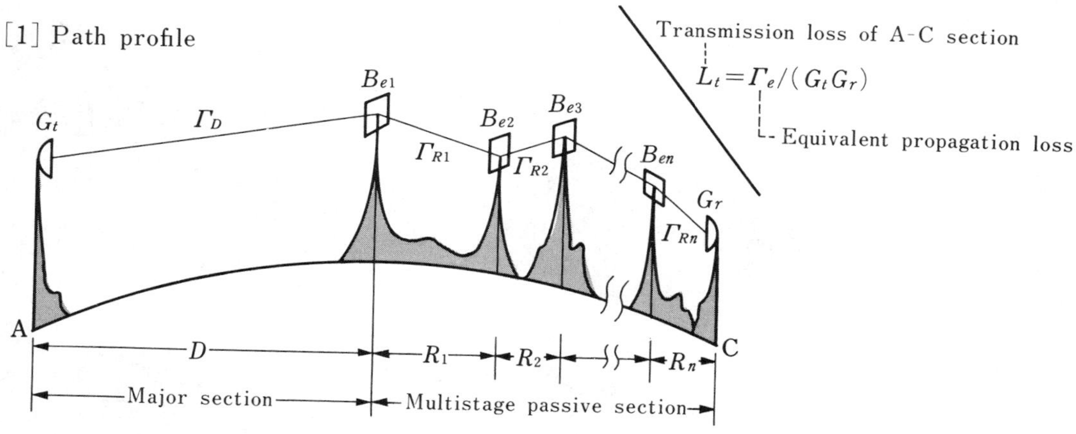

[2] Basic Equations for equivalent propagation loss Γ_e

	Basic equations		Condition
$\Gamma_e =$	$\Gamma_D \cdot L_{p1} L_{p2} \cdots\cdots L_{pn}$	①	Gereral expression L_p : Passive relay additional loss G_p : Unidirectional gain of reflector B_e : Effective area of reflector
	$\Gamma_D \cdot \Gamma_{R1} \Gamma_{R2} \cdots\cdots \Gamma_{Rn} / (G_{p1} G_{p2} \cdots G_{pn})^2$	②	
	$\Gamma_D \cdot \Gamma_{R1} \Gamma_{R2} \cdots\cdots \Gamma_{Rn} \{\lambda^{2n} / (B_{e1} B_{e2} \cdots B_{en} (4\pi)^n)\}^2$	③	
	$\Gamma_D \{(R_1 R_2 \cdots\cdots R_n \cdot \lambda^n) / (B_{e1} B_{e2} \cdots\cdots B_{en})\}^2$	④	Passive section is treated as free space
	$\{4\pi D \cdot R_1 R_2 \cdots\cdots R_n \cdot \lambda^{n-1} / (B_{e1} B_{e2} \cdots B_{en})\}^2$	⑤	Both major and passive sections are treated as free space

[3] Two-stage passive relay ($n=2$)

Far-field type	$\Gamma_e =$	$\Gamma_D \cdot L_{p1} L_{p2} = \Gamma_D \Gamma_{R1} \Gamma_{R2} / (G_{p1} G_{p2})^2 = \Gamma_D \Gamma_{R1} \Gamma_{R2} \{\lambda^4 / B_{e1} B_{e2} (4\pi)^2)\}^2$
		$\Gamma_D \{R_1 R_2 \cdot \lambda^2 / (B_{e1} B_{e2})\}^2$...... to be applied when the passive relay section is treated as free space
		$\{4\pi D \cdot R_1 R_2 \cdot \lambda / (B_{e1} B_{e2})\}^2$.... to be applied when both the major and the passive sections are treated as free space

Near-field type	Both R_1 and R_2 are of near-field type	Ignore the loss of section R_1 and treat section R_2 as the near-field type path
	R_2 is of near-field type	Add the additional loss (dB) in the near-field of section R_2 to Γ_e for section R_1
	R_1 is of plane-wave near-field type (Double reflector system).	Ignore the loss of section R_1 and the section is treated as a single-stage passive relay propagation path (Single relay by means of double plane reflector system)

Graph to Determine Type of Reflector—Far Field or Near Field

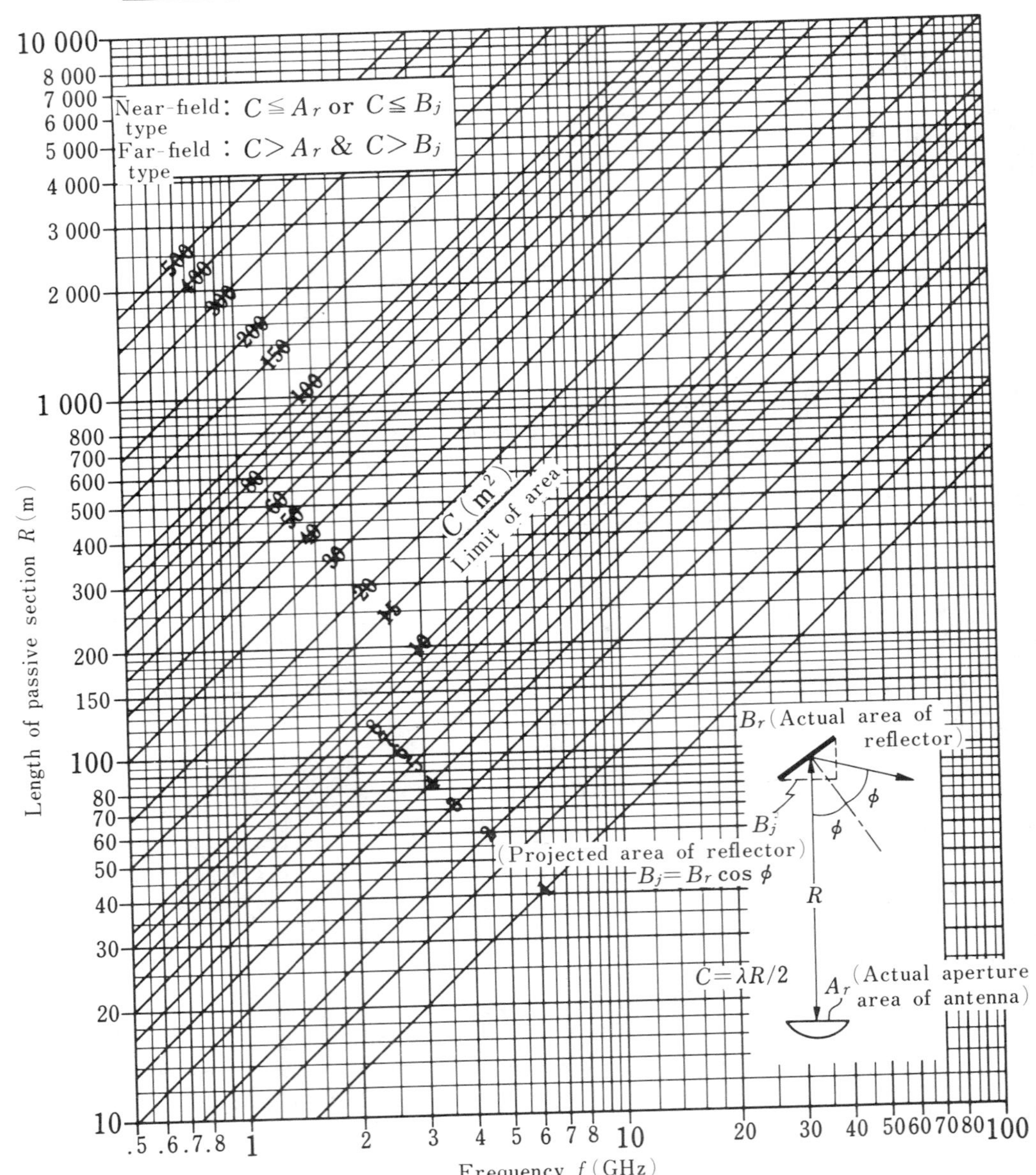

Reflector (Bidirectional) Gain (1)

Actual Area Gain

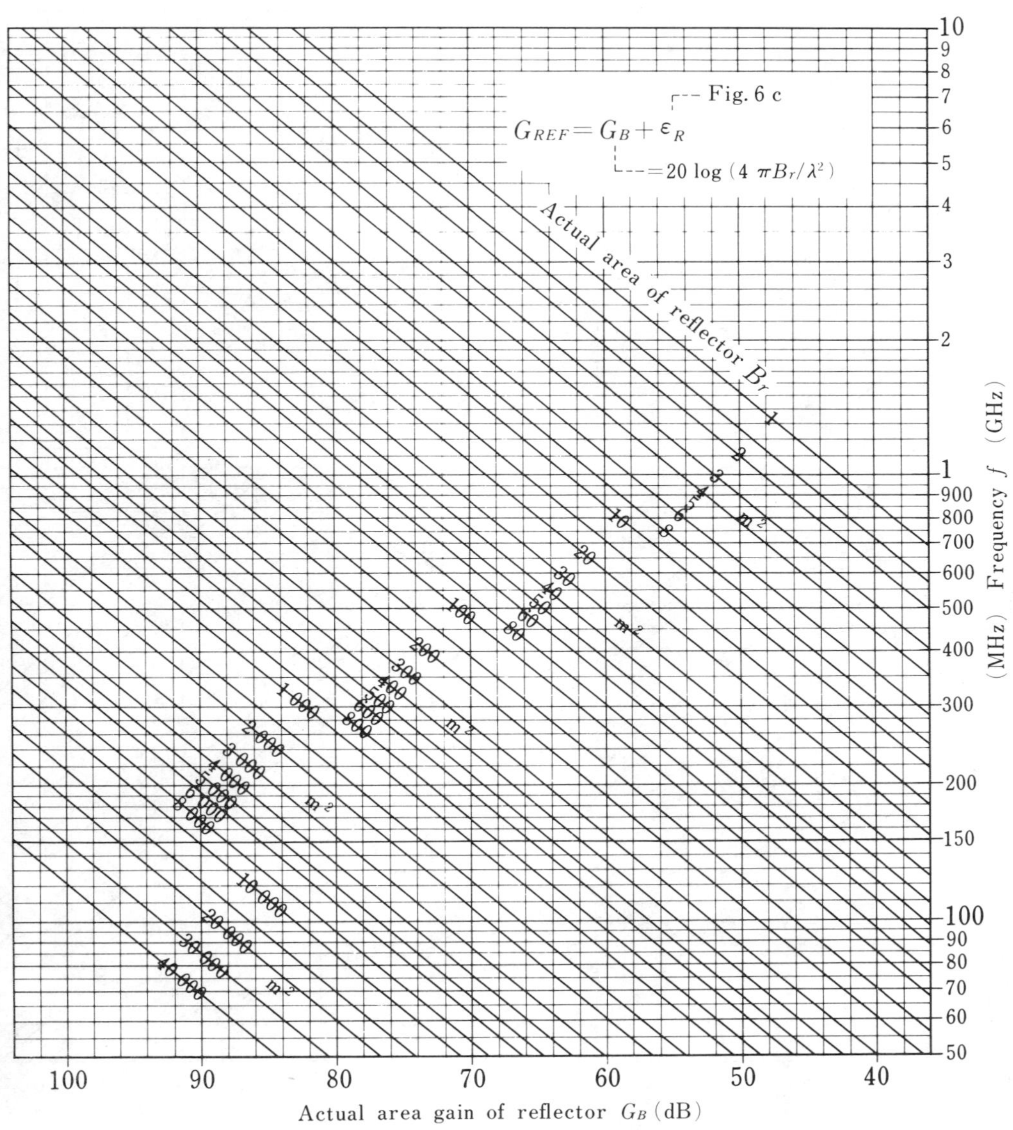

Reflector (Bidirectional) Gain (2)

Actual Area Gain

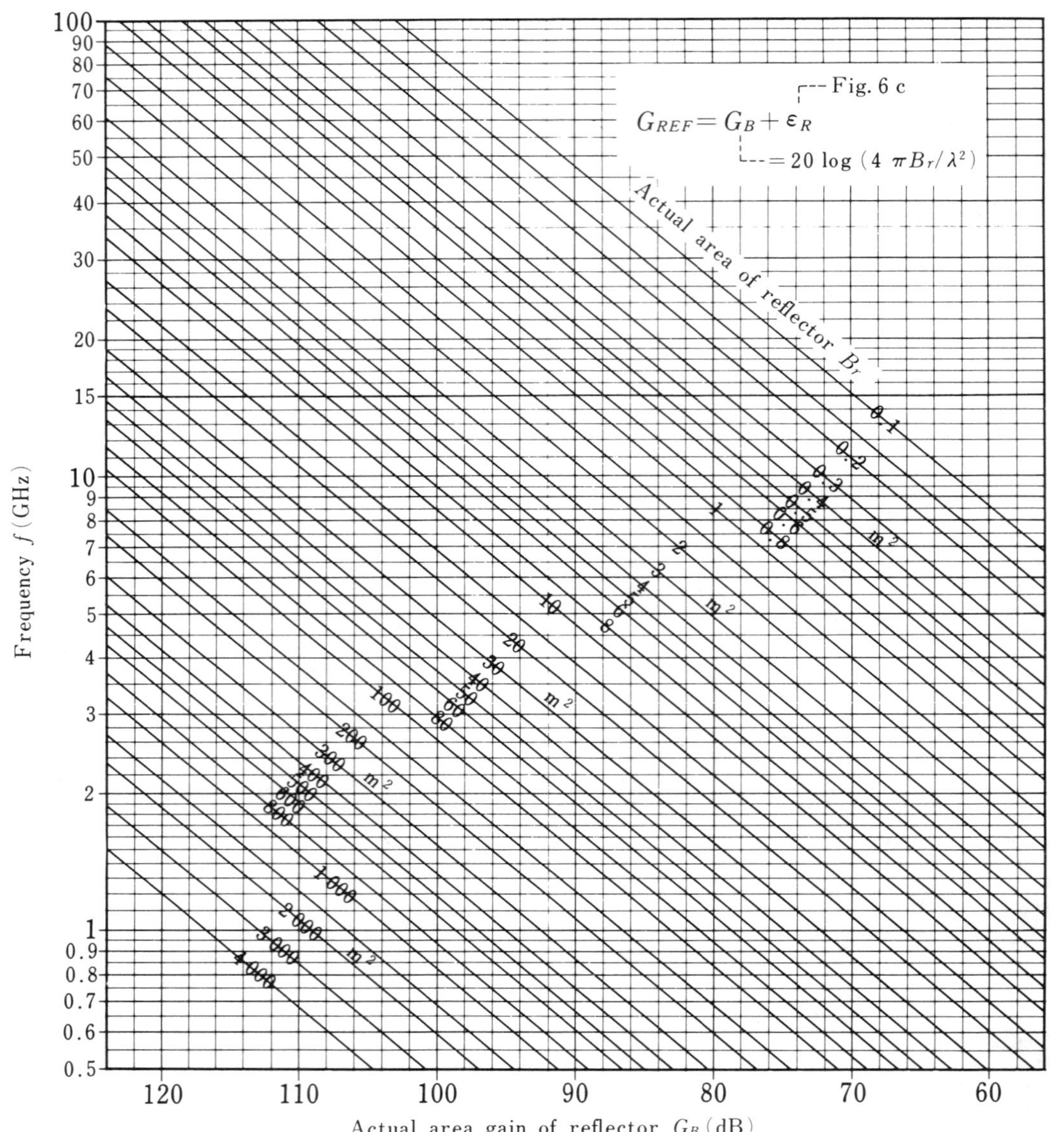

Reflector (Bidirectional) Gain (3)

Performance Efficiency

Actual area gain
Reflector gain | Performance efficiency

$$G_{REF} = G_B + \varepsilon_R$$

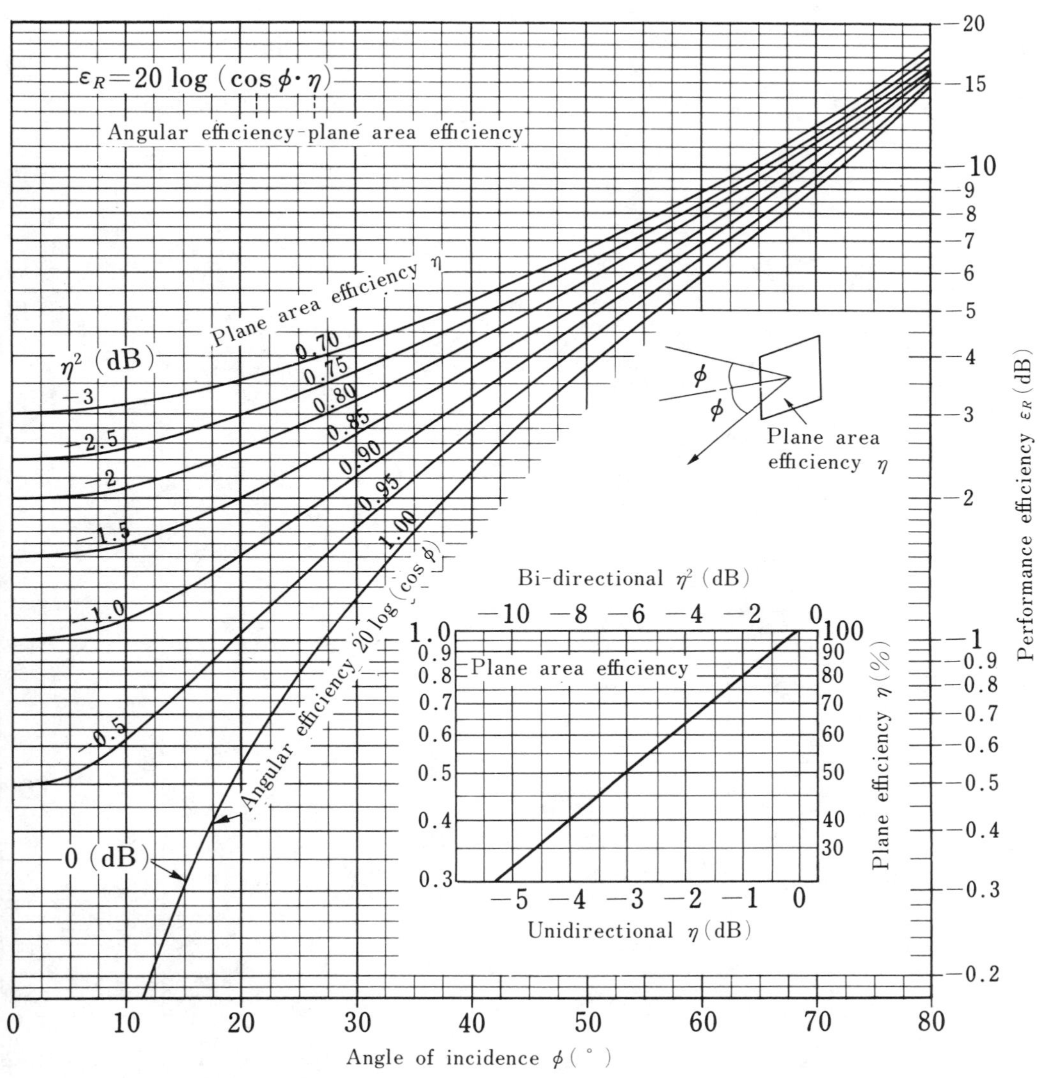

Passive-Relay Additional Loss (Far-Field Type) (1)

Additional Loss per Unit Area

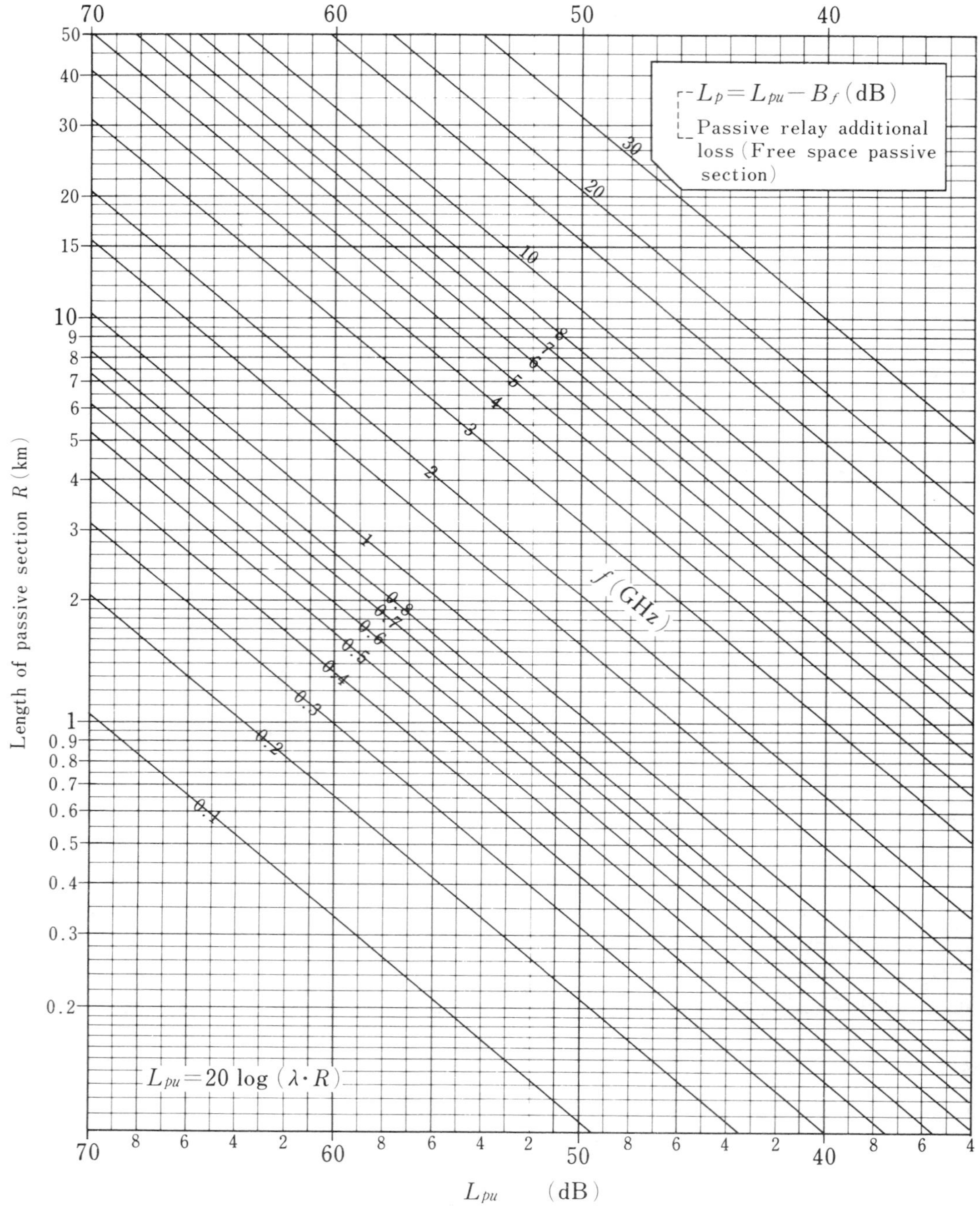

Passive-Relay Additional Loss (Far-Field Type) (2)

Additional Loss per Unit Area

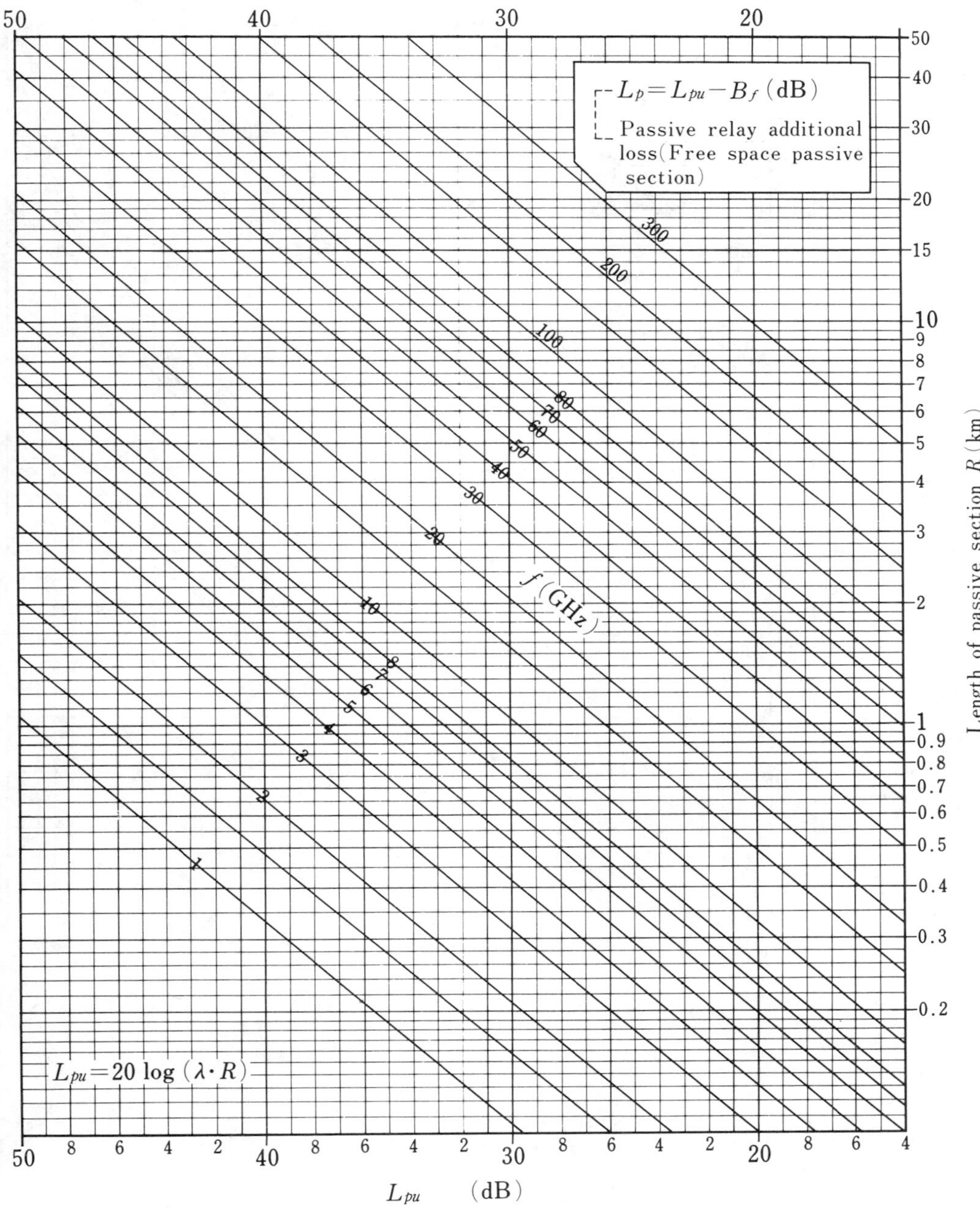

Passive-Relay Additional Loss (Far-Field Type) (3)

Dual-Area Parameter

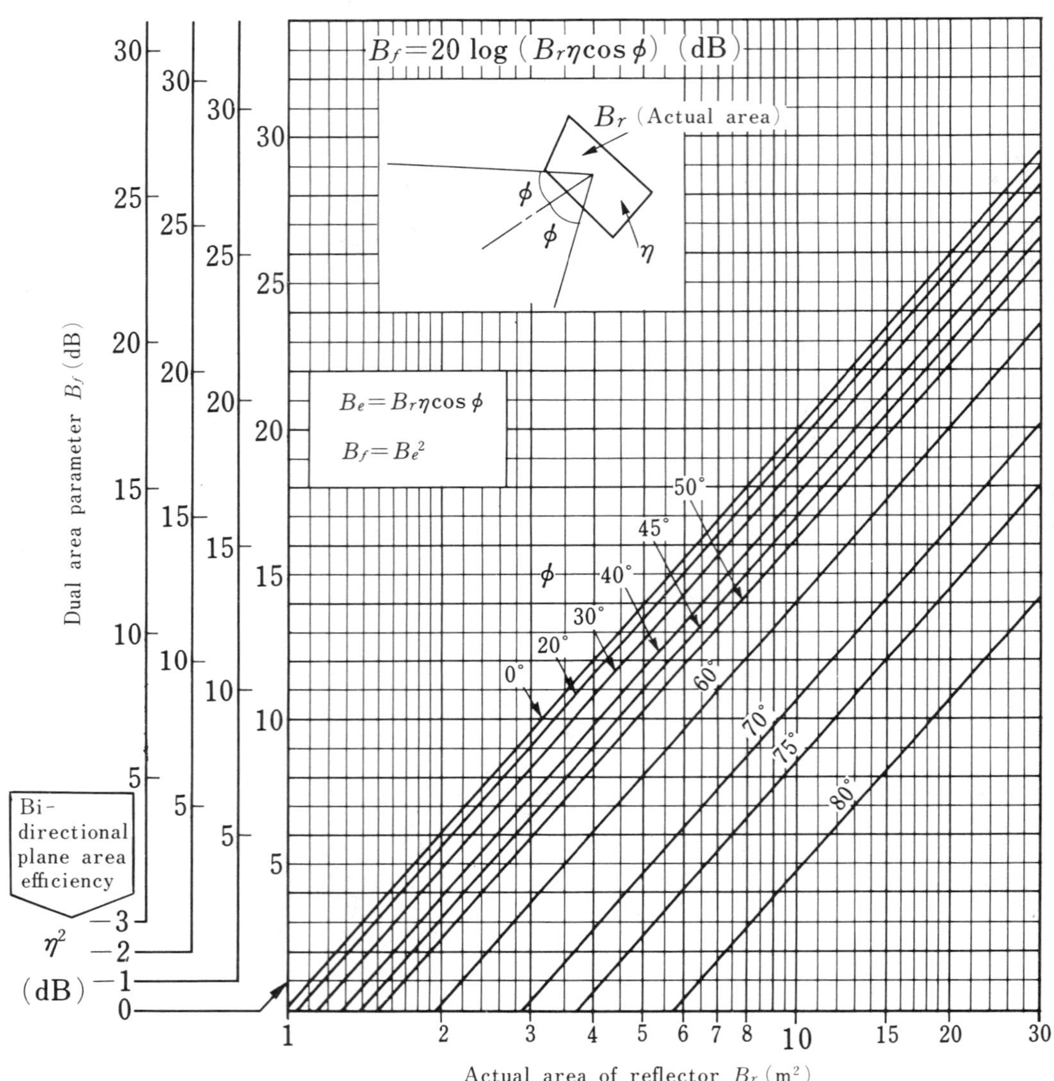

Passive-Relay Additional Loss (Far-Field Type) (4)

Dual-Area Parameter

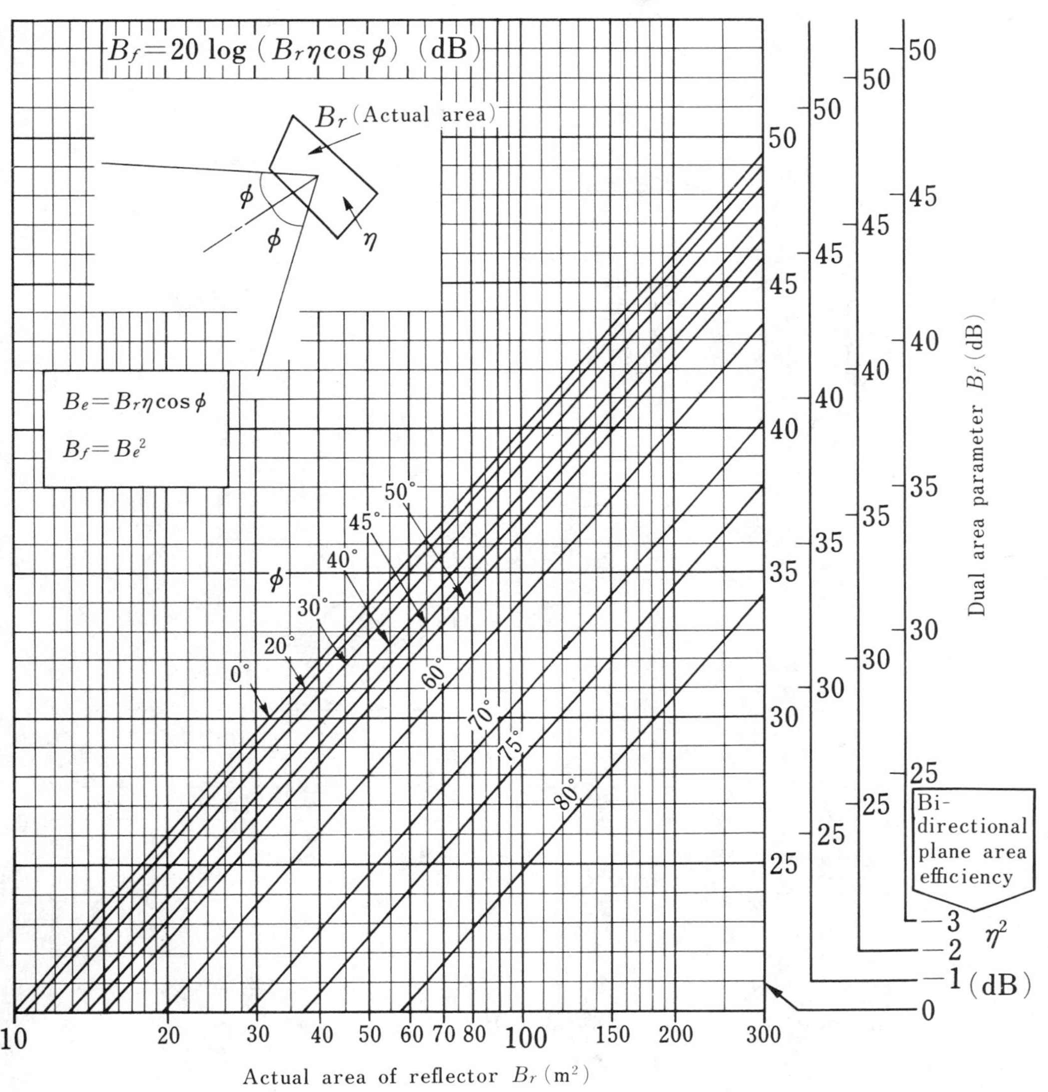

Passive-Relay Additional Loss (Far-Field Type) (5)

Interreflector Loss

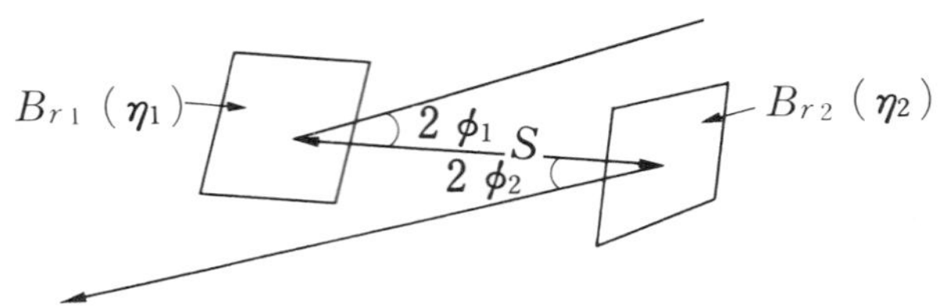

Interreflector loss X_R	Dual area parameter B_f (Fig. 7 c, d)
① $X_R = X_u - B_f$ (dB) ② If $X_R < 0$, $X_R = 0$ (dB)	$B_f = 20 \log B_e$ $B_e = \begin{cases} B_{r1}\ \eta_1 \cos\ \phi_1 \\ B_{r2}\ \eta_2 \cos\ \phi_2 \end{cases}$ Whichever is on the side of major section

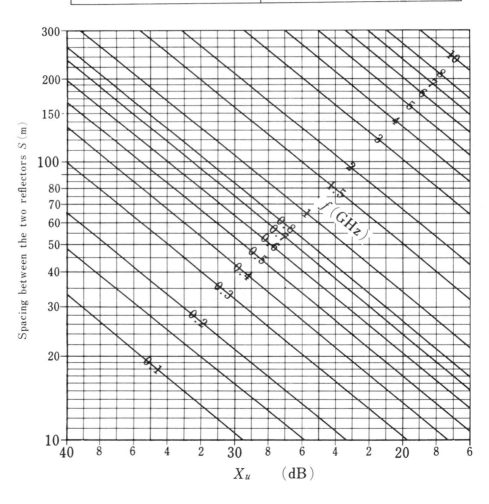

Spacing between the two reflectors S (m)

300, 200, 150, 100, 80, 70, 60, 50, 40, 30, 20, 10

f (GHz): 0.1, 0.2, 0.3, 0.4, 0.5, 0.6, 0.7, 0.8, 1, 1.5, 2, 3, 4, 5, 6, 7, 8, 10

40 8 6 4 2 30 8 6 4 2 20 8 6

X_u (dB)

Passive-Relay Additional Loss (Far-Field Type) (6)

Interreflector Loss

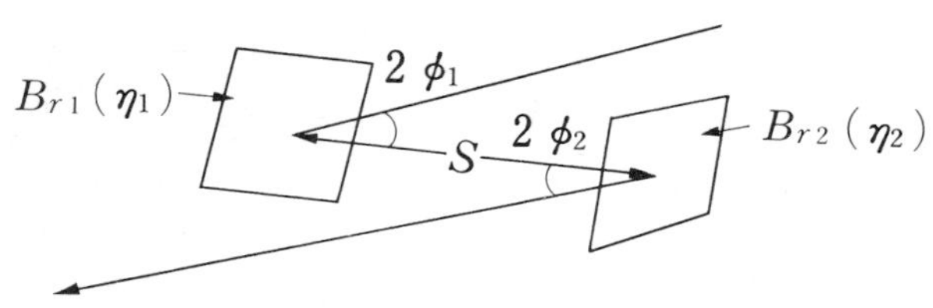

Interreflector loss X_R	Dual area parameter B_f (Fig. 7 c, d)
① $X_R = X_u - B_f$ (dB) ② If $X_R < 0$, $X_R = 0$ (dB)	$B_f = 20 \log B_e$ $B_e = \begin{cases} B_{r1}\,\eta_1 \cos\phi_1 \\ B_{r2}\,\eta_1 \cos\phi_2 \end{cases}$ Whichever is on the side of the major section

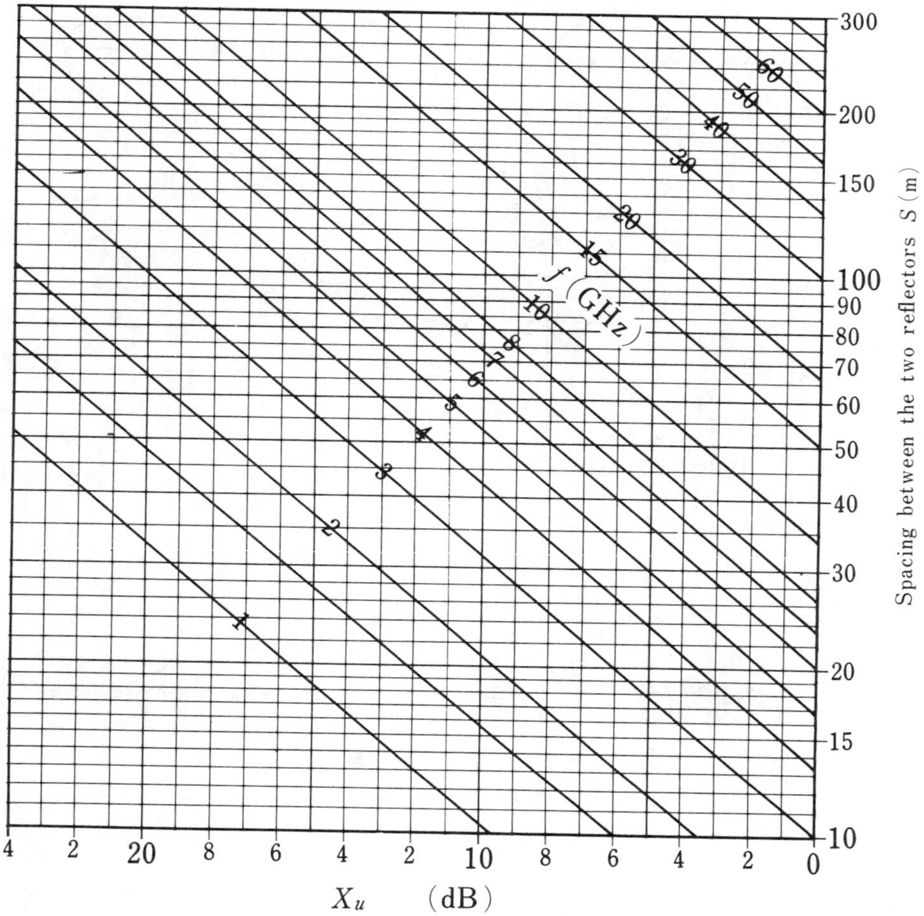

f (GHz): 1, 2, 3, 4, 5, 6, 7, 8, 10, 15, 20, 30, 40, 50, 60

Spacing between the two reflectors S (m): 10, 15, 20, 30, 40, 50, 60, 70, 80, 90, 100, 150, 200, 300

X_u (dB): 4, 2, 20, 8, 6, 4, 2, 10, 8, 6, 4, 2, 0

Passive-Relay Additional Loss (Near-Field Type) (1)

Parameter u

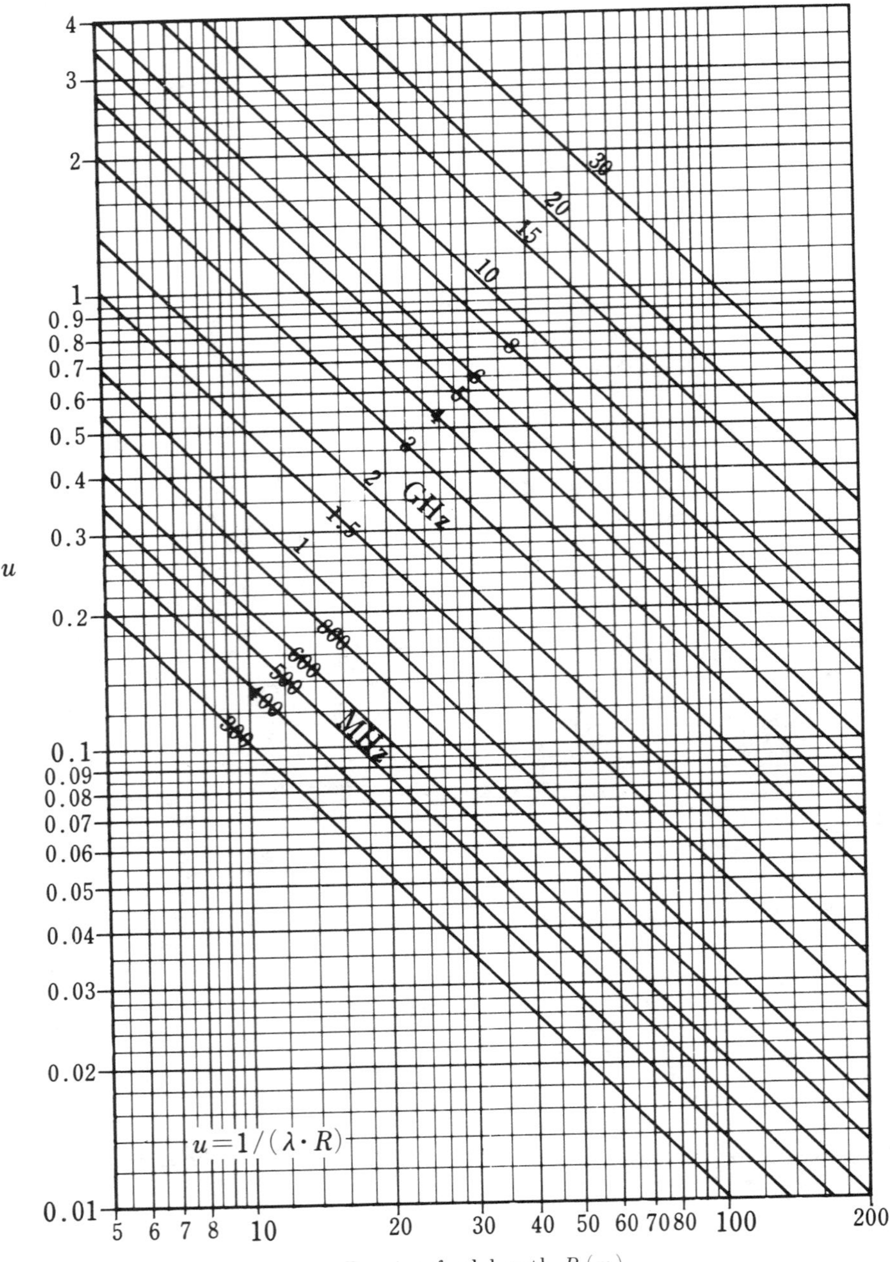

Passive-Relay Additional Loss (Near-Field Type) (2)

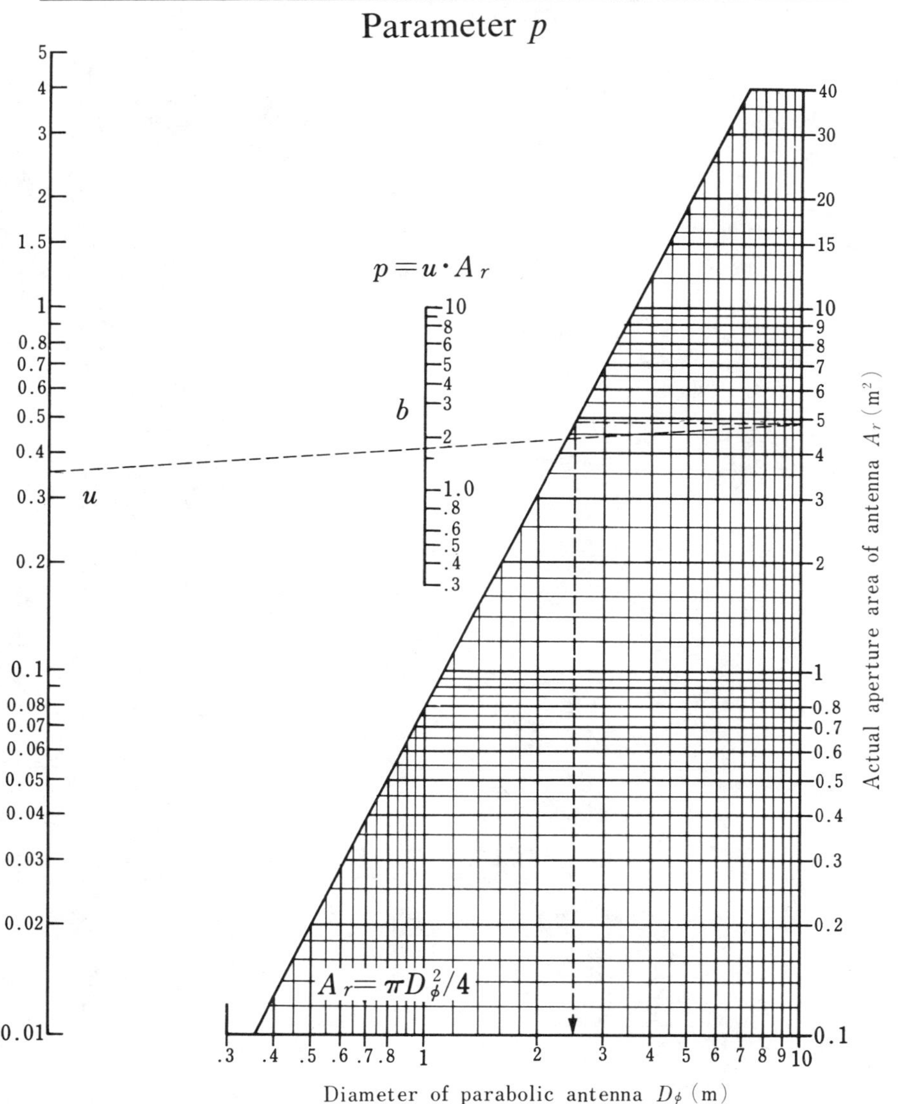

Passive-Relay Additional Loss (Near-Field Type) (3)

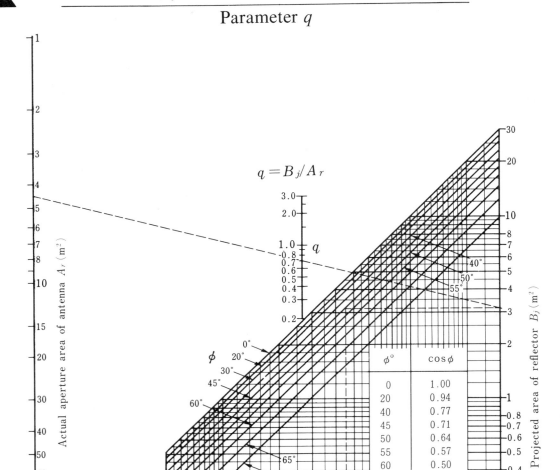

Parameter q

$q = B_j / A_r$

$B_j = B_r \cos\phi$

Actual aperture area of antenna A_r (m²)

Projected area of reflector B_j (m²)

Actual area of reflector B_r (m²)

$\phi°$	$\cos\phi$
0	1.00
20	0.94
40	0.77
45	0.71
50	0.64
55	0.57
60	0.50
65	0.42
70	0.34
75	0.26
80	0.17

Passive-Relay Additional Loss (Near-Field Type) (4)

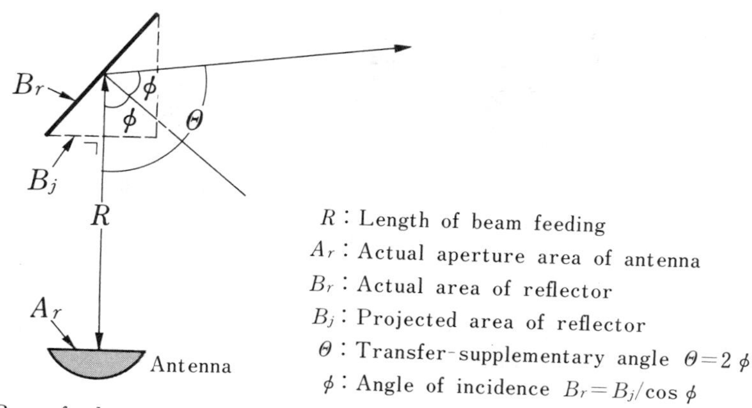

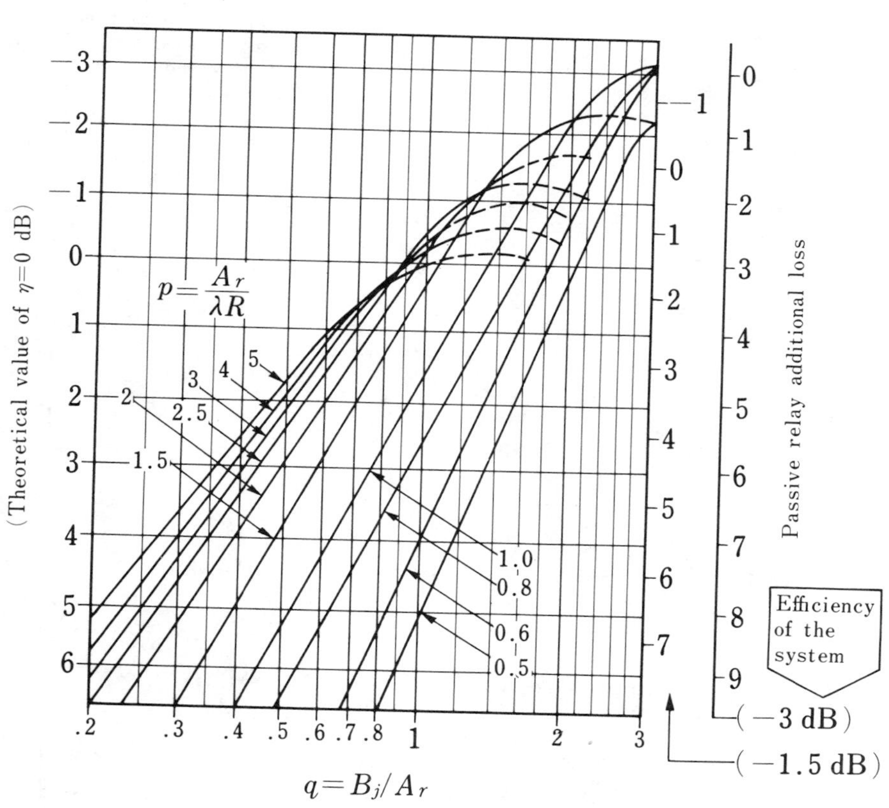

Equivalent Propagation Loss in Totally Free Space (1)

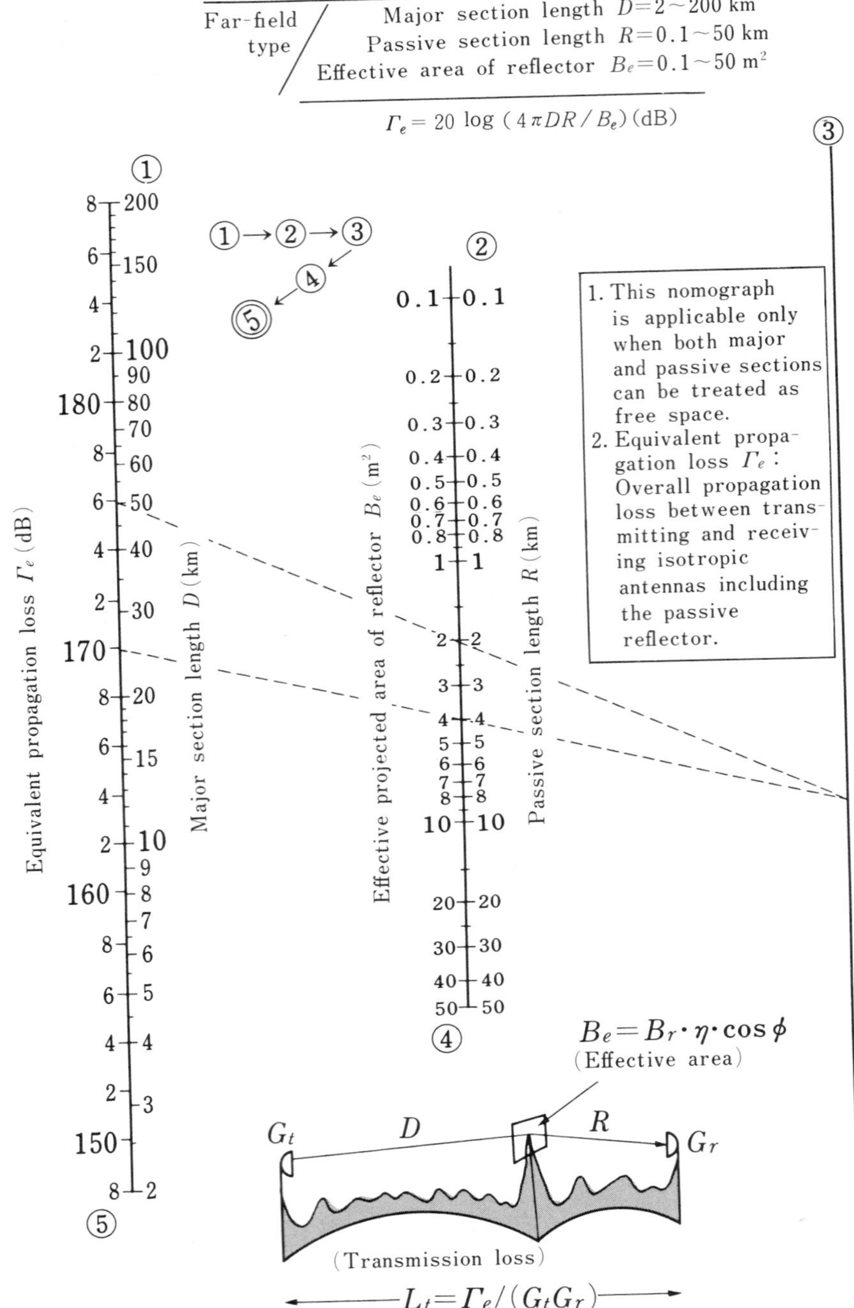

Equivalent Propagation Loss in Totally Free Space (2)

Far-field type	Major section length $D=2\sim200$ km Passive section length $R=0.1\sim50$ km Effective area of reflec. $B_e=1\sim500$ m^2

$$\Gamma_e = 20\log(4\pi DR/B_e)\ (\text{dB})$$

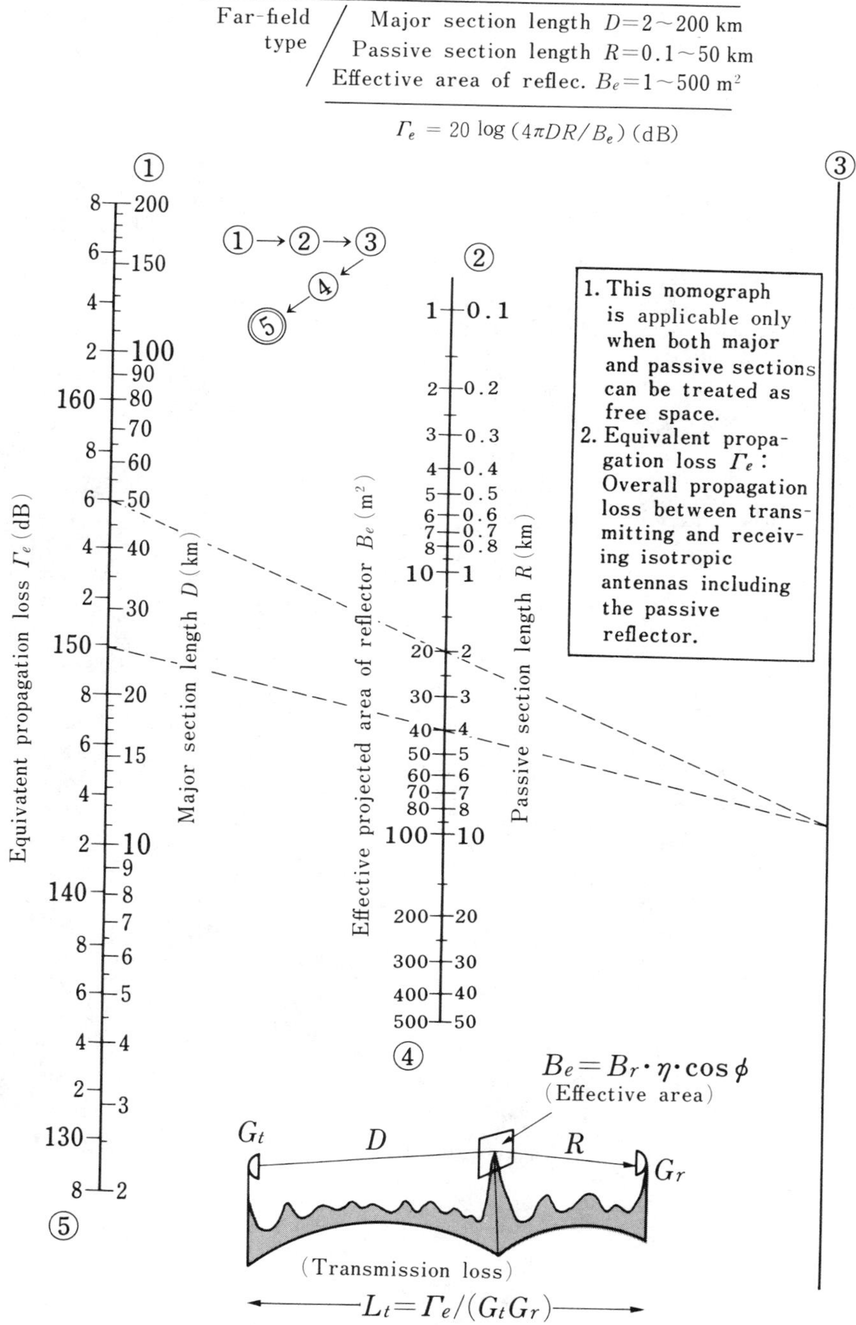

1. This nomograph is applicable only when both major and passive sections can be treated as free space.
2. Equivalent propagation loss Γ_e: Overall propagation loss between transmitting and receiving isotropic antennas including the passive reflector.

Transfer-Supplementary Angle and Angle of Incidence (1)

Intersection Type ($90° \leq \theta \leq 120°$)

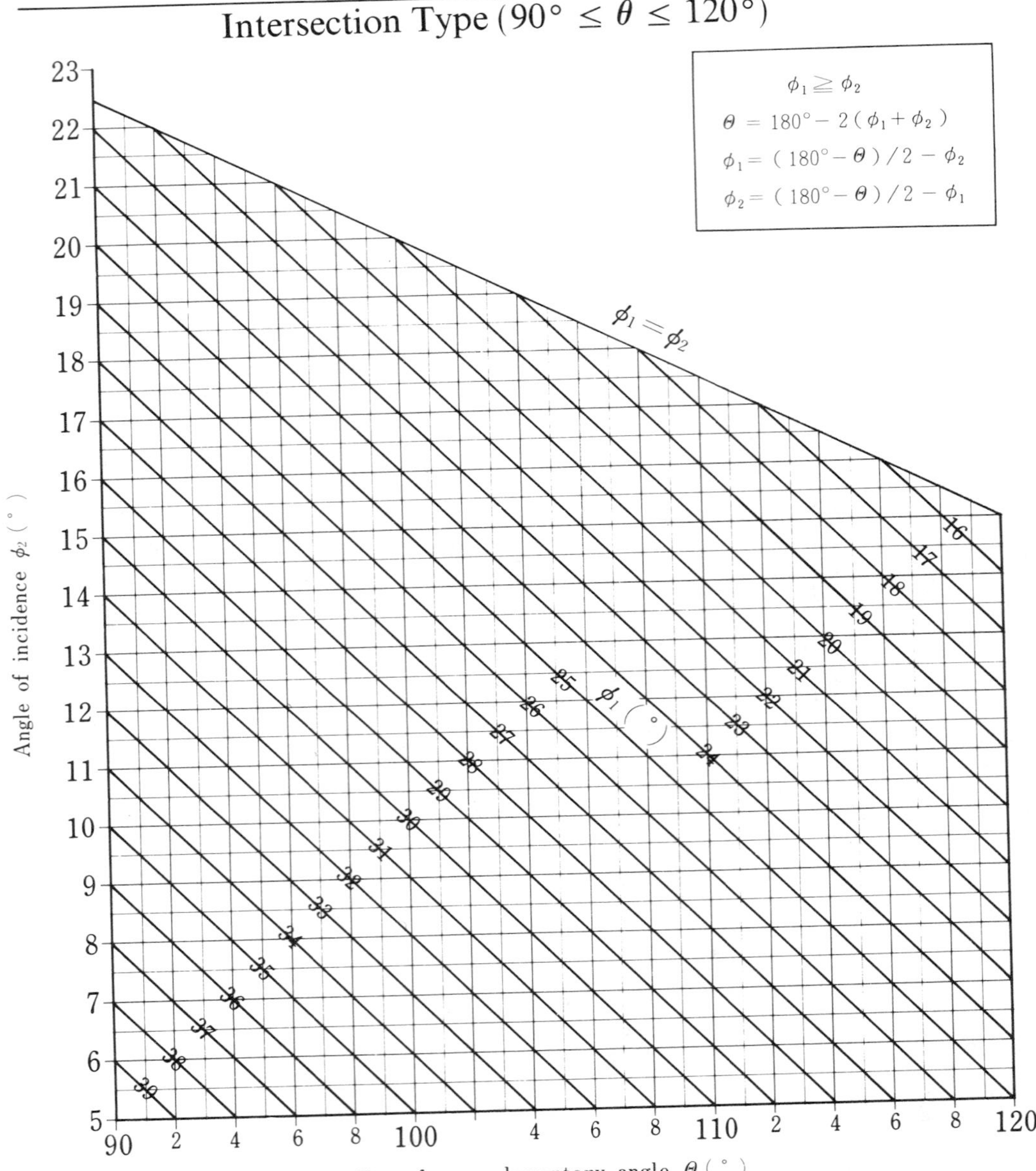

$\phi_1 \geqq \phi_2$
$\theta = 180° - 2(\phi_1 + \phi_2)$
$\phi_1 = (180° - \theta)/2 - \phi_2$
$\phi_2 = (180° - \theta)/2 - \phi_1$
$\phi_1 = \phi_2$
ϕ_1 (°)
Angle of incidence ϕ_2 (°)
Transfer-supplementary angle θ (°)

Transfer-Supplementary Angle and Angle of Incidence (2)

Intersection Type ($120° \le \theta \le 170°$)

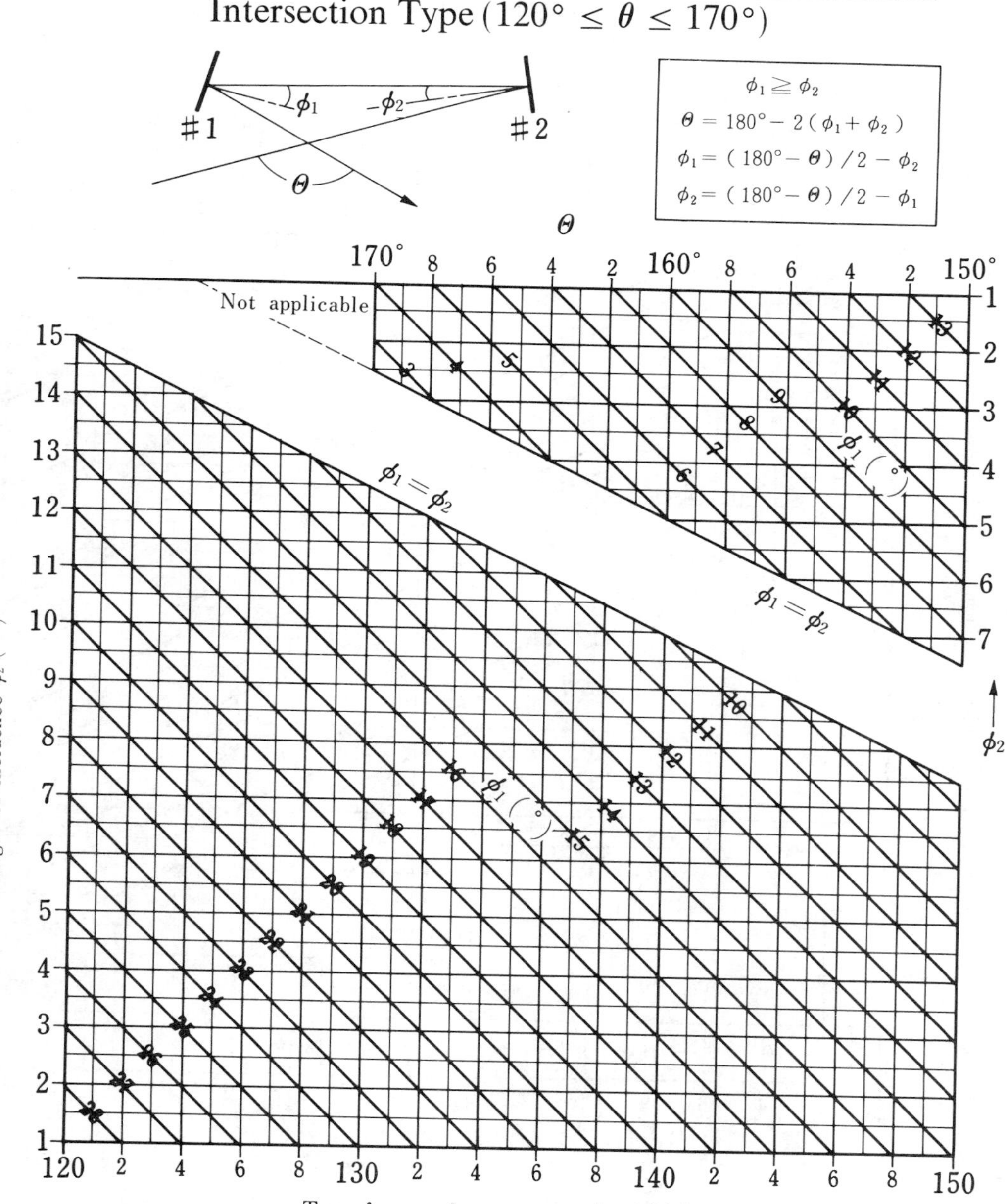

Transfer-Supplementary Angle and Angle of Incidence (3)

Parallel Type ($120° \le \theta \le 150°$)

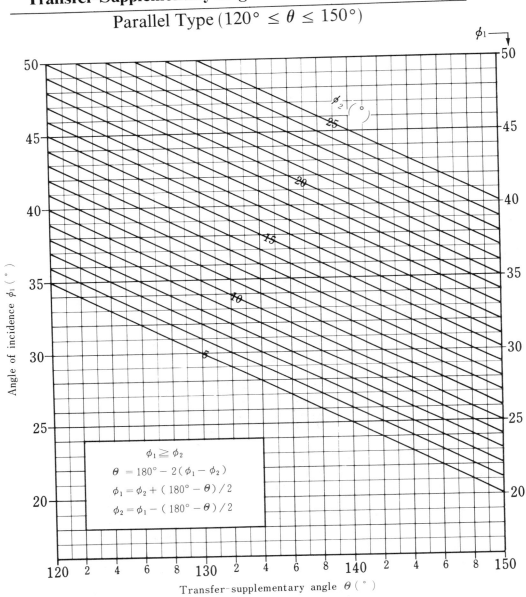

Transfer-Supplementary Angle and Angle of Incidence (4)

Parallel Type ($150° \leq \theta \leq 180°$)

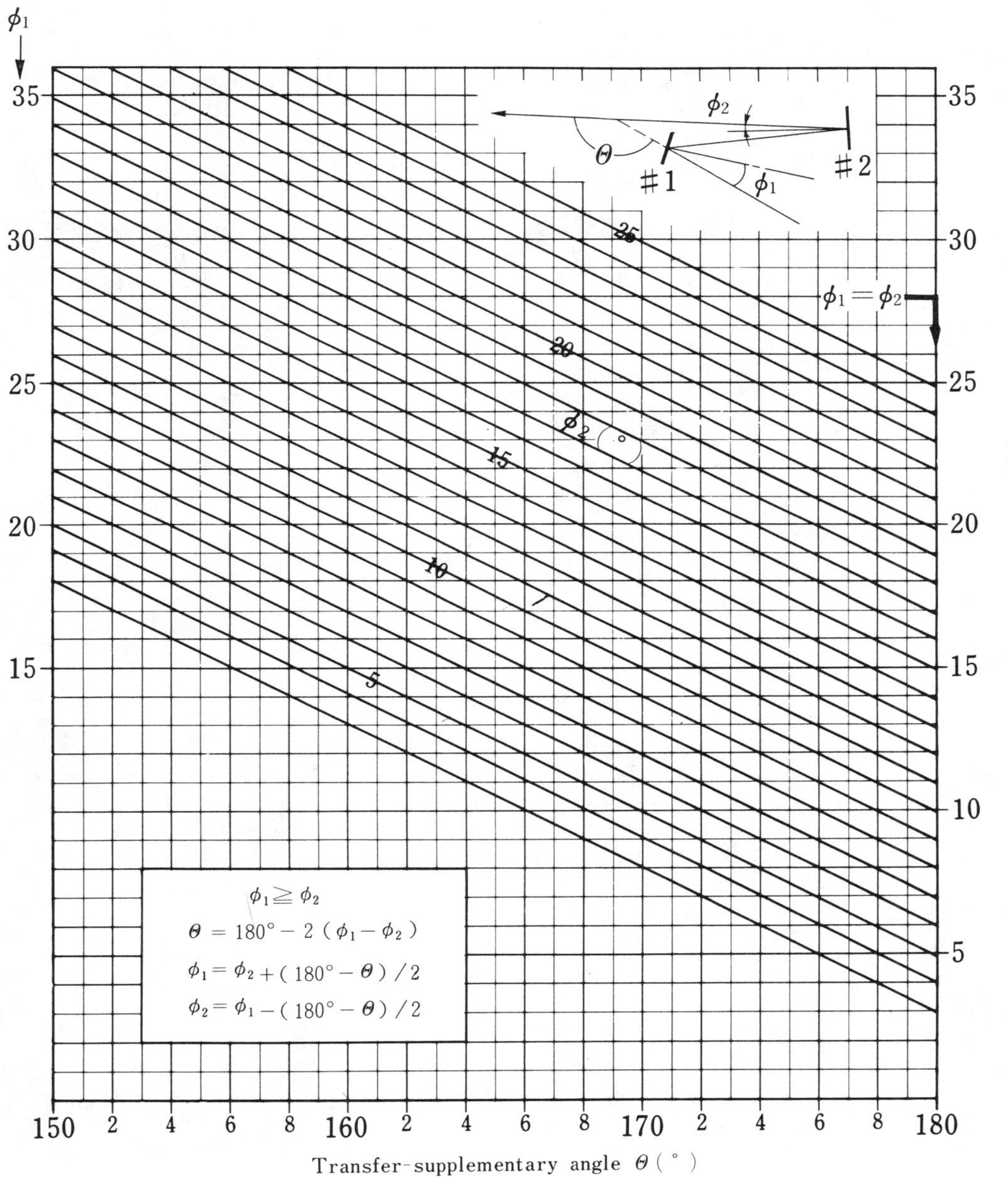

6

11a

Minimum Spacing (1)

Parameter p_s

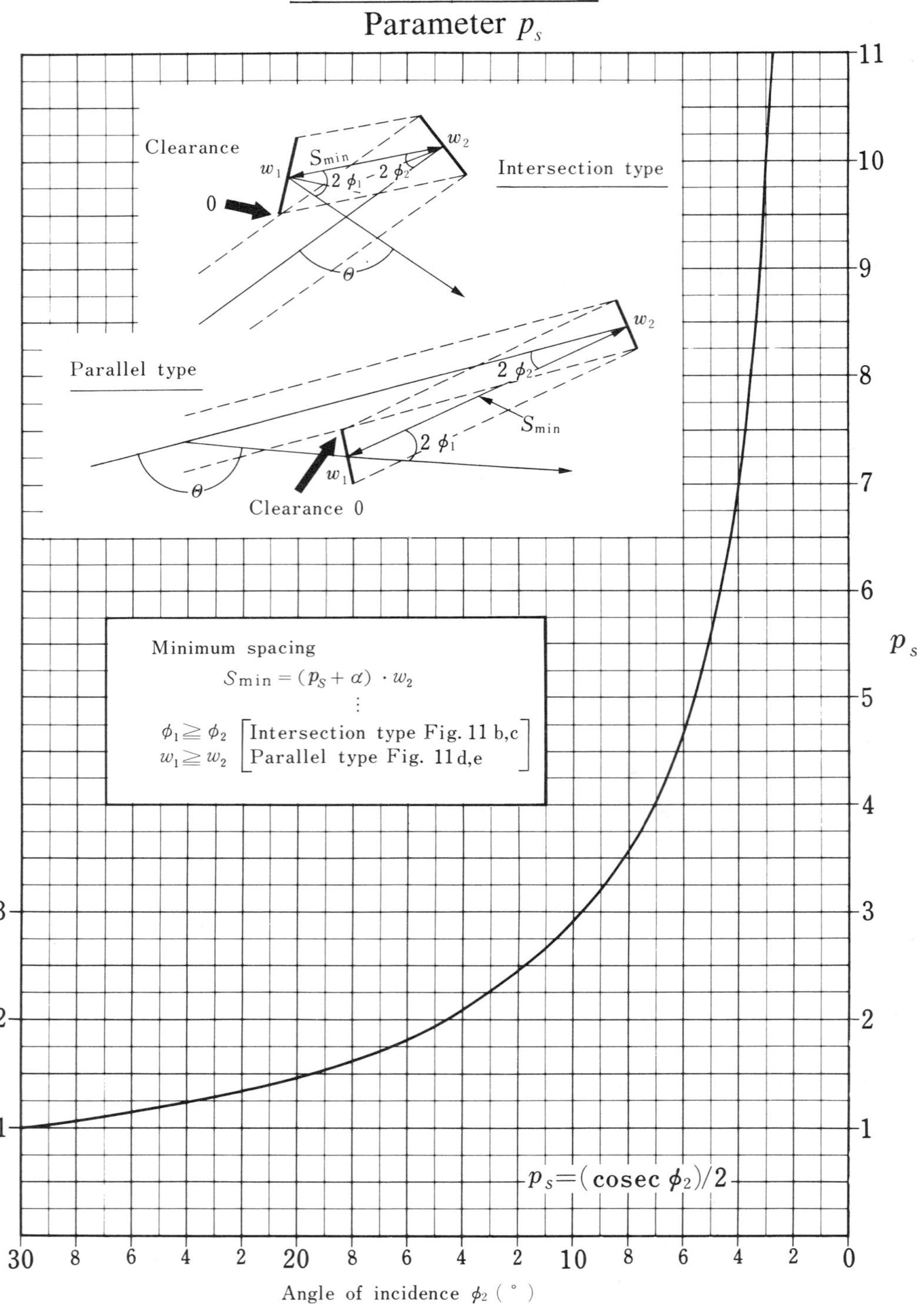

Minimum Spacing (2)

Parameter α, Intersection Type

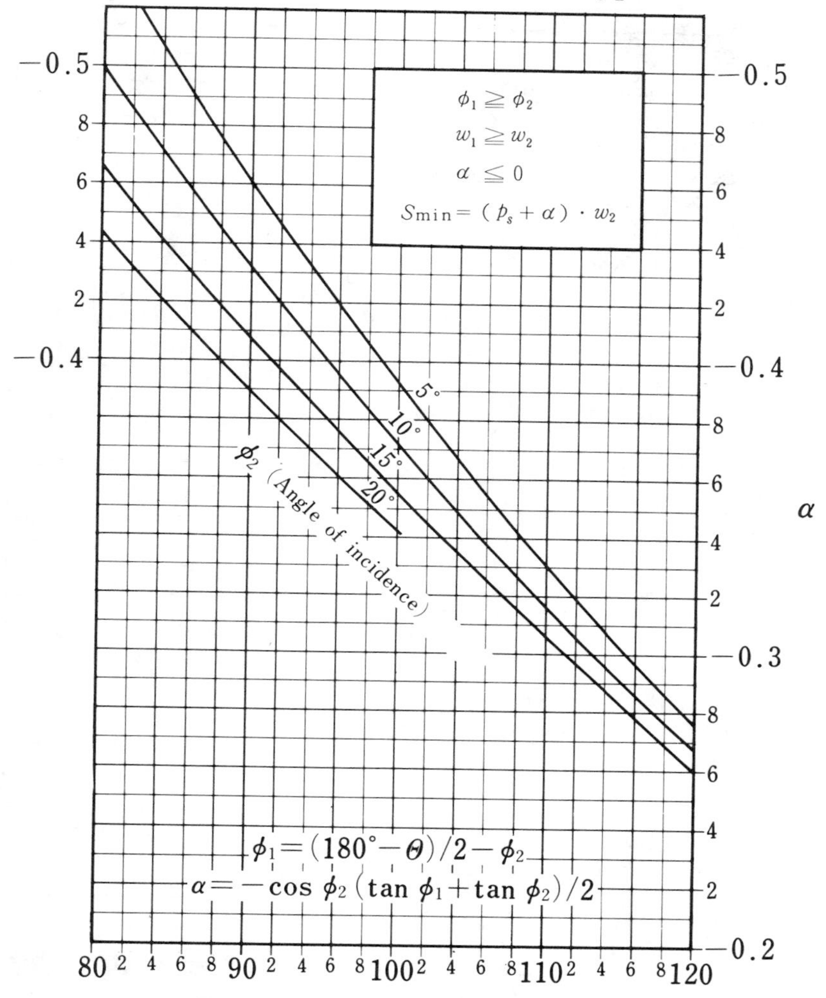

Minimum Spacing (3)

Parameter α, Intersection Type

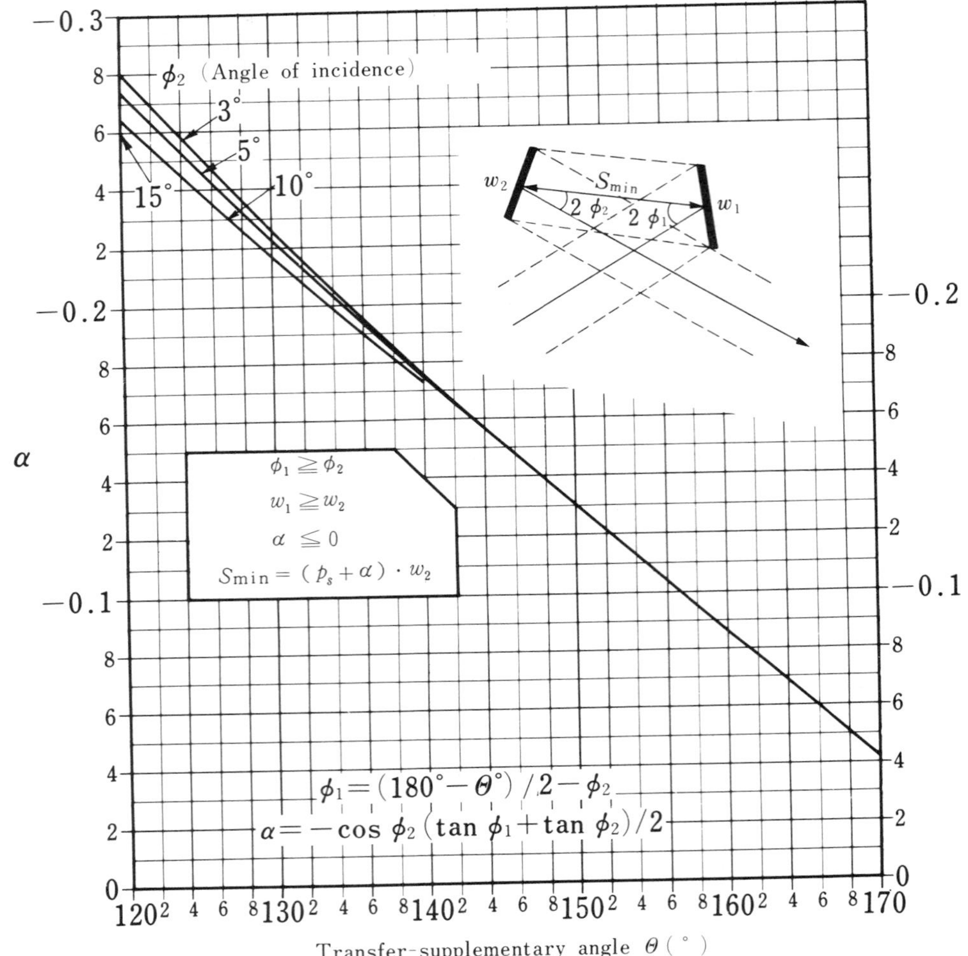

Minimum Spacing (4)

Parameter α, Parallel Type

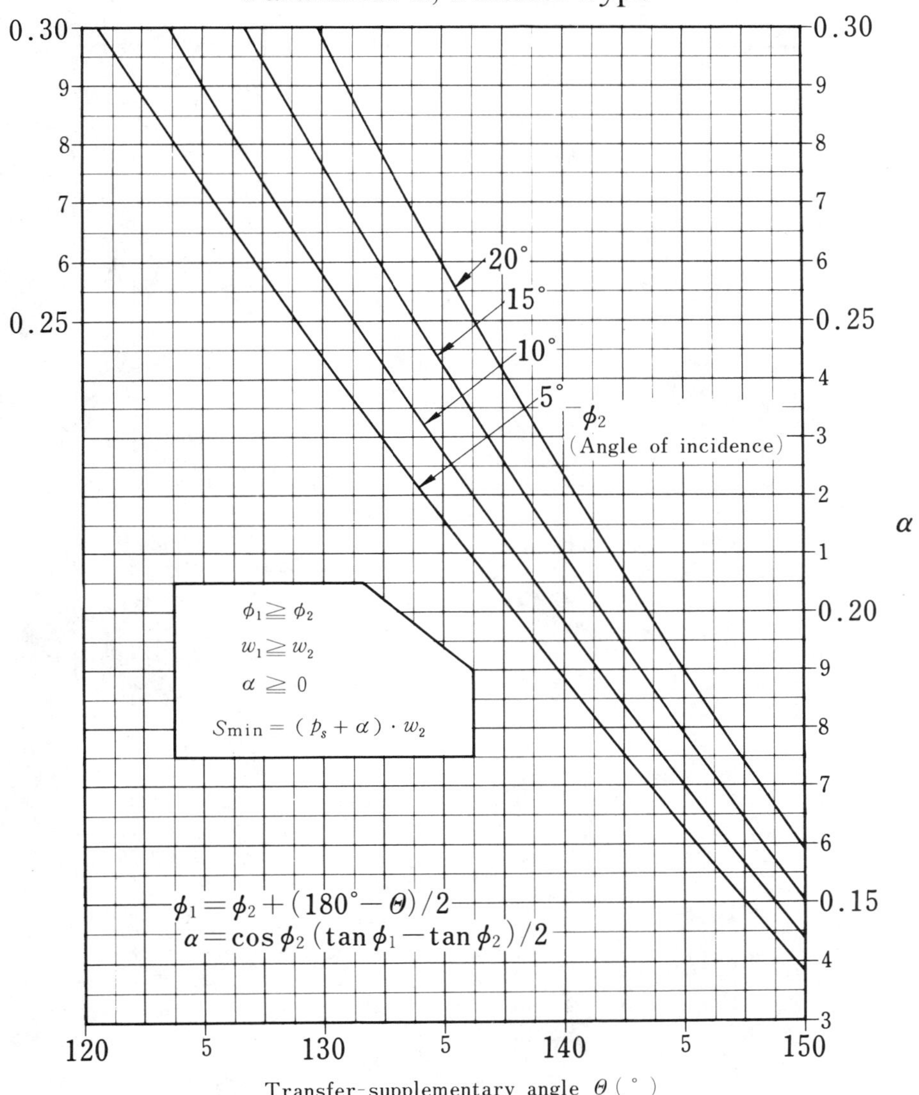

Minimum Spacing (5)

Parameter α, Parallel Type

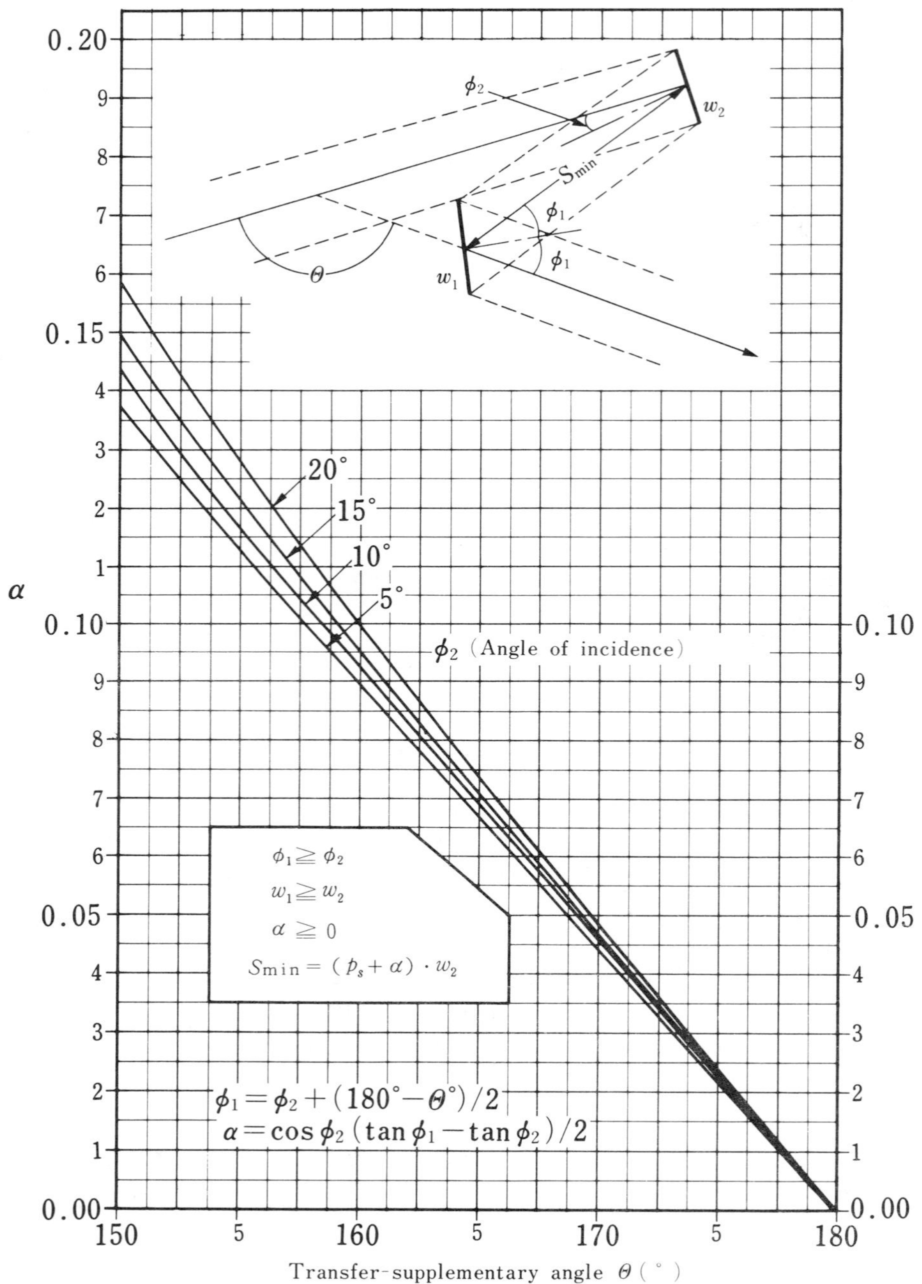

Minimum Spacing in Symmetrical-Intersection-Type Reflector

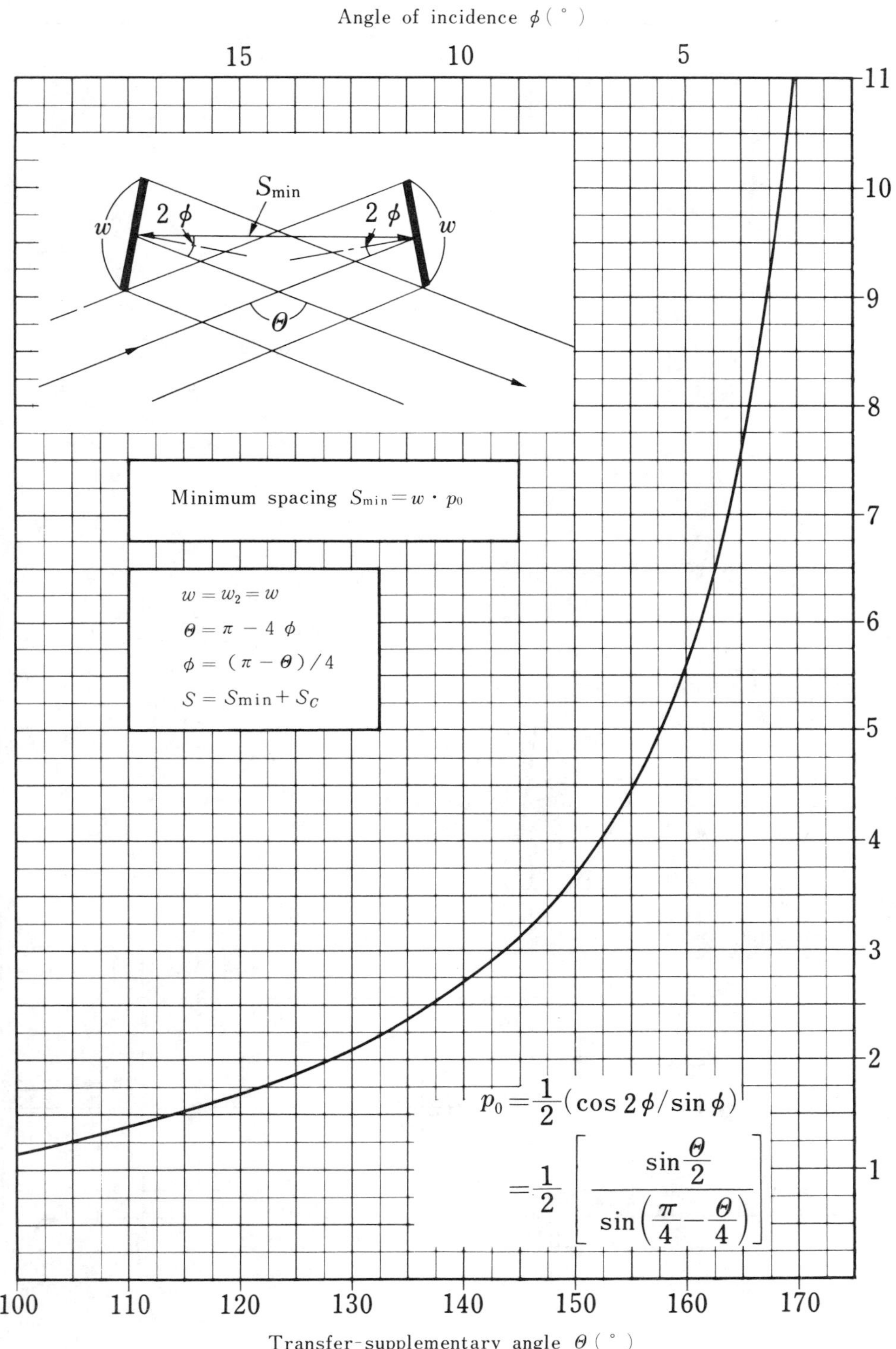

Additional Spacing to Give a Clearance

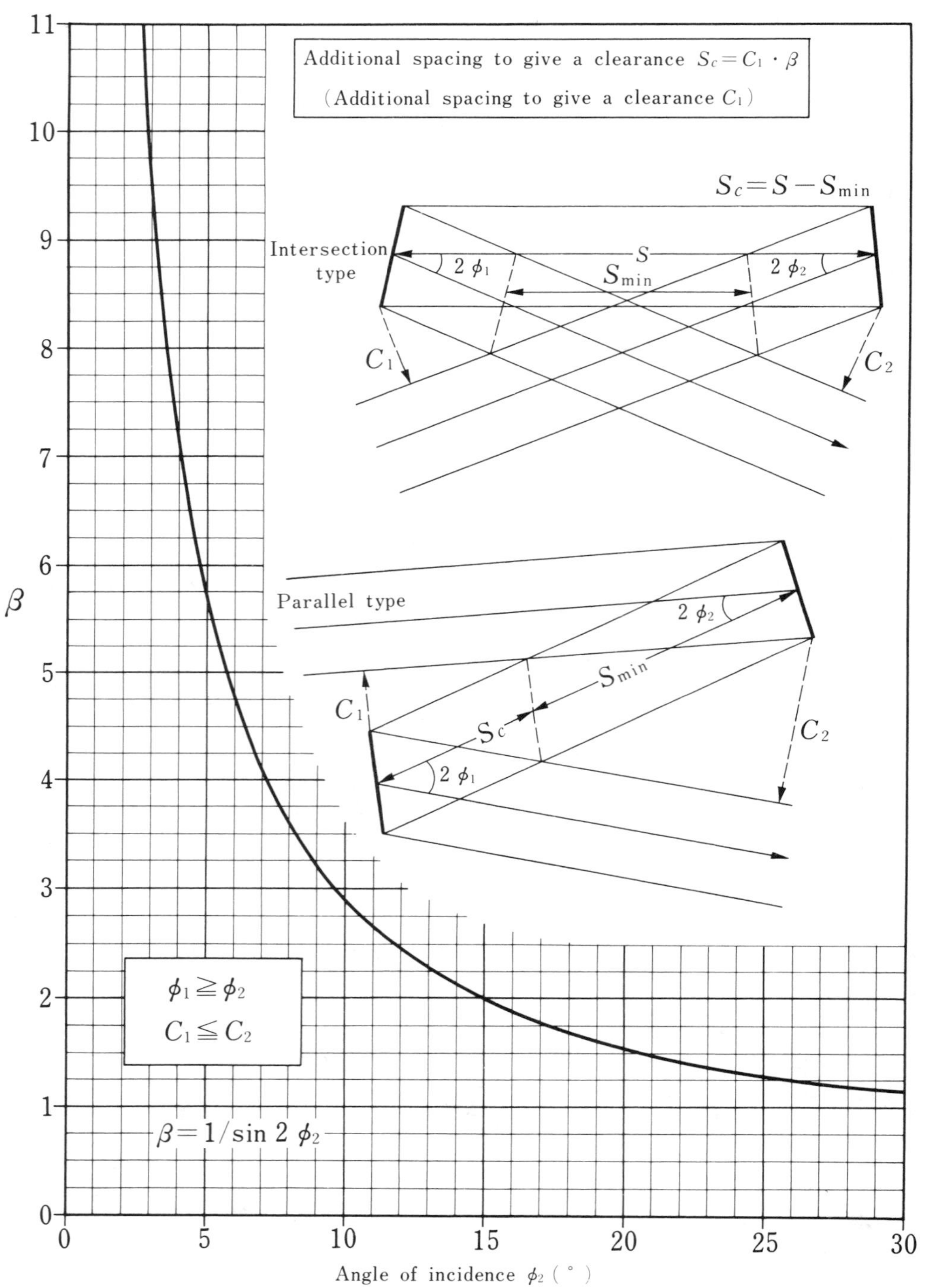

Width Ratio of Reflectors (1)

Intersection Type (100° ≤ θ ≤ 130°)

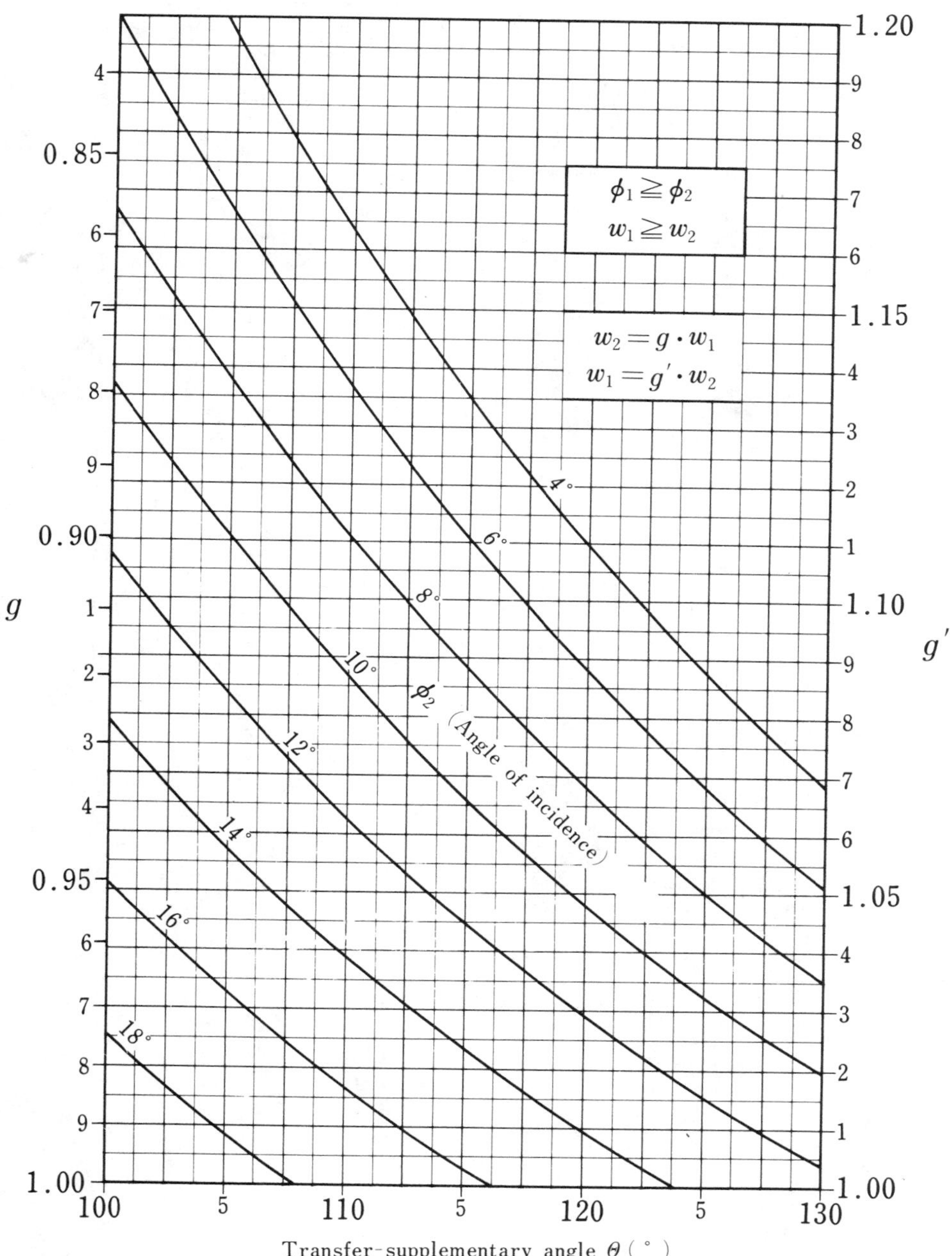

Width Ratio of Reflectors (2)

Intersection Type (130° ≤ θ ≤ 160°)

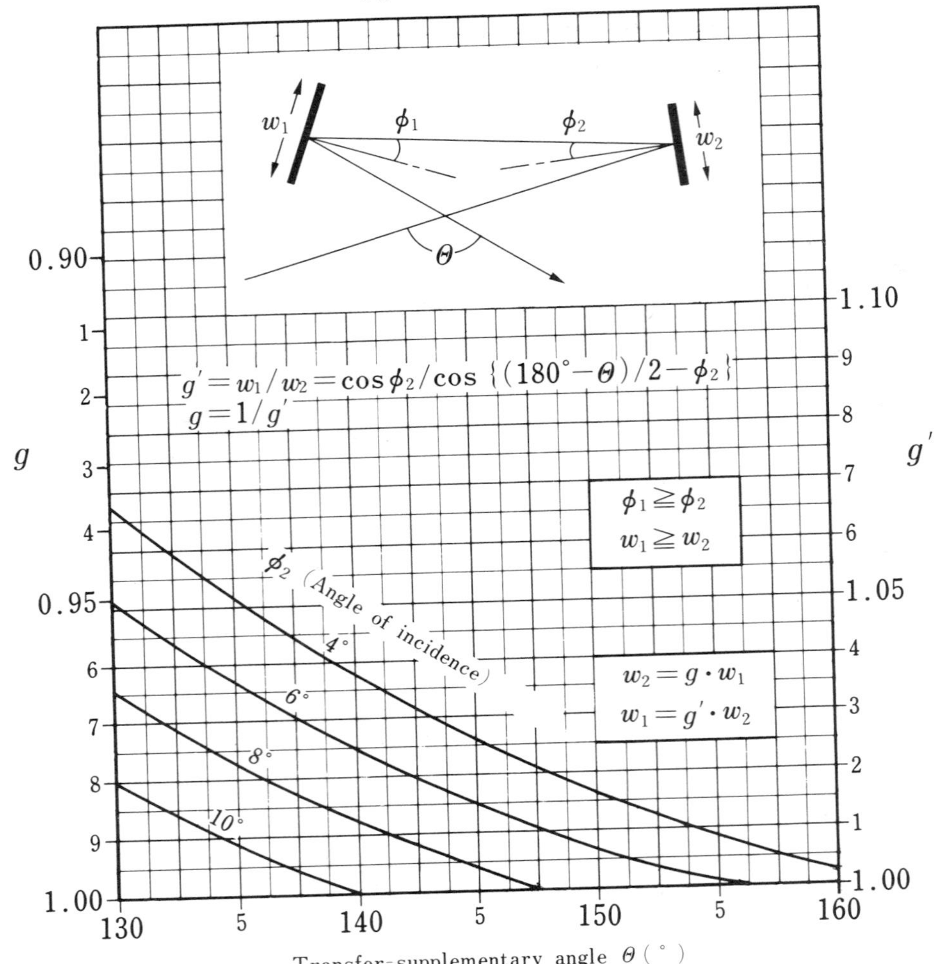

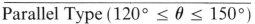

Width Ratio of Reflectors (3)

Parallel Type (120° ≤ θ ≤ 150°)

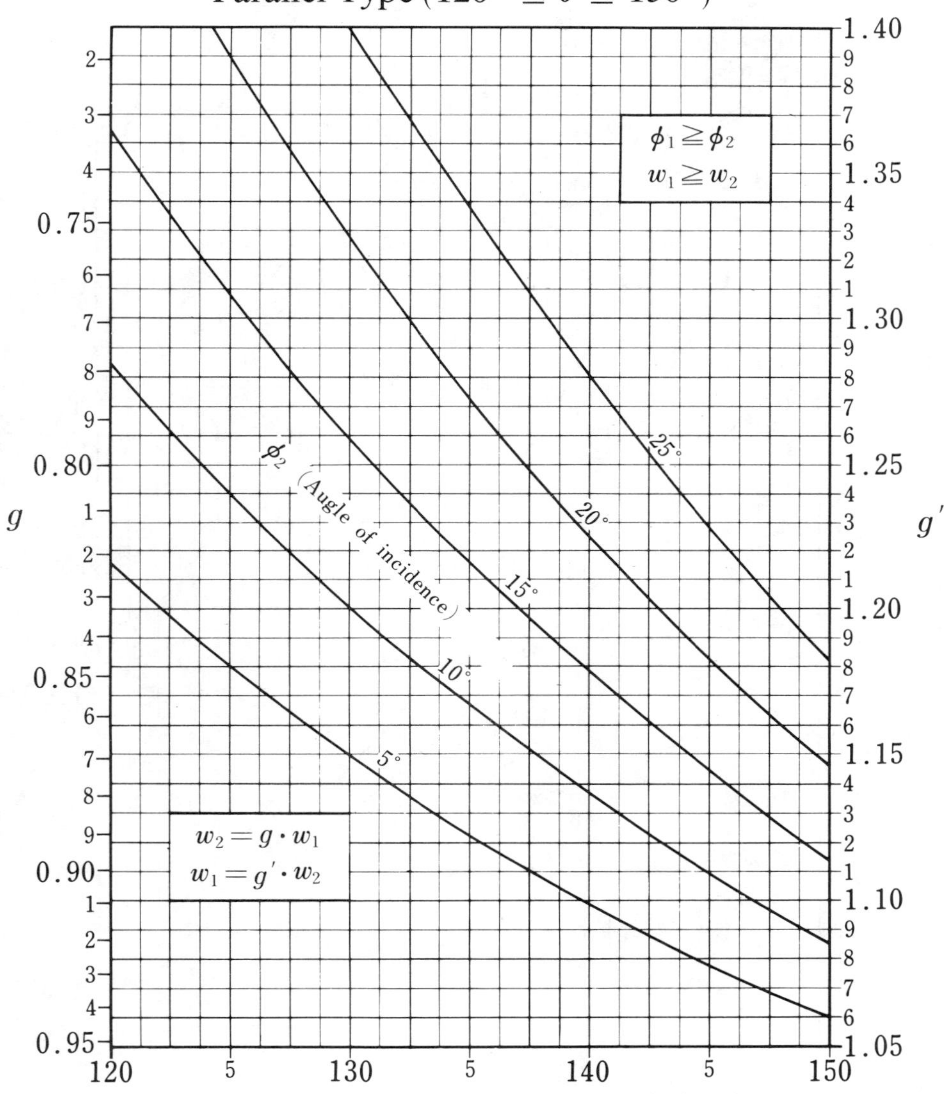

Width Ratio of Reflectors (4)

Parallel Type ($150° \leq \theta \leq 180°$)

$g' = w_1/w_2 = \cos\phi_2/\cos\{\phi_2 + (180° - \Theta)/2\}$

$g = 1/g'$

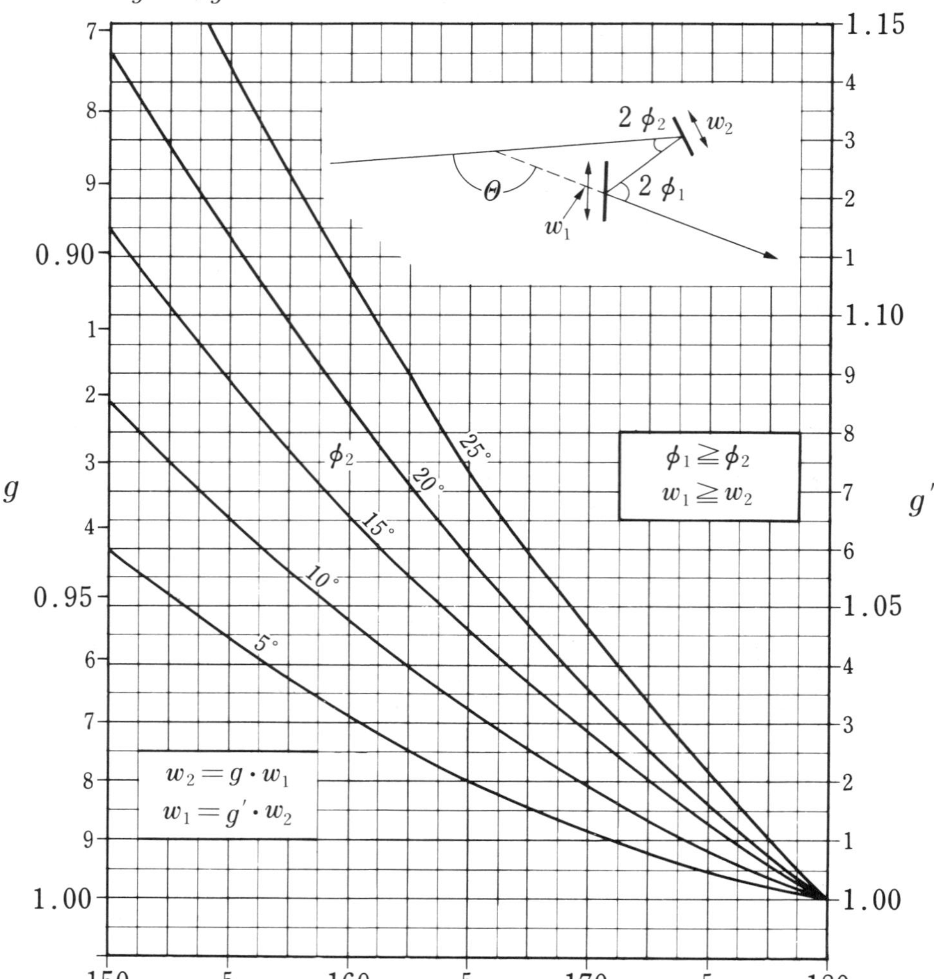

Geometry of Intersection-Type Arrangement

Equations

$$S\begin{cases} = \{w_2/\sin\phi_2-(w_1\sin\phi_1+w_2\sin\phi_2)\}/2+C_1/\sin 2\phi_2 & \text{①} \\ = \{2\cos\phi_1/\sin 2\phi_2-\cos\phi_1(\tan\phi_1+\tan\phi_2)\}w_1/2+C_1/\sin 2\phi_2 & \text{②} \\ = \{\mathrm{cosec}\,\phi_2-\cos\phi_2(\tan\phi_1+\tan\phi_2)\}w_2/2+C_1/\sin 2\phi_2 & \text{③} \end{cases}$$

$$g=w_2/w_1=\cos\phi_1/\cos\phi_2 \quad \text{④}$$

$$\Theta=\pi-2(\phi_1+\phi_2) \quad \text{⑤}$$

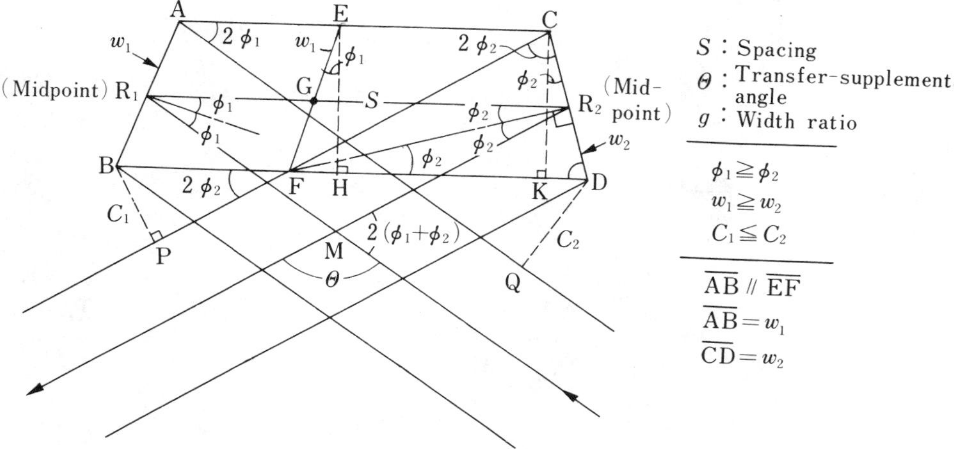

Derivation of equations

$\triangle R_1MR_2$: $\angle R_1=2\phi_1$, $\angle R_2=2\phi_2$, $\angle M=\pi-(\angle R_1+\angle R_2)\equiv\Theta$

$\therefore \Theta=\pi-2(\phi_1+\phi_2)$ ⑤

Ordinary-intersection type

$\triangle R_2FD$: $FD=w_2/(2\sin\phi_2)$
$\triangle EFH$: $FH=w_1\sin\phi_1$
$\triangle CKD$: $KD=w_2\sin\phi_2$

$EC=FD-(FH+KD)$
$GR_2=(FD+EC)/2$
$=FD-(FH+KD)/2$ …… S_{min}

$\therefore S_{min}=\{w_2/\sin\phi_2-(w_1\sin\phi_1+w_2\sin\phi_2)\}/2$ ①

$\square ABFE$: $BF=C_1/\sin 2\phi_2$, $BF=R_1G\equiv S_c$

$\therefore S_c=C_1/\sin 2\phi_2$, $S=S_{min}+S_c$ ①

$\square EHKC$: $EH=w_1\cos\phi_1$, $CK=w_2\cos\phi_2$, $EH=CK$

$\therefore g=w_2/w_1=\cos\phi_1/\cos\phi_2$ ④

②, ③ : Substitute $w_2=w_1\cos\phi_1/\cos\phi_2$ obtained from ④ into ① and eliminate w_2 …②

: Substitute $w_1=w_2\cos\phi_2/\cos\phi_1$ obtained from ④ into ① and eliminate w_1 …③

Symmetrical-intersection type

$\phi_1=\phi_2=\phi$, $w_1=w_2=w$, $g=1$

①′ → $(w/\sin\phi-2w\sin\phi)/2=w(\cos 2\phi/\sin\phi)/2$

$\therefore S=w(\cos 2\phi/\sin\phi)/2+C/\sin 2\phi$ …… Separation ①

⑤ → $\pi-2(2\phi)=\pi-4\phi$

$\therefore \{\Theta=\pi-4\phi,\ \phi=\pi/4-\Theta/4\}$ …… Angular relationship ⑤

Geometry of Parallel-Type Arrangement

Equations

$$S\begin{cases} =\{w_2/\sin\phi_2+(w_1\sin\phi_1-w_2\sin\phi_2)\}/2+C_1/\sin 2\phi_2 \cdots\cdots ① \\ =\{2\cos\phi_1/\sin 2\phi_2+\cos\phi_1(\tan\phi_1-\tan\phi_2)\}w_1/2+C_1/\sin 2\phi_2 \cdots\cdots ② \\ =\{\operatorname{cosec}\phi_2+\cos\phi_2(\tan\phi_1-\tan\phi_2)\}w_2/2+C_1/\sin 2\phi_2 \cdots\cdots ③ \end{cases}$$

$$g=w_2/w_1=\cos\phi_1/\cos\phi_2 \cdots\cdots ④$$

$$\theta=\pi-2(\phi_1-\phi_2) \cdots\cdots ⑤$$

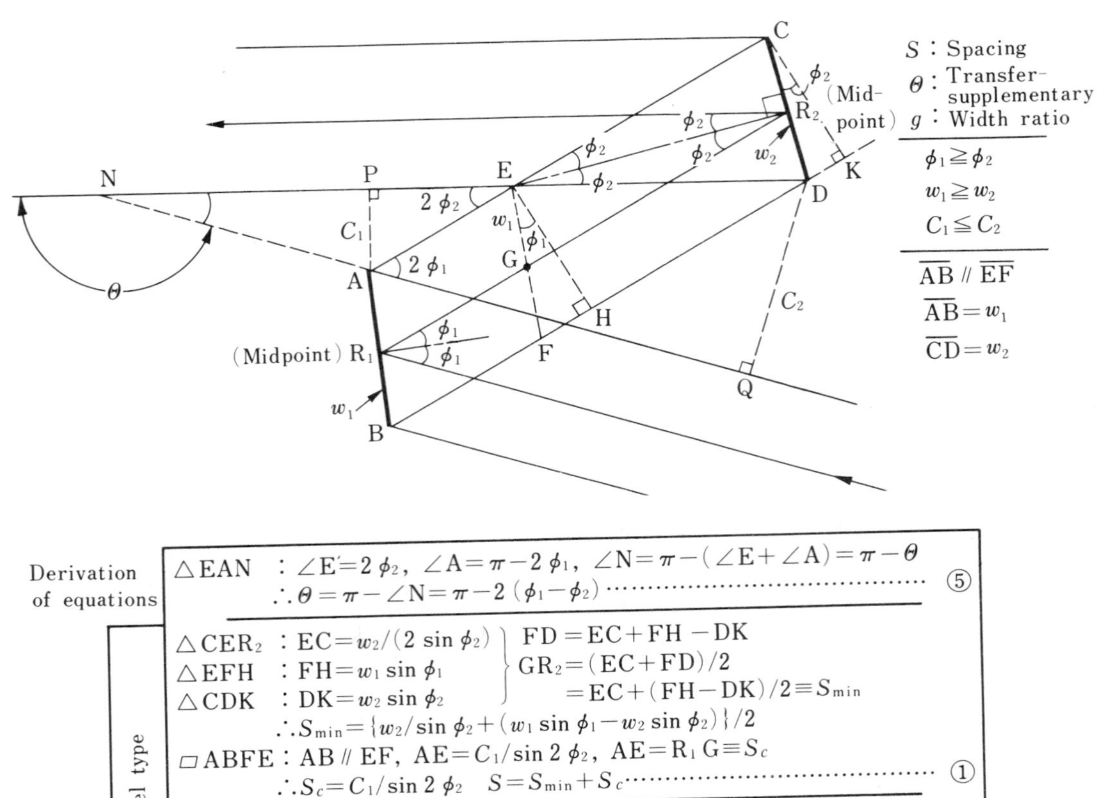

Derivation of equations

△EAN : $\angle E=2\phi_2$, $\angle A=\pi-2\phi_1$, $\angle N=\pi-(\angle E+\angle A)=\pi-\theta$

$\therefore \theta=\pi-\angle N=\pi-2(\phi_1-\phi_2)$ ⋯⋯ ⑤

Parallel type

△CER_2 : $EC=w_2/(2\sin\phi_2)$

△EFH : $FH=w_1\sin\phi_1$

△CDK : $DK=w_2\sin\phi_2$

$FD=EC+FH-DK$

$GR_2=(EC+FD)/2$

$=EC+(FH-DK)/2\equiv S_{min}$

$\therefore S_{min}=\{w_2/\sin\phi_2+(w_1\sin\phi_1-w_2\sin\phi_2)\}/2$

▱ABFE : $AB /\!/ EF$, $AE=C_1/\sin 2\phi_2$, $AE=R_1G\equiv S_c$

$\therefore S_c=C_1/\sin 2\phi_2 \quad S=S_{min}+S_c$ ⋯⋯ ①

□EHKC : $EH=w_1\cos\phi_1$, $CK=w_2\cos\phi_2$, $EK=CK$

$\therefore g=w_2/w_1=\cos\phi_1/\cos\phi_2$ ⋯⋯ ④

②, ③ : Substitute $w_2=w_1\cos\phi_1/\cos\phi_2$ obtained from ④ into ① and eliminate w_2 ⋯ ②

: Substitute $w_1=w_2\cos\phi_2/\cos\phi_1$ obtained from ④ into ① and eliminate w_1 ⋯ ③

7

NOISE AND SIGNAL-TO-NOISE RATIO

7.1 CLASSIFICATION OF NOISE

The word "noise" originally meant any sound that was disturbing to a listener. However, it is now used to mean everything contained in the receiving signal, excluding the desired signal (intentionally transmitted information), but including visual disturbances such as ghosts or obscurity in television and stripes in facsimile.

Ambience or a background scene picked up via a microphone or a television camera, including background conversation, noises caused by a jet plane or a train, a car horn, an electric light pole in the scene of a classic play, are called background noise and surrounding noise.

These types of noise sources appear also on the receiving side. For example, often it is difficult to converse on a public telephone located on a train platform or to watch television in a sunny room. Since these noise sources occur only at the two ends of the electronic communication transmission system, we call them terminal environmental noise.

The noise in the physical system, that is, in the electronic communication equipment and the associated transmission media, such as transmission lines and radio propagation paths, is usually called transmission system noise or circuit noise.

Thus noise is classified into two kinds—internal noise produced in equipment components and external noise. External noise includes interference noise.

Figure 1 classifies the noise involved in a radio system in terms of its causes. This text does not intend to analyze and describe in detail individual types of noise, but it does describe briefly the nature of various kinds of noises and the influence on the transmission system operated over VHF, UHF, and SHF bands.

7.2 TERMINAL ENVIRONMENTAL NOISE AND INTERSYSTEM NOISE

A. Terminal Environmental Noise

Noise present at the system input is difficult to eliminate because it must be treated in the same manner as the desired signal. This holds until the noise reaches the human ear or eye. The noise present at the output reduces intelligibility and impairs speech quality. This type of noise is independent of the radio frequencies employed or the type of transmission system. An effective approach to minimize this kind of noise lies in improving the environment and the handling (software) at the junction of humans and the communication system (man–machine interface).

B. Noise Produced within a System

The noise produced within a system may be classified into two groups—internal noise, generated in such facilities as feeders, receivers (repeaters), modulators, demodulators, and control and supervisory equipment, and propagation system noise, involved in the propagation path and also in the performance of antennas.

Internal Noise

Thermal Noise (White Noise). Resistors and any circuit components generate noise due to the random movement of free electrons unless the absolute temperature is at 0 K (= −273°C). The magnitude of the noise voltage is proportional to the square root of the resistance R, the absolute temperature T, and the square root of the frequency band B. Consequently, this kind of noise is called thermal noise, thermal agitation noise, or white noise. Sometimes it may be called Johnson noise after the scientist who defined it. It is called white noise because it exhibits a uniformly distributed power spectrum like white light. Noise generated in a thermionic tube and diode also exhibits white noise that can be treated as the noise mentioned here, despite its different noise-generating mechanism.

In the succeeding paragraphs these two kinds of noise are treated in the same manner. This unavoidable equivalent thermal noise makes up a major part of the system internal noise as described in detail in Sections 7.4 to 7.10.

Intermodulation Noise (Unintelligible Crosstalk Noise). This is not linear and intelligible crosstalk, but unintelligible crosstalk of distorted signals, as the name indicates. Intermodulation noise, whose magnitude and distribution depend on the modulation type, the number of speech channels multiplexed, and the repeating system configuration, occurs as the result of imperfect fidelity of the transmission equipment and system. Disturbances of the desired signal are worsened through accumulation of the distortion. This type of noise is particularly important in FM multichannel communication and may exceed thermal noise in magnitude.

The main causes and sources of such intermodulation noise are

1. Nonlinear distortion in a modulator and a demodulator.
2. Phase distortion (linear distortion) in a transmission system, such as a repeater and an amplifier.
3. Echo in a feeder.

Crosstalk. There are two types of crosstalk—intelligible and unintelligible. In a multichannel communication system, both types of crosstalk originate mainly between the multiplex carrier equipment and the terminal equipment.

Quantum Noise. This noise arises in the quantization process of a pulse-code modulated (PCM) signal. Here the amplitude of a sampled pulse is approximated to one of the discrete levels, and when quantizing the analog multichannel signal, the noise exhibits a form similar to the intermodulation noise due to the modulator nonlinear distortion in an analog FM system.

Impulsive Noise. This noise occurs due to the switchover of equipment or a circuit in the case of equipment failure or deep fading. It arises also in telephone exchange equipment. As a result, an error may occur in telegraph and data transmission.

Propagation System Noise

Propagation Echo Noise. The ground reflection waves (interference waves, echo) arriving at a receiving point behind the direct (desired) wave may cause phase distortion in a broadband supermultichannel communication system.

An echo caused by a man-made structure, such as a building, usually produces interference rather than distortion due to its considerably larger delay time. In the lower frequency region of the VHF band the reflection at an ionized layer also produces interference.

Delay Distortion Caused by Aperture-to-Medium Coupling in Troposcatter Propagation. The troposcatter propagation resulting from component waves generated by an irregularity of the refractive indices within a large volume of the atmosphere, produces a very large phase difference due to a large difference between the shortest and the longest paths connecting the transmitter with the receiver. This imposes a limitation upon the communication capacity.

By sharpening the antenna directivity (increasing the antenna gain by enlarging the antenna aperture), the coupling volume and, accordingly, the above-mentioned phase difference will be reduced. In addition, this reduces the virtual antenna gain (increase of the antenna-to-medium coupling loss). In addition the reduction of the antenna gain will give rise to a corresponding increase of thermal noise.

Interference Noise Due to Backward Radiation of an Antenna. When the two-frequency plan is applied to a radio relay system, the eastbound (up-bound) and the westbound (down-bound) transmitting radio frequencies at a certain relay station are the same. Therefore the co-channel interference to the adjacent station occurs in the opposite direction of transmission when the antenna characteristic, that is, the directivity, is not perfect.

Interference Noise Due to Forward or Backward Reflection by a Nearby Object. In the radio relay system using a two-frequency plan, the receiving signal may be disturbed by the signal wave coming from behind and reflected by any man-made object such as a building, billboard, or tower located in front of the receiving antenna. If a building is located within an antenna main beam, the radio system may suffer serious interference.

Since such interference is observed frequently in medium-sized and larger cities, it may be called interference due to reflection by a nearby object in an urban area.

Overreach Interference Noise. When the same radio frequency is used repeatedly within a radio relay route, as in the two-frequency plan, a receiver may receive a signal from other than the facing transmitting station under particular topographical and meteorological conditions. The signal reaches the receiver usually by passing by the two intermediate repeater stations. This is called overreach interference noise.

REMARK. A distinction must be made between the propagation system noise mentioned above and the system propagation noise described in Chapter 8.

7.3 EXTERNAL NOISE

External noise is classified as natural noise caused by natural phenomena, and man-made noise produced by human activities in a civilized society.

A. Natural Noise

It is well known that radio static or the absorption by an ionized layer prevails in the MF, HF, and lower bands. However, natural noise will be almost insignificant at frequencies above 800 MHz, except for a satellite communication system using a very weak radio signal.

Noise Radiated from outside the Earth

These noise sources are the sun, the planets of our solar system, our galaxy, other galaxies, and celestial bodies outside our galaxy.

Sun. The sun normally radiates radio noise over a very wide frequency range of between 15 MHz and 70 GHz. The strength of this radiation sometimes reaches more than 10 times the normal over a period of several seconds to several days. Such an abnormal condition occurs when the number of sun spots increases or an eruption occurs on the sun's surface. Such a phenomenon is called a noise storm or burst, but it is not harmful unless an antenna with a minimum high gain of more than 30 dB is directed toward the sun.

Planets. The noise level depending upon the planet temperature and secondary radiation from the sun is considered to be negligibly low. However, since Venus exhibits a very high level of radiation, and the available data over the frequencies up to the microwave region are considerable, Venus may be used as a reference radiation point source.

Galactic Systems. Many galaxies exist in outer space, including the one to which our solar system belongs. The galaxy to which the sun belongs is usually called the Milky Way.

Most of the groups of stars viewed in the celestial sphere under ideal conditions belong to our galactic system. The Milky Way, which runs almost along the great circle of the celestial sphere, corresponds to an area where the fixed stars radiate light and electromagnetic waves similar to that of the sun.

A strong radiation belt is found along the Milky Way, which is strongest in the direction toward the center of the galactic system (around Sagittarius). Radio waves from the galactic system are called galactic radio waves.

A portion of the galactic radio wave spectrum, whose radiation is stable and strong, is used as an electric field strength reference to measure antenna gain at an earth station for satellite communications.

There are many other sources of radio waves in the galactic system. One of the most outstanding are pulsars, which radiate very periodic pulses with a stability on the order of between 10^{-12} and 10^{-15}. So far about 350 pulsars have been found, but their radiation level is too low to interfere with the communication system.

Celestial Bodies Outside Our Galaxies. Radio galaxies, which are galaxies outside our own galaxy and radiate radio waves of markedly high strength, and quasars are harmless as far as noise disturbance is concerned although interesting astronomically.

Earth-Based Noise

Atmospheric Noise. The most conspicuous and popular among various types of noise generated in the atmosphere surrounding the earth is radio static (generated in the process of lightning discharge).

The static noise interferes more seriously with broadcasting or long-distance point-to-point communication in the lower frequency band. At frequencies in the VHF band or higher, the effect rapidly becomes less serious as the frequency increases.

Noise caused by a collision of charged sand particles, rain droplets, snow, or hail with any part of an antenna system and absorption by an ionized layer may be ignored at VHF or higher frequencies.

The resonant absorption by oxygen or water vapor in the atmosphere is great in the vicinity of 60 and 22 GHz and causes an increase in the equivalent noise temperature. (Refer to Chapter 5 for the absorption of the desired wave.) However, the noise thus produced is masked by other types of noise in an ordinary communication system without the use of low-noise receiving techniques. Figure 2 shows the noise temperature of the atmosphere for a satellite communication system.

Ground Noise. When the direction of the antenna of the satellite earth station is lowered toward the earth's horizon, the noise increases. This is called ground noise, since the ground at a normal temperature generates thermal noise. This is not harmful in an ordinary communication system, since the equivalent noise temperature is usually between 200 and 300 K.

B. Man-Made Noise

As a result of the diversification of science and civilization, the frequency range of man-made noise has expanded. It extends from the super-low-frequency band to the microwave frequency band.

In regions with extensively developed communication systems, mutual interference between various systems may become a problem.

Man-Made Noise Generated from a Source Other than a Communication System

Possible noise sources are high-frequency appliances, automobile ignition valves, thermostats, vibrators, motor brushes, power transmission lines, discharge tubes, etc. The direct causes of the noise are a leakage of harmonic waves, sparks, corona discharge, glow discharge, etc.

In the lower frequency region, induction can also be a cause of noise. The above-mentioned man-made noise will usually be greater in the lower frequency band and may be as harmful to the quality of communication using the VHF band as urban noise.

Interference Noise in a Communication System

Co-channel interference that arises when a common frequency is used in two or more sections is important.

Spurious Interference Noise. This is due to a spurious emission by parasitic oscillation, intermodulation, harmonics, leakage of local oscillator frequency, etc. When several pieces of radio equipment using radio frequencies in the VHF and the lower UHF bands are installed in the same room, an interference problem may be produced. However, this is not the case for equipment in the SHF band or higher due to the use of three-dimensional microwave circuitry, one-way antennas, etc.

Co-Channel Interference to or from a Different Radio Relay System. For effective use of frequencies, a frequency usually has been assigned to two or more different areas simultaneously. Consequently co-channel interference may occur if the frequency allotment, station siting, and equipment installation plan are carried out inadequately or the climatic conditions become abnormal. The most typical cases are as follows.

1. For a radio mobile communication system the interference is observed between zones where the same radio frequency channel is allotted.
2. For television broadcasting the interference is caused by abnormally long-distance propagation of the transmitted wave.
3. For a microwave radio relay system the interference occurring to or from a different route and the mutual interference between the satellite and the terrestrial communication systems share the same radio frequency band. The satellite communication system operated by an extraordinarily weak signal is apt to undergo interference, and a satellite earth station transmitting at a higher power frequently may interfere with a nearby terrestrial microwave station.

Interference Noise Due to High-Power Radiation. An excessively higher field strength may occur in the first stage of the receiver of the radio relay system. It may also produce nonlinear distortion, that is, phase distortion due to an intermodulation noise, if in the neighborhood of the station there is an FM broadcasting station, a television station, a transhorizon path station (by tropospheric scattering or mountain diffraction), or a satellite earth station, even when any one of these stations does not use the same frequency as the radio system of interest. (In the worst case the solid-state components used in the first stage may be burned out.)

In particular, radar radiates high-power impulses in all directions. As a result, nearby radio stations are required to insert a filter in the feeder system to reject the interference.

7.4 THERMAL NOISE

A. Propagation Design and Thermal Noise

Of the various types of noise already mentioned, those most closely associated with propagation design are propagation system noise, external noise, and thermal noise.

External noise is discussed separately in terms of its origin; however, it is hard to describe in general. The internal noise of the equipment is classified into two parts, in general—thermal noise and intermodulation noise. The magnitudes of these depend largely on equipment design. The intermodulation noise power appearing at the output terminal of the communication channel remains almost constant even when the receiver input level changes. The thermal noise is closely determined according to the radio receiver input.

For a modulation system other than PCM, normally the output thermal noise is inversely proportional to the radio receiver input. In other words, increasing the radio receiver input results in suppressing the output thermal noise and thereby improving the signal-to-noise (S/N) ratio of the radio channel.

In the case of a PCM system, the S/N ratio remains almost constant even when the receiver input level changes, but exceeds a certain level. The noise is composed mainly of quantization noise, but if the input is decreased gradually from the threshold level, the S/N ratio is impaired rapidly, and finally the signal will be masked by the noise. In other words, the S/N ratio is not independent of thermal noise, even in a PCM system.

B. Available Power of Thermal Noise Source

The thermal noise voltage produced by the sum of a large number of component voltages caused by the random movement of free electrons differs considerably from that of a single sinusoidal wave.

The effective value of the thermal noise e_{rmg} is defined as the square root of the mean of the sum of the instantaneous voltages squared. In regard to resistance R, absolute temperature T, and bandwidth B, it is given by

$$e_{\text{rms}} = \sqrt{4KTBR} \qquad \text{(effective value of thermal noise)} \tag{7.1}$$

where K is Boltzmann's constant,s

$$K = 1.38 \times 10^{-23} \text{ J/K} \tag{7.2}$$

Since the source of thermal noise may be considered a power source with the internal resistance R and the electromotive force e_{rms}, the available power (matched maximum power) P_a is obtained when the load resistance is equal to R. Accordingly,

$$P_a = \begin{pmatrix} \text{terminal voltage} \\ \text{developed} \end{pmatrix}^2 R = (e_{\text{rms}}/2)^2/R$$

If Eq. (7.1) is substituted into Eq. (7.2),

$$P_a = (\sqrt{4KTBR})^2/4R = KTB \tag{7.3}$$

It is obvious from this that the available power of the resistance noise is independent of the resistance R.

C. Thermal Noise Power at an Ambient Temperature

When the temperature of the input circuit T is equal to an ambient normal temperature T_0 (= 290 K as a standard), the available power of the noise P_a is

$$\begin{aligned} P_a &= KT_0B \\ &= 1.38 \times 290 \times 10^{-23} \times B \\ &= 4 \times 10^{-21} \times B \end{aligned} \tag{7.4}$$

For the bandwidth $B = 1$ Hz,

$$\begin{aligned} P_a &= 4 \times 10^{-21}\ \text{W} \\ &= 4 \times 10^{-9}\ \text{pW} \end{aligned} \tag{7.5}$$

or in decibels,

$$[P_a] = -174\ \text{dBm} \tag{7.6}$$

For an arbitrary value of bandwidth B,

$$[P_a] = -174 + 10 \log B \quad \text{(dBm)} \tag{7.7}$$

Frequently these figures are used to calculate the S/N ratio, etc.

REMARKS. The ambient temperature to be used in the calculation is frequently 15 or 20°C, which corresponds to 288 or 293 K absolute temperature. However, it is not so rare for the temperature to rise above 30°C (303 K) in a tropical zone, or even in a temperate zone during the summer. Although there might be disagreement on which value of the temperature is to be used in calculating the radio system, the values of 290 or 300 K are often used, since they are convenient for the calculation, and the subsequent error in decibels is very small. (The difference between 288 and 303 K is only 0.22 dB.)

D. Peak Factor of Thermal Noise

Thermal noise is composed of a set of very small and random voltage components. Thus it is difficult to define its peak factor because peak points of the voltage are not as distinct as in a sinusoidal wave.

Since the thermal noise follows a Gaussian (normal) distribution, the equivalent peak factor is defined as the ratio of the noise voltage exceeded for a certain time rate to the effective voltage. The time rate usually employed is 0.01%.

The 0.01% value for the Gaussian distribution is 3.89σ. Here σ denotes the standard deviation, which is equal to the square root of the mean of the sum of the squared instantaneous voltages, and implies the effective voltage of the thermal noise. Accordingly it is summarized as

$$\text{peak factor, measured at 0.01\% time rate} = 3.89\ (= 11.8\ \text{dB}) \tag{7.8}$$

Usually the value of 4 (= 12 dB) is used in place of the figure obtained in the equation.

The peak factor for any time rate other than 0.01% can be found using the table for a Gaussian distribution (e.g., approximately 13 dB for a 0.001% time rate).

7.5 NOISE TEMPERATURE

A. Definition of Noise Temperature

The available power P_a of a resistance noise source is given by KTB as in Eq. (7.3), where T is the noise temperature and is equal to the physical temperature of the resistor.

Any noise source other than the resistance components, such as an antenna or a diode, can be treated as resistance noise if the generating noise source is white noise. The available noise power of a noise source P_a' is expressed in the same form as that of the resistance noise,

$$P_a' = KTB$$

Then

$$T = P_a'/KB \tag{7.9}$$

Here T denotes an effective noise temperature and is not necessarily equal to the physical temperature of the noise source.

The preceding paragraphs define the bases on which the noise performance of a system with amplifiers can be expressed, generally in terms of effective noise temperature. In the following, thermal noise is assumed to mean white noise, for convenience, and equivalent noise temperature is simply noise temperature.

B. Overall System Equivalent Noise Temperature (See Fig. 3)

The available noise power P_{no} at the output of a receiving facility connected to an antenna system with noise temperature T_a is given for the available gain g,

$$P_{no} = (KT_aB)g + P_{nr} \tag{7.10}$$

The first and second terms of Eq. (7.10) denote antenna system noise and receiver noise, respectively. The second term can be converted into power at the receiving antenna output terminal P_n in line with the antenna system available noise KT_aB. Then P_n is given by

$$\begin{aligned} P_n &= KT_aB + P_{nr}/g \\ &= K(T_a + T_r)B \\ &= KT_tB \end{aligned} \tag{7.11}$$

where T_t = overall system equivalent noise temperature,

$$T_t = T_a + T_r \tag{7.12}$$

T_r = receiver equivalent noise temperature,

$$T_r = P_{nr}/gKB \tag{7.13}$$

A comparison of Eqs. (7.13) and (7.9) shows that the thermal noise generated in receiving facilities is expressed by T_r resulting from $1/g$ output-to-input conversion and $1KB$ power-to-noise temperature conversion.

The main part of the receiving facilities determining the noise temperature is located between the antenna terminal and the limiter input, that is, the area comprising the feeder system, preamplifier, mixer, and intermediate-frequency amplifier in cascade.

Let $g_f, g_1, g_2, \ldots, g_n$ be the available gains of these components and $T_f, T_1, T_2, \ldots, T_n$ the corresponding equivalent noise temperatures. Then

$$T_r = T_f + T_1/g_f + T_2/g_fg_1 + T_3/g_fg_1g_2 + \cdots + T_n/g_fg_1g_2 \cdots g_{n-1} \tag{7.14}$$

In Eq. (7.14) g_f represents the feeder system gain that is less than 1.0, but each $g_1, g_2, \ldots, g_{n-1}$ is greater than 1.0.

Accordingly the older the stage is, the less it affects the determination of T_r. That is, it is essential to minimize the noise and increase the gain of the initial-stage amplifier and also to lower the feeder loss in order to lower the equivalent noise temperature.

C. Influence of Feeder System on Noise Temperature

The equivalent noise temperature of the feeder system T_f and the available gain as given in Eq. (7.14) are derived differently from those for the amplifier stages as follows.

Assuming T_0 = 290 K,

$$\begin{aligned} T_f &\doteqdot 290(l_f - 1) \\ g_f &= 1/l_f \end{aligned} \tag{7.15}$$

where L_f is the loss of the feeder system, expressed in decibels, L_f (dB) = $10 \log l_f$.

Figure 4 shows a plot of L_f versus T_f and g_f prepared for a low-noise (receiving) system for satellite communication.

D. Antenna Equivalent Noise Temperature T_a

The equivalent noise temperature of antennas used in VHF, UHF, and SHF terrestrial links is greater than the normal ambient temperature of 290 K. This is due to considerably lower directivity of the antenna and the foreground terrain or to man-made noise, and may exceed 3000 K in an extreme case.

When an antenna at a satellite earth station is directed toward the horizon, the main beam picks up the noise, thereby enhancing the noise temperature to 200 to 250 K. However, when the elevation angle of the antenna is 20 degrees or more, and kept away from the sun and from the strong galactic radio-wave belt, the noise temperature will be below approximately 30 to 10 K (4 K at the lowest).

Figure 2 presents graphs giving measured values of the noise temperature. It is noted that T_a is determined by the external noise temperature and not by the physical temperature of the antenna.

7.6 NOISE FIGURE $\overline{\text{NF}}$ (SEE FIG. 3)

A. Definition

In general the ratio of the available output noise power, referred to the input terminals of a four-terminal network, to the available noise power KT_0B of the resistor at an ambient temperature is called the noise figure (NF) or the internal noise figure,

$$\overline{\text{NF}} = P_{no}/gKT_0B = P_n/KT_0B \tag{7.16}$$

where $\overline{\text{NF}}$ = noise figure

P_{no} = available noise power at output

P_n = equivalent available noise power at output referred to input terminals

g = available gain

K = Boltzmann's constant = 1.38×10^{-23} J/K

T_0 = ambient temperature (standard 290 K)

B = frequency bandwidth

The noise figure may be defined as follows. Assuming that the noise temperature of an input power source is at an ambient temperature T_0, the noise figure is defined as the ratio of the total noise power at the output terminals of the four-terminal network P_{no} (i.e., the total noise power contained in the input power source and generated in the network) to the output noise power attributable to the noise temperature of the input power source.

This definition clearly agrees with the definition shown in Eq. (7.16).

Assuming that the noise temperature of the input source is equal to the ambient temperature T_0, the noise figure is defined as the ratio of the S/N ratio at the input terminals of a four-terminal network $(S/N)_i$ to the S/N ratio at the output terminals $(S/N)_o$; S and N denote the signal and the noise levels before demodulation.

Proof. Let S_i be the available input power and g the network gain. Then the S/N ratio at the input is

$$(S/N)_i = S_i/KT_0B$$

and the S/N ratio at the output,

$$(S/N)_o = gS_i/P_{no}$$

Accordingly, the noise figure is given by

$$\overline{\mathrm{NF}} = (S/N)_i/(S/N)_o = P_{no}/gKT_0B \tag{7.17}$$

This expression agrees with the first expression of Eq. (7.16).

B. Noise Figure of Cascaded Four-Terminal Networks

The term noise figure indicates the relative amount of noise generated in amplifiers, etc., and is used widely.

In general the noise figures of VHF, UHF, and SHF receivers, excluding low-noise receivers, are between 3 and 30 (5 and 15 dB). This range represents the overall noise figure taking into account the effects caused by all the stages connected in cascade in a receiver. It is derived from Eq. (7.14),

$$\overline{\mathrm{NF}} = \overline{\mathrm{NF}}_1 + \frac{(\overline{\mathrm{NF}}_2 - 1)}{g_1} + \frac{(\overline{\mathrm{NF}}_3 - 1)}{g_1 g_2} + \cdots + \frac{(\overline{\mathrm{NF}}_n - 1)}{g_1 g_2 \cdots g_{n-1}} \tag{7.18}$$

where $\overline{\mathrm{NF}}$ = noise figure of four-terminal networks in cascade in a receiver, etc.

NF_n = noise figure of nth-stage networks, such as amplifiers and mixers

g_n = available gain of nth-stage networks, such as amplifiers and mixers

The reason why all the terms except the first have a numerator in which 1 is subtracted from $\overline{\mathrm{NF}}$ is that each $\overline{\mathrm{NF}}$ includes the noise contribution (in the amount of KT_0) from the input power source.

C. Feeder System Loss and Noise Figure

When the antenna noise temperature is at the ambient temperature or higher and the noise figure is greater than unity, the effect of the feeder system can simply be calculated separately as in the case of the low-noise receiver.

Noise Figure Referred to Receiver Input Terminals

Here the feeder system is regarded as an attenuator to weaken both the signal and the external noise by an amount equivalent to the feeder system loss. To calculate the noise figure, the signal and the external noise power at the receiver input are employed.

Equivalent Noise Figure Referred to Antenna Output Terminals

When the feeder system with a loss L_f is regarded as a four-terminal network, the equivalent gain g_f is

$$g_f = 1/L_f < 1 \tag{7.19}$$

Since the available noise power at the output is always KT_0B, the noise figure $\overline{\text{NF}}_f$ according to its definition is

$$\overline{\text{NF}}_f = KT_0B/g_fKT_0B = l_f \qquad \text{(noise figure of feeder system)} \tag{7.20}$$

Thus it agrees with the loss factor L_f. The noise figure of receiving facilities including a feeder system $\overline{\text{NF}}_t$ is given using Eq. (7.18),

$$\begin{aligned}\overline{\text{NF}}_t &= \overline{\text{NF}}_f + (\overline{\text{NF}} - 1)/g_f \\ &= \overline{\text{NF}}_f + (\overline{\text{NF}} - 1)l_f \qquad \text{[using Eq. (7.19)]} \\ &= \overline{\text{NF}}_f + (\overline{\text{NF}} - 1)\overline{\text{NF}}_f \qquad \text{[using Eq. (7.20)]} \\ &= \overline{\text{NF}}_f\overline{\text{NF}}\end{aligned} \tag{7.21}$$

When the value in decibels is expressed in brackets,

$$[\overline{\text{NF}}_t] = \underbrace{[\overline{\text{NF}}_f]}_{\substack{\text{feeder}\\ \text{system}}} + \underbrace{[\overline{\text{NF}}]}_{\text{receiver}} \quad \text{(dB)} \tag{7.22}$$

Thus the noise figure of a receiving system is equal to the sum of the receiver noise figure and the feeder loss in decibels.

7.7 EXTERNAL NOISE FIGURE $\overline{\text{EN}}$

A 0 K antenna noise temperature is not realizable, since the antenna picks up a certain amount of external noise, such as ground noise, atmospheric noise, cosmic noise, and man-made noise. The presence of such noise requires defining the external noise figure $\overline{\text{EN}}_i$, for the sake of convenience, similarly to the receiver noise figure.

A. External Noise Figure (Isotropic Antenna)

If an isotropic antenna is employed and the available power due to external noise is represented by P_{ei}, then the external noise figure is given by

$$\overline{\text{EN}}_i = P_{ei}/KT_0B \tag{7.23}$$

This equation shows numerically the noise environment for the entire solid angle viewed from the receiving antenna. However, it is not very useful since an actual antenna has a certain directivity and the received noise is not very uniform for various directions.

B. External Noise Figure (Ordinary Antenna)

Let P_e denote the available power due to external noise received through an antenna, and let $\overline{\text{EN}}$ denote the external noise figure. Then

$$\overline{\text{EN}} = P_e/KT_0B \tag{7.24}$$

This may be called the effective external noise figure. A frequently used noise figure is that defined by noise generated at the antenna (output) terminals or the receiver input terminals. It may be called the internal noise figure to distinguish it from the external noise figure. Although the internal noise figure $\overline{\text{NF}}$ should never be less than 1.0 because the noise temperature is equal to the ambient temperature of an input circuit, the external noise figure $\overline{\text{EN}}$ can be less than 1.0 (negative in decibel value), that is,

$$\overline{\text{NF}} \geq 1, \qquad \overline{\text{EN}} \geq 0 \tag{7.25}$$

7.8 SYSTEM OVERALL NOISE FIGURE (OPERATING NOISE FIGURE) *F*

The amount of thermal noise in the operating condition where an antenna is connected to the receiver is divided into the external noise figure $\overline{\text{EN}}$ and the (internal) noise figure $\overline{\text{NF}}$.

The equivalent thermal noise power P_n referred to the antenna output terminals or the receiver input terminals, depending upon how the feeder system is treated, is expressed as

$$P_n = \left(\overline{\text{EN}} + \overline{\text{NF}} - 1\right)KT_0B = FKT_0B \tag{7.26}$$

where F is the overall system noise figure given by

$$F = \overline{\text{EN}} + \overline{\text{NF}} - 1 \tag{7.27}$$

The term -1 in the parentheses of Eq. (7.26) is needed because the definition of $\overline{\text{NF}}$ assumes the presence of the input noise source of KT_0B equivalent to $\overline{\text{NF}} = 1$. However, in this case it is easy to understand if it is considered that KT_0B is to be replaced by the input noise from the antenna.

F in the Terrestrial Microwave System

In ordinary terrestrial microwave systems using frequencies of 1 GHz or higher, man-made noise can be ignored, and the antenna noise temperature is approximately equal to the ambient temperature (i.e., $\overline{\text{EN}} = 1$). Then

$$F = 1 + \overline{\text{NF}} - 1 = \overline{\text{NF}} \tag{7.28}$$

Thus the overall noise figure agrees with the internal noise figure. This relation will be used widely for calculation of the S/N ratio.

F in the Low-noise Receiving System

In ordinary receiving systems, generally, $\overline{\text{EN}} \geq 0$ and $\overline{\text{NF}} \geq 1$, and the values of F lie between 3 and 50 (5 and 17 dB). However, a typical satellite earth station uses a low-noise receiving system in which

$$\overline{\text{EN}} = 0.05 \text{ to } -13 \text{ dB}, \qquad \overline{\text{NF}} = 1.1 \text{ to } 0.41 \text{ dB}$$

Accordingly,

$$F = 0.05 + 1.1 - 1 = 0.15$$

If expressed in decibels,

$$[F] = 10 \log 0.15 = -8.24 \text{ dB}$$

REMARK. In the low-noise receiving system the term equivalent noise temperature (of antenna and systems) is used as a direct expression rather than using $\overline{\text{NF}}$, $\overline{\text{EN}}$, or F as indirect expressions. Equivalent noise temperature or noise figure referred to the antenna output terminals differs from that referred to the receiver input terminals. However, the S/N ratios estimated from either parameter will yield the same result. Therefore any one of these two reference points may be used.

7.9 THRESHOLD INPUT LEVEL (LOWEST INPUT LEVEL FOR REQUIRED S/N RATIO)

A. S/N Ratio Improvement Factor

The S/N ratio at an intermediate-frequency stage of the receiving system, that is, the carrier-to-noise ratio C/N, differs from the demodulated message S/N ratio.

In general the latter is greater than the former. Their ratio

$$I = (S/N)/(C/N) \tag{7.29}$$

is called the S/N ratio improvement factor and depends on the methods of modulation and conditions of operation.

It should be noted that the N in S/N differs from the N in C/N, although the same symbol N is used customarily.

From Eq. (7.29),

$$(S/N) = (C/N)I \tag{7.30}$$

This C/N ratio may be expressed, using the noise figure, as

$$C/N = P_r/P_n = P_r/KT_0BF \tag{7.31}$$

where P_r = receiver input level
K = Boltzmann's constant
T_0 = ambient temperature (290 K as a reference value)
B = effective noise bandwidth, which is normally equal to -6 dB bandwidth in the IF stage
P_n = equivalent noise power referred to receiver input terminals

Thus the S/N ratio at the demodulator output can be calculated employing the S/N ratio improvement factor I.

B. Threshold Input

In most modulation systems, except digital modulation systems such as PCM and PNM, the S/N ratio of a system is proportional to the receiver input P_r under normal operating conditions. However, the S/N ratio in systems including PCM and PNM, but excluding AM, will deteriorate rapidly, and the signal will be masked by noise if the receiver input decreases below a certain level. This limiting level is called the threshold, and the input to this receiver, is called the threshold input P_{th}.

Threshold Input of Continuous Signal

The threshold occurs where the modulated carrier peak and the white noise peak are at the same level, that is, at the threshold,

$$\text{peak of continuous signal} = \text{peak of white noise} \tag{7.32}$$

Since the peak factor of white noise is about 4 (= 12 dB) and that of the sinusoidal wave is $\sqrt{2}$ (= 3 dB), Eq. (7.32) may be expressed by the threshold input P_{th} and the noise power referred to the input terminals as

$$P_{\text{th}}(\sqrt{2})^2 = P_n(4)^2$$

Then

$$P_{\text{th}} = P_n(4/\sqrt{2})^2 = P_n \cdot 8 \tag{7.33a}$$

and, expressed in decibels,

$$[P_{\text{th}}] = [P_n] + 9\text{ dB} \tag{7.33b}$$

The ratio of the threshold input P_{th} to the noise power P_n is called the threshold coefficient C_{th}. Then the C_{th} for analog signals is 8 (= 9 dB).

Threshold Input for Pulse Signal

Although the PAM-AM system and the PCM system are to be treated differently as described later, systems such as PFM-AM, PPM-AM, and PWM-AM, where the frequency, position, and width of pulses are used as main modulating parameters, can be analyzed as follows.

The time of the leading or the trailing edges of pulses is determined by the points corresponding to the instances for half the peak amplitude of the pulse composed of a series of sinusoidal waves. Thus the threshold point occurs at the point where half the pulse peak amplitude equals the peak value of white noise,

$$P_{\text{th}}(\sqrt{2})^2(1/2)^2 = P_n(4)^2 \tag{7.34}$$

Then

$$P_{\text{th}} = P_n \cdot 32 \tag{7.35a}$$

and expressed in decibels,

$$[P_{\text{th}}] = [P_n] + 15 \text{ dB} \tag{7.35b}$$

Thus C_{th} of the pulse signal is 32 (= 15 dB).

A unit pulse can be broken down into a large number of high-frequency sinusoidal components, and P_{th} represents the threshold power. However, when the threshold power is expressed by $P_{\text{th}\,m}$, which is equivalent to the mean power, where the power for the with-pulse periods and the power for the without-pulse periods are averaged, the duty factor D_f is introduced,

$$P_{\text{th}\,m} = P_{\text{th}} D_f \tag{7.36}$$

where

$$D_f = \tau/T_p$$

with τ being the width of the pulse and T_p the pulse period.

Although the PAM-AM system as well as the single-channel AM system where the threshold is ambiguous has no such restriction imposed by Eq. (7.34), the same threshold level as in the continuous signal may be assumed as a lower limit of the S/N ratio for convenience.

If it is to be expressed through the mean power $P_{\text{th}\,m}$, the duty factor D_f is considered.

In digital pulse systems such as PNM-AM and PCM-AM, the S/N ratio is in principle maintained above a certain value, if it is possible to detect the presence or absence of pulses and to distinguish them from noise.

Consequently the condition given by Eq. (7.34) is not essential and P_{th} can take a smaller value while the system design requires a greater allowance than in the other system to prevent loss or false recognition of pulses, leading to a large error.

Figure 6 summarizes the threshold factors for various modulation systems. Here both PNM-AM and PCM-AM systems have been treated in the same manner as PPM-AM. Figure 7 is used to determine the noise power referred to the receiver input terminals, the threshold input P_{th} for the overall noise figure F, and the effective noise bandwidth B. P_n is obtained by replacing B by Δf in the figure, where Δf is the message or video bandwidth.

7.10 GRAPH FOR ESTIMATION OF S/N RATIO

Formulas to calculate the S/N ratio for white noise differ according to the modulation system, as described below. Figures 8 to 12 show the calculation graphs for a single-channel FM system and for a single-sideband multiplex carrier system [FM(SS-FM)].

A. S/N Ratio for Ordinary FM System

$$\frac{S}{N} = 10\log\frac{P_r}{KT_0F}\left(\frac{3}{2}\frac{S_0^2}{f_m^3}\right) \quad \text{(dB)} \tag{7.37}$$

where P_r = receiver input
P_{th} = threshold receiver input, $P_r \geq P_{\text{th}}$
K = Boltzmann's constant
T_e = normal temperature, Kelvin (290 to 300 K)
F = overall noise figure
B = effective noise bandwidth
S_0 = rms frequency deviation
f_m = highest modulating frequency

Equation (7.37) can be rewritten as

$$\frac{S}{N} = 10\log P_r - 10\log F + 10\log\{(3/2KT_0)(S_0^2/f_m^3)\}$$

$$= [P_r]\text{ (dB)} - [F]\text{ (dB)} + [G_{\text{FM}}]\text{ (dB)} \tag{7.38}$$

Figure 8 presents the graph for $[G_{\text{FM}}]$, and the S/N ratio can be estimated directly for $[G_{\text{FM}}]$, P_r, and F.

B. S/N Ratio for FM Television Transmission System

$$\frac{S}{N} = 10\log\frac{P_r}{KT_0F}\left(\frac{3S_p^2}{f_m^3}\right) \text{ (dB)} \tag{7.39}$$

or

$$\frac{S}{N} = 10\log P_r - 10\log F + 10\log\left\{\frac{3}{KT_0}\left(\frac{S_p^2}{f_m^3}\right)\right\}$$

$$= [P_r]\text{ (dB)} - [F]\text{ (dB)} + [G_{\text{TV}}]\text{ (dB)} \tag{7.40}$$

where $P_r \geq P_{\text{th}}$
S_p = peak-to-peak frequency deviation of video signal, corresponding to level difference between base and peak white level

For other parameters, see Eq. (7.37). Figure 9 shows $[G_{\mathrm{TV}}]$. It also shows the relationship between S_p, f_m, and $[G_{\mathrm{TV}}]$ for a broad application range so that it may be useful for designing not only ordinary television, but also industrial television, special-purpose television with broader or narrower bandwidth, and facsimile.

REMARK. The signal and the noise levels to be used for calculating the television transmission S/N ratio are usually expressed in peak-to-peak and rms, respectively.

C. S/N Ratio for Arbitrary Channel of Single-Sideband System

$$\frac{S}{N} = 10\log\frac{P_r}{KT_0F}\left(\frac{S_0^2}{\Delta f \cdot f_{\mathrm{ch}}^2}\right) \tag{7.41}$$

or

$$\frac{S}{N} = 10\log P_r - 10\log F + 20\log(S_0/f_{\mathrm{ch}}) - 10\log(KT_0\,\Delta f)\quad(\mathrm{dB})$$

$$= [P_r]\,(\mathrm{dB}) - [F]\,(\mathrm{dB}) + [G_v]\,(\mathrm{dB}) + [J_v]\,(\mathrm{dB}) \tag{7.42}$$

where $P_r \geq P_{\mathrm{th}}$
$\Delta f \ll f_{\mathrm{ch}}$
$[G_v] = 20\log(S_0/f_{\mathrm{ch}})$
$[J_v] = -10\log(\Delta f KT_0)$
Δf = bandwidth of channel of interest
f_{ch} = frequency of channel of interest within single-sideband bandwidth
S_0 = rms frequency deviation of channel

Figures 10 and 11 show $[G_v]$ and $[J_v]$, respectively. Using these figures, the S/N (power) ratio in an arbitrary channel with an arbitrary bandwidth may be estimated easily.

D. Graph for Quick Estimation of S/N Ratio in Single-Sideband FM System (According to CCIR Recommendation)

Figure 12 shows $[G_{\mathrm{TP}}]$ for the channel bandwidth of 3.1 kHz (bandwidth between 0.3 and 3.4 kHz required for telephone transmission),

$$\frac{S}{N} = \frac{P_r}{KT_0F}\left(\frac{S_0^2}{\Delta f f_{\mathrm{ch}}^2}\right) \tag{7.43}$$

$$\frac{S}{N} = 10\log P_r - 10\log F + 10\log\left\{\frac{S_0^2}{f_{\mathrm{ch}}^2 KT_0\,\Delta f}\right\}$$

$$= [P_r]\,(\mathrm{dB}) - [F]\,(\mathrm{dB}) + [G_{\mathrm{TP}}]\,(\mathrm{dB}) \tag{7.44}$$

where $P_r \geq P_{th}$
$\Delta f = 3.1$ kHz

The table in Fig. 12 shows the CCIR recommended standard parameters and the notations to distinguish the curves. Using these curves, $[G_{TP}]$ denotes the top channel where the S/N ratio is worst within the multiplexed single-sideband signal.

E. Graphs of Receiver Input versus S/N Ratio for Quick Estimation (Single-Sideband FM System)

Figures 13 to 24 present the results of calculations of top-channel S/N ratios over the $\Delta f = 3.1$-kHz frequency bandwidth for various channel capacities from 12 to 2700, using parameters of the receiver input P_r and the overall noise figure F. Substituting the top-channel frequency f_m for f_{ch} in Eq. (7.44), the top-channel S/N ratios of 5 to 95 dB Eq. (7.44), can be read off the graphs. However, it should be noted that a receiver input level P_r that is lower than the threshold input P_{th} as found in Fig. 7 is to be excluded.

F. White Noise Generated in a Radio Relay System

In Figs. 13 to 24 the values of N (pW) written alongside of the values of the S/N ratios are the actual power corresponding to the white noise level, referred to a test point at which the demodulated signal level is 1 mW (= 0 dB).

The graphs are used to find the overall noise power generated in the radio relay system.

Single-Section System (without Repeater)

The S/N ratios (dB) and N (pW) can be found directly on the graphs.

The m-Section Relay System

The white noise generated in a multisection relay system will follow a power-addition law. Then the overall noise power N_{sys} and the resultant S/N_{sys} are given by

$$N_{sys} = N_1 + N_2 + \cdots + N_m \quad \text{(pW)} \tag{7.45}$$

$$S/N_{sys} = 10 \log\left(10^9/N_{sys}\right) = 90 - 10 \log N_{sys} \quad \text{(dB)} \tag{7.46}$$

where $N_1, N_2, \ldots, N_m$ = noise power generated in each section
S = signal power (= 1 mW, or 10^9 pW)

System Comprising Through Repeaters

The white noise generated in a radio relay system without modulator and demodulator accumulates in every relay process. Therefore the total white noise generated in a section between a modulator and a demodulator can be treated as described in the preceding section. Accordingly, for a portion of the system with no demodulator the equivalent S/N ratio per hop can be defined.

System Consisting of Homogeneous Sections, Each of Which Generates an Equal Amount of Noise

If the S/N ratio for each hop is assumed to be the same, although this is not realistic, then

$$N_{\text{sys}} = mN \quad \text{(pW)}$$

$$S/N_{\text{sys}} = S/N - 10\log m \quad \text{(dB)} \tag{7.47}$$

where m = number of relay sections
N = white noise power per hop
S/N = S/N ratio per hop

REMARK. The overall system noise N_{sys} with radio fading and the resultant S/N_{sys} shall be evaluated carefully, and in practice some correction to the above equations becomes necessary. For details see Chapter 8.

EXAMPLES

EXAMPLE 7.1

Find the effective temperatures T_x and T_y at the antenna feed point X and the receiver input Y, respectively, when the antenna noise temperature T_a = 15 K.

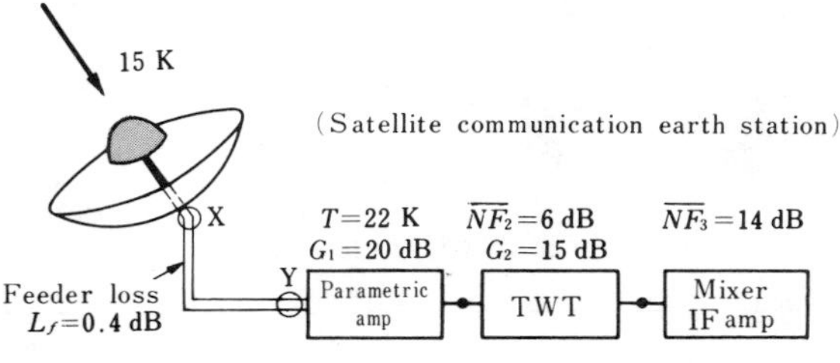

Solution

1. Convert G (dB) into g, the antilogarithm.

$$G_1 = 20 \text{ dB} \rightarrow 100 \cdots g_1 \qquad \text{(parametric amplifier)}$$

$$G_2 = 15 \text{ dB} \rightarrow 32 \cdots g_2 \qquad \text{(TWT amplifier)}$$

2. Calculate the feeder system equivalent gain g_f and the feeder system equivalent noise temperature T_f,

$$L_f = 0.4 \text{ dB}, \qquad g_f = 0.912, \qquad T_f = 28 \text{ K} \qquad \text{(Fig. 4)}$$

3. Equivalent noise temperature for each stage in the above diagram,

parametric amplifier	$T_1 = 22$ K	
TWT amplifier ($\overline{\text{NF}}_2 = 6$ dB)	$T_2 = 870$ K	(use curves or equations
mixer, IF amplifier ($\overline{\text{NF}}_3 = 14$ dB)	$T_3 \doteqdot 7000$ K	in Fig. 5)

4. Equivalent noise temperature of receiving system at point X, T_{rx},

$$\begin{aligned} T_{rx} &= T_f + T_1/g_f + T_2/(g_f g_1) + T_3/(g_f g_1 g_2) \\ &= 28 + 22/0.912 + 870/(0.912 \times 100) + 7000/(0.912 \times 100 \times 32) \\ &= 28 + 24.1 + 9.5 + 2.4 = 64 \text{ K} \end{aligned}$$

5. Equivalent noise temperature of receiving system at point Y, T_{ry} and converted noise temperature of the antenna input noise T_a',

$$\begin{aligned} T_{ry} &= T_1 + T_2/g_1 + T_3/(g_1 g_2) \\ &= 22 + 870/100 + 7000/(100 \times 32) = 32.9 \text{ K} \\ T_a' &= (T_a + T_f) g_f = (15 + 28) \times 0.912 = 39.2 \text{ K} \end{aligned}$$

6. Overall equivalent noise temperature,

$$\begin{aligned} T_x &= T_a + T_{rx} = 15 + 64 = 79 \text{ K} \\ T_y &= T_a' + T_{ry} = 39.2 + 32.9 = 72.1 \text{ K} \end{aligned}$$

Example 7.2

Calculate the overall noise figures F (dB) at points X and Y in Example 7.1.

Solution A

1. True number of $\overline{\text{NF}}$,

parametric amplifier	$T_1 = 22\ K$,	$\overline{\text{NF}}_1 = 1.076$ (0.32 dB)	
TWT amplifier	$\overline{\text{NF}}_2 = 6 \text{ dB} = 4$		(Fig. 5*a*)
mixer, IF amplifier	$\overline{\text{NF}}_3 = 14 \text{ dB} = 25$		

2. Antilogarithm of gain

$$L_f = 0.4\ \text{dB} \qquad l_f = 1.096, \qquad g_f = 0.912 \qquad \text{(Fig. 4)}$$

$$G_1 = 20\ \text{dB}, 100 \cdots g_1$$

$$G_2 = 15\ \text{dB},\ \ 32 \cdots g_2$$

3. Internal noise figure $\overline{\text{NF}}_y$ referred to point Y (receiver),

$$\overline{\text{NF}}_y = \overline{\text{NF}}_1 + (\overline{\text{NF}}_2 - 1)/g_1 + (\overline{\text{NF}}_3 - 1)/(g_1 g_2)$$

$$= 1.076 + (4 - 1)/100 + (25 - 1)/(100 \times 32)$$

$$= 1.076 + 0.03 + 0.008 = 1.114$$

4. External noise figure $\overline{\text{EN}}_y$ referred to point Y. The converted antenna noise temperature T_a' and the feeder system equivalent noise temperature T_f are

$$T_a' = (T_a + T_f)g_f, \qquad T_f \text{ to be read off Fig.4}$$

$$= (15 + 28)0.912 = 39.2\ \text{K}$$

Then

$$\overline{\text{EN}}_y = T_a'/290 = 0.135, \qquad \text{(Fig. 5}b)$$

5. Overall noise figure F_y referred to point Y,

$$F_y = (\overline{\text{EN}}_y + \overline{\text{NF}}_y - 1) = (0.135 + 1.114 - 1) = 0.249$$

$$[F_y]\ \text{dB} = 10 \log F_y \doteqdot -6\ \text{dB}$$

6. Overall noise figure F_x at point X. The internal noise figure $\overline{\text{NF}}_x$ (receiving system) at point X is

$$\overline{\text{NF}}_x = l_f\ \text{(feeder system)} \times \overline{\text{NF}}_y\ \text{(receiver)} = 1.096 \times 1.114 = 1.221$$

The external noise figure $\overline{\text{EN}} \doteqdot 0.052$ from Fig. 5b. Hence,

$$F_x = (\overline{\text{EN}} + \overline{\text{NF}}_x - 1) = 0.273$$

$$[F_x]\ \text{dB} = 10 \log F_x = -5.6\ \text{dB}$$

Solution B (Employing the Solution to Example 7.1)

1. Overall noise temperature F_x at point X. Applying $T_x = 79$ K to Fig. 5c,

$$F_x \doteqdot -5.6\ \text{dB}$$

2. Overall noise temperature F_y at point Y. Applying $T_y = 72.1$ K to Fig. 5c,

$$F_y = -6 \text{ dB}$$

EXAMPLE 7.3

Let the receiving signal level at the antenna output in Example 7.1 be $P_r = -90$ dBm. Calculate the carrier-to-noise ratios C/N at points X and Y assuming that the effective noise bandwidth $B = 2.5$ MHz.

Solution A (Employing the Overall Noise Temperature as Given in the Solution of Example 7.1)

1. Signal level C_x at point X,

$$C_x = P_r = -90 \text{ dBm}$$

2. Equivalent noise level N_x at point X,

$$N_x = KT_xB$$

$$= (1.38 \times 10^{-23})(79)(2.5 \times 10^6) \times 10^3 = 2.73 \times 10^{-12} \text{ mW}$$

When expressed in decibels,

$$10 \log(2.73 \times 10^{-12}) = -115.6 \text{ dBm}$$

3. $[C_x/N_x]$ dB at point X,

$$[C_x/N_x] \text{ dB} = -90 - (-115.6) = 25.6 \text{ dB}$$

4. Signal level C_y at point Y,

$$C_y = P_r - L_f$$

$$= -90 - 0.4 = -90.4 \text{ dBm}$$

5. Equivalent noise level N_y at point Y. Since the noise level is proportional to the overall noise temperature,

$$N_y \text{ (dB)} = N_x \text{ (dB)} + 10 \log(T_y/T_x)$$

$$= -115.6 + 10 \log(72.1/79)$$

$$= -115.6 - 0.4 = -116 \text{ dBm}$$

6. $[C_y/N_y]$ at point Y,

$$[C_y/N_y] = -90.4 - (116) = 25.6 \text{ dB}$$

REMARKS. $[C_x/N_x] = [C_y/N_y]$. It is obvious that the equivalent C/N ratio at the antenna output coincides with the equivalent C/N ratio at the receiver input because T_x or T_y corresponds to the amount of noise (comprising external and internal noises) generated in a total system but referred to point X or Y. The same explanation shall apply to the overall noise figure F. It follows that the evaluation of a system may be made at either point, antenna output or receiver input if the overall noise temperature or the overall noise figure is employed.

Solution B (Employing the Overall Noise Figure Obtained in the Solution of Example 7.2)

1. Signal level C_x at point X,

$$C_x = P_r = -90 \text{ dBm}$$

2. Equivalent noise level N_x at point X,

$$N_x = KT_0BF_x$$

where

$$KT_0 = (1.38 \times 10^{-23})(290) \times 10^3 = 4 \times 10^{-18} \text{ mW}$$

and

$$\begin{array}{lr} [KT_0] \text{ dBm} = 10 \log(KT_0) & \doteqdot -174 \\ [B] \text{ dB} = 10 \log(2.5 \times 10^6) \doteqdot & 64 \\ [F_x] \text{ dB} & = \quad -5.6 \\ \hline [N_x] = [KT_0BF_x] & = -115.6 \text{ dBm} \end{array}$$

3. $[C_x/N_x]$ at point X,

$$[C_x/N_x] \text{ dB} = -90 - (-115.6) = 25.6 \text{ dB}$$

4. The calculation of $[C_y/N_y]$ at point Y is made for confirmation. C_y is lower than C_x by 0.4 dB, corresponding to the feeder loss of 0.4 dB, while N_y is also lower than N_x by 0.4 dB ($= F_x - F_y$). Therefore $[C_y/N_y] = [C_x/N_x]$.

EXAMPLE 7.4

Assume that the modulation system for Example 7.3 is FM. Find the margin in decibels to the threshold input level.

Solution A

1. Noise power referred to input terminals,

$$N_x = -115.6 \text{ dBm}$$
$$N_y = -116 \text{ dBm} \qquad \text{(from Example 7.3)}$$

2. Threshold factor,

$$C_{\text{th}} = 9 \text{ dB} \qquad \text{(continuous wave, FM)}$$

3. Threshold input level,

$$P_{\text{th}\,x} = -115.6 + 9 = -106.6 \text{ dBm}$$

$$P_{\text{th}\,y} = -116 + 9 = -107.0 \text{ dBm}$$

4. Signal input,

$$S_x = -90 \text{ dBm}$$

$$S_y = -90.4 \text{ dBm}$$

5. Margin,

$$S_x - P_{\text{th}\,x} = -90 - (-106.6) = 16.6$$

$$S_y - P_{\text{th}\,y} = -90.4 - (107.) = 16.6$$

REMARK. The same margin is obtained for both the antenna output and the receiver input.

Solution B

1. Threshold input level $P_{\text{th}\,0}$ for $F = 0$ dB,

$$P_{\text{th}\,0} = -100.8 \text{ dBm} \qquad (B = 2.5 \text{ MHz, continuous wave, Fig. 7}c)$$

Make a correction to the value by -0.15 dB [$= 10\log(300/290)$] for the assumed temperature $T_0 = 290$ K since the curves in Fig. 7*c* are for $K =$ 300 K.

2. $P_{\text{th}\,x}$ and $P_{\text{th}\,y}$ for $F_x = -5.6$ dB and $F_y = -6$ dB, respectively,

$$P_{\text{th}\,x} = P_{\text{th}\,0} + F_x = -101 + (-5.6) = -106.6 \text{ dBm}$$

$$P_{\text{th}\,y} = P_{\text{th}\,0} + F_y = -101 + (-6) = -107 \text{ dBm}$$

3. Margin. To be obtained as in Solution A, steps 4 and 5.

EXAMPLE 7.5

Calculate the S/N ratio of a 400-MHz FM mobile radio system under the following conditions: field strength $E = 20$ dBμV/m, gain of Braun antenna for reception $G_r = -2$ dB (referred to a half-wave dipole), feeder system loss $L_f = 1$ dB, receiver noise figure $\overline{\text{NF}} = 8$ dB, IF bandwidth $B = 12$ kHz, frequency deviation (rms) $S_0 = 4$ kHz, and telephone channel bandwidth $\Delta f = 3$ kHz.

Solution

1. Basic input P_{rb}. Field strength $E \rightarrow$ power density $P_u \rightarrow$ basic input (from isotropic antenna) P_{rb},

$$P_u = E^2/120\,\pi$$

$$P_{rb} = P_u A_t = (E^2/120\pi)(\lambda^2/4\pi) = E^2\lambda^2/480\pi^2 \qquad \text{(Chapter 1)}$$

$$P_{rb}\ (\text{dBm}) = E\ (\text{dB}\mu\text{V/m}) + 20\log\lambda\ (\text{m}) - 10\log(480\pi^2 \times 10^9)$$

$$= 20 - 2.5 - 126.8 = -109.3\ \text{dBm}$$

Here P_{rb} can be read directly from the conversion graph for field strength, power density, and basic input given in Chapter 0.

2. Receiver input P_r,

$$P_r = P_{rb}\ (\text{dB}) + \text{receiving (isotropic) antenna gain} - \text{feeder system loss } L_f$$

$$= -109.3 + (-2 + 2.2) - 1.0 = -110.1\ \text{dBm}$$

where the gain of a half-wave dipole antenna is assumed to be 2.2 dB.

3. Threshold input P_{th}. $F = \overline{\text{NF}}$. Then

$$P_{\text{th}} = -116\ \text{dBm} \qquad \text{(Fig. 7}b\text{)}$$

4. S/N ratio. Figure 8 is used to calculate the S/N ratio when $P_r > P_{\text{th}}$.

$$G_{\text{FM}} = 143.5\ \text{dB} \qquad \text{(Fig. 8}c\text{)}$$

$$S/N = P_r - F + G_{\text{FM}} \quad (\text{dB})$$

$$= -110.1 - 8 + 143.5 = 25.4\ \text{dB}$$

EXAMPLE 7.6

Calculate the S/N ratio of a television link using FM under the following assumptions:

receiver input $P_r = -55$ dBm, overall noise figure $F = 12$ dB, frequency deviation for video signal $S_p = 8$ (MHz, peak-to-peak), IF bandwidth $B = 20$ MHz, and maximum modulating frequency $f_m = 4$ MHz.

Solution

1. Threshold input P_{th},

$$P_{\text{th}} \doteqdot -80\ \text{dBm} \qquad \text{(Fig. 7}d\text{)}$$

2. S/N ratio. Figure 9 is used to calculate the S/N ratio when $P_r > P_{th}$,

$$G_{TV} = 119 \text{ dB} \qquad (\text{Fig. } 9c)$$

$$S/N = P_r - F + G_{TV} \quad (\text{dB})$$

$$= -55 - 12 + 119 = 52 \text{ dB}$$

Example 7.7

A single-sideband FM system with the parameters given below is transmitting a high-speed facsimile channel within the baseband accommodated. Calculate the S/N ratio.

Channel frequency of interest $f_{ch} = 220$ kHz, channel bandwidth $\Delta f = 15$ kHz, frequency deviation $S_0 = 200$ kHz, receiver input $P_r = -50$ dBm, and overall noise figure $F = 10$ dB.

Solution

1. Calculate the S/N ratio assuming that the receiver input is higher than the threshold level.
2.

$$S/N = P_r + G_V + J_V \quad (\text{dB})$$

Then,

$$G_V = -0.8 \text{ dB} \qquad (\text{Fig. } 10c)$$

$$J_V = 122 \text{ dB} \qquad (\text{Fig. } 11a)$$

$$P_r = -50 \text{ dBm}$$

$$S/N \doteqdot 71 \text{ dB}$$

Example 7.8

Design a system for an international communication route composed of 44 relay sections using a frequency division multiplex (FDM) single-sideband FM system capable of transmitting 600 telephone channels. Standard receiver input level P_r for each hop is assumed to be at -30 dBm, taking into account the path length, size of antenna, and transmitter output of the equipment as well as the overall noise figure $F = 9$ dB. Estimate roughly the mean S/N_{sys} and the mean noise power N_{sys} (unweighted) of the overall system under nonfading conditions.

Solution

1. Figure 12 is used for rough estimation.
2. CCIR 600-channel system is assumed.
3. S/N ratio per hop (for worst channel),

$$G_s \doteqdot 116.6 \text{ dB} \qquad \text{(Fig. 12}a\text{)}$$

$$S/N = P_r - F + G_s = -30 - 9 + 116.5 = 77.5 \text{ dB}$$

4. System overall S/N_{sys} (number of hops $m = 44$). If all the hops provide the same S/N ratio,

$$S/N_{\text{sys}} = S/N - 10 \log m = 77.5 - 16.4 \fallingdotseq .61 \text{ dB}$$

5. Mean noise N_{sys} (pW)

$$N_{\text{sys}} = 800 \text{ pW} \qquad \text{(Fig. 25)}$$

EXAMPLE 7.9

Calculate the mean S/N ratio of the worst channel for sections AB and AC. The intermodulation noise and the interference noise are ignored.

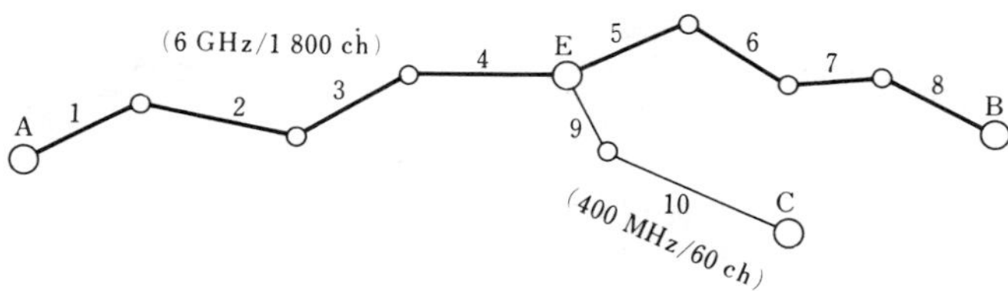

Conditions

Design Parameter	6 GHz/1800 Channels	400 MHz/60 Channels
Noise figure $\overline{\text{NF}}$	10 dB	8 dB
Frequency deviation per channel S_0	140 kHz	100 kHz
Channel bandwidth Δf	3.1 kHz	3.1 kHz
Top channel frequency f_m	8120 kHz	300 kHz

Section	1	2	3	4	5	6	7	8	9	10
Mean receiver input P_r (dBm)	−20	−24	−18	−20	−23	−20	−20	−26	−52	−65
Frequency	6 GHz								400 MHz	

Solution

1. Find a suitable chart for estimating the S/N ratio for each hop through Auxiliary Indexes B6 and B7,

Fig. 23a, b for 6-GHz/1800-channel system

Fig. 15 c, d for 400-MHz/60-channel system

2. S/N ratio and noise power in the top channel (worst channel) for each hop:

Hop	1	2	3	4	5	6	7	8	9	10
S/N	73.5	69.5	71.5	73.5	70.5	73.5	73.5	67.5	69	56
N (pW)	45	112	36	45	90	45	45	180	125	2500
N_t (pW)		(AE) 238				(EB) 360			(EC) 2625	

3. S/N (AB) and S/N (AC), using Fig. 25,

$$N_t\ (AB) = 238 + 360 = 598, \qquad S/N\ (AB) = 62.2 \text{ dB}$$

$$N_t\ (AC) = 238 + 2625 = 2863, \quad S/N\ (AC) = 55.4 \text{ dB}$$

REMARKS. For overall noise and S/N ratio in a radio relay system with fading see Chapter 8.

Chart Index,
Noise Level and S/N Ratio

Auxiliary Index, Effective Noise and Threshold Input

Effective (white) noise input --- $P_n = 10 \log (KT_0\ F \cdot B)$ (dB)
Threshold input --- $P_{th} = P_n + C_{th}$ (dB)

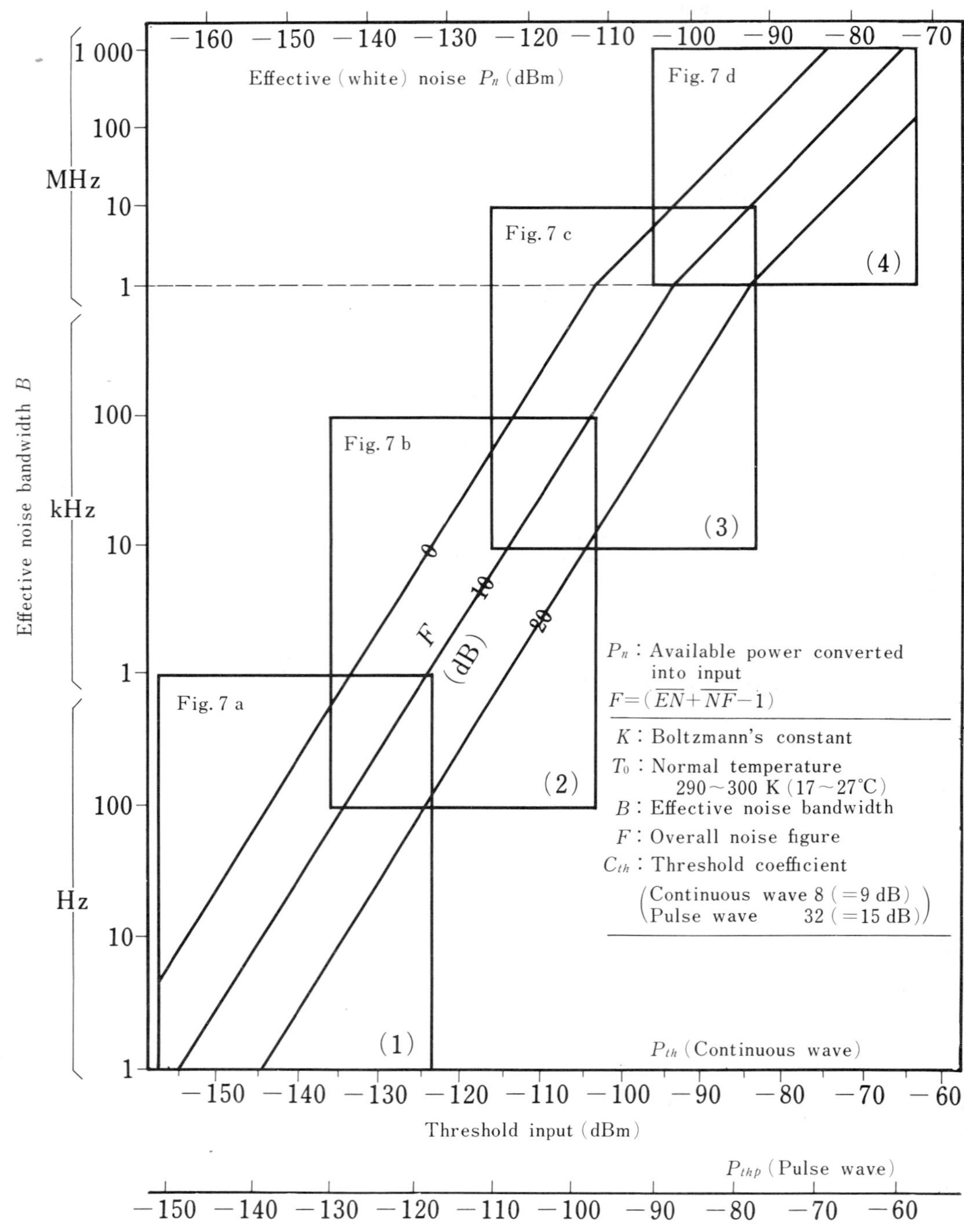

Auxiliary Index, S / N Ratio for Conventional FM System

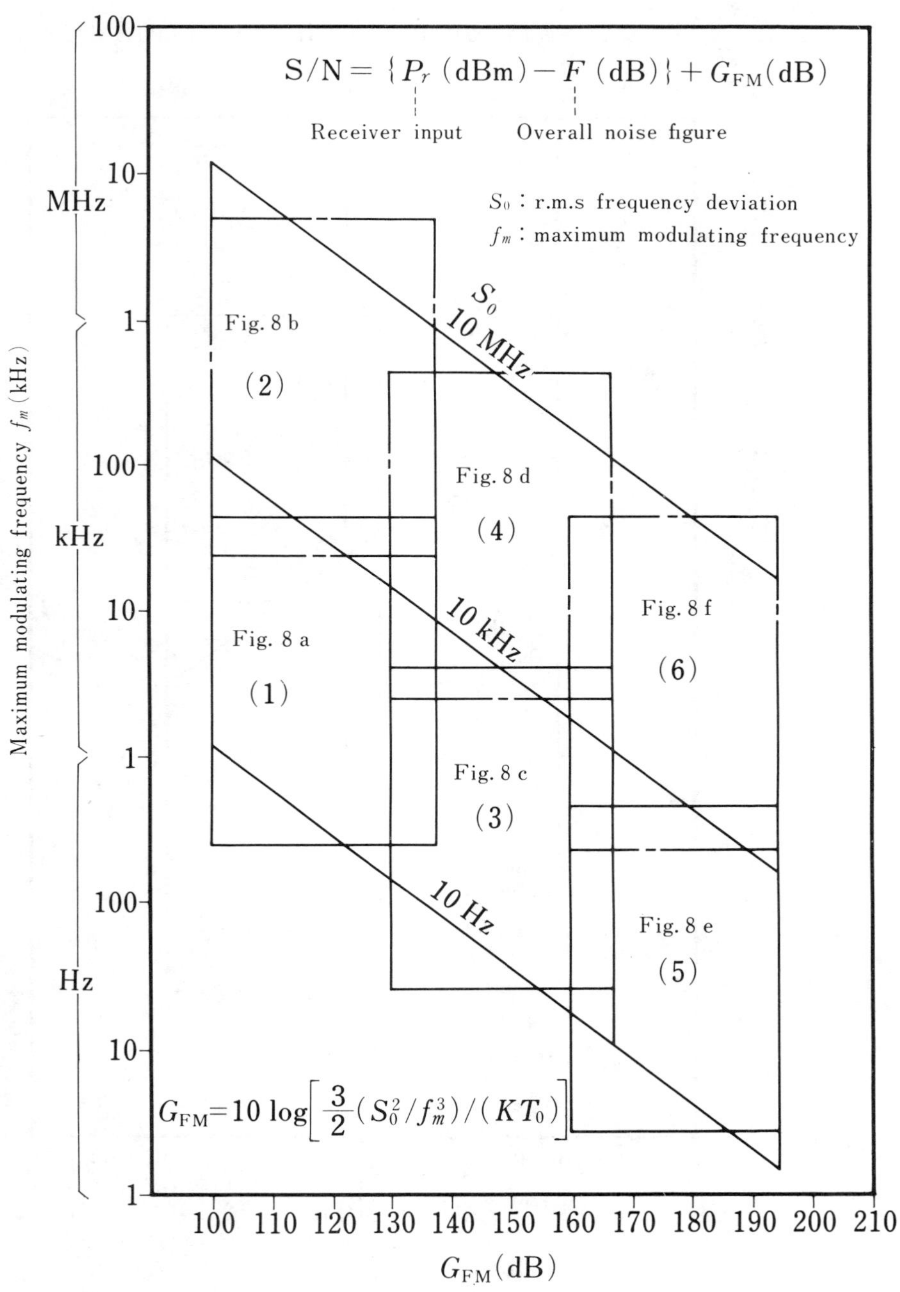

Auxiliary Index, S / N Ratio for (TV Transmission) FM System

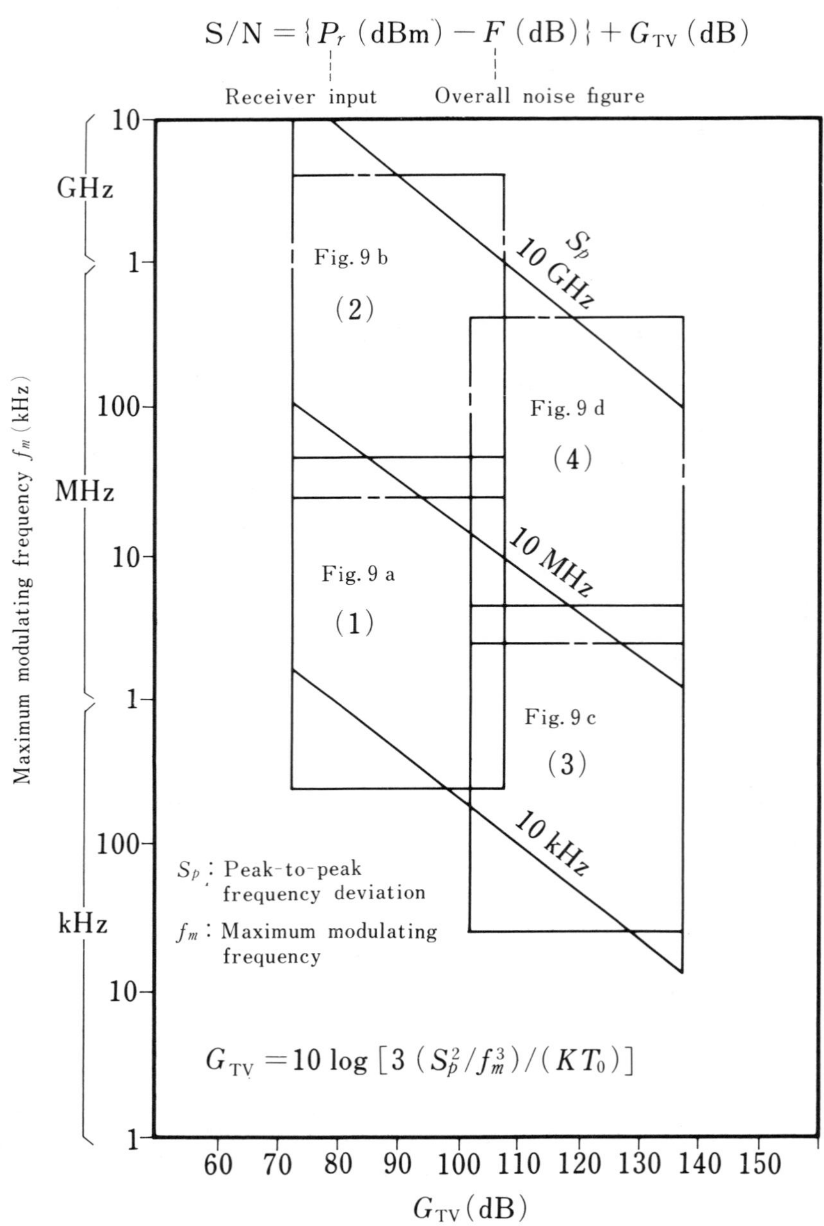

Auxiliary Index, S / N Ratio for Arbitrary Channel of SS-FM System

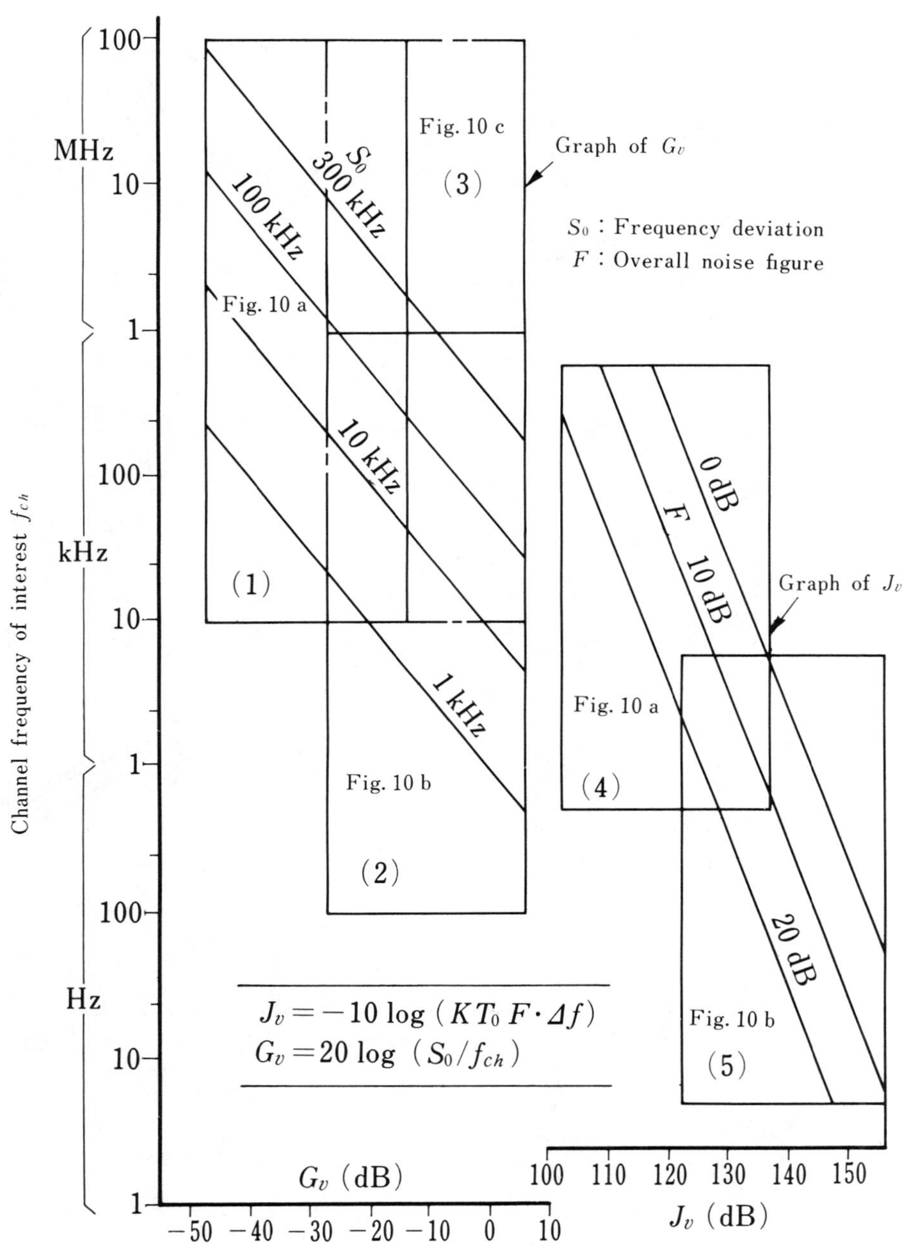

Auxiliary Index, Graph for Quick Estimation of S / N Ratio for SS-FM System (CCIR Recommendation)

$$S/N\,(ch) = \{P_r\,(dBm) - F\,(dB)\} + G_{TP}\,(dB)$$

P_r: Receiver input F: Overall noise figure

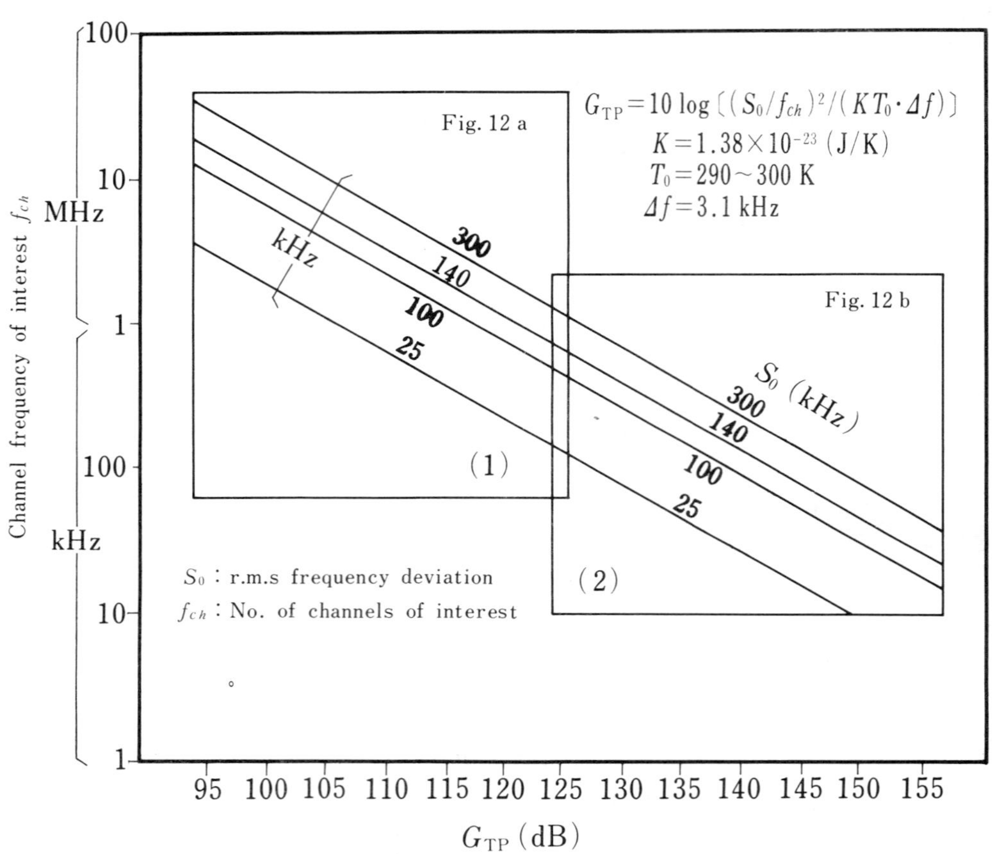

Auxiliary Index, Receiver Input versus S / N Ratio (SS-FM System) (1)

n : No. of channels
S_0 : r.m.s frequency deviation (kHz/ch)
f_m : Maximum modulating frequency (kHz)

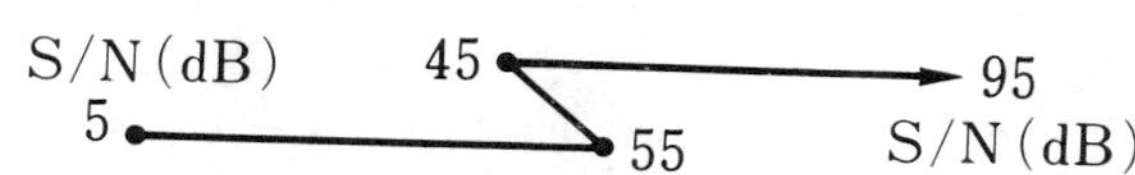

n ⟨f_m⟩	S_0	Receiver input P_r (dBm) (−120 … −10)
12 ⟨60⟩	CCIR 35	Fig. 13 a → Fig. 13 b
24 ⟨108⟩	CCIR 35	Fig. 14 a → Fig. 14 b
60 ⟨252 / 300⟩	CCIR 50	Fig. 15 a → Fig. 15 b
	CCIR 100	Fig. 15 c → Fig. 15 d
	CCIR 200	Fig. 15 e → Fig. 15 f
120 ⟨552⟩	CCIR 50	Fig. 16 a → Fig. 16 b
	CCIR 100	Fig. 16 c → Fig. 16 d
	CCIR 200	Fig. 16 e → Fig. 16 f
	280	Fig. 16 g → Fig. 16 h
240 ⟨1 054⟩	200	Fig. 17 a → Fig. 17 b
	280	Fig. 17 c → Fig. 17 d
300 ⟨1 300⟩	CCIR 200	Fig. 18 a → Fig. 18 b
	280	Fig. 18 c → Fig. 18 d

Auxiliary Index, Receiver Input versus S / N Ratio (SS-FM System) (2)

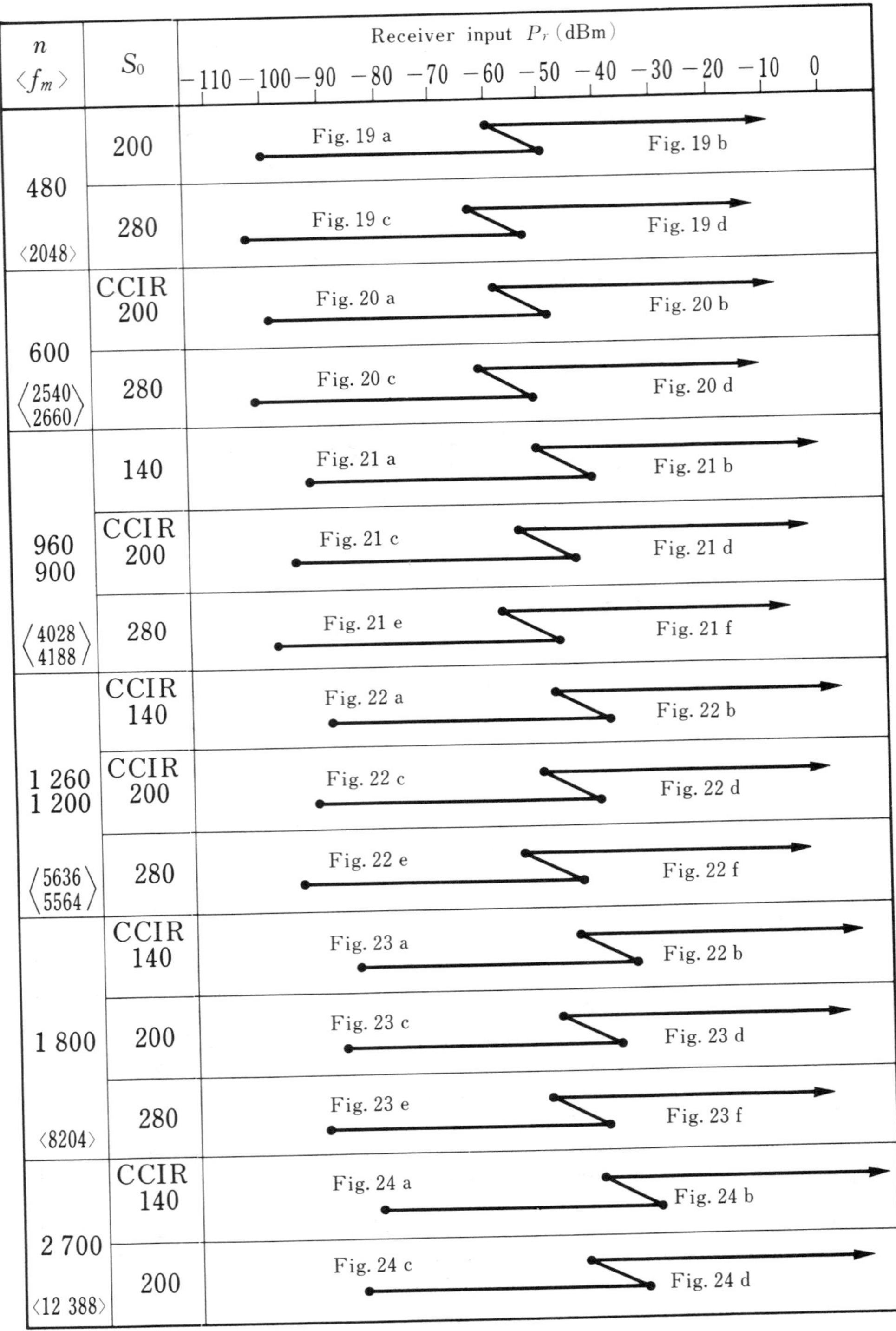

n ⟨f_m⟩	S_0	Receiver input P_r (dBm) −110 −100 −90 −80 −70 −60 −50 −40 −30 −20 −10 0	
480 ⟨2048⟩	200	Fig. 19 a	Fig. 19 b
	280	Fig. 19 c	Fig. 19 d
600 ⟨2540 / 2660⟩	CCIR 200	Fig. 20 a	Fig. 20 b
	280	Fig. 20 c	Fig. 20 d
960 / 900 ⟨4028 / 4188⟩	140	Fig. 21 a	Fig. 21 b
	CCIR 200	Fig. 21 c	Fig. 21 d
	280	Fig. 21 e	Fig. 21 f
1 260 / 1 200 ⟨5636 / 5564⟩	CCIR 140	Fig. 22 a	Fig. 22 b
	CCIR 200	Fig. 22 c	Fig. 22 d
	280	Fig. 22 e	Fig. 22 f
1 800 ⟨8204⟩	CCIR 140	Fig. 23 a	Fig. 22 b
	200	Fig. 23 c	Fig. 23 d
	280	Fig. 23 e	Fig. 23 f
2 700 ⟨12 388⟩	CCIR 140	Fig. 24 a	Fig. 24 b
	200	Fig. 24 c	Fig. 24 d

Radio System Noise

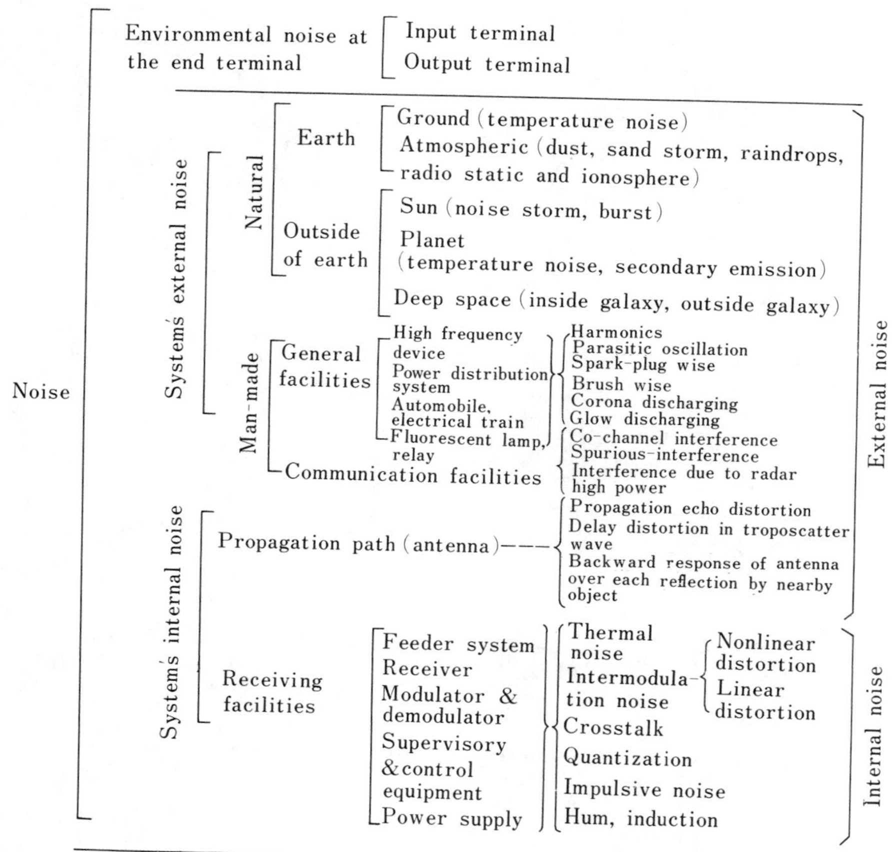

〔Note.〕 (1) Noises inherent in the switching equipment and in the wire transmission system have been excluded.

(2) The noises can be classified by their physical nature.

Random noise	Impulsive noise	White noise
Periodic noise	Continuous noise	Filter noise (filtration noise)

(3) White noise : Noises with a uniform frequency spectrum (within the bandwidth concerned)

White noise
- Thermal noise (resistor)
- Shot noise (electron tube, diode, transistor)
- Some of the intermodulation noise and of the interference noise in broadband transmission system.

(4) White noise and thermal noise have frequently the same meaning unless clear distinction between them is necessary.

Noise Temperature of Atmosphere

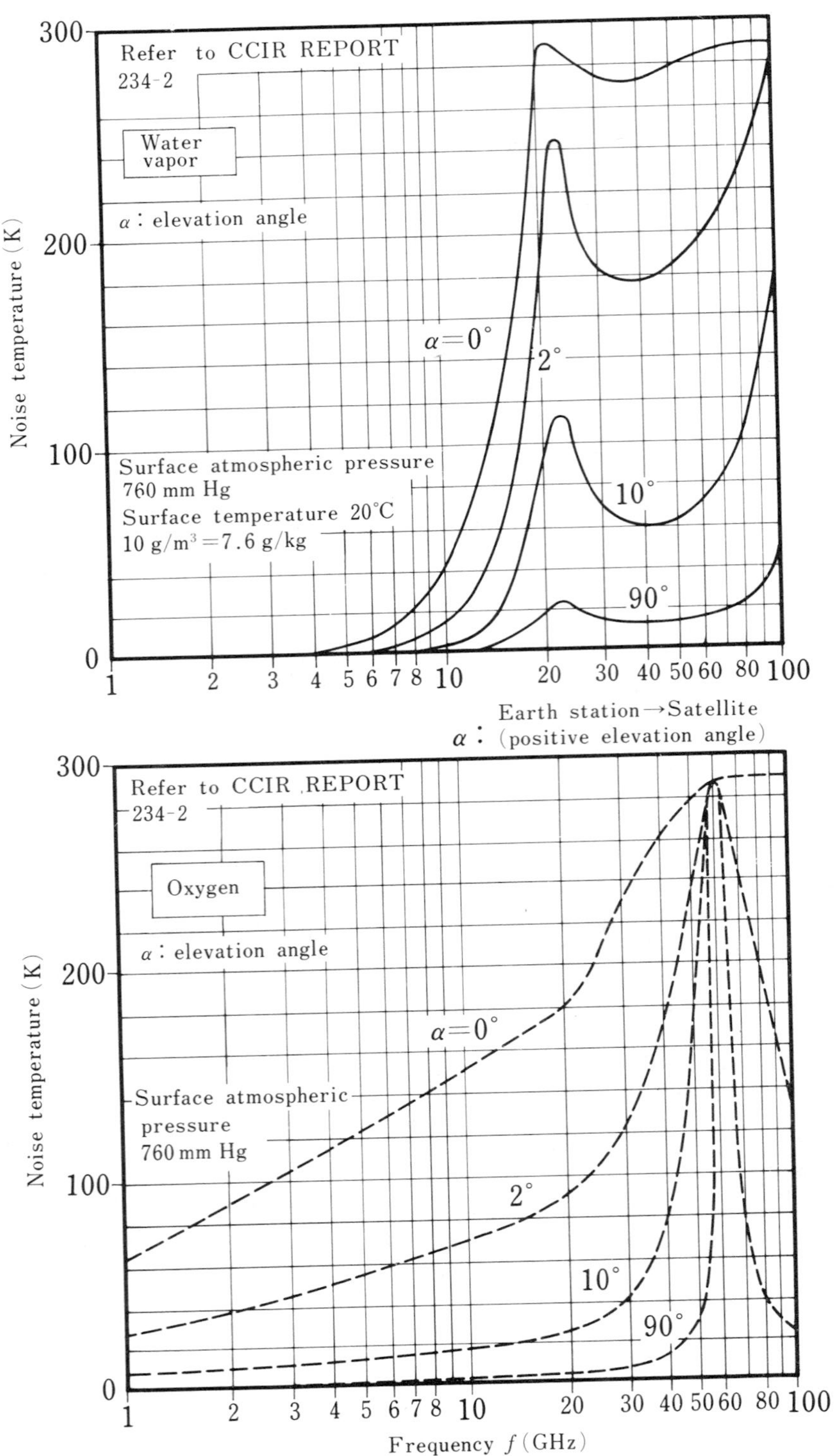

Representation of Thermal Noise

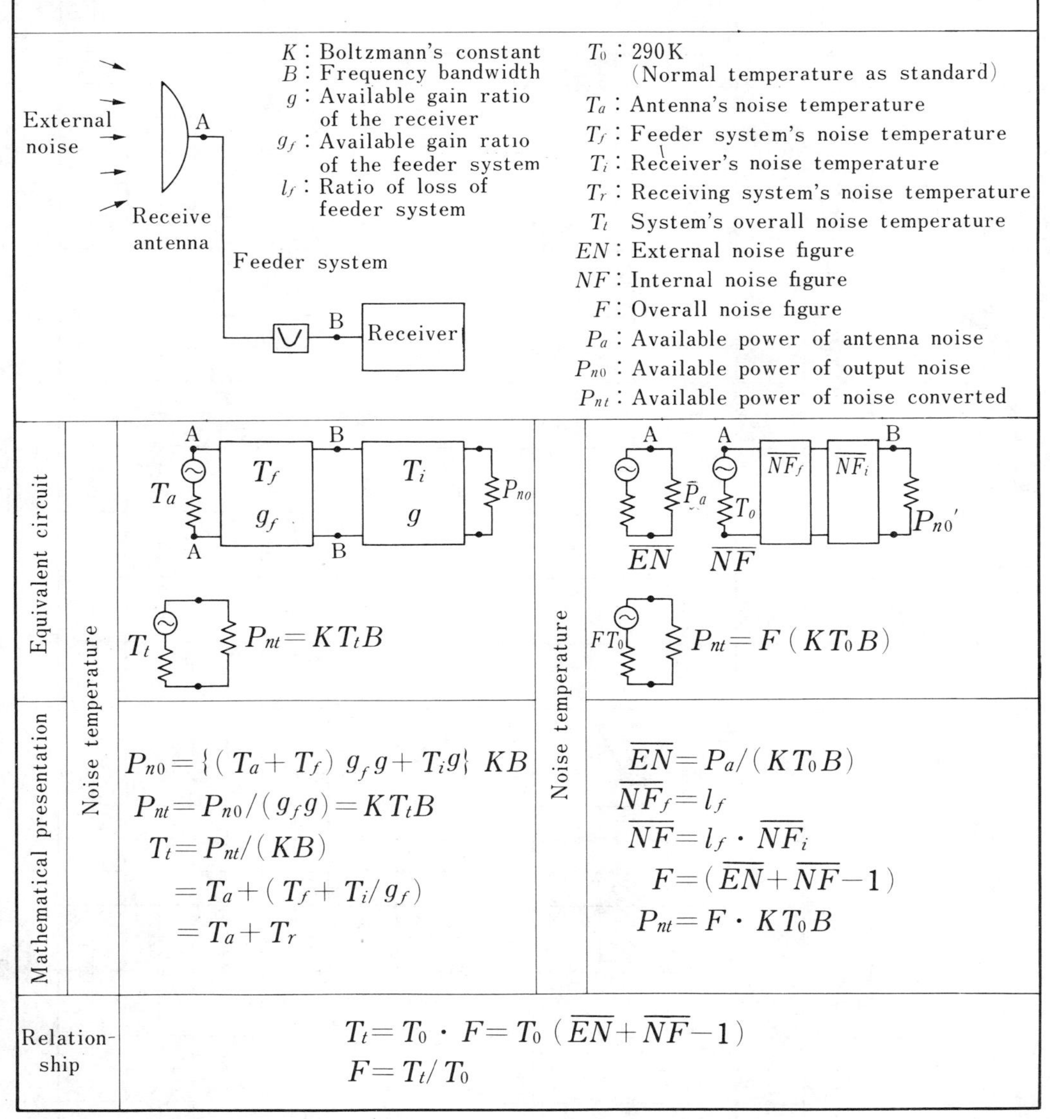

(1) Above equations use antilogarithm instead of logarithm in dB

(2) When $\overline{EN}$, $\overline{NF}$, g and l_f are given in dB convert them into real numbers

(3) Unit of temperature : kelvin (K), $T_0=290\text{K}=17°\text{C}$

Influence of Feeder System upon Low-Noise Receiving System

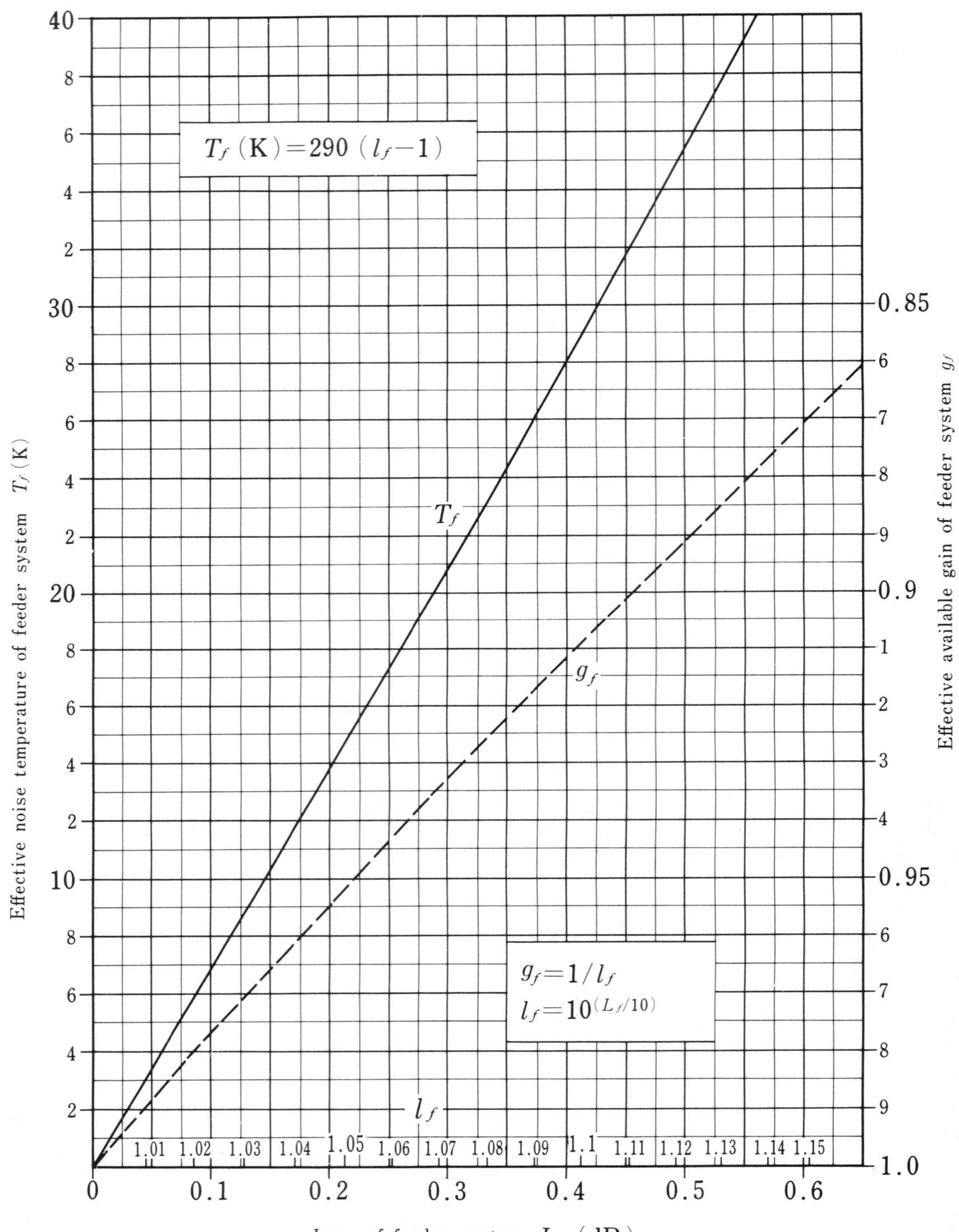

Effective Noise Temperature T_i versus Internal Noise Figure $\overline{NF}$

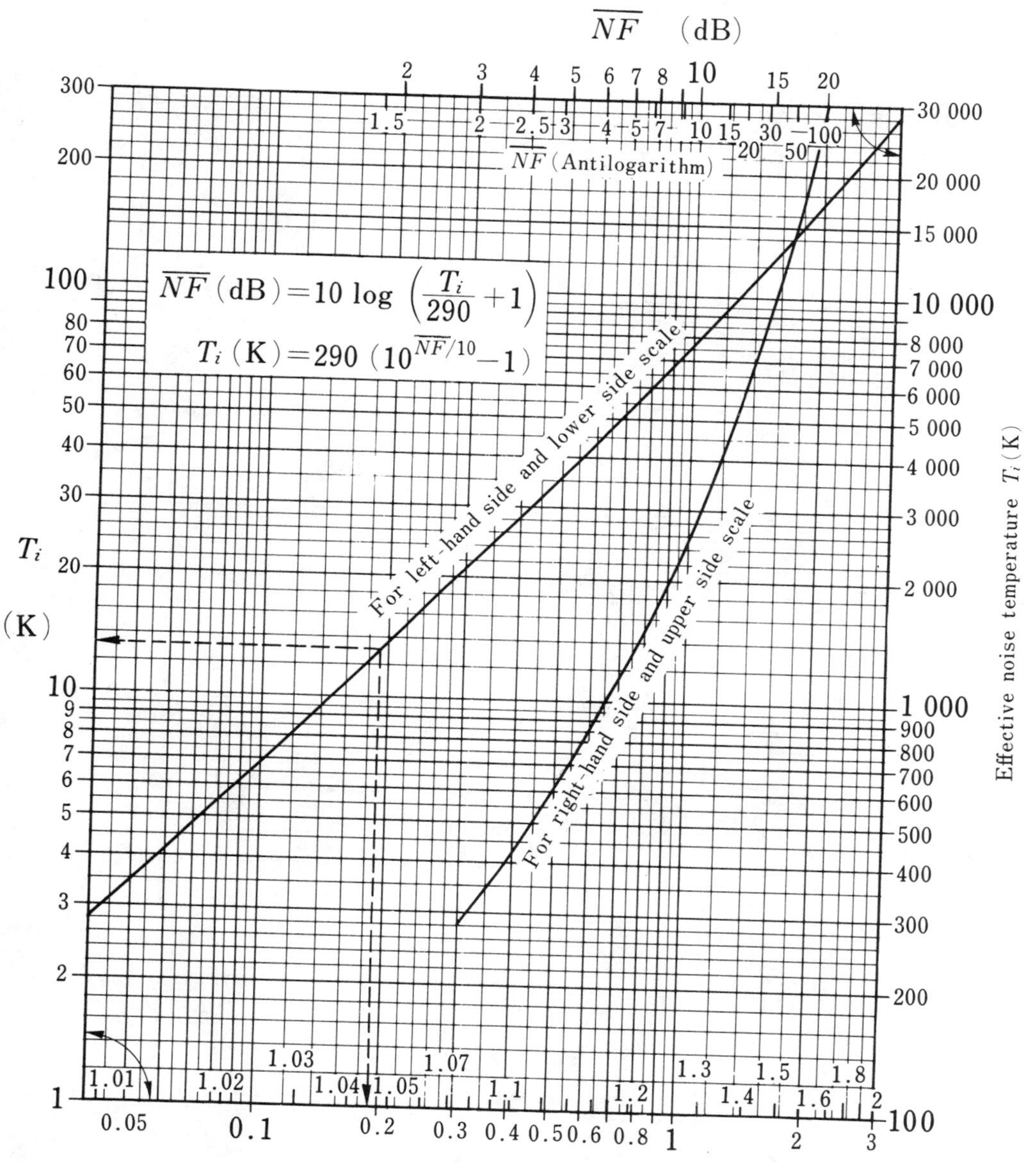

Antenna Noise Temperature T_a versus External Noise Figure $\overline{EN}$

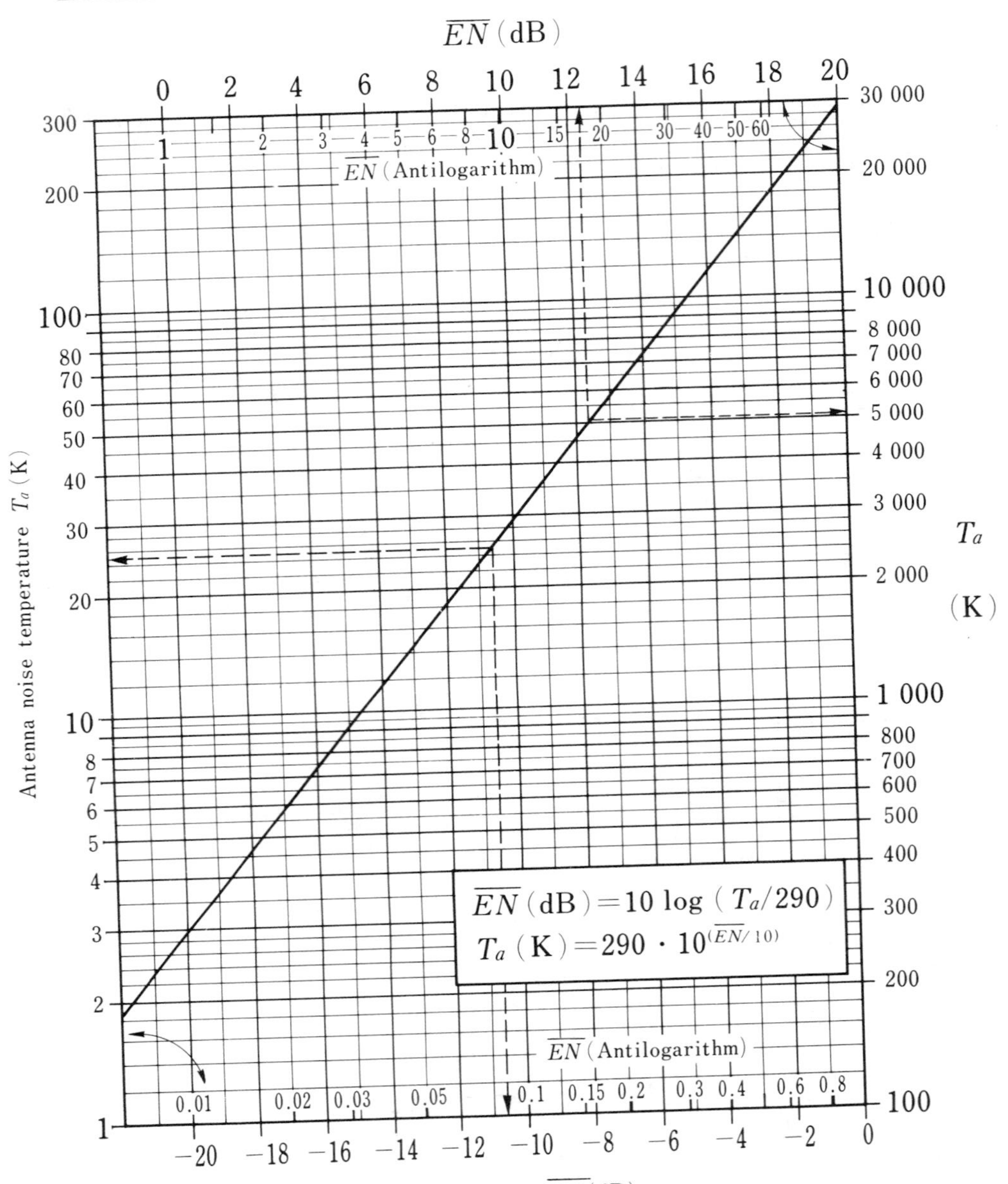

Overall Noise Temperature T_t versus Overall Noise Figure F

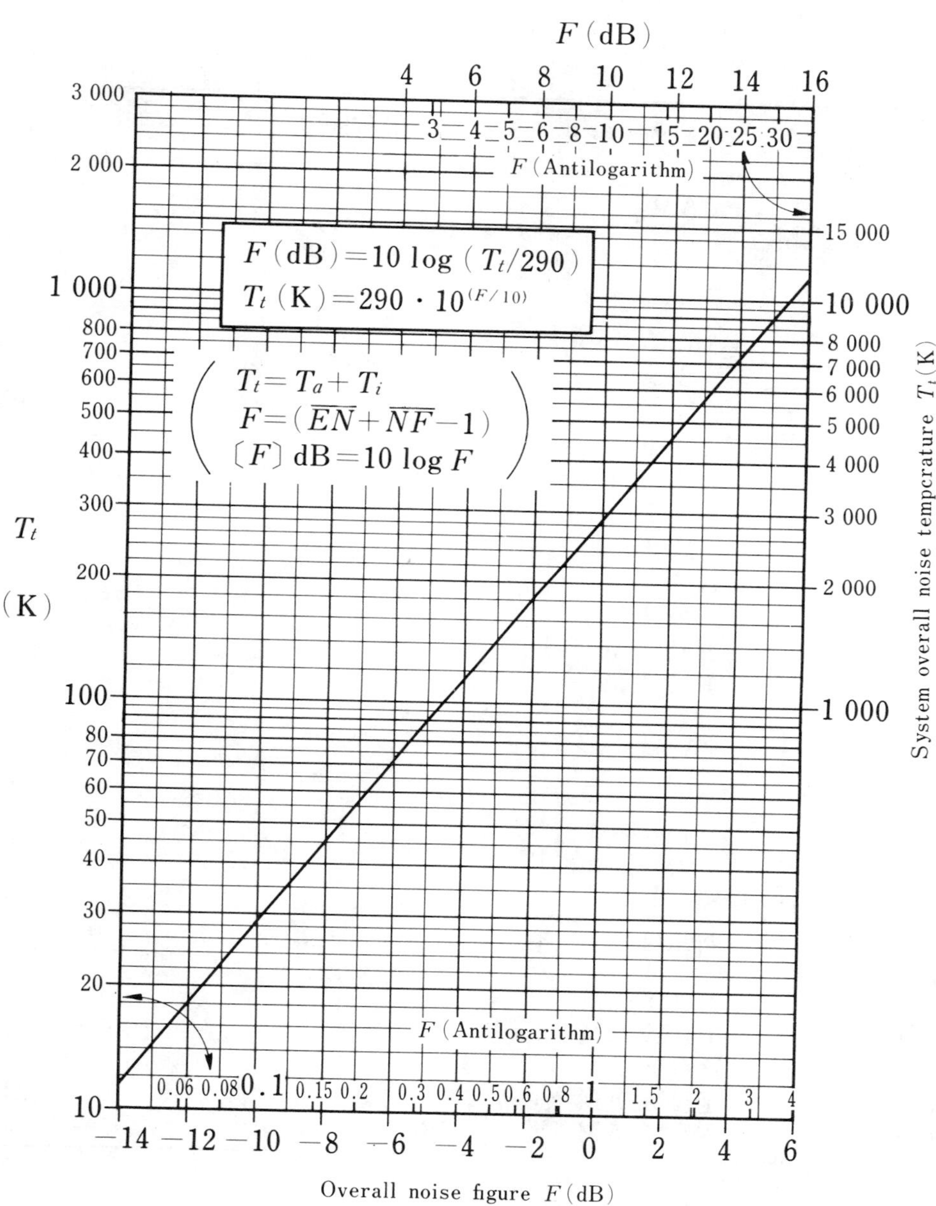

Threshold Coefficient

Threshold input $P_{th} = KTBF \cdot C_{th}$

(If P_{th} is a mean power, P_{th} (mean) $= P_{th} \cdot D_f$)

—White noise—

Type of radio wave	Method of modulation	Threshold coefficient C_{th}	Duty factor D_f
Analog wave	(AM) FM PM	8 (=9 dB)	1 (=0 dB)
	(AM-AM) (SS-AM) (SS-SS) AM-FM SS-FM FM-FM PM-FM AM-PM SS-PM FM-PM PM-PM		
	PAM-FM PFM-FM PPM-FM PWM-FM PNM-FM PCM-FM		
Pulse wave	(PAM-AM)		D_f (=10 log D_f)
	PPM-AM PFM-AM PWM-AM PNM-AM PCM-AM	32 (=15 dB)	

(1) AM : Amplitude modulation
FM : Frequency modulation
PM : Phase modulation
SS : Single-sideband
PAM : Pulse-amplitude modulation
PFM : Pulse-frequency modulation
PPM : Pulse-position modulation
PWM : Pulse-width modulation
(PAM, PFM, PPM, PWM: Analog pulse modulation)
PNM : Pulse-number modulation
PCM : Pulse-code modulation
(PNM, PCM: Digital pulse modulation)

(2) $D_f = \tau / T_p$ τ : pulse width, T_p : period of pulse

(3) Method of modulation given in the parenthesis presents no clear threshold point

Effective Noise and Threshold Input (1)

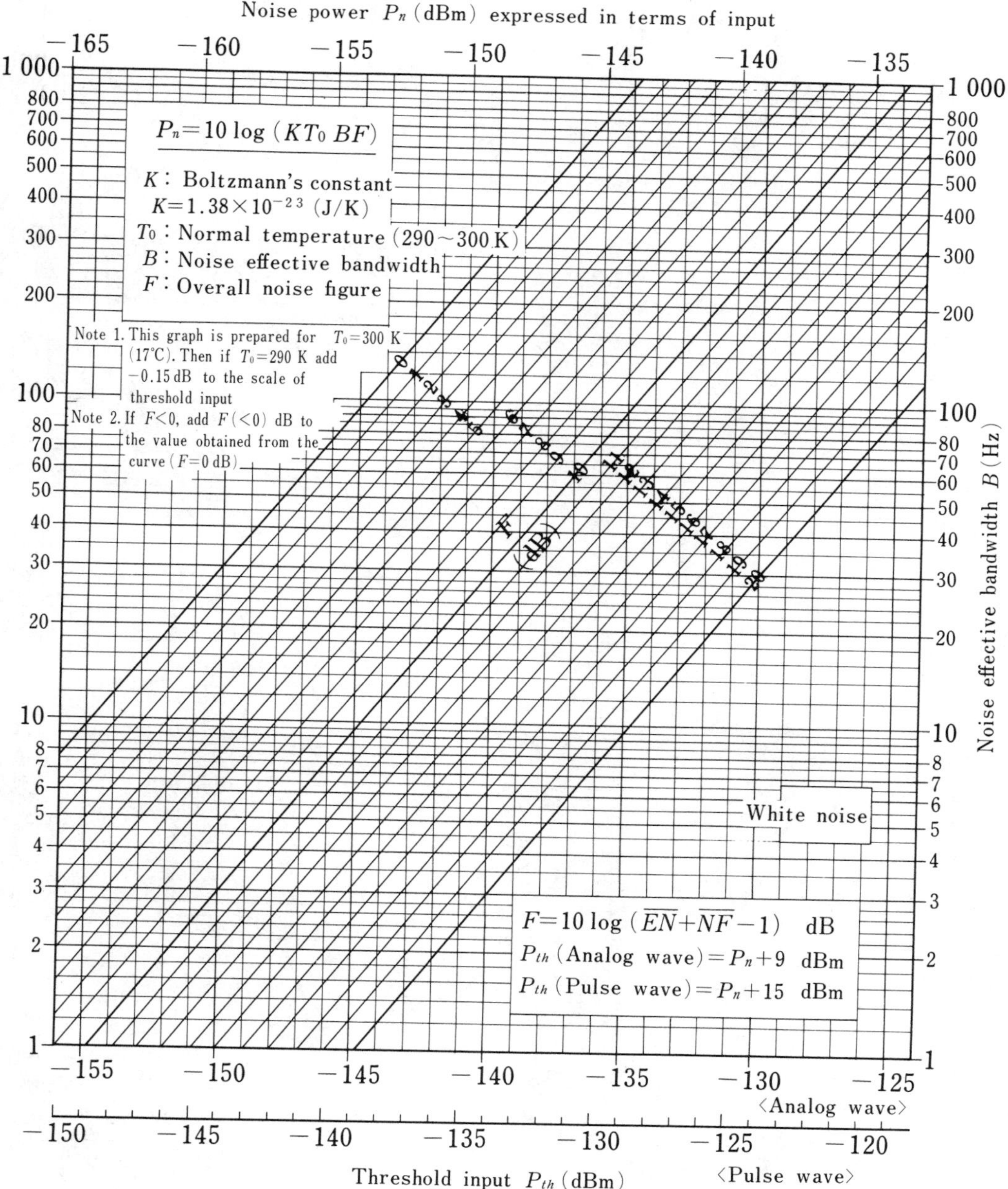

Effective Noise and Threshold Input (2)

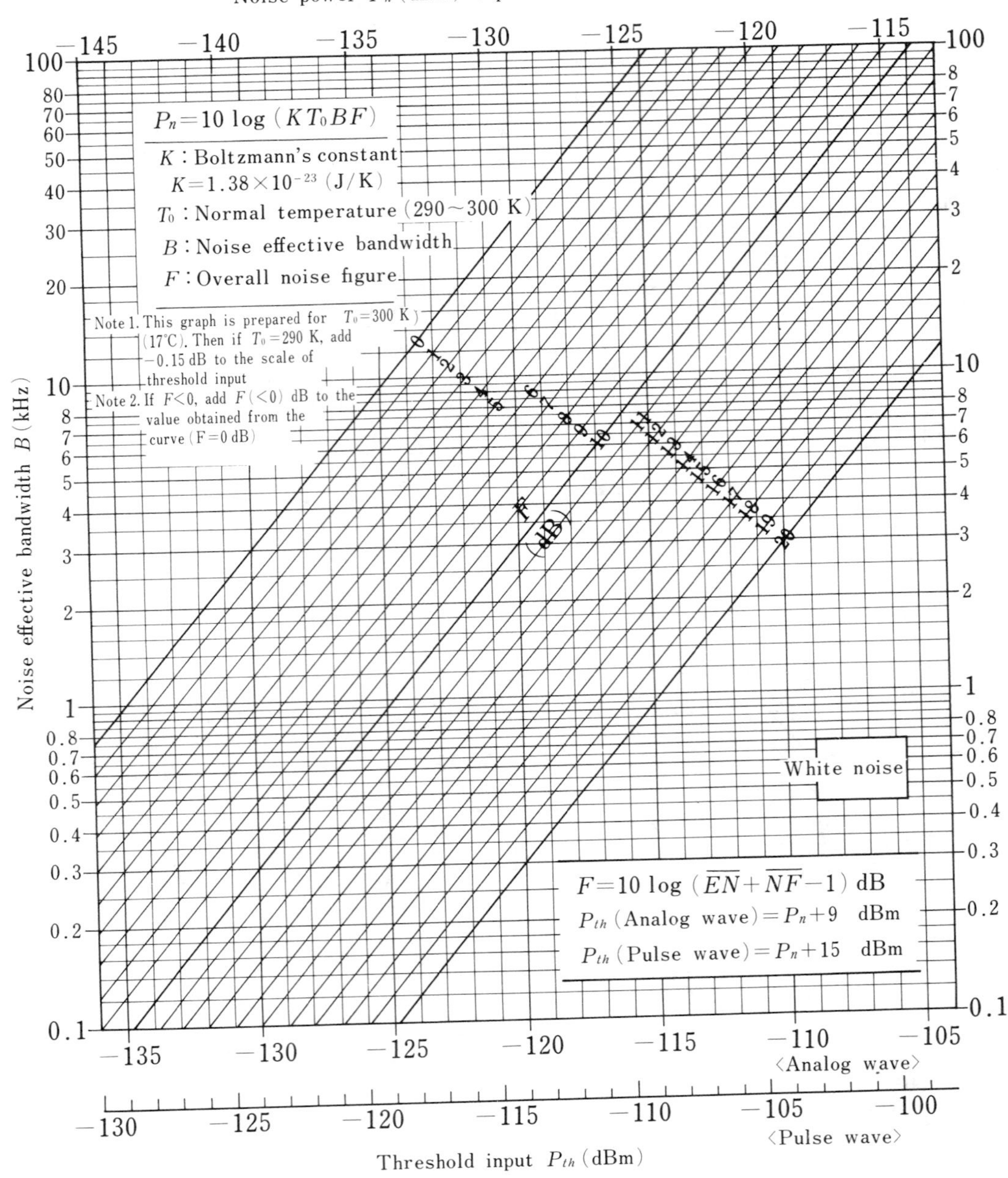
Noise power P_n (dBm) expressed in terms of input
−145 −140 −135 −130 −125 −120 −115
$P_n = 10 \log (KT_0BF)$
K : Boltzmann's constant
$K = 1.38 \times 10^{-23}$ (J/K)
T_0 : Normal temperature (290 ~ 300 K)
B : Noise effective bandwidth
F : Overall noise figure
Note 1. This graph is prepared for $T_0 = 300$ K (17°C). Then if $T_0 = 290$ K, add −0.15 dB to the scale of threshold input
Note 2. If $F < 0$, add $F (< 0)$ dB to the value obtained from the curve (F = 0 dB)
F (dB)
White noise
$F = 10 \log (\overline{EN} + \overline{NF} - 1)$ dB
P_{th} (Analog wave) $= P_n + 9$ dBm
P_{th} (Pulse wave) $= P_n + 15$ dBm
Noise effective bandwidth B (kHz)
−135 −130 −125 −120 −115 −110 −105
⟨Analog wave⟩
−130 −125 −120 −115 −110 −105 −100
⟨Pulse wave⟩
Threshold input P_{th} (dBm)

Effective Noise and Threshold Input (3)

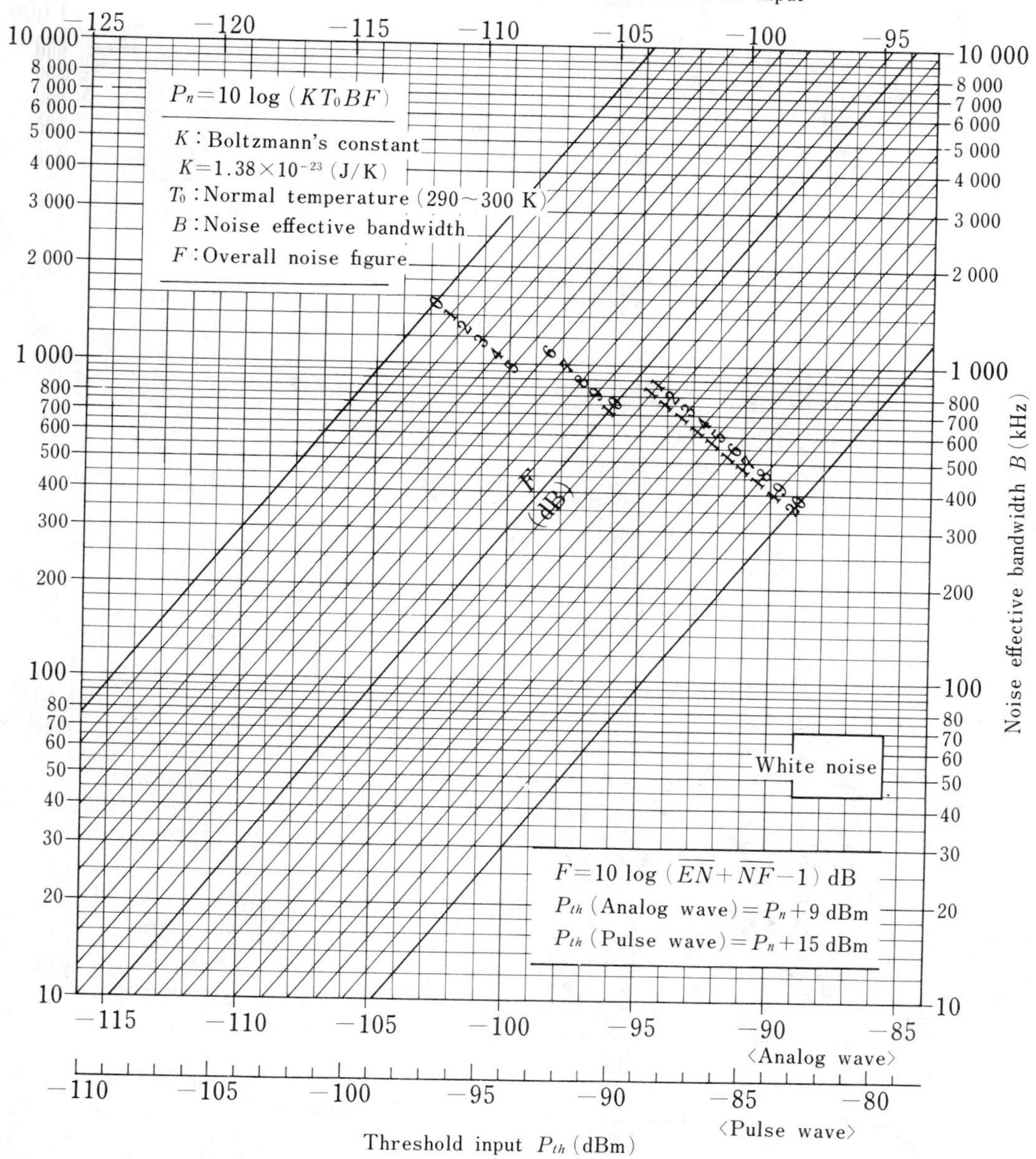

Note 1. This graph is prepared for $T_0 = 300$ K (17°C). Then if $T_0 = 290$ K, add −0.15dB to the scale of threshold input.

Notc 2. IF $F < 0$, add $F\ (< 0)$ dB to the value obtained from the curve ($F = 0$dB).

Effective Noise and Threshold Input (4)

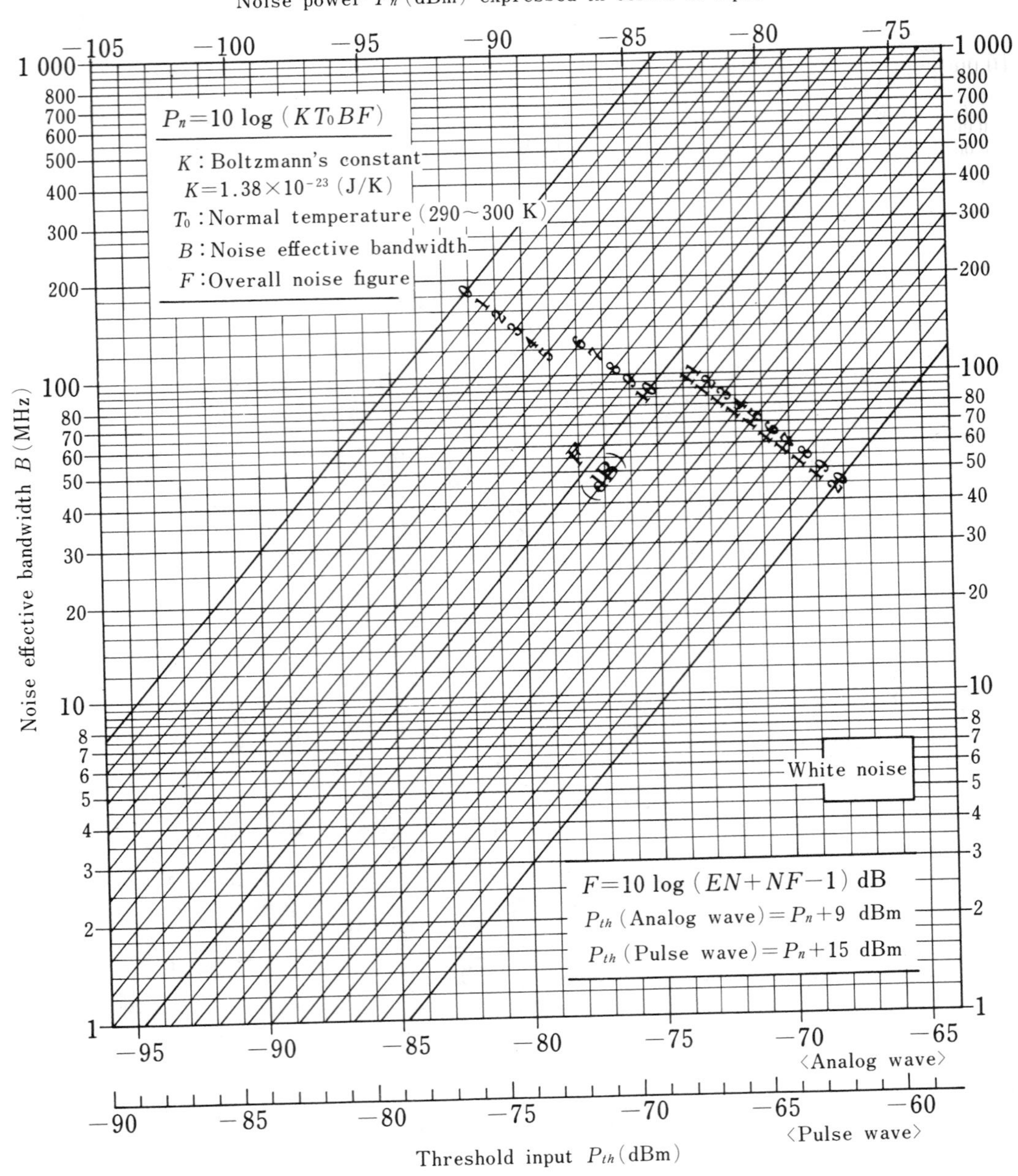

Note 1. This graph is prepared for $T_0 = 300$ K (17°C). Then if $T_0 = 290$ K, add −0.15dB to the scale of threshold input.

Note 2. If $F < 0$, add F (< 0) dB to the value obtained from the curve ($F = 0$dB).

S / N Ratio for (Conventional) FM System (1)

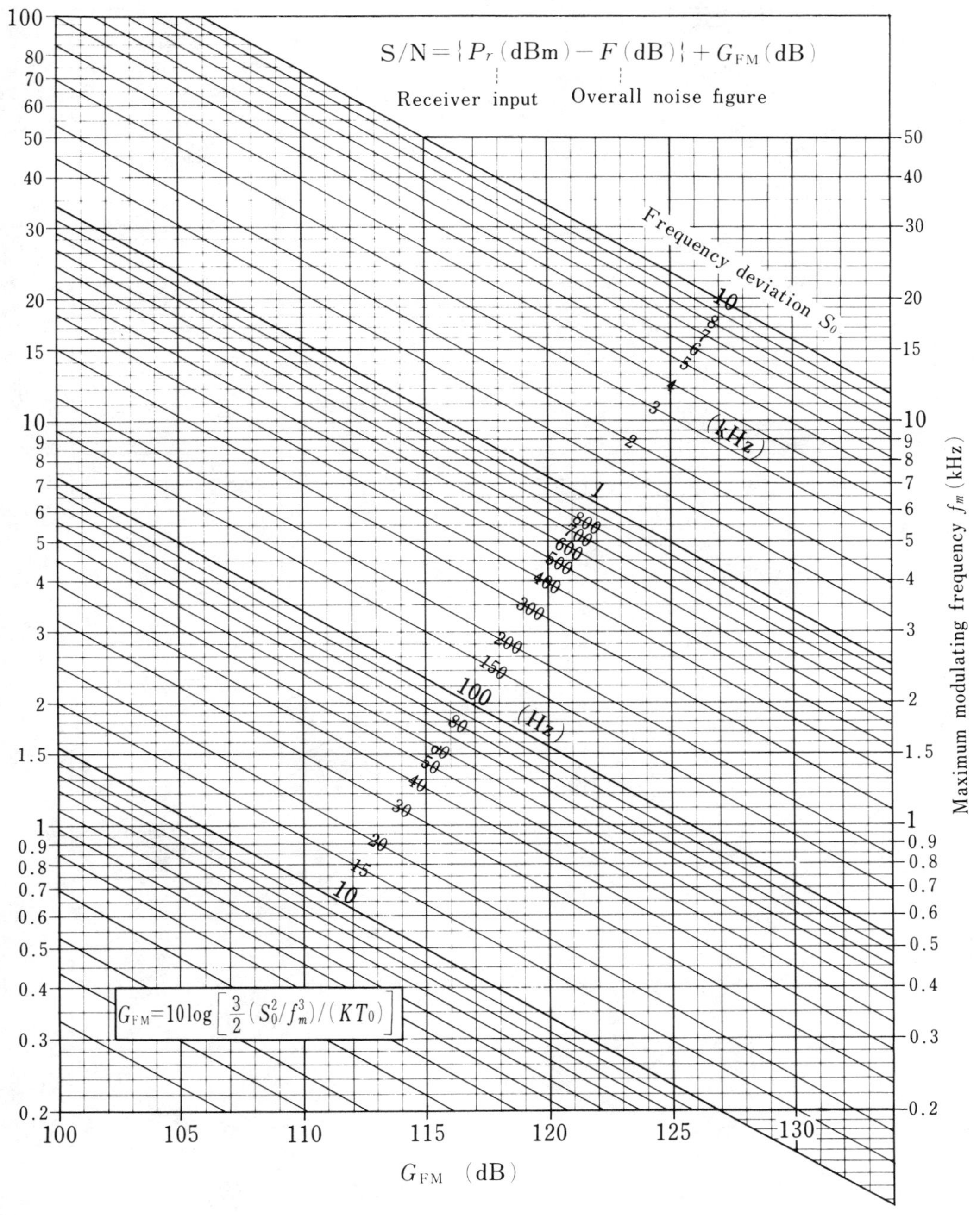

S / N Ratio for (Conventional) FM System (2)

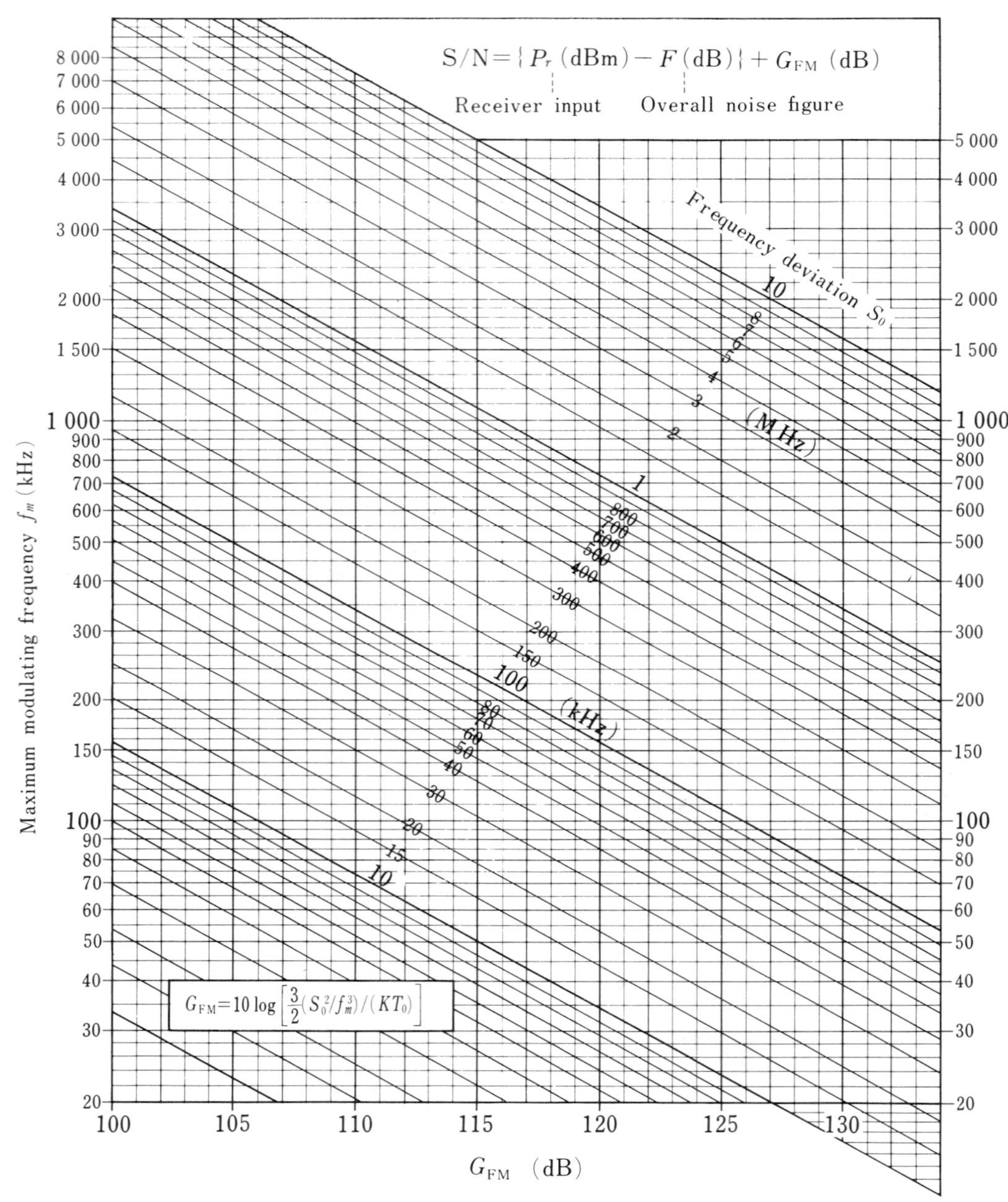

S / N Ratio for (Conventional) FM System (3)

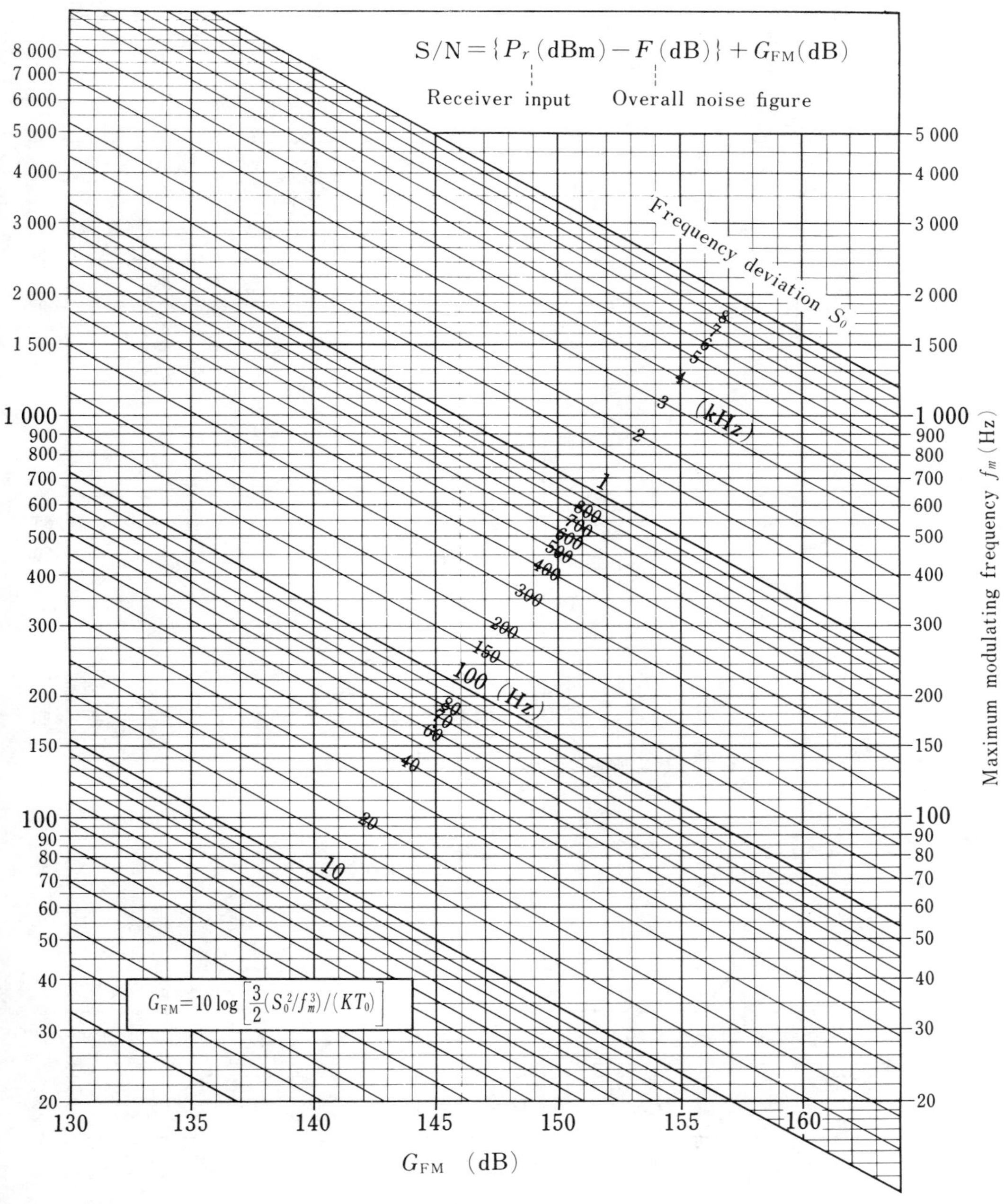

S / N Ratio for (Conventional) FM System (4)

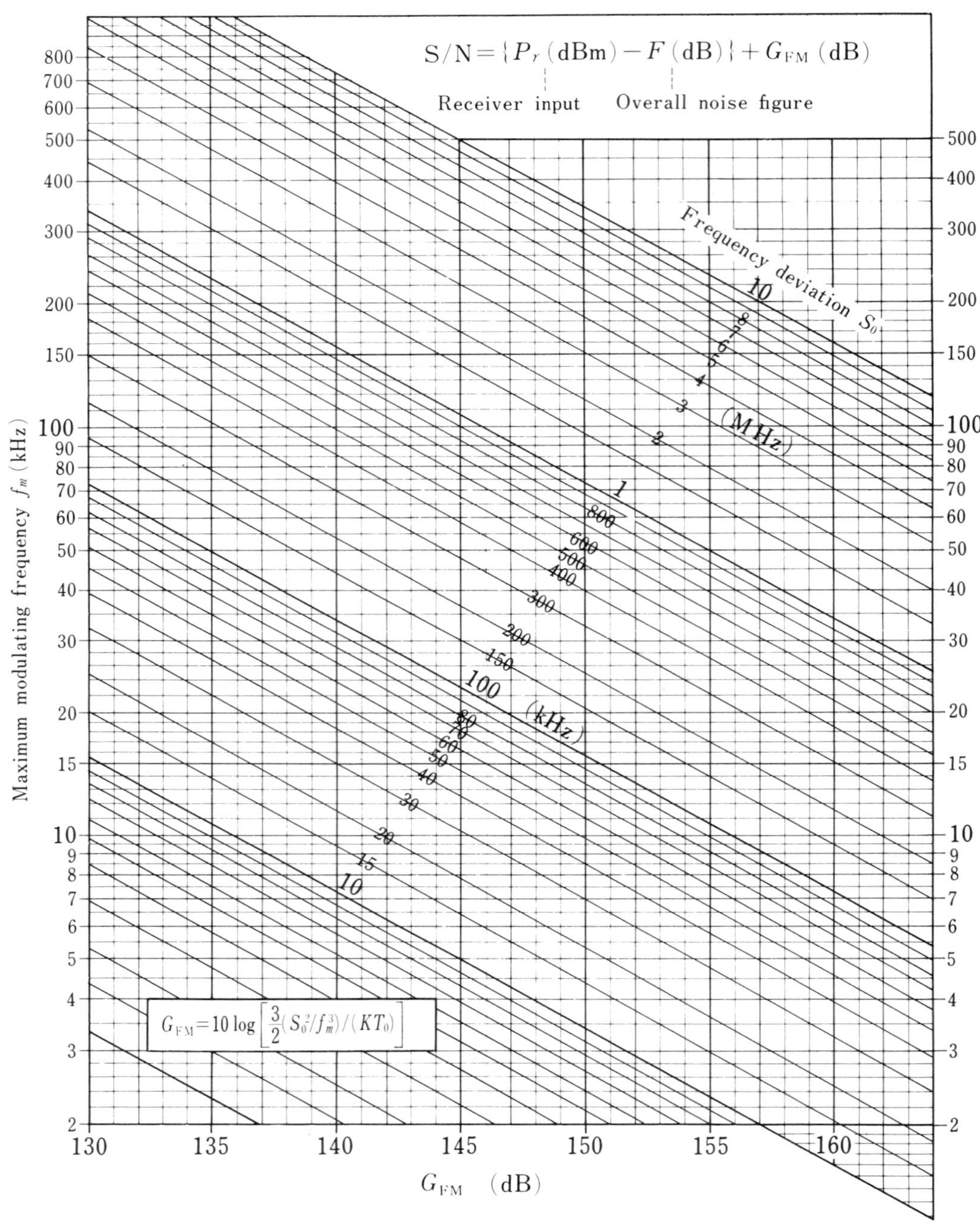

S / N Ratio for (Conventional) FM System (5)

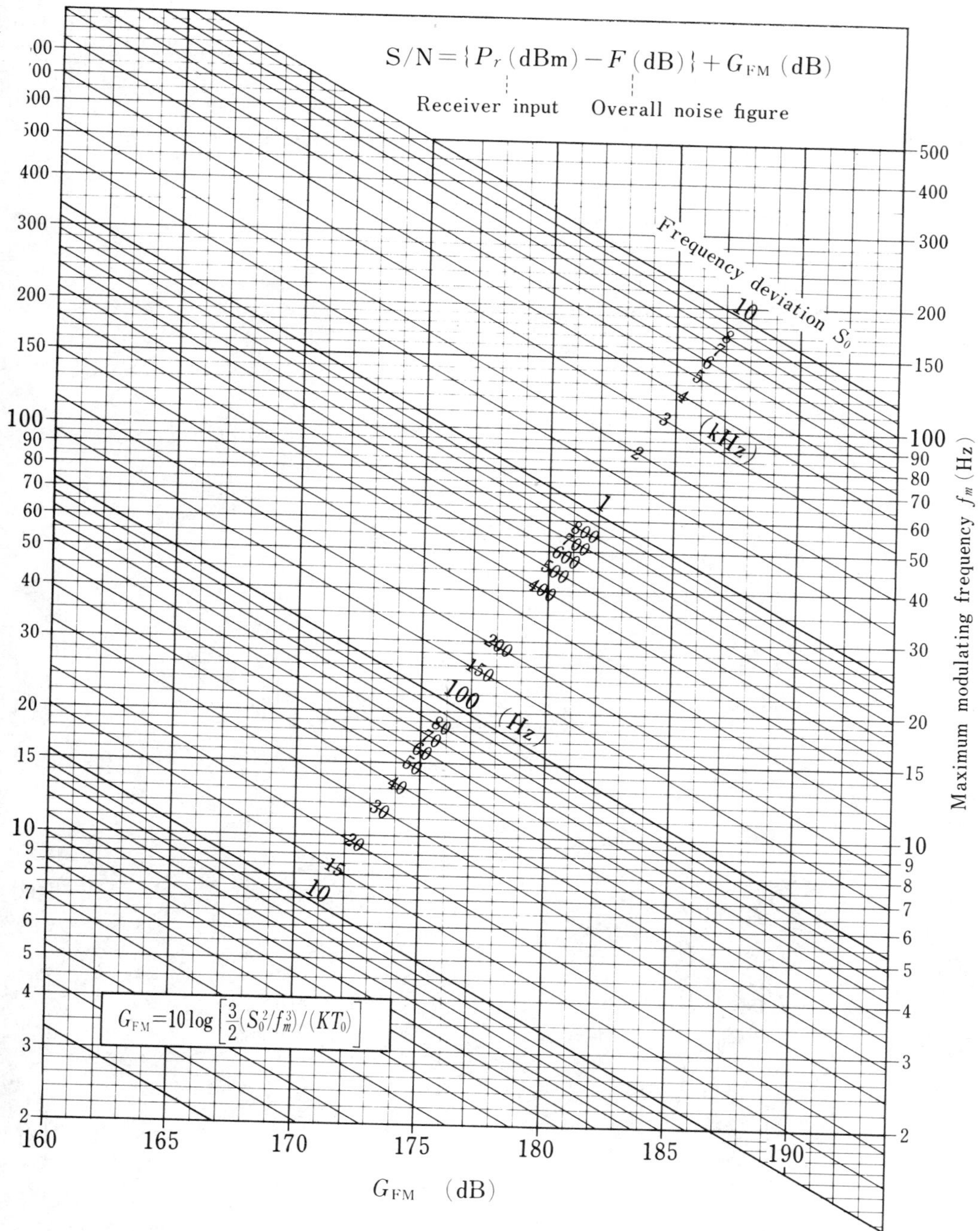

S / N Ratio for (Conventional) FM System (6)

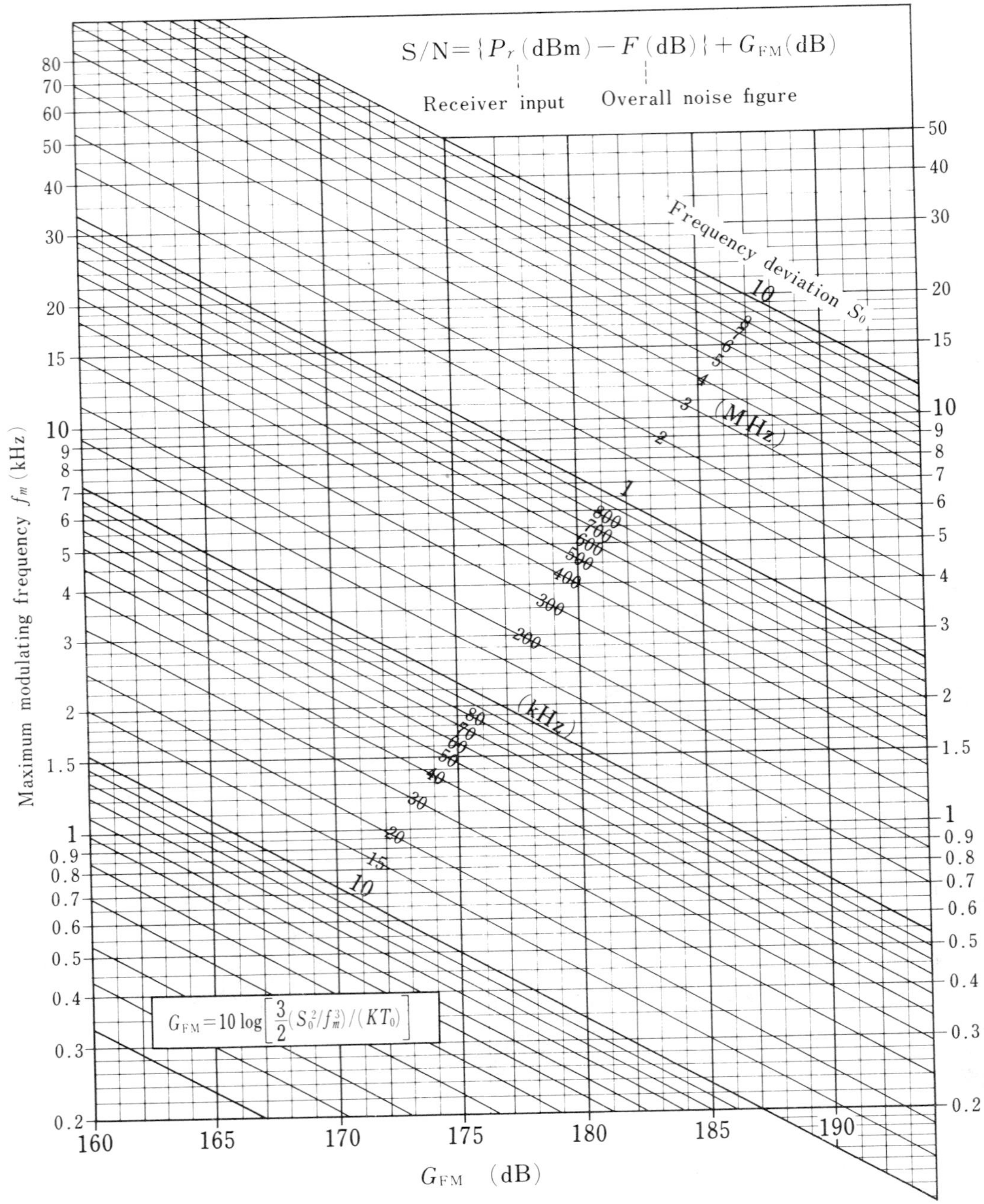

S / N Ratio for (TV Transmission) FM System (1)

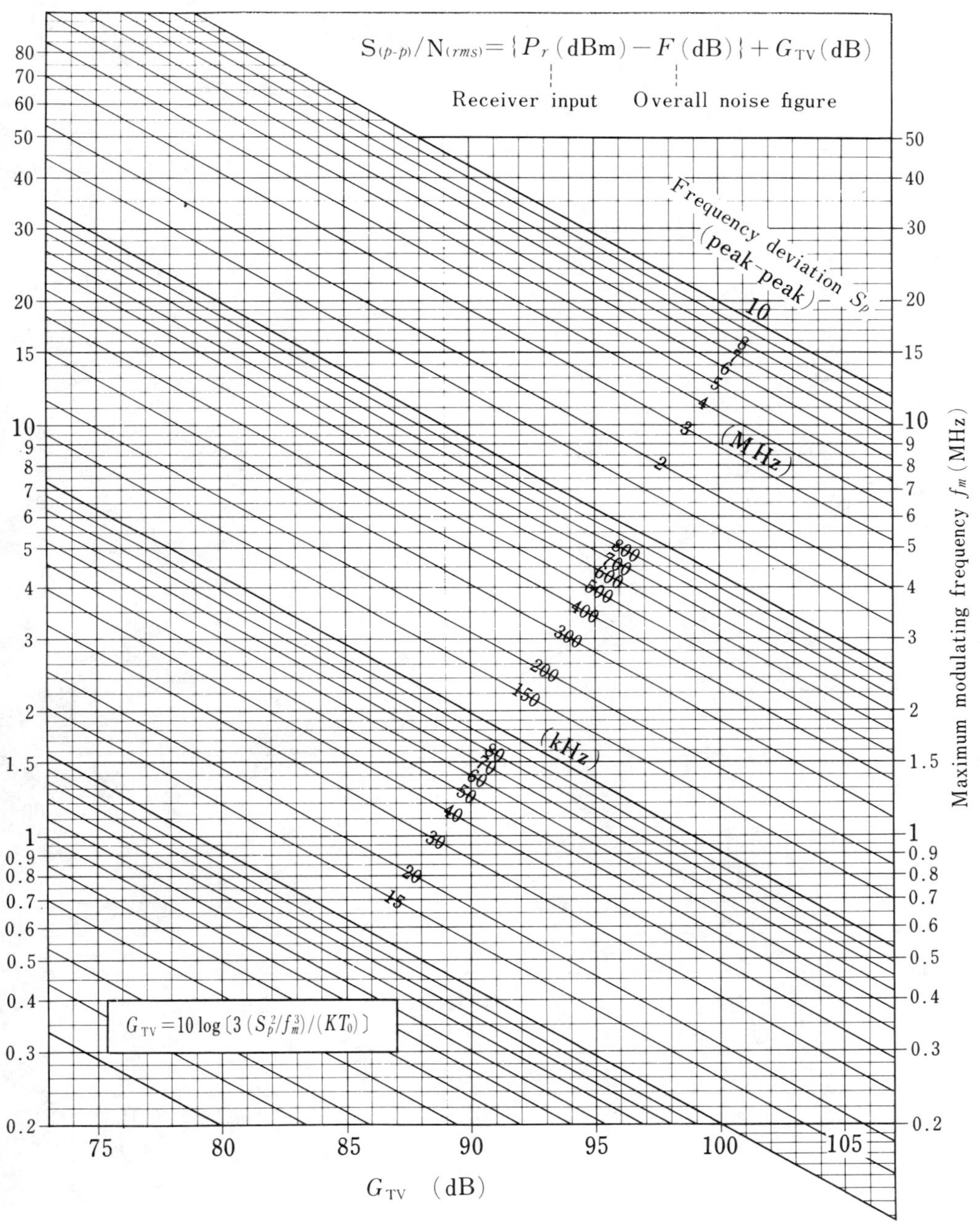

S / N Ratio for (TV Transmission) FM System (2)

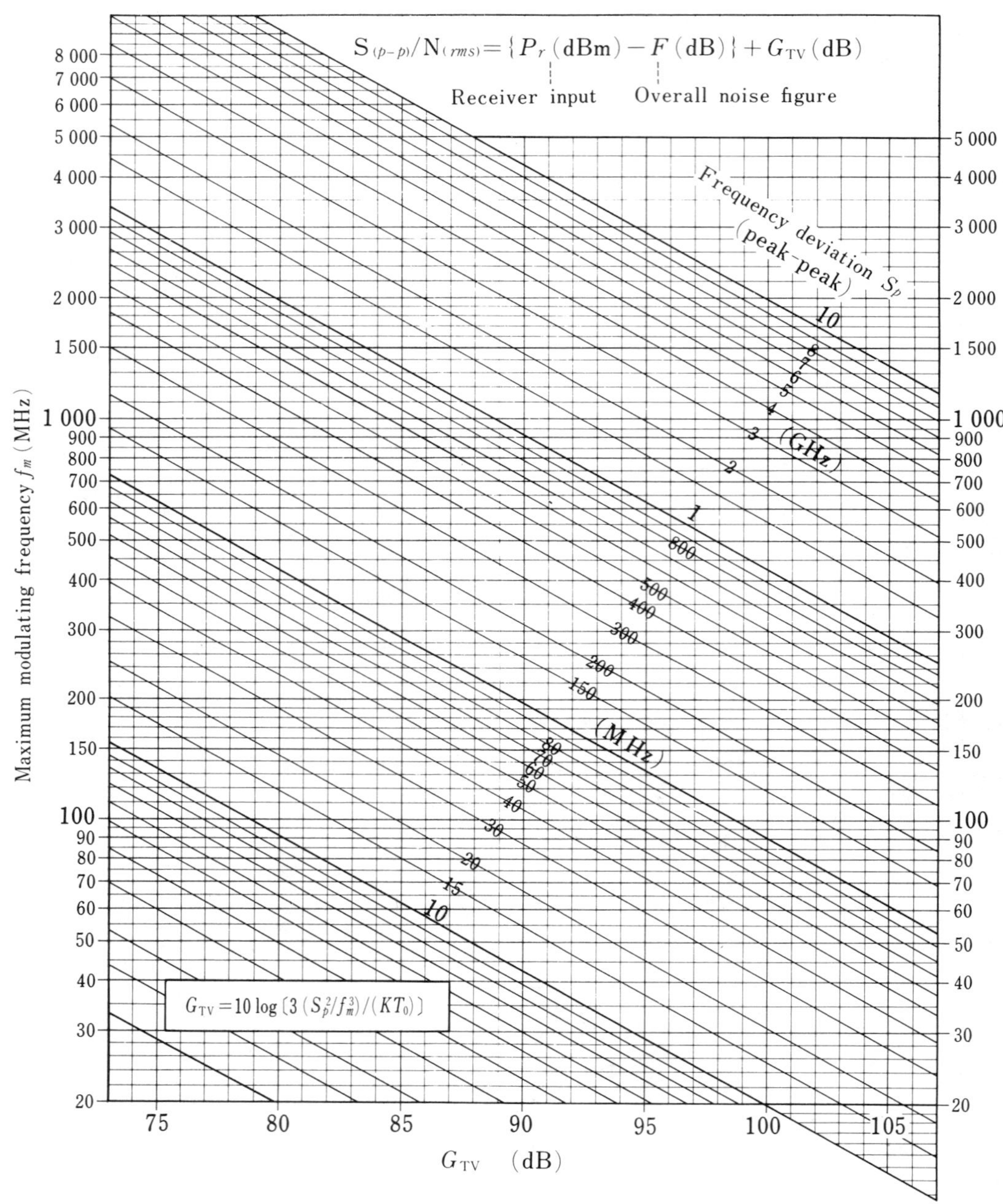

S / N Ratio for (TV Transmission) FM System (3)

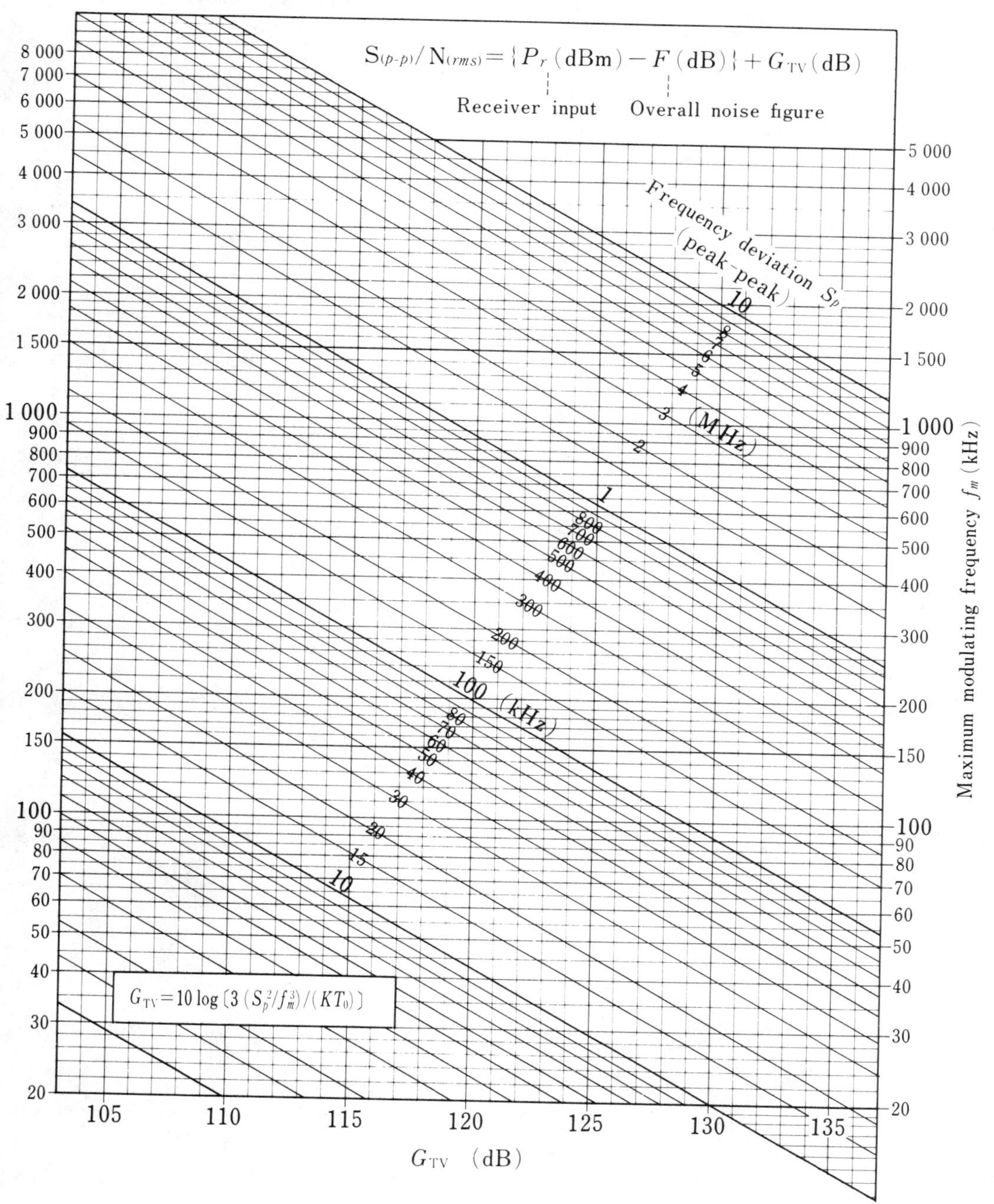

S / N Ratio for (TV Transmission) FM System (4)

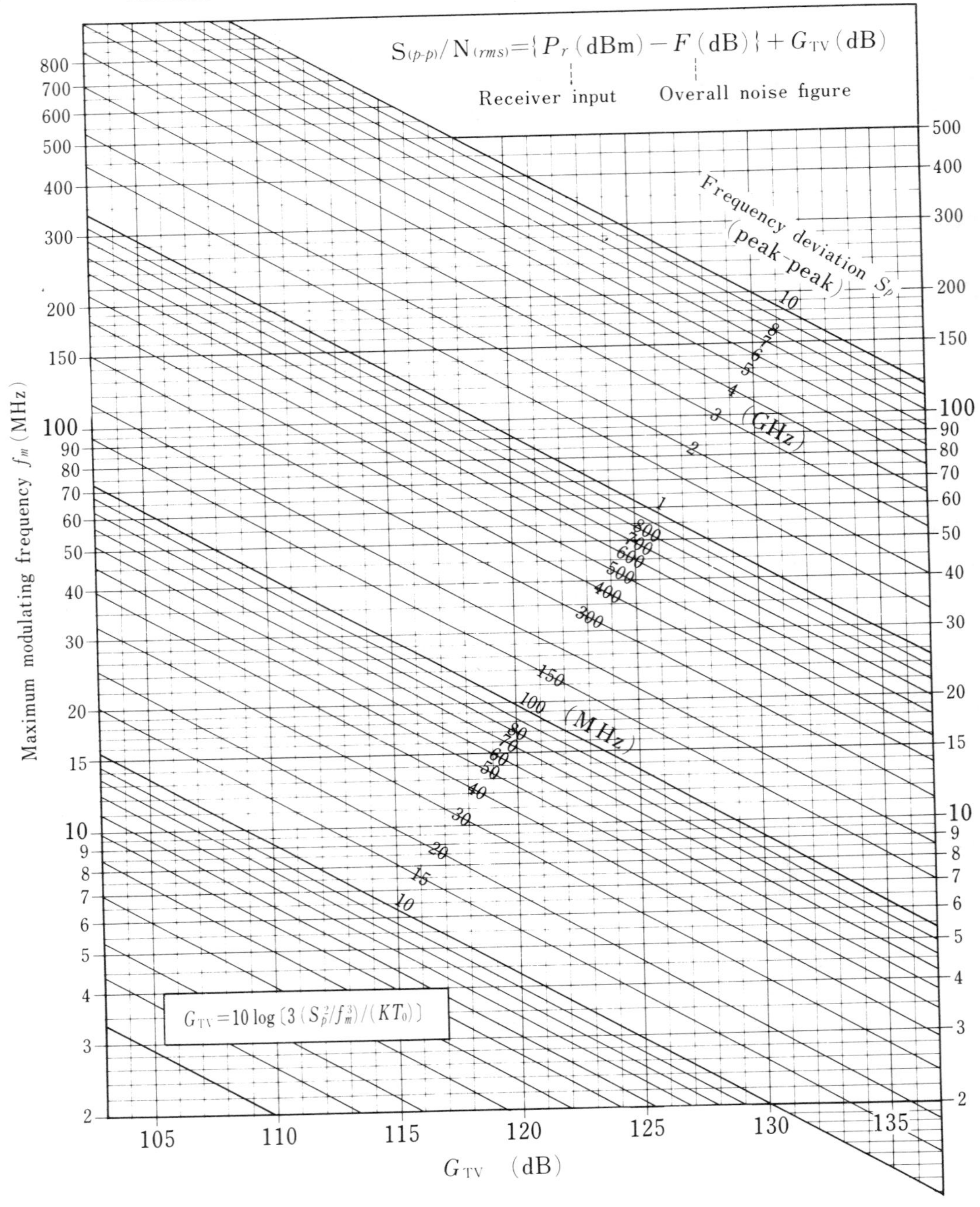

S / N Ratio for Arbitrary Channel of SS-FM System (1)

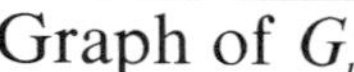

Graph of G_v

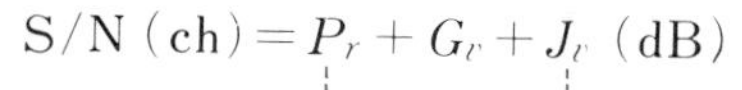

$$S/N\,(ch) = P_r + G_v + J_v \ (dB)$$

Receiver input　　Fig. 11 a, b

Frequency deviation S_0 (ch) (kHz): 3 000, 2 000, 1 000, 900, 800, 700, 600, 500, 400, 300, 200, 100, 90, 80, 70, 50, 40, 30, 20, 10, 9, 8, 7, 6, 5, 4, 3, 2, 1.5, 1, 0.9, 0.8, 0.7, 0.6, 0.5, 0.4, 0.3, 0.2, 0.1

Frequency of interest f_{ch} (MHz): 0.01 – 100

G_v (dB): −45, −40, −35, −30, −25, −20, −15

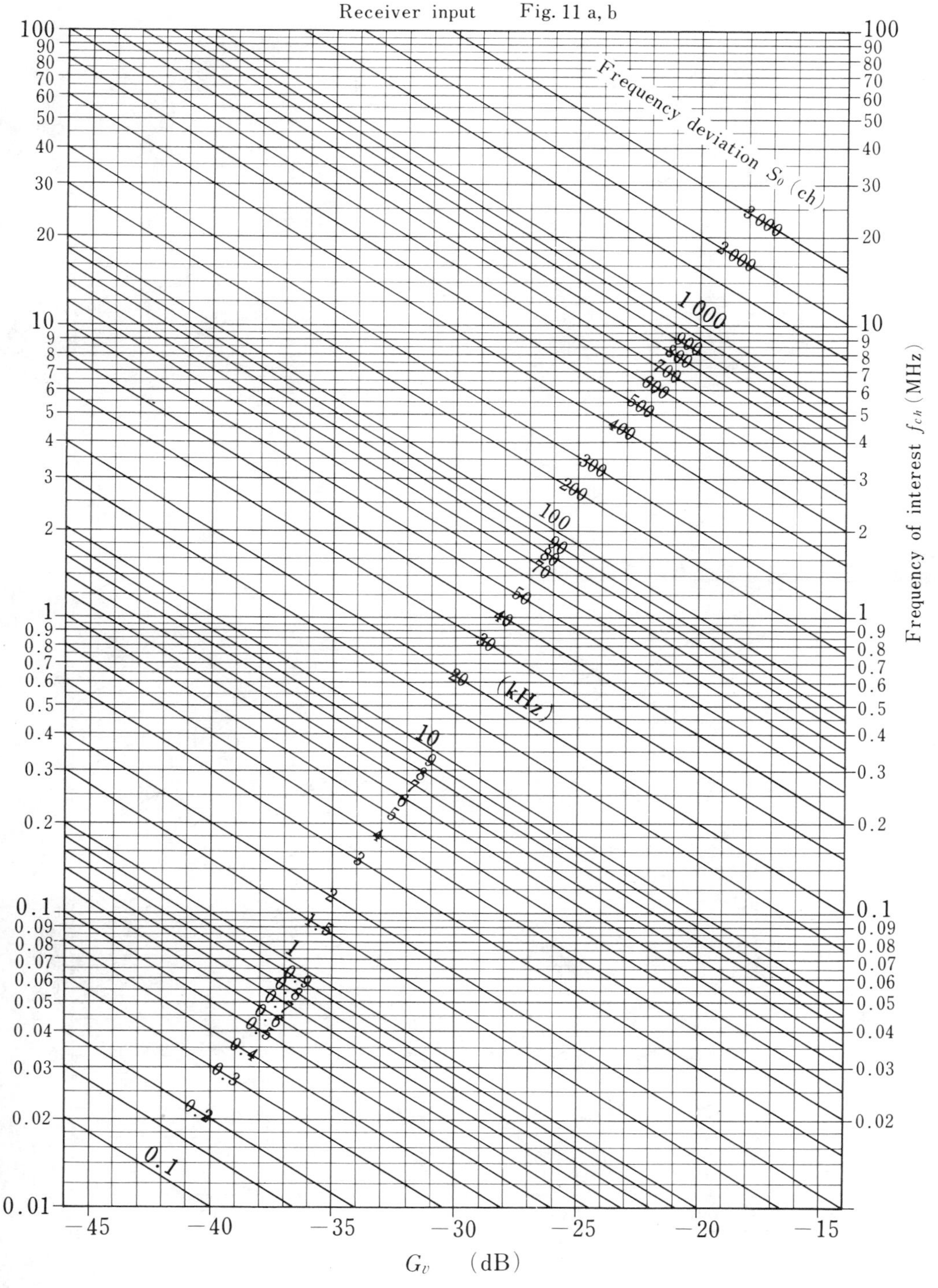

S / N Ratio for Arbitrary Channel of SS-FM System (2)

Graph of G_v

$$S/N\,(ch) = P_r + G_v + J_r \ (dB)$$

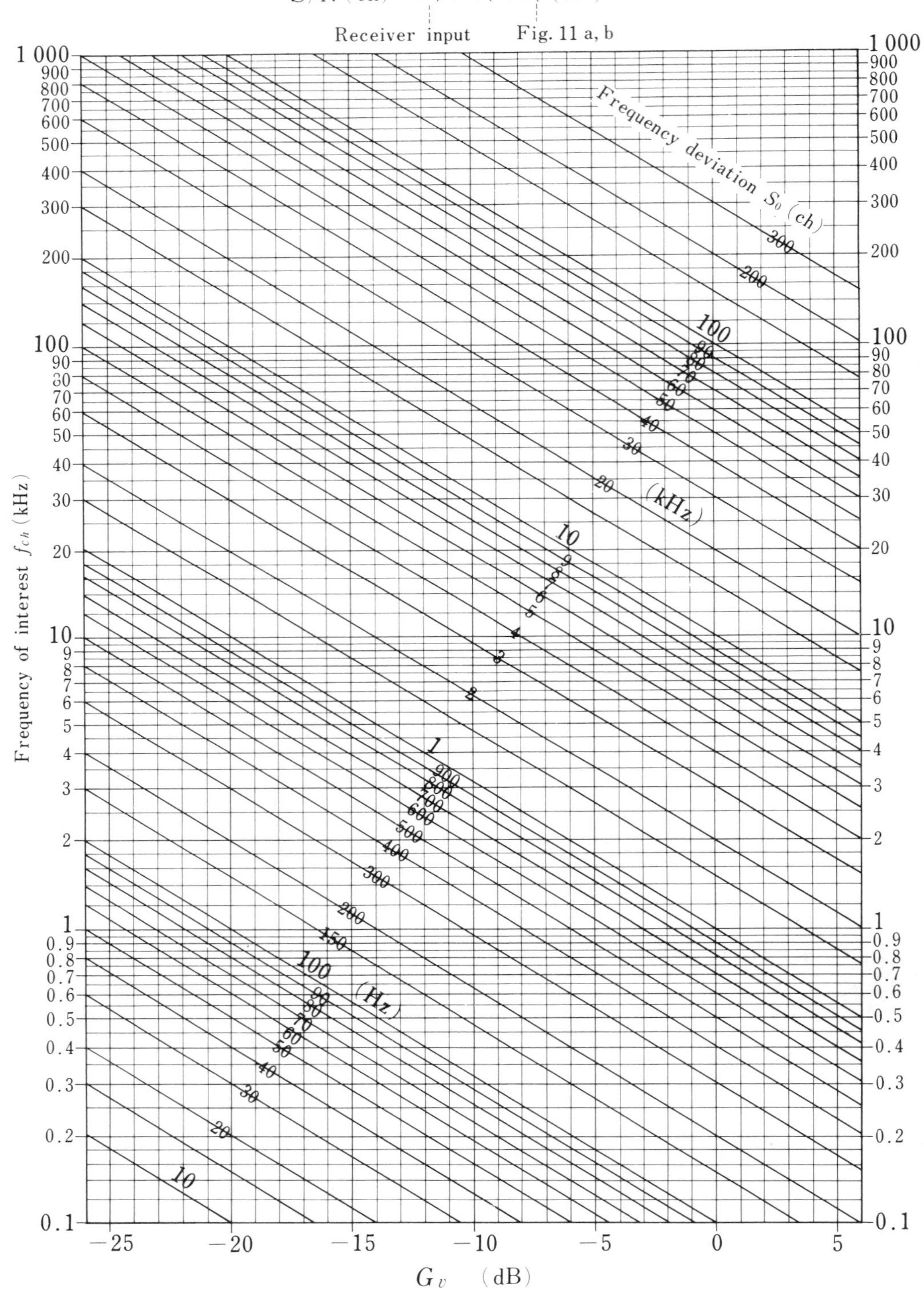

S / N Ratio for Arbitrary Channel of SS-FM System (3)

Graph of G_v

$$S/N\,(ch) = P_r + G_v + J_v\ (dB)$$

Receiver input　　Fig. 11 a, b

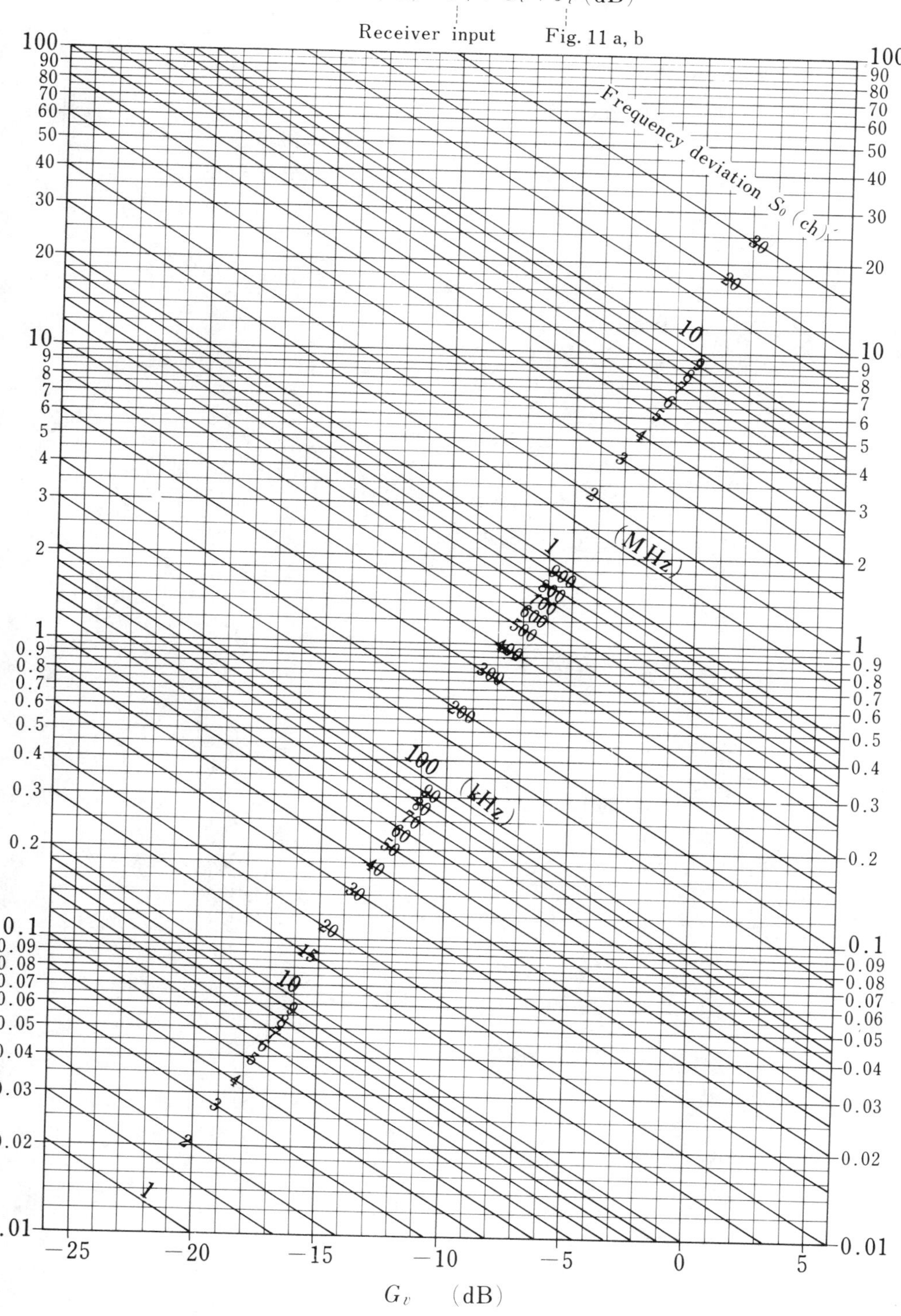

S / N Ratio for Arbitrary Channel of SS-FM System (4)

Graph of J_v

$$S/N\,(ch) = P_r\,(dBm) + G_v + J_v$$

Receiver input — Fig. 10 a,b,c

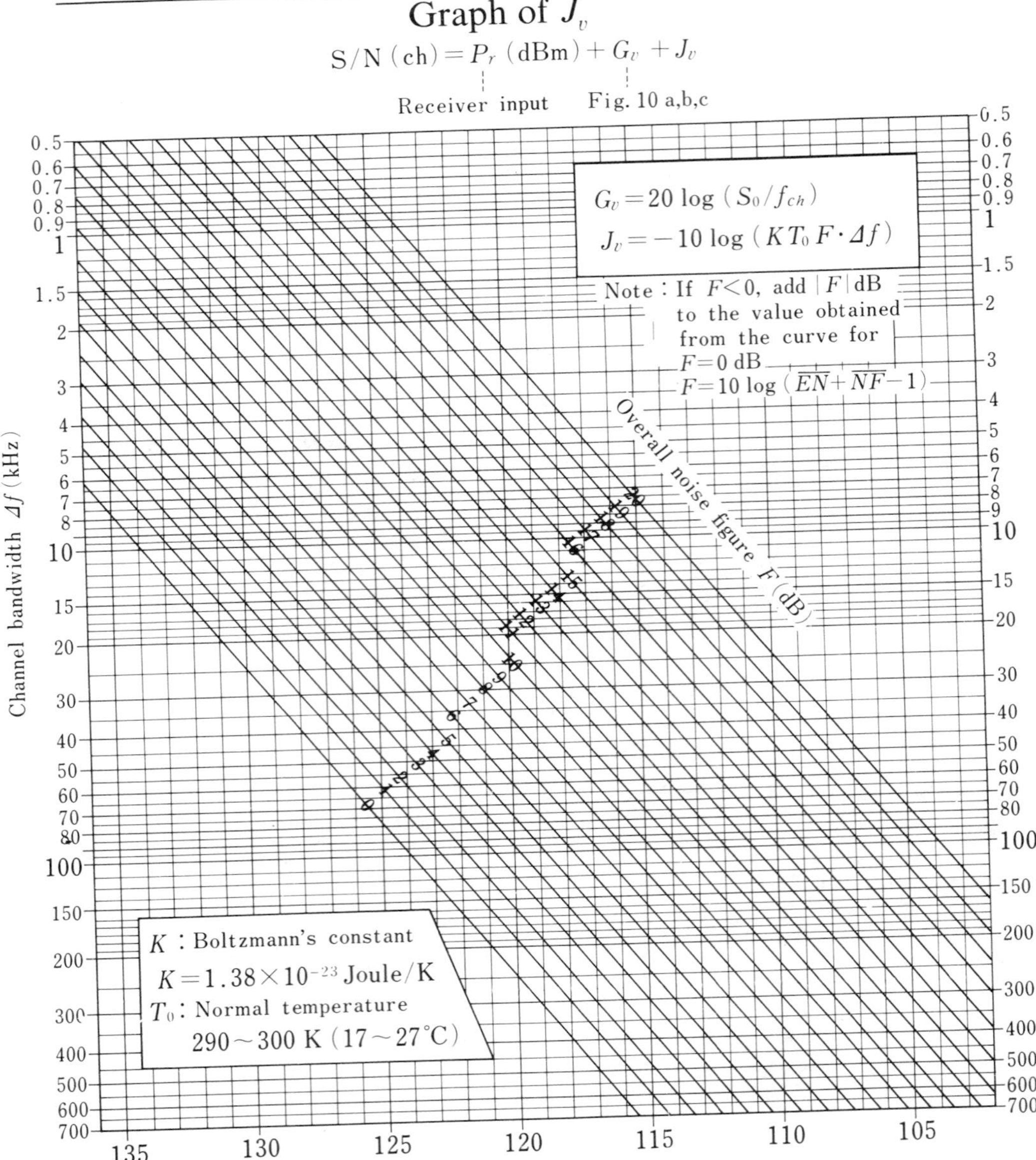

S / N Ratio for Arbitrary Channel of SS-FM System (5)

Graph of J_v

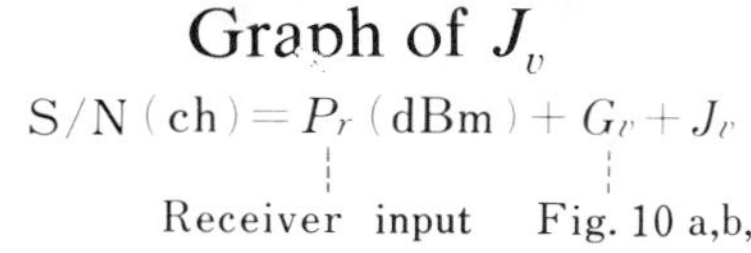

$G_r = 20 \log (S_0/f_{ch})$

$J_r = -10 \log (K T_0 F \cdot \Delta f)$

Note : If $F<0$ add $|F|$ dB to the value obtained from the curve for $F=0$ dB

$F = 10 \log (\overline{EN} + \overline{NF} - 1)$

Overall noise figure F (dB)

0 1 2 3 4 5 6 7 8 9 10 12 13 14 15 17 18 19

K : Boltzmann's constant

$K = 1.38 \times 10^{-23}$ Joule/K

T_0 : Normal temperature

290 ~ 300 K (17 ~ 27°C)

Channel bandwidth Δt (kHz)

5, 6, 7, 8, 9, 10, 15, 20, 30, 40, 50, 60, 70, 80, 90, 100, 150, 200, 300, 400, 500, 600, 700, 800, 1 000, 1 500, 2 000, 3 000, 4 000, 5 000, 6 000, 7 000

155 150 145 140 135 130 125

J_r (dB)

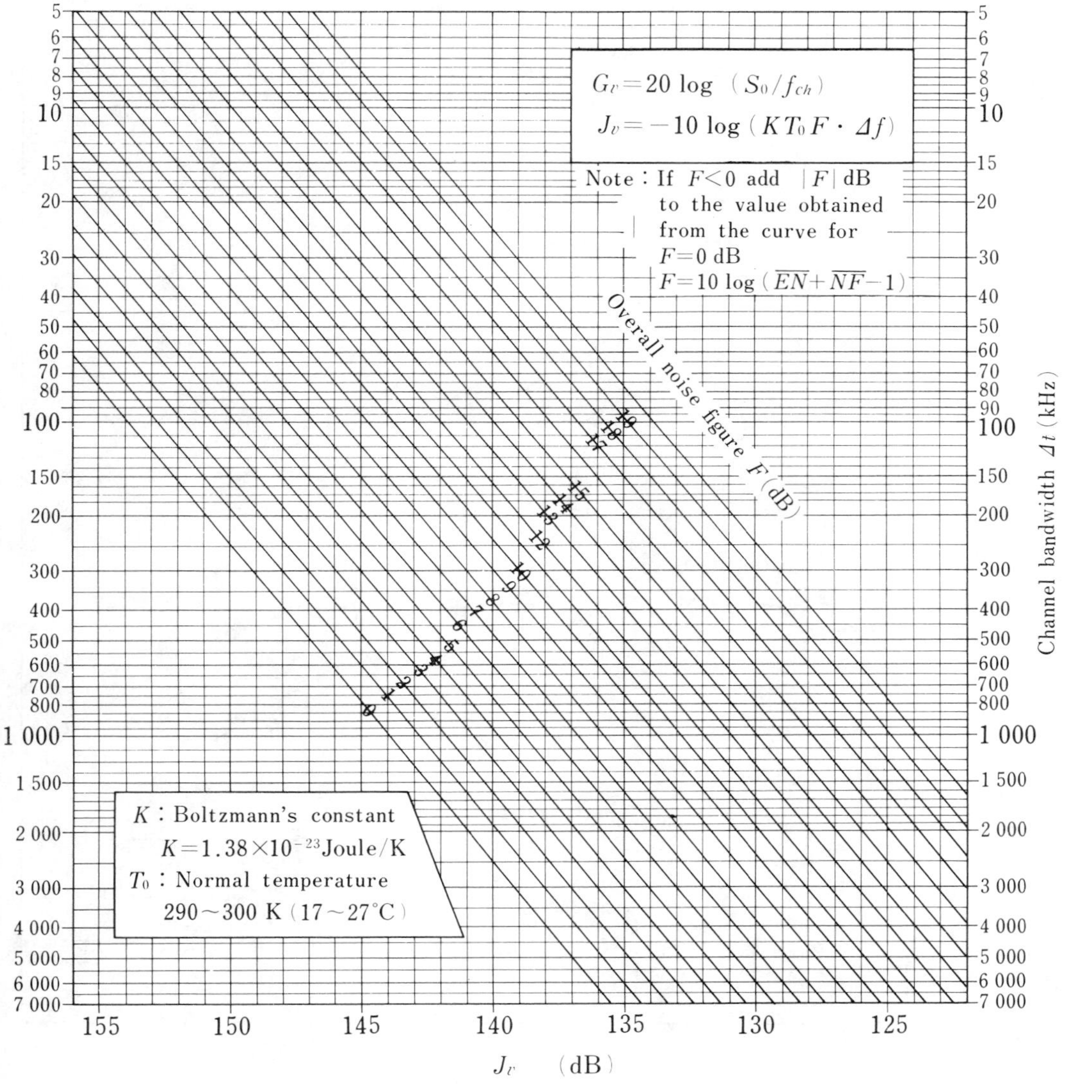

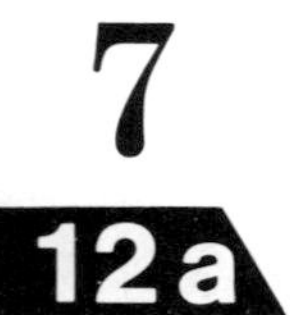

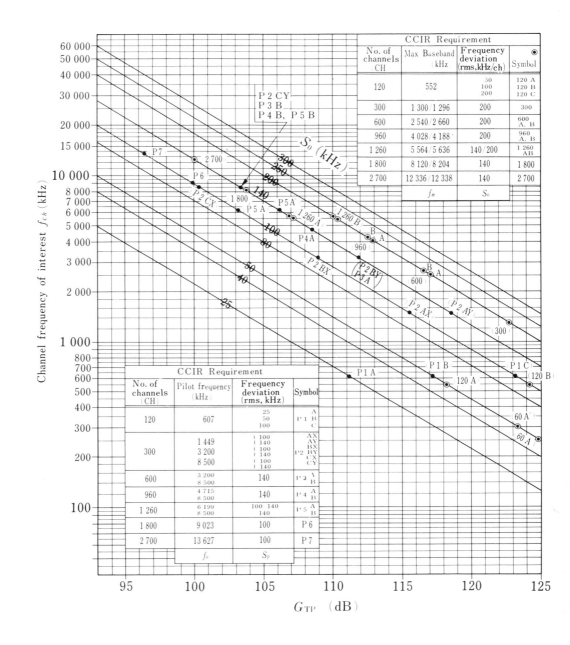

Graph for Quick Estimation of S / N Ratio for SS-FM System (CCIR Recommendation) (1)

CCIR Requirement			
No. of channels (CH)	Max Baseband (kHz)	Frequency deviation (rms, kHz/ch)	⊙ Symbol
120	552	50 100 200	120 A 120 B 120 C
300	1 300/1 296	200	300
600	2 540/2 660	200	600 A, B
960	4 028/4 188	200	960 A, B
1 260	5 564/5 636	140/200	1 260 AB
1 800	8 120/8 204	140	1 800
2 700	12 336/12 338	140	2 700
	f_m	S_0	

CCIR Requirement			
No. of channels (CH)	Pilot frequency (kHz)	Frequency deviation (rms, kHz)	Symbol
120	607	25 50 100	P1 A B C
300	1 449 3 200 8 500	100 140 100 140 100 140	P2 AX AY BX BY CX CY
600	3 200 8 500	140	P3 A B
960	4 715 8 500	140	P4 A B
1 260	6 199 8 500	100 140 140	P5 A B
1 800	9 023	100	P6
2 700	13 627	100	P7
	f_0	S_p	

Channel frequency of interest f_{ch} (kHz)

60 000, 50 000, 40 000, 30 000, 20 000, 15 000, 10 000, 8 000, 7 000, 6 000, 5 000, 4 000, 3 000, 2 000, 1 000, 800, 700, 600, 500, 400, 300, 200, 100

S_0 (kHz): 300, 250, 200, 140, 100, 80, 50, 40, 25

P2 CY, P3 B, P4 B, P5 B, P7, 2 700, P6, P2 CX, 1 800, P5 A, P5 A', 1 260 A, 1 260 B, B, A, P4 A, 960, P2 BX, P2 BY, P3 A, B, A, 600, P2 AX, P2 AY, 300, P1 A, P1 B, 120 A, P1 C, 120 B, 60 A', 60 A

95 100 105 110 115 120 125

G_{TP} (dB)

Graph for Quick Estimation of S / N Ratio for SS-FM (CCIR Recommendation) (2)

Channel's S/N (white noise) Receiver input Overall noise figure

$$S/N = \{P_r\,(\text{dBm}) - F\,(\text{dB})\} + G_{TP}\,(\text{dB})$$

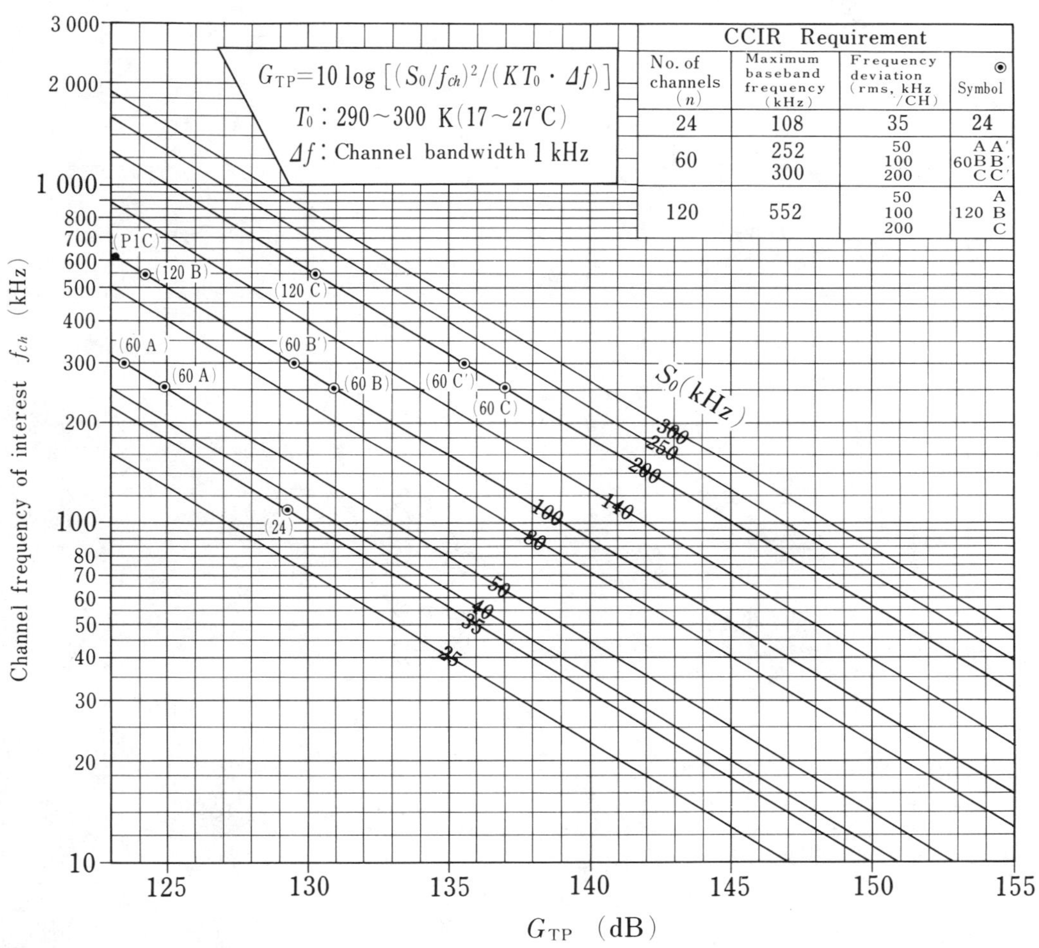

CCIR Requirement

No. of channels (n)	Maximum baseband frequency (kHz)	Frequency deviation (rms, kHz/CH)	⊙ Symbol
24	108	35	24
60	252 300	50 100 200	60 A A' B B' C C'
120	552	50 100 200	120 A B C

Receiver Input versus S / N Ratio (SS-FM System)

No. of channels	n	12 ch
Top channel frequency	f_m	60 kHz
Frequency deviation	S_0	35 rms, kHz/ch
Channel bandwidth	Δf	3.1 kHz
Overall noise figure	F	0～20 dB

12 CH
$S_0 = 35$ kHz

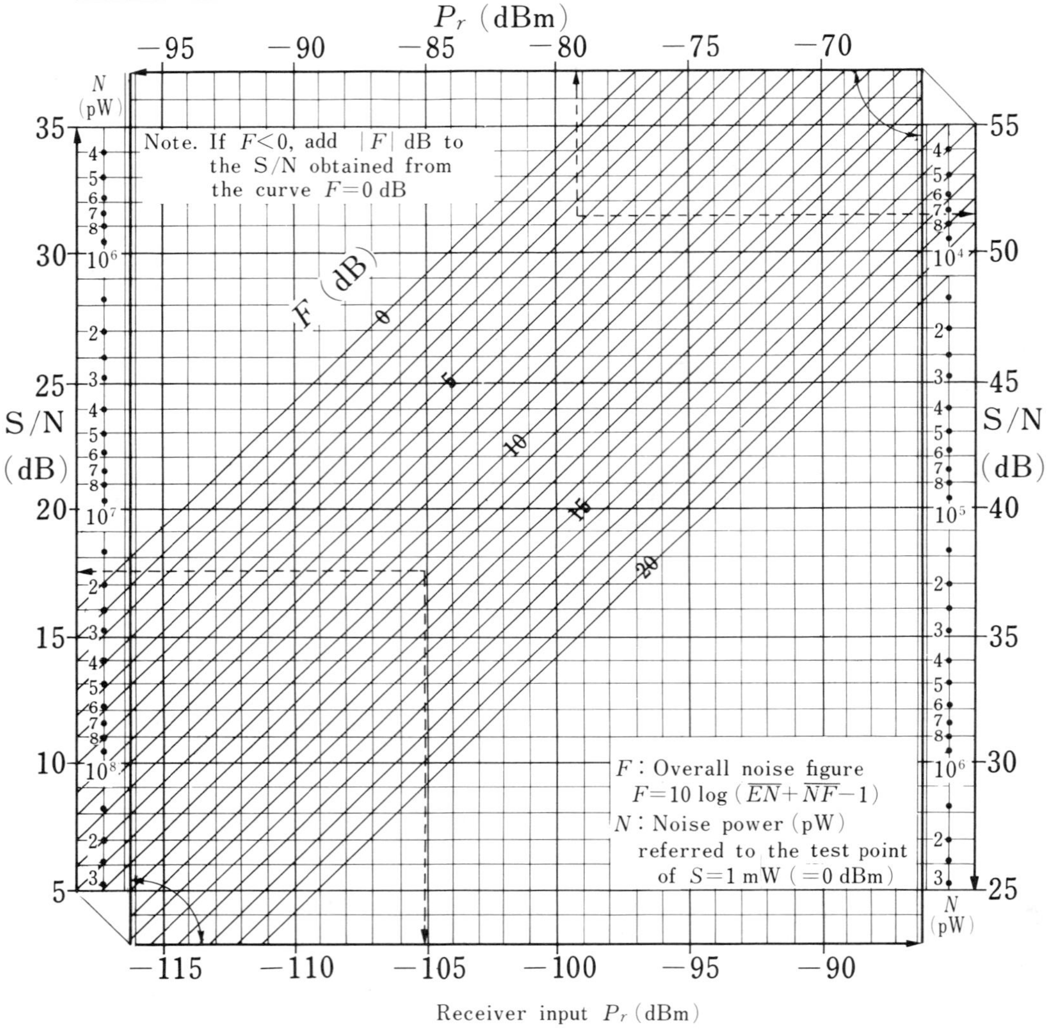

P_r (dBm)

−95 −90 −85 −80 −75 −70

Note. If $F<0$, add $|F|$ dB to the S/N obtained from the curve $F=0$ dB

F (dB)

0 5 10 15 20

S/N (dB): 5 10 15 20 25 30 35 (left); 25 30 35 40 45 50 55 (right)

N (pW): 10^6, 10^7, 10^8 (left); 10^4, 10^5, 10^6 (right)

F : Overall noise figure
$F = 10 \log(\overline{EN} + \overline{NF} - 1)$
N : Noise power (pW) referred to the test point of $S = 1$ mW (= 0 dBm)

−115 −110 −105 −100 −95 −90

Receiver input P_r (dBm)

Receiver Input versus S / N Ratio (SS-FM System)

No. of channels	n	12 ch	**12 CH**
Top channel frequency	f_m	60 kHz	$S_0 = 35$ kHz
Frequency deviation	S_0	35 rms, kHz/ch	
Channel bandwidth	Δf	3.1 kHz	
Overall noise figure	F	0 ~ 20 dB	

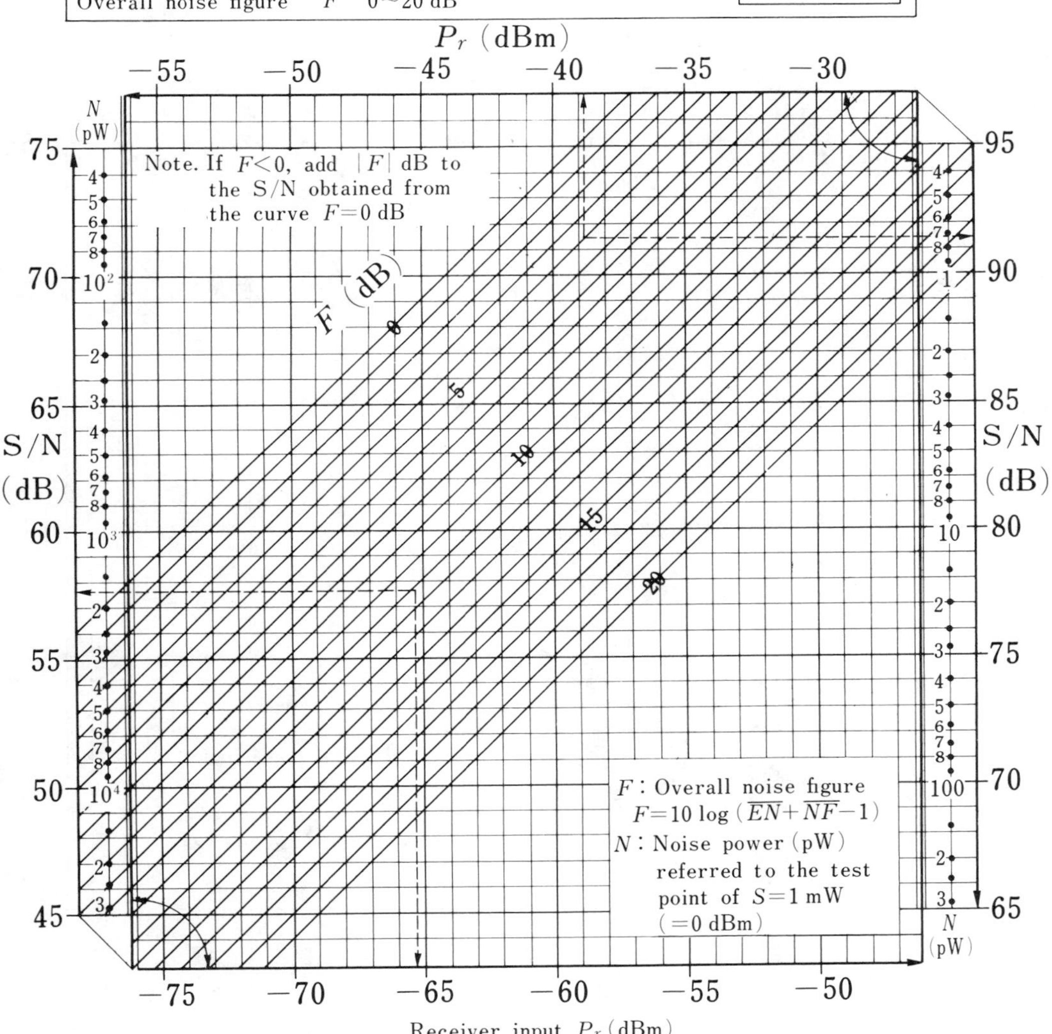

Receiver Input versus S / N Ratio (SS-FM System)

No. of channels	n	24 ch	
Top channel frequency	f_m	108 kHz	
Frequency deviation	S_0	35 rms, kHz/ch	
Channel bandwidth	Δf	3.1 kHz	
Overall noise figure	F	0∼20 dB	

24 CH
$S_0 = 35$ kHz

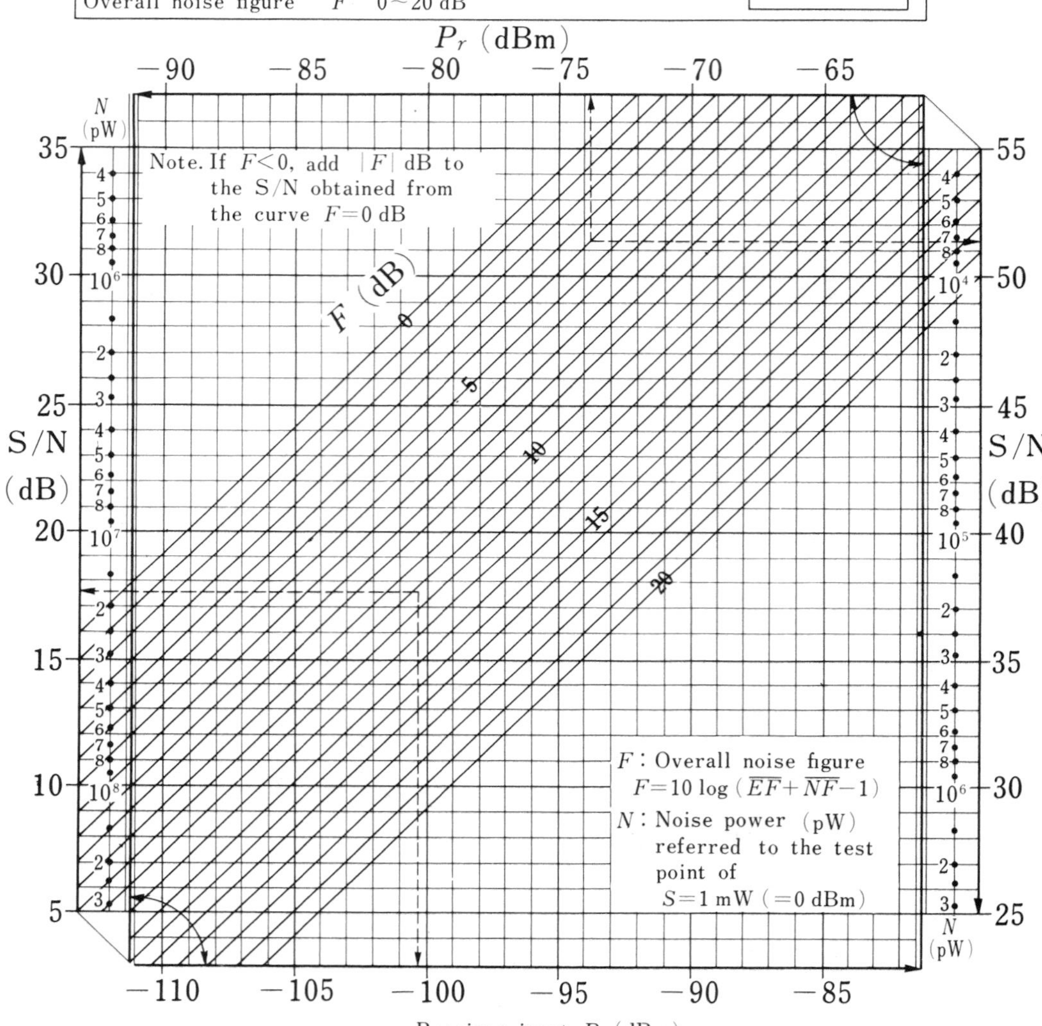

Receiver Input versus S / N Ratio (SS-FM System)

No. of channels	n	24 ch
Top channel frequency	f_m	108 kHz
Frequency deviation	S_0	35 rms, kHz/ch
Channel bandwidth	Δf	3.1 kHz
Overall noise figure	F	0~20 dB

24 CH
$S_0 = 35$ kHz

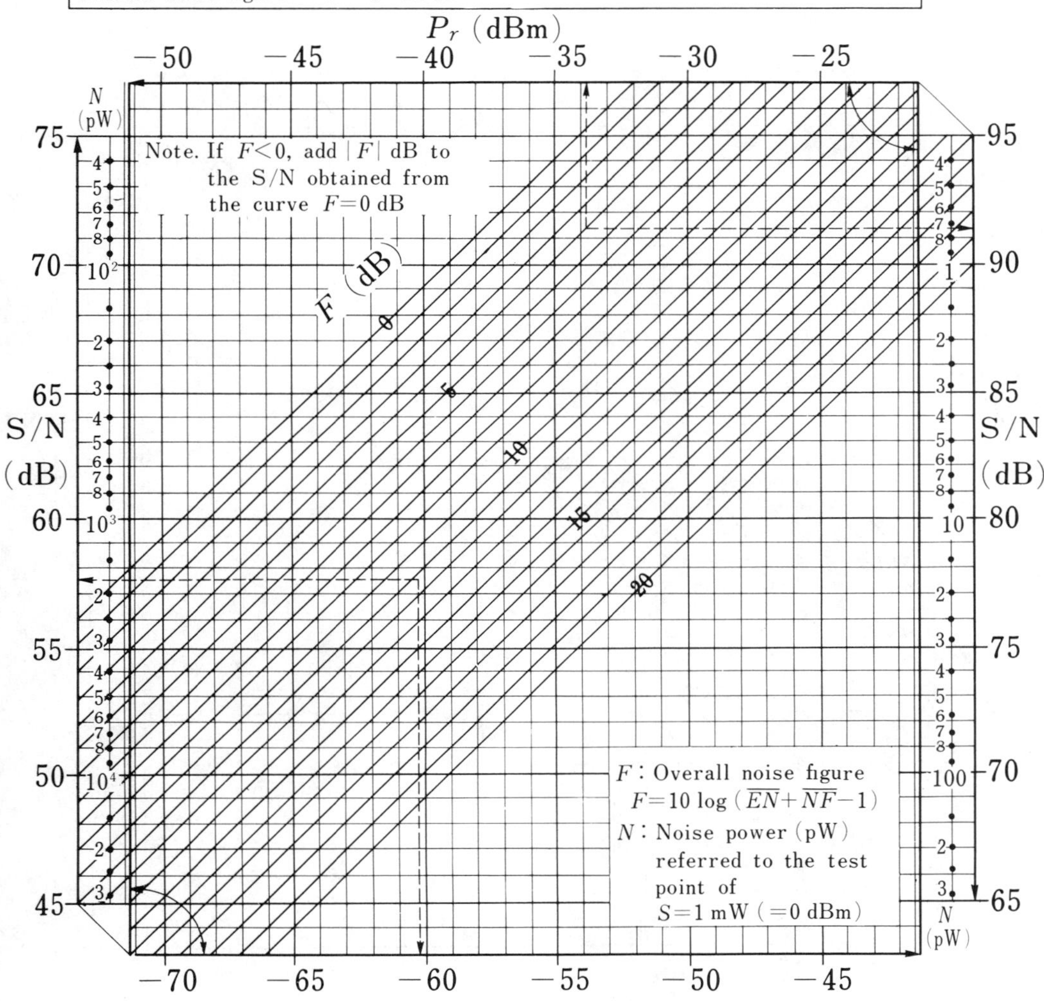

Receiver Input versus S / N Ratio (SS-FM System)

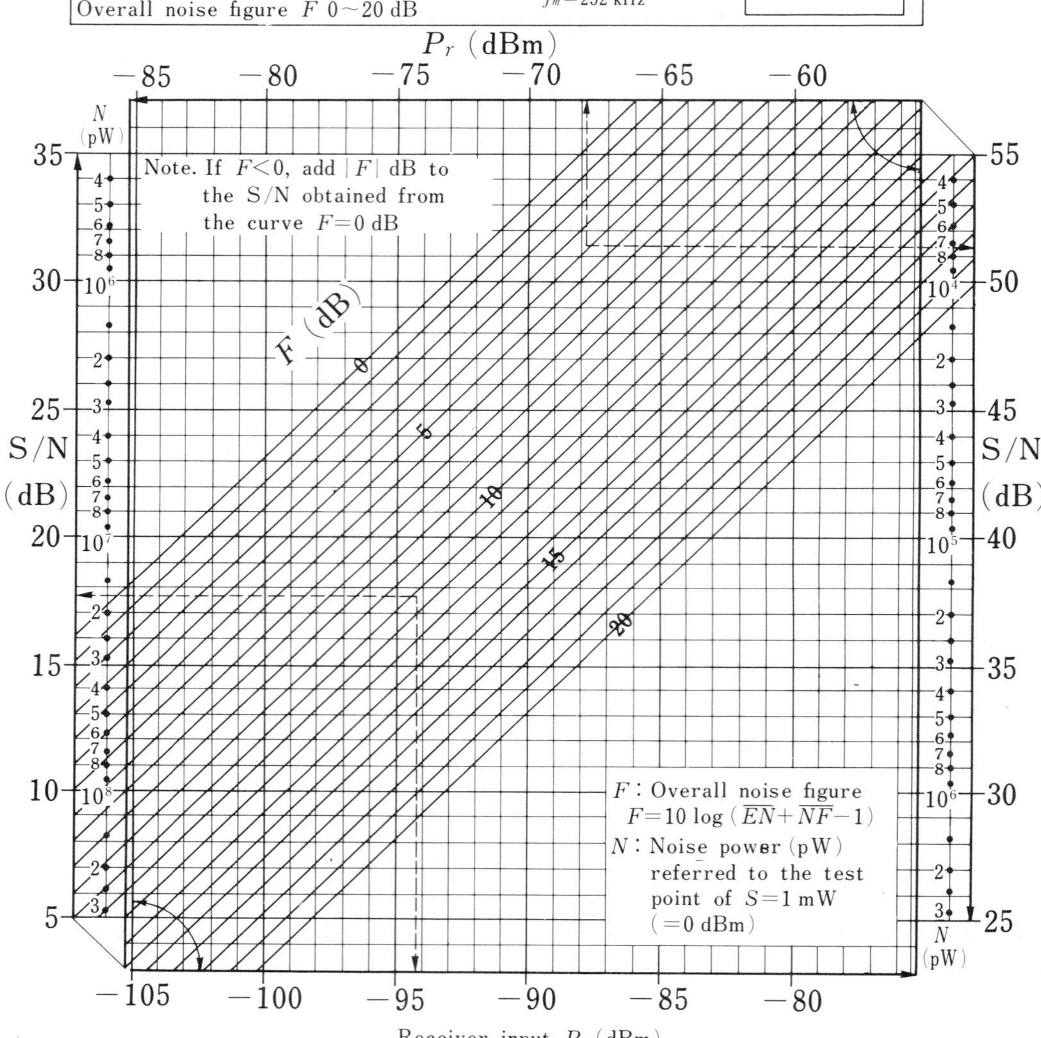
No. of channels n 60 ch
Top channel frequency f_m 300 kHz
Frequency deviation S_0 50 rms, kHz/ch
Channel bandwidth Δ_f 3.1 kHz
Overall noise figure F 0~20 dB
Note. this graph is prepared for f_m=300 kHz, then add +1.5 dB to the obtained value when f_m=252 kHz
60 CH
S_0=50 kHz
P_r (dBm)
−85 −80 −75 −70 −65 −60
N (pW)
Note. If F<0, add |F| dB to the S/N obtained from the curve F=0 dB
F (dB)
0
5
10
15
20
S/N (dB)
35 30 25 20 15 10 5
55 50 45 40 35 30 25
10^6 10^7 10^8
10^4 10^5 10^6
F : Overall noise figure
F=10 log ($\overline{EN}$+$\overline{NF}$−1)
N : Noise power (pW) referred to the test point of S=1 mW (=0 dBm)
−105 −100 −95 −90 −85 −80
Receiver input P_r (dBm)

Receiver Input versus S / N Ratio (SS-FM System)

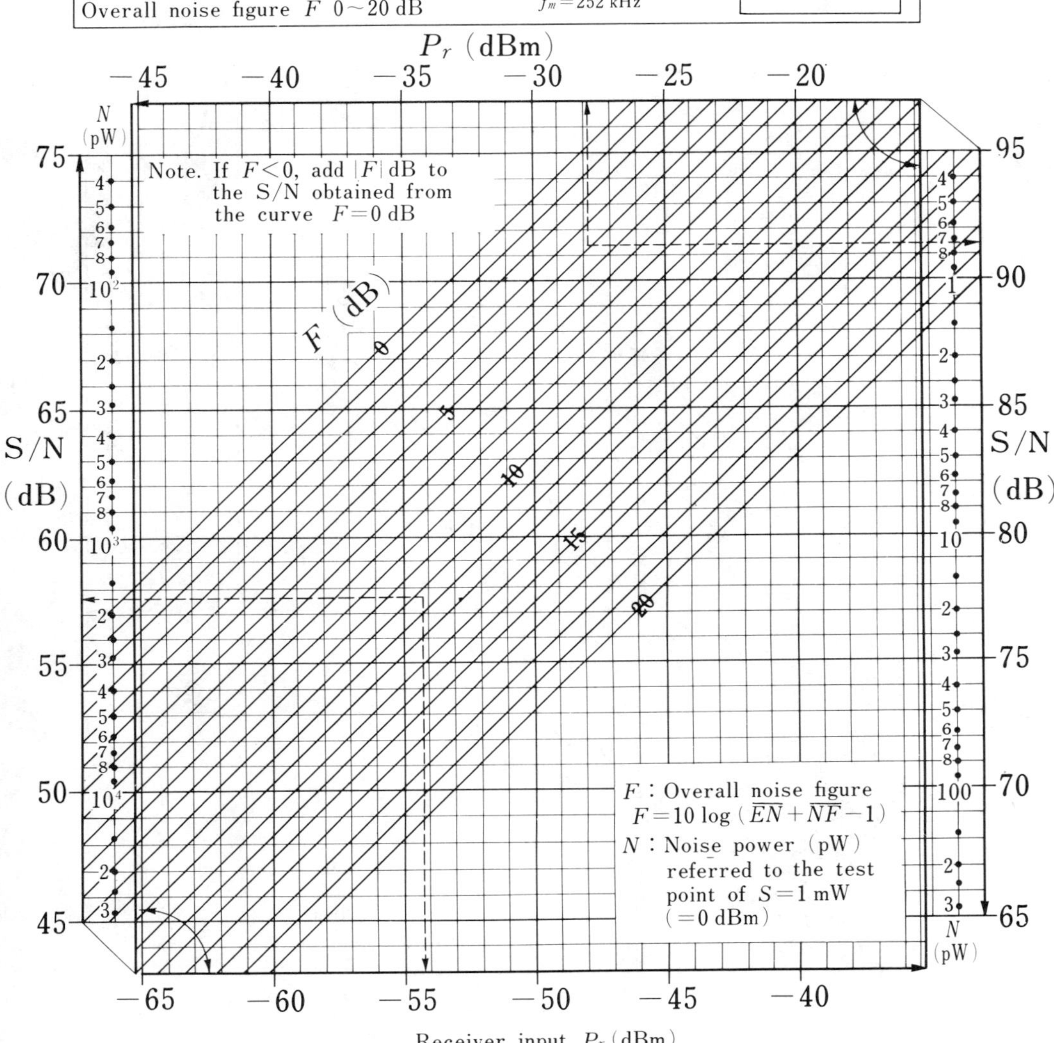
No. of channels n 60 ch
Top channel frequency fm 300 kHz
Frequency deviation S0 50 rms, kHz/ch
Channel bandwidth Δf 3.1 kHz
Overall noise figure F 0~20 dB
Note. this graph is prepared for fm=300 kHz then add +1.5 dB to the obtained value when fm=252 kHz
60 CH
S0=50 kHz
Pr (dBm)
Note. If F<0, add |F| dB to the S/N obtained from the curve F=0 dB
F (dB)
S/N (dB)
N (pW)
F : Overall noise figure
F=10 log (EN+NF−1)
N : Noise power (pW) referred to the test point of S=1 mW (=0 dBm)
Receiver input Pr (dBm)

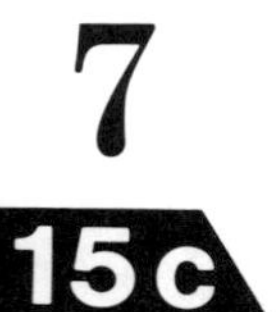

15c

Receiver Input versus S / N Ratio (SS-FM System)

No. of channels n 60 ch
Top channel frequency f_m 300 kHz
Frequency deviation S_0 100 rms, kHz/ch
Channel bandwidth Δf 3.1 kHz
Overall noise figure F 0～20 dB

Note : This graph is prepared for $f_m = 300$ kHz, then add +1.5 dB to the obtained value when $f_m = 252$ kHz

60 CH
$S_0 = 100$ kHz

P_r (dBm)

−90 −85 −80 −75 −70 −65

S/N (dB): 35 30 25 20 15 10 5

S/N (dB): 55 50 45 40 35 30 25

N (pW)

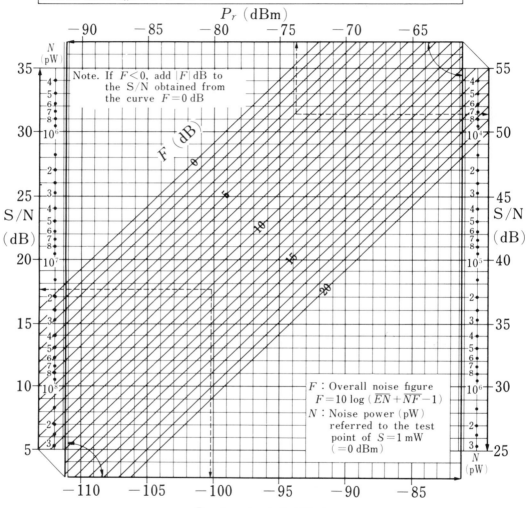

Note. If $F < 0$, add $|F|$ dB to the S/N obtained from the curve $F = 0$ dB

F (dB): 0, 5, 10, 15, 20

F : Overall noise figure
$F = 10 \log(\overline{EN} + \overline{NF} - 1)$
N : Noise power (pW) referred to the test point of $S = 1$ mW (= 0 dBm)

−110 −105 −100 −95 −90 −85

Receiver input P_r (dBm)

Receiver Input versus S / N Ratio (SS-FM System)

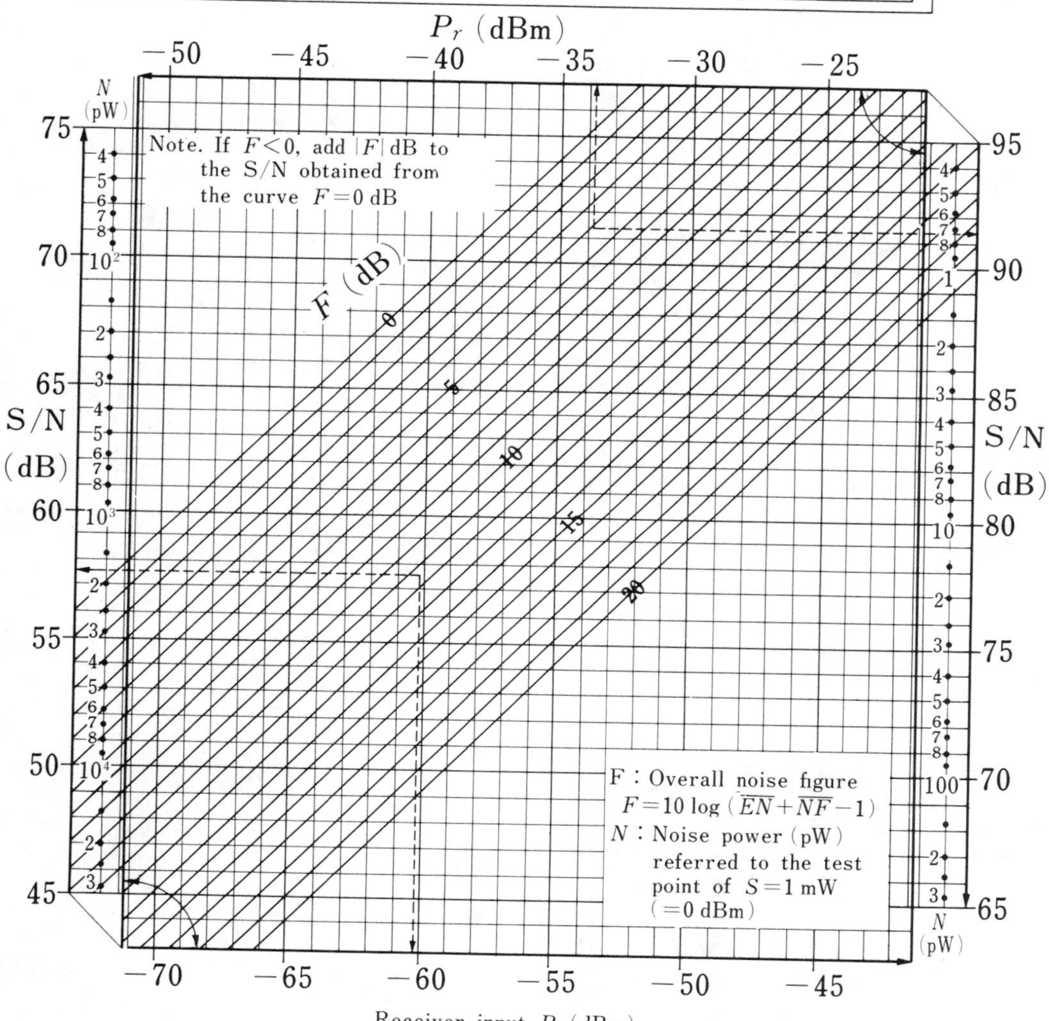
No. of channels n 60 ch
Top channel frequency f_m 300 kHz
Frequency deviation S_0 100 rms, kHz/ch
Channel bandwidth Δf 3.1 kHz
Overall noise figure F 0~20 dB
Note: This graph is prepared for f_m=300 kHz, then add +1.5 dB to the obtained value when f_m=252 kHz
60 CH
S_0=100 kHz
P_r (dBm)
Note. If $F<0$, add $|F|$ dB to the S/N obtained from the curve $F=0$ dB
F (dB)
S/N (dB)
N (pW)
F : Overall noise figure
$F=10\log(\overline{EN}+\overline{NF}-1)$
N : Noise power (pW) referred to the test point of $S=1$ mW (=0 dBm)
Receiver input P_r (dBm)

Receiver Input versus S / N Ratio (SS-FM System)

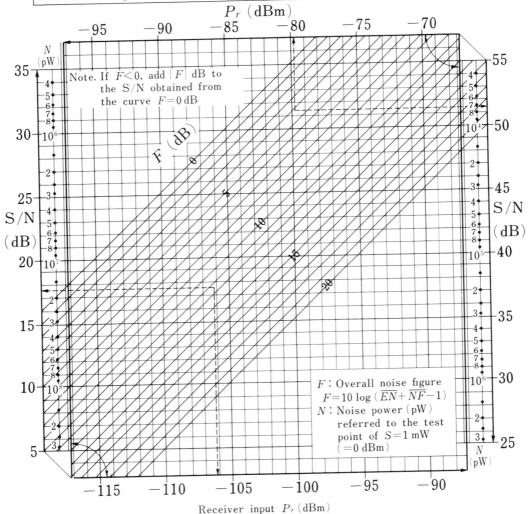

No. of channels n 60 ch
Top channel frequency f_m 300 kHz
Frequency deviation S_0 200 rms, kHz/ch
Channel bandwidth Δf 3.1 kHz
Overall noise figure F 0~20 dB

Note : This graph is prepared for f_m=300 kHz, then add +1.5 dB to the obtained value when f_m=252 kHz

60 CH
S_0=200 kHz

P_r (dBm)

−95 −90 −85 −80 −75 −70

N (pW)

S/N (dB)

35 30 25 20 15 10 5

55 50 45 40 35 30 25

Note. If $F<0$, add $|F|$ dB to the S/N obtained from the curve F=0 dB

F (dB)

0 5 10 15 20

F : Overall noise figure
$F=10 \log (\overline{EN}+\overline{NF}-1)$
N : Noise power (pW) referred to the test point of S=1 mW (=0 dBm)

−115 −110 −105 −100 −95 −90

Receiver input P_r (dBm)

Receiver Input versus S / N Ratio (SS-FM System)

No. of channels n 60 ch Top channel frequency f_m 300 kHz Frequency deviation S_0 200 rms, kHz/ch Channel bandwidth Δf 3.1 kHz Overall noise figure F 0∼20 dB	Note: This graph is prepared for $f=300$ kHz, then add +1.5 dB to the obtained value when $f_m=252$ kHz	60 CH $S_0=200$ kHz

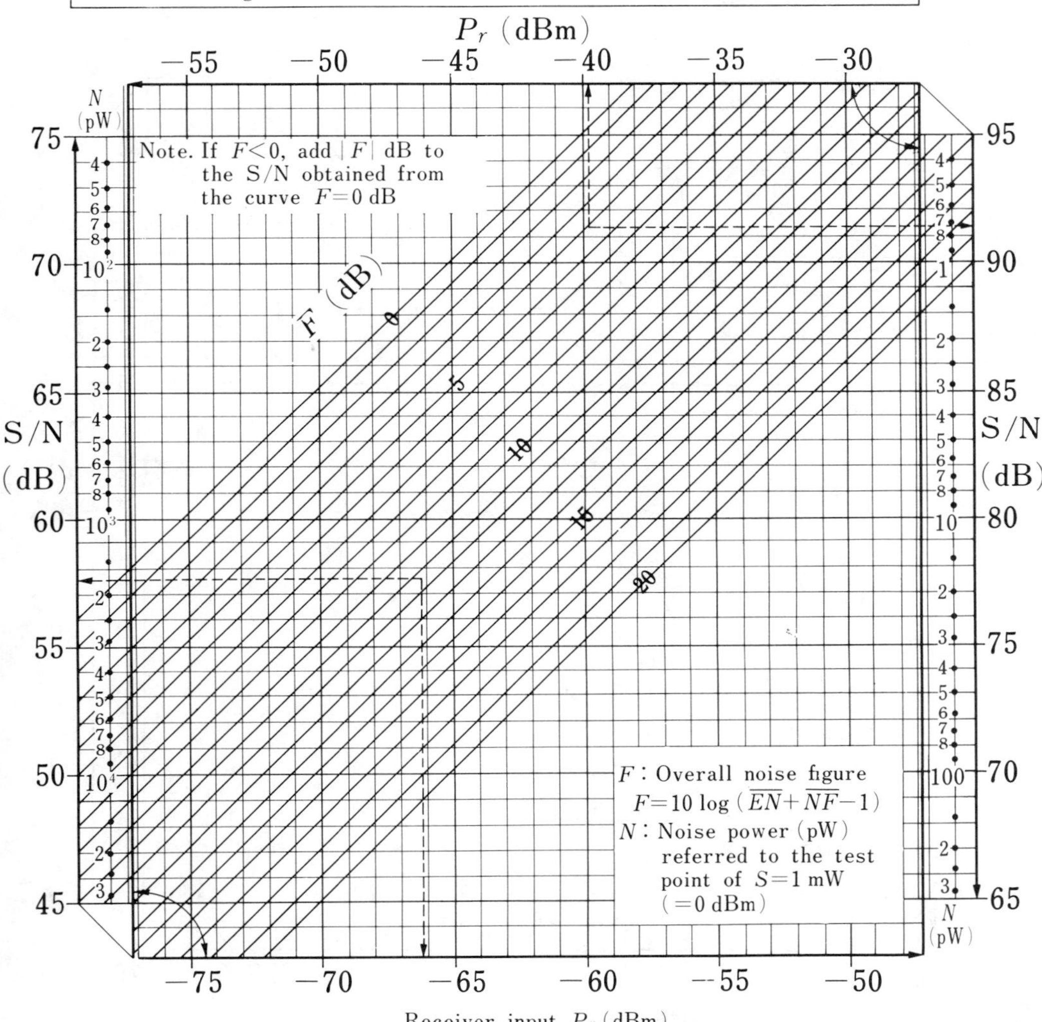

Receiver Input versus S / N Ratio (SS-FM System)

No. of channels	n	120 ch
Top channel frequency	f_m	552 kHz
Frequency deviation	S_0	50 rms, kHz/ch
Channel bandwidth	Δf	3.1 kHz
Overall noise figure	F	0～20 dB

120 CH
$S_0 =$ 50 kHz

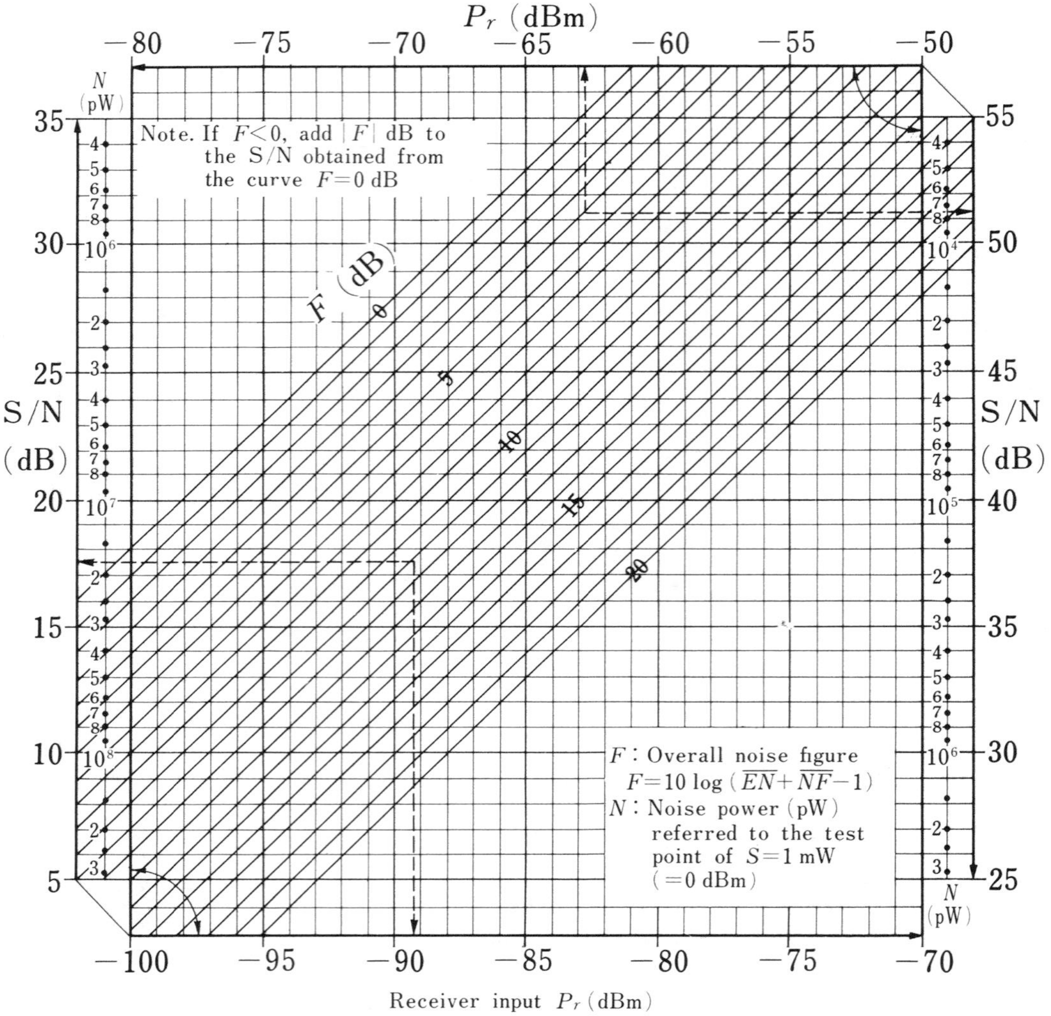

Receiver Input versus S / N Ratio (SS-FM System)

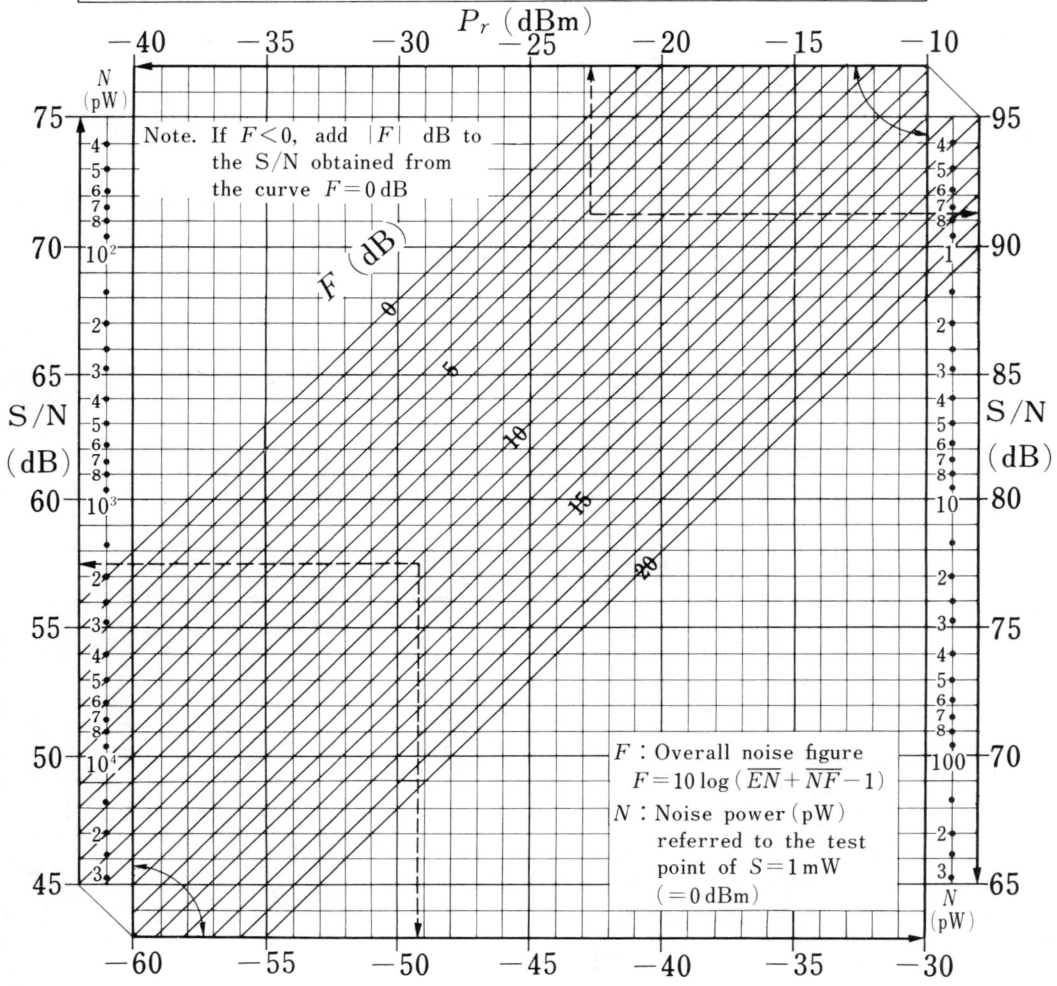
No. of channels n 120 ch
Top channel frequency f_m 552 kHz
Frequency deviation S_0 50 rms, kHz/ch
Channel bandwidth Δf 3.1 kHz
Overall noise figure F 0～20 dB
120 CH
S_0=50 kHz
P_r (dBm)
−40 −35 −30 −25 −20 −15 −10
N (pW)
Note. If $F<0$, add $|F|$ dB to the S/N obtained from the curve $F=0$ dB
F (dB)
0
5
10
15
20
S/N (dB)
75 70 65 60 55 50 45
95 90 85 80 75 70 65
10^2
10^3
10^4
1
10
100
F : Overall noise figure
$F=10\log(\overline{EN}+\overline{NF}-1)$
N : Noise power (pW) referred to the test point of $S=1$ mW (=0 dBm)
−60 −55 −50 −45 −40 −35 −30
Receiver input P_r (dBm)

Receiver Input versus S / N Ratio (SS-FM System)

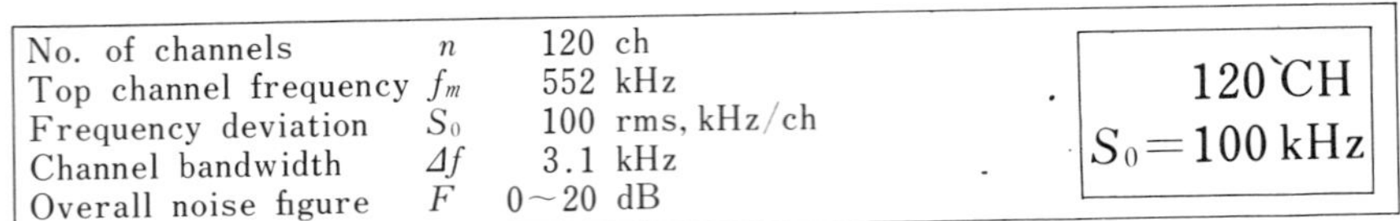

No. of channels	n	120 ch
Top channel frequency	f_m	552 kHz
Frequency deviation	S_0	100 rms, kHz/ch
Channel bandwidth	Δf	3.1 kHz
Overall noise figure	F	0 ~ 20 dB

120CH
$S_0 = 100$ kHz

P_r (dBm)

−85 −80 −75 −70 −65 −60

S/N (dB)

N (pW)

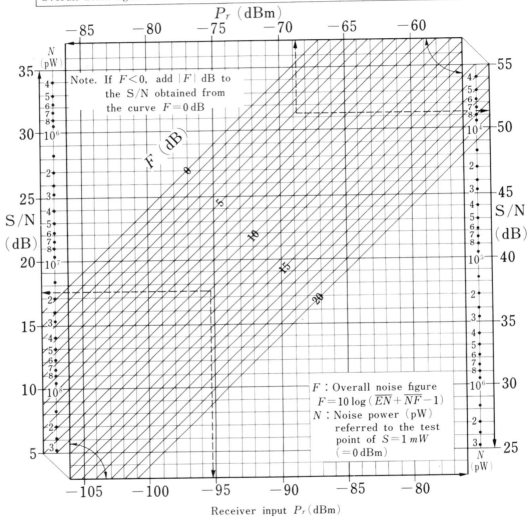

Note. If $F<0$, add $|F|$ dB to the S/N obtained from the curve $F=0$ dB

F (dB)

0 5 10 15 20

F : Overall noise figure
$F = 10 \log(\overline{EN} + \overline{NF} - 1)$
N : Noise power (pW) referred to the test point of $S = 1\,mW$ (= 0 dBm)

−105 −100 −95 −90 −85 −80

Receiver input P_r (dBm)

Receiver Input versus S / N Ratio (SS-FM System)

No. of channels	n	120 ch
Top channel frequency	f_m	552 kHz
Frequency deviation	S_0	100 rms, kHz/ch
Channel bandwidth	Δf	3.1 kHz
Overall noise figure	F	0~20 dB

120 CH
$S_0 = 100$ kHz

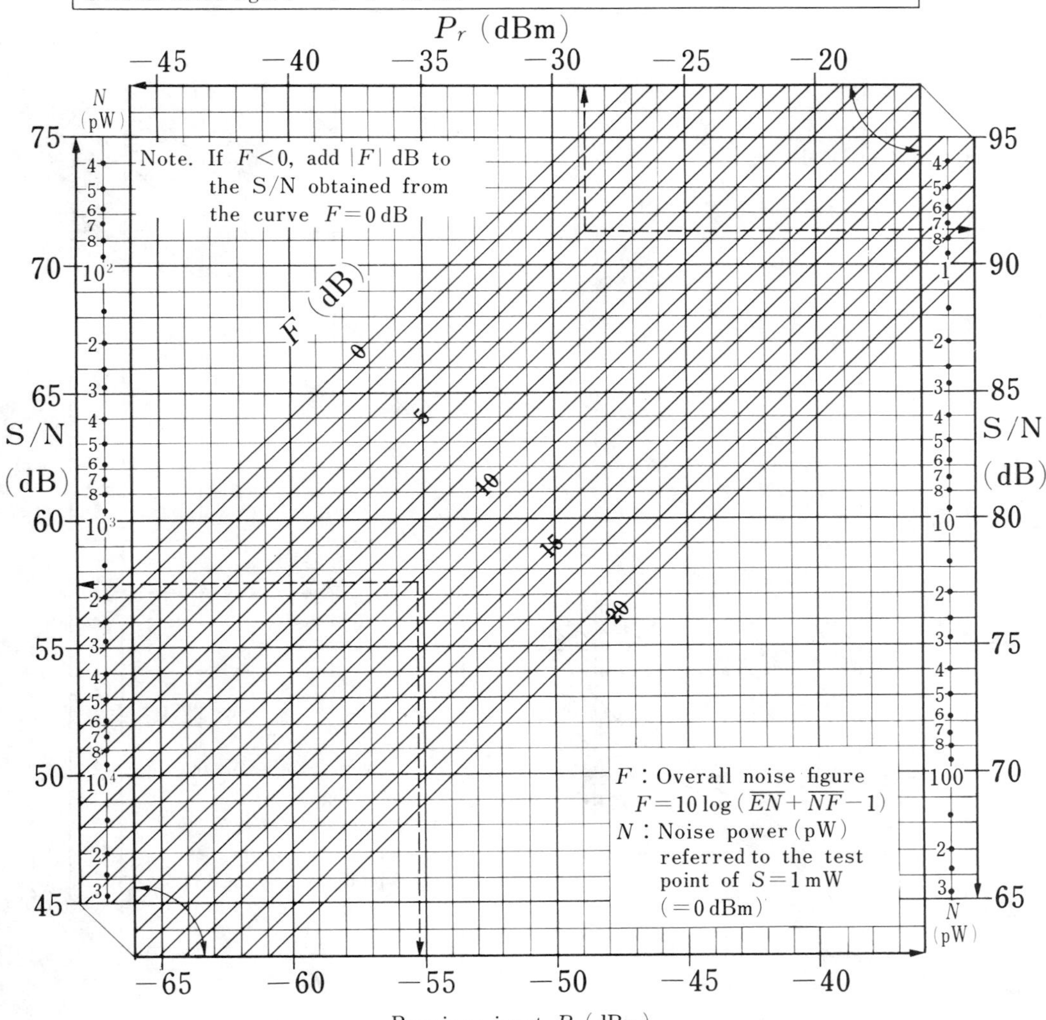

Receiver Input versus S / N Ratio (SS-FM System)

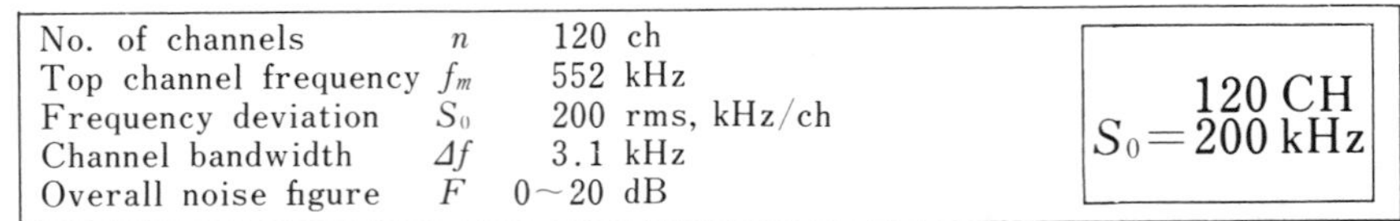

No. of channels	n	120 ch
Top channel frequency	f_m	552 kHz
Frequency deviation	S_0	200 rms, kHz/ch
Channel bandwidth	Δf	3.1 kHz
Overall noise figure	F	0~20 dB

120 CH
$S_0 = 200$ kHz

P_r (dBm)

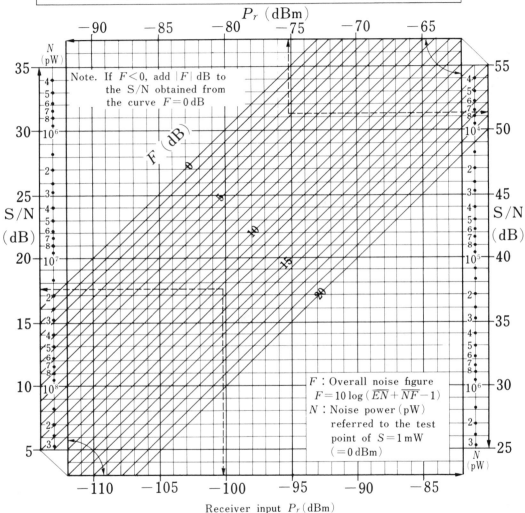

Note. If $F<0$, add $|F|$ dB to the S/N obtained from the curve $F=0$ dB

F (dB)

F : Overall noise figure
$F = 10 \log(\overline{EN} + \overline{NF} - 1)$
N : Noise power (pW) referred to the test point of $S = 1$ mW (= 0 dBm)

S/N (dB)

N (pW)

Receiver input P_r (dBm)

Receiver Input versus S / N Ratio (SS-FM System)

No. of channels	n	120 ch	
Top channel frequency	f_m	552 kHz	
Frequency deviation	S_0	200 rms, kHz/ch	
Channel bandwidth	Δf	3.1 kHz	
Overall noise figure	F	0~20 dB	

120 CH
$S_0 = 200$ kHz

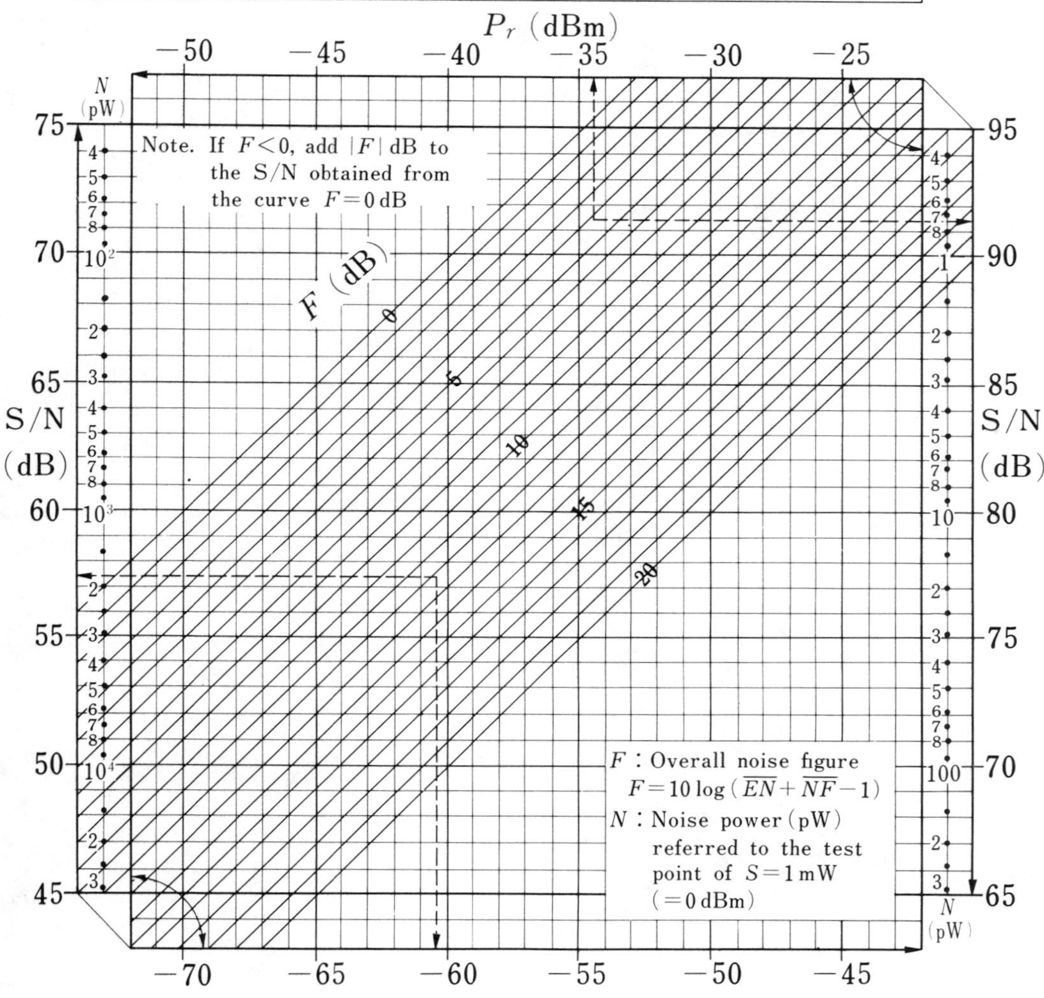

Receiver Input versus S / N Ratio (SS-FM System)

No. of channels	n	120 ch	120 CH
Top channel frequency	f_m	552 kHz	$S_0=280$ kHz
Frequency deviation	S_0	280 rms, kHz/ch	
Channel bandwidth	Δf	3.1 kHz	
Overall noise figure	F	0～20 dB	

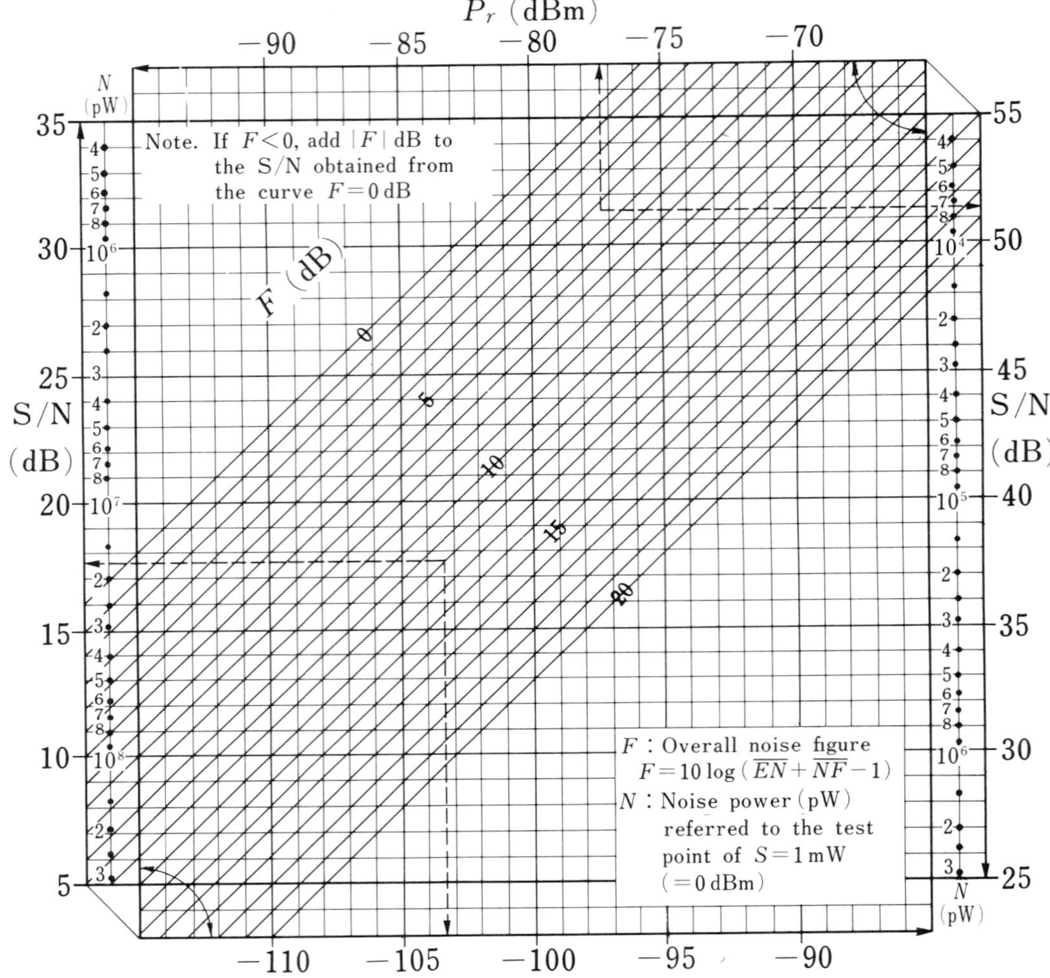

Receiver input P_r (dBm)

Receiver Input versus S / N Ratio (SS-FM System)

No. of channels	n	120 ch
Top channel frequency	f_m	552 kHz
Frequency deviation	S_0	280 rms, kHz/ch
Channel bandwidth	Δf	3.1 kHz
Overall noise figure	F	0~20 dB

120 CH
$S_0 = 280$ kHz

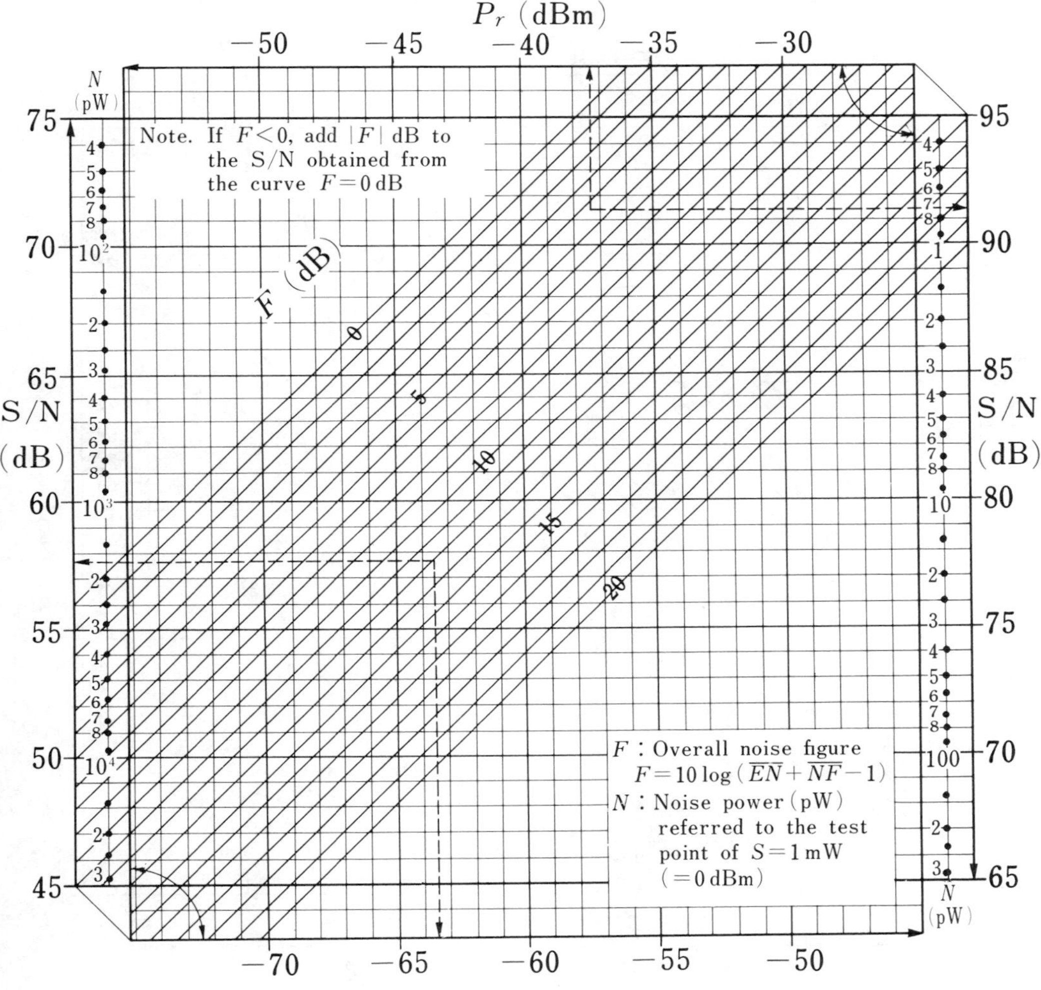

Receiver Input versus S / N Ratio (SS-FM System)

No. of channels	n	240	ch
Top channel frequency	f_m	1 054	kHz
Frequency deviation	S_0	200	rms, kHz/ch
Channel bandwidth	Δf	3.1	kHz
Overall noise figure	F	0 ~ 20	dB

240 CH
$S_0 = 200$ kHz

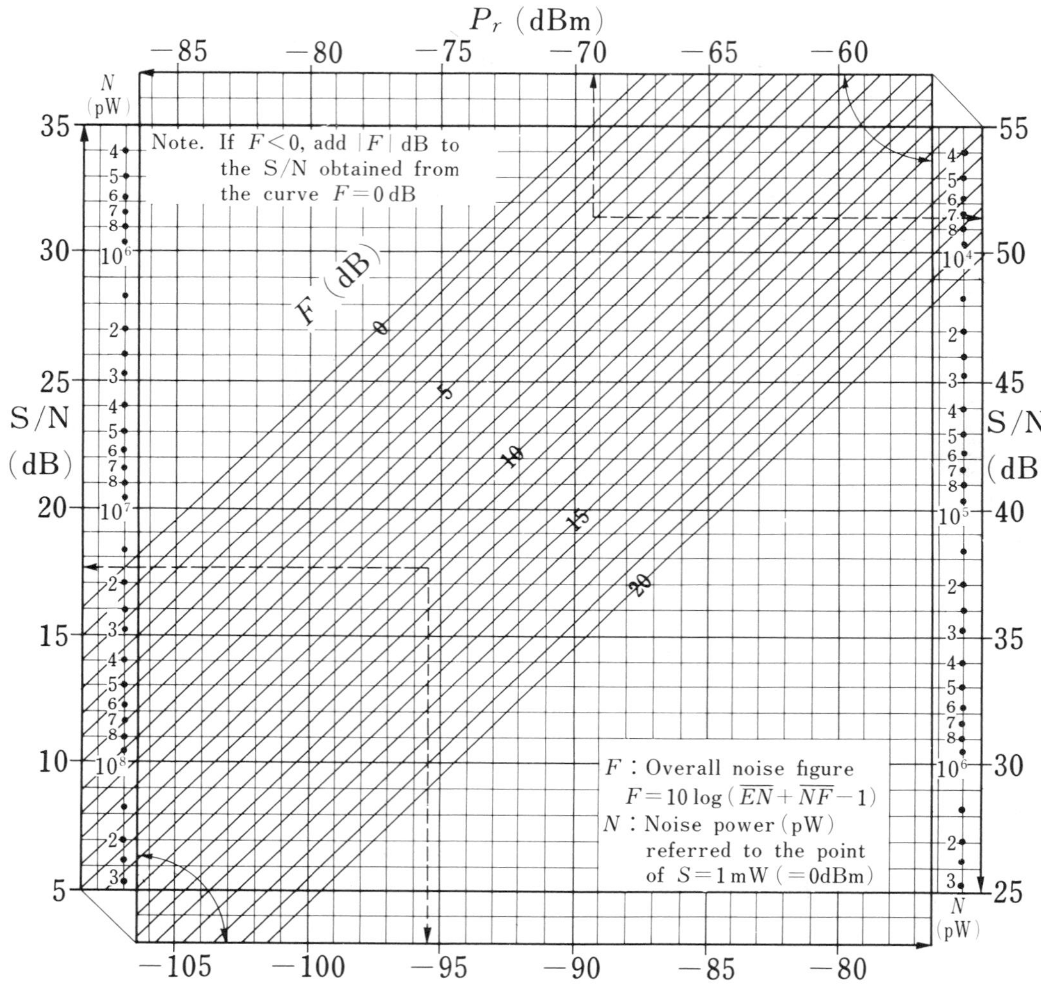

Receiver input P_r (dBm)

Receiver Input versus S / N Ratio (SS-FM System)

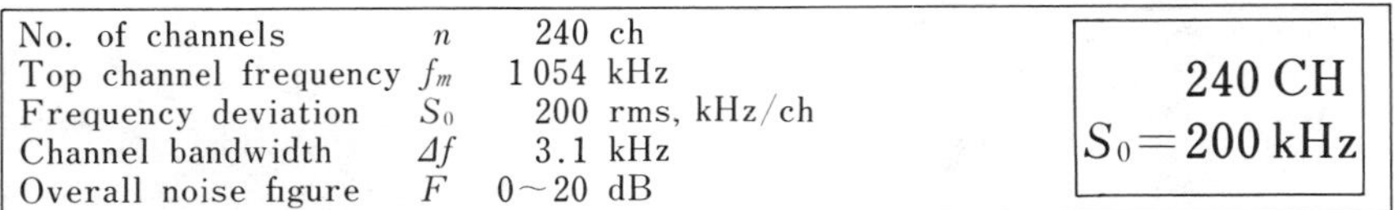

No. of channels	n	240	ch
Top channel frequency	f_m	1 054	kHz
Frequency deviation	S_0	200	rms, kHz/ch
Channel bandwidth	Δf	3.1	kHz
Overall noise figure	F	0～20	dB

240 CH
$S_0 = 200$ kHz

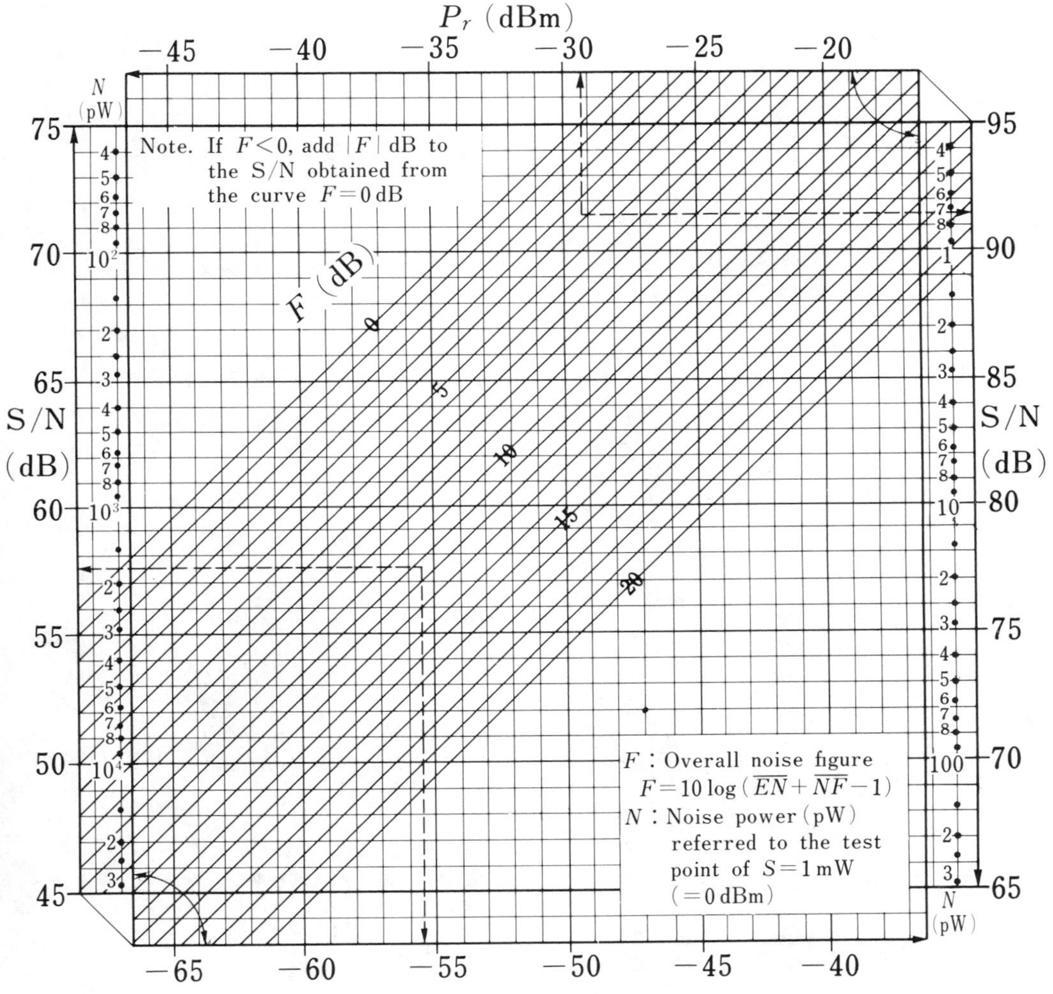

P_r (dBm)

−45 −40 −35 −30 −25 −20

S/N (dB)

Note. If $F<0$, add $|F|$ dB to the S/N obtained from the curve $F=0$ dB

F (dB)

0 5 10 15 20

F : Overall noise figure
$F = 10 \log (\overline{EN} + \overline{NF} - 1)$
N : Noise power (pW) referred to the test point of $S = 1$ mW (= 0 dBm)

N (pW)

−65 −60 −55 −50 −45 −40

Receiver input P_r (dBm)

Receiver Input versus S / N Ratio (SS-FM System)

No. of channels	n	240 ch
Top channel frequency	f_m	1 054 kHz
Frequency deviation	S_0	280 rms, kHz/ch
Channel bandwidth	Δf	3.1 kHz
Overall noise figure	F	0 ~ 20 dB

240 CH

$S_0 = 280$ kHz

P_r (dBm)

−85 −80 −75 −70 −65 −60

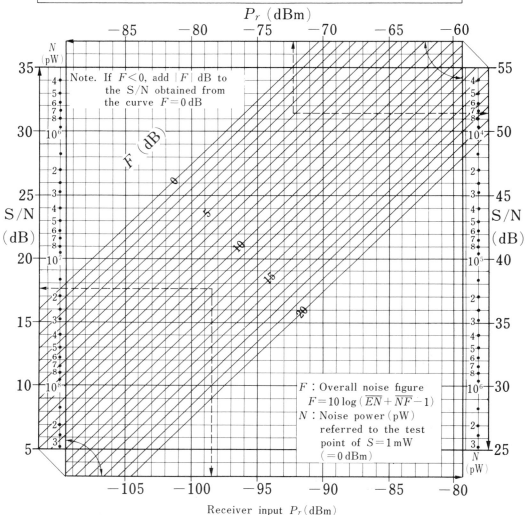

Note. If $F < 0$, add $|F|$ dB to the S/N obtained from the curve $F = 0$ dB

F (dB)

0 5 10 15 20

S/N (dB): 35 30 25 20 15 10 5 (left scale); 55 50 45 40 35 30 25 (right scale)

N (pW)

F : Overall noise figure
$F = 10 \log(\overline{EN} + \overline{NF} - 1)$
N : Noise power (pW) referred to the test point of $S = 1$ mW (= 0 dBm)

−105 −100 −95 −90 −85 −80

Receiver input P_r (dBm)

Receiver Input versus S / N Ratio (SS-FM System)

No. of channels	n	240 ch
Top channel frequency	f_m	1 054 kHz
Frequency deviation	S_0	280 rms, kHz/ch
Channel bandwidth	Δf	3.1 kHz
Overall noise figure	F	0 ~ 20 dB

240 CH
$S_0 = 280$ kHz

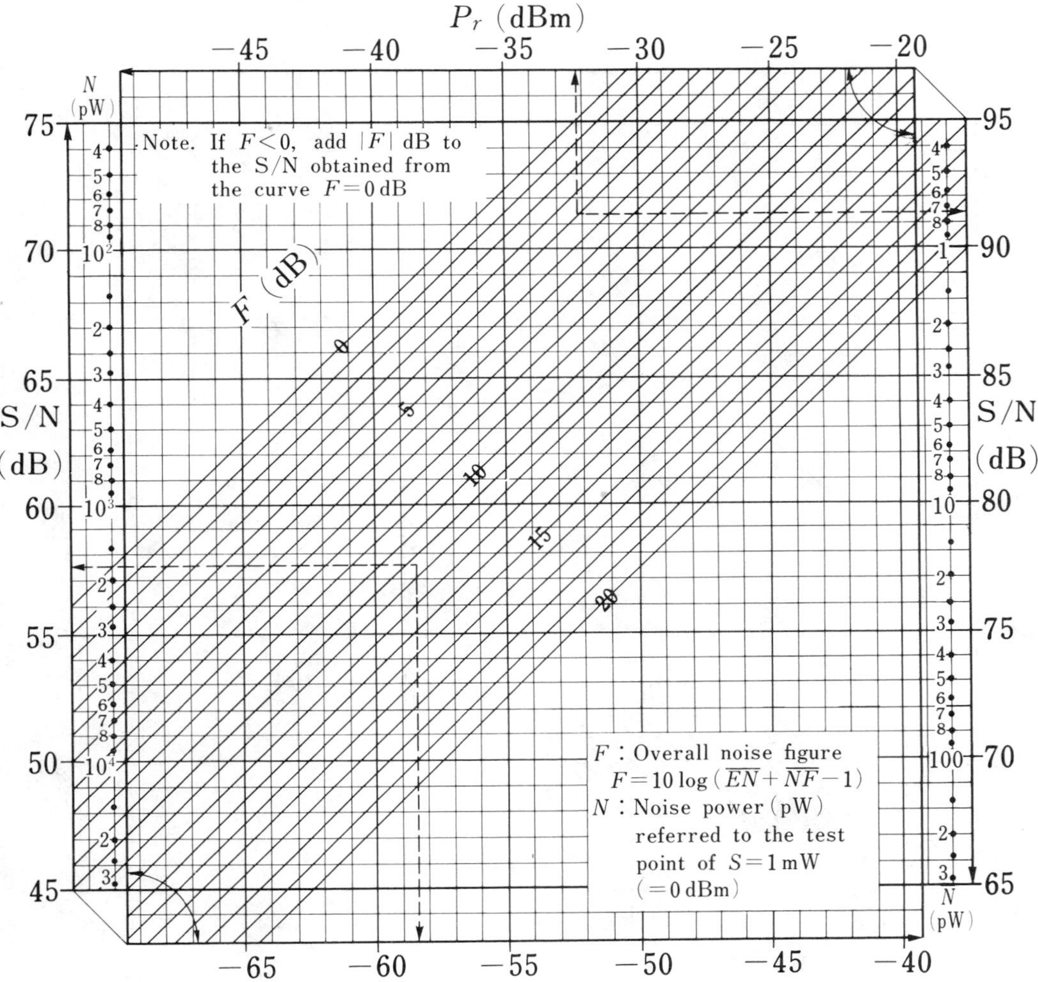

Receiver Input versus S / N Ratio (SS-FM System)

No. of channels	n	300 ch
Top channel frequency	f_m	1 300/1 296 kHz
Frequency deviation	S_0	200 rms, kHz/ch
Channel bandwidth	Δf	3.1 kHz
Overall noise figure	F	0～20 dB

300 CH
$S_0 = 200$ kHz

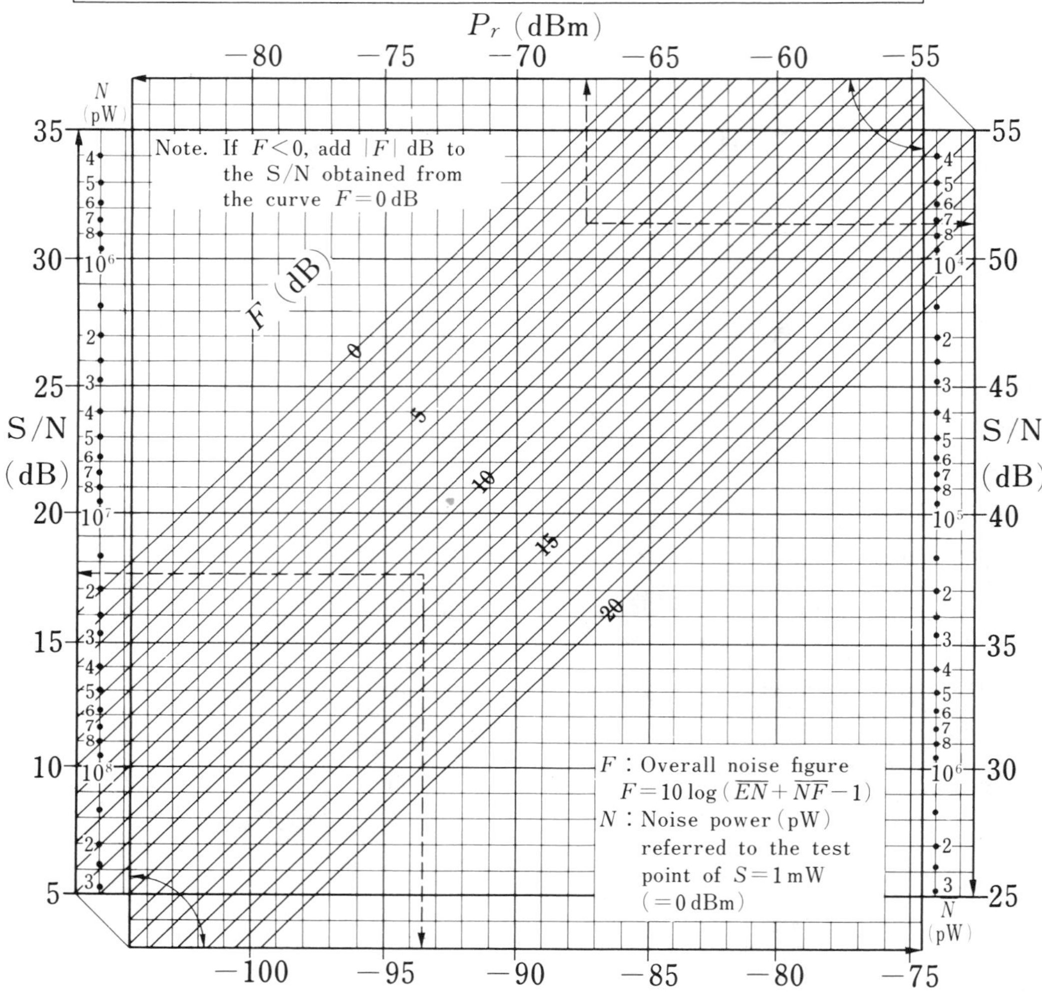

Receiver Input versus S / N Ratio (SS-FM System)

No. of channels	n	300 ch
Top channel frequency	f_m	1 300/1 296 kHz
Frequency deviation	S_0	200 rms, kHz/ch
Channel bandwidth	Δf	3.1 kHz
Overall noise figure	F	0~20 dB

300 CH
$S_0 = 200$ kHz

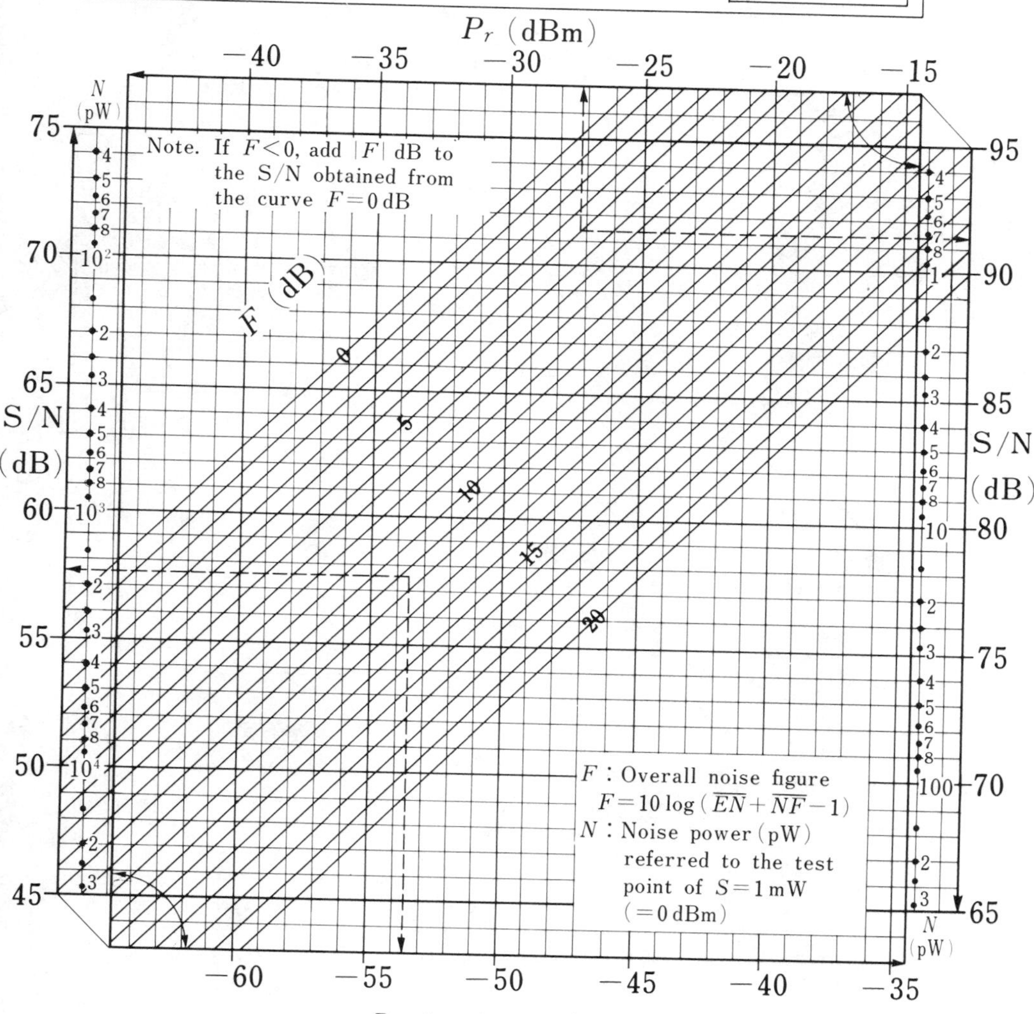

Receiver input P_r (dBm)

Receiver Input versus S / N Ratio (SS-FM System)

No. of channels	n	300 ch
Top channel frequency	f_m	1 300/1 296 kHz
Frequency deviation	S_0	280 rms, kHz/ch
Channel bandwidth	Δf	3.1 kHz
Overall noise figure	F	0~20 dB

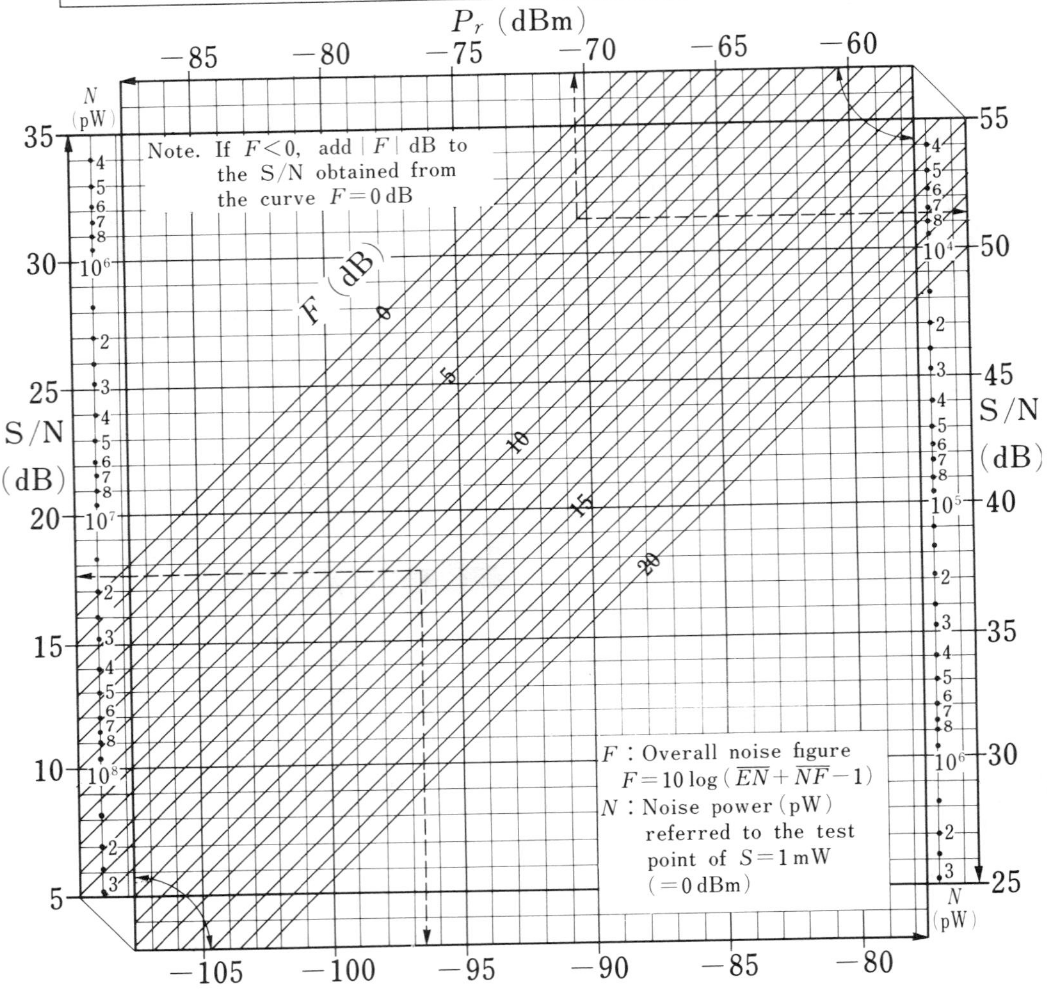

300 CH
$S_0 = 280$ kHz

P_r (dBm)

−85 −80 −75 −70 −65 −60

Note. If $F<0$, add $|F|$ dB to the S/N obtained from the curve $F=0$ dB

F (dB): 0, 5, 10, 15, 20

S/N (dB): 35, 30, 25, 20, 15, 10, 5

N (pW): 4, 5, 6, 7, 8, 10^6, 2, 3, 4, 5, 6, 7, 8, 10^7, 2, 3, 4, 5, 6, 7, 8, 10^8, 2, 3

S/N (dB): 55, 50, 45, 40, 35, 30, 25

N (pW): 4, 5, 6, 7, 8, 10^4, 2, 3, 4, 5, 6, 7, 8, 10^5, 2, 3, 4, 5, 6, 7, 8, 10^6, 2, 3

F : Overall noise figure
$F = 10 \log(\overline{EN} + \overline{NF} - 1)$
N : Noise power (pW) referred to the test point of $S = 1$ mW (= 0 dBm)

−105 −100 −95 −90 −85 −80

Receiver input P_r (dBm)

Receiver Input versus S / N Ratio (SS-FM System)

No. of channels	n	300 ch
Top channel frequency	f_m	1 300/1 296 kHz
Frequency deviation	S_0	280 rms, kHz/ch
Channel bandwidth	Δf	3.1 kHz
Overall noise figure	F	0~20 dB

300 CH
$S_0 = 280$ kHz

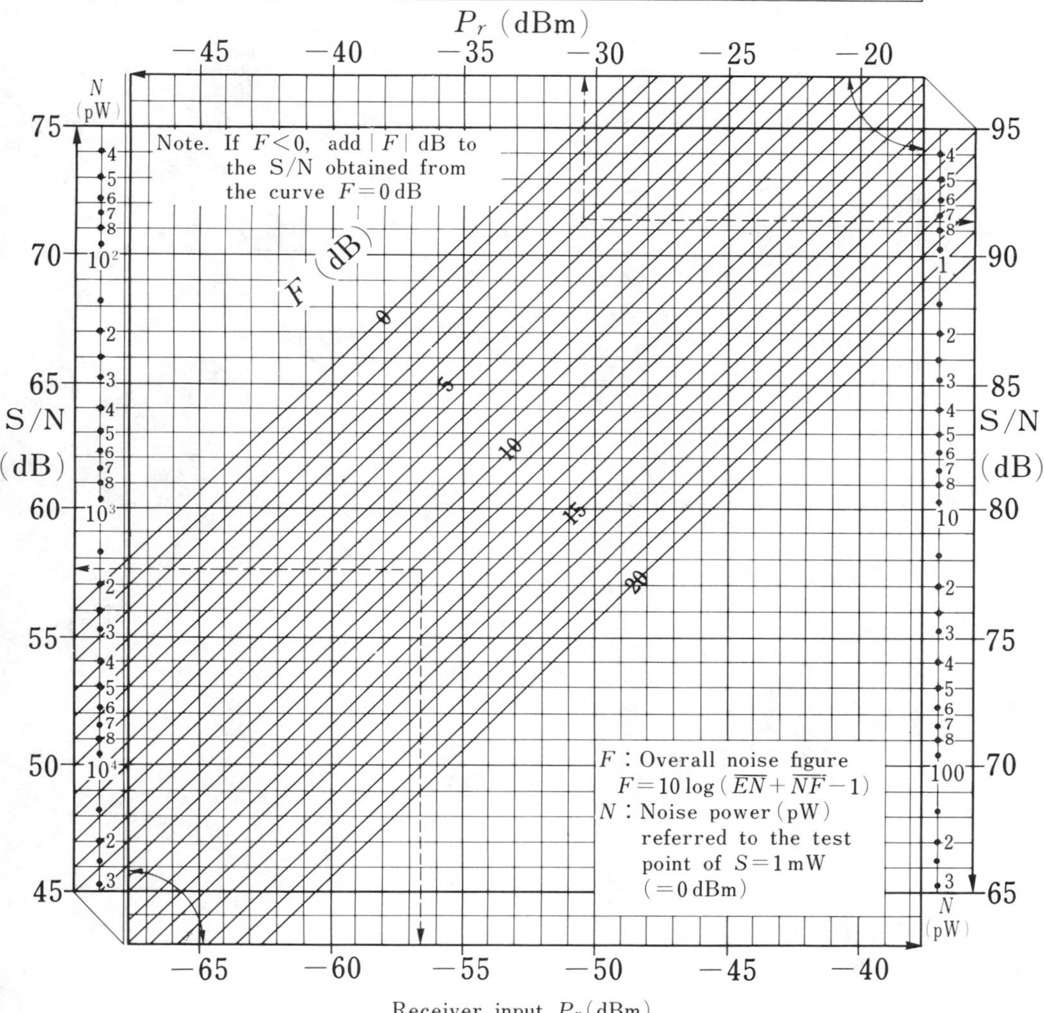

Receiver Input versus S / N Ratio (SS-FM System)

No. of channels	n	480 ch
Top channel frequency	f_m	2 048 kHz
Frequency deviation	S_0	200 rms, kHz/ch
Channel bandwidth	Δf	3.1 kHz
Overall noise figure	F	0~20 dB

480 CH
$S_0 = 200$ kHz

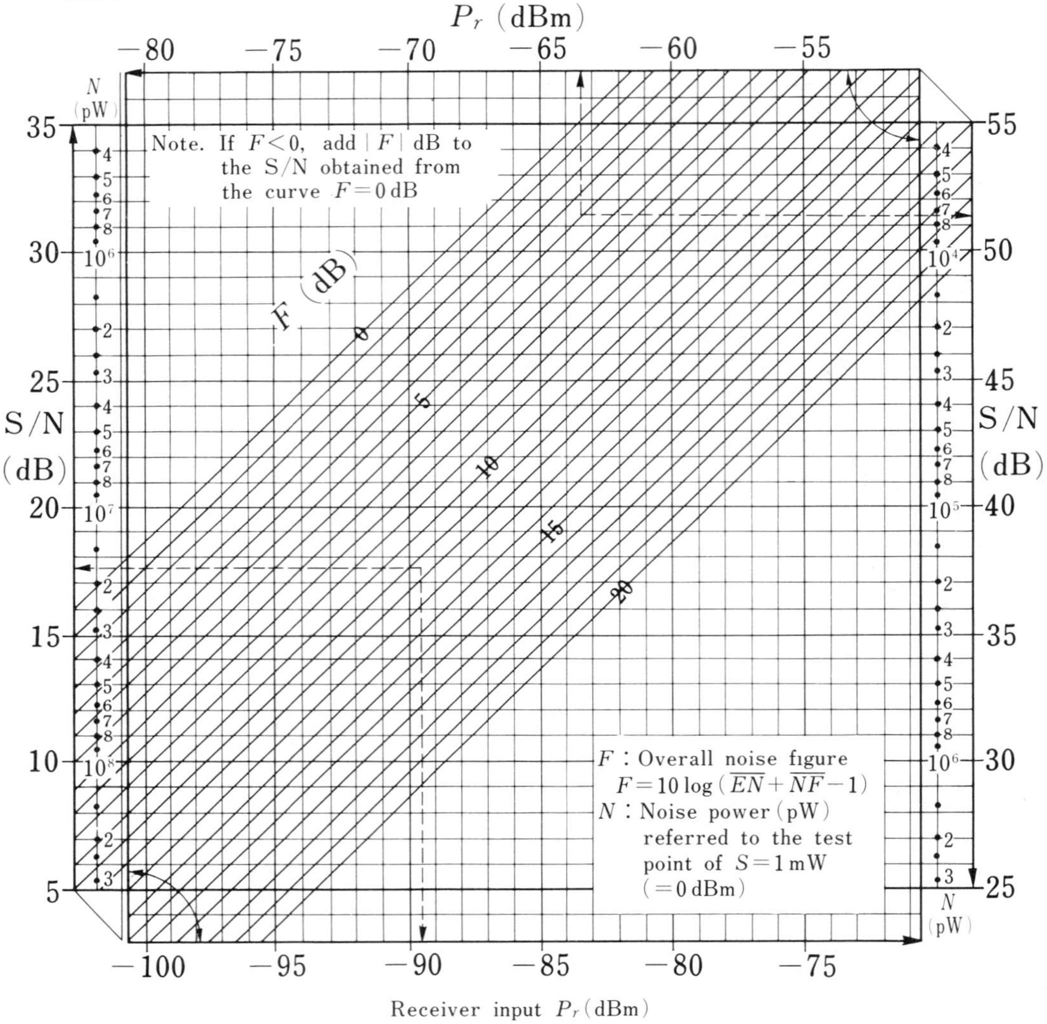

Receiver input P_r (dBm)

Receiver Input versus S / N Ratio (SS-FM System)

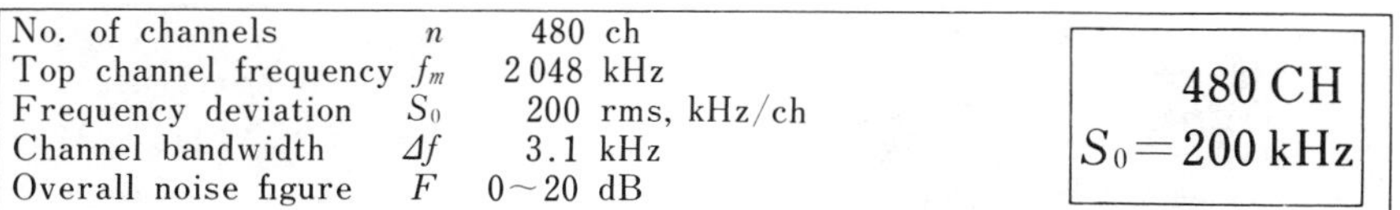

No. of channels	n	480 ch
Top channel frequency	f_m	2 048 kHz
Frequency deviation	S_0	200 rms, kHz/ch
Channel bandwidth	Δf	3.1 kHz
Overall noise figure	F	0 ~ 20 dB

480 CH
$S_0 = 200$ kHz

P_r (dBm)

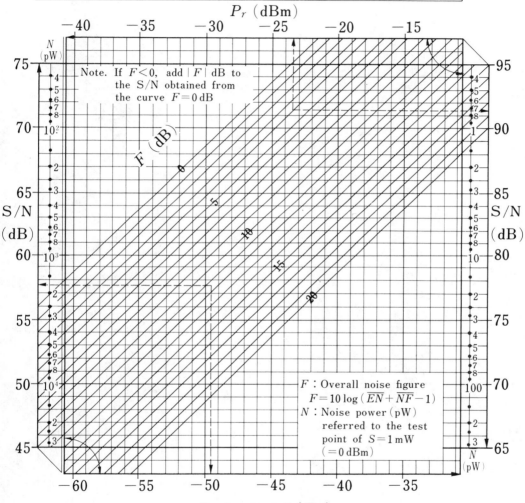

Note. If $F < 0$, add $|F|$ dB to the S/N obtained from the curve $F = 0$ dB

F (dB)

S/N (dB)

N (pW)

F : Overall noise figure
$F = 10 \log(\overline{EN} + \overline{NF} - 1)$
N : Noise power (pW) referred to the test point of $S = 1$ mW ($= 0$ dBm)

Receiver input P_r (dBm)

Receiver Input versus S / N Ratio (SS-FM System)

No. of channels	n	480	ch
Top channel frequency	f_m	2 048	kHz
Frequency deviation	S_0	280	rms, kHz/ch
Channel bandwidth	Δf	3.1	kHz
Overall noise figure	F	0~20	dB

480 CH
$S_0 = 280$ kHz

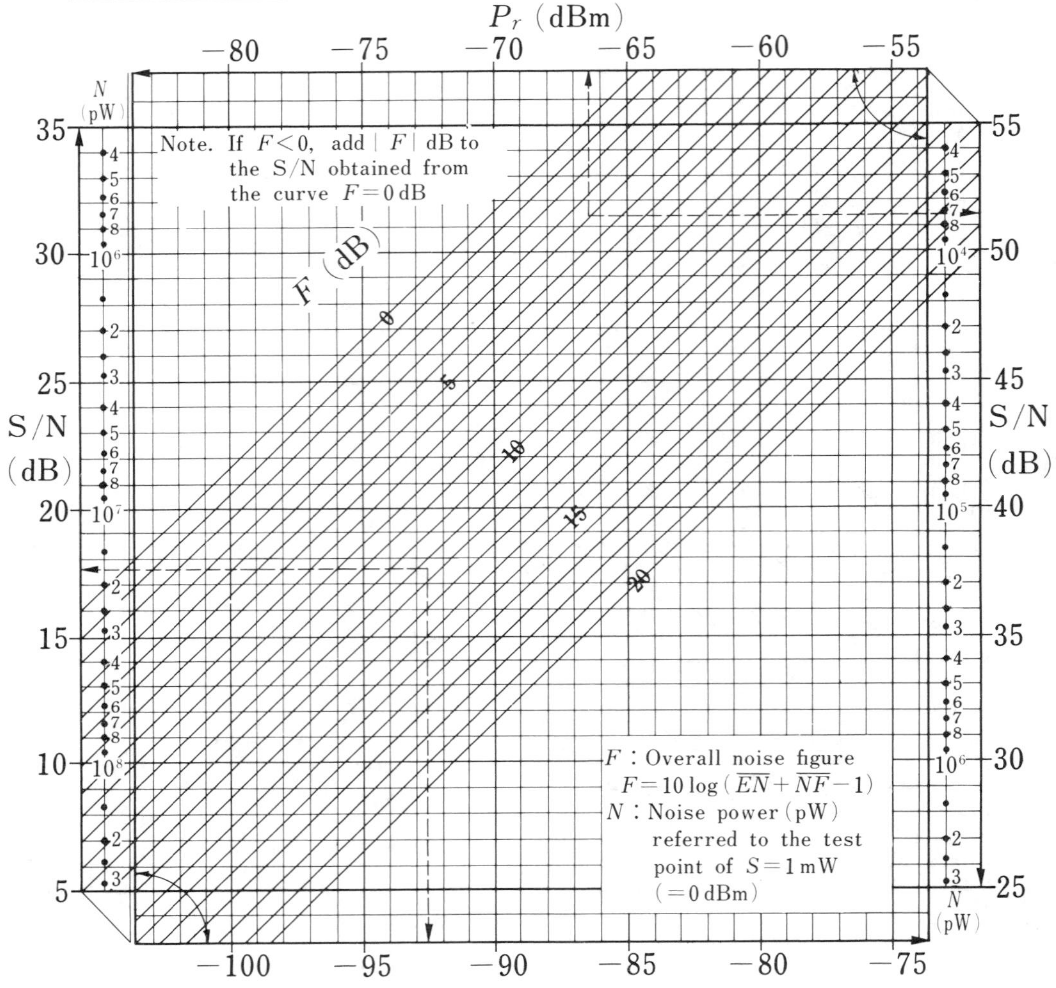

Receiver input P_r (dBm)

Receiver Input versus S / N Ratio (SS-FM System)

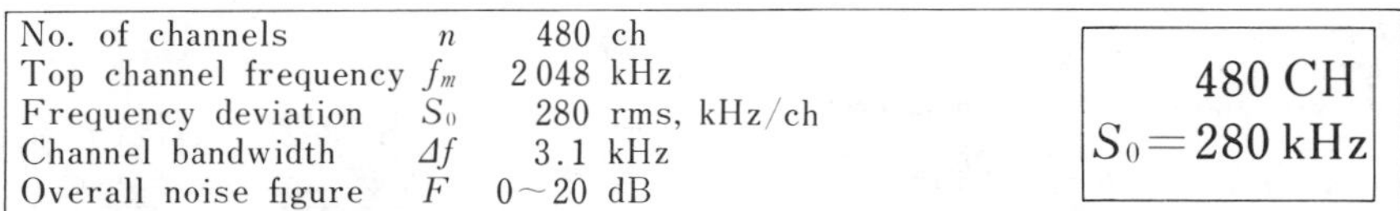

No. of channels	n	480	ch
Top channel frequency	f_m	2 048	kHz
Frequency deviation	S_0	280	rms, kHz/ch
Channel bandwidth	Δf	3.1	kHz
Overall noise figure	F	0~20	dB

480 CH
$S_0 = 280$ kHz

P_r (dBm)

−40 −35 −30 −25 −20 −15

N (pW)

S/N (dB)

75 70 65 60 55 50 45

95 90 85 80 75 70 65

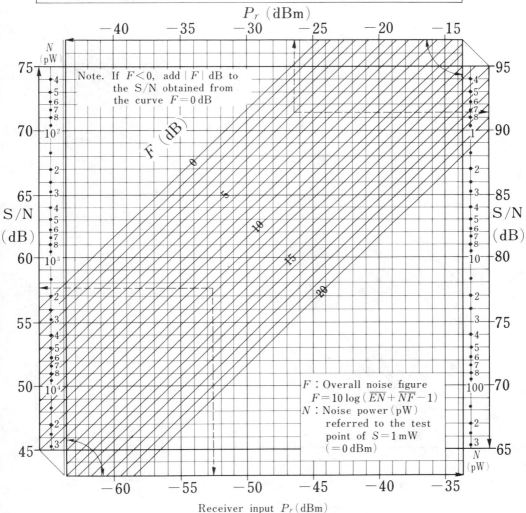

Note. If $F<0$, add $|F|$ dB to the S/N obtained from the curve $F=0$ dB

F (dB)

0 5 10 15 20

F : Overall noise figure
$F = 10 \log(\overline{EN} + \overline{NF} - 1)$
N : Noise power (pW) referred to the test point of $S = 1$ mW (= 0 dBm)

−60 −55 −50 −45 −40 −35

Receiver input P_r (dBm)

Receiver Input versus S / N Ratio (SS-FM System)

No. of channels n 600 ch Top channel frequency f_m 2 540/2 660 kHz Frequency deviation S_0 200 rms, kHz/ch Channel bandwidth Δf 3.1 kHz Overall noise figure F 0～20 dB	Note : This graph is prepared for f_m = 2 540 MHz, then add −0.4 dB to the obtained value when f_m = 2 660 MHz	600 CH S_0 = 200 kHz

P_r (dBm)

−75 −70 −65 −60 −55 −50

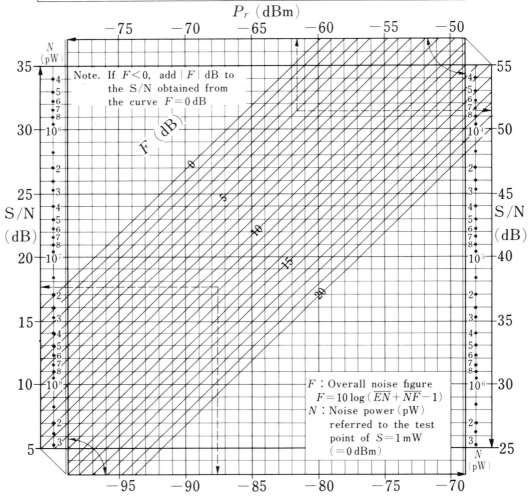

Note. If $F<0$, add $|F|$ dB to the S/N obtained from the curve F = 0 dB

F (dB)

0 5 10 15 20

S/N (dB)

35 30 25 20 15 10 5

55 50 45 40 35 30 25

N (pW)

F : Overall noise figure
$F = 10 \log (\overline{EN} + \overline{NF} - 1)$
N : Noise power (pW) referred to the test point of S = 1 mW (= 0 dBm)

−95 −90 −85 −80 −75 −70

Receiver input P_r (dBm)

Receiver Input versus S / N Ratio (SS-FM System)

No. of channels n 600 ch Top channel frequency f_m 2 540/2 660 kHz Frequency deviation S_0 200 rms, kHz/ch Channel bandwidth Δf 3.1 kHz Overall noise figure F 0～20 dB	Note : This graph is prepared for f_m = 2 540 MHz, then add −0.4 dB to the obtained value when f_m = 2 660 MHz	600 CH S_0 = 200 kHz

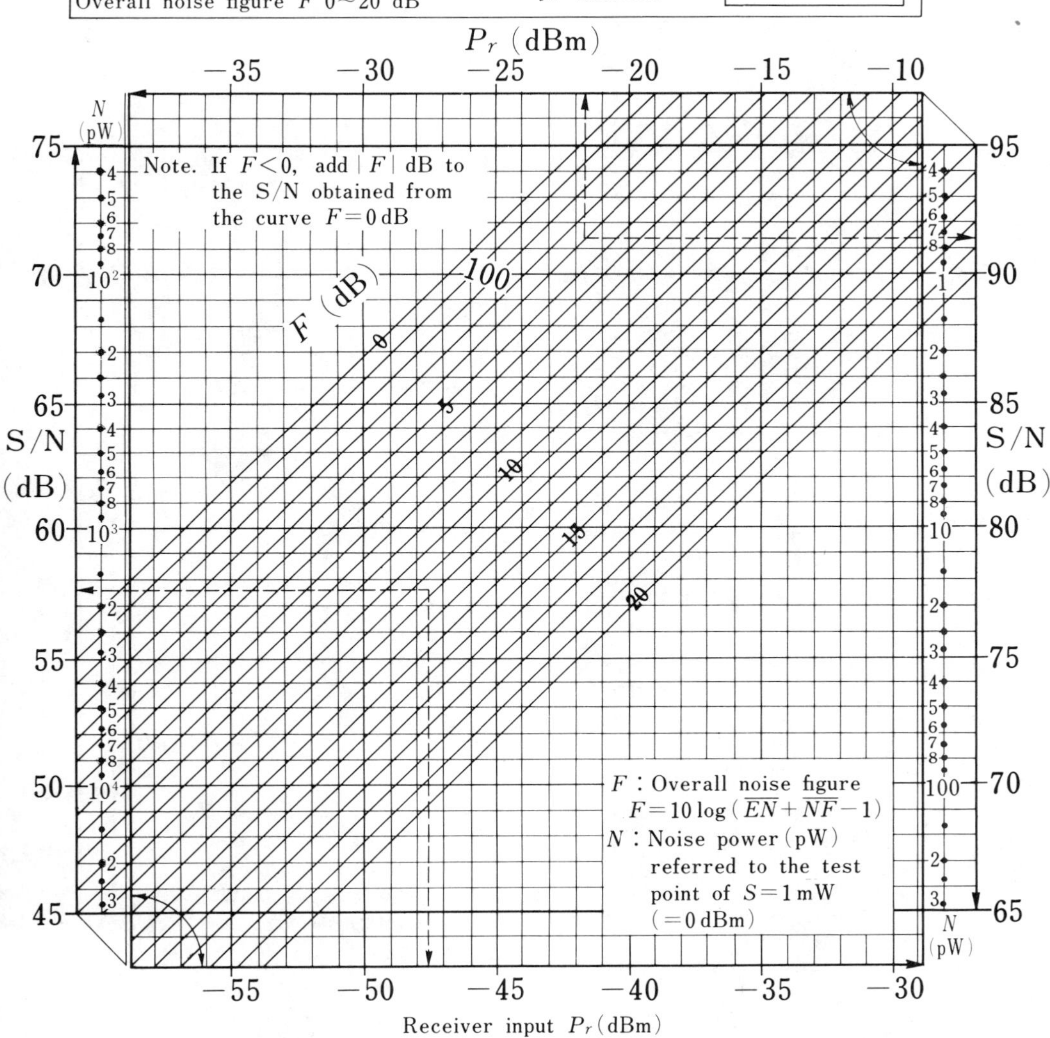

Receiver Input versus S / N Ratio (SS-FM System)

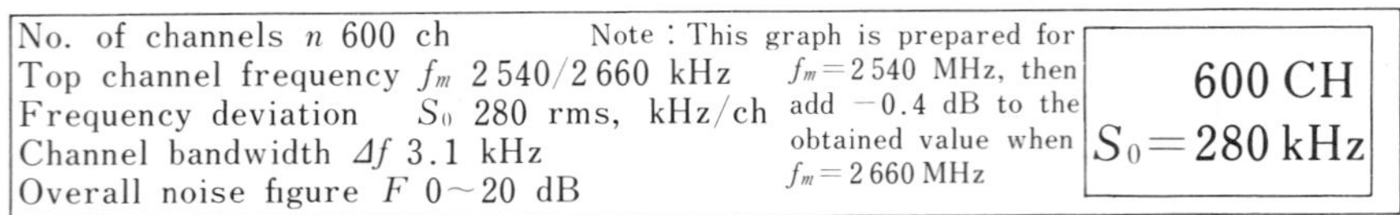
No. of channels n 600 ch
Top channel frequency f_m 2 540/2 660 kHz
Frequency deviation S_0 280 rms, kHz/ch
Channel bandwidth Δf 3.1 kHz
Overall noise figure F 0～20 dB

Note : This graph is prepared for f_m = 2 540 MHz, then add −0.4 dB to the obtained value when f_m = 2 660 MHz

600 CH
S_0 = 280 kHz

P_r (dBm)

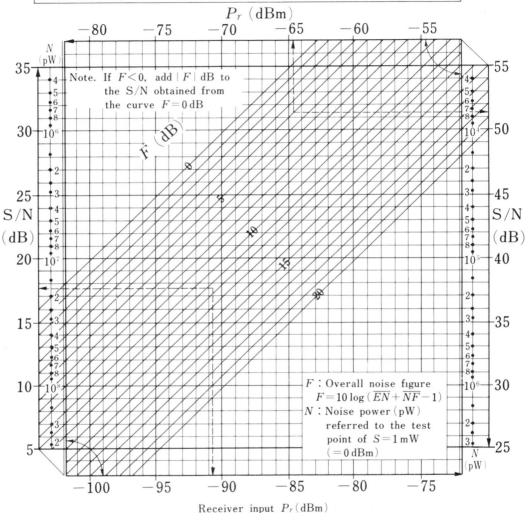

Note. If $F<0$, add $|F|$ dB to the S/N obtained from the curve $F=0$ dB

F (dB)

F : Overall noise figure
$F=10\log(\overline{EN}+\overline{NF}-1)$
N : Noise power (pW) referred to the test point of $S=1$ mW (= 0 dBm)

S/N (dB)

N (pW)

Receiver input P_r (dBm)

Receiver Input versus S / N Ratio (SS-FM System)

No. of channels n 600 ch
Top channel frequency f_m 2 540/2 660 kHz
Frequency deviation S_0 280 rms, kHz/ch
Channel bandwidth Δf 3.1 kHz
Overall noise figure F 0~20 dB

Note : This graph is prepared for f_m=2 540 MHz, then add −0.4 dB to the obtained value when f_m=2 660 MHz

600 CH
S_0=280 kHz

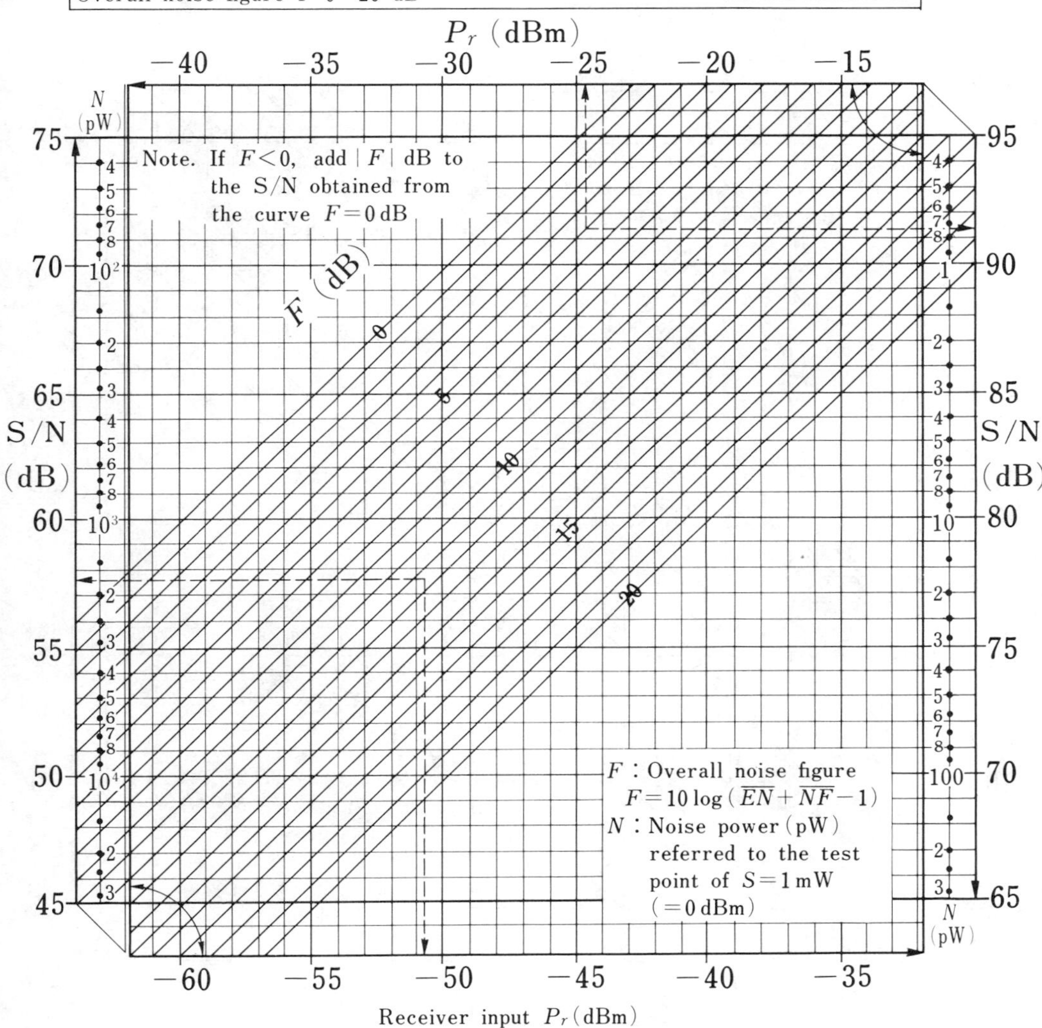

Receiver Input versus S / N Ratio (SS-FM System)

No. of channels n 960 ch Top channel frequency f_m 4 028/4 188 kHz Frequency deviation S_0 140 rms, kHz/ch Channel bandwidth Δf 3.1 kHz Overall noise figure F 0~20 dB	Note: This graph is prepared for f_m=4 028 kHz, then add −0.3 dB to the obtained value when f_m=4 188 kHz

960 CH
900 CH
S_0=140 kHz

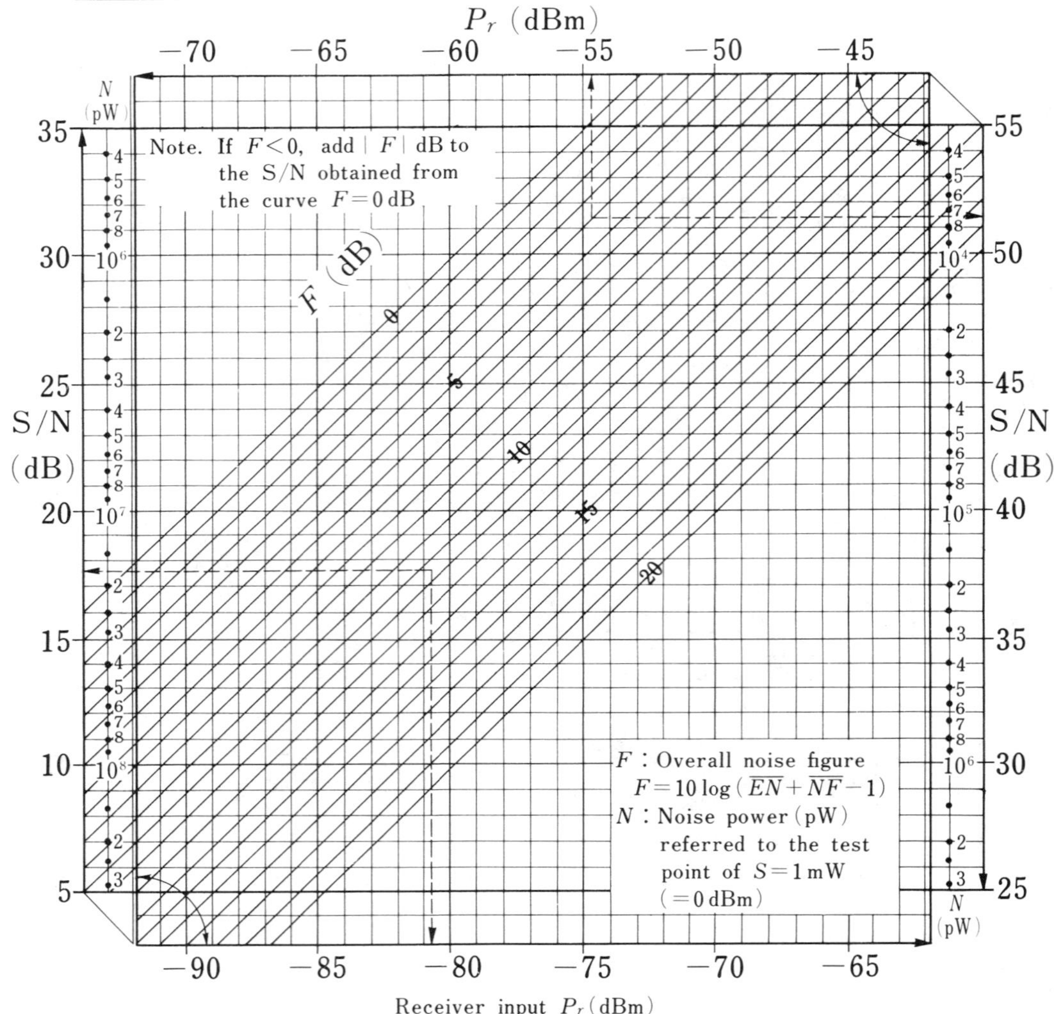

Receiver Input versus S / N Ratio (SS-FM System)

No. of channels n 960 ch Top channel frequency f_m 4 028/4 188 kHz Frequency deviation S_0 140 rms, kHz/ch Channel bandwidth Δf 3.1 kHz Overall noise figure F 0~20 dB	Note : This graph is prepared for f_m=4 028 kHz, then add −0.3 dB to the obtained value when f_m=4 188 kHz	960 CH 900 CH S_0=140 kHz

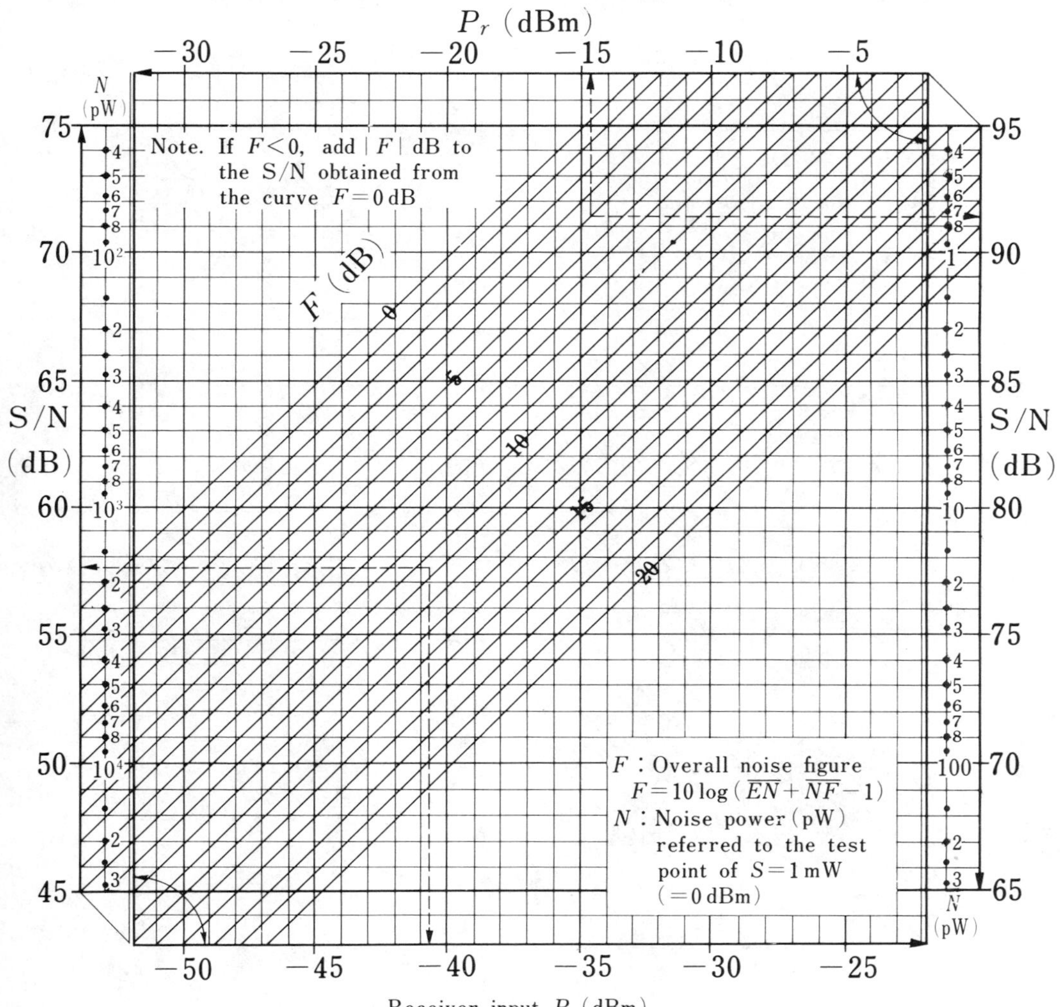

21c Receiver Input versus S / N Ratio (SS-FM System)

No. of channels n 960 ch Top channel frequency f_m 4 028/4 188 kHz Frequency deviation S_0 200 rms, kHz/ch Channel bandwidth Δf 3.1 kHz Overall noise figure F 0~20 dB	Note : This graph is prepared for f_m=4 028 kHz, then add −0.3 dB to the obtained value when f_m=4 188 kHz	960 CH 900 CH S_0=200 kHz

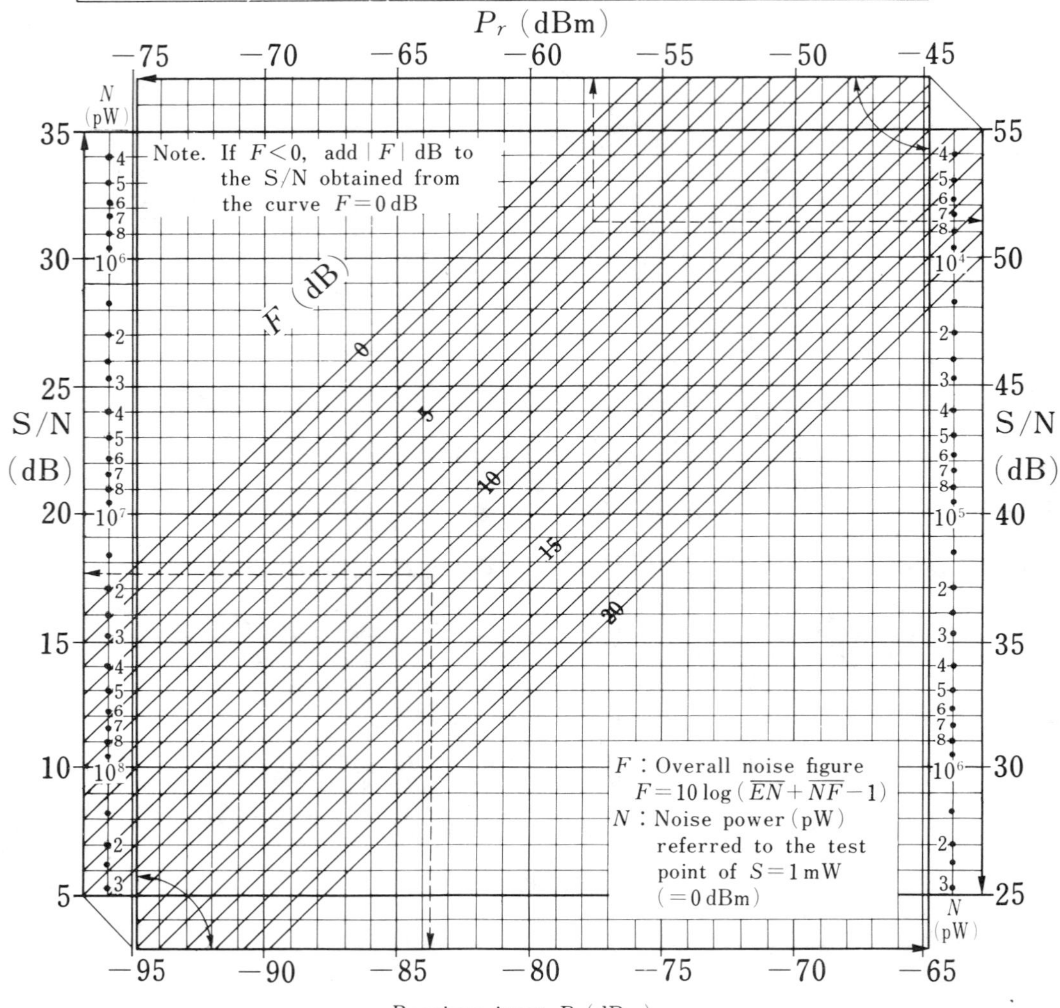

Receiver Input versus S / N Ratio (SS-FM System)

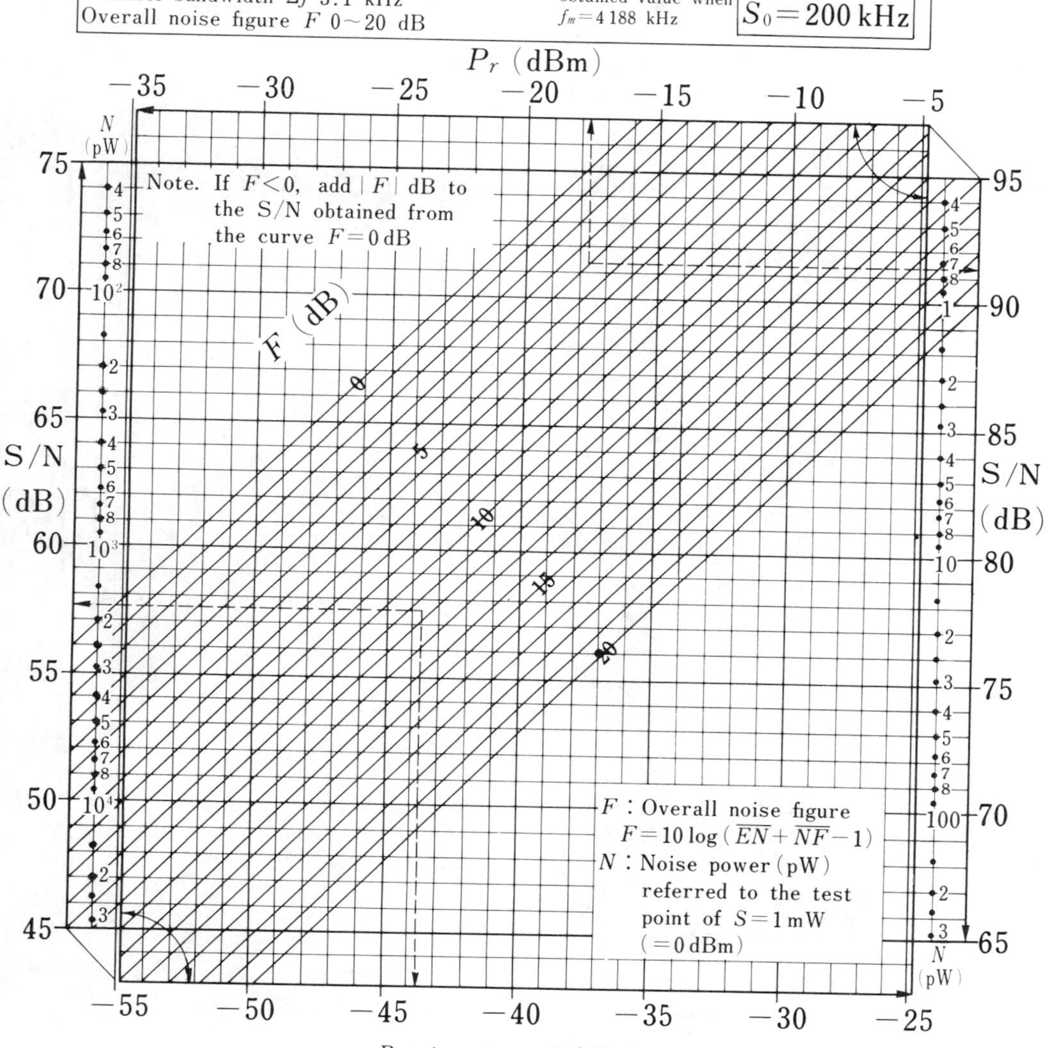

No. of channels n 960 ch
Top channel frequency f_m 4 028/4 188 kHz
Frequency deviation S_0 200 rms, kHz/ch
Channel bandwidth Δf 3.1 kHz
Overall noise figure F 0~20 dB
Note : This graph is prepared for f_m=4 028 kHz, then add −0.3 dB to the obtained value when f_m=4 188 kHz
960 CH
900 CH
S_0=200 kHz
P_r (dBm)
−35
−30
−25
−20
−15
−10
−5
N (pW)
Note. If $F<0$, add $|F|$ dB to the S/N obtained from the curve $F=0$ dB
F (dB)
0
5
10
15
20
S/N (dB)
75
70
65
60
55
50
45
95
90
85
80
75
70
65
F : Overall noise figure
$F=10\log(\overline{EN}+\overline{NF}-1)$
N : Noise power (pW) referred to the test point of $S=1$ mW (=0 dBm)
−55
−50
−45
−40
−35
−30
−25
Receiver input P_r (dBm)

Receiver Input versus S / N Ratio (SS-FM System)

No. of channels n 960 ch Top channel frequency f_m 4 028/4 188 kHz Frequency deviation S_0 280 rms, kHz/ch Channel bandwidth Δf 3.1 kHz Overall noise figure F 0~20 dB	Note : This graph is prepared for f_m=4 028 kHz, then add −0.3 dB to the obtained value when f_m=4 188 kHz	960 CH 900 CH S_0=280 kHz

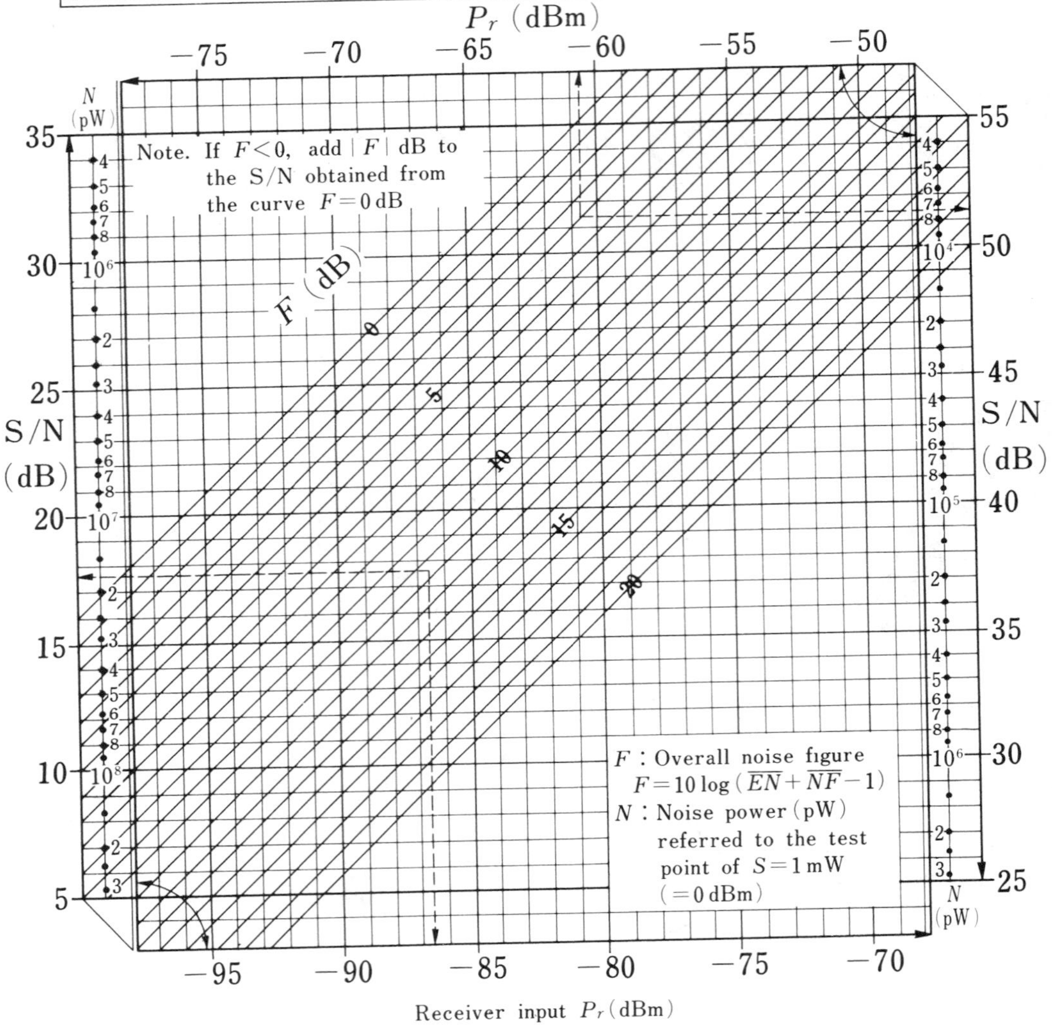

Receiver input P_r (dBm)

Receiver Input versus S / N Ratio (SS-FM System)

No. of channels n 960 ch
Top channel frequency f_m 4 028/4 188 kHz
Frequency deviation S_0 280 rms, kHz/ch
Channel bandwidth Δf 3.1 kHz
Overall noise figure F 0~20 dB

Note : This graph is prepared for f_m = 4 028 kHz, then add −0.3 dB to the obtained value when f_m = 4 188 kHz

960 CH
900 CH
S_0 = 280 kHz

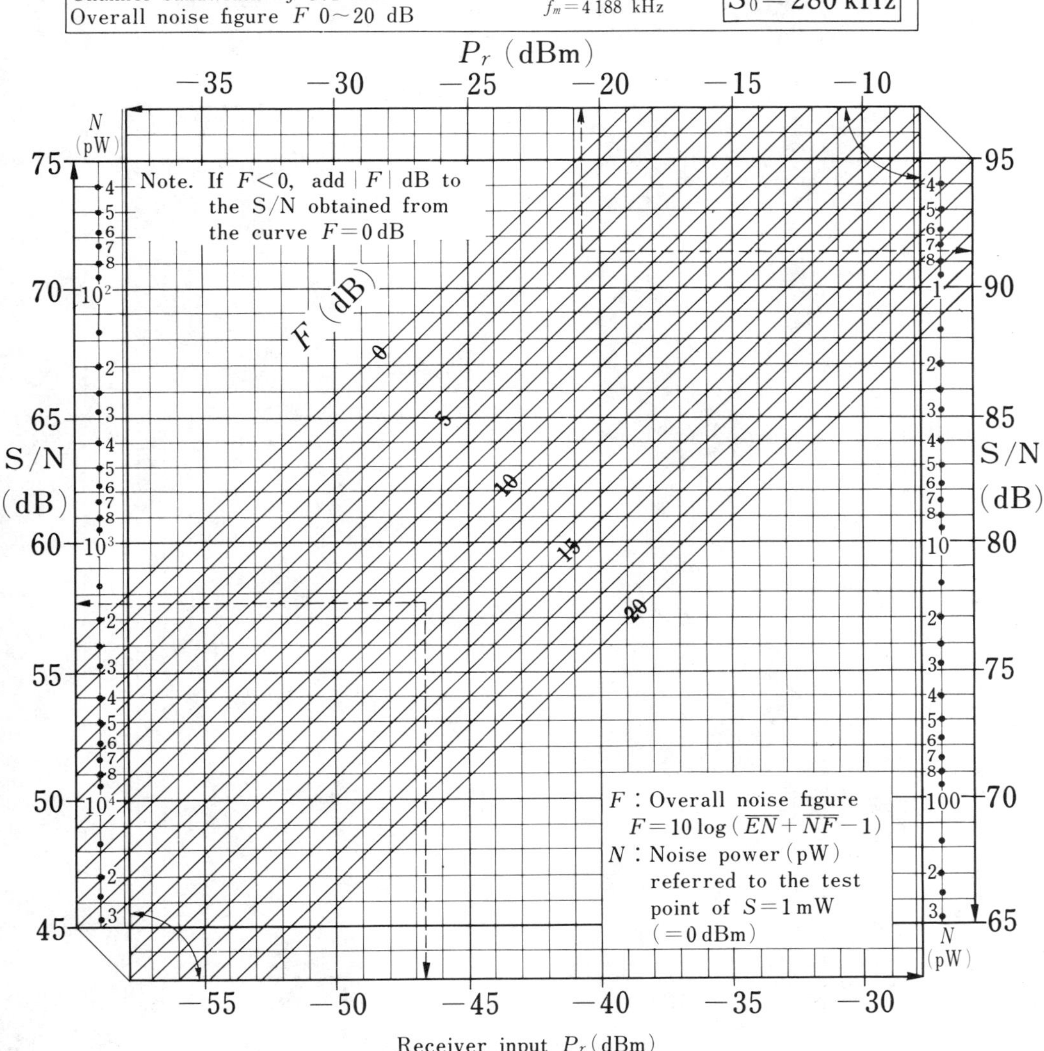

Receiver Input versus S / N Ratio (SS-FM System)

No. of channels	n	1 200 ch
Top channel frequency	f_m	5 636/5 564 kHz
Frequency deviation	S_0	140 rms, kHz/ch
Channel bandwidth	Δf	3.1 kHz
Overall noise figure	F	0~20 dB

1 260 CH
1 200 CH
S_0=140 kHz

P_r (dBm)

−65 −60 −55 −50 −45 −40

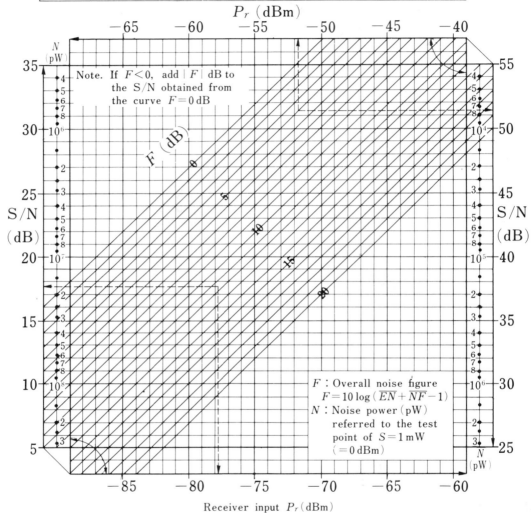

Note. If $F<0$, add $|F|$ dB to the S/N obtained from the curve $F=0$ dB

F (dB)

0 5 10 15 20

S/N (dB)

35 30 25 20 15 10 5

S/N (dB)

55 50 45 40 35 30 25

N (pW)

F : Overall noise figure
$F=10\log(\overline{EN}+\overline{NF}-1)$
N : Noise power (pW) referred to the test point of $S=1$ mW (=0 dBm)

−85 −80 −75 −70 −65 −60

Receiver input P_r (dBm)

Receiver Input versus S / N Ratio (SS-FM System)

No. of channels	n	1 200 ch
Top channel frequency	f_m	5 636/5 564 kHz
Frequency deviation	S_0	140 rms, kHz/ch
Channel bandwidth	Δf	3.1 kHz
Overall noise figure	F	0~20 dB

1 260 CH
1 200 CH
$S_0 =$ 140 kHz

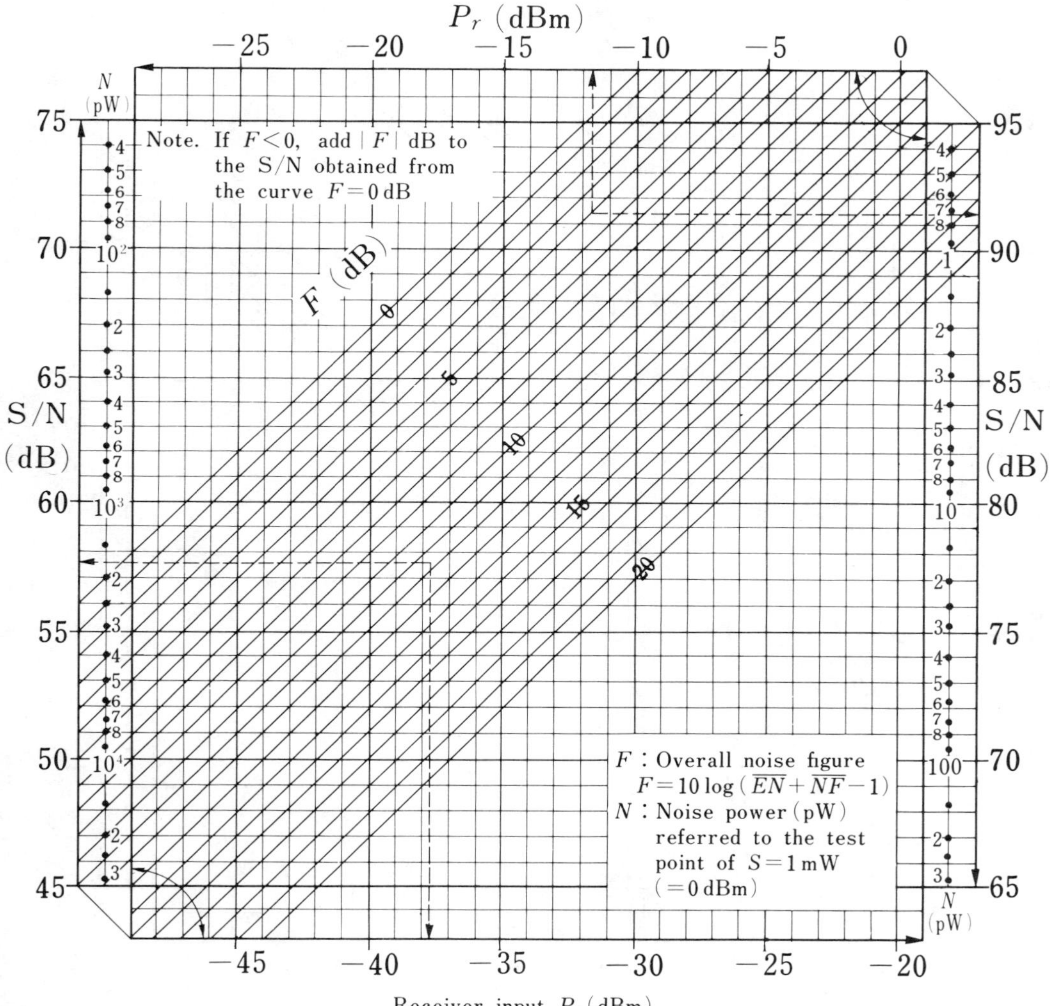

Receiver Input versus S / N Ratio (SS-FM System)

No. of channels	n	1 200 ch
Top channel frequency	f_m	5 636/5 564 kHz
Frequency deviation	S_0	200 rms, kHz/ch
Channel bandwidth	Δf	3.1 kHz
Overall noise figure	F	0~20 dB

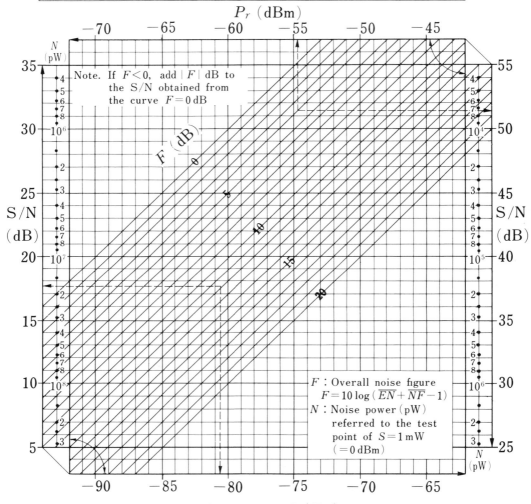

Receiver input P_r (dBm)

Receiver Input versus S / N Ratio (SS-FM System)

No. of channels	n	1 200 ch
Top channel frequency	f_m	5 636/5 564 kHz
Frequency deviation	S_0	200 rms, kHz/ch
Channel bandwidth	Δf	3.1 kHz
Overall noise figure	F	0～20 dB

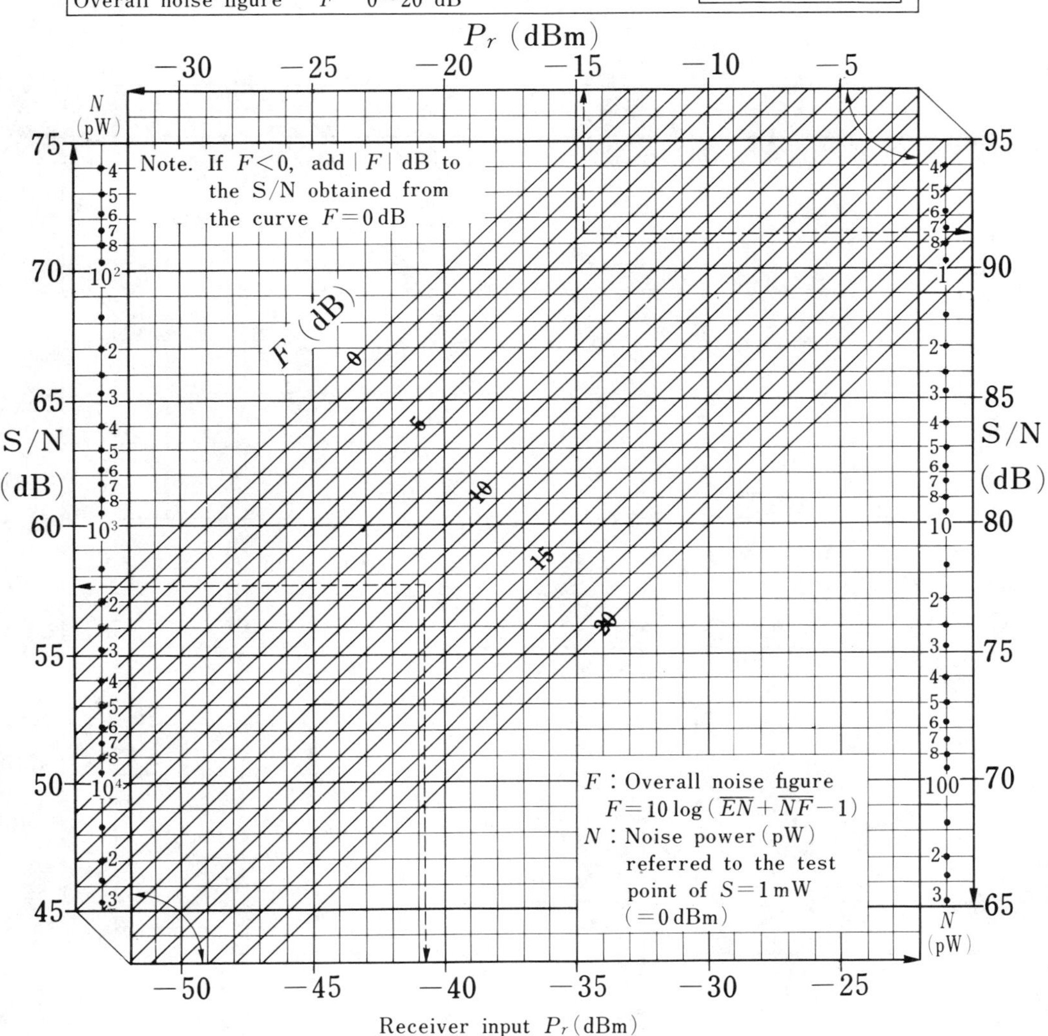

Receiver Input versus S / N Ratio (SS-FM System)

No. of channels	n	1 200 ch
Top channel frequency	f_m	5 636/5 564 kHz
Frequency deviation	S_0	280 rms, kHz/ch
Channel bandwidth	Δf	3.1 kHz
Overall noise figure	F	0～20 dB

1 260 CH
1 200 CH
$S_0 =$ 280 kHz

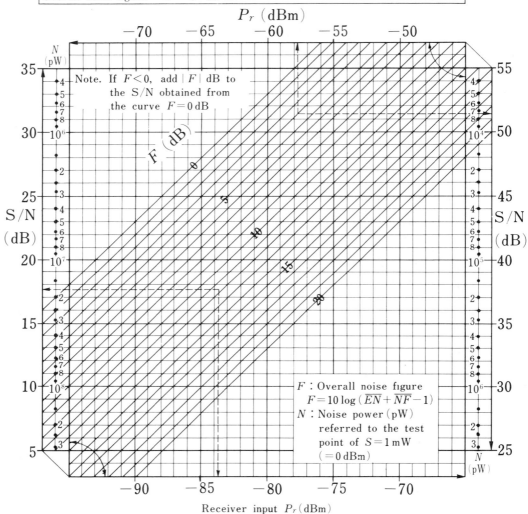

Receiver Input versus S / N Ratio (SS-FM System)

No. of channels	n	1 200 ch
Top channel frequency	f_m	5 636/5 564 kHz
Frequency deviation	S_0	280 rms, kHz/ch
Channel bandwidth	Δf	3.1 kHz
Overall noise figure	F	0~20 dB

1 260 CH
1 200 CH
$S_0 =$ 280 kHz

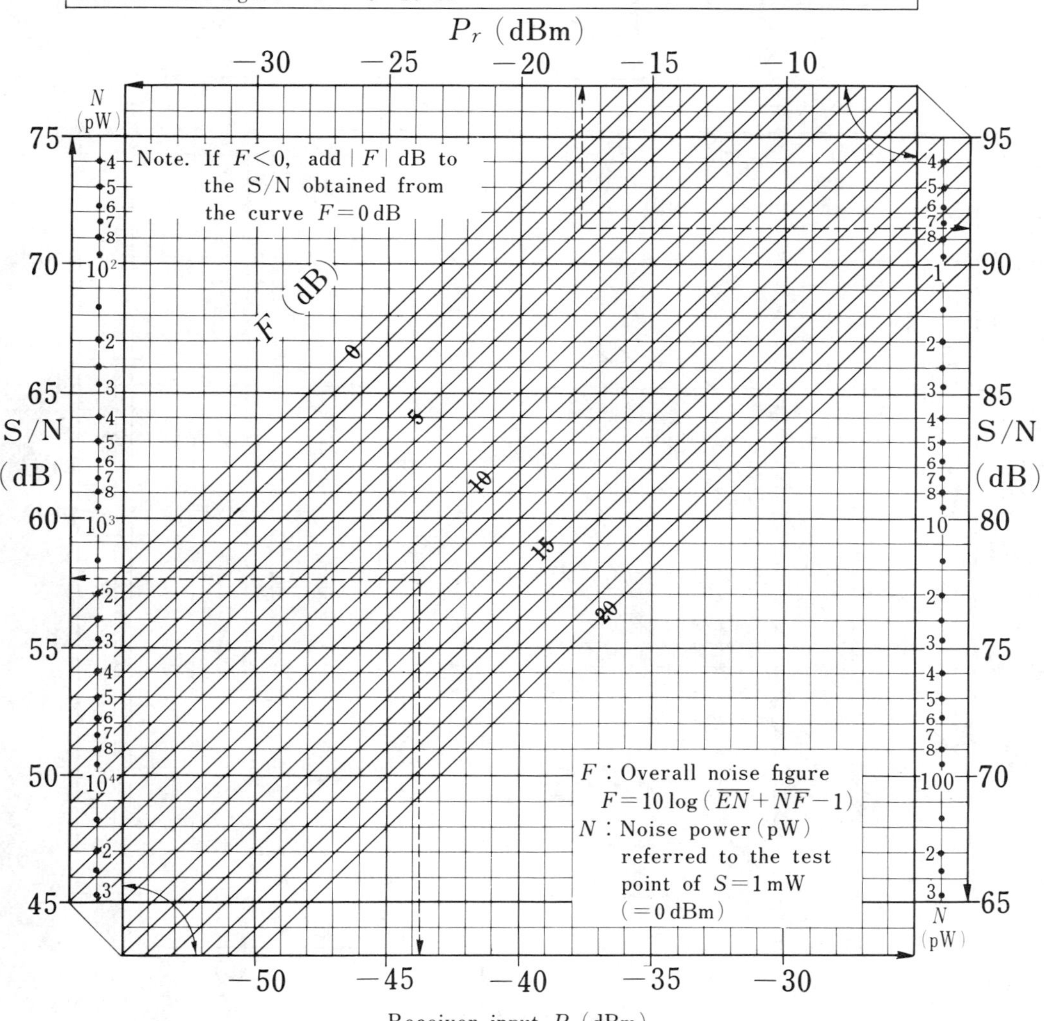

Receiver Input versus S / N Ratio (SS-FM System)

No. of channels	n	1 800 ch
Top channel frequency	f_m	8 204/8 120 kHz
Frequency deviation	S_0	140 rms, kHz/ch
Channel bandwidth	Δf	3.1 kHz
Overall noise figure	F	0~20 dB

1 800 CH
$S_0 =$ 140 kHz

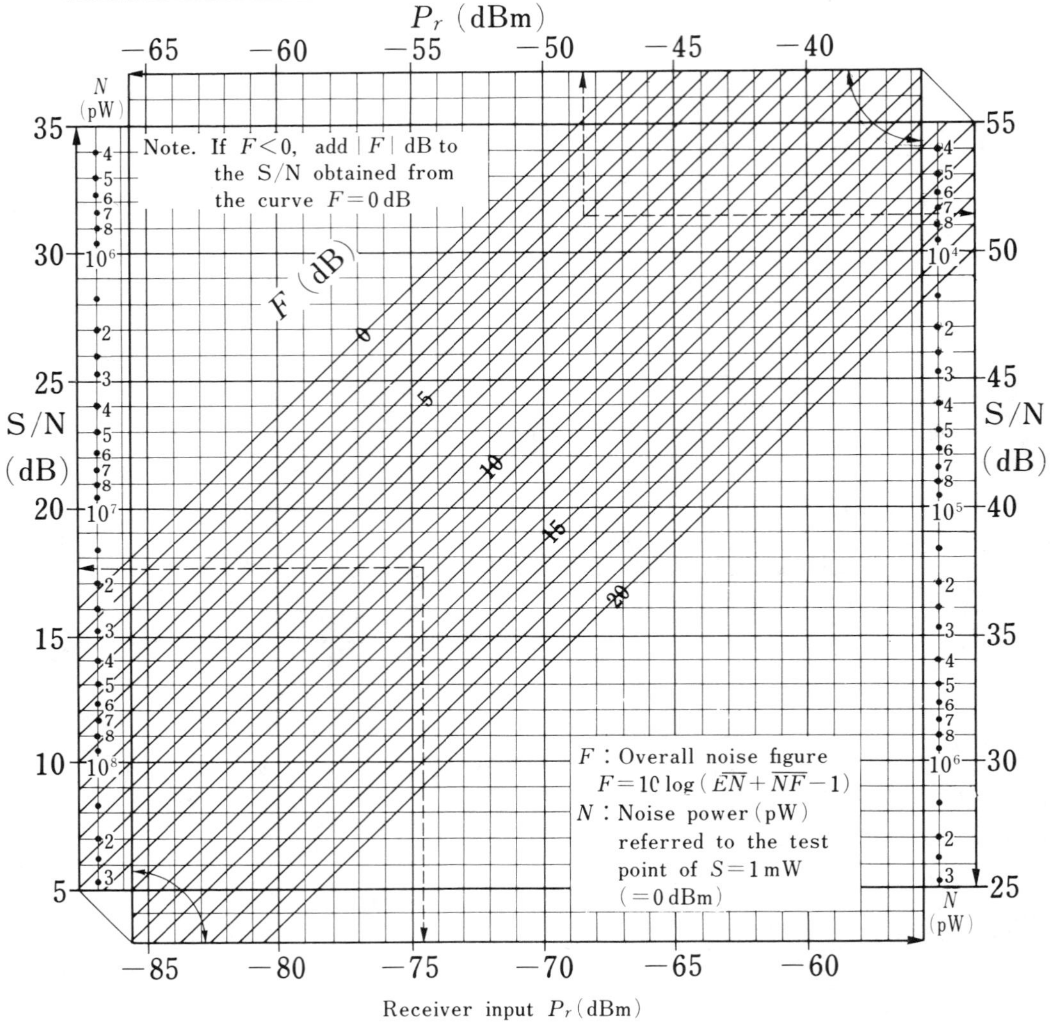

Receiver Input versus S / N Ratio (SS-FM System)

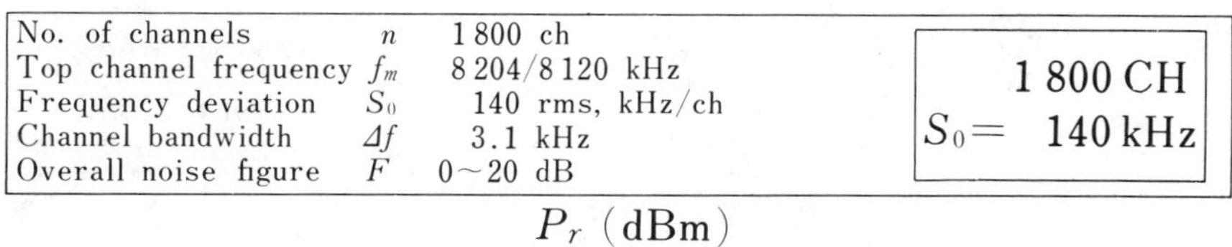

No. of channels	n	1 800 ch
Top channel frequency	f_m	8 204/8 120 kHz
Frequency deviation	S_0	140 rms, kHz/ch
Channel bandwidth	Δf	3.1 kHz
Overall noise figure	F	0~20 dB

1 800 CH
$S_0 =$ 140 kHz

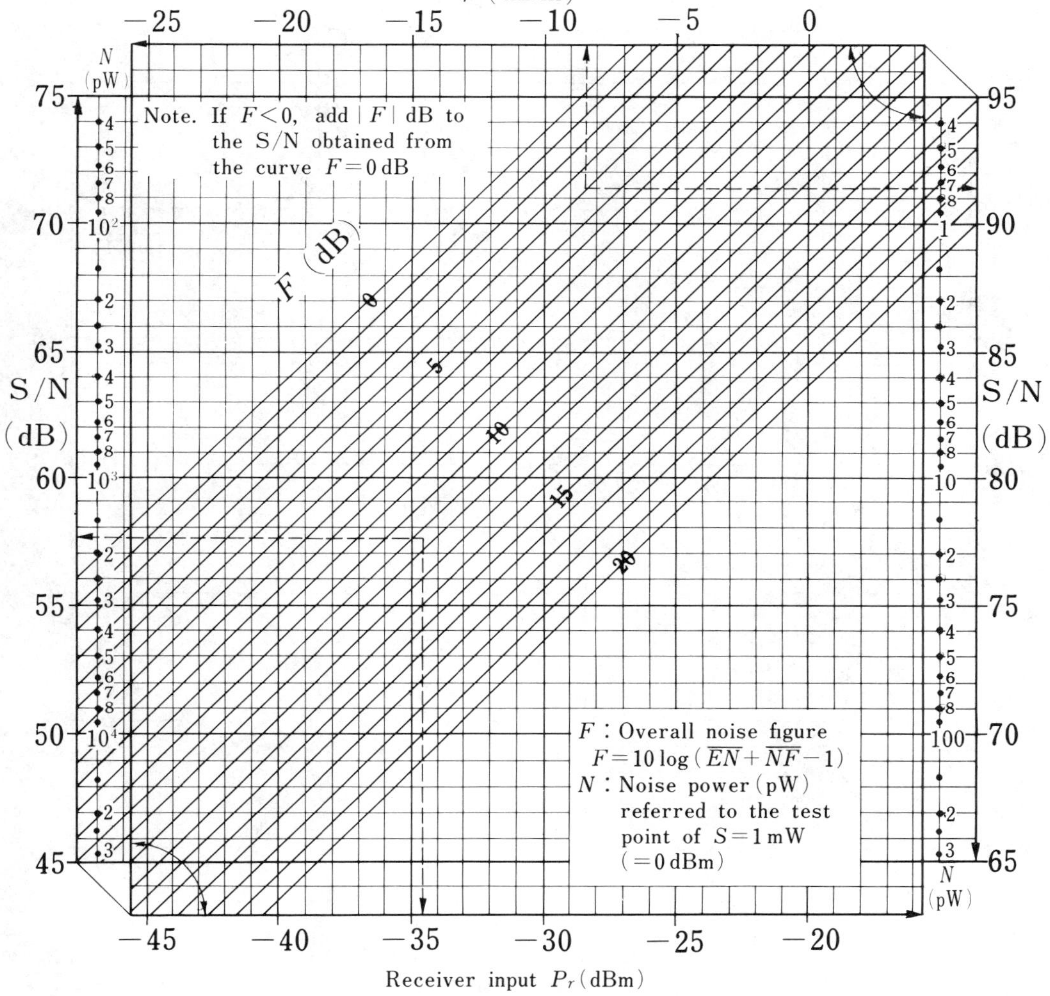

P_r (dBm)
−25 −20 −15 −10 −5 0
S/N (dB)
75 70 65 60 55 50 45
95 90 85 80 75 70 65
N (pW)
Note. If $F<0$, add $|F|$ dB to the S/N obtained from the curve $F=0$ dB
F (dB)
0 5 10 15 20
F : Overall noise figure
$F = 10 \log(\overline{EN} + \overline{NF} - 1)$
N : Noise power (pW) referred to the test point of $S = 1$ mW (=0 dBm)
−45 −40 −35 −30 −25 −20
Receiver input P_r (dBm)

Receiver Input versus S / N Ratio (SS-FM System)

No. of channels	n	1 800 ch
Top channel frequency	f_m	8 204/8 120 kHz
Frequency deviation	S_0	200 rms, kHz/ch
Channel bandwidth	Δf	3.1 kHz
Overall noise figure	F	0∼20 dB

1 800 CH
$S_0 =$ 200 kHz

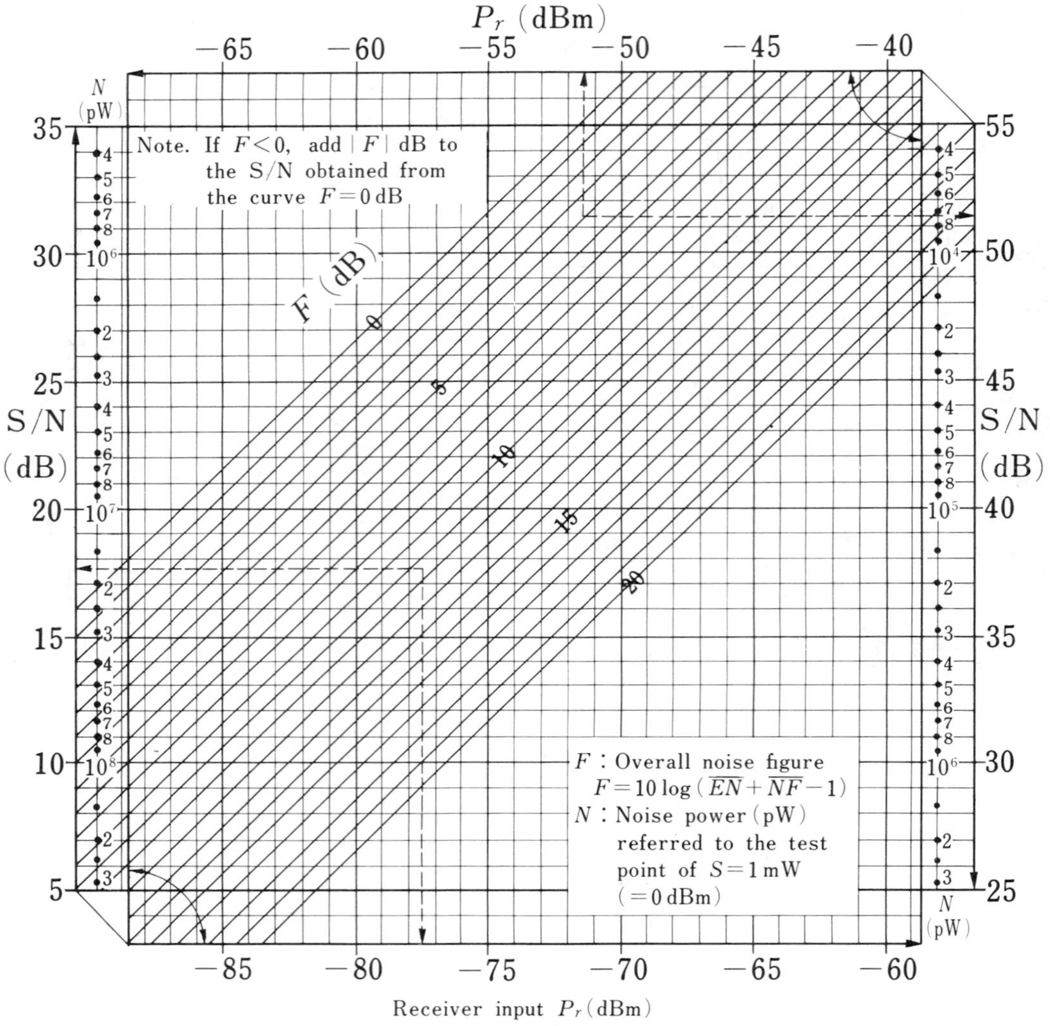

Receiver Input versus S / N Ratio (SS-FM System)

No. of channels	n	1 800 ch
Top channel frequency	f_m	8 204/8 120 kHz
Frequency deviation	S_0	200 rms, kHz/ch
Channel bandwidth	Δf	3.1 kHz
Overall noise figure	F	0 ~ 20 dB

1 800 CH
$S_0 =$ 200 kHz

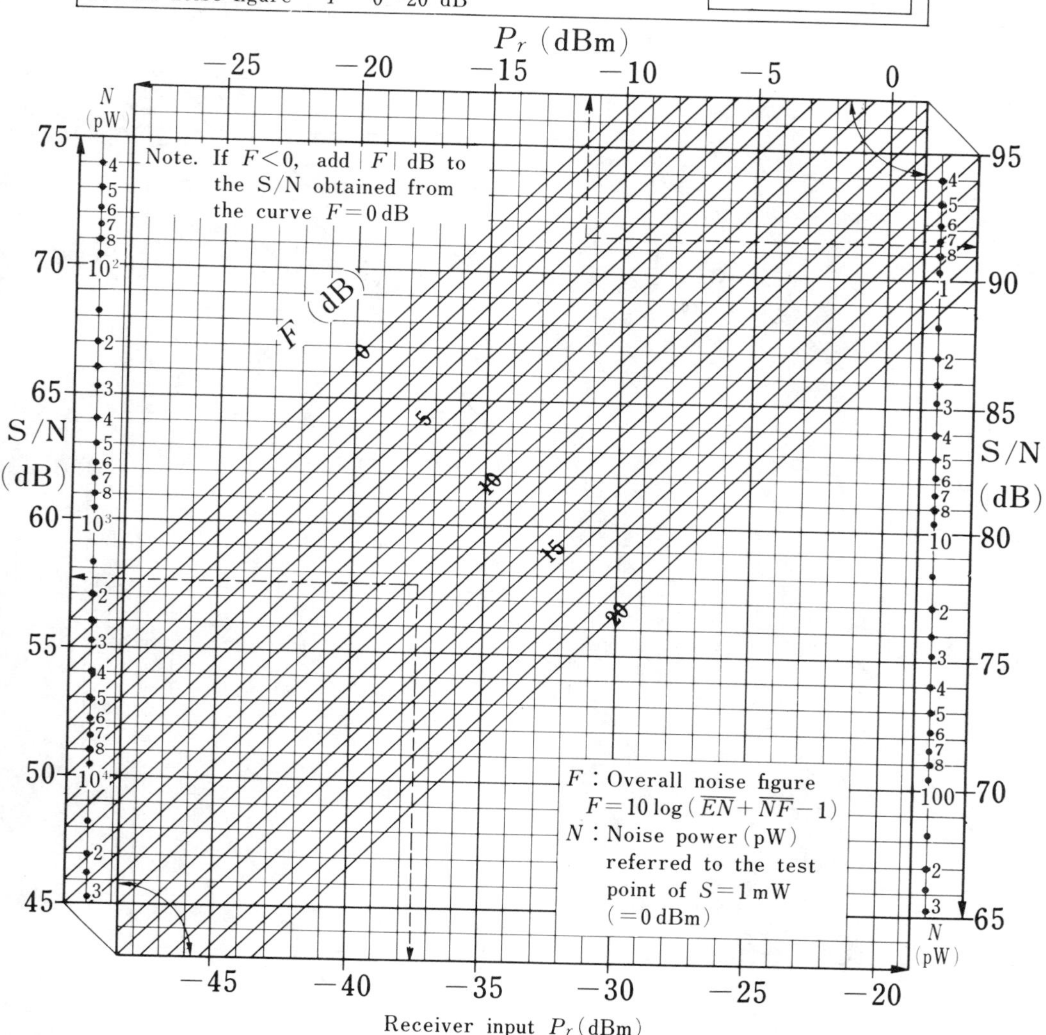

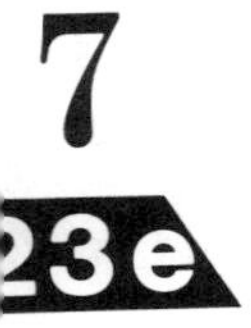

Receiver Input versus S / N Ratio (SS-FM System)

No. of channels	n	1 800 ch
Top channel frequency	f_m	8 204/8 120 kHz
Frequency deviation	S_0	280 rms, kHz/ch
Channel bandwidth	Δf	3.1 kHz
Overall noise figure	F	0 ~ 20 dB

1 800 CH
$S_0 =$ 280 kHz

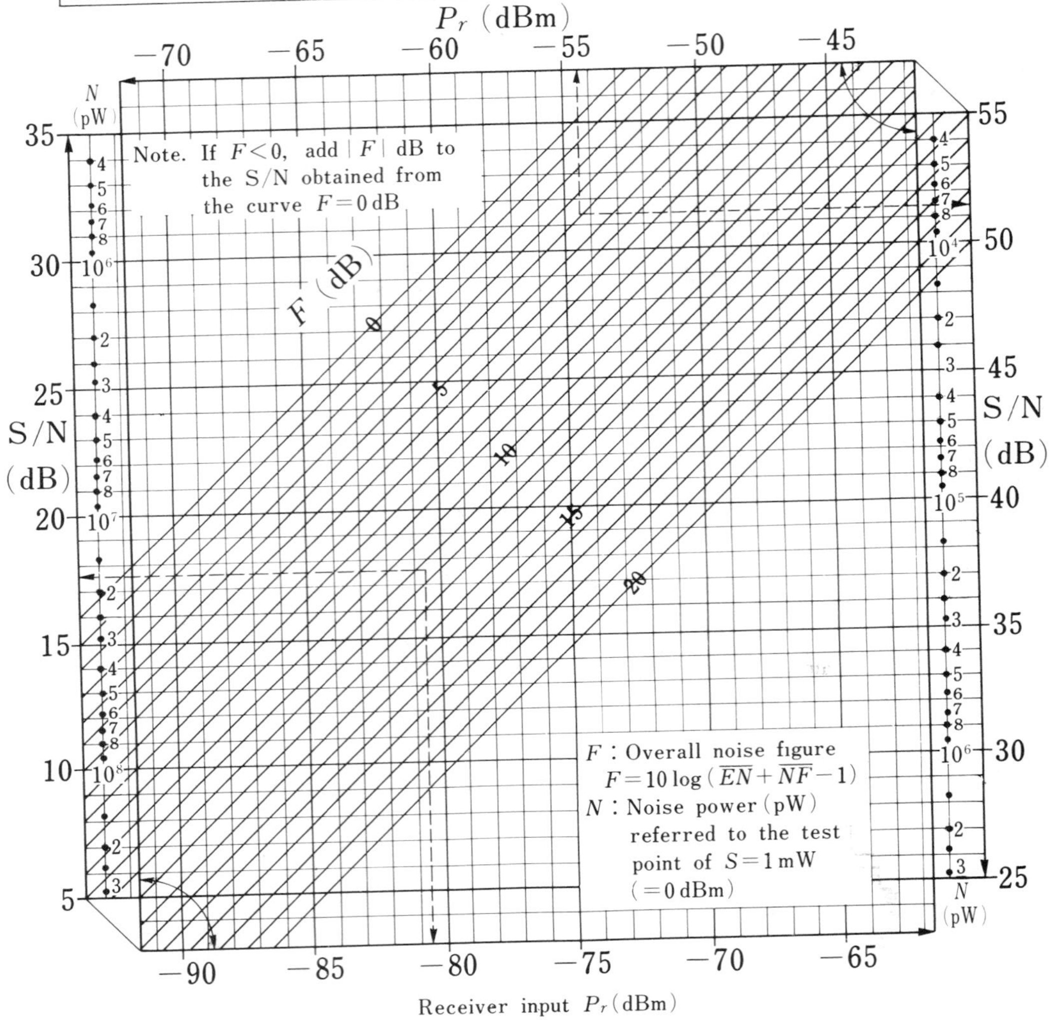

P_r (dBm)
−70
−65
−60
−55
−50
−45
N (pW)
35
30
25
20
15
10
5
S/N (dB)
55
50
45
40
35
30
25
Note. If $F<0$, add $|F|$ dB to the S/N obtained from the curve $F=0$ dB
F (dB)
0
5
10
15
20
F : Overall noise figure
$F = 10 \log(\overline{EN} + \overline{NF} - 1)$
N : Noise power (pW) referred to the test point of $S = 1$ mW (= 0 dBm)
−90
−85
−80
−75
−70
−65
Receiver input P_r (dBm)

Receiver Input versus S / N Ratio (SS-FM System)

No. of channels	n	1 800 ch
Top channel frequency	f_m	8 204/8 120 kHz
Frequency deviation	S_0	280 rms, kHz/ch
Channel bandwidth	Δf	3.1 kHz
Overall noise figure	F	0~20 dB

1 800 CH
$S_0 =$ 280 kHz

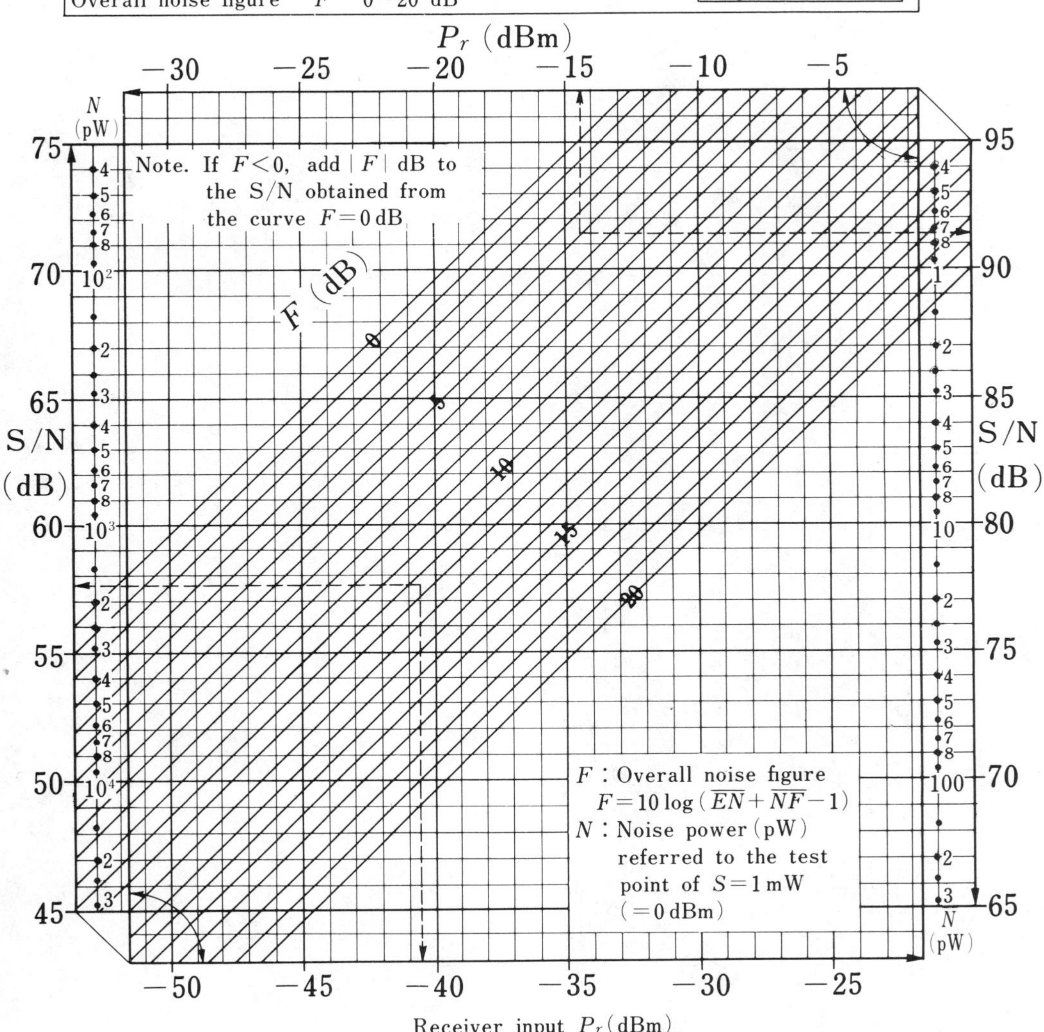

Receiver Input versus S / N Ratio (SS-FM System)

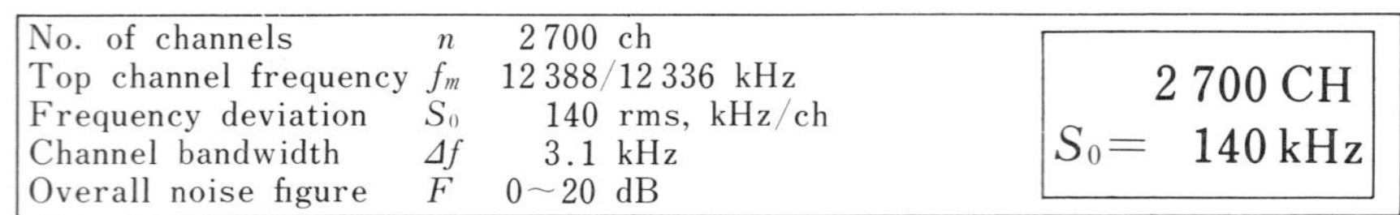

No. of channels	n	2 700 ch
Top channel frequency	f_m	12 388/12 336 kHz
Frequency deviation	S_0	140 rms, kHz/ch
Channel bandwidth	Δf	3.1 kHz
Overall noise figure	F	0～20 dB

2 700 CH
$S_0 =$ 140 kHz

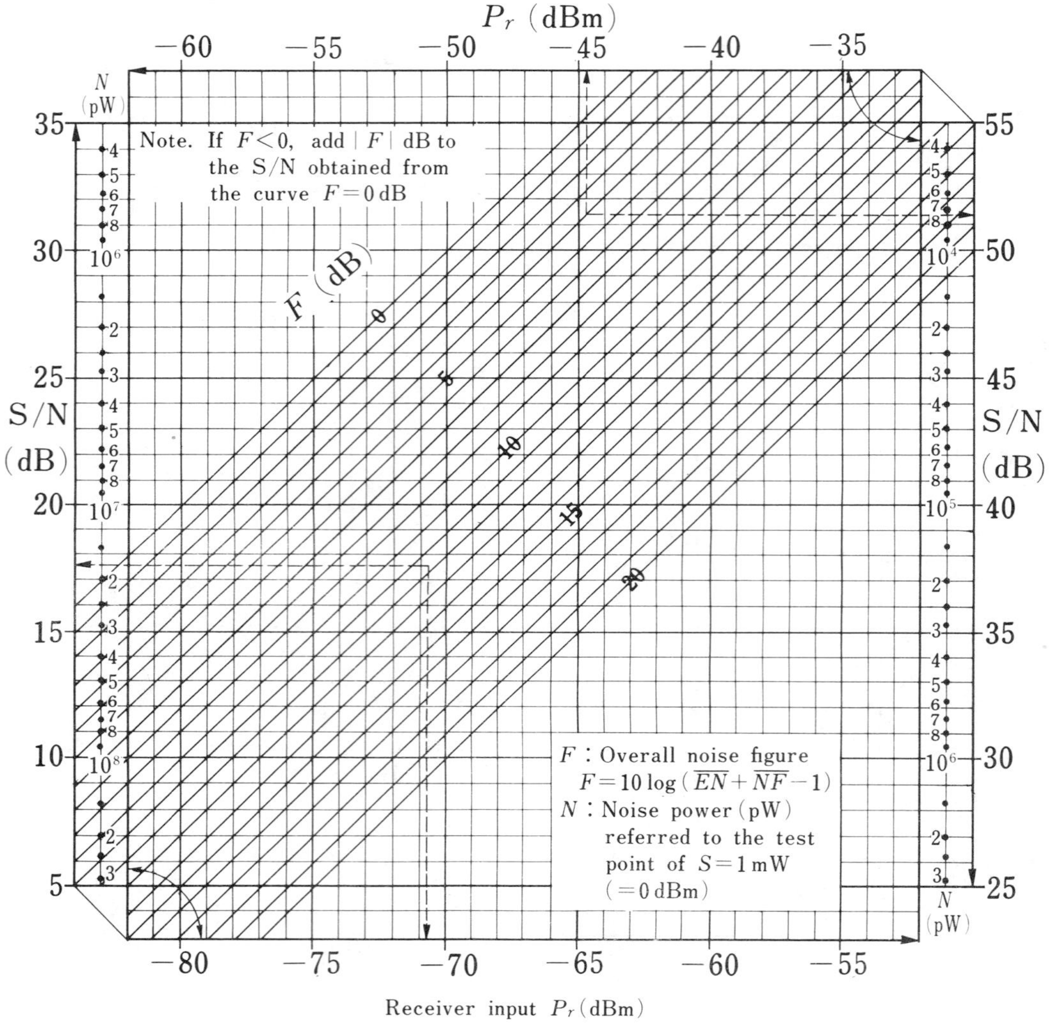

P_r (dBm)
−60 −55 −50 −45 −40 −35
N (pW)
Note. If $F<0$, add $|F|$ dB to the S/N obtained from the curve $F=0$ dB
F (dB)
0
5
10
15
20
S/N (dB)
35 30 25 20 15 10 5
55 50 45 40 35 30 25
10^6 10^7 10^8 10^4 10^5 10^6
F : Overall noise figure
$F=10\log(\overline{EN}+\overline{NF}-1)$
N : Noise power (pW) referred to the test point of $S=1$ mW (=0 dBm)
−80 −75 −70 −65 −60 −55
Receiver input P_r (dBm)

Receiver Input versus S / N Ratio (SS-FM System)

No. of channels	n	2 700 ch
Top channel frequency	f_m	12 388/12 336 kHz
Frequency deviation	S_0	140 rms, kHz/ch
Channel bandwidth	Δf	3.1 kHz
Overall noise figure	F	0~20 dB

2 700 CH
$S_0 =$ 140 kHz

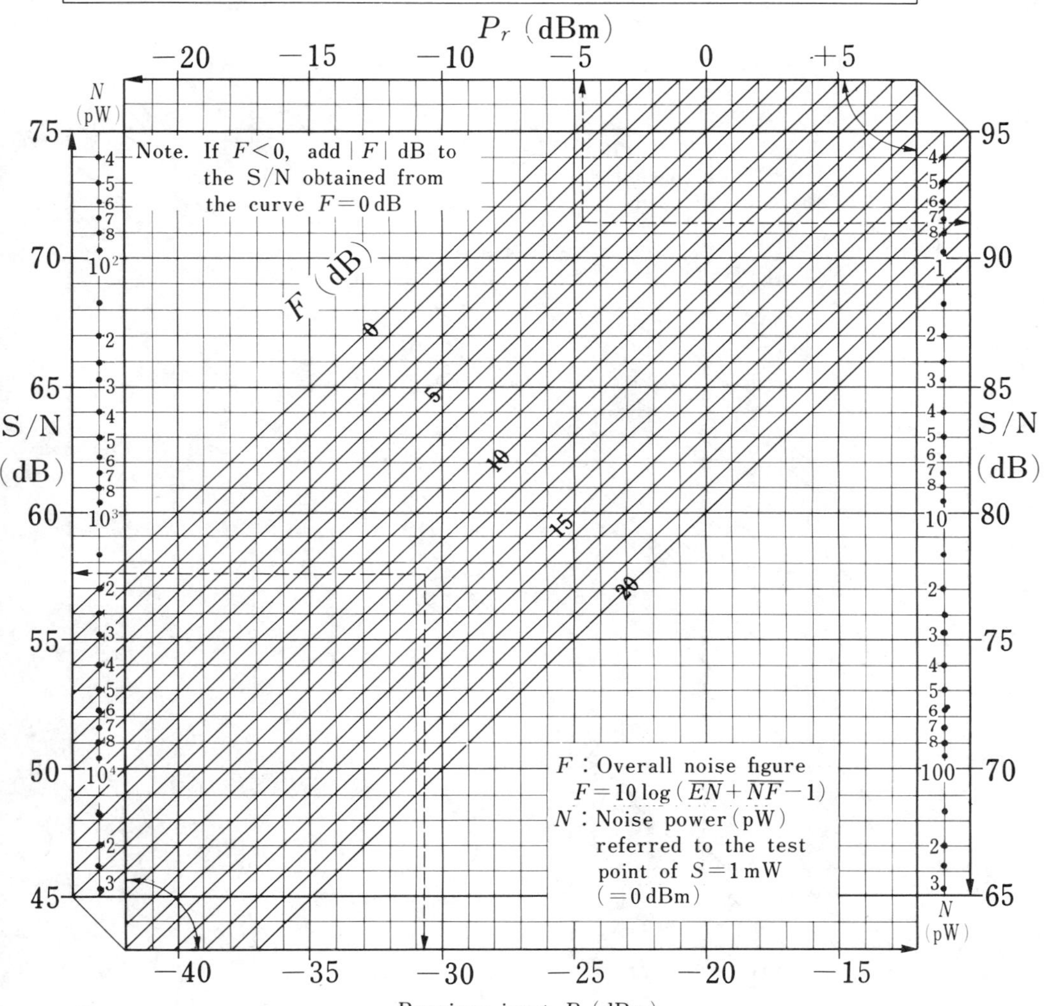

Receiver Input versus S / N Ratio (SS-FM System)

No. of channels	n	2 700 ch
Top channel frequency	f_m	12 388/12 336 kHz
Frequency deviation	S_0	200 rms, kHz/ch
Channel bandwidth	Δf	3.1 kHz
Overall noise figure	F	0～20 dB

2 700 CH
$S_0 =$ 200 kHz

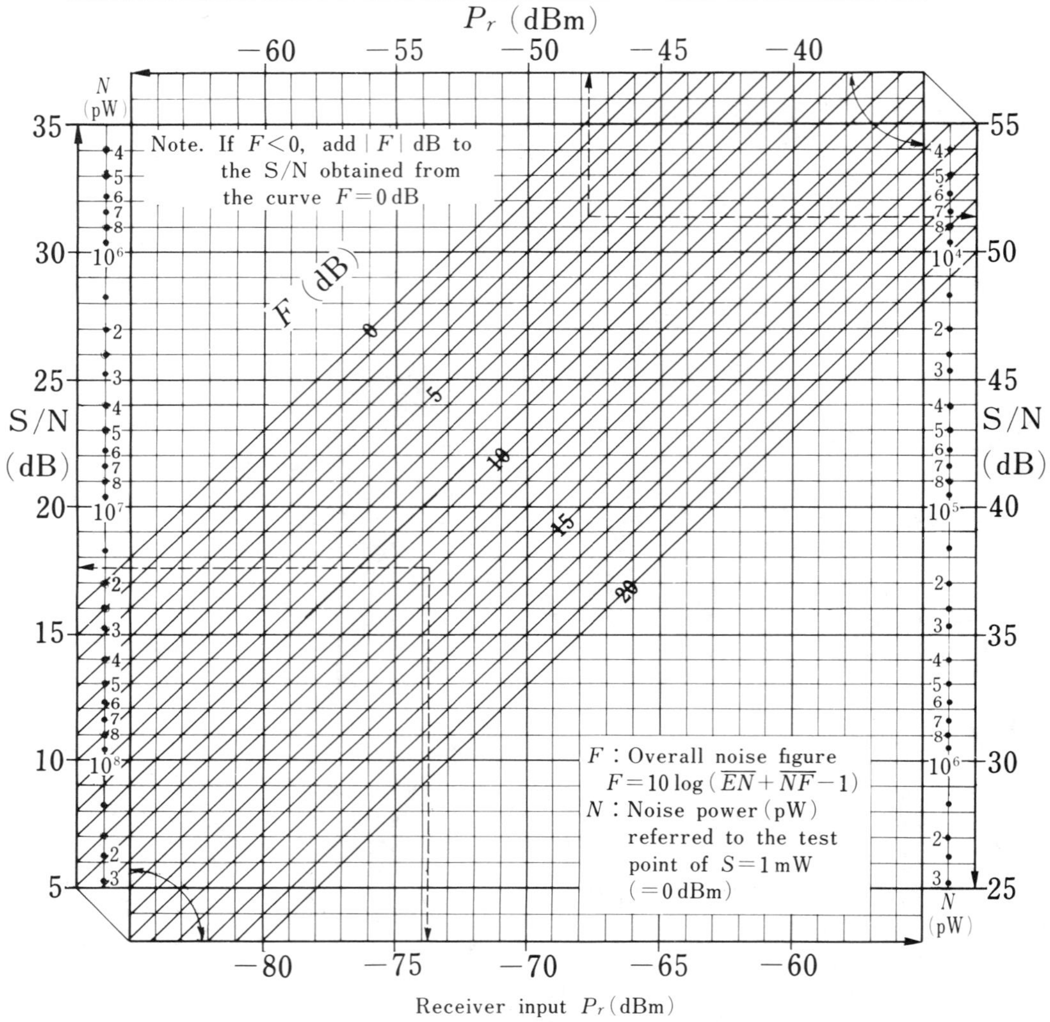

P_r (dBm)

−60 −55 −50 −45 −40

N (pW)

Note. If $F<0$, add $|F|$ dB to the S/N obtained from the curve $F=0$ dB

F (dB)

0 5 10 15 20

S/N (dB)

35 30 25 20 15 10 5

55 50 45 40 35 30 25

F : Overall noise figure
$F = 10 \log(\overline{EN} + \overline{NF} - 1)$
N : Noise power (pW) referred to the test point of $S = 1$ mW (= 0 dBm)

−80 −75 −70 −65 −60

Receiver input P_r (dBm)

Receiver Input versus S / N Ratio (SS-FM System)

No. of channels	n	2 700 ch
Top channel frequency	f_m	12 388/12 336 kHz
Frequency deviation	S_0	200 rms, kHz/ch
Channel bandwidth	Δf	3.1 kHz
Overall noise figure	F	0~20 dB

2 700 CH
$S_0 =$ 200 kHz

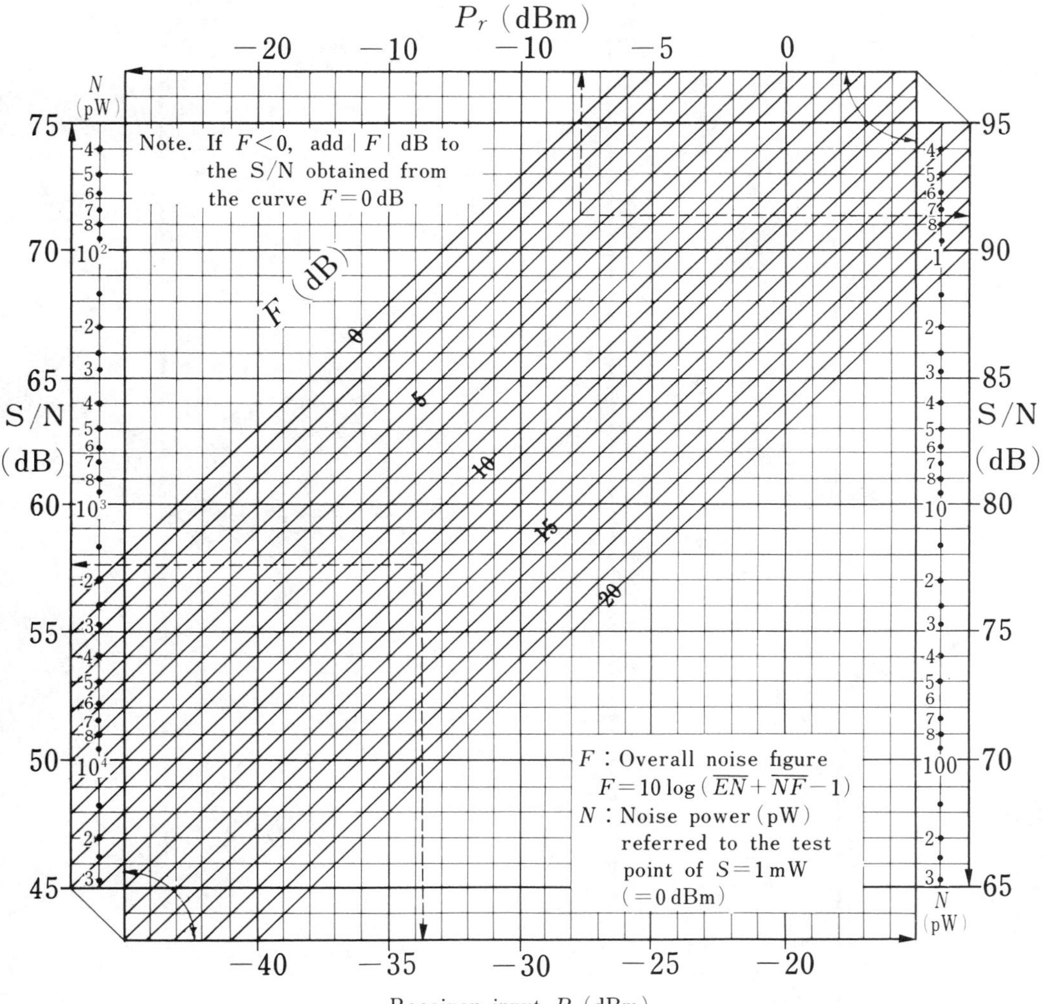

Graph for S/N (dB) to N (pW) Conversion

Value of N(pW) is a noise power expressed in picowatts referred to the test point of $S=1$ mW (dBm)

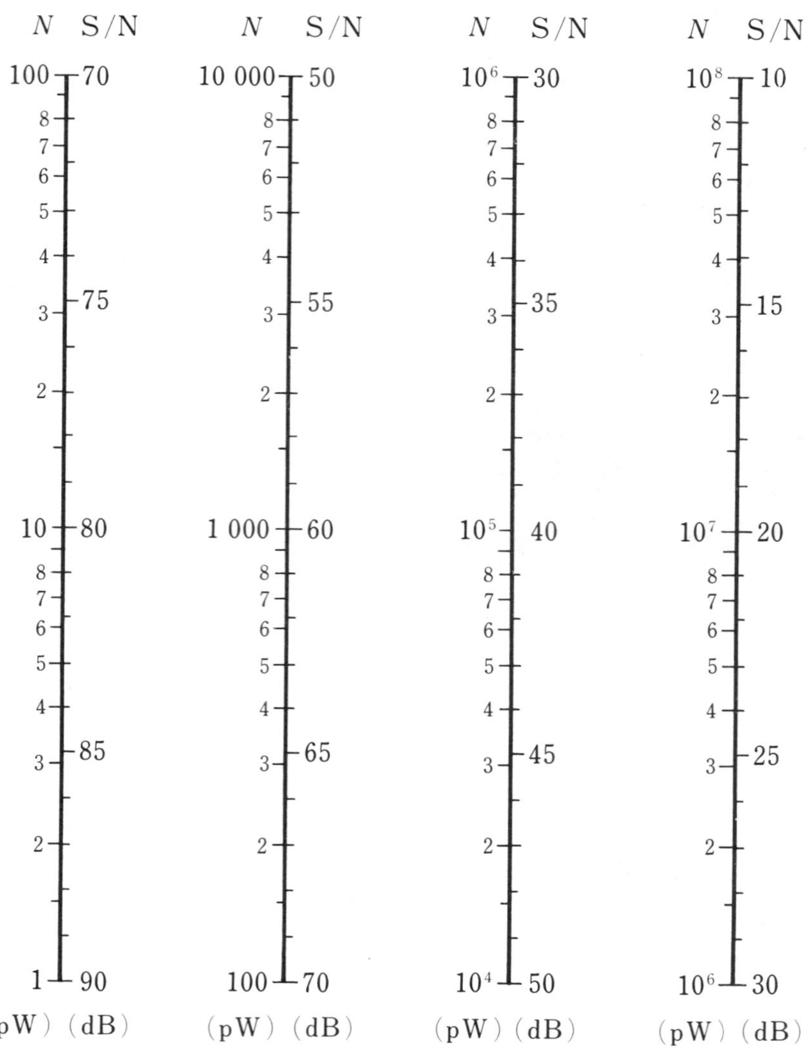

8

FADING ESTIMATION AND SYSTEM EVALUATION

8.1 ESTIMATION LIMIT AND APPROACH TO ENGINEERING DESIGN

A. Fading and Radio Meteorology

Radio waves propagating in a troposphere undergo variations due to changes in meteorological and ground surface conditions.

The field strength variation with time is called fading. A drop in the receiver input signal level causes a corresponding rise of the internal noise of the receiving system, thereby impairing communication quality. A further decrease below a limiting input level (i.e., the threshold of the squelch operation level) will disable the communication transmission system.

A rise in a receiver input signal level following fading will reduce the internal noise and lead accordingly to an enhancement of the system performance through an improvement in the S/N ratio. On the other hand, an excessively strong radio wave may reach a point too far away, thereby causing harmful interference to a system using the co-frequency radio channel. Consequently, unless the fading phenomenon is considered, it is difficult to evaluate any radio system engineering design.

Fading that occurs in and above the VHF band is caused directly by refraction, reflection, diffraction, absorption, scattering of radio waves in a tropospheric atmosphere, tidal movement, snow accumulation, icing, deicing, and the growing and withering of vegetation on a reflection surface or shielding obstacle. The fading behavior is affected greatly by weather conditions of the area (such as local climate, microclimate, surface climate) along and surrounding the radio path. The climatic condition related to fading is called radio meteorology, to distinguish it from a so-called climate sensed directly through human eyes, ears, and skin.

For example, rainfall or strong wind, which can usually be considered rough weather, is a calm climate for radio propagation at frequencies below approximately 10 GHz, with little fading. Conversely, good weather with a gentle wind is an unfavorable radio climate, and very severe fading may occur frequently between evening and midnight.

One factor used widely in radio meteorology is the vertical structure of the atmospheric refractive index perpendicular to the radio path. For example, the refractivity N or the refractive modulus M versus height curve and the main parameters that affect fading obtained experimentally and statistically from the curves are the intensity and probability of duct, the variation distribution of the effective earth radius factor K, etc.

However, in practice the distribution measurement of M and N with height is not so easy; it is in fact almost impossible for an individual radio path. Therefore the relationship between these factors and general meteorological elements on the ground surface have not yet been clarified sufficiently.

Accordingly it is essential to analyze the direct correlation between fading and ordinary meteorological parameters which are processed from available and reliable statistical data covering a period of 10 to 30 years or longer, and also to extract and systematize these parameters, which affect fading.

The method presented herein addresses this problem, but it should be noted that so far the results obtained in estimating fading are very insignificant compared with the task ahead.

B. Estimation of Fading

Experimental studies on radio meteorology and fading have been reported in Japan, the United States, and Europe since 1950, and much progress has been made. Although some of these studies have given academic solutions to important problems, no universal method of estimating fading has been established. This is due to the complexity of the fading behavior and to disclosure restrictions by governments or organizations. Some reports have insisted that it is possible to estimate fading with high reliability. However, close examination has revealed a sort of dogmatic oversimplification of the fading phenomena and the placing of too much credibility on a small amount of data, that is, presuming that such data represent the population by means of statistical mathematics.

Although meteorological observation dates back to the prehistoric age, a scientific method of measuring the principal meteorological parameters has only been developed in the last 100 years or so.

At present short- and long-term weather forecasts are being performed routinely by fixed-point observation from ground and ship weather stations, and also via the three-dimensional observation network of radiosondes, aircraft, rockets, and meteorological satellites, as well as by control and data processing by computer.

Nevertheless, sufficiently reliable weather forecasts, such as "tomorrow's weather in area A" as a daily forecast, and "precipitation in August," or "the first frost of this year" as a statistical forecast, are not yet ensured.

Considering the fact that the reliability of daily weather forecasts is still at the level mentioned, it will be natural to consider estimating fading a superperceptive phenomenon due to the weather. In other words, the uncertainty inherent in fading estimations must always be borne in mind.

It is urged that data collected on fading and climate over the next 10 years be made available and publicized worldwide for scientific and objective analysis to improve the estimation of fading.

C. Procedures to Accomplish a Radio Relay System

All subsequent actions taken to complete a relay system depend to a great extent on the thoroughness of the fading estimation. If the reliability of a design parameter is overestimated, serious damage may result later. On the other hand, if the reliability of a parameter is underestimated, it is also not safe to apply it in the system design. The provision of an excessive margin in the system design, however, is not a sound economic solution.

The methods used in this book for the estimation and system design assume 90% parameter reliability, with engineering covering the remaining 10%. The design may match the requirement or may result in a 10% penalty or bonus. The 10% deficiency is to be made up in the system evaluation and subsequent engineering action for improvement.

Figure 1*a* places radio system noise into three categories. The allowable noise indicates the target according to type and purpose of the transmission system; the designed noise is the noise estimated based on the fading probability of each propagation path and according to the composition of the facilities; and the achieved noise is the actual noise measured after the installation has been completed.

In the planning stage the required performance of the facilities is first specified, and then the locations of the station sites are determined. In this way the designed noise will not exceed the allowable noise.

By considering the limited accuracy achieved in estimating fading and overall noise types it is possible to advance to the implementation stage at an appropriate time before arriving at a 100% credible plan.

After installation has been completed, the overall system noise is measured over as long a period as possible to include fading with seasonal variations. The data thus obtained are evaluated and compared with the allowable noise and the designed noise, as mentioned above. If the measured value exceeds the allowable noise, determine the cause of failure and find the appropriate remedy to improve the system.

Since such an improvement requires redesigning, providing additional facilities, etc., the plan–do–see cycle as shown in Fig. 1*a* will be repeated a number of times from a large circle in the beginning, up to a small circle in the end to complete a perfect system.

Figure 1*b* shows the five steps of the installation procedure for the radio relay system. The essential points are summarized in the figure.

Figure 1*c* is a flowchart, indicating steps to be taken in propagation design, system evaluation, and system improvement. For details, see Sections 8.6 to 8.9.

8.2 BASIC CLASSIFICATION OF FADING

A. Basic Causes and Classification of Fading

The generation of fading in VHF and higher frequencies is attributable in principle to the presence of the ground (surface of earth) and the tropospheric atmosphere. A radio communication system between artificial satellites in space, outside the troposphere, is able to maintain stable free-space propagation without being affected by fading because of the absence of terrain and atmosphere if the antenna orientation is controlled properly.

The line-of-sight and the transhorizon radio systems interconnecting the base stations on the lunar surface are considered quite stable, because the radio path along which the radio waves propagate has the ground but no atmosphere over the lunar terrain. This makes the variation of the reflection coefficient and the diffraction loss over the lunar surface negligibly small, although the temperatures vary greatly between day and night.

The high-altitude and over-mountain propagation path of the earth is considered as an example of the "with-atmosphere-but-without-ground" propagation path. The radio path between two peaks in a mountainous area is almost free from the influence of ground reflection or diffraction. Moreover, it can maintain the almost free-space condition ensuring stable radio propagation since the radio path is well separated from the average ground surface and the atmosphere around the path is effectively mixed due to the air turbulence caused by a rolling mountainous terrain.

Severe fading frequently occurs on a radio path over an area with both ground and atmosphere, especially on flat ground (sea, lake, swamp, plain, etc.). The lower the altitude of the radio path, the more pronounced is this tendency.

The discussion so far has indicated that the influences of the ground and the atmosphere upon fading are not produced independently, but are correlated closely with each other. Also, the presence of both ground and atmosphere is necessary to cause fading.

Fading may be classified according to

1. Cause of generation.
2. Feature of propagation mode.
3. Time sequential mode of variation.
4. Type of statistical distribution.
5. Phenomenon observed in system quality.

Various types of fading may occur occasionally on the same propagation path due to the complexity of interaction between climate, topography, and environmental conditions. A classification of fading by basic types according to direct causes and the variation mode is given in Fig. 2*a*.

B. Atmospheric Fading

The main cause of this type of fading is the tropospheric atmosphere. Another cause is the presence of ground. Fading types A through D for the line-of-sight propagation path as well as their variation modes are shown in Fig. 2*b*.

Type F fading, which normally can be observed in transhorizon long-distance paths, results from interference between the forward-traveling components of radio waves partially refracted and/or reflected when passing through a turbulent atmosphere when several layers or air masses with various refractivity index values develop, vanish, or move.

Type A (Absorbing)

Fading of this type is caused by radio-wave absorption and scattering accompanied by such precipitation as rain, mist, clouds, snow, sleet, etc., or by the absorption through water vapor and oxygen. The resultant receiving signal level fluctuation occurs only below its normal level. The most influential precipitation is rainfall, and its influence can be detected at frequencies above 6 GHz; its influence is most serious at frequencies above 10 GHz.

Water vapor presents the first peak attenuation at approximately 22 GHz, and oxygen shows an outstanding absorption spectrum at approximately 60 GHz. See Chapter 5 for details.

The *M* curve during precipitation is approximately similar to the standard type as shown in Fig. 2*b* (left-hand side).

Since no fading except type A will occur during precipitation, the design margin for fading should be sufficient to cover the estimated attenuation by type A fading or another type of fading, whichever is greater.

Type B (Birefracted Path)

When the *M* curve is not a straight line, its inflection point is apt to move up or down, and its average inclination, or the shape of the curve, changes. Fading occurs as a result of the change of phase difference between waves passing through various portions of the *M* curve. According to the actual measurement of the refractive index versus height distribution obtained via radiosonde, the *M* curve is subject to bending, as shown in Fig. 2*b* (or in the opposite direction) and to interference of the birefracted-path type caused by the interaction between paths 1 and 2, leading to fading.

However, the presence of an inflection point on the *M* curve does not necessarily produce fading. This type of fading occurs only when the inflection point of the *M* curve is situated within or near the first Fresnel zone and is changing its level and the inclination of the *M* curve.

Type B fading is fading with a medium degree of variation ranging from several decibels to 10 dB, and the period of the variation is between several tens of seconds and several tens of minutes. Cancellation by two waves is not clearly observed in this type of fading.

Type C (Scintillation)

If an air mass whose refractivity index differs from that of the surrounding air passes a radio path and its nearby area transversally, the received radio-wave level undergoes a small variation around its mean value. This is due to abnormal refraction taking place on a part of the radio path.

In the case of scintillation fading, the variation is about several decibels, and its period is between a fraction of a second and several seconds. The higher the radio frequency, the more rapid the variation.

The small air mass, which may be warm air, supercooled cloud, dried land breeze, wet sea breeze, etc., is observed frequently even on a windy day, and it is apt to form a duct.

Fading of this type alone can never be detrimental to obtaining a satisfactory system design. However, type C fading may cause an undesirable effect if it is superimposed upon fading of types D, G, K, etc.

The interference propagation path experiences scintillation fading more frequently than other paths because the reflected waves passing through the atmospheric surface layer, whose refractive index is very unstable, are scattered irregularly and superimposed upon the direct wave. Sea waves may become a cause of scintillation fading.

Type D (Duct)

The term radio duct is used analogously with its use denoting a covered channel because in this particular case it describes the propagation of a radio wave within a confining atmospheric layer of abnormal refractivity. The duct in radio meteorology is defined when a portion of the *M* curve has a negative gradient with altitude as shown in Fig. 2*b*. A duct whose lower end is determined by the ground surface is a surface duct, otherwise it is an elevated duct.

If the duct, even when it is recognized in the *M* curve, is not very close to the radio path, no fading occurs. On the other hand, if the first Fresnel zone of a radio path is intercepted by the duct, most of the radio-wave energy may be blocked and prevented from reaching the receiver. (The received signal level may drop from 1/100 to 1/10,000 in power ratio.)

The main wave undergoing fading of this type is decomposed into a number of component waves whose phases and amplitudes take random values. The resulting vectorial combination of these waves at the receiving end produces severe fading within one to several tens of seconds.

The variation mode observed in such a process resembles the theoretical Rayleigh distribution in statistics. Therefore type D fading, which has a large depth and a short period of variation, is often called Rayleigh fading.

Type D fading in the line-of-sight propagation path has such a wide range of amplitude variations as to significantly affect communication. Characteristics of type D fading are summarized below.

1. Type D fading occurs when the *M* curve showing the vertical distribution of *M* includes the negative-gradient portion (*M* duct) within the first Fresnel zone of a radio path. It follows that the *M* curve with a duct portion does not always generate fading. Consequently care must be taken in studying the relationship between radio meteorology and fading. It should also be noted that the probability of fading changes if the radio path profile under consideration changes, even if it is in the same local area.

2. It is almost impossible for identical types of *M* duct to form over the entire radio path. The *M* duct cannot remain unchanged for a long time in the atmosphere near the radio path due to its intrinsic instability.

3. The *M* duct is influenced macroscopically by the mesoclimate (assuming horizontal extension of 200 km or less, within 6000 m or less in altitude). However, the influence by microclimate (extending over 10 km or less within 1000 m in altitude) is considered to be predominant. The formation of the surface duct, which is one type of *M* duct, is closely related to the surface climate (within 20 m in altitude). It follows that the estimation of type D fading or the discussion of the correlation between fading and the *M* duct requires analysis of the climatic environment and the topography of the radio path, as well as of the nature of the terrain under the radio path. Since the radiosonde data collected at a fixed point and a fixed

time are related only to the macroclimate and the mesoclimate, it is pointless to discuss a correlation between them.

4. It will be more reasonable to explain a heavy signal fade experienced during type D fading (Rayleigh fading) to be produced as a result of more than one *M* duct traversing the radio path at a time rather than by a single duct over the entire radio path, as explained conventionally.

5. The frequency of occurrence of the *M* duct and its intensity, depending on the negative gradient and the duct thickness of the *M* curve, respectively, have a climatic locality in the topography and the season that are apt to cause the *M* duct. According to meteorological data, even in a cold region covered by snow and ice, a temperature inversion layer occurs so often that an *M* duct will not always result. Wind speed and humidity are both significant influences on the *M* duct.

6. An *M* duct causing type D fading accompanied by a heavy drop of the receiving signal is formed more frequently (1) in the hotter season than in the colder season; (2) in the lower altitude region than in the high-altitude region; (3) over flat land, the coast, or seawater; and (4) over mountainous terrain.

7. Type D fading occurs more frequently over the coast, seawater, or flat land becuase the radio path is traversed by an elevated duct which develops from an almost ever-occurring surface duct and ascends to an altitude of 300 to 500 m over the ground surface. In a mountainous area, where the sun is obstructed by undulating ground, vegetation, or mountain ridge, the surface duct rarely occurs, and the radio propagation is very stable. An air stream ascending or descending along the mountain slope prevents an elevated duct from being formed. On a rainy or windy day the duct rarely develops even over flat land or seawater (refer also to type G fading in Section 8.2E.

8. In summary, the most essential site selection design is one in which local climatic conditions, radio path profiles, and the nature of the surrounding terrain lying under the radio path are taken into account. It should be noted that the application of diversity (especially space diversity) reception can improve the performance remarkably.

Type F (Forward-Scatter)

Type F fading may be called forward-scatter type or troposcatter type, and its outstanding feature is in its mode of variation. In a transhorizon long-distance propagation path comprising a transhorizon portion of 50 to 100 km or longer and using frequencies higher than the VHF band, neither a direct wave nor a reflected wave exists, and the surface wave and the diffraction wave can be ignored due to extremely large attenuation along the ground surface.

The receiving signal on such a propagation path is affected by air turbulence in a high-altitude region. If, as found on a lunar surface, this region were a vacuum, the receiving signal would be at such a very low level as calculated by diffraction theory.

A similar phenomenon may be observed if there is no air turbulence (the *M* curve is perfectly linear or remains unchanged even when it is bent or curved) even if it is not in a vacuum.

The radio energy emitted into an atmospheric layer with air turbulence is scattered (partially refracted and/or partially reflected). This results in the generation of groups of minute waves having random phases and amplitudes. Part of the energy reaches the destination point.

The minute waves are collected by a large-diameter antenna and combined vectorially to produce a receiving signal.

The type F fading thus formed is very severe. Normally short-term fading of 10 dB or more is superimposed upon the continuously but slowly changing mean level. There is a tendency for the mean level to decrease as the variation becomes smaller.

The troposcatter propagation path is realized only when type F fading occurs.

C. Ground-Based Fading

Ground-based fading is caused by a change in electrical characteristics or in the physical construction of the ground (refer to the typical examples in Fig. 2*c*).

Type P (Path Length)

The most frequently observed type of fading affecting the receiving signal intensity is caused by a change of the path-length difference between direct and reflected waves when the level of the reflection surface changes due to tidal movement. However, it is not correct to conclude that any interference propagation path over seawater always produces the adverse effect mentioned.

The following conditions may be cause for lowering the receiving signal by more than 3 dB below the free-space level.

1. The propagation echo ratio ρ_θ is greater than 0.3 ($= -10$ dB).
2. The path-length difference against the tide is greater than about $\lambda/2$ ($= 180°$ in phase difference).
3. The phase difference between direct and reflected waves at mean sea level is approximately 180° if the path-length difference caused by the tide is very small.

A typical example of long-term type P fading is that caused by a change of the reflection surface level during a period of snow accumulation or thawing.

Type Q (Quality Shifting)

A road surface flooded with rainwater will conceal its concrete or asphalt characteristics. The electrical characteristics of the ground surface will change when a lake surface is frozen over or when buds sprout all over the surface of a withered savanna. For an ordinary propagation path, such a change has little effect on the radio-wave reflection of horizontal polarization (HP), while in the case of vertical polarization characteristics, the effect may be considerable.

Type R (Roughness)

Snowfall can make the surface of a field and bush entirely level in a single night. It may also turn a rice field into a naked field and a sugar or cotton field into a smooth plain over which the horizon may be viewed.

Whether the ground surface is rough or smooth has a great effect on the magnitude of the reflection coefficient. This tendency will become more conspicuous as the wavelength becomes shorter.

Consequently the propagation path design shall be made using the smoothest terrain surface, taking into account a possible change in roughness of the ground surface.

A site selection survey or a radio propagation test performed in a season when the ground is covered by dense bush leads to erroneous system design. Hence interference due to reflection may be found after installation. This is the result of the intensified effect of type G fading in addition to type R fading through enhancement of the equivalent reflection coefficient.

Type S (Shadow)

If an adjacent station that could be clearly viewed over a mountain ridge the previous winter cannot be seen in the summer, it may be necessary to check the leaf growth of trees on the mountain ridge and the presence of clouds and fog. Since the radius of the Fresnel zone is proportional to the square root of the wavelength and also to the radio hop length, the radius r_s at a point on the mountain ridge very near to the station can be on the order of up to 10 m. Therefore varying the clearance within several meters ($\approx r_s/2$) on a grazing radio path may invite a large variation in the propagation loss, that is, from zero up to a large loss relative to the free-space condition.

In general the penetration loss of a radio wave through a defoliated forest is rather small. However, the loss may reach up to 10 dB or more, causing a corresponding reduction of the receiving signal level when the trees become leafy again.

To avoid such an additional propagation loss, it is essential to determine first whether the vegetation on the mountain ridge is from evergreen trees or from deciduous trees. Second, anticipate the growth of these and then design the radio system in such a manner that the station site or the antenna height is determined to ensure the necessary clearance, or that a sufficient margin is provided in radio span engineering to compensate for losses caused by that tree growth.

Type S fading may become a serious problem in determining the service area for a television broadcast or for a mobile radio system. According to experience, the reception at a vehicle moving in the suburbs or on a mountainside is rather good in winter but poor in the period from early summer to autumn. For television reception beyond deciduous trees, a distinct change of television picture quality may be observed when combined with the seasonal variations of atmospheric fading (less serious in winter and very serious in summer). Since the propagation loss may increase by 10 to 20 dB in the summer over that in the winter, the service area should be determined with seasonal factors in mind.

D. Man-Made Fading (Type U or V)

This type of fading is not serious at frequencies below 100 MHz because of its very large diffraction coefficient and lower antenna directivity (except for particular cases). Generally speaking, the diffraction coefficient of the radio wave is proportional to the square root of the wavelength. Also, the antenna beamwidth is proportional to the wavelength and inversely proportional to the aperture diameter. Accordingly this type of fading cannot be ignored for a large-aperture antenna used in a high-frequency band.

Type U (Caused by Moving Object)

An example of type U fading is found in the decrease of field intensity of a radio link installed near an airport or harbor. This is due to the radio path being obstructed by aircraft or ships. However, the resultant loss will be less than approximately 6 dB provided that no more than half the first Fresnel zone is obstructed.

If the main beam (equivalent to approximately 2.4 times the half-power beamwidth) of an antenna is obstructed completely, the resultant loss will reach 20 dB or more, causing serious impairment.

In the engineering design of a radio system for navigation, beacon, and safety, special attention should be paid to such large moving objects as jumbo jet aircraft and supertankers, which are used extensively at present. In an obstructed radio path the reflection due to moving objects may cause serious interference.

Type U fading may also occur between moving objects. A patrol boat or a tugboat may sometimes by obstructed by another vessel. However, such difficulty can be avoided by establishing more than one base station.

Type V (Variable Antenna)

The main beam axis of an antenna mounted on top of a high tower is deflected by wind. The tolerable deflection of the antenna main-beam axis with respect to the center axis of the emission should be within half the half-power (−3 dB) beam width in both the horizontal and the vertical planes. By increasing the deflection further from the tolerable value, the transmitting (receiving) level falls rapidly, and the resultant loss of the signal level reaches 20 to 30 dB.

Ordinary towers connected by bolts and nuts tilt toward the direction of wind when a storm wind blows, because of the disposition of the gap between the hole and the bolt at each connecting point. Such deflection changes its direction as the wind direction changes and combines with the intrinsic vibration of the tower. Thus, to avoid excessive signal attenuation, the tower structure design and construction should be made so that the maximum deflection of the antenna direction against the estimated wind velocity may be kept within a tolerable limit.

However, since such an extremely stringent performance may have an unrealistically high cost, some trade-off between performance and cost will become necessary. Several cases have been reported in which the receiver input decreased after the system was commissioned and severe fading occurred during a period of strong wind (no other type of fading occurred during this period). Such fading is attributable to incorrect initial antenna orientation or insufficient adjustment after having made the correct orientation. (This is particularly important for a multiantenna system.)

It is desirable that the antenna be installed with great care, that is, that sufficient time be allowed for correct antenna orientation. Thereafter fix the position with lock pins. Then perform a careful final inspection during installation work, followed by periodical inspections (at least once a year).

E. Composite Fading (Types G, K, and T)

Although several combinations of types A, B, C, and D are possible, the typical and probable combination types are G, K, and T (see Fig. 2*b*).

Type G (Ground Duct Interference)

When the ground is heated by the sun, the temperature of the air above the ground surface rises rapidly and water evaporates. Immediately after sunset, radiant cooling starts, and temperature inversion occurs in the air near the ground surface.

On the other hand, the atmospheric layer close to the seawater surface may produce temperature or humidity inversions. This is because such a layer is always maintained in a state of humidity saturation through evaporation by solar heating and seawater splashing. After sunset the wind from the cooled land flows onto the sea surface. In all cases, that is, whether it is over the sea, over land, in daytime, or at nighttime, the surface atmospheric layer over the earth, including both sea and land, always presents an anomalous M curve which is clearly different from that for the standard refraction gradient. This tendency is particularly pronounced in the atmosphere within 50 m above the earth, and it can easily lead to the formation of a surface duct.

Reflected waves on an interference propagation path undergo irregular refraction and reflection upon entering and leaving the surface duct. Hence many waves with random phases and amplitudes thus formed interfere with the direct wave, thereby producing severe fading. This is called type G fading. Although the fading on the interference propagation path has been explained so far as the result of the overlapping of duct type D fading with type K fading, this explanation is not sufficient to describe the phenomenon. It is more accurate to explain it as the result of the combined effect of type G, type D, and type T fading (i.e., G + D + K). Type G fading will become very harmful if the propagation echo ratio ρ_θ is large. Even where fading is caused only by type G, the effect of ρ_θ is serious. Fading of this type will be minimized when the equivalent reflection coefficient ρ_θ is made sufficiently small through a rough reflection surface or through antenna directivity (suppression of the reflected wave using the multiantenna system).

Type K (Interference K)

No fading occurs in the free-space propagation path for which the M curve for the atmosphere over the path is essentially a straight line (the effective earth radius is defined as the slope of the straight curve), even when K or the slope of the M curve varies. However, in the interference propagation path with a reflected ray, varying the M curve slope changes the phase difference between direct and reflected waves and produces a fading whose magnitude depends upon the phase shift $\Delta\Theta$ and the propagation echo ratio ρ_θ. This type of fading is called type K and is observed most frequently in propagation paths with pronounced reflection. The characteristics of type K fading are described below.

1. As mentioned in detail in Chapter 2, the maximum and minimum receiving input levels obtained by combining the direct wave with the reflected wave, whose amplitude is ρ_θ times the amplitude of the direct wave, are given relative to the direct wave as

$$\begin{aligned} &\text{maximum value} \quad 20\log(1+\rho_\theta) \quad (\text{dB}) \\ &\text{minimum value} \quad 20\log(1-\rho_\theta) \quad (\text{dB}) \end{aligned} \tag{8.1}$$

The range of variations of the receiving input levels is obtained during type K fading where the phase shift exceeds 2π rad (corresponding to λ), which agrees with the difference between the two extreme values given above. Since the theoretical maximum of ρ_θ is unity, the receiving level will never increase more than 6 dB for type K fading, but it will drop rapidly, approaching the theoretical minimum $-\infty$ dB as ρ_θ approaches -1. The effective reflection coefficient of a water surface ρ_e (equal to ρ_θ in this case) is between 0.7 and 0.95, depending on the roughness of the water surface, the radio frequency to be used, the grazing angle, and the polarization, neglecting antenna directivity. Assuming that the maximum ρ_θ obtained in a radio propagation path is 0.95, the range of the type K fading will be between $+6$ dB and -26 dB, or 32 dB in total.

2. In the microwave frequency band a slight change of the M curve slope, and therefore a slight change in the K factor, may easily increase the range of variations of the phase difference between direct and reflected waves $\Delta\Theta$, and the range often exceeds 2π rad (equivalent to λ). The resultant receiving level oscillates between the maximum and the minimum. In longer wavelengths, as in the VHF and UHF bands, the geometrical conditions of the radio path, such as antenna height and distance, and the variation range of the K factor will determine whether or not the receiving level varies while passing through the above-mentioned maxima and/or minima.

However, in general the fading range also depends on the relative phase between the direct and reflected waves at the K factor (usually equal to 4/3 under normal conditions), depending on whether the relative phase is around the in-phase region or in the inverse-phase region.

3. Usually the radio path where the phases of the reflected and the direct waves nearly cancel each other (close to the minimum receiving field intensity) is unstable in terms of propagation. This is because a slight change of the K factor easily causes a substantial drop in the receiving input, bringing the receiving level to the minimum if ρ_θ is large. If the state of a radio path is close to the in-phase condition, the minimum point can be kept sufficiently away from the normal state and the radio propagation of the path is quite stable. The phase relationship between these waves is found using the method in Chapter 2. In this respect, it is most important to know the probable statistical range of variations of the K factor. It is almost certain that the variation of the K factor defined by a slope of the M curve (CCIR Recommendation 310-2) will be between 1 and 2 at most, and the value of the K factor can hardly fall below 1 or rise above 2. Radio propagation in an atmosphere where the variation of the K factor can usually be defined as $1 < K < 2$ is called normal propagation. The propagation with fading (excluding slight fading such as scintillation fading) in an atmosphere whose M curve has a point of inflection is called anomalous propagation.

4. Estimating type K fading is not very difficult since it is sufficient to make the estimation only for normal propagation. Let its range of variations be F_k (dB). Then

$$F_k = |A_{k1} - A_{k2}| \quad (\text{dB}) \tag{8.2}$$

$$\begin{aligned} A_{k1} &= 10\log\left[1 + \rho_\theta^2 + 2\rho_\theta\cos(\Delta\Theta_{k1})\right] \\ A_{12} &= 10\log\left[1 + \rho_\theta^2 + 2\rho_\theta\cos(\Delta\Theta_{k2})\right] \end{aligned} \tag{8.3}$$

where A_{k1} = maximum level of receiving wave relative to direct wave
A_{k2} = minimum level of receiving wave relative to direct wave
$\Delta\Theta_{k1}$ = angle that is the phase difference between direct and reflected waves, and that is farthest from 180° (π rad) when the K factor varies between 1 and 2 and the phase difference is expressed as an angle between 0° (0 rad) and 360° (2π rad)
$\Delta\Theta_{k2}$ = same as above except that the angle is the one nearest to 180°

To find these angles, Fig. 15 in Chapter 2 is used by substituting $\Delta\Theta_k$ for Θ and A_k for A.

Several views have been given on the range of variations of the K factor. For example, one investigation reports that $K \geq 2/3 \approx 0.8$ and $-\infty < K < -4$ occur 99.9 and 0.1% percent of the time, respectively. However, several negative views have also been reported. It is indispensible to determine the effective range of variations of K for radio propagation design, and this problem is treated in Chapter 0.

5. If the propagation echo ratio $\rho_\theta \leq 0.1$, the signal fade caused by type K fading will never exceed 1 dB. Hence the radio path with $\rho_\theta \leq 0.1$ and free from any other influence such as diffraction or absorption, is regarded as a free-space propagation path.

Type T (Tangential, Ridge K)

When a mountain obstructs the line of sight of a direct ray or is lying underneath and closely to the direct ray, diffraction loss is produced. If K varies, the obstruction height above or the clearance below the line of sight changes accordingly, leading to fading due to diffraction loss.

1. Fading of this type can be observed most clearly when the radio path is nearly tangential to the mountain ridge. This is why it is called type T fading.

2. The range of variations is approximately proportional to $\sqrt{f}$, where f is the frequency. However, the drop in the signal level is not as large as in type K fading, barely exceeding 8 dB.

3. The greater the ridge height above the line of sight (poor line of sight or large diffraction parameter U), the smaller the ranges of variations. For example, for a 50-km radio path using 8 GHz or lower frequency and having an obstructive mountain at its center, at a height more than three times that of the first Fresnel zone radius ($U \geq 3$), the range of variations will be within 3 dB.

4. The range of variations of fading decreases as visibility is improved at $K = 4/3$ or as the clearance is increased (or as the clearance parameter U_c becomes greater). Assuming the same frequency and the same ridge as in item 3, and also assuming a clearance greater than the Fresnel zone radius ($U_c \geq 1$), the range of variations will be within approximately 2 dB, and the drop in the receiving signal will be within 1 dB relative to the free-space level.

5. Therefore any radio path to the ridge with a clearance of the first Fresnel zone radius or greater, and free from the influence of reflection and/or absorption, is regarded as a free-space propagation path.

6. In general the radio path of the type in item 5 is very stable since it eliminates the influence of such severe fading as types G and K, and also reduces type B fading.

8.3 FADING ON ACTUAL RADIO PATHS

Radio paths are classified into a variety of types according to topography, social situation, and economy. Figure 3 summarizes several types of fading encountered in typical radio paths. Each of these basic types of fading may occur independently timewise or in combination with another type on the same radio path.

Several features of radio paths are described. Since type A fading becomes significant only when the frequency is raised to 10 GHz or higher, any other type of significant fading rarely occurs during precipitation. Also, since type A fading can be found on any path, no particular explanation will be given.

A. Free-Space Radio Path

Criteria for the free-space radio path are that the propagation echo ratio $\rho_\theta \leq 0.1$ and that the clearance be more than half the first Fresnel zone radius. However, even when these conditions are satisfied, the frequency and the depth of the fading will vary greatly depending on the radio meteorological and regional environment, the relation between radio path and topography, the proportion of the plain, and the characteristics of the ground. Although the birefracted path-type fading is observed most frequently, a heavy drop in the receiving signal occurs in duct-type fading as well as in types B and D combined. In general this type of propagation path is considered most stable compared to the other types.

B. Interference Radio Path

Duct-type fading, or type D fading, for direct rays, type G (grounded duct interference) fading mainly for reflected rays, and type K fading interact with each other to worsen the fading. The propagation echo ratio ρ_θ determines the degree of fading. The outstanding feature will appear when ρ_θ becomes larger and approaches about 0.3. The frequency of such fading to produce a circuit interruption increases rapidly when ρ_θ exceeds 0.6. Type B (birefracted path) fading, type C (scintillation) fading, and type P, Q, and R (topographical) fading are less harmful, giving only a secondary effect if no other fading occurs simultaneously. However, when these types of fading are superimposed on type D, type G, or type K fading, the resultant fading becomes quite serious.

C. Ridge Diffraction Grazing Radio Path

The most significant deep fading is type D, followed by type B and then type T inherent in this type of radio path. In addition, type S fading caused by forest or accumulated snow on top of a ridge is included. However, usually the effects of type T and type S fading are insignificant. Then the variation caused by this fading may be assumed to be similar to that in the free-space propagation path.

By shielding the lower half of the space contributing to radio propagation, the effects due to the lower portion of the *M* curve or to the lower layer duct leading to improvement of the stability of propagation may be eliminated.

D. Mountain Diffraction Radio Path

In a deep diffraction region the range of variations due to type T (diffraction K) fading falls within approximately 2 dB and is negligible. A single-ridge diffraction propagation path in the fading analysis of this path can be treated as a combination of two free-space propagation paths connected at the mountain ridge located between the transmitting and the receiving points. In general this type of radio path is stable. However, the multiridge propagation path is considered unstable because of its complicated propagation mechanism.

E. Mountain Diffraction Interference Radio Path

When there is significant ground reflection ($\rho_\theta \gg 0.1$) between the transmitting (receiving) point and a mountain ridge, the effects of type K and type G (reflected-ray surface duct) fading combine to increase the signal variation. This increase depends on the propagation echo ratio ρ_θ, which represents the overall effect expected between transmitting and receiving points. The diffraction loss by mountain ridges is greater for the reflected ray than for the direct ray, and consequently, the equivalent reflection coefficient is reduced as well as ρ_θ (refer to Chapter 3).

If both sections, that is, the transmitting side and the receiving side, have reflection, several radio ray reflections are possible. However, any radio ray undergoing two or more time reflections can be ignored in the radio frequency band or UHF or higher. Therefore it suffices to analyze only the interference occurring between a main radio ray and two other radio rays with a single reflection. If the radio path contains two or more mountain ridges, the reflection between ridges may be negligible.

F. Tropospheric Scatter Propagation Path

Since the air turbulence as seen in type F fading is a principal means of conveying the signal to the receiver, severe short-term fading occurs continuously. The addition of type D (duct) fading, type G (reflection surface duct) fading, type B (birefracted path) fading, and type C (scintillation) fading to type F fading may produce a very complicated variation mode. Features of this fading are given in the following.

1. Short-term fading of 10 dB or more occurs continuously.
2. Yearly variation is observed everywhere except for a part of the tropical region. In the medium-latitude zone the mean receiving signal level is low in the winter and high in the summer.
3. At sunrise and sunset a large variation is observed.
4. The range of variations is greater on a radio path over seawater than over land.
5. The radio meteorological condition is different for different localities along a single radio section.
6. The statistical distribution of a mean level during a period from 10 min to 1 hour is approximated by a decibel normal distribution, and that for an instantaneous value within a few minutes by a Rayleigh distribution. In such a radio path it is necessary to estimate statistically the mean level as well as the range of variations since no stationary main ray is found there. Methods to do this are based on actually obtained data as given in Chapter 4.

G. Spherical-Ground Diffraction Radio Path

At frequencies in and above the VHF band, the diffraction loss over the spherical ground and the attenuation of the surface wave are very large. Hence the tropo-scatter wave component will become dominant if the length of the transhorizon portion is greater by about 50 km. Thus it is no longer correct to call the path a diffraction propagation path. Therefore this type of propagation path is limited in application, and very severe fading occurs because the main rays pass very close to the earth. General characteristics of this type of path are as follows.

1. Fading occurs almost continuously since the main path passes the surface duct region over a long distance.
2. Particularly large variations are observed at sunrise and sunset on a clear day. It is not unusual to observe an increase of electric field intensity of 20 to 30 dB over the mean value.
3. The electric field intensity, when fading ceases, falls approximately to a diffraction propagation level.

This type of propagation path exhibits the largest fading range and the lowest propagation stability of all the types mentioned.

8.4 STATISTICAL TREATMENT OF FADING

The fading that is a temporal variable requires the use of a statistical method for quantitative presentation. The components described in the following are considered essential for collecting and analyzing the data and for the estimation of fading.

A. Variables

Fading inherently implies a variation of electric field intensity, power density, received voltage, and received (input) power or propagation loss. However, it may be regarded as a variation of noise power or S/N ratio, determined by the parameters mentioned above. It should be noted that the latter depends on the type of modulation and the performance of the receiving devices. The noise appearing at the output of the demodulator comprises both distortion noise and interference noise.

B. Sampling Time Interval

For practical fading measurement, and for data processing, the minimum time interval required for the measurement of fading must be determined.

To ensure accuracy of measurement, a smaller interval is preferred. However, decreasing the interval will increase the amount of data, thereby making data processing difficult. Thus this interval is determined taking into account the minimum period of fading. In general the troposcatter propagation path requires intervals of 5 to 50 ms to measure short-term fading; the other type of propagation path requires intervals of 1 second or longer.

In the line-of-sight propagation path the duration of fading decreases with increasing fading depth. Rarely does the period of the fade fall below 1 second. Here 1 second is used as a reference unit in the estimation of fading, as in Section 8.8.

C. Statistical Period

The statistical period differs depending on the use of the data, as shown in Fig. 4. Usually it agrees with a commonly used time unit; 5 ms or 1 s is used as the sampling unit to directly obtain the statistical distribution. The long-term statistical distribution is obtained through 1-min or 1-hour mean values. The radio meteorological condition is repeated once every year, which corresponds to the revolution of the earth around the sun.

However, it may be very irregular, as indicated by such characterizations as cool year, warm year, or dry year. The fact that a statistical period exceeding 1 year is not justified in terms of economy suggests that 1 year be recommended as the maximum for the statistical period.

When the yearly variation is to be taken into account, some arrangements must be made such as the following.

1. A certain allowance is made for data taken in a specific year.
2. Statistical values obtained from the meteorological data for a period of 5 to 10 years or longer are used.

The table in Fig. 4 shows time rates of various kinds of statistical periods for a specific unit period. Integers shown indicate the quantity of samples for the statistical distribution at a higher level.

D. Reliability and Fade Probability

To process the sampled statistical data of various propagation parameters pertaining to fading, the cumulative distribution curve in Fig. 5 is used. The graph has a scale for a normal (Gaussian) distribution. When the samples follow the normal distribution, the distribution curve becomes a straight line. The mean and the median (50% value) of the normal distribution are equal.

A relationship exists between the deviation from the median to the sample value at an arbitrary percentage and the standard deviation σ as shown in Fig. 5[B].

The long-term distribution (e.g., 1 month) of the receiver input or noise power level expressed in decibels, and caused by fading, almost follows a normal distribution. However, this does not apply for a portion of an extremely small or an extremely large percentage, that is, saturation occurs in a portion of the higher receiver input while a decrease continues further in a portion of the lower receiver input.

The percentage of time is expressed in the reliability q or the fade probability p in Fig. 5,

$$\begin{aligned} p &= 1 - q, & (q &= 1 - p) \\ p &= 100 - q \quad (\%), & q &= 100 - p \quad (\%) \end{aligned} \tag{8.4}$$

Whether p or q is to be used depends on the purpose. Thus to avoid confusion, it is recommended that the following be defined clearly.

Reliability q

The cumulative probability expressed as the percentage of time that a receiver input level exceeds P_r (dB) for the total period of interest, that is, the reliability q (%) at which the receiver input exceeds P_r, is expressed as

$$q(P_r\,\mathrm{dB}) \quad (\%) \tag{8.5}$$

The receiver input P_r corresponding to a certain reliability q (%) is expressed as

$$P_r\,(q\,\%) \quad (\mathrm{dB}) \tag{8.6}$$

denoting by q (%) the reliability of the receiver input P_r.

As to the noise power, the reliability that the noise power is below N (pW), and the noise power N that will give q (%) reliability are expressed as

$$\begin{aligned} &q\,(N\,\mathrm{pW}) \quad (\%) \\ &N\,(q\,\%) \quad (\mathrm{pW}) \end{aligned} \tag{8.7}$$

Fade Probability p

This is the opposite of the reliability mentioned above, and these two factors are interchangeable. However, it should be noted that this may produce a slight error depending on the grading interval for which a sample value is given.

The notation and the definitions of the related parameters are as follows.

$p\ (P_r\,\text{dB})$ (%)	(fade probability that receiver input falls below P_r dB)	(8.8)
$p\ (N\ \text{pW})$ (%)	(fade probability that noise power exceeds N pW)	
$P_r\ (p\ \%)$ (dB)	(receiver input corresponding to fade probability p %)	(8.9)
$N\ (p\ \%)$ (pW)	(noise power corresponding to fade probability p %)	

REMARKS

1. When it is necessary to distinguish p from q numerically, the following expressions are used:

$$P_r\ (q = 0.01\%), \qquad N\ (q = 99.992\%), \qquad N\ (p = 0.008\%)$$

2. Customarily q is used to denote the reliability in electric field intensity, receiver input, and propagation loss; p is used to denote the fade probability with respect to noise voltage and noise power.
3. The reason p is used rather than q with respect to noise power is that the percentage of time a salient drop of the receiver input (corresponding to a small P_r or a large N) due to fading occurs is expressed simply and conveniently by p, for example, $p = 0.006\%$ as compared with $q = 99.994\%$. Moreover, the total percentage of time for m-time repeating can be expressed simply by $\Sigma_1^m p$ (%) if p is very small.
4. Allowable noise objectives according to CCIR recommendations are specified using p.
5. It is required to define the field intensity according to CCIR Recommendation 311-1 (Presentation of Data in Studies of Tropospheric Propagation) for a certain reliability and uniformity of investigation. Figure 5[B] has been drawn with this in mind.
6. *Fade probability that the noise power exceeds a certain amount* should not be confused with *time rate at which the noise power falls below a certain level.* Fade probability should be interpreted as *time rate at which the noise power exceeds a certain level* because fade probability has been defined as *time rate at which an unfavorable condition occurs* as a result of the decrease of the receiver input.

E. Estimated Loss due to Fading (See Fig. 16)

Some decrease of field intensity or receiver input due to fading with respect to the median (50% value) is called estimated loss due to fading. For the receiver input,

$$\begin{aligned} F_e\ (q\ \%) &= P_r\ (50\%) - P_r\ (q\ \%) \quad (\text{dB}) \\ F_e\ (p\ \%) &= P_r\ (50\%) - P_r\ (p\ \%) \quad (\text{dB}) \end{aligned} \tag{8.10}$$

where $q \geq 50\%$ and $p \leq 50\%$.

Accordingly, this corresponds to a difference in decibels between the receiver input and the median, with an arbitrary value of reliability, or fade probability. Also, it may be interpreted as the decrement of the receiver input due to fading at an arbitrary time rate.

Conventional definitions dating back to the early period of VHF technology pertain to the variation distribution of field intensity and may be described as follows for reliability:

Fading range	Difference in decibels between 1 and 99% values	
Fading rise	Difference in decibels between 1 and 50% values	(8.11)
Fading depth	Difference in decibels between 50 and 99% values	

However, the concept of fading depth may be applied to an arbitrary percentage value to evaluate the decrease in receiver input as in the following.

1. Although the estimated loss due to fading is expressed as F_e (q %) or F_e (p %), depending on the results of the statistical processing of the collected data, it is a function of the fade probability p (%) and the fading parameter χ. It is expressed as F_e (χ, p %) in estimating the loss due to fading, as described and illustrated in section 8.8 and Fig. 17.
2. If there is no need to refer to χ and p (%), a simple expression F_e is used. The same shall apply to p (%), q (%), P_r, N, etc.

8.5 SYSTEM STANDARD COMPOSITION AND ALLOWABLE NOISE

A. Allowable Noise

The better the communication system, the less the amount of noise produced. In an ideal system, transmitted information is delivered to a recipient instantly and in perfect form. However, any communication system will have a number of noise sources along its transmission path. Moreover, thermal noise originating from the propagation space and from the circuits of equipment cannot be avoided.

Therefore elimination of all noise conflicts with the installation cost for the transmission system. Specifically, the reduction of noise to an unnecessarily low level has an adverse effect on the economics of the system.

Generally speaking, as long as the quality of information transmission is maintained within a certain limit, the noise appearing at the receiver end of a communication system is tolerable. Noise tolerance is determined taking into account the aim and method of transmission and its economy. For example, in an emergency radio system, priority is given to the capability of receiving a signal with low field intensity over high fidelity. A marginal S/N ratio is tolerated near the threshold point.

On the other hand, in a broadcasting relay system that transmits a symphony concert maximum high fidelity is required under the present state of the art, and also a high S/N ratio at which the noise is hardly noticeable.

However, even when the purpose of transmission is the same, from the viewpoints of technology and economy, the tolerable noise level may differ according to the system. In the international radio telephone system, for example, there are four different types of systems—line-of-sight radio relay system, transhorizon radio system, satellite communication system, and HF radio system. When two or more systems are connected in cascade, the noise requirement is different based on which international agreement, such as CCITT or CCIR recommendations, is adopted.

The term noise power compromises unweighted noise power and psophometrically weighted noise power.

1. Unweighted noise power frequently is used to express a flat (white) noise, such as thermal noise.
2. Psophometrically weighted noise power is measured with a psophometer, which comprises a psophometric weighting network. Using the psophometer, the flat noise power measured in a 3.1-kHz band (0.3 to 3.4 kHz) is 2.5 dB lower than the noise power measured without the psophometer. This value is derived from an evaluation of the responsiveness of the telephone receiver and the human ear to the flat noise.

B. Criteria for Allowable Noise Power Applied to a Specific Country or Region

Criteria for allowable noise power applied to a radio system originating from and also terminating in any country or region may be determined autonomously according to an administrative or regional agreement. Such criteria are established for the following items.

1. Mobile radio service in an urban area.
2. National television broadcasting service.
3. Allowable noise power for private microwave relay systems for petroleum pipeline or electric power companies, for railway or highway control radio systems, and for radio systems to collect data sent from weather satellites.

In general the criteria for the allowable noise power are specified in terms of the S/N ratio and the temporal reliability (usually in percent) at which the obtained S/N ratio at least is ensured. Figure 25 shows, in the form of a table, the application standards in Japan for domestic fixed services, and Fig. 26 gives the graphs to be used for an arbitrary reliability value.

C. Criteria for Allowable Noise Power of Systems for International Connection

For an international circuit connecting two adjacent countries, the allowable noise power may only be agreed upon by these two countries. For interconnection between three or more countries, however, agreement may be more difficult to establish because of increased changes of conflict of interest over differences in the conditions of a country and the technical requirements.

To avoid such inconsistencies and to provide a rational means for setting up the quality of international long-distance circuits, the CCIR has made the following recommendations.

1. An imaginary reference system containing a radio circuit 2500 km in length for a terrestrial system or an earth–satellite–earth link for a satellite communication system is proposed. It is called a hypothetical reference circuit useful in the engineering design of equipment and facilities.

2. Allowable noise is specified for a hypothetical reference circuit.

3. The allowable noise power in a real circuit is specified. Here consideration is given to the fact that the real circuit usually differs appreciably from the hypothetical reference circuit with respect to geographical conditions, such as restrictions imposed by mountains, water, towns, cities, or borders, and to maintenance conditions.

4. The allowable noise power for the real circuit of arbitrary length within 2500 km (minimum 50 km) is specified. Any portion of an international circuit passing through a country, if such portion is designed to satisfy the stated noise objectives, can constitute or be part of a qualified international circuit linking more than one country.

The recommended hypothetical reference circuits for FDM terrestrial systems and for satellite communication systems are given in the following figures:

Figure 6*a*	12- to 60-channel line-of-sight relay system
Figure 6*b*	Line-of-sight relay system exceeding 60 channels
Figure 8	Transhorizon relay system
Figure 9	Satellite communication system

The allowable noise power is given in the following figures:

Figure 7*a*	Line-of-sight hypothetical reference circuit
Figure 7*b*	Line-of-sight real circuit
Figure 8	Transhorizon relay system
Figure 9	Satellite communication system

These figures give the details of the hypothetical reference circuits and the allowable noise power.

D. Prevention of Mutual Interference between Satellite Communication System and Terrestrial Radio System

Since the satellite communication system was developed after most of the frequency bands had been allocated to terrestrial radio systems (fixed station, mobile station, beacon station, etc.), the subsequent reservation of a frequency band for exclusive use by the satellite system was difficult. However, since effective use of frequency bands is essential, some of the radio frequency bands have been assigned for the satellite communication system for joint use with the terrestrial radio system.

To prevent mutual interference to both systems sharing a common frequency band, several recommendations have been adopted as given in the following figures:

Figure 10	Allowable interference noise
Figure 11*a*	Restriction on radiated power of terrestrial microwave station with respect to direction of radiation and to effective radiated power
Figure 11*b*	Restriction on earth station with respect to effective radiated power and on satellite station with respect to power density on earth surface

Such restrictions are indispensable for the further development of both the satellite and the terrestrial systems in the future. For terrestrial systems installed before such restrictions were imposed, it is suggested that appropriate modifications (suppression of radiated power, change of frequency, change of station site, etc.) be made to satisfy the requirements at the time of equipment replacement.

E. Graphs to Determine the Allowable Noise

Figure 12 shows the allowable noise power for three different kinds of systems. Also, a given temporal probability is shown under the assumption that 1-min mean noise power expressed in decibels for the worst month follows a normal distribution.

Figure 13 shows graphs to be used for the system design of a line-of-sight real circuit. Descriptions of Fig. 13 are also given in Section 8.4F. The 2500-km circuit length specified in the CCIR recommendation is considered appropriate for Europe or Japan. In an intercontinental communication system, two to ten 2500-km links are interconnected in tandem. The result is an accumulation of noise power and fade probability. (Note that along the shortest path, the distance between two points separated furthest from the earth is approximately 20,000 km since a quadrant of the earth corresponds to 10,000 km.) However, this problem has been solved in practice by using the high-performance satellite communication system, but the problem of delay time due to its long propagation path still remains.

F. Allowable Noise Power of Line-of-Sight FDM Relay System

Graphs showing the allowable noise power assigned for the L-km system are given in Fig. 13. The significance and the application of requirements I, II, III, and IV in Fig. 13 are described in this section. These requirements apply to the baseband section, that is, the radio transmission system that includes the radio propagation path, the radio transmitter and receiver, and the radio modulator and demodulator. Multiplex carrier equipment is not included. This is distinguished from the total system noise as shown in Fig. 14*a*.

Requirement I (Mean Noise Power)

$$\text{The mean noise power in any hour should not exceed } N_{AL} \text{ (pW)} \tag{8.12}$$

"Any" hour seems to imply "an unconditionally selected" hour. However, considering that fading is a natural phenomenon, this is difficult to accept without reservations. Here the preference would be to interpret it as "any hour within a limited period," with the limited period being at least 1 year. The same consideration should apply to requirements II, III, and IV.

Compared with this requirement is the worst hourly mean noise power in a year of the total instantaneous noise power (including distortion noises, interference noises, and propagation noises) as a result of the power sum of the individual radio relay section noise.

$$N_{AL} = 13{,}350 \text{ pW } (7500 \text{ pW}) \qquad [\text{unweighted (weighted) value}]$$

This requirement according to CCIR Recommendation 393-1 (1956-59-63-66) and Recommendation 395-1 (1966) was deleted in Recommendation 393-3 and Recommendation 395-2 in 1978. Hence this requirement will be treated as a reference requirement.

Requirement II (Noise Specified for p = 20%)

The 1-min noise power should not exceed N_{BL} (pW) for more than 20% in any month (8.13)

Compared with this requirement, the total 1-min mean noise power represents 20% of the fade probability as a power sum of the instantaneous level of noise produced in all the radio sections during the worst month of the year.

It is obvious that N_{AL} differs statistically from N_{BL}, although $N_{AL} = N_{BL}$.

N_{AL} and N_{BL} for an arbitrary length of the system L (km) can be estimated using Figs. 13 and 14.

Requirement III (Fade Probability or Percentage of Time the Noise Exceeds a Specified Value)

The 1-min mean noise power exceeds the specified noise power N_0 for more than p_0 % of any month (8.14)

In other words, the fade probability or the percentage of time the 1-min mean noise power exceeds the specified noise power N_0 should not be greater than p %.

$$N_0 = 84{,}550 \text{ pW } (47{,}500 \text{ pW}) \qquad [\text{unweighted (weighted) value}]$$

Figures 13 and 14 are used to estimate p_0.

Requirement IV (Outage Factor at a Specified Noise Level)

The noise power measured with 5-ms integration should not exceed the specified value N_c for more than p_c % of any month (8.15)

In other words, the fade probability or the time the noise power measured with 5-ms integration exceeds a specified noise power N_c should not be greater than p_c %.

N_c is specified as 10^6 pW; unweighted it corresponds to 30 dB of S/N ratio. The S/N ratio below this value is considered to be insufficient for international connections. Hence such a low S/N ratio is regarded as being equivalent to disability or outage of the system. p_c is called outage factor.

This requirement is specified for the hypothetical reference circuit. No similar requirement exists for the real circuit; however, it is important for determining an allowable limit of the system disability or outage.

Figures 13 and 14 are used for estimating p_c.

REMARKS

1. Requirements I to IV were proposed from the viewpoint of the user before the fading behavior or the noise variation mechanism at the end of a repeating system was elucidated sufficiently. Accordingly, these requirements should be provisional recommendations.
2. The question still remains of justifying these requirements, and a proposition for their modification exists. It is sufficient to satisfy only two of the requirements, that is, requirement I or II, and requirement III or IV. Hence in applying these requirements to the actual system, the most expedient treatments (such as alteration or deletion of a requirement) should be made taking into account the potency and economy of the system.
3. Requirements I to IV are treated here uniformly by selecting a second interval for the statistical sampling unit for fading in each hop of a radio relay system.

G. Allowable Propagation Noise

The allowable noise of a radio transmission system should be allocated to distortion noise generated in transmitters, receivers, radio modulators, and demodulators, and also to interference noise caused by overreach propagation and antenna backward radiation as well as to interfering waves from external sources. Therefore the allowable propagation noise to be allocated to the fading in a propagation path shall be the total amount of allowable noise of a radio transmission system less the interference noise.

The noise allocation principle is to be determined taking into account the modulation system, equipment performance, system composition, possible interference to and from another radio system, etc.

Typical noise allocation is given in Fig. 14a, where $+\alpha$, as specified in the CCIR recommendation for real circuits, is allocated to the distortion noise and the interference noise. This makes propagation path design somewhat more stringent.

Compared to the concept analogous to the term propagation noise, the source of the thermal noise is not in the propagation path itself but in the receiver. Nevertheless the fade-dependent internal receiver noise, accumulated at every repeater, reaches the terminal. Although the allowable limit of the propagation noise is specified for an overall system, the amount of noise generated in each radio hop must be examined in the engineering design process. It follows that

1. The propagation noise power, or S/N ratio, in a hop is the noise power expressed in pW or dBm (or S/N dB) at a point of zero transmission level at the end of a radio hop telephone channel. This comprises the ideal equipment for the noise-free radio modem multiplex carrier equipment.

2. In the FDM system the highest telephone channel generating, in general, the highest noise power, is analyzed.
3. The relationship between the receiving signal input power and the noise power (S/N ratio) is to be clarified.
4. The propagation noise of each hop accumulates power additively to produce an overall propagation noise at the end of a relaying system.
5. Since the propagation noise is a statistical variable depending on fading, the requirements are defined for various time rates, as given in Fig. 14*a*, I, II, III and IV.

8.6 PROPAGATION DESIGN

The method of propagation design varies according to the type of system, such as line of sight or transhorizon, satellite communication, mobile communication, or broadcasting, domestic or international, relay system, or direct system (without repeating). However, the following description is given for a typical system, the internationally connected line-of-sight FDM relay system (total length $L = 50$ to 2500 km) whose performance requirements are considered most stringent. Some modification or simplification may be made when this method is applied to different systems, services, or propagation modes. The design procedures and a brief explanation are given (refer to Fig. 1*c*).

A. Route Engineering

After the two terminal points that are to be connected to each other have been established, an appropriate route of the radio relay system is determined with macroscopic consideration given to radio propagation, type of modulation, system composition, ease and economy of installation, and maintenance.

Route Engineering A

1. Prepare as many crude routing plans as possible using a topographical map drawn to a scale of 1–1,000,000 to 1–2,000,000.

2. Plot the above plans together with the existing routes and the proposed trunk routes on a map drawn to a scale 1–500,000 to 1–200,000 and reject plans considered clearly unsuitable.

3. Select the relay station sites that belong to a baseband section and plot them on the map scaled 1–500,000 to 1–200,000. In general several route plans are prepared based on extensive studies of the frequency allotment plan, interference due to overreach propagation, interference to or from other routes, and interference to or from branching routes.

Field Route Survey

1. Organize a field survey team under an experienced field engineer. Dispatch it to the field to evaluate the previously prepared plans and to obtain or arrange to obtain the basic information necessary for the system design.

2. Stations to be surveyed, in general, are all the terminal stations with multiplex carrier equipment, radio modem terminal stations (baseband stations), attended stations, route-branching stations, and existing stations where new equipment is accommodated. Since this is a macroscopic survey, the field survey of through repeater stations is omitted at this stage.

3. Investigate particular conditions concerning the station site and the radio path which previously could not be confirmed on a map. These include weather conditions (such as snow, the possibility of landslides, and river flooding), vegetation conditions (such as forests, woods, and dense jungle) local diseases and hygienic control, public order, and local power supply.

4. Obtain or arrange to obtain the meteorological data required to estimate fading.

Route Engineering B

1. Repeat the work for route engineering A based on the data obtained in the field route survey above.

2. Determine a route plan using a map drawn on a scale of 1–200,000 or less. If more than one route plan is available, a preference order is to be established.

3. Find the total length L (km) of the radio relay system, and the number of radio relay hops m included in the single baseband section (two or more baseband sections when the propagation noise is to be estimated for these sections combined).

Calculation of Allowable Propagation Noise (Figs. 13 and 14a and b)

1. Find the allowable noise for a radio transmission system for the baseband section according to CCIR recommendations (N_{AL}, N_{BL}, p_0, p_c)

2. Find the distribution of allowable propagation noises (N_A, N_B, $p'_0 p_c$).

B. Propagation Design for Radio Hop Locations

This work determines the site for terminal stations at both ends and intermediate relay stations. Also, propagation characteristics of the radio hops may be estimated and called site selection design in a narrow sense. Figures 1*c* and 15 show the design procedure.

Site Selection (Survey on a Map and in the Field)

1. Select the sites of transmitting and receiving points on a topographical map scaled 1–100,000 to 1–200,000.

2. Prepare profile charts (path profile) showing the cross section of the ground surface along the path connecting both the transmitting and the receiving points. Also prepare a strip map from the above map covering the radio path.

3. Conduct a field survey (preliminary site survey) of the relay station sites, reflection areas, and obstacles if time and budget permit.

Propagation Path Parameters and Fading Margins

1. Calculate or estimate the following parameters: elevations (above mean sea level) of transmitting and receiving station sites, propagation distance, antenna height, clearance between a radio path and the peak of a ridge, reflection point, path-length difference, equivalent reflection coefficient, propagation echo ratio, a half-pitch in the height pattern of the antenna, clearance parameter, effective low path length, underlying terrain factor (of radio path).

2. Calculate the allowable propagation noise (provisionally assigned value) of a section (N_{Ai}, N_{Bi}, p_{0i}, and p_{ci}; Fig. 15).

3. Find the propagation loss and a mean receiving input power P_r (refer to Chapters 1 to 3, 5, and 6).

4. Calculate the receiving input power P_0 corresponding to 84,550 pW (unweighted value), and the specified noise power according to requirement III (Chapter 7).

5. Calculate the receiving input power P_r corresponding to 10^6 pW (unweighted value), with the specified noise power according to requirement IV or the receiving input power P_0 being the squelch operating level, threshold level, or noise switchover level, as the case may be.

6. Calculate the fading margin for requirement III and the outage margin for requirement IV,

$$\begin{aligned} &\text{fading margin} \quad & M_0 = P_r - P_0 \quad (\text{dB}) \\ &\text{outage margin} \quad & M_c = P_r - P_c \quad (\text{dB}) \end{aligned} \tag{8.16}$$

Collection of Radio Meteorological Data

1. Obtain as much past climatic data as possible from meteorological observatories, schools, laboratories, governmental agencies, port authorities, companies, etc., situated near the proposed route.

2. Obtain essential meteorological parameters, that is, statistical values of monthly data during a period of 5 to 10 years or longer: monthly mean temperature, monthly mean of the daily highest and lowest temperatures, monthly averages of the number of fairly clear, clear, cloudy, and nonsunny days, average number of days in which there are strong winds, monthly mean of humidity.

3. Investigate local specific climate (microclimate, surface climate).

4. Summarize the data from items 1 to 3 to facilitate the propagation path study.

Estimate the Fading Parameter χ

1. Find the frequency-distance coefficient K_{fD} from Fig. 18,

$$K_{fD} = g_f g_D \tag{8.17}$$

2. Determine the radio meteorological coefficient K_A from Fig. 19*b* using parameter a_1 (monthly mean temperature and humidity, see Fig. 19*a*) and parame-

ter a_2 (monthly mean of the difference between the daily highest and lowest temperatures and the monthly rate of strong-wind days, see Fig. 19*b*).

3. Determine the coefficient of surrounding terrain conditions K_S using the map of surrounding terrain conditions (Fig. 20)

4. Obtain the path profile coefficient K_P by selecting an appropriate model for the actual path profile of the radio path under study (Fig. 21*a* and *b*) and determining the related parameters:

Parameter b_1	Propagation echo radio	(Fig. 21*c*)
Parameter b_2	Effective low path length	(Fig. 21*d*)
Parameter b_3	Underlying terrain factor	(Fig. 21*c*)

Then estimate K_P from Fig. 21*d*.

5. Calculate the fading parameter χ in order to know the fading evaluation loss and the statistical value of noise power,

$$\chi = K_{fD} K_A K_S K_P \tag{8.18}$$

Mean Noise Power, 20% Noise Power (per Hop)

1. Estimate the mean noise power N_{Axi} knowing the mean receiving input power P_r (refer to Chapter 7).

2. Estimate the loss due to fading F_e (20%) for the fade probability $p = 20\%$ (refer to Fig. 17*a*).

3. Estimate the receiving input power P_B for the fade probability $p = 20\%$,

$$P_B = P_r - F_e\,(20\%) \quad (\text{dB}) \tag{8.19}$$

4. Estimate the noise power for $p = 20\%$ using P_B (refer to Chapter 7).

Specified Noise Power Fade Probability, Outage Factor (per Hop)

1. Estimate the fade probability p_{0xi} (%) using the fade margin M_0 and χ (refer to Fig. 17*b* and *c*).

2. Estimate the outage factor p_{cxi} (%) using the outage margin M_c and χ (refer to Fig. 17*b* and *c*).

Tentative Judgment for Section Noise Performance

$$\begin{array}{ll} \text{mean noise power} & N_{Ai} \geq N_{Axi} \text{ (reference)} \\ 20\% & N_{Bi} \geq N_{Bxi} \end{array} \tag{8.20}$$

$$\begin{array}{ll} \text{fade probability} & p_{0i} \geq p_{0xi} \\ \text{outage factor} & p_{ci} \geq N_{cxi} \end{array} \tag{8.21}$$

Note the estimated results that do not satisfy the required conditions in order to facilitate design modifications to achieve better system performance.

C. Assessment of Design Propagation Noise Power

Overall Propagation Noise Power of a Relay System

After completion of the steps for m sections ($i = 1$ to m) as described in section 8.6B, calculate the total noise power using the following formulas.

1. Hourly mean noise power (value for requirement I) for reference only,

$$N_{Ax} = \left(2 \sum_{i=1}^{m} N_{Axi}\right) 10^{-E/10} \quad \text{(pW)} \tag{8.22}$$

2. 20% value of 1-min mean noise power for any month (value for requirement II),

$$N_{Bx} = \left(2 \sum_{i=1}^{m} N_{Bxi}\right) 10^{-E/10} \quad \text{(pW)} \tag{8.23}$$

3. Fade probability of receiver input where 1-min mean noise power exceeds a specified value for any month (value for requirement III),

$$p_{0x} = \sum_{i=1}^{m} p_{0xi} \quad (\%) \tag{8.24}$$

4. Outage factor to a specified value of noise power measured in 5-ms integration time,

$$p_{cx} = \sum_{i=1}^{m} p_{cxi} \quad (\%) \tag{8.25}$$

Evaluation of Estimated Propagation Noise Power

The following conditions are to be satisfied.

1. Requirement I,

$$N_A \geq N_{Ax} \tag{8.26}$$

2. Requirement II,

$$N_B \geq N_{Bx}$$

3. Requirement III,

$$p_0 \geq p_{0x} \tag{8.27}$$

4. Requirement IV,

$$p_c \geq p_{cx}$$

The propagation design will be complete when all the above requirements are satisfied.

D. Modification of Designed Value

If N_{Bx} does not meet the requirements, the following modification may be made.

1. Improvement in the transmitting power, antenna gain, and feeder system loss and change of modulation method.
2. Reduction of mean propagation loss of long-distance sections within a route.
3. Modification of routing design.

If either or both p_{0x} and p_{cx} do not meet the requirements, the following modifications may be made.

1. Improvement in the propagation echo ratio ρ_θ of a section undergoing an interference caused by reflected rays.
   (a) Change of antenna height or position (shifting the reflection point).
   (b) Improvement in antenna directivity in the vertical plane.
   (c) Suppression of reflected ray through composite directivity (dual antenna combining system).
   (d) Minor change in station siting.
2. Use of a diversity system (Figs. 23 and 24).
   (a) Apply the diversity system to the worst section first, then the second worst, and so on, reviewing the results of estimation (see Section 8.6B, Tentative Judgment for Section Noise Performance).
   (b) Take into consideration future radio frequency allotment plans if the use of frequency diversity (FD) is necessary.
   (c) Pay particular attention to the construction of towers, if the use of space diversity (SD) is necessary, with respect to how to mount two antennas and to install the additional feeders.
   (d) Consider the use of both frequency and space diversity if necessary. The number of hops m_v to apply the diversity is made as low as possible. For example,

$$m_v \leq m/5 \qquad \text{(target in design stage)} \tag{8.28}$$

3. Change of radio route.

E. Station Site Selection Survey and Measurement

The results presented in Sections 8.6A to D, that is, the field survey of the sites for all stations, including both terminal stations, and of the associated radio paths were obtained by several survey teams. This survey aims primarily at ensuring the accuracy of propagation design. Thus a number of specialists in the design of access roads, station sites, station buildings, power lines, and so on, may, if necessary, join one of the teams or be organized into separate teams.

Suitability of Sites Selected on Map

1. Perform on-the-spot field surveys of station sites to confirm their suitability or select an alternative site near the original site if necessary.

2. Pay particular attention to such obstacles as mountains, buildings, and woods situated on and near the radio path toward the adjacent station of the main route and also toward the direction of a branching radio link. This is because confirmation of topographical conditions in nearby terrain is rather difficult on the map.

3. Investigate the sites and the access roads for the possibility of such natural disasters as avalanches, landslides, falling rocks, floods, high tides, billows, tidal waves, earthquakes, and forest fires, and determine countermeasures. Also study salt damage, discharged gases, and air pollution.

4. Investigate the availability of vehicles and aircraft, the distance between the site and the road or heliport, the length of new access roads and power lines, and the presence of rough routes.

5. Investigate the available site and soil conditions for station buildings and towers, and the possibility for future expansion.

Conditions of Reflection Surface and Nearby Ridge

1. Perform a detailed survey of the reflection surface on a radio path that is subject to interference with reflected rays. The reflection surface is a long and narrow area where the first Fresnel zone intersects the earth surface at $K = 4/3$ (refer to Chapter 2).

2. Conduct an on-the-spot field survey and aerial photosurvey to determine details of the reflection surface such as trees, grasses, houses, banks, snowfall, and flood conditions which could not be confirmed on a map.

3. If a ridge exists that may be included within the first Fresnel zone of a direct ray at $K = 4/3$, determine the shape of the ridge below the direct ray path and estimate the tree height, including possible growth.

Optical Test by Mirror and Measurement (to Be Performed in Principle for all Hops)

1. A mirror is used to confirm the line of sight between two facing station sites (to be the locations of the antennas) as far as possible from each other. This mirror test measures the azimuth and elevation angles of a facing station site by using a mirror to reflect sunlight rays with a theodolite to a distant site.

2. If the mirror test is difficult to perform because of topographical conditions, such as tall buildings in an urban area or a dense forest area, conduct indirect measurements.

3. Indirect measurements are made between points where the mirror test and the associated measurements are possible. However, it is preferable that these points be selected on the radio path or as close to the path as possible.

4. If necessary, apply triangulation and astronomical observation (of the sun or a fixed star).

5. Results of indirect measurements must be converted into the figures for the antenna-to-antenna radio path.

F. Radio Propagation Test

The radio propagation test is to be conducted for a radio hop where

1. An abnormal propagation is anticipated due to a specific weather condition in the region.
2. The propagation echo ratio is expected to be very large but is difficult to estimate.
3. Line-of-sight clearance is doubtful due to the presence of a nearby ridge or ridges.
4. Predicting the pitch of the spatial interference pattern or the location of the in-phase point is difficult.

Height (Height-Gain) Pattern

1. Estimate the equivalent reflection coefficient ρ_e given the difference between the peak and the dip of the height-gain pattern. The height difference between the peak and the dip of the pattern corresponds to a half-pitch p_h.

2. The test transmitter and receiver are to be positioned as closely as possible to the sites where the antennas for the radio relay system are to be installed. It is suggested that the height pattern measurement be made so that two pitches (two peaks and two dips) may be included in the measured pattern. However, this is not always possible due to limited available tower height.

3. The height pattern measurement is made during a calm period where the received signal amplitude variation is within 1 dB. Use the mean value of measured data obtained at or around the planned antenna mounting height since the reflection point moves as the antenna height changes.

4. In general, very high accuracy cannot be expected in this test. Therefore care must be taken in determining the statistical distribution of ρ_e or K from the obtained data, even when there are a considerable number of data. In other words, the data obtained in an actual measurements may lead to an incorrect conclusion.

5. Since radio propagation is reciprocal in nature, we can choose either a lower or a higher site as a transmitting or a receiving point. Usually, however, the lower site is preferred for the location of the test equipment, with a lift available for the height pattern measurement. This is because the pitch in the height pattern is smaller at the lower site than at the higher site.

6. The geometrical main-beam axis of the test antenna is determined in advance through the preliminary measurement of its directivity pattern in the vertical plane, made in a free-space model span. Then the main axis of the antenna is oriented to the distant site via the mirror test after the antenna has been mounted on the tower for testing. If the antenna main beam in the vertical plane has been determined by measurements made in a radio path with a reflected ray, the height pattern data obtained using such an antenna may be inaccurate.

Fading

1. Conduct the fading measurement during the time or season when adverse weather conditions are apt to occur. This is because the test aims at elucidating the

degree of distortion in radio propagation resulting from specific meteorological conditions.

2. The longer the period in which the measurement is performed, the higher the reliability of the obtained results. However, usually there are budget limitations. Generally it will be difficult to estimate annual fading through measurements performed during a 1-month period.

3. It is necessary to make meteorological observations or to collect the climate data for the test period. Then those data should be analyzed to determine whether or not they represent the expected results from a certain region during a season of an ordinary year.

4. Short-term data may be used to estimate fading in the range of approximately 10 to 90% values, or 5 to 95% values including the median. In other words, care must be taken in estimating extreme values (below 1% or above 99%) from the fading data obtained in only a 1- and 2-week measurement period. The significance of such a short-term test is in analyzing the data and applying the results to the propagation design from the standpoint described in item 3 on the general tendency of abnormal propagation for the period concerned.

5. For interference paths, the variation pattern of the receiving signal changes depending upon the antenna height (the relative position of the antenna in the height pattern, i.e., at the peak, the dip, or somewhere in between). Therefore it is necessary to note the measurement conditions together with the data on the variation of the height pattern with time.

Ridge Loss

1. If a ridge is located close to one of two adjacent sites or if the path length or the wavelength is short, the influence due to change of clearance between a ridge and the radio path will be significant. This is because the first Fresnel zone radius at the ridge r_s is very small.

2. Such a measurement is possible only when the movable range of the antenna with height is sufficient to cover the possible change of clearance at the ridge r_s, preferably to include the zero clearance found on a grazing path.

The measurement accuracy of the field intensity is limited. Even a careful measurement employing present state-of-the-art measurement techniques may still yield an error on the order of 3 to 6 dB for the absolute value of the field intensity, even when the median or the mean is calculated from a large number of data obtained in a long-term measurement. Hence the measurements outlined in this section will be significant only when a conclusion has to be drawn from the relative values. If the measured median differs by more than 6 dB from the theoretical value, it will be necessary to recheck the antenna orientation or the test procedures.

G. Additional Corrections and Reevaluation of Propagation Design

Referring to the results of the field survey for site selection and a radio propagation test, site rerouting or reselection may become necessary, and in this case procedures B to D are to be repeated.

Upon completion of the propagation design, proceed to the next step, as shown in Fig. 1*b*, and begin installation.

8.7 ESTIMATION OF FADING

A. Numerical Expression of Fading

It is most important in radio system design to estimate the numerical magnitude of fading that occurs on a radio propagation path. Such expressions as "fading is severe" or "probability of surface duct is high" cannot be used in system design, unless they are accompanied by a corresponding numerical figure to define the magnitude, frequency or time probability of fading. Although fading is treated statistically, its probability of occurrence cannot be estimated simply. For example, on the same line-of-sight propagation path, under almost identical climatic conditions, and for days when fading as well as no fading occurs within the same 24-hour period variations will sometimes be found.

Even in a troposcatter propagation path, where fading occurs continuously and where the mean value of the received radio signal is determined by the fading, the mode of variation of the received signal shows differences between day and night, between fine days and rainy days, and a large difference in mean values between summer and winter. However, if fading is observed during a period of one to several months and is analyzed statistically, a certain variation tendency may be noted. Although it has to be estimated statistically, the estimation reliability of the probability that signal fading is between 34 and 36 dB is not very high. However, it is very high for the probability that signal fading is 35 dB or more.

It is most essential in system design to know the probability of the received power, S/N ratio, or noise power exceeding or falling below a certain level as mentioned above, rather than the probability for specifically fixed levels.

The most typical presentation of such an estimate is the use of a cumulative percentage distribution curve, as described in Section 8.4 and Fig. 5. It determines the numerical relationship in terms of probability between the ordinate (received power, noise power, etc.) and the abscissa (reliability q or fade probability p) in Fig. 5 for each propagation path with its individual characteristics. The same relationships for troposcatter propagation paths and absorption-type propagation paths have been described in Chapters 4 and 5, respectively.

The present section describes a method of fading estimation generally applicable to both domestic and international connection systems, which employs the most commonly used line-of-sight propagation path.

B. Basic Philosophy of Fading Estimation

1. To estimate the fading it is necessary to know the type and characteristic of the fading that may occur on a propagation path and then to select an appropriate method of estimation. Several types of fading for which the system is to be designed

on the line-of-sight path are shown in the following:

Type of Propagation Path	Possible Type*	Types for Which System Is to Be Designed
Free space	A, B, C, D	B, D
Interference	A, B, C, D, G, K	B, D, G, K
Grazing ridge diffraction	A, B, C, D, T	B, D, T

*NOTE

Type A	Fading due to absorption caused by rainfall, etc. (Effect is significant at a frequency of 10 GHz or above. For details refer to Chapter 5.)
Type B	Birefracted path type fading.
Type C	Scintillation fading. (Effect caused by this type of fading is insignificant; therefore the effect is to be added to the other types of fading, if any.)
Type D	Duct-type fading.
Type G	Surface duct for reflected ray type fading.
Type K	Interference K type fading.
Type T	Diffraction K type fading.

2. Each type of fading may occur time sequentially as well as independently, or it may combine with other types to intensify the fading effect. Accordingly the total effect by multiple fading cannot be estimated by simply adding individually estimated values. In general, type B and T fading does not produce a heavy drop of the received signal level. On the other hand, types D or G accompany a large variation, and the associated heavy drop of the received signal or variation will be intensified further if type K fading is added.

3. Factors to determine the frequency of fading and the amount of drop or rise of the received power are found in the atmospheric and the underlying ground conditions, and the interaction between them. Such factors are found by analyzing climatic data, radio path profiles, and the surrounding terrain under the radio path.

4. The climate exhibits specific characteristics depending upon the time of day or night and the season. Similar features are observed in diurnal and yearly fading patterns.

5. The period of signal level variation due to fading is irregular, and usually the deeper the drops of the signal level, the shorter the period (barely shorter than 1 second). Thus it is practical to fix the minimum time unit for measurement and system design at 1 second (for average power).

6. It is essential to system design, taking into account fading, to know the relationship between the signal level and the time probability for the signal to fall below the given level in the worst month in a certain time span, such as 1 year. The term, cumulative percentage of time curve is used frequently to show such a relationship.

7. It has been elucidated empirically that the statistical distribution of radio field intensity expressed in decibels follows approximately the normal distribution. In other words, the logarithmic normal distribution for the field intensity is expressed in watts. Such a cumulative percentage of time curve for a decibel value

following the normal distribution and drawn on a normal probability chart is a straight line. The statistical distribution of received signal power during severe fading differs considerably according to fading behavior with respect to the types of fading and their combination. Thus it tends to indicate that the time probability for the drop in received power is greater than that for the rise in received power from the free-space value. Therefore the shape of the curve becomes asymmetrical and irregular to the median (50% value).

8. If the type of fading distribution agrees with such known statistical distributions as normal distribution, exponential distribution, gamma distribution, or Pearson's distribution, then the mathematical treatment of fading will become easy. So far considerable efforts have been made to fit the fading distribution to one of the preceding distributions to facilitate numerical analysis. However, excessive adherence to this method causes problems.

What is needed is a statistical distribution type that represents the situation more accurately, and on which the system design is carried out. However, if a suitable graph is available, it may fit the purpose whether it agrees with any of the known distribution types or not.

Signal variations due to fading well approximated by known distribution types, such as log-normal type, gamma (Pearson III) type, and Nakagami–Rice type, have been observed. However, in general fading frequently results from a combination of several variation types of fading, each of which belongs to a different population. Such fadings exhibit a distribution whose mathematical expression for more general application is rather difficult. From the data of distribution types observed during a 5- to 10-day period, including a day of severe fading, a bimodal-type distribution containing a couple of distribution modes is often observed.

C. Radio Meteorological Coefficient

This coefficient is a constant representing factors related to fading and extracted from long-term meteorological data on a middle- and microclimate. It is used to determine macroscopically the amount of fading on the propagation path of interest.

The following radio meteorological factors and intermediate parameters are used to determine the radio meteorological coefficient:

Radio Meteorological Factors	Intermediate Parameters	Remarks
Monthly average temperature T (°C) Monthly average relative humidity H (%)	a_1(Fig. 19a)	Use an average value obtained from the data for 10 years or longer if possible
Monthly average of daily difference between highest and lowest temperatures ΔT (°C) Monthly rate of strong-wind days W	a_2(Fig. 19b)	

In the preceding list,

$$a_1 = 0.765 + T[1.95 + 0.0983(H - 40)]/1000 \tag{8.29}$$

$$a_2 = (0.86 + 0.0056\,\Delta T)(0.97 - W^{0.8}/10) \tag{8.30}$$

where $H = 40$ for $H < 40$.

The radio meteorological coefficient K_A is given by

$$K_A = a_1 a_2 + 0.04\sqrt{a_1 a_2} \tag{8.31}$$

REMARKS

1. Although the radio meteorological elements mentioned above are limited to those that are easy to obtain, and even if such information is not available in a country or region, it is necessary to estimate them from daily weather information or from data available in another similar location.
2. The data should be collected from weather stations located as closely as possible to the radio path under consideration. It might be rather difficult but ideal to use the average value of the data taken at the transmitting point, the receiving point, and the midpoint.
3. Since this coefficient is macroscopic, it is sufficient if it represents general climatic characteristics expected in a region near the radio path. If a similar climatic condition is applicable to an adjacent radio path, a common radio meteorological coefficient may be used.

The radio meteorological coefficient is derived from surface meteorological data but not from upper air meteorological data. A radiosonde is used to observe upper air meteorological conditions while most radio propagation paths pass through much lower altitudes. The climate in such a lower altitude atmosphere is most important to the generation of fading. Abnormal refraction observed in an atmosphere of medium and high altitudes is believed to be dependent upon the surface climate. Therefore direct use of ample data on the ground and in the surface layer is desirable in view of the reliability.

D. Coefficient of Surrounding Terrain Conditions

The coefficient of surrounding terrain conditions is a constant indicating the effect of the microclimate or local climate according to the topographic condition of the area surrounding the radio path. It is used to provide a certain adjustment to the radio meteorological coefficient, which is macroscopic.

A horizontal area within 10 km of the projected line-of-sight radio path is examined and studied on a map. This area is classified into four different topographic areas, as seen in the following table, according to the undulation characteristics of the earth surface (called terrain). Then the terrain factor is estimated for each type of terrain included in the entire area under consideration. Figure 20 shows practical examples and steps of analysis.

Classified Topography	Topographic Factor	Description
Watery area	r_W	Sea, lake, swamp, damp ground, rice paddy, ice or snow field in the area, and during thawing season
Flat land	r_P	Plain, plateau, permanent ice or snow field, savanna, thick forest, desert, town or city in a flat land
Sloping land	r_C	Sloped lands such as the skirt of a large mountain and fan-shaped land
Mountainous	r_M	Land where undulation is clearly perceived even on a 1–200,000 scale map

The coefficient of the surrounding terrain conditions is given by

$$K_S = C_1 r_W + C_2 r_P + C_3 r_C + C_4 r_M \leq 1 \tag{8.32}$$

where $C_1, C_2, \ldots, C_4$ are constant, ≤ 1. In practice, the following equation is used,

$$\begin{aligned} K_S &= r_W + 0.95 r_P + 0.9 r_C + 0.8 r_M \\ &= S_W + W_P + S_C + S_M, \qquad 1 \geq K_S \geq 0.8 \end{aligned} \tag{8.33}$$

The presence of sloping land tends to assist cooled air in flowing down into the lower atmospheric layer below the radio path and produces temperature inversion after sunset.

E. Path Profile Coefficient

This is a constant determined by fading parameters extracted as the result of an analysis of the electrical and geometrical relationships between the radio path and the shape of the terrain immediately below the radio path, based on the path profile chart for the reflected rays, and also determined by the effect of the surface climate upon both direct and reflected rays. It provides a microscopic correction to the radio meteorology.

The necessary propagation path profile parameters and intermediate parameters are obtained diagrammatically or through calculation referring to Fig. 21*a* and *b*.

1. Use the equation in Fig. 21*a* and *b* to estimate the following path profile parameters:

   Propagation echo ratio ρ_θ
   Effective low path length L_e
   Underlying terrain factor T_p

2. Use Fig. 21*c* and *d* to estimate the intermediate parameters b_1, b_2, and b_3.

The following relationship exists between the path profile parameters and intermediate parameters:

Path Profile Parameters	Intermediate Parameters	Remarks
Propagation echo ratio ρ_θ	b_1	$\rho_\theta = \rho_e D_{\theta 1} D_{\theta 2}$ where ρ_e = equivalent reflection coefficient D_θ = directivity
Effective low path length L_e	b_2	$L_e = L_{de} + L_{re}\rho_\theta$ where L_{de} = low path length of direct ray L_{re} = low path length of reflected ray
Underlying terrain factor T_p	b_3	where $T_p = (d_w + d_p/2)/D$ d_w = length of water area under radio path d_p = length of flat land under radio path D = propagation path length

The path profile coefficient K_P is given by

$$K_P = (1.04b_1 + b_2)b_3 \tag{8.34}$$

F. Frequency-Distance Coefficient

The propagation of a radio wave with shorter wavelength is apt to be influenced by a smaller radio duct or a smaller change of refractivity since the first Fresnel zone becomes smaller for both direct and reflected rays. It also tends to increase the possibility of multiray interference since a slight change in the path-length difference between rays or between direct and reflected rays of shorter wavelength produces a large phase shift.

The larger the propagation distance is, the more often the duct crosses the radio path. Convexity due to the curvature of the earth surface reduces the radio path height, thereby enhancing the influence of the surface duct. The tendency for the magnitude and frequency of fading to increase with the radio frequency has been confirmed empirically. Part of this effect has already been included indirectly in the path profile coefficient. However, the major portion of this effect is obtained from Fig. 18.

The frequency-distance coefficient K_{fD} is given by

$$K_{fD} = g_f g_D \tag{8.35}$$

where f = radio frequency (GHz)
D = distance (km)

$$g_f = \log(4.6f^{1/2.4})$$

$$g_D = [\log(D/5)]^{1.3}$$

Equation (8.35) has been introduced empirically as were the other coefficients. Since the reliability of this equation may become worse in the frequency region below approximately 1 GHz and at a distance greater than approximately 100 km due to an insufficient amount of available data, a certain adjustment in the estimation is necessary.

G. Fading Parameter (See Figs. 16, 18, 19, 20 and 21)

This parameter is used to define the magnitude and probability of fading. It is given by the product of the radio meteorological coefficient K_A, the coefficient of surrounding terrain conditions K_S, the path profile coefficient K_P, and the frequency-distance coefficient K_{fD},

$$\chi = K_A K_S K_P K_{fD} \tag{8.36}$$

where χ is the fading parameter. Assume $\chi = 1.5$ if $K_A K_S K_P K_{fD} > 1.5$.
Graphs for fading estimation are given in Fig. 17.

1. The graphs presented in Fig. 17 for estimating the fading parameter χ were prepared as the result of an analysis of the relationship between the actual fading data and the climatic element, the path profile, and the topographical locality.
2. The samples to be taken are of 1-second mean values measured for a 1-month period and the graph may be used for the relative levels expressed in decibels of receiving input, field intensity, power flux density, propagation loss, S/N ratio, and noise power.
3. Substituting the estimated loss due to fading F_e with one of various kinds of margins M to squelch operation, tolerable and minimum S/N ratios according to system design, etc., the fade probability can be estimated.

The deviation of the received power from the median, that is, the estimated loss due to fading F_e (dB), and the occurrence time probability (fade probability) p (%) are estimated using the parameter χ as follows.

1. Range $p \leq 1\%$ or time reliability $q \geq 99\%$ (Fig. 17*b* and *c*). The following equations are used

$$\begin{aligned} F_e(\chi, p\,\%)\ (\text{dB}) &= \chi[-10\log(p\,\%/100)] \\ p(\chi, F_e\,\text{dB})\ (\%) &= 100(10^{-Fe/10\chi}) \end{aligned} \tag{8.37}$$

2. Range $p > 1\%$ or time reliability $q \leq 99\%$ (Fig. 17*a*).
   (a) For computation assume the normal distribution for each range, that is, $p = 1$ to 10%, 10 to 50%, or 50 to 99.99%, in which the cumulative percentage distribution curve agrees with a straight line, and estimate the values through interpolation. Let F_e be the estimated loss due to fading

(fading depth) and F_r the fading rise. Then

$$\begin{aligned} F_e &= 20\chi \text{ (dB) at the point } p = 1\%, \quad q = 99\% \\ F_e &= 10\chi \text{ (dB) at the point } p = 10\%, \quad q = 90\% \\ F_r &= 20\chi \text{ (dB) at the point } p = 99.99\%, \quad q = 0.01\% \end{aligned} \tag{8.38}$$

(b) $\chi = 0$ and $\chi = 1.5$ correspond to the states of no fading and of the severest fading likely to occur, respectively. (The period in which the fading is to be observed is any one month.)

(c) Rayleigh fading caused by randomly interfering multirays may be expressed approximately by $\chi = 1$ when F_e is very large. Hereafter the distribution corresponding to $\chi = 1$ will be called Rayleigh distribution for the sake of convenience.

The following important remarks on the application of graphs of Fig. 17 should be noted.

1. The medians (50% values) in Fig. 17 and the graphs for fading estimation are assumed to agree with the calculated field intensity at $K = 4/3$ of the propagation path under consideration.

2. The standard value of the coefficient of the effective earth radius K is to be used for preparing the path profile and for calculating the propagation parameters, that is,

$$\begin{aligned} &K = 4/3 \text{ or } 1.333 \\ &\text{effective earth radius } Ka = 8500 \text{ km} \end{aligned} \tag{8.39}$$

A unified use of $K = 4/3$ may simplify the design and prevent confusion that could be caused by using an unnecessarily wider range of K values. The median of the receiver input level may be called mean receiver input or normal receiver input.

H. Comparison with Conventional Estimating Methods

Most of the methods reported in the past to estimate the amount of fading have been inferred from the results of propagation tests carried out in specific model sections. However, very few methods have been reported based on experiences in real systems. Even the present method is based on statistics obtained for each relaying section, and therefore no clear instructions have been given for this method to cope with the actual condition of the individual propagation path. In the following tables the features and associated problem of the conventional methods and the method proposed herein are summarized.

Summary of Conventional Methods

Method of Estimation	Feature	Application and Associated Problems
1. Ordinary meteorological elements	Establishment of an empirical formula according to such parameters as temperature and vapor pressure in addition to such path parameters as distance and antenna height.	There are so few parameters that it is difficult to express the radio path characteristics with a variety of features.
2. Upper air meteorological element	Using mainly statistics on the difference of M between specific pressure levels, the refractivity characteristics of the medium and lower altitude atmospheric layers are estimated indirectly.	It is not suitable for a line-of-sight propagation path which is affected by the local climate, since the relation between climatic conditions of the upper air and of the medium and lower altitude atmospheric layers has not yet been elucidated.
3. Distribution of M curves	Statistics of M curves actually obtained close to the radio path are used. Variations of M may be substituted by variations of the apparent K value.	Measurement of M or acquisition of M data is difficult in practice. Apparent K leads to an exaggerated interpretation of the concept of K and to dangerous misunderstandings of the fading phenomenon.
4. Past data on duct formation	Correlation between fading and duct parameters, such as frequency of occurrence, thickness, and magnitude of duct.	Measurement of M or acquisition of M data is difficult in practice. It is doubtful whether the observed data obtained at one or two spots on the radio path can represent the general characteristics of the expected variations.
5. Approximation to an appropriate statistical distribution model	Fading distribution is approximated to log-normal distribution, Nakagami–Rice distribution, gamma distribution, etc., according to the radio path characteristics. Using this method, an extensive distribution range may be approximated subject to an appropriate choice of distribution type and variable parameters.	For typical radio paths, such as the free-space type or the interference type, the empirical approximation may be possible, but it has not yet been clarified how to apply the method to a radio path with arbitrary topographical and climatic conditions.

Method of Estimation	Feature	Application and Associated Problems
6. Probability of Rayleigh fading	Assuming Rayleigh fading to be the worst possible fading, all fading is substituted by Rayleigh fading, and the probability of Rayleigh fading is estimated for each topographic condition, such as over seawater, coasts, plains, and mountains.	It seems unreasonable to interpret different kinds of fading, including slight or minor fading and fading more severe than Rayleigh fading, as reflecting the characteristics of an individual path as a single type, that is, Rayleigh fading. This method is not very appropriate for designing the radio path individually, and the establishment of countermeasures to improve the system becomes complicated.
7. Data of radio propagation (direct estimation)	Data obtained in many radio sections actually in use and in several test sections are classified according to area, length, and frequency, and the cumulative percentage curve is drawn. Estimation is made by this curve.	This method is considered most practical if further analysis of the cause of fading is carried out and systematic classification is realized. However, at present there are insufficient data for route-by-route propagation design.

Features of New Method for Estimation

This method, which has been obtained through further development of method 7 above and through studying the advantages and disadvantages of conventional methods, is only a temporary solution. The reason for this are limited available data. (Some designers or clients tend not to disclose unsuccessful design data which would be very valuable for subsequent system design.) However, the method proposed here has been prepared according to actual data obtained from about 440 line-of-sight radio sections.

1. Parameters to determine the fading in a line-of-sight system, if they are chosen from a variety of easy to acquire data, may be classified into four kinds of natural constituents for which 9 parameters are defined as follows:

Location of Cause of Fading	Name of Parameter	Symbol
Atmosphere (meteorology)	Monthly average temperature Monthly average humidity Monthly average of difference between daily highest and lowest temperatures Monthly rate of days of strong wind	K_A
Radio path and terrain	Propagation echo ratio Effective low path length	K_P

	Underlying terrain factor	
Environment of radio path	Coefficient of surrounding terrain conditions	K_S
Distance and radio frequency	Frequency-distance coefficient	K_{fD}

2. To find a radio meteorological effect related closely to propagation on the radio path, the climate is analyzed in three steps:

(a) Macroscopic climate, that is, observed data at a fixed point (radio meteorological coefficient K_A)

(b) Microclimate, that is, surrounding terrain data (coefficient of surrounding terrain conditions K_S)

(c) Local-surface climate, that is, path profile data (path profile coefficient K_P)

3. Variation distribution types are constructed so as to reflect the actual situation to the maximum extent possible and to facilitate a mathematical and graphical presentation instead of adhering to approximation of a known distribution type.

4. Estimation procedures have been prepared so that the fading characteristics of an individual propagation path can be represented by a numerical value through system analysis performed automatically in the course of finding the fading parameters.

5. Although the reliability of this method for the estimated absolute value is not always sufficient depending upon the quantity and quality of the data available, the relative value, for comparison of a certain radio path with another one, is believed to be more definite and practicable in this method than in any other method as far as is known. System design and performance improvement are relative in principle and cannot be discussed without having any lead figure for comparison.

The terrains along the radio paths have been classified as over seawater, coasts, plains, and mountains. However, such a classification may be difficult in actual cases, and the design may become ambiguous unless rules for classification and numerical evaluation are defined clearly.

The radio meteorological coefficient K_A varies according to the location of the local meteorological observatory and has a direct effect on the fading parameters. This coefficient is defined macroscopically, and it raises the question of balance between this and a microscopically estimated value. However, it is acceptable for a long-distance system made up of many repeater stations where the estimated values are averaged.

I. Alternative Approximation

In some developing countries, a sparsely populated region, or areas along a radio link route via an isolated island, meteorological data are often unavailable or difficult to collect. The methods for estimating fading as mentioned in the foregoing involve very complex procedures. Therefore depending on the purpose, a more simplified method is desirable. Fig. 22 shows an alternative approximation.

Radio Meteorological Coefficient K_A

An approximate value of K_A is obtained by using the monthly average temperatures in a macroscopically classified region. If such temperature data are not available or are insufficient, the only alternative is to use estimated values.

Coefficient of Surrounding Terrain Conditions K_S

Approximate values of K_S are determined using the ratio u of the water area to the entire area within the extended rectangle drawn along the radio path, as shown at the bottom of Fig. 22a, and also the ratio of the mountainous area to the entire land area.

Path Profile Coefficient K_P

As shown at the top of Fig. 22b, the parameter b is obtained employing L_u, a portion of the direct-wave path passing within an altitude of 200 m over a flat land, and ρ_θ, the propagation echo ratio, or ρ_e, the effective reflection coefficient (when $D_{\theta 1} D_{\theta 2} \doteqdot 1$, or $D_{\theta 1}$ and $D_{\theta 2}$ may be ignored). Then the parameter t is determined by D/D_p, where D is the propagation distance and D_p the length of the flat land portion (over water and land),

$$t = 0.9 + 0.1 D_p/D$$

Thus

$$K_P = bt$$

Fading parameter χ

As mentioned in Section G, above.

$$\chi = K_{fD} K_A K_S K_P$$

where K_{fD} is obtained using Fig. 18.

J. Approximate Fading Estimation Applicable to Preliminary Design Stage

As is often the case with rough route selection in a preliminary stage, a quick estimation of the fading is required in the absence of any meteorological data and path profiles. Also, sometimes a system exists where such an approximate estimation suffices for the engineering design. In such cases the margin for fading in Fig. 26 or the graph for fading estimation to be used in the initial planning stage, given in Fig. 27, can be applied.

Figure 27 represents an annual reliability of 99.9%. However, if values for a different level of reliability are necessary, the value F_e (q = 99.9%) obtained in Fig. 27 is applied to the appropriate table in Fig. 26 for the corresponding frequency band. Then a percentage value other than 99.9% can be determined, making the slope of the curve coincide with or assume a log-normal distribution.

8.8 ESTIMATION OF TOTAL NOISE IN RADIO RELAY SYSTEM

CCIR recommendations specify the allowable noise power for real line-of-sight FDM radio relay systems whose total length L is 50 to 2500 km. The engineering design for each radio hop must be made so that the total noise power may not exceed the limit when the average hop distance D is 35 to 60 km. Accordingly the number of radio sections connected in cascade m is 42 to 72 at maximum. If no variation due to fading occurs and, accordingly, no variation occurs in the produced noise power, then the total noise power of the radio transmission system N_T is given by

$$N_T = \sum_{i=1}^{m} N_i$$

where N_i is the noise power produced in the ith section. Moreover, if all the radio sections are homogeneous, the equation is simplified further as

$$N_T = mN_i$$

Most distortion and interference noise of radio relay and modulator–demodulator equipment follows a power-additive law as given in the preceding equation. However, this is not always true for radio propagation noise because it undergoes temporal variation.

The relationship between the propagation noise of an individual hop and the total propagation noise of the end-to-end relay system and its application to the system design are described in the following sections.

A. Mean Propagation Noise (Allowable Value N_A, Designed Value N_{AX})

According to the old CCIR recommendation, the allowable mean noise is specified by the mean power in any 1 hour. Hence both the allowable mean propagation noise N_A and the designed mean propagation noise N_{AX} are defined as 1-hour mean power, as shown in the following table:

Requirement I	Specification	Remarks	
CCIR recommendation (allowable mean noise)	Mean noise power for any 1 hour shall not exceed $3L + \alpha$ (pW)	Total noise in radio transmission system	Weighted value
Allowable mean propagation noise N_A	Mean noise power for any 1 hour shall not exceed $1.78L$ (pW)	According to noise distribution plane as shown in Fig. 14	Unweighted value
Designed mean propagation noise N_{AX}	$\left(2 \sum_{i=1}^{m} N_{AXi}\right) 10^{-E/10}$ (pW)	E = preemphasis improvement E = 3 to 5 dB	Unweighted value

NOTES. α = 200 pW ($50 \le L \le 840$ km); 400 pW ($840 < L \le 1670$ km); 600 pW ($1670 < L \le 2500$ km)

E = 4 dB with preemphasis according to CCIR recommendation; = 0 dB without preemphasis

N_{AXi} = median of propagation noise for ith section in the worst month in a year

The conditions to satisfy the propagation design are

$$N_A > N_{AX} = \left(2 \sum_{i=1}^{m} N_{AXi}\right) 10^{-(E/10)} \tag{8.40}$$

The following relationship holds between N_{AX} and N_{AXi}.

1. N_{AXi} is the median of the 1-second mean propagation noise power distribution in the worst month.
2. Denote the noise power produced in each hop at an arbitrary time by N_i. Then N_{AX} provides the 1-hour mean of the instantaneous noise power appearing at the end of the relay system with m relaying sections, where the instantaneous value of N_i is the accumulated power. Also, N_{AX} varies according to the number of relay hops m and the mean receiver input and fading condition in each hop.
3. If no fading occurs in all the hops for 1 hour, then the mean value agrees with

$$N_{AX} = N_T = \sum_{i=1}^{m} N_i = \sum_{i=1}^{m} N_{AXi} \tag{8.41}$$

4. The allowable noise is defined as the mean noise power in any 1 hour which is the same as the hour with the worst fading. In such cases N_{AX} cannot, in principle, be expressed in a formula. This is because it is determined by the state of fading in the individual hop, and a correlation between hops pertaining to the time in which the fading occurs. Nevertheless, a means must be found for estimation in order to proceed with the system design. An empirical equation has been introduced as given by Eq. (8.40).

In the frequency modulation system, so-called triangular noise impairs the S/N ratio of the higher frequency channel appreciably, and the overall distortion noise generated in the equipment shows a similar tendency. Consequently it is necessary to suppress this noise in the higher frequency channel to equalize the S/N ratio in the baseband. Conversely, the S/N ratio in the lower frequency channel is reduced by this preemphasis so that the total baseband signal power may be maintained at almost the same level as that without preemphasis.

The purpose of preemphasis is to improve the S/N ratio in the higher frequency channel in the baseband by increasing the frequency deviation for the higher frequency channel. In propagation design the preemphasis effect is defined as the improvement of noise in the top channel of the baseband, E in decibels when the preemphasis circuit is employed.

The preemphasis effect is expressed by a ratio of the noise power,

$$\text{preemphasis effect} \qquad 10^{-E/10} \qquad \text{(power ratio)} \tag{8.42}$$

Since the preemphasis is applied for each baseband section, the 1-hour mean noise produced in a radio relay system comprising the k baseband sections is given by

$$N_{AX} = \left(2\sum_{1}^{m_1} N_{AXi}\right)10^{-E_1/10} + \left(2\sum_{m_1+1}^{m_2} N_{AXi}\right)10^{-E_2/10} + \cdots + \left(2\sum_{m_{k-1}+1}^{m} N_{AXi}\right)10^{-E_k/10} \tag{8.43}$$

Assume $E = 0$ for the baseband section without preemphasis. If the preemphasis effect is the same for all the baseband sections, Eq. (8.43) is simplified and agrees with Eq. (8.40).

B. 20% Propagation Noise (Allowable Value N_B, Designed Value N_{BX})

This is a value specified by CCIR recommendations for 1-min mean noise power for 20% probability in the worst month. However, it is difficult to find a direct relationship with the 1-hour mean noise power because each of these two values is defined on a different basis. The following table summarizes the requirements and the designed values:

Requirement II	Specification	Remarks	
CCIR recommendation (allowable 20% noise)	Less than $3L + \alpha$ (pW) 1-min mean power for more than 20% of any month	Total noise of radio relay system	Psophometrically weighted value
Allowable 20% propagation noise N_B	Less than $1.78L$ (pW) 1-min mean power for more than 20% of any month	Noise distribution plan as in Fig. 14 is assumed	Unweighted value
Designed 20% propagation noise N_{BX}	$\left(\sum_{i=1}^{m} N_{BXi}\right) \Big/ 10^{-E/10}$ (pW)	$E = 3$ to 5 dB; $E = 0$ without preemphasis	Unweighted value

NOTE. N_{BXi} = propagation noise power at $P = 20\%$ for ith hop in the worst month of a year

The requirement for propagation design is

$$N_B \geq N_{BX} = \left(\sum_{i=1}^{m} N_{BXi} \right) 10^{-E/10} \tag{8.44}$$

The following relationship holds between N_{BX} and N_{BXi}. The estimated loss due to fading for $p = 20\%$, that is, F_e (20%) is well within the range of slight fading since it is only 6.4 dB even for severe fading where $\chi = 1$. Within this range the receiver input levels of an individual hop in a radio relay system go up or down at random, thereby canceling each other. Therefore the noise power at the end of a radio relay system N_{BX} can safely be assumed to be the sum of N_{BXi} for all the radio hops. The noise power N_{BX} itself is considered to include the effect of deep fading.

This requirement may be insignificant considering that the simultaneous satisfaction of requirement I for 1-hour mean noise power, and of Requirements III and IV for a small-percentage-of-time probability, normally is sufficient in view of the system performance. Accordingly, only a brief description has been given here.

C. Fade Probability and Outage Factor for a Specified Noise Power Level

The fade probability and the outage factor for a specified noise power level in the worst month of the year have been given in CCIR recommendations. These two requirements differ from each other in their prerequisites, and the designed values are summarized in the following tables to show the relationship between requirements and designed values:

Requirement III	Specification	Remarks	
CCIR recommendation (fade probability at a specified noise power level)	Percentage of time the 1-min mean noise power exceeds 47,500 pW in any month should be less than $0.1 \times L/2500$ (%) where L is a length (km) of the radio system.	Total noise power of a radio transmission system	Psophometrically weighted value
Allowable fade probability p_0	Percentage of time the 1-min mean noise power exceeds 84,500 pW in any month should be less than $0.1 \times L/2500$ (%) where L is a length (km) of the radio system.		Unweighted value
Designed fade probability p_{0x}	$\sum_{i=1}^{m} p_{0xi}$ (%)	Radio propagation noise	Unweighted value

NOTE. p_{0xi} = fade probability that the 1-sec mean propagation noise power in the ith hop exceeds 84,550 pW in the worst month of a year

Requirement IV	Specification	Remarks	
CCIR recommendation (outage factor at a specified noise power level)	Percentage of time the 5-ms noise power exceeds 10^6 pW in any month should be less than $0.01 \times L/2500$ (%)	Total noise power of a radio transmission system	Unweighted value
Allowable outage factor p_c			Unweighted value
Designed outage factor p_{cx}	$\sum_{i=1}^{m} p_{cxi}$ (%)	Radio propagation noise	Unweighted value

NOTE. p_{cxi} = outage factor in percent that the 1-second radio propagation noise exceeds 10^6 pW, either the squelch operating level or the noise switching level, in the worst month

The requirement for propagation design is

$$\begin{aligned} p_0 &\geq p_{0x} \\ p_c &\geq p_{cx} \end{aligned} \tag{8.45}$$

REMARKS

1. Specified allowable noise power given in the table outlining Requirement III is the total noise power, which includes the influence of deep fading produced in any propagation path upon the output appearing at the end of a relay system. On the other hand, all the specified noise power is allocated to the propagation noise for design purposes. The resultant deficiency for the other type of noise may be omitted from the allocation that includes a margin for the instantaneous value of the 1-min mean noise power specified and the 1-second mean noise power for design.
2. Variation in the fading which lasts for less than 1 second is observed rarely in a line-of-sight system. Thus is it permissible to use an estimation graph that is prepared based on 1-second mean noise power, even when the allowable specified noise power is determined in 5 ms. It may be considered that the variation within 1 second is well represented by the 1-second mean value.
3. The preemphasis effect is ignored since it is reserved as a design margin for unknown factors.

EXAMPLES

In the following examples the allowable noise power requirement I and the related descriptions are to be considered only as reference.

Example 8.1

Route planning of the international radio relay system between Paris (France) and Rome (Italy) passing through Lyons, Milan, and the Alps indicated a requirement of 24 relay hops over the total route length $L = 1204$ km. Determine the allowable noise power in the total system and the radio transmission system. The system is of the FDM (frequency division multiplex), 1800-channel line-of-sight type in the 4-GHz band and differs from the CCIR hypothetical reference circuit.

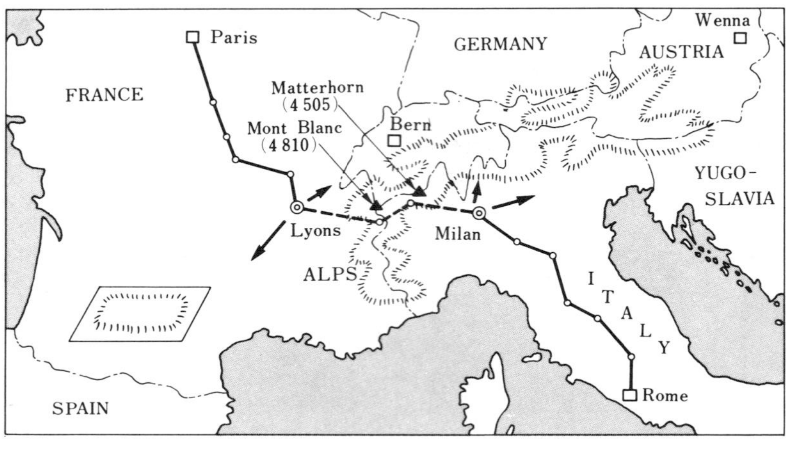

Solution

1. Check items of the allowable noise power on the Chart Index.

Allowable noise to line-of-sight system for real link	Fig. 7*b*
Graph for allowable noise (for real link)	Fig. 13
Allowable, allocated, and achieved noise power	Fig. 14*a*

2. Allowable noise power in radio transmission system. If $L = 1204$ km, then $840 < L < 1670$ km, and $\alpha = 400$ pW,

Requirement I	$N_{AL} = 3L + \alpha = 3 \times 1204 + 400 = 4012$ pW	
Requirement II	$N_{BL} = N_{AL}$	
Requirement III	$p_0 = 0.1 \times L/2500 = 0.0482\%$	(allowable fade probability that noise exceeds 47,500 pW)
Requirement IV	$p_c = 0.1 \times p_0 = 0.00482\%$	(allowable outage factor if noise exceeds 10^6 pW)

Note. Requirements I to III are psophometrically weighted values, Requirement IV is an unweighted value.

3. Allowable total noise power for overall system,

Requirement I	$N_{ALL} = 4L + \alpha = 4 \times 1204 + 400 = 5216$ pW	
Requirement II	$N_{BLL} = N_{ALL}$	
Requirement III	$p_{OLL} = 0.0482\%$	(allowable fade probability that noise exceeds 50,000 pW)
Requirement IV	$p_{cLL} = 0.00482\%$	(allowable outage factor if noise exceeds 10^6 pW)

Requirements I to III are psophometrically weighted values, Requirement IV is an unweighted value.

4. Summary:

Requirement	Allowable Noise Power (weighted/unweighted) (pW): Total System	Radio Transmission System	Allowable Fade Probability/ Reliability (%)	Statistical Unit
I	5216/9285	4012/17, 141	Any hour in a year	1-hour mean value
II	5216/9285	4012/17, 141	20/80	1-min mean value in worst month
III	50,000/89,000	47,500/84,550	0.0482/99.9518	1-min mean value in worst month
IV	$5.6 \times 10^5/10^6$	$5.6 \times 10^5/10^6$	0.0048/99.9952	5-ms value in worst month

EXAMPLE 8.2

Indicate the allowable propagation noise in picowatts and the S/N ratio in decibels (both in unweighted values) to be allocated to the Paris–Rome radio relay system in its radio propagation design. The total length of the radio system is assumed to be $L = 1204$ km. Apply the allowable noise allocation as shown in Fig. 14*a*, although it may change according to the purpose and composition of the system.

Solution

1. According to Fig. 14*a* the allocation of allowable noise power of a radio transmission system N_{AL} and N_{AB} is as follows, although it is slightly

stringent:

$$N_{AL} = N_{AB} \begin{Bmatrix} \text{distortion noise } N_E = L + \alpha/2 \\ \text{interference noise } N_I = L + \alpha/2 \\ \text{propagation noise } N_A = N_B = L \end{Bmatrix} = 3L + \alpha$$

2. Using the charts in Fig. 14*b*, where the power is expressed in unweighted values, we obtain

$$N_A = N_B = 1.78L = 1.78 \times 1204 = 2143 \text{ pW}$$

$$p_0 = 0.1 \times L/2500 = 0.0482\% \qquad \text{(allowable fade probability that noise exceeds 84,550 pW)}$$

$$p_c = 0.1 p_0 = 0.0048\% \qquad \text{(allowable outage factor if noise exceeds } 10^6 \text{ pW)}$$

3. Determine the S/N ratio from the preceding. The power obtained in step 2 gives the noise power at zero relative level point, that is, at a point where the test zone level is 1 mW = 10^9 pW. Then S/N (dB) for the test tone is given by

$$S/N \text{ (dB)} = 10 \log(10^9/N) = (90 - 10 \log N \text{ pW}) \quad \text{(dB)}$$

Thus,

$$S/N_A = S/N_B = 90 - 10 \log 2143 = 56.7 \text{ dB}$$

$$S/N_0 \text{ (corresponding to } S/N \text{ for Requirement III)}$$

$$= 90 - 10 \log 84{,}550 = 40.7 \text{ dB}$$

$$S/N_c \text{ (corresponding to } S/N \text{ for Requirement IV)}$$

$$= 90 - 10 \log 10^6 = 30 \text{ dB}$$

4. Summary (for a radio propagation path of length $L = 1204$ km between Paris and Rome):

	Allowable Propagation Noise (unweighted value)			
Requirement	Noise Power (pW)	S/N (dB)	Allowable Fade Probability/Reliability (%)	Statistical Unit
I	2143	56.7	Any hour in a year	1-hour mean value
II	2143	56.7	20/80	1-min mean value in worst month
III	84,550	40.7	0.0482/99.9518	1-min mean value in worst month
IV	10^6	30	0.0048/99.9952	5-ms value in worst month

EXAMPLE 8.3

Among the route plans of the Paris–Rome international microwave system is one that includes the longest hop between point A (2720 m above mean sea level) located at the south shoulder of the Matterhorn in the Alps and point B (130 m above mean sea level) in the city of Milan. Analyze this propagation path and examine its economic feasibility.

System Parameters	Propagation Parameters
Radio frequency $f = 4$ GHz	Total system length $L = 1204$ km
Top-channel frequency per number of channels $f_m = 8120$ kHz/1800	Propagation distance $D = 140$ km
Frequency deviation $S_0 = 140$ kHz ch	Antenna height above ground. $A_1 = 10$ m, $A_2 = 120$ m
Transmitter output $P_t = 40$ dBm	Antenna height above mean sea level $h_{s1} = 2730$ m,
Overall noise figure $F = 8$ dB	$h_{s2} = 250$ m
IF bandwidth $B = 25$ MHz	Radio meteorological coefficients
Antenna gain (3.3-m diameter) $G_t = G_r = 40$ dB	$K_{A1} = 0.61$, $K_{A2} = 0.79$
Feeder system loss $L_t = L_r = 1.5$ dB	

NOTE. Subscript 1 = point A, subscript 2 = point B.

For a path profile and strip map see the following figures.

Solution

The analysis is carried out according to Fig. 15 after completion of a macroscopic study according to the steps shown in Fig. 1*c*.

1. Calculation of propagation path parameters.
   (a) Antenna heights h_1 and h_2. The estimated height above mean sea level of the presumed reflection surface area is $h_r = 130$ m, and

$$h_1 = h_{s1} - h_r = 2730 - 130 = 2600 \text{ m}$$

$$h_2 = h_{s2} - h_r = 250 - 130 = 120 \text{ m}$$

   (b) Reflection points d_1 and d_2,

$$\text{parameter } C = (h_1 - h_2)/(h_1 + h_2) = 2480/2720 = 0.912$$

$$m = 0.03D^2/(h_1 + h_2)$$

$$= 0.03 \times 140^2/2720 = 0.216$$

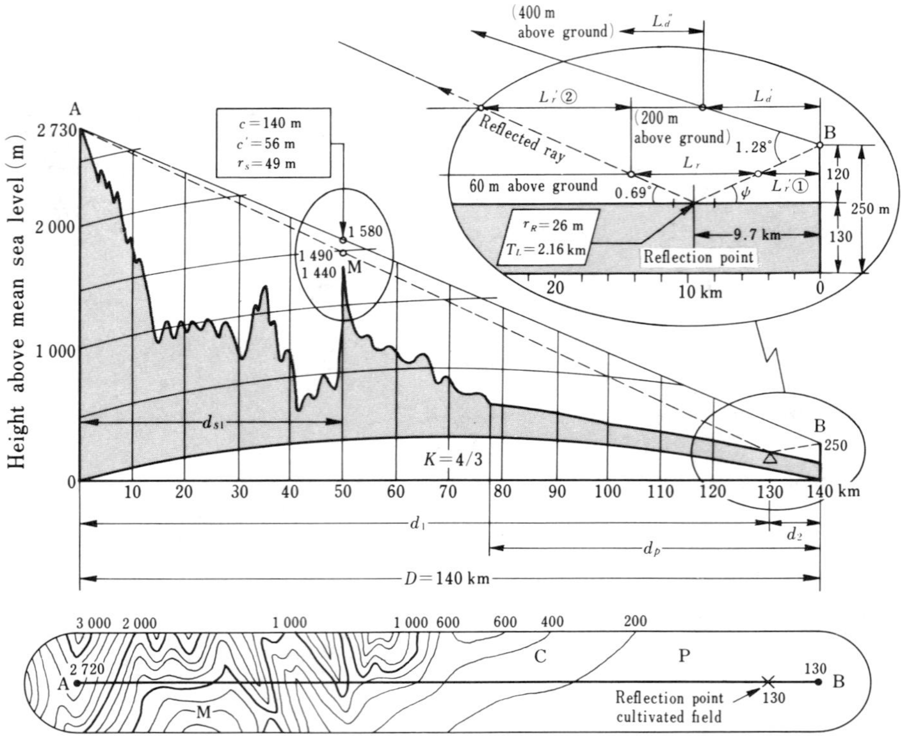

According to Fig. 25a in Chapter 2, we obtain $y = 0.931$ and

$$d_1 = yD = 0.93 \times 140 = 130.3$$

$$d_2 = D - d_1 = 140 - 130.3 = 9.7$$

(c) Equivalent antenna heights h_1' and h_2',

$$h_1' = h_1 - u_1 = 2600 - 1000 = 1600$$

$$h_2' = h_2 - u_2 = 120 - 5.7 = 114.3$$

Refer to Chapter 2 for u_1 and u_2.

(d) Path-length difference Δ and path-length difference to a half-wave-length ratio $\Delta/(\lambda/2)$,

$$\Delta = 2h_1'h_2'/D = 2 \times 1600 \times 114.3/(140 \times 10^3) = 2.61 \text{ m}$$

$$\Delta/(\lambda/2) = 261/(7.5/2) = 69.6 > 5$$

(e) Grazing angle ψ,

$$\psi_1 \doteqdot h_1'/d_1 = 1600/(130.3 \times 10^3) = 12.3 \text{ mrad}$$

$$\psi_2 \doteqdot h_2'/d_2 = 114.3/(9.7 \times 10^3) = 11.8 \text{ mrad}$$

$$\psi = (\psi_1 + \psi_2)/2 = 12 \text{ mrad} = 0.69°$$

The error is within 10%.

(f) Angles included by a direct ray and a reflected ray θ_1 and θ_2,

$$\theta_1 = h_1/d_1 - (h_1 - h_2)/D - d_2/2Ka$$

$$= [2600/130.3 - 2480/140 - 9.7/17]10^{-3} = 1.7 \text{ mrad}$$

$$= 0.1°$$

$$\theta_2 = h_2/d_2 - (h_2 - h_1)/D - d_1/2Ka$$

$$= [120/9.7 + 2480/140 - 130.3/17]10^{-3} = 22.4 \text{ mrad}$$

$$= 1.28°$$

(g) Half-pitch of height pattern p_{h1} and p_{h2},

$$p_{h1} = \lambda/(2 \sin \theta_1) = 0.075/(2 \sin 0.1°) = 21.5 \text{ m}$$

$$p_{h2} = \lambda/(2 \sin \theta_2) = 0.075/(2 \sin 1.28°) = 1.7 \text{ m}$$

(h) Major radius of effective reflection area T_L,

$$T_L = r_R/\sin \psi = \sqrt{\lambda d_1 d_2/D}\,/\sin \psi$$

$$= \sqrt{7.5 \times 130.3 \times 9.7 \times 10/140}\,/\sin 0.69° = 2.16 \text{ km}$$

2. Studies of the diffraction ridge.

(a) An examination of path profile and its associated map indicates the presence of a ridge 50 km from point A. This may cause a diffraction loss to the direct and the reflected rays. The on-the-spot field survey and measurement have confirmed that the height h_M of ridge M, including trees, is 1440 m above mean sea level. Here d_{s1} denotes the distance between points A and M, while $d_{s2} = D - d_{s1}$.

(b) Path height of direct ray h_p and path height of reflected ray h_p',

$$h_p = (h_{s1}d_{s2} + h_{s2}d_{s1})/D - d_{s1}d_{s2}/2Ka$$

$$= (2730 \times 90 + 250 \times 50)/140 - 50 \times 90/17 = 1580 \text{ m}$$

$$h_p' = h_1(d_1 - d_{s1})/d_1 - d_{s1}(d_1 - d_{s1})/Ka + h_r$$

$$= 2600 \times (130.3 - 50)/130.3 + 50(130.3 - 50)/17 + 130$$

$$= 1496 \text{ m}$$

As is clearly shown, h'_p is obtained by adding h_r to the path height calculated assuming that the distance between point A and the reflection point is d_1, and the elevation of the reflection point is 0 m.

(c) Clearance C and C',

$$C = h_p - h_M = 1580 - 1440 = 140 \text{ m}$$

$$C' = h'_p - h_M = 1496 - 1440 = 56 \text{ m}$$

(d) First Fresnel zone radius r_s at the ridge,

$$r_s = \sqrt{\lambda d_{s1} d_{s2}/D}$$

$$= \sqrt{0.075 \times 50 \times 90 \times 10^3/140} = 49.1 \fallingdotseq 49 \text{ m}$$

(e) Clearance parameters U_c and U'_c,

$$U_c = C/r_s = 140/49 = 2.86$$

$$U'_c = C'/r_s = 56/49 = 1.14$$

(f) Effect of diffraction. Since $U_c > 1$ and $U'_c > 1$, the path clearance is sufficiently large, and the influence of the ridge diffraction may be ignored for both the direct and the reflected rays.

3. Studies on interference effects due to ground reflection.

(a) Since the radio path is of the high–low type, the reflection point approaches the lower side near point B. In the range of 5 to 15 km from point B toward point A on the map, indicate a flat land of approximately 130 m. Then the elevation above mean sea level $h_r = 130$ m is assumed for the reflection point as in step 1, items (a) to (h).

(b) Reflection area and equivalent reflection coefficient ρ_e. The position is 9.7 km (at center) from point B. The projection area of the main reflection is an area enclosed by the ellipse with $\pm T_L = 2.16$ km along the radio path and $\pm r_R = 26$ m transverse from the reflection point. The flat-plane reflection coefficient $R > 0.9$ for both horizontal and vertical polarizations, and the spherical-ground divergence coefficient $D_v \gtrsim 0.9$. Hence

$$RD_v > 0.8$$

According to the results of an on-the-spot field survey, it has been inferred that the main projection area of reflection is a rough ground formed by a cultivated field and a less dense forest, and the equivalent reflection coefficient was determined, with a safety margin from Fig. 11*b* in Chapter 2, as $\rho_e = 0.65$.

(c) Effect of antenna directivity $D_{\theta 1}$ and $D_{\theta 2}$. The angle θ_1 included by a direct ray and a reflected ray at point A is only 0.1°. Then no improvement in suppressing the reflected ray by antenna directivity is expected; $D_{\theta 1} = 1$ (0 dB). In the case of the included angle $\theta_2 = 1.28°$ and using a parabolic antenna of 3.3-m diameter ($\eta = 55\%$) at point B, calculate $D_{\theta 2}$. The half-power beam width angle $\theta_{1/2}$ and the directivity effect $D_{\theta 2}$ are estimated from Fig. 10 [2] in Chapter 2 as follows, assuming $k = 1.4$, including an antenna orientation error:

$$\theta_{1/2} = 57k\lambda/D_{\phi} = 57 \times 1.4 \times 7.5/330 = 1.8$$

$$\theta_2/\theta_{1/2} = 1.28/1.8 = 0.7$$

$$D_{\theta} = -6.9 \text{ dB} \quad \text{(Antilogarithm: 0.45)}$$

(d) Propagation echo ratio ρ_{θ},

$$\rho_{\theta} = \rho_e D_{\theta 1} D_{\theta 2} = 0.65 \times 1 \times 0.45 = 0.29$$

NOTE. Some of the figures obtained in steps 1 to 3 are shown for clarity in the path profile presented at the beginning of this example, although these figures are not usually known at the outset.

4. Effective low path length L_e.
   (a) Estimate the path length of the direct and reflected rays passing through the low path zone using the enlarged path profile for the vicinity of point B.

Low Path Zone	Direct Ray	Reflected Ray
ZI (0 to 60 m)	$L_d = 0$	$L_r = 9$
ZII (60 to 200 m)	$L'_d = 8.7$	$L'_r = L'_r$ ① L'_r ② $= 17$
ZIII (200 to 400 m)	$L''_d = 24.5$	$L''_r = 17$

   (b)

$$\begin{aligned} L_e &= (L_d + L'_d/3 + L''_d/5) + (L_r + L'_r/3 + L''_r/5)\rho_{\theta} \\ &= L_{de} + L_{re}\rho_{\theta} \\ &= (0 + 8.7/3 + 24.5/5) + (9 + 17/3 + 17/5)0.29 \\ &= 7.8 + 18.1 \times 0.29 = 13 \end{aligned}$$

5. Allowable propagation noise (provisional allocation).
   (a) Requirement I_i $\quad N_{Ai} = D/2 = 140/2 = 70$ pW
   (b) Requirement II_i $\quad N_{Bi} = D = 140$ pW
   (c) Requirement III_i $\quad p_{0i} = 0.1 \times D/2500 = 140/25{,}000 = 0.0056\%$
   (d) Requirement IV_i $\quad p_{ci} = 0.1 p_{0i} = 0.00056\%$

6. S/N ratio and noise power N_{AXi} for standard receiver input P_r.
   (a) Standard receiver input P_r. Let an average receiver input by only the direct ray be the standard input because the interference effect due to reflected rays is to be taken into account in the fading analysis (free-space propagation loss for 4 GHz and 140 km is $\Gamma_0 = 147.4$ dB);

$$P_r = P_t + (G_t + G_r - L_t - L_r) - \Gamma_0$$

$$= 40 + (40 \times 2 - 1.5 \times 2) - 147.4$$

$$= -30.4 \text{ dBm}$$

   (b) S/N ratio and noise power N_{AXi} (Requirement I_i) for standard receiver input (see Chapter 7). The parameters used here are $F = 8$ dB, $S_0 = 140$ kHz, $\Delta f = 3.1$ kHz, $P_r = -30.4$ dBm, and $f_m = 8204$ kHz (1800 channels). Calculation proceeds as follows:

$$S/N = P_r - F + 10\log\left[S_0^2/(f_m^2\,\Delta f)\right] + 174$$

$$= -30.4 - 8 + 10\log\left[140^2/(8120^2 \times 3.1 \times 10^3)\right] + 174$$

$$= 65.4 \text{ dB}$$

   Assuming that the noise power at a point of 0 dBm by the test tone ($S = 10^9$ pW at zero relative level point) is N_{AXi} (pW), we obtain

$$10\log N_{AXi} = 24.6 \text{ dB pW}$$

$$N_{AXi} = 10^{24.6/10} = 10^{2.46} = 290 \text{ pW}$$

7. Fading margin M_0 for Requirement III_i.
   (a) The input level P_0 to produce 84,550 pW of noise power in the region where the noise power is inversely proportional to the receiver input is given by

$$P_0 = P_r - 10\log(84{,}550/N_{AXi})$$

$$= -30.4 - 10\log(84.550/290) = -55 \text{ dBm}$$

   (b)

$$M_0 = P_r - P_0$$

$$= 10\log(84{,}550/N_{AXi}) = 24.6 \text{ dB}$$

8. Outage margin M_c for requirement IV_i.
   (a) The input level P_c to produce 10^6 pW of the noise power is given by

$$P_c = P_r - 10\log\left(10^6/N_{AXi}\right)$$

$$= -30.4 - 10\log(10^6/290) = -65.8 \text{ dBm}$$

(b) Threshold input P_{th},

$$P_{th} = 10\log(KTBFC_{th})$$
$$= 10\log(KTC_{th}) + 10\log B + 10\log F$$
$$= 10\log(1.38 \times 10^{-20} \times 290 \times 8) + 10\log(2.5 \times 10^{7}) + 8$$
$$= -165 + 74 + 8 = -83 \text{ dBm}$$

(c) If $P_c > P_{th}$, then adopt P_c as a limiting level.

(d)

$$M_c = P_r - P_c = 10\log(10^6/N_{AXi})$$
$$= 10\log(10^6/290) = 35.4 \text{ dB}$$

9. Fading parameter χ.

(a) Frequency-distance coefficient K_{fD},

$$K_{fD} = g_f g_D = 1.915 \times 1.67 = 1.53 \qquad \text{(Fig. 18)}$$

(b) The radio meteorological index K_A is derived from the mean values of K_{A1} at point A and K_{A2} at point B,

$$K_A = (K_{A1} + K_{A2})/2 = 0.7$$

(c) The coefficient of surrounding terrain conditions K_S is calculated using the topographic factor r found on the map of surrounding terrain conditions as $r_W = 0$, $r_P = 0.35$, $r_C = 0.20$, $r_Z = 0$, and $r_M = 0.45$. Then

$$K_S = 0.95 r_P + 0.9 r_C + 0.8 r_M$$
$$= 0.95 \times 0.35 + 0.9 \times 0.2 + 0.8 \times 0.45 = 0.873$$

(d) Path profile coefficient K_P. Find the parameters b_1, b_2, and b_3 (Fig. 21),

$$b_1 = 1.0 \qquad \rho_\theta = 0.29$$
$$b_2 = 0.14 \qquad L_e = 13 \text{ km}$$
$$b_3 = 0.875 \qquad T_p = (d_p/2)/D = 31/140 = 0.22$$

Then $K_P = 1.033$.

(e) Fading parameter χ,

$$\chi = K_{fD} K_A K_S K_P$$
$$= 1.53 \times 0.7 \times 0.873 \times 1.033 = 0.97$$

10. Receiver input P_B and noise power N_{BXi} corresponding to Requirement II_i.
   (a) Estimated loss due to fading $F_e(20\%)$. Using Fig. 17*a*,

$$F_e(20\%) = 6.2 \text{ dB}$$

   (b)

$$P_B = P_r - F_e(20\%) = -30.4 - 6.2 = -36.6 \text{ dBm}$$

   (c)

$$N_{BXi} = N_{AXi} 10^{F_e(20\%)/10}$$

$$= 290 \times 10^{0.62} = 290 \times 4.18 = 1212 \text{ pW}$$

11. Fade probability p_{0xi} and outage factor p_{cxi} at a specified noise power level.
   (a) Fade probability p_{0xi} at the specified noise power (84,550 pW), corresponding to Requirement III_i. Applying $M_0 = 24.6$ dB and $\chi = 0.97$ to Fig. 17*b*,

$$p_{0xi} = 0.28\%$$

   (b) Outage factor p_{cxi} at the specified noise power (10^6 pW) corresponding to Requirement IV_i. Applying $M_c = 35.4$ dB and $\chi = 0.97$,

$$p_{cxi} = 0.023\%$$

12. Comparison and comments. The provisional requirements are compared with the estimated design values in the following table:

Requirement	Provisional Allotment	Estimated Design Value	Satisfied (○) or Not Satisfied (×)	Ratio of Deficiency
I_i	$N_{Ai} = 70$ pW	$N_{AXi} = 290$ pW	×	6.2 dB
II_i	$N_{Bi} = 140$ pW	$N_{BXi} = 1212$ pW	×	9.4 dB
III_i	$p_{0i} = 0.0056\%$	$p_{0xi} = 0.28\%$	×	50 times
IV_i	$p_{ci} = 0.00056\%$	$p_{cxi} = 0.023\%$	×	41 times

   (a) All figures given in the table as estimated design values exceed the provisional limits. However, improvement may not be so difficult.
   (b) Since the total system length is 1204 km and the designed number of hops is 24, the average hop length is approximately 50 km, and the average noise power per hop is less than 40 pW, which is a sufficient margin. Therefore the CCIR specifications concerning Requirements I and II appear to be satisfied, unless the number of long-distance hops is particularly large.
   (c) For Requirements III and IV the designed values have already exceeded the overall requirement of the system. As in Example 8.1, an appropriate improvement will be required for this section.

(d) In view of the specific features of this section, an application of space diversity is considered most effective and permits realization of the noise performance within the provisional allotment for Requirements III and IV.

REMARKS. It should be noted that in general, stable radio propagation (small x) along a higher altitude mountainous path or along a sloped path, as in this example, can be achieved. This is because of such features as ease of preventing reflected rays (small ρ_θ), effective low path length (small L_e), small radio meteorological coefficient K_A, etc. Therefore the requirement for international connection specified by CCIR may be satisfied even for a path of more than 100 km.

EXAMPLE 8.4

There is a 2-GHz FDM broadband multiplex telephone and television transmission system for international connection across a channel in the sea, for which the time reliability for dropping out is assumed to be 99.9998%. Produce an engineering guideline of the facilities provided for the site conditions given as follows.

Conditions

1. Radio meteorological data (worst month: August in the northern hemisphere).

Monthly mean temperature	$T = 27.5°C$
Monthly mean of temperature difference between daily highest and daily lowest	$\Delta T = 11°C$
Monthly mean humidity	$H = 82\%$
Monthly rate of strong-wind days (for maximum wind velocity of 10 m/s or higher)	$W = 0.2$

2. Path profile.

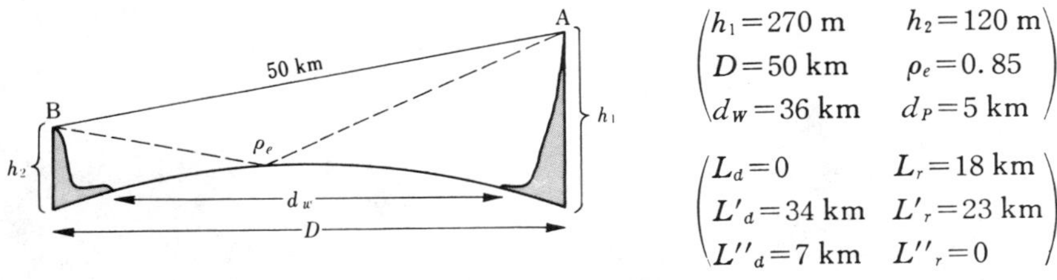

$$\begin{pmatrix} h_1 = 270 \text{ m} & h_2 = 120 \text{ m} \\ D = 50 \text{ km} & \rho_e = 0.85 \\ d_W = 36 \text{ km} & d_P = 5 \text{ km} \end{pmatrix}$$

$$\begin{pmatrix} L_d = 0 & L_r = 18 \text{ km} \\ L'_d = 34 \text{ km} & L'_r = 23 \text{ km} \\ L''_d = 7 \text{ km} & L''_r = 0 \end{pmatrix}$$

3. Map of surrounding terrain conditions.

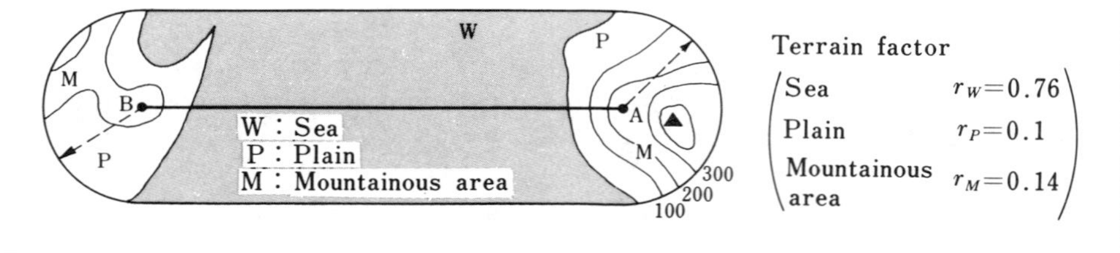

Solution

1. First calculate the fading parameter and then the estimated loss due to fading $F_e(0.0002\%)$ for a reliability factor $q = 99.9998\%$ (corresponding to a fade probability $p = 0.0002\%$).
2. Radio meteorological coefficient K_A,

$$\text{Figure 19}a\ (T, H) \qquad a_1 = 0.93$$

$$\text{Figure 19}b\ (W, \Delta T) \qquad a_2 = 0.87$$

$$a_1 a_2 = 0.81$$

$$K_A = 0.86 \qquad (\text{Fig. 19}b)$$

3. Coefficient of surrounding terrain conditions, K_S

$$K_S = r_W + 0.95 r_P + 0.8 r_M$$

$$= 0.76 + 0.95 \times 0.1 + 0.8 \times 0.14 = 0.97 \qquad (\text{Fig. 20}b)$$

4. path profile coefficient K_P,

$$L_e = (L_d + L'_d/3 + L''_d/5) + (L_r + L'_r/3 + L''_r/5)\rho_\theta$$

$$= (34/3 + 7/5) + (18 + 23/3)0.85$$

$$= 12.7 + 25.7 \times 0.85 = 34.5 \text{ km} \qquad (\text{Fig. 21}b)$$

$$T_p = (d_W + d_p/2)/D = (36 + 5/2)/50 = 0.77$$

where $\rho_e = \rho_\theta$ is assumed.

$$\text{Figure 21}e\ (\rho_\theta) \qquad b_1 = 1.29$$

$$\text{Figure 21}f\ (L_e) \qquad b_2 = 0.43$$

$$\text{Figure 21}f\ (T_p) \qquad b_3 = 0.98$$

$$K_P = 1.74$$

5. Frequency-distance coefficient K_{fD},

$$\text{Figure 18 } (f, D) \qquad g_f = 0.79, \qquad g_D = 1.0$$

$$K_{fD} = 0.79$$

6. Fading parameter χ,

$$\chi = K_A K_S K_P K_{fD}$$

$$= 0.86 \times 0.97 \times 1.74 \times 0.79 = 1.15$$

7. Estimated loss due to fading F_e. From Fig. 17*c*, F_e for $p = 0.0002$ and $\chi = 1.15$ is given as

$$F_e(0.0002\%) \doteqdot 65.5 \text{ dB}$$

8. Guidance in engineering the facilities.
   (a) The system design is to be carried out in such a manner that the margin of the normal receiver input level compared to a limiting level (threshold input level or squelch operating input level) is 65.5 dB or more.
   (b) The free-space value used as the normal receiver input for F_e is a value estimated in consideration of interference due to reflected rays. Free-space propagation loss $\Gamma_0 = 132.5$ dB for 2 GHz, 50 km (Chapter 1).
   (c) Requirement for normal receiver input P_r,

$$P_r = P_t + (G_t + G_r - L_t - L_r) - 132.5 \text{ dBm}$$

$$P_r \geq P_c + 65.5 \text{ dBm}$$

where P_t = transmitter output
G_t, G_r = transmitting or receiving antenna gain
L_t, L_r = transmitting or receiving feeder system loss
P_c = threshold input

   (d) System design objective,

$$P_t + (G_t + G_r - L_t - L_r) - P_c \geq 198 \text{ dBm}$$

NOTE. The allowable fade probability $p = 0.0002\%$ corresponds to 5.3 s per month. If such fading occurs independently in all the relaying hops of a 2500-km radio relay system containing 50 hops, where each hop is 50 km long, the allowable dropout period (CCIR Recommendation 393-3) corresponds to $p = 0.01\%$, or 264 s (approximately 4.4 min).

EXAMPLE 8.5

Assume the use of a 3-m diameter parabolic antenna with aperture efficiency $\eta = 55\%$ in Example 8.4. Discuss the possibility of designing an economical system and then present an improved alternative design. The limit of the receiver input is -75 dBm, where the squelch is set to initiate, and the transmitting and receiving feeder losses L_t and L_r are assumed to be 2 dB each.

Solution

1. Comments on the previous plan.
   (a) The solution of the previous example is used.
   (b) From step 8(d) of Example 8.4,

$$P_t + (G_t + G_r - L_t - L_r) - P_c \geq 198 \text{ dBm}$$

$$\text{antenna gain } G_t = G_r = 33.4 \text{ dB}$$

$$\text{limit of receiver input } P_c = -75 \text{ dBm}$$

   (c) Then

$$P_t + (33.4 \times 2 - 2 \times 2) - (-75) \geq 198 \text{ dBm}$$

   Accordingly,

$$P_t \geq 60.2 \text{ dBm} = 30.2 \text{ dBW} = 1050 \text{ watts}$$

   (d) Problems on economical design. The transmitter output on the order of 1 kW at 2 GHz, although not technically impossible, is not preferable because of the extreme cost of operation and installation of the facilities, including the power plant. An improved alternative plan is given next.
2. Improved alternative plan.
   (a) Increase the antenna gain using a parabola of a larger diameter.
   (b) Lower the limit of receiver input.
   (c) Suppress the propagation echo ratio by sharpening the antenna directivity.
   (d) Apply a diversity system.
   (e) Improvement can be achieved by changing the station site.

EXAMPLE 8.6

Using a comparison of various plans for improvement (including resiting plan), determine a method to design and erect an economical facility with the transmitter output P_t reduced to 10 watts or less. It is also assumed that the overall noise figure of the receiving system $F = 9$ dB and the IF bandwidth $B = 25$ MHz.

Solution

1. Increase of antenna gain.
   (a) A reduction of the required transmitter output by higher antenna gains can be achieved as shown in the following table:

Diameter of Parabola D'_ϕ Station A × Station B	Improvement (Power) Ratio $I_\phi = (D'_\phi/D_\phi)^2$	Required Transmitter Output (W) $P'_t = P_t/I_\phi$
4 × 4	1.78 × 1.78	332
6 × 6	4 × 4	68
6 × 8	4 × 7.11	37
8 × 8	7.11 × 7.11	20

NOTE. Previous values of antenna diameter and transmitter output were $D_\phi = 3$ m and $P_t = 1050$ watts.

   (b) The 8-m parabolic antenna is required for both stations *A* and *B* in order to reduce the transmitter output to approximately 20 watts.
   (c) If in addition to the preceding, reduction of the propagation echo ratio ρ_θ due to the antenna directivity effect is made, the fading parameter χ is improved and, accordingly, $P_t < 10$ watts may be achievable.
2. Reduction of the limiting receiver input level.
   (a) If the limiting receiver input level where the circuit outage occurs is assumed at the threshold point P_{th}, then $P_{\text{th}} = -82$ dBm (refer to Chapter 7), leading to a 7-dB improvement against the squelch operating point (-75 dBm).
   (b) Improvement of noise figure $F = 9$ dB by more than 5 dB makes the receiving facilities costly because use of a low-noise receiver of a particular type, such as a parametric amplifier, becomes necessary.
   (c) A reduction of the IF bandwidth B is not desirable in view of television signal transmission, and a reduction of more than 3 dB (corresponding to $B/2$) is difficult to attain.
   (d) To maintain the transmitter output power P_t at 20 watts or less, a total improvement amounting to 20 dB or more is required. However, this is practically unachievable, as mentioned in items (b) and (c).

Releasing the squelch operation during system operation is not advised since it produces a noise burst which may interfere with other RF channels within the same system.

3. Application of diversity system.
   (a) Frequency diversity (FD). Although the required improvement will be attained for $\Delta f/f \geq 2\%$, the frequency diversity system cannot be considered an economical approach when a future addition of an RF channel is needed. This is explained by the required provision of another transmitter and receiver, and the difficulty in RF channel frequency allotment.

(b) Space diversity (SD). The space diversity using two antennas installed with 4- to 5-m spacing provides an improvement ratio of 20 or more in the outage factor. However, it requires duplication of the receiving facilities.

4. Improvement of propagation echo ratio ρ_θ by the dual-antenna combining method.

   (a) Mounting two antennas separated vertically by P_h which equals one-half the pitch of the height pattern, and combining the two antenna outputs, each of which is fed to the receiver through a feeder of equal length, will result in in-phase addition of direct waves and cancellation of the reflected waves, leading to an improvement (reduction) of ρ_θ.

   (b) This method can be applied more advantageously to station B in view of the tower construction since station B, whose antenna elevation above sea level is lower than that of station A, gives the smaller P_h.

   (c) The diversity effect is achieved using two antennas (whose diameters may be smaller than designed originally) at only one side of two stations, requiring no adjustment after installation and no provision of additional in-station facilities.

   (d) It is considered to be most stable and economical compared with other systems because of its effectiveness in suppressing the sea-surface reflected waves, thereby improving the fading.

   (e) Whether or not the spacing P_h is practical economically is to be examined, because a greater P_h requires a greater tower construction cost.

   (f) For a given station site plan,

$$P_h \approx 7 \text{ m}$$

   which is the optimum condition at station B, and the transmitter output may be reduced to 5 watts or less.

5. Comparison and evaluation of alternate plans. The symbols ⊙, ○, △, and × are used in the table below in the order good → not good.

   (a) Evaluation table.

Alternate Plans	Order	Contribution to: Reduction of P_t	Contribution to: Improvement of Ordinary Fading	Construction Cost	Ease of Maintenance and Further Expansion
Use of antenna of larger diameter	2	○	△	○	○
Lowering the lower limit of the receiver input	4	×	×	×	×
Application of diversity system	3	⊙	△	△	△
Dual-antenna combining method	1	⊙	○	⊙	○

(b) Conclusion. The dual-antenna combining method, with the simplest arrangement and lowest cost, produces the greater improvement.

EXAMPLE 8.7

Present an engineering example to improve the fading by means of the dual-antenna combining method under the conditions given in Examples 8.4 and 8.5 and describe why the required transmitter output may be reduced to 5 watts or less.

Solution

First an engineering example is presented.

1. Determine the configuration and the aperture diameter D'_ϕ of the parabola in the dual-antenna system, where the total aperture area of the two antennas is equal to or greater than the aperture area of a single parabola in the original plan,

$$\left(D'_\phi\right)^2 \times 2 \geq 3^2$$

$$D'_\phi \geq 2.12 \text{ m}$$

2. Use of a standard size parabola of 2.5-m diameter, which provides a certain margin, is proposed.
3. When configured as the figure below, the reflected waves cancel each other to be ideal at the feeding point F, making $\rho'_\theta \to 0$, and the direct waves produce in-phase (+3 dB) addition. However, the reduction factor of the reflected wave in the dual-antenna system J_p with reference to the reflected wave in the original plan is assumed conservatively to be $J_p \geq 15$ dB (5.6 in field intensity ratio), taking into account the residual reflection wave due to the estimation error of the pitch, deviation of the performance of the parabola, mounting error of the parabola, variation of the K factor, etc.

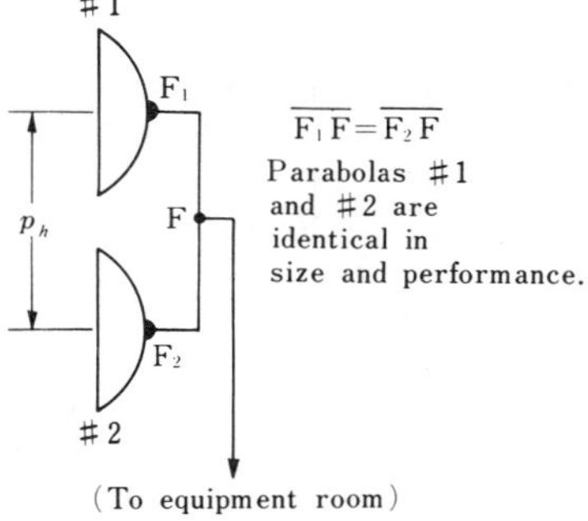

4. Expected propagation echo ratio ρ'_θ after improvement, to where $\rho_\theta = \rho_e$,

$$\rho'_\theta = \rho_\theta / J_p = 0.85/5.6 = 0.15$$

5. Fading parameter χ' expected after improvement.
   (a) Figure 21*c* (ρ'_θ) $\quad b'_1 = 0.95$
   (b) Effective low path length L_e,

$$L_e = 12.7 + 25.7 \times 0.15 = 16.6 \text{ km}$$

$$\text{Fig. } 21d\ (L_e) \qquad b'_2 = 0.175$$

   (c) T_p (as in the original plan) $\quad b_3 = 0.98$

$$K'_P = 1.13$$

   (d) The radio meteorological coefficient $K_A = 0.86$, the coefficient of surrounding terrain conditions $K_S = 0.97$, and the frequency-distance coefficient $K_{fD} = 0.79$ are the same as in the original plan (Example 8.4).
   (e) Fading parameter χ' after improvement ($\chi = 1.15$ in the original plan),

$$\chi' = K_A K_S K_P K_{fD}$$

$$= 0.86 \times 0.97 \times 1.13 \times 0.79 = 0.74$$

6. Estimated loss due to fading after improvement F'_e ($F_e = 65.5$ dB in the original plan). From Fig. 17*c*,

$$F'_e(0.0002\%) = 42 \text{ dB}$$

   for the given (χ, p).
7. Spacing between two parabolas P_h (at station B).
   (a) Determine reflection points d_1 and d_2 at $K = 4/3$ in reference to the profile chart given in Example 8.4,

$$d_1 = 33.3 \text{ km}, \qquad d_2 = 16.7 \text{ km}$$

   (b) Angle included by the direct and reflected rays θ at station B (refer to Chapter 2),

$$\theta = 0.64$$

   (c) One-half of the pitch in the height pattern (refer to Chapter 2),

$$P_h = 6.7 \text{ m}$$

Next the transmitter output P'_t required after improvement is calculated ($P_t =$ 1050 watts in the original plan).

1. Combined gain of dual-antenna system G_T at the feeding point F.
   (a) Gain of 2.5-m diameter parabolic antenna with $\eta = 0.55\%$,

$$G' = 31.8 \text{ dB per single antenna}$$

   (b) Loss of dual-antenna feeder system L_f',

$$L_f' = 1 \text{ dB}$$

   (c) Combined gain for dual-antenna system,

$$G_r = 31.8 + 3 - 1 = 33.8 \text{ dB}$$

2. Improvement ratio I_p in reference to the original plan

$$I_p \text{ (dB)} = (G_r - G) + (F_e - F_e')$$

$$= (33.8 - 33.4) + (65.5 - 42) = 23.9$$

$$I_p' \text{ (power ratio)} = 245$$

3. Required transmitting output expected after improvement P_t',

$$P_t' = P_t/I_p' = 1050/245 = 4.3 \text{ watts} < 5 \text{ watts}$$

To improve the fading, use of the dual-antenna combining method at any one side, that is, the transmitting or the receiving side, is sufficient. Therefore for the system employing a common transmitting and receiving antenna, addition of only one parabola per hop formed by stations A and B is sufficient to achieve the purpose.

EXAMPLE 8.8

There is a project to install an east–west trunk microwave system, which at its center will pass through Havana, Cuba. In addition to the present submarine cable system, it is requested that, if feasible, a microwave link be used to connect the proposed trunk system to the Isle of Pines, Cuba's largest island situated about 100 km south of the mainland. The purpose is to meet the demand for a network of television signal transmission, national telephone automation, as well as industrial and administrative data transmission. Present the approach to the site plan under the following conditions and indicate problems if any.

Conditions

1. Radio frequency band, 2 GHz; transmission capacity, telephony (300 channels) or television signal (3 channels).

Item	Telephony	Television
Mean S/N ratio per hop due to propagation noise (dB)	75	70 (peak-to-peak/rms)
Frequency deviation S (kHz)	$S_0 = 200$ rms	$S_p = 8000$ (peak-to-peak)
Channel bandwidth f (kHz)	3.1	6000
Transmitter output P_t (dBm)	40	40
Antenna gain G_t, G_r (dB)	35.8 (4 m)	35.8 (4 m)
Feeder system loss L_t, L_r	2	2
Overall noise figure F (dB)	7	7
Squelch operation input level level P_c (dBm)	−74	−74

2. Reliability. 5-min signal dropout is allowable during the month of worst fading.
3. The radio meteorological coefficient $K_A = 0.86$ is assumed from the data of other areas with similar conditions since long-term actual observation data in the area near the site are unavailable.
4. The station site selection design is to take into account the future addition of an 800-MHz telemetry system as well as a 4-GHz aeronautical radar link system.

Solution

1. Station site selection.
   (a) Guideline. Study the possibility of direct paths linking any site of the trunk microwave system in the mainland with a certain site in the northern part of the Isle of Pines. If it is found unsuitable, consider the use of a relay station.
   (b) Route plan (see Fig. A). In plan 1 the first relay station A on the western part of the trunk microwave link is connected to station D in the northern part of the Isle of Pines through the 120-km overseas propagation path. In plan 2 the spur terminal station C on the western route is connected to station D. This plan gives the shortest path length, that is, an 85-km overseas propagation path. However, the erection of a 100-m tower is intended to overcome the lower elevation of station C. In plan 3 the second relay station B of the western route is connected to station E in the northern part of the Isle of Pines through a 103-km overseas propagation path. In this path, one or two of the intermediate islands of the Mangles archipelago serve as a shielding ridge for reflected waves to suppress the sea-surface reflection. The location of point E is determined so that the path may go over the island and the antenna height is fixed at 200 m above sea level (180 to 190 m of site elevation

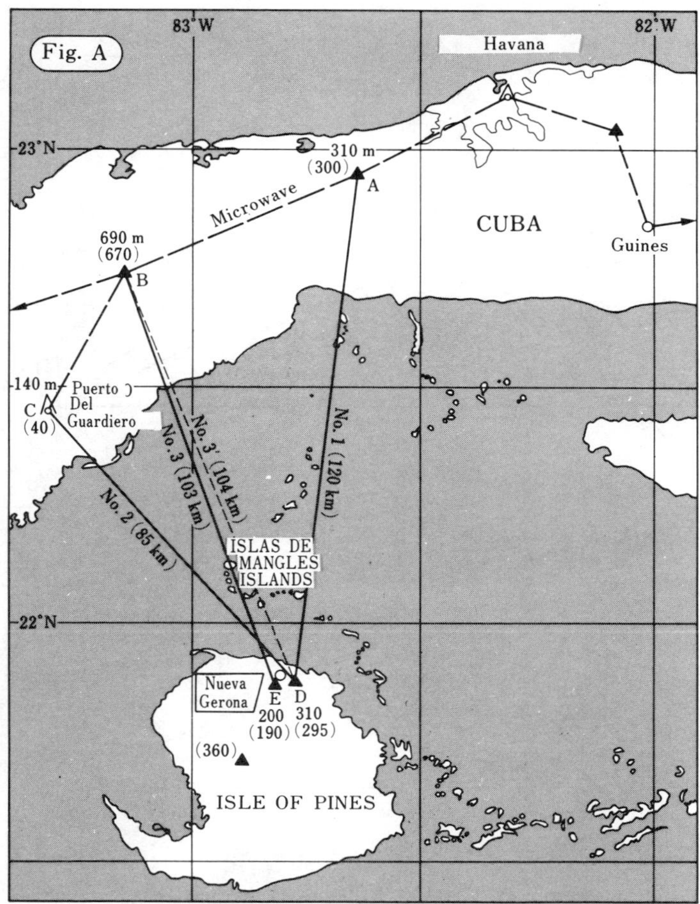

plus 20 to 10 m of antenna height above ground) at point E. Thus the center of the reflection area may approach the island.

(c) Path profile (see Fig. B).

(d) The strip map showing the surrounding terrain condition is to be prepared separately but is not shown here.

2. Design calculation.

(a) Propagation path parameters:

Item	Equation	Plan 1	Plan 2	Plan 3	Remarks
D (km)		120	85	103	Original data
h_1 (m)		310	310	690	
h_2 (m)		310	140	200	
C	$(h_1 - h_2)/(h_1 + h_2)$	—	0.38	0.55	Chapter 2,
m	$0.03D^2/(h_1 + h_2)$	—	0.324	0.358	Figs. 1, 24, 25
y		0.5	0.647	0.713	

Fig. B

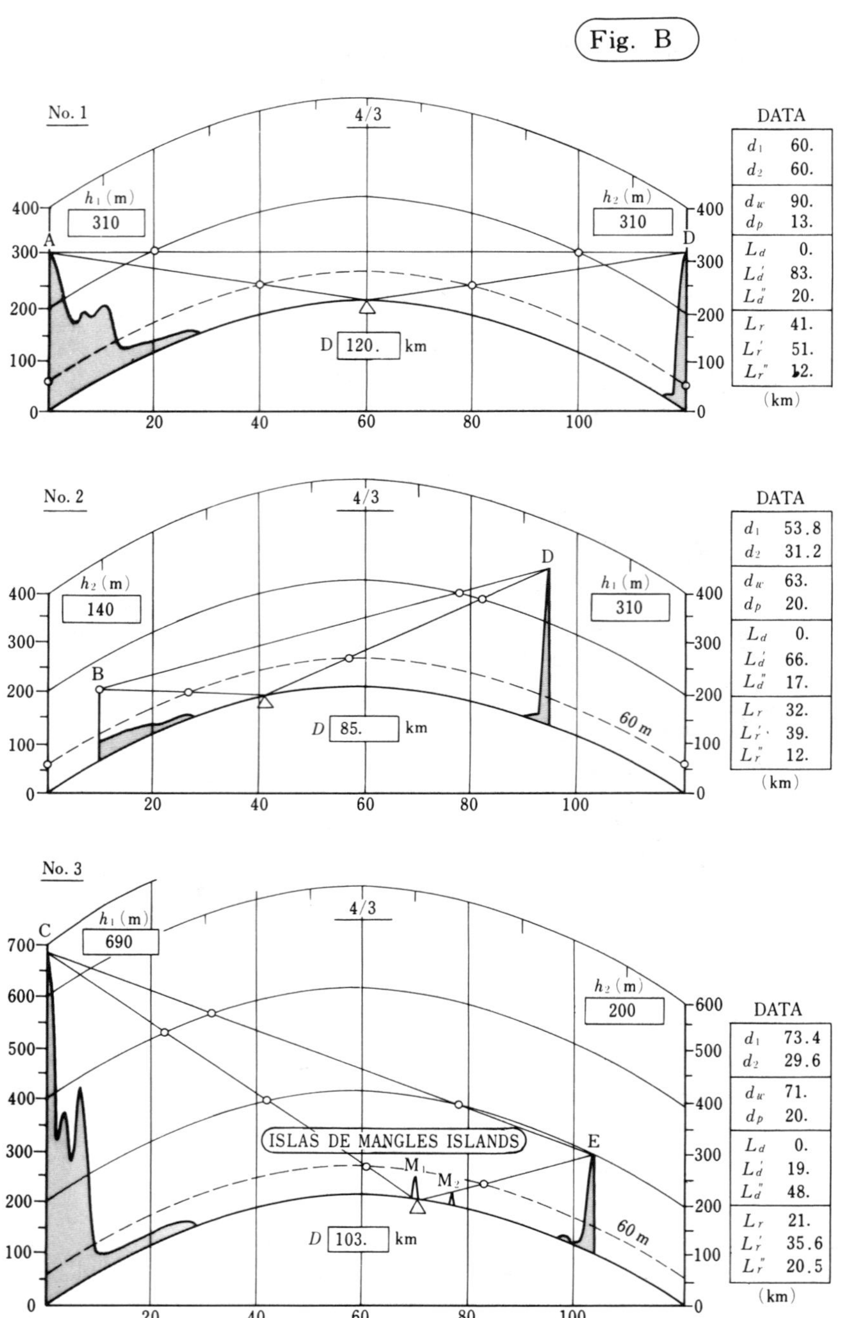

No. 1
4/3
h_1 (m) 310
h_2 (m) 310
A
D
D 120. km
DATA
d_1 60.
d_2 60.
d_w 90.
d_p 13.
L_d 0.
L_d' 83.
L_d'' 20.
L_r 41.
L_r' 51.
L_r'' 12.
(km)
No. 2
4/3
h_2 (m) 140
h_1 (m) 310
B
D
D 85. km
60 m
DATA
d_1 53.8
d_2 31.2
d_w 63.
d_p 20.
L_d 0.
L_d' 66.
L_d'' 17.
L_r 32.
L_r' 39.
L_r'' 12.
(km)
No. 3
4/3
C
h_1 (m) 690
h_2 (m) 200
ISLAS DE MANGLES ISLANDS
M_1
M_2
E
D 103. km
60 m
DATA
d_1 73.4
d_2 29.6
d_w 71.
d_p 20.
L_d 0.
L_d' 19.
L_d'' 48.
L_r 21.
L_r' 35.6
L_r'' 20.5
(km)

Item	Equation	Plan 1	Plan 2	Plan 3	Remarks
d_1 (km)	yD	60	67.3	73.4	
d_2 (km)	$D - d_1$	60	36.7	29.6	
d_{t1} (km)	$4.12\sqrt{h_1}$	0.75	108.2	108.2	Chapter 2,
P	d_1/dt_1	0.83	0.62	0.68	Fig. 23
ψ_1 (°)	$0.057(h_1/d_1 - 0.06d_1)$	—	0.149	0.291	Chapter 2,
ψ_2 (°)	$0.057(h_2/d_2 - 0.06d_2)$	—	0.152	0.287	Figs. 1, 31, 32
ψ (°)	$(\psi_1 + \psi_2)/2$	0.09	0.150	0.289	
θ_2 (°)		0.09	0.187	0.408	
R		≒ 1	0.94	0.91	Chapter 2,
D_v		0.48	0.67	0.715	Figs. 8a, 9, 12, 13,
Z		—	—	0.32	
ρ_e	RD_vZ	0.48	0.63	0.21	
$D_{\theta 1}D_{\theta 2}$		≒ 1	≒ 1	≒ 1	
ρ_θ	$\rho_e D_{\theta 1} D_{\theta 2}$	0.48	0.63	0.21	
u_1	$d_1^2/17$	212	266	317	Chapter 2,
u_2	$d_2^2/17$	212	72	52	Figs. 1, 23, 26, 30
h_1'	$h_1 - u_1$	98	44	373	
h_2'	$h_2 - u_2$	98	68	148	
Δ	$(2h_1'h_2'/D)10^{-3}$	0.16	0.07	1.07	
$\Delta/(\lambda/2)$	$(2\Delta/\lambda)$	2.13	0.47	7.13	
P_{h2}	$\lambda/(2\sin\theta_2)$	47.7	23.0	10.5	

(b) Fading parameter χ:

	Equation	Plan 1	Plan 2	Plan 3	Remarks
L_{de}	$L_d + L_d'/3 + L_d''/5$	31.7	25.4	15.9	Fig. B, Fig. 21*b*
L_{re}	$L_r + L_r'/3 + L_r''/5$	60.4	47.4	37	
L_e	$L_{de} + L_{re}\rho_\theta$	58.3	55.3	23.5	
T_p	$(d_W + d_p/2)/D$	0.804	0.86	0.786	
b_1		1.09	1.15	0.96	
b_2		0.816	0.75	0.23	
b_3		0.98	0.995	0.875	Fig. 21*c*
K_P		1.911	1.935	1.119	Fig. 21*d*
g_f		0.79	0.79	0.79	$f = 2$ GHz, Fig. 18
g_D		1.52	1.31	1.44	
K_{fD}	$g_f g_D$	1.201	1.035	1.138	
r_W	(Water surface ratio)	0.68	0.64	0.56	To be found from
r_P	(Plain area ratio)	0.15	0.31	0.22	strip map of
r_M	(Mountain area ratio)	0.17	0.05	0.22	surrounding terrain

	Equation	Plan 1	Plan 2	Plan 3	Remarks
K_S	$r_W + 0.95r_P + 0.8r_M$	0.956	0.978	0.945	conditions (figure is omitted here).
K_A		0.86	0.86	0.86	
χ	$K_P K_{fD} K_S K_A$	1.887	1.684	1.035	

(c) Standard receiver input P_r and dropout margin M_c:

	Equation	Plan 1	Plan 2	Plan 3
P_t		40	40	40
$G_t + G_r$		71.6	71.6	71.6
$L_t + L_r$		4	4	4
Γ_0		140.1	137.1	138.7
L	$\Gamma_0 - (G_t + G_r - L_t - L_r)$	72.5	69.5	71.1
P_r	$P_t - L$	−32.5	−29.5	−31.1
P_c	$(P_c = P_{SQ})$	−74	−74	−74
M_C	$P_r - P_c$	41.5	44.5	42.9

3. Mean S/N ratio (dB):

Equation	Plan 1	Plan 2	Plan 3	Specified Value
Telephony (Top channel rms/rms)				
$S/N = P_r - F + 10\log[(S_0/f_m)^2/(KT_0\,\Delta f)]$	83.3	86.3	84.7	75
Television (peak-to-peak/rms)				
$S/N = P_r - F + 10\log[3(S_p^2/f_m^3)/KT_0]$	74.0	77.0	75.4	70

All the estimated values agree with the specified values.

4. Percentage of dropout time:

	Plan 1	Plan 2	Plan 3	Remarks
Estimated drop out time percentage				
($\chi > 1.5$ for plans 1 and 2)	> 0.2	> 0.1	0.001	Fig. 17*b*
Specified value				
5 min/(60 × 24 × 31 min) × 100%		0.0112		

5. Comments and associated problems.
   (a) Plans 1 and 2 cannot be adopted because of the possibility of deep fading; plan 3 almost satisfies the specification.
   (b) Although plan 3 estimated the shielding effect by the Mangles Islands against the reflected waves as 10 dB ($Z = 0.32$), it is preferable, if

possible, to confirm the exact propagation path on a larger scale (1:50,000 to 1:25,000) map, take the field survey and measurements, conduct the radio propagation test in sections B to E, and perform the measurement of the height pattern (in reference to pitch and amplitude) at point E.

(c) For the equivalent reflection coefficient $\rho_e > 0.2$ it is necessary to carry out the shielding of the reflected waves by means of the dual-antenna combining method at point E. Spacing of these two antennas is equal to one-half the pitch P_h, that is, for a 2- to 2.5-m diameter parabola,

$$P_{h2} = \begin{cases} 26.3 \text{ m}, & \text{for 800 MHz} \\ 10.5 \text{ m}, & \text{for 2 GHz} \\ 5.3 \text{ m}, & \text{for 4 GHz} \end{cases}$$

Accordingly, this plan is easier to carry out.

(d) In case of plans 1 and 2, P_{h2} for 800 MHz is large (120 m for plan 1 and 60 m for plan 2). Then an economical system cannot be attained. If space diversity is adopted, the vertical spacing between the upper and lower antennas is to be more than $P_{h2}/2$, taking into consideration type K fading.

(e) Adoption of dual-antenna combining will permit fixing the site of the station at point D on the Isle of Pines (plan 3′ on the propagation path as shown in Fig. A).

EXAMPLE 8.9

There is an installation plan for a 6-GHz international-grade microwave multirelay system passing through the savanna area of Argentina, Paraguay, and Brazil. Considering that the propagation echo ratio ρ_θ in the savanna is within the range of 0.3 to 0.7 and the types of propagation paths encountered are almost similar, it is intended to provide a standard-type tower on which an

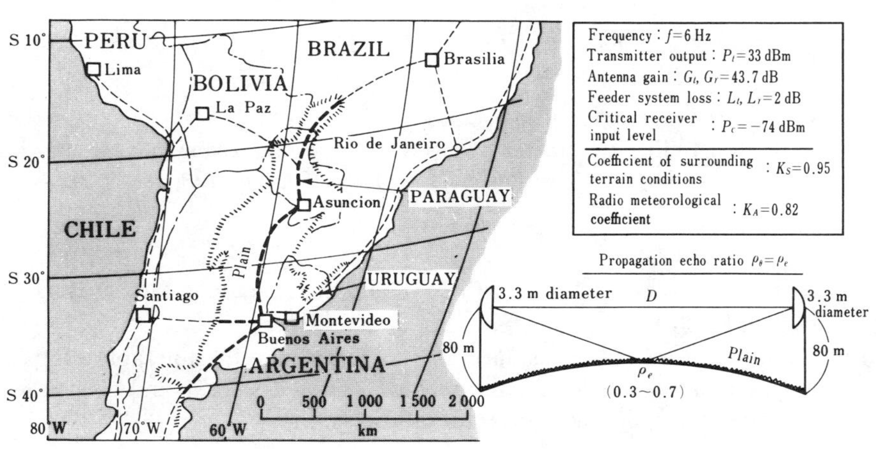

antenna can be mounted 80 m above ground, and to seek uniformity for station building, relay equipment, etc. Use of diversity is restricted, in principle, to particular hops in the initial stage. Further addition of the diversity will be decided some time after commencement of service, based on the actual fading data, to avoid unnecessary investment for the diversity arrangement. Calculate the allowable propagation path length for various values of ρ_θ when a CCIR requirement is applied to the allowable outage factor.

Solution

First the procedure of the analysis is outlined.

1. Tolerable value of fading parameter χ_i.
   (a) Calculate the tolerable outage (dropout) factor p_{ci} (%) for the propagation distance D = 30 to 50 km.
   (b) Calculate the outage margin M_c (dB) for the propagation distance of 30 to 50 km for the performance of the given facilities.
   (c) Find the tolerable value for this fading parameter χ_i (%) using M_c and p_{ci} from Fig. 17.
2. Analysis of path profile.
   (a) Draw a path profile for distances D = 30, 35, 40, 45, and 50 km.
   (b) From this profile estimate the effective low path length L_e for $\rho_\theta = 0.1$ to 0.7.
   (c) Calculate the underlying terrain factor T_p,

$$T_p = (d_p/2)/D = 0.5 \qquad \text{(flat land for whole length)}$$

3. Path profile coefficient K_P.
   (a) Each $\rho_\theta \to b_{11}, b_{12}$
   ($\Delta/(\lambda/2) > 5$ is assumed.)
   (b) Each $L_e \to b_{22}$
   (c) $T_p = 0.5 \to b_3 = 0.925$
   (d) $K_P = (b_{11} \cdot b_{12} + b_{22})b_3$

   } $\rho_e = 0.1, 0.2$ to 0.7; D = 30, 35 to 50 km (35 combinations)
4. Estimated fading parameter χ. Calculate χ employing the already known parameters, that is, $K_A = 0.82$, $K_S = 0.95$, K_P, and K_{fD},

$$\chi = K_A K_S K_P K_{fD}$$

5. Estimated outage factor p_{cxi} (%). Estimate the outage factor using χ and M_c from Fig. 17c.
6. Tolerable propagation distance $D_{\max}$. Determine the tolerable propagation distance on a graph where the relation between steps 1 and 5 is shown for distance D and propagation echo ratio ρ_θ.

The design calculations can now be carried out.

1. Tolerable value of fading parameter χ:

D (km) p_{ci} (%)	30 0.00012	35 0.00014	40 0.00016	45 0.00018	50 0.0002	Propagation Distance Outage Factor Specified by CCIR
P_t	33	33	33	33	33	Dropout margin
$G_t + G_r$	87.4	87.4	87.4	87.4	87.4	
$-(L_t + L_r)$	-4	-4	-4	-4	-4	
$-\Gamma_0$	-137.6	-138.9	-140.1	-141.1	-142	
P_r	-21.2	-22.5	-23.7	-24.7	-25.6	
$-P_c$	74	74	74	74	74	
M_C (dB)	52.8	51.5	50.3	49.3	48.4	
χ_i	0.89	0.88	0.87	0.86	0.85	Tolerable value of fading parameter

NOTE. p_{ci} (%) $= 0.01 \times D/2500$

$$M_c \text{ (dB)} = P_r - P_c = [P_t + (G_t + G_r - L_t - L_r) - \Gamma_0] - p_c$$

2. Fading parameter:

D ρ_θ	30	35	40	45	50	Remark
0.1	0.533	0.619	0.758	0.931	1.072	
0.2	0.576	0.672	0.832	1.007	1.160	
0.3	0.619	0.727	0.894	1.099	1.264	
0.4	0.656	0.779	0.951	1.164	1.343	$K_A K_S K_P K_{fD}$
0.5	0.679	0.834	1.021	1.244	1.430	
0.6	0.740	0.889	1.085	1.319	1.524	$(K_A = 0.82$
0.7	0.784	0.948	1.155	1.407	1.618	$K_S = 0.95)$

3. Tolerable propagation distance. From the table in step 2 of the design calculations the graph showing the relationship between distance D and the estimated fading parameter χ can be drawn for each value of the propagation echo ratio ρ_θ. The relationship between distance D and the tolerable fading parameter according to the CCIR objective, as mentioned in step 1 of the design calculations, can be drawn with a number of curves in the same graph (Figure A). From this graph the conditions to satisfy the CCIR objectives can be determined. The tolerable propagation distance $D_{\max}$ can be found at the intersection of the curves corresponding to the CCIR objective in Fig. A and

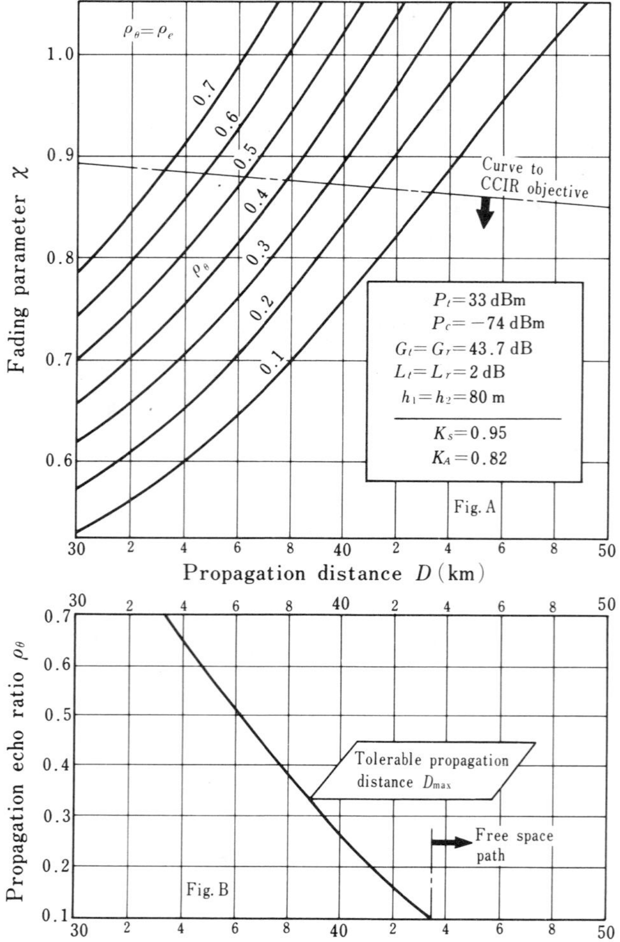

the vertical line for the propagation echo ratio ρ_θ. The curve of D_{max} versus ρ_θ thus obtained is shown in Fig. B.

4. Conclusion.

(a) The propagation distance D_{max} versus the propagation echo ratio ρ_θ is shown in Fig. B.

(b) Roughly estimated values are

$$\rho_\theta = 0.3 \qquad D_{max} = 39 \text{ km}$$

$$\rho_\theta = 0.5 \qquad D_{max} = 36 \text{ km}$$

$$\rho_\theta = 0.7 \qquad D_{max} = 33 \text{ km}$$

(c) Since a smaller ρ_θ can produce a greater D_{max} and more stable propagation due to decreased χ, as shown in Fig. A, it is preferable to choose the station site in such a manner that the center of the reflection area falls in a rolling terrain, a forest, or a village.

(d) If $\rho_\theta < 0.1$ (free-space type), then $D_{max} > 44$ km and $\chi < 0.75$. Such conditions are hardly expected in a savanna, but may be satisfied when the propagation path is designed in such a manner that the reflection point is located at a hill or in woods having a height equal to or greater than the first Fresnel zone radius r_R at the reflection point. The guideline figures of r_R are

$$r_R = 19 \text{ m}, \qquad D = 30 \text{ km}$$

$$r_R = 22 \text{ m}, \qquad D = 40 \text{ km}$$

$$r_R = 25 \text{ m}, \qquad D = 50 \text{ km}$$

(e) The savanna extends over more than 3000 km and the radio meteorological conditions and surrounding terrain conditions are not likely to be uniform. Although the radio meteorological coefficient $K_A = 0.85$ and the coefficient of surrounding terrain conditions $K_S = 0.95$ (corresponding to uniformly flat land) are assumed in the example, these figures have to be corrected at the lower altitude (northern) region or in a region near the sea or a lake, thereby making the conditions more stringent.

Remarks

1. If K_A is corrected and a new calculation is made according to step 2 of the design calculations, the preceding method can apply generally to other prairie types, such as the Australian plain, the Libyan desert, Saudi Arabia, or the Mongolian plain.
2. If the facility design condition undergoes modification, step 1 of the design calculation is to be recalculated.
3. If the antenna height is modified, recalculate it from step 2 of the design calculations.
4. Although the path profile has been omitted here, the measurement of the length of the low path portion has to be made carefully.
5. This preliminary study will be useful in eliminating blind application of space diversity to lower layer paths over flat land and will assist in the effective use of diversity.

Example 8.10

The preferred route of the Peking–Hanoi international system should be the shortest in view of maintaining the required transmission quality, even in the extended long-distance connection. The figure shows such a route plan: Peking–Chengchow–Hankow–Nanking–Hanoi. However, in the execution of the project to establish China's nationwide microwave network, the planning and installation of the Peking–Tsinan–Nanking–Shanghai route and the

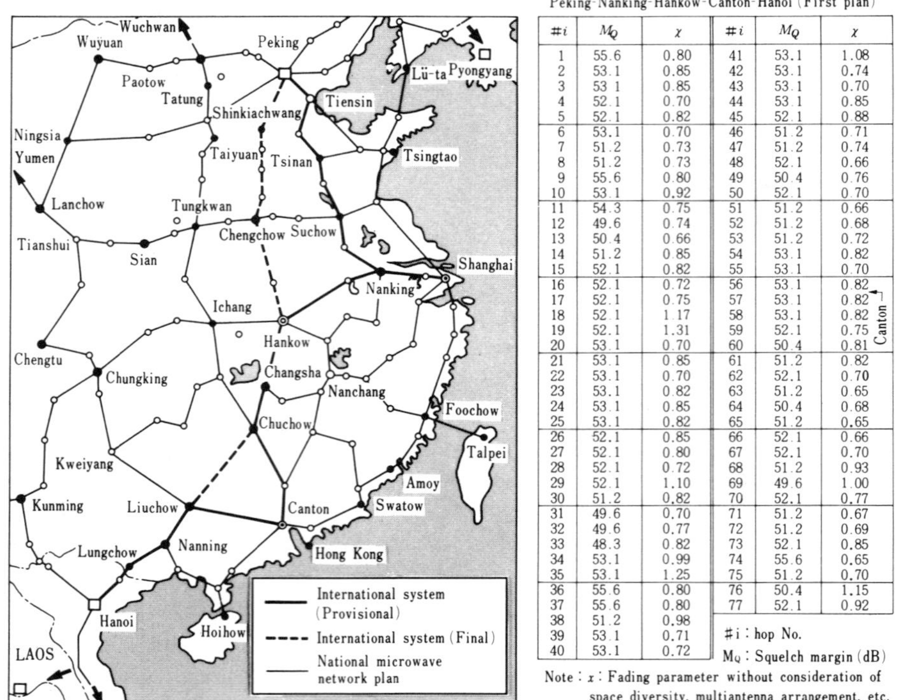

Peking-Nanking-Hankow-Canton-Hanoi (First plan)

#i	M_Q	χ	#i	M_Q	χ
1	55.6	0.80	41	53.1	1.08
2	53.1	0.85	42	53.1	0.74
3	53 1	0.85	43	53.1	0.70
4	52.1	0.70	44	53.1	0.85
5	52.1	0.82	45	52.1	0.88
6	53.1	0.70	46	51.2	0.71
7	51.2	0.73	47	51.2	0.74
8	51.2	0.73	48	52.1	0.66
9	55.6	0.80	49	50.4	0.76
10	53.1	0.92	50	52.1	0.70
11	54.3	0.75	51	51.2	0.66
12	49.6	0.74	52	51.2	0.68
13	50.4	0.66	53	51.2	0.72
14	51.2	0.85	54	53.1	0.82
15	52.1	0.82	55	53.1	0.70
16	52.1	0.72	56	53.1	0.82 ← Canton
17	52.1	0.75	57	53.1	0.82
18	52.1	1.17	58	53.1	0.82
19	52.1	1.31	59	52.1	0.75
20	53.1	0.70	60	50.4	0.81
21	53.1	0.85	61	51.2	0.82
22	53.1	0.70	62	52.1	0.70
23	53.1	0.82	63	51.2	0.65
24	53.1	0.85	64	50.4	0.68
25	53.1	0.82	65	51.2	0.65
26	52.1	0.85	66	52.1	0.66
27	52.1	0.80	67	52.1	0.70
28	52.1	0.72	68	51.2	0.93
29	52.1	1.10	69	49.6	1.00
30	51.2	0.82	70	52.1	0.77
31	49.6	0.70	71	51.2	0.67
32	49.6	0.77	72	51.2	0.69
33	48.3	0.82	73	52.1	0.85
34	53.1	0.99	74	55.6	0.65
35	53.1	1.25	75	51.2	0.70
36	55.6	0.80	76	50.4	1.15
37	55.6	0.80	77	52.1	0.92
38	51.2	0.98			
39	53.1	0.71			
40	53.1	0.72			

#i : hop No.

M_Q : Squelch margin (dB)

Note : x : Fading parameter without consideration of space diversity, multiantenna arrangement, etc.

Shanghai–Hankow–Canton route should precede the aforementioned plan. Then it is intended that the route plan in which the latter two are combined, that is, the Peking–Tsinan–Hankow–Canton–Hanoi route, be used as a provisional international route. Estimate the S/N ratio and the outage factor for the Peking–Hanoi system assuming that the data given in the table have been prepared as a result of the first routing plan and calculation of propagation. Give an appropriate approach to satisfy CCIR objectives I and IV when applied to a system with more than 2500 km length. System total length $L = 3485$ km, number of hops $N = 77$, and average hop length = 45.3 km are assumed.

Assumed Conditions

Frequency $f = 5$ GHz, transmitter output $P_t = 33$ dBm, antenna gain $G_t = G_r = 43.8$ dB (4-m diameter), feeder system loss $L_t = L_r = 2$ dB, overall noise figure $F = 9$ dB, squelch operation level $P_Q = -75$ dBm. Telephony: Required number of channels $n = 60$ (utilizing a part of the 1260 channels for the national multiplex system), frequency deviation $S_0 = 200$ kHz, maximum modulating frequency $f_m' = 5636$ kHz. Color television: Frequency deviation $S_p = 8$ MHz peak-to-peak maximum modulating frequency $f_m' = 6$ MHz, required average S/N ratio (peak-to-peak/rms) = 55 dB, CCIR objectives I and IV.

Solution

Allowable noise.

1. Conditions for required circuit performance.
   (a) International terminal stations along the route are assumed to be Peking, Shanghai, Hankow, and Canton. Accordingly, the section Hanoi–Canton within the provisional route Peking–Hanoi is treated as a permanent plan.
   (b) CCIR performance objectives are specified for 2500 km. A system exceeding 2500 km in length is treated as a combined system where several routes of 2500 km or less are connected in tandem, that is, the noise powers and the outage factors are simply added.
   (c) The length of the Peking–Hanoi system is 3485 km. The objectives and the performances are examined for three baseband sections as given in the following:

Baseband Section	Route Length	Number of Relaying Hops
Peking–Canton (provisional)	2485	56 (#1 to #56)
Canton–Hanoi (permanent)	1000	21 (#57 to #77)
Peking–Hanoi (provisional)	3485	77 (#1 to #77)

   (d) Although the number of telephone channels initially planned is 60, these channels are to be accommodated at any part within the FDM 1260-channel system, and the study will be made for the top telephone channel (the worst channel). It follows that the national system is to satisfy CCIR objectives I and IV for the above baseband sections.
   (e) For color television transmission the separately specified S/N ratio is distributed in proportion to the length; noise power is added proportionately to length.
2. Noise power distribution and evaluation.
   (a) The distribution plan of the designed noise power is made according to Fig. 14.
   (b) If the estimated propagation noise satisfies the assigned objective, the total noise power is also deemed to be satisfactory. (The distortion and interference noises are not explained here.)
3. Allowable propagation noise and S/N ratio:

	Color Television	Telephony (1260 Channels-Top)		
		Propagation noise (unweighted)		
	S/N (dB)	I (Hourly Mean Noise)		IV (Outage Factor)
	(peak-to-peak/rms)	N_A (pW)	S/N (dB)	p_c (%)
Peking–Canton	56.5	4423	53.5	0.00994
Canton–Hanoi	60.4	1780	57.5	0.00400
Peking–Hanoi	55	6203	52.1	0.01394
		$1.78 \times L$ (pW)		$0.01 \times L/2500$

Total noise power (for telephony, unweighted), corresponding to $3L + \alpha$ (pW) of the radio transmission system, is as follows:

Peking–Canton	Canton–Hanoi	Peking–Hanoi
8050 pW	3400 pW	11,450 pW

Estimated noise power for each section of propagation.

1. Relation of S/N ratios for telephony and television transmission systems (quantitative relation if the same transmission path is employed),

$$(S/N)_{\mathrm{TP}} = 10\log\frac{P}{KTF}\left(\frac{S_0^2}{\Delta f f_m^2}\right) \quad \text{(dB)}$$

$$(S/N)_{\mathrm{TV}} = 10\log\frac{P_r}{KTF}\left(\frac{3S_p^2}{f_m'^3}\right) \quad \text{(dB)}$$

Then

$$(S/N)_{\mathrm{TV}} = (S/N)_{\mathrm{TP}} + 10\log\left(\frac{3f_m^2\,\Delta f S_p^2}{f_m'^3 S_0^2}\right)$$

$$= (S/N)_{\mathrm{TP}} + 10\log\left(\frac{3\times 5.636^2\times 3.1\times 8^2\times 10^{12}}{6^3\times 2^2\times 10^{13}}\right)$$

$$= (S/N)_{\mathrm{TP}} + 3.4\ \text{dB}$$

Various calculations on the telephony system are done first, and then the summary of the results, including those for television, will be given.

2. Propagation noise of each hop N_{AXi}.
   (a) From the given parameters, the squelch operating input P_Q, and the squelch margin M_Q $(= P_r - P_Q)$, calculate first $(S/N)_Q$ at P_Q, and then $(S/N)_i$ and the noise power N_{AXi} of each hop.

$$(S/N)_Q = P_Q - F + G_s$$

$$= -75 - 9 + 110 = 26\ \text{dB} \qquad (< 30\ \text{dB})$$

   using Fig. 12 in Chapter 7, graph for the quick estimation of the S/N ratio for a single-sideband FM system employing CCIR parameters.
   (b) $(S/N)_i$ for receiver input P_r,

$$(S/N)_i = (S/N)_Q + M_Q = M_Q + 26\ \text{dB}$$

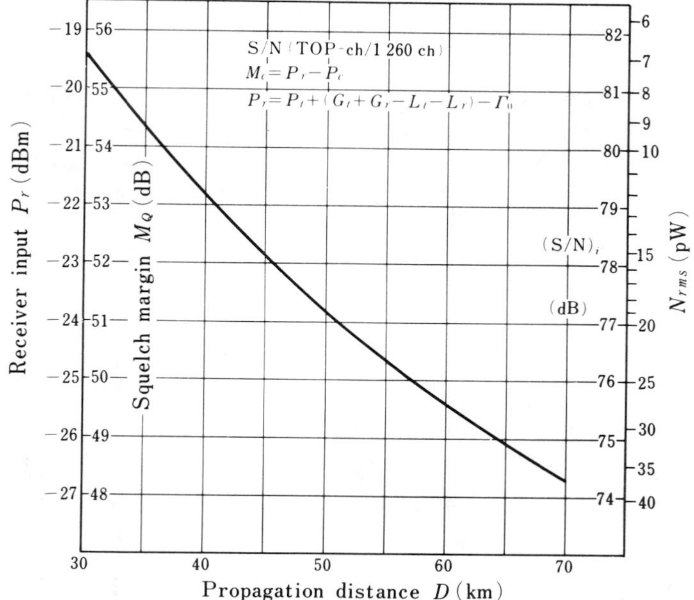

(c)

$$N_{AXi} = 10^{[90-(S/N)_i]/10} \quad \text{pW} \qquad \text{(Fig. 25, Chapter 7)}$$

(d) Since the number of hops for which the calculation is to be made reaches 77, it is better to present the preceding relationship on a graph for the given range to save labor.

The following figure has been prepared for distance D, since P_r, M_Q, etc., have been derived for the basic variable D. This figure is convenient when some modification in route selection design is needed.

3. Outage factor of each hop p_{cxi}.
   (a) The outage factor is defined as the probability that the amount of the generated noise exceeds 10^6 pW at a point where the test tone signal level is 0 dBm ($= 10^9$ pW). To estimate the outage factor with a very small figure, the assumption is made that the fadings occurring in individual hops will not be correlated with each other.
   (b) The outage factor for each hop p_{cxi} is estimated from the fading parameter χ and the outage margin defined as a margin of a 30-dB S/N ratio, corresponding to the S/N ratio for 10^6 pW noise.
   (c) The outage margin M_c is easily found by subtracting 4 dB from the known squelch margin M_Q since the S/N ratio at the squelch operating receiver input is 26 dB and the outage margin is defined at $S/N =$ 30 dB, which is 4 dB higher than 26 dB.
   (d) The outage factor for each hop is thus estimated from M_c and χ using Fig. 17.

4. The results obtained in the preceding steps 2 and 3 for each propagation path are summarized in the following table:

Peking–Canton						Canton–Hanoi		
# i	N_{AXi} (pW)	p_{cxi} (10^{-5} %)	# i	N_{AXi} (pW)	p_{cxi} (10^{-5} %)	# i	N_{AXi} (pW)	p_{cxi} (10^{-5} %)
1	7	3	31	27	3	57	13	10
2	13	16	32	27	12	58	13	10
3	13	16	33	37	60	59	16	4
4	16	2	34	13	* 115	60	23	20
5	16	14	35	13	* 1200	61	19	18
6	13	1	36	7	3	62	16	1
7	19	3	37	7	3	63	19	—
8	19	3	38	19	* 160	64	23	2
9	7	3	39	13	11	65	19	—
10	13	45	40	13	11	66	16	—
11	10	2	41	13	* 290	67	16	1
12	27	7	42	13	3	68	19	80
13	23	1	43	13	1	69	27	* 280
14	19	26	44	13	16	70	16	6
15	16	14	45	16	35	71	19	1
16	16	2	46	19	3	72	19	2
17	16	4	47	19	4	73	16	22
18	16	* 790	48	16	—	74	7	—
19	16	* 2100	49	23	8	75	19	2
20	13	1	50	16	1	76	23	* 950
21	13	16	51	19	—	77	16	64
22	13	1	52	19	1			
23	13	11	53	19	3			
24	13	16	54	13	11	$\sum_{57}^{77}$	748	1473
25	13	11	55	13	1			
26	16	22	56	13	11	Peking–Hanoi		
27	16	10						
28	16	2						
29	16	* 420	$\sum_{1}^{56}$	1778	5546	$\sum_{1}^{77}$	2526	7019
30	19	18						

NOTE.* $p_{cxi} > 10^2 \times 10^{-5}$ %

Estimated Noise for Each Base-band Section (Preliminary Examination).

1. Estimated value of hourly mean noise power (worst value) N_{AX} and associated S/N ratio,

$$N_{AX} = 2\sum N_{AXi}\ (\text{pW}) \quad (\text{all sections})$$

$$(S/N)_{\text{Tp}} = 10\log(10^9/N_{AX})$$

$$(S/N)_{\text{TV}} = (S/N)_{\text{TP}} + 3.4\ \text{dB}$$

2. Estimated outage factor p_{cx},

$$p_{cx} = \sum p_{cxi}\ (\%) \qquad (\text{all sections})$$

3. Comparison of the estimated values to the base-band section with the specified values (preliminary examination):

	Mean Noise		Outage Factor (%)	
Route	N_{AX}/N_A	$(S/N)_{\text{TV}}$ Specified Value	p_{cx}/p_c	Excess
Peking–Canton	1778/4423	60.9/56.5	* 0.05546/0.00994	4552×10^{-5}
Canton–Hanoi	748/1780	64.7/60.4	* 0.01473/0.00400	1073×10^{-5}
Peking–Hanoi	2526/6203	59.4/55.0	* 0.07019/0.01394	5625×10^{-5}
Judgment	Good	Good	* No good	

NOTE. N_A, p_c = specified values; N_{AX}, p_{cx} = Designed values.

Improvement of Estimated Outage Factor

1. Hops that require improvement.
   (a) Reviewing the two preceding tables, notice that unsatisfactory (not good) performance may be attributed to significantly large outage factors of the outstanding 9 hops from a total of 77 hops, as indicated by asterisks. This does not necessarily indicate defective engineering as a whole, but the existence of a small number of difficult hops in view of the site selection design.
   (b) Hops requiring improvement shall be selected carefully so that the performance of both base-band sections, that is, Peking–Canton and Canton–Hanoi, may be satisfactory. It follows that the performance of the Peking–Hanoi section will automatically be satisfied since both the specified and the expected values of the outage factor may follow the simple addition of percentage law.
   (c) The hops requiring improvement are to be determined one by one in the order from the worst to the best, until the required total performance is

obtained. If the achievable improvement is assumed to be 20, implying the improvement of the outage factor by the amount of 1/20, the hops requiring improvement are determined as shown in the following table:

Hops Not Good (Note)	p_{cxi} (10^{-5} %)	Amount of Improvement	Resultant Total Excess
	Peking–Canton (Excess 4552 × 10^{-5} % Total)		
#19	2100	1995	2557
#35	1200	1140	1417
#18	790	750	667
#29	420	399	268
#41	290	275	−7
	4800		
#38 #34	(Further improvement is required only when the above improvement is insufficient.)		
	Canton–Hanoi (Excess 1073 × 10^{-5} % Total)		
#76	950	902	171
#69	280	266	−95
	1230		

NOTE. Written order corresponds to order of priority.

2. Causes of not-good performance and possible remedies. The outline of improvement is as follows:

Hop Not Good	Cause of Defect	Outline of Remedies
#18 #19	Large ρ_θ Low path, large ρ_θ	Increase antenna height at the station located between #18 and #19 paths, shift reflection point to land, and use dual-antenna combining method for #19 path.
#29	Large ρ_θ	Lower one of the antennas to suppress ray reflected by a ridge.
#35	Low path, large ρ_θ	Raise both antennas and use dual-antenna combining method.
#41	Large ρ_θ	Use dual-antenna combining method.
#69	Large ρ_θ	Raise one antenna to shift reflection point to land.
#76	Low path, long distance, large ρ_θ	Use space diversity.

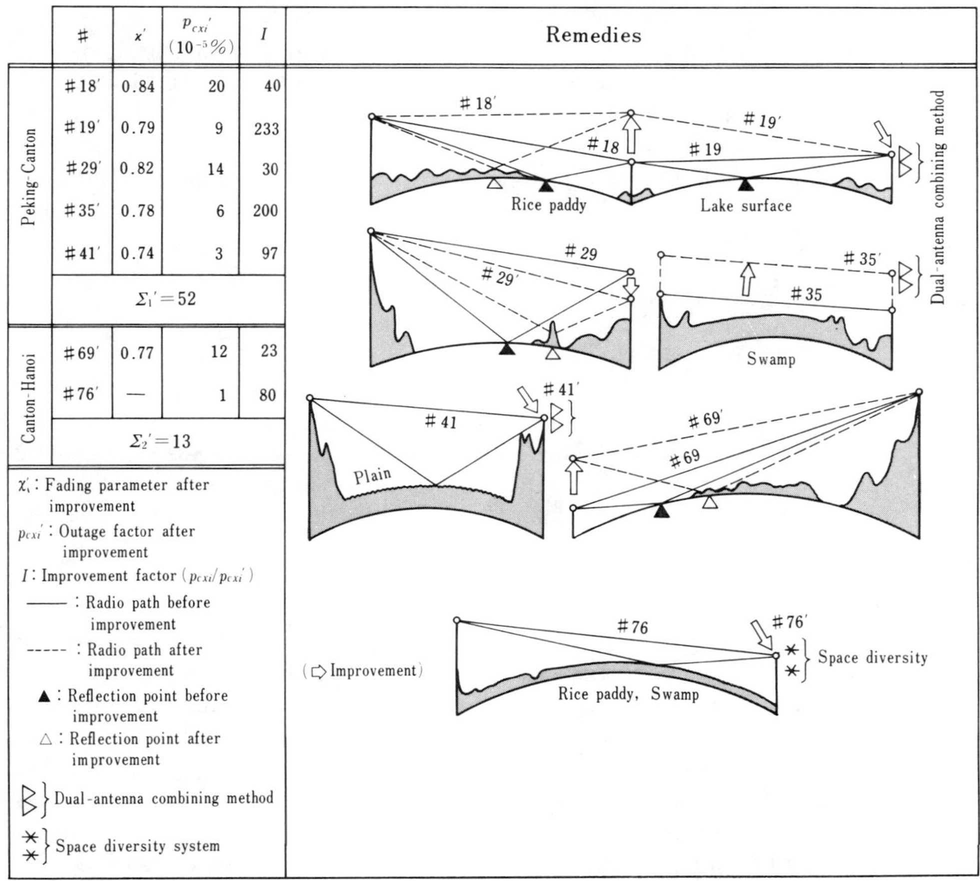

	#	x'	p_{cxi}' ($10^{-5}\%$)	I
Peking-Canton	#18′	0.84	20	40
	#19′	0.79	9	233
	#29′	0.82	14	30
	#35′	0.78	6	200
	#41′	0.74	3	97
	$\Sigma_1' = 52$			
Canton-Hanoi	#69′	0.77	12	23
	#76′	—	1	80
	$\Sigma_2' = 13$			

3. Estimated improvement achieved by the remedies:

Evaluation of system design.

1. Propagation noise:

After Improvement	Mean S/N Ratio and Noise Power: Telephony (1260 channels, Top)		Television (Color, 6 MHz)	Outage Factor
Route	$(S/N)_{TP}$ (dB)	N_{AX}/N_A (pW)	$(S/N)_{TV}$ (dB) Specified Value	p_{cx}/p_c (%)
Peking–Canton	57.5	1778/4423	60.9/56.5	0.00798/0.009
Canton–Hanoi	61.3	748/1780	64.7/60.4	0.00256/0.004
Peking–Hanoi	56.0	2526/6203	59.4/55.0	0.01054/0.013
Judgment	Good	Good	Good	Good

2. Total system noise and radio transmission system noise.
   (a) The specified values are expected to be satisfied if no erroneous design, manufacturing, adjustment, or installation work is found. This is because the distortion and interference noise is sufficiently low in the designed noise allotment.
   (b) For site selection in a plain, particularly between Peking and Nanking, the overreach interference should be borne in mind, and appropriate protection, such as cross-polarization or frequency change, to restrict the interference to within a tolerable limit should be made.

REMARKS

1. The routing design of this system has been established in such a manner that the system passes through as many intermediate cities as possible. The purpose of this is to handle the domestic traffic demand and also to facilitate future system expansion and the establishment of spur routes. However, if the forementioned design philosophy is not followed, the total length of the Peking–Hanoi system could be shortened by approximately 200 km, thereby omitting six to eight hops. The design to be adopted depends on a comparison of overall economy as well as on future plans.
2. Radio meteorology in China is almost unknown. Especially the vast low-plain area tends to give rise very frequently to a strong duct in the summer, and thus it is difficult to predict in which section and to what extent the duct will be produced.
3. To obtain a functional radio system, it is necessary to start earlier with an experiment to depict the relationship between the radio-wave propagation and meteorological conditions, and to collect data systematically and continually. The first long-distance microwave system to be installed may be pertinent to study. Therefore it is advisable to perform the observation during a period of more than one year after completion of the installation.
4. Estimation of fading mentioned in this solution may be of poor reliability with respect to radio meteorology for reasons given in item 2. It is expected that there will be disagreement between designed and measured values, and improvements will be made based on data collected during the first year after installation (at least six months, with the worst month included). It should be noted that the use of the diversity system is reserved, in principle, for this purpose, and accordingly, the application of a diversity system should not necessarily be determined at the time of initial installation.
5. Use of the diversity system is very effective, particularly in minimizing the outage during deep fading. Therefore it may appear that use of the diversity can prolong the propagation distance and delete a certain number of relay stations along the route, leading to system economy. However, it should also be noted that each time the system is expanded or a new radio relay system is installed using the higher frequency band, addition of a diversity arrangement is required, thereby raising installation, operation, and maintenance costs.
6. While application of the diversity system has simplified the site selection plan, several unsatisfactory examples of the diversity system have also been noted.

Diversity is not the final solution for minimizing fading. For example, it is not effective for fading caused by trapping, which is observed very often in the low-altitude path, as found in flat water land in the area between Peking and Nanking. Hence it is essential to design the station to satisfy requirements without using diversity, except for cases where the topographical conditions and local climate necessitate the use of diversity for propagation paths between islands, over the strait, lake, etc.

Chart Index (1), Fading Estimation and System Evaluation

Chart Index (2), Fading Estimation and System Evaluation

Pattern of Noise Performance Design of Radio System

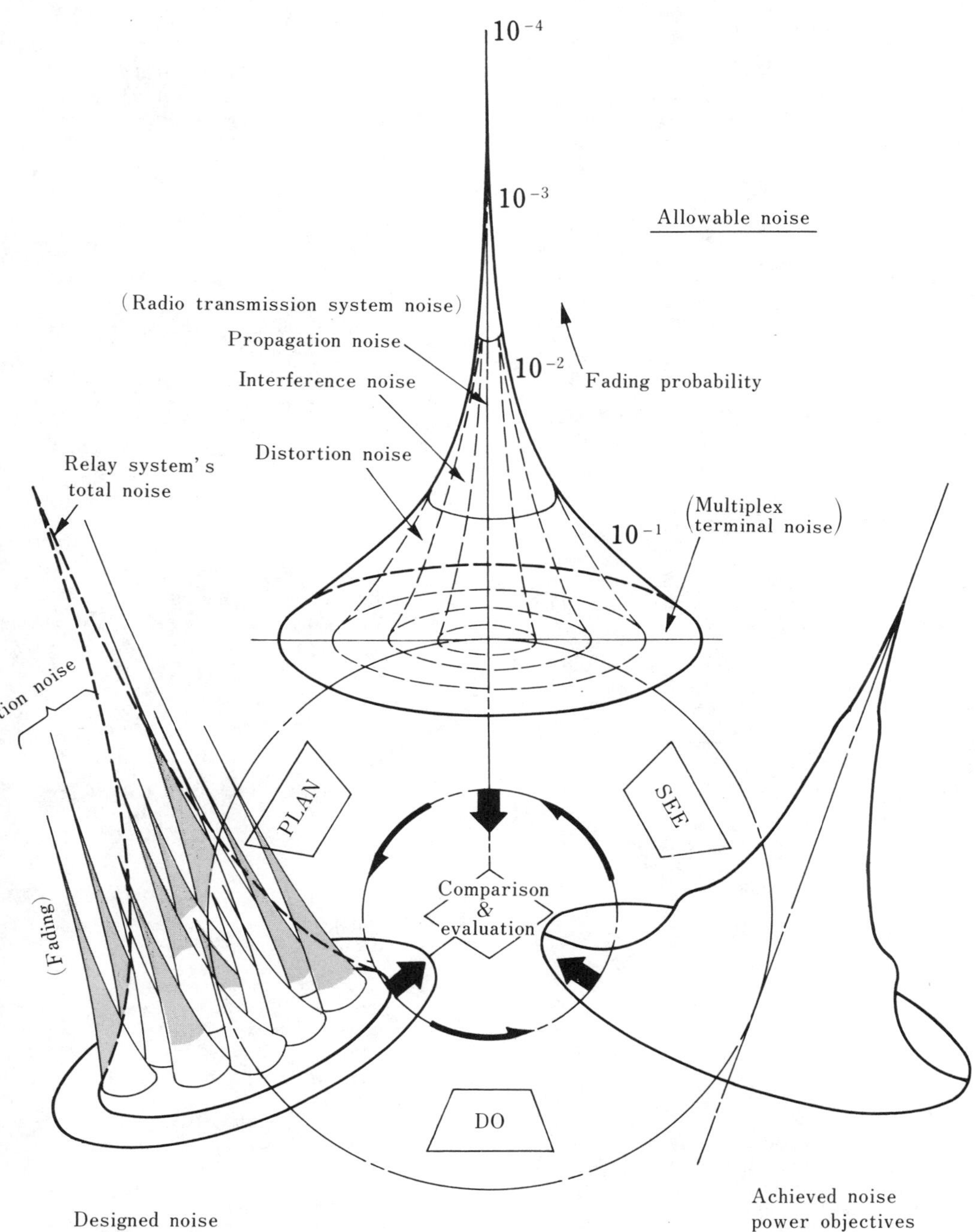

Installation Steps of Radio System

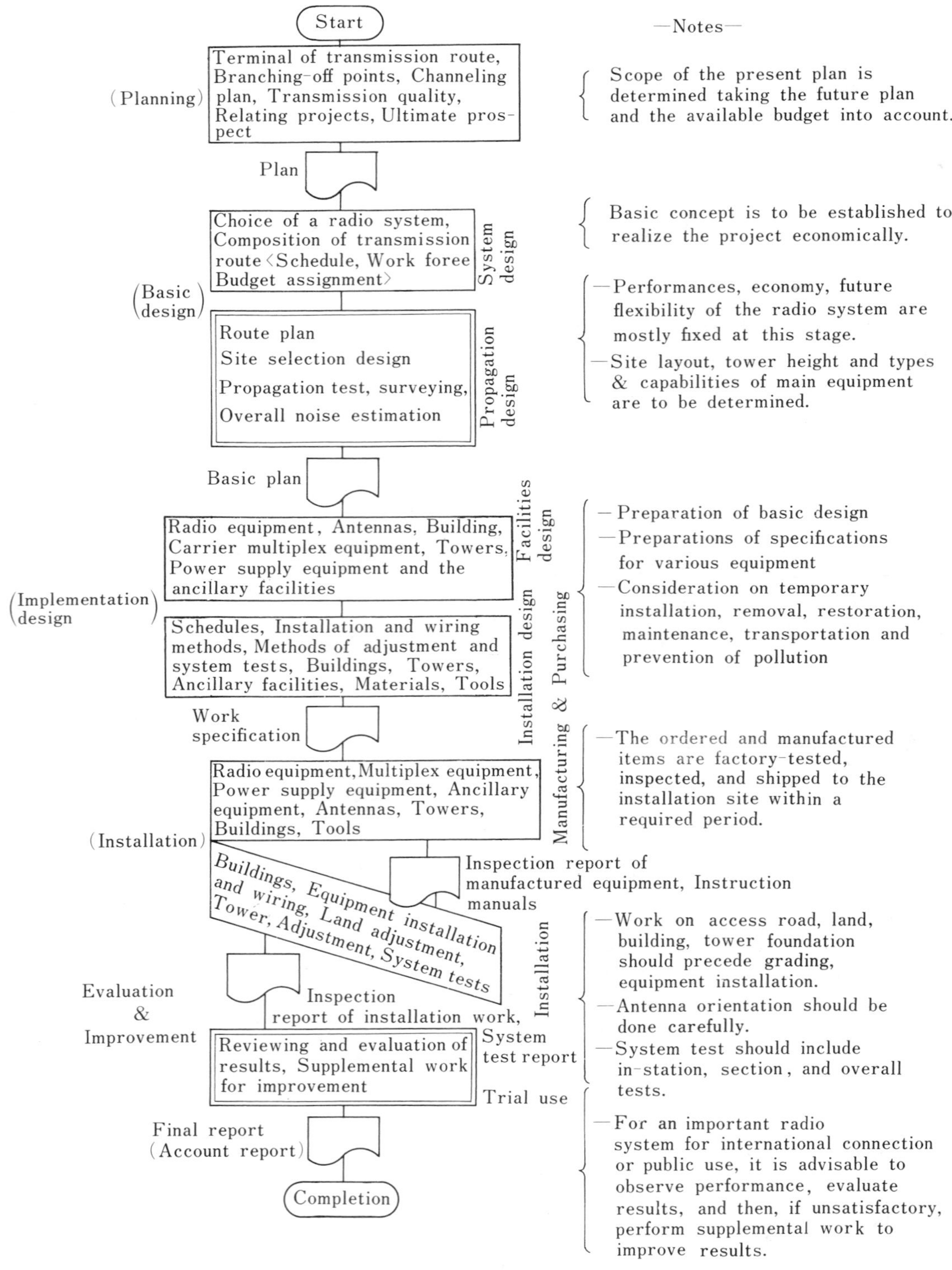
Start
(Planning)
Terminal of transmission route, Branching-off points, Channeling plan, Transmission quality, Relating projects, Ultimate prospect
Plan
(Basic design)
Choice of a radio system, Composition of transmission route <Schedule, Work foree Budget assignment>
System design
Route plan
Site selection design
Propagation test, surveying,
Overall noise estimation
Propagation design
Basic plan
(Implementation design)
Radio equipment, Antennas, Building, Carrier multiplex equipment, Towers, Power supply equipment and the ancillary facilities
Facilities design
Schedules, Installation and wiring methods, Methods of adjustment and system tests, Buildings, Towers, Ancillary facilities, Materials, Tools
Installation design
Work specification
Manufacturing & Purchasing
(Installation)
Radio equipment, Multiplex equipment, Power supply equipment, Ancillary equipment, Antennas, Towers, Buildings, Tools
Inspection report of manufactured equipment, Instruction manuals
Buildings, Equipment installation and wiring, Land adjustment, Tower, Adjustment, System tests
Installation
Inspection report of installation work,
System test report
Evaluation & Improvement
Reviewing and evaluation of results, Supplemental work for improvement
Trial use
Final report (Account report)
Completion
—Notes—
Scope of the present plan is determined taking the future plan and the available budget into account.
Basic concept is to be established to realize the project economically.
—Performances, economy, future flexibility of the radio system are mostly fixed at this stage.
—Site layout, tower height and types & capabilities of main equipment are to be determined.
—Preparation of basic design
—Preparations of specifications for various equipment
—Consideration on temporary installation, removal, restoration, maintenance, transportation and prevention of pollution
—The ordered and manufactured items are factory-tested, inspected, and shipped to the installation site within a required period.
—Work on access road, land, building, tower foundation should precede grading, equipment installation.
—Antenna orientation should be done carefully.
—System test should include in-station, section, and overall tests.
—For an important radio system for international connection or public use, it is advisable to observe performance, evaluate results, and then, if unsatisfactory, perform supplemental work to improve results.

Flowchart of Propagation Design and System Evaluation

Line-of-Sight FDM Radio Relay System

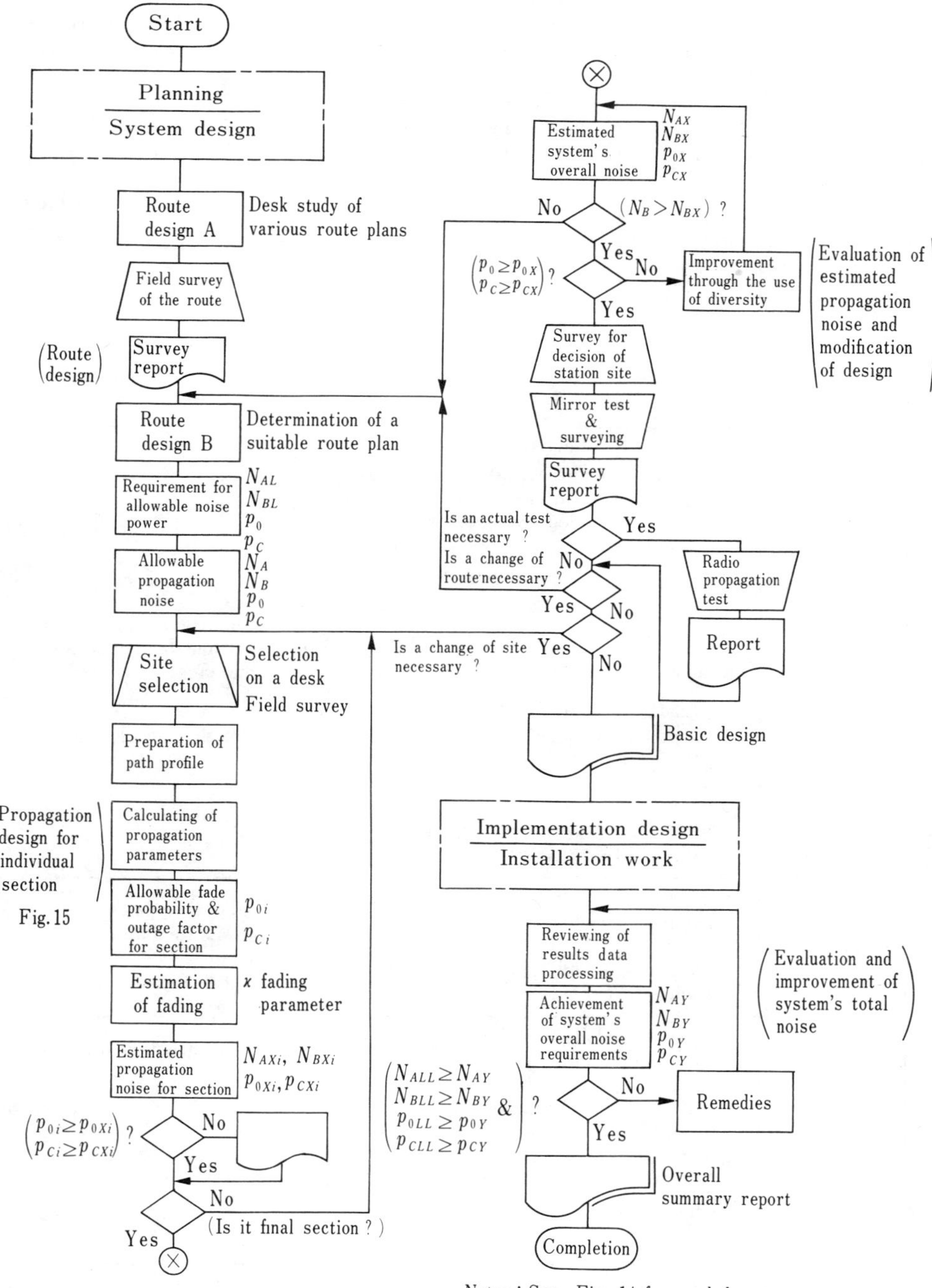

Notes : See Fig. 14 for symbol.

8

2a

Basic Classification of Fading

According to Cause of Fading over VHF to SHF Bands

	Nature	Direct cause	Type of fading	Abbreviation
I	Atmosphere	Absorption (Absorption or scattering due to rain, fog, water vapor, oxygen, etc. at frequency of 8 GHz or higher)	Absorbing type (Attenuation type)	A
		Refraction — Bent M-curve	Birefracted path type	B
		Refraction — Slightly wavy M-curve	Scintillation type	C
		Refraction — M-curve including a portion of duct	Duct type	D
		Turbulence (Forward scatter)	Forward-scatter type (Troposcatter type)	F
II	Ground-based	Path-length difference (Change of path-length difference between the direct and the reflected rays, produced by tide, snow accumulation, etc.)	Path length type	P
		Geological feature (Change of reflection coefficient caused by freezing, grown moss, etc.)	Quality transition of nature of reflection surface	Q
		Roughness (Change of equivalent reflection coefficient caused by snow accumulation, vegetation, etc.)	Rough-smooth type	R
		Shielding (Change of screening effect due to trees, etc.)	Screen type	S
III	Man-made objects	Screening or reflection by vessels, airplanes, temporary buildings, etc.	U.F.O type (Mobile object type)	U
		Displacement and vibration of antenna or reflector	Vital antenna type	V
IV	Composite	Ground-based duct and ground-reflected rays	Grounded duct interference type	G
		Variation of K (Variation of gradient of M-curve and ground-reflected ray)	K type	K
		Variation of K and ridge-diffracted rays	Tangential ridge K type	T

Atmospheric Fading

Basic Types in Line-of-Sight System

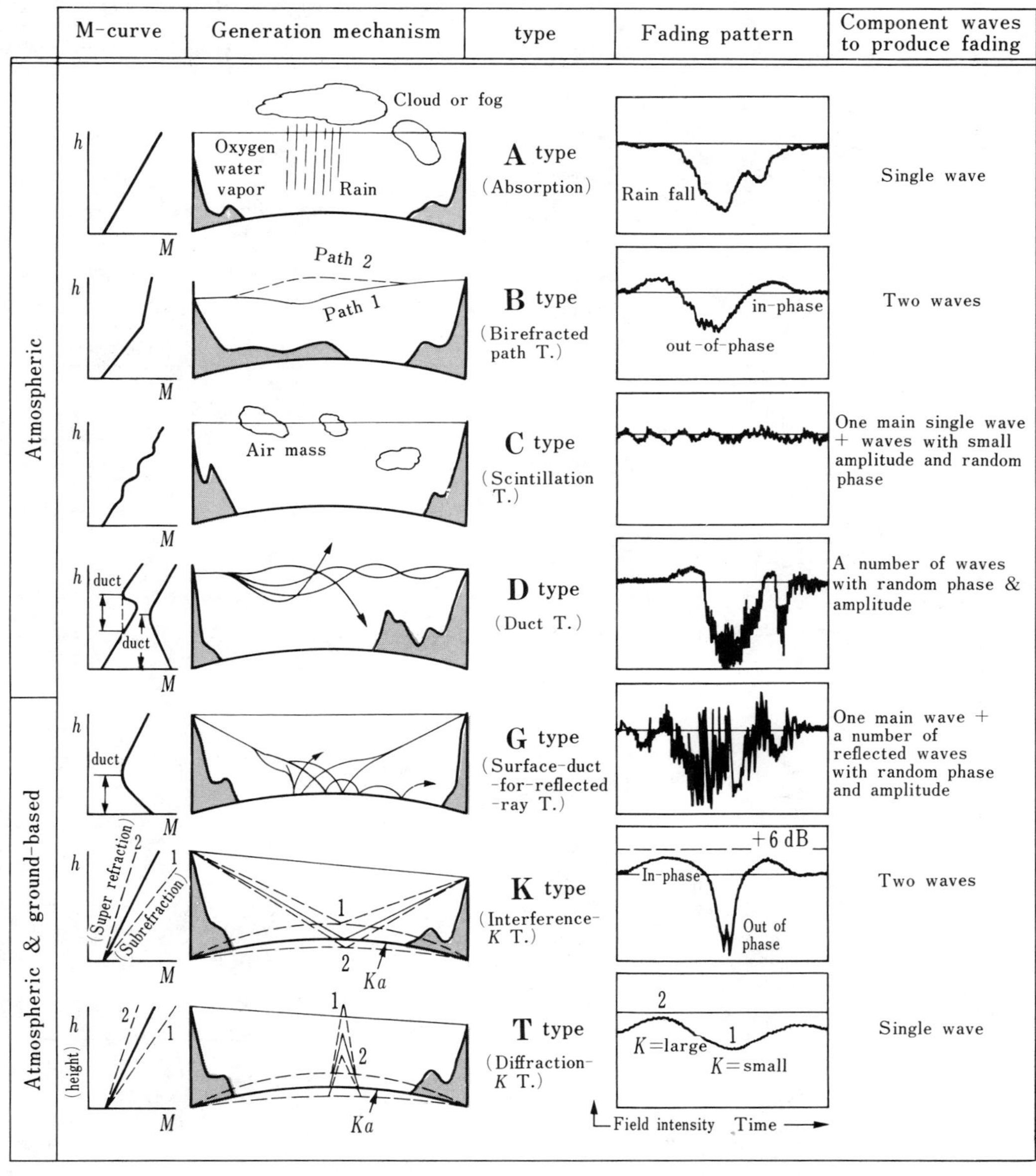

M-curve
Generation mechanism
type
Fading pattern
Component waves to produce fading
Atmospheric
Atmospheric & ground-based
h
M
Cloud or fog
Oxygen water vapor
Rain
A type
(Absorption)
Rain fall
Single wave
Path 2
Path 1
B type
(Birefracted path T.)
in-phase
out-of-phase
Two waves
Air mass
C type
(Scintillation T.)
One main single wave + waves with small amplitude and random phase
duct
D type
(Duct T.)
A number of waves with random phase & amplitude
G type
(Surface-duct-for-reflected-ray T.)
One main wave + a number of reflected waves with random phase and amplitude
(Super refraction) 2
1 (Subrefraction)
K type
(Interference-K T.)
+6 dB
In-phase
Out of phase
Ka
Two waves
(height)
T type
(Diffraction-K T.)
K=large
K=small
Single wave
Field intensity
Time

Ground-Based Fading

Typical Examples of Basic Form

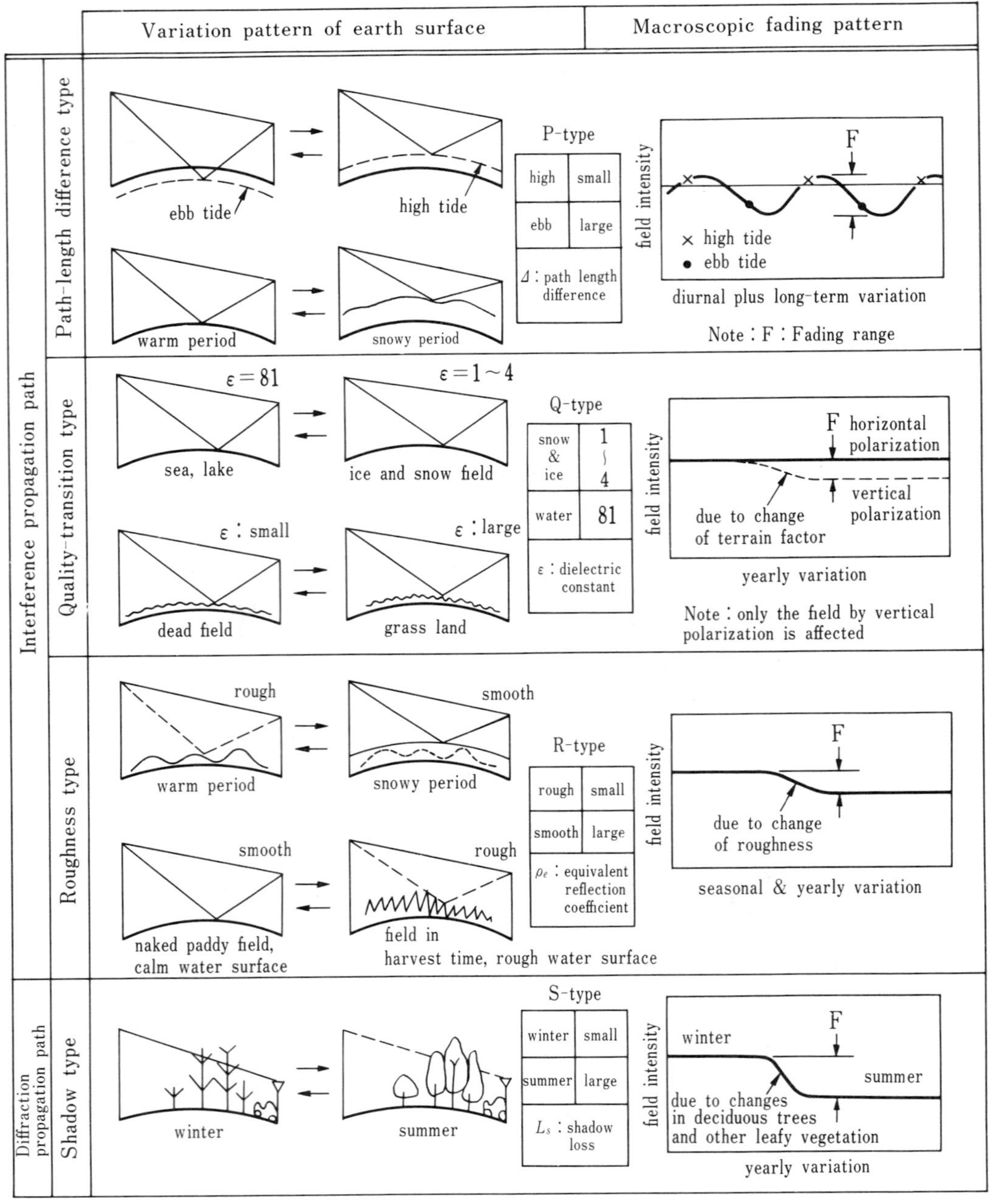

Fading on Actual Radio Path

Type of propagation path		Atmospheric A	Atmospheric B	Atmospheric C	Atmospheric D	Atmospheric F	Composite G	Composite K	Composite T	Ground based P	Ground based Q	Ground based R	Ground based S	Typical path profile
Line of sight propagation	Free space	⊗	⊙	△	●									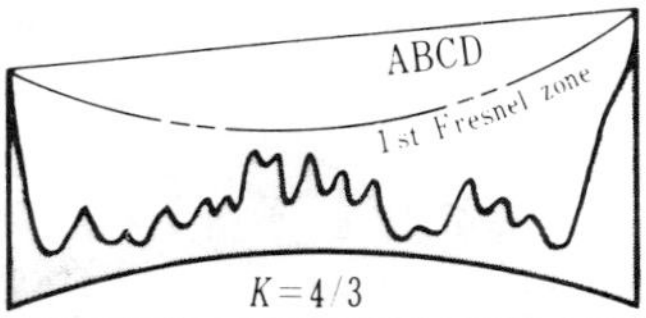 ABCD; 1st Fresnel zone; $K=4/3$
Line of sight propagation	Interference	⊗	⊙	△	●		●	●		△	△	△		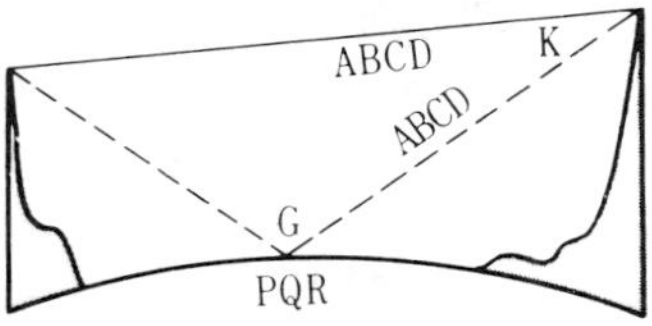 ABCD; K; ABCD; G; PQR
Line of sight propagation	Diffraction by grazing ridge	⊗	⊙	△	●				⊙				△	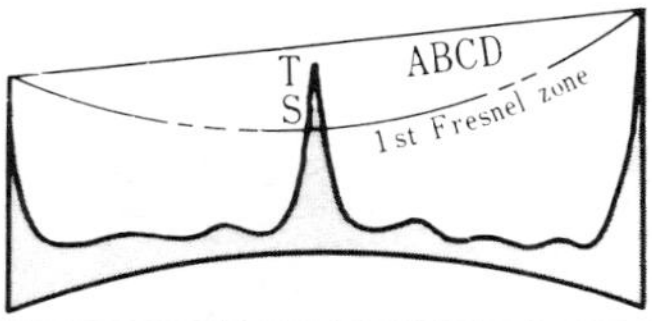 T; S; ABCD; 1st Fresnel zone
Transhorizon propagation	Mountain diffraction	⊗	⊙	△	●				△					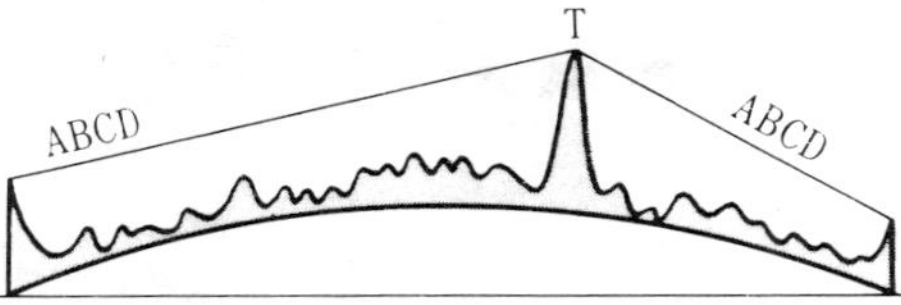 T; ABCD; ABCD
Transhorizon propagation	Interference due to mountain diffraction	⊗	⊙	△	●		⊙	△	△					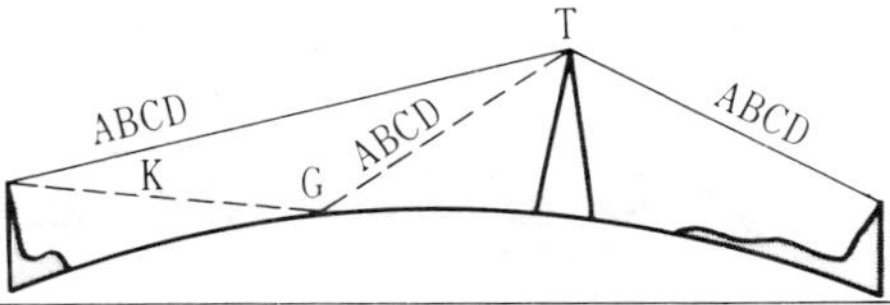 T; ABCD; K; G; ABCD; ABCD
Transhorizon propagation	Troposcatter dispersion	⊗	⊙	⊙	⊙	●	⊙							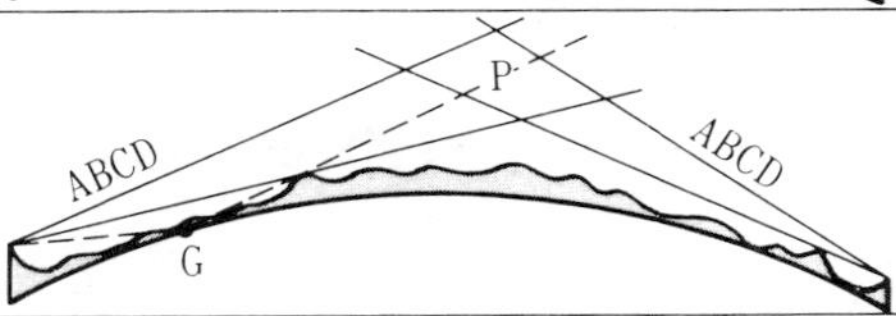 P; ABCD; ABCD; G
Transhorizon propagation	Diffraction over the spherical earth	⊗	⊙	△	●		⊙				△	△	△	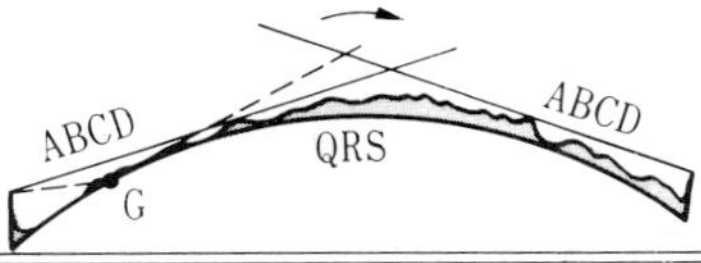 ABCD; ABCD; QRS; G
		A	B	C	D	F	G	K	T	P	Q	R	S	Profile

Remarks : Degree of contribution to fading is expressed by the following symbols.

● Great contribution
⊙ Medium contribution
△ Small contribution
⊗ Contribution only to the system using approx. 10 GHz or above

Units for Processing Time Series Data of Fading

Common to Field Intensity, Propagation Loss, Noise and S/N Ratio

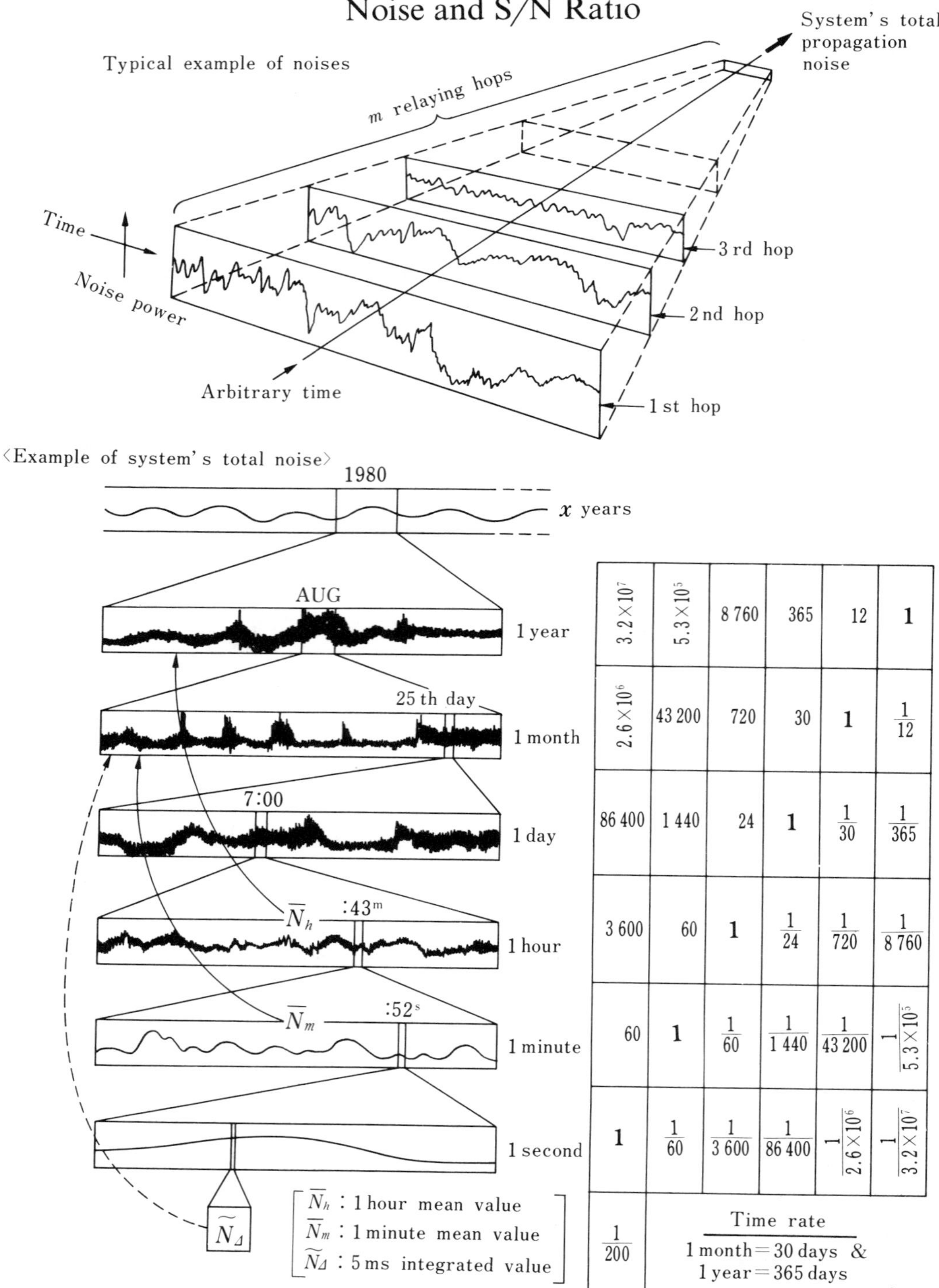

1 year	3.2×10^7	5.3×10^5	8 760	365	12	**1**
1 month	2.6×10^6	43 200	720	30	**1**	$\frac{1}{12}$
1 day	86 400	1 440	24	**1**	$\frac{1}{30}$	$\frac{1}{365}$
1 hour	3 600	60	**1**	$\frac{1}{24}$	$\frac{1}{720}$	$\frac{1}{8\,760}$
1 minute	60	**1**	$\frac{1}{60}$	$\frac{1}{1\,440}$	$\frac{1}{43\,200}$	$\frac{1}{5.3\times10^5}$
1 second	**1**	$\frac{1}{60}$	$\frac{1}{3\,600}$	$\frac{1}{86\,400}$	$\frac{1}{2.6\times10^6}$	$\frac{1}{3.2\times10^7}$
$\widetilde{N}_\Delta$	$\frac{1}{200}$					

Definition of Reliability, Fading Probability, and Estimated Fading Loss

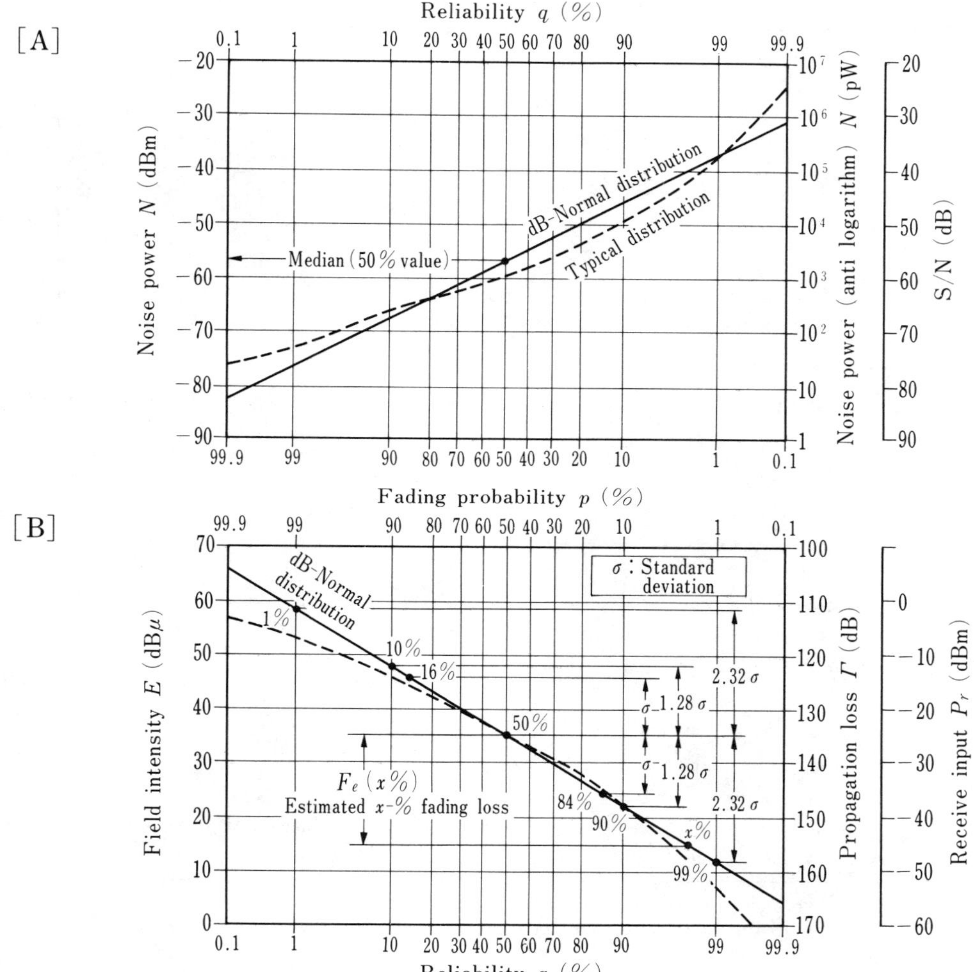

Hypothetical Reference Circuit in Line-of-Sight Radio System (1)

FDM 12 to 60 Channels (CCIR Rec. 391)

Hypothetical reference circuit

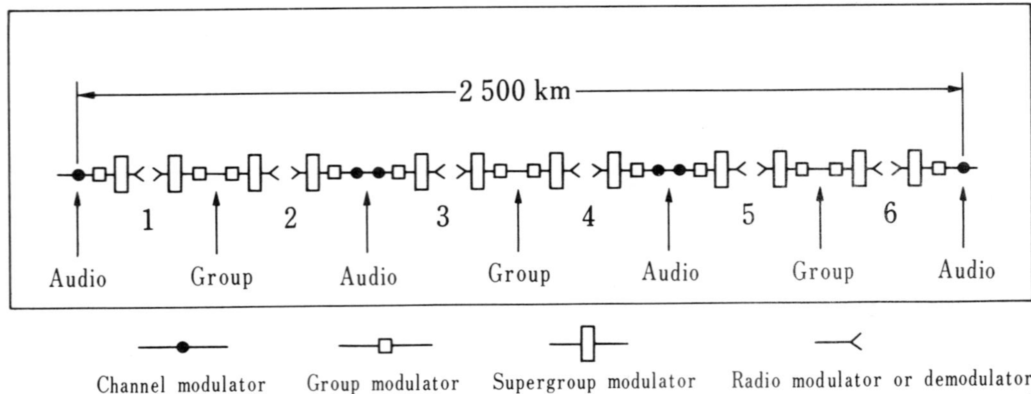

(1) Hypothetical reference circuit for FDM multisection radio relay system 12～60 telephony channels, 2500 km (for international circuit connection)

(2) The single direction circuit is composed of :

- 3 sets of channel modulators
- 6 sets of group modulators
- 6 sets of supergroup modulators
- 6 sets of radio modulators

} Each modulator is composed of a modulator and a demodulator

(3) Each of 6 sections divided by radio modulators and demodulators is assumed to be of equal length.

The hypothetical reference circuit presents a guideline for system planning or the facility design concept but will not impose a restriction on planning and designing of a real link that is subject to geographical and social conditions.

Hypothetical Reference Circuit in Line-of-Sight Radio System (2)

FDM 60 Channels or More (CCIR Rec. 392)

Hypothetical reference circuit

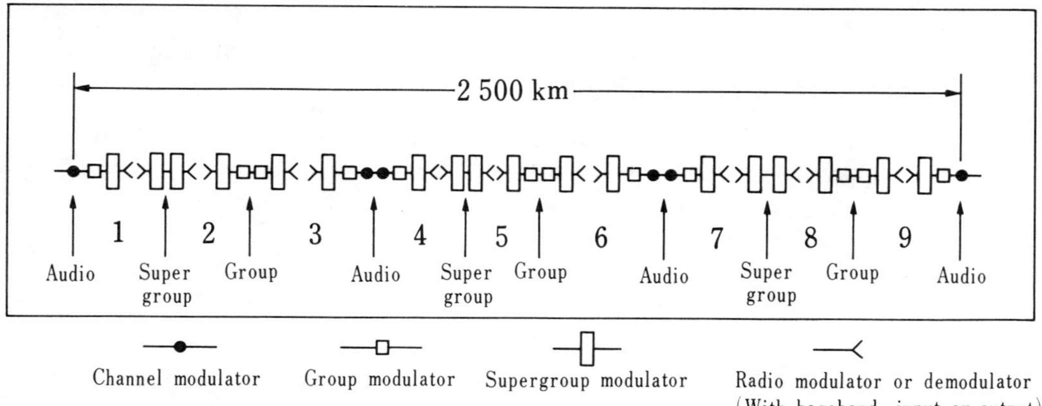

(1) Hypothetical reference circuit for FDM multisection radio relay system 60 telephony channels or more, 2 500 km (for international circuit connection)

(2) The single direction circuit is composed of

- 3 sets of channel modulators
- 6 sets of group modulators
- 9 sets of supergroup modulators
- 9 sets of radio modulators & demodulators

} Each modulator is composed of a modulator and a demodulator.

(3) Each of 9 sections divided by the radio modulators and demodulators is assumed to be of equal length.

The hypothetical reference circuit presents a guideline for system planning or the facility design concept but will not impose a restriction on planning and designing of a real link that is subject to geographical and social conditions.

Allowable Noise in Line-of-Sight System (1)

Hypothetical Reference Circuit (CCIR Rec. 393-1)

Allowable noise power measured at a zero reference point (where the test signal level is at 0 dBm=1 mW=10^9 pW) of a channel at the receiving end of the 2 500 km hypothetical reference circuit is specified as below.
However it does not include 2 500 pW of the total hourly mean noise allocated by CCITT, to various frequency-division multiplex modulators and demodulators.

(1) 7 500 pW psophometrically weighted mean power in any hour
(2) 7 500 pW psophometrically weighted one-minute mean power for more than 20 percent of any month
(3) 47 500 pW psophometrically weighted one-minute mean power for more than 0.1 percent of any month
(4) 1 000 000 pW unweighted and measured in 5 ms integration time for more than 0.01 percent of any month

Summary of allowable noise and required S/N ratio (* Converted figures)

Spec. item	Allowable noise power (pW)		S/N ch (dB)		Time conditions
	Weighted value	Unweighted value	Weighted value	Unweighted value	
I	7 500	* 13 350	* 51.25	* 48.75	for mean noise power in any hour
II	7 500	* 13 350	* 51.25	* 48.75	for one-minute mean noise power for more than 20% of any month
III	47 500	* 84 550	* 43.23	* 40.73	for one-minute mean noise power for more than 0.1% of any month
IV	* 561 800	1 000 000	32.50	30.0	for noise power measured with 5 ms integrating time for more than 0.01% of any month

Note 1: Total power of the noise distributed uniformly over the bandwidth of 3.1 kHz is reduced by 2.5 dB when it passes through the psophometrically weighting circuit. (2.5 dB corresponds to a power ratio of 1.78, and −2.5 dB to a power ratio of 0.562).

2: It should be noted that the requirements as given in (1) and I (specified item) have been deleted in CCIR Rec. 393-3 (1978) (This atlas uses as these reference)

Allowable Noise in Line-of-Sight System (2)

Real Circuit (CCIR Rec. 395-1)

Radio relay systems actually planned and installed are barely similar to the hypothetical reference circuit and are likely to differ appreciably from it. The allowable noise power measured at the zero point of any telephone channel excluding the noise allocated for the FDM translation equipment (2 500 pW in any hour for 2 500 km link according to CCITT) is given below for the real link. The actual link can be either similar to the hypothetical reference circuit or very different from it.

Length of real link L (km)	Allowable noise power Weighted value (unweighted value) (pW) Real link similar to HRC (Note 1)	Real link differing appreciably from HRC	Time conditions	Spec. item
$280 \leqq L \leqq 2\,500$	$3L$ ($5.34L$)		for the mean noise power in any hour and for one-minute mean noise power for more than 20 % of any month	I
$50 \leqq L \leqq 840$	—	$3L+200$ ($5.34L+356$)		
$840 < L \leqq 1\,670$	—	$3L+400$ ($5.34L+712$)		II
$1\,670 < L \leqq 2\,500$	—	$3L+600$ ($5.34L+1\,068$)		
$L < 280$	—	47 500 S/N = 43.23 dB (84 550 S/N = 40.73 dB)	for one-minute mean noise power for more than (280/2 500) × 0.1 % of any month	III
$280 \leqq L \leqq 2\,500$	47 500 S/N = 43.23 dB (84 550 S/N = 40.73 dB)		for one-minute mean noise power for more than (L/2 500) × 0.1 % of any month	

Note 1. HRC denotes Hypothetical Reference Circuit

2. The allowable percentage of time, i.e., the outage factor as in spec. item IV that the noise exceeds 10^6 pW, though it has not been specified in this recommendation, should preferably be specified in conformity with the CCIR Recommendation 393-1 as given in Fig. 7 a, if a highly reliable radio relay system is to be installed. As regards weighted and unweighted please refer to the note on Fig. 7 a.

3. The spec. item I for one-hour mean noise power not to be exceeded has been deleted by CCIR Rec. 395-2 (1978), however, this atlas uses these as reference.

Reference Composition and Allowable Noise of Transhorizon Radio System

[A] Hypothetical reference circuit (CCIR Rec. 396-1)

(1) It seems to be difficult to define the hypothetical reference circuit for the transhorizon radio system as in the case of line-of-sight radio relay systems because the performance of the former is more severely affected by geographical conditions than the latter.

(2) If a section under study is L km long, the hypothetical reference circuit may comprise 2 500/L sections in tandem and apply the nearest whole number to the value 2 500/L.

(3) Such hypothetical reference circuit of 2 500 km length is assumed to comprise FDM modulators as follows :

- 3 sets of channel modulators
- 6 sets of group modulators
- 6 sets of supergroup modulators

} Each modulator comprises a modulator and a demodulator.

[B] Allowable noise (CCIR Rec. 397-3)

(1) Transhorizon systems operating between points capable of linkage by line-of-sight radio relay without excessive technical or economic difficulty must meet the requirements for the line-of-sight systems as specified in CCIR Rec. 393-3.

(2) Transhorizon systems operating under conditions precluding alternative means of communication must satisfy the requirement given in the table below.

* converted figure

Allowable noise power (pW)		S/N ch (dB)		Time conditions
Weighted value	Unweighted value	Weighted value	Unweighted value	
25 000	* 44 500	46.0	* 43.5	for one-minute mean noise power for more than 20% of any month
63 000	* 112 100	42.0	* 39.5	for one-minute mean noise power for more than 0.5% of any month
* 561 800	1 000 000	* 32.5	30.0	for noise power measured with 5-ms integrating time for more than 0.05% of any month

Note : All the figures applied to a point of zero relative level.
Include intermodulation and propagation distortion noises for the radio system portion, but exclude the noise (2 500 pW/2 500 km according to CCITT) generated in FDM modulators. As regards weighted and unweighted, please refer to the note on Fig. 7 a.

Reference Composition and Allowable Noise of Satellite Communication System

[A] Hypothetical reference circuit (CCIR Rec. 352-1)

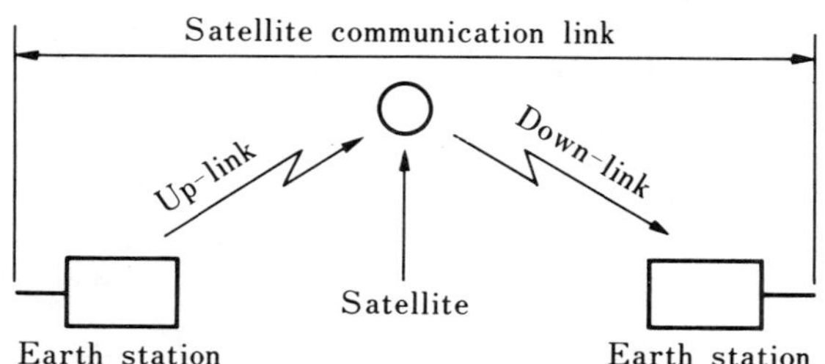

(1) Hypothetical reference circuit is defined as the one-way link, i.e., earth-satellite-earth communication path.

(2) A set of modulator and demodulator for conversion between the baseband frequency and the radio frequency is included.

(3) Terrestrial links between the earth station and other earth stations and the same between the earth station and the switching center are not included.

[B] Allowable noise (CCIR Rec. 353-3)

(1) This requirement presents a guideline for the actual system design due to a lack of data on fading, etc.

(2) The allowable noise shall include the interference noise from a terrestrial microwave system and any other satellite sharing the same frequency band.

(3) It shall also include noise power caused by atmospheric absorption and rainfall.

Summary of allowable noise and required S/N ratio

* Converted figures

Allowable noise power (pW)		S/N ch (dB)		Time conditions
Weighted value	Unweighted value	Weighted value	Unweighted value	
10 000	* 17 800	50.0	* 47.5	for one-minute mean noise power for more than 20% of any month
50 000	* 89 000	43.0	* 40.5	for one-minute mean noise power for more than 0.3% of any month
* 561 800	1 000 000	* 32.5	30	for noise power measured with 5-ms integrating time for more than 0.03% of any month

Note : All the figures are applied to a point of zero relative level.
As regards weighted and unweighted, please refer to the note on Fig. 7 a.

Mutual Interference between Satellite Communication System and Terrestrial Radio System

Interference paths

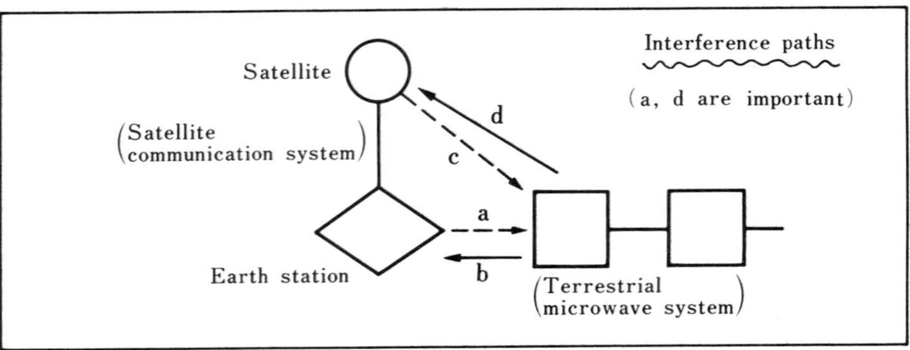

Allowable interference noise

* Converted figure

Terrestrial microwave system → Satellite communication system			Satellite communication system → Terrestrial microwave system		
Allowable noise (pW))		Time conditions	Allowable noise (pW)		Time conditions
Weighted value	Unweighted value		Weighted value	Unweighted value	
1 000	* 1 780	for one-minute mean noise power for more than 20% of any month	1 000	* 1 780	for one-minute mean noise power for more than 20% of any month
50 000	* 89 000	for one-minute mean noise power for more than 0.03% of any month	50 000	* 89 000	for one-minute mean noise power for more than 0.01% of any month
CCIR Rec. 356-4			CCIR Rec. 357-3		

Note : All the figures are applied to the point of zero relative level.
As regards weighted and unweighted please refer to the note on Fig. 7 a.

Restriction on Radiated Power from Terrestrial Microwave Station

Interference in Sharing Frequency Bands for the Uplink of Satellite Communication

Various types of restrictions have been specified by CCIR document and ITU's Radio regulation to prevent the interference in sharing frequency bands for the satellite up-link

Sharing frequency for up-link	Antenna feed power P_t	(e.i.r.p.) P_e	Angle separating the direction of maximum radiation taking into account the atmospheric refraction, from geostationary satellite orbit δ
1 ~ 10 (GHz)	13 dBW or less	55 dBW or less	• $\delta \geqq 2°$ when $P_e > 35$ dBW If not possible, the following must be satisfied. • $P_e \leqq 47$ dBW when $\delta \leqq 0.5°$ • $47 < P_e \leqq 55$ dBW when $0.5 < \delta < 1.5°$ where the relationship 8 dB per one degree exists.
10 ~ 15 (GHz)	10 dBW or less		$\delta \geqq 1.5°$ when $P_e > 45$ dBW
Above 15 (GHz)			No restriction

Note : e.i.r.p. (Effective isotropically radiated power) :
Antenna feed power × antenna gain (as referred to isotropic antenna)

8

11b

Restriction on Radiated Power from Satellite Communication System

Prevention of Interference with Terrestrial Microwave Station

[A] Restriction on e.i.r.p. of earth station

Frequency band (GHz)	Upper limit of e.i.r.p.		Bandwidth to define e.i.r.p.
	$\theta = 0°$	$0° < \theta \leq 5°$	
1 ~ 15	40	$40 + 3\theta$	any 4 kHz
Above 15	64	$64 + 3\theta$	any 1 MHz

θ : Elevation angle of the antenna beam axis toward the satellite at the earth station

[B] Restriction on power flux density at the surface of the earth produced by satellites

Frequency band (GHz)	Upper limit of earth surface flux density (dBW/m^2)			Bandwidth to define the flux density
	$0 \leqq a < 5°$	$5° < a \leqq 25°$	$25° < a \leqq 90°$	
1.5~2.5	−154	$-154 + (\frac{a-5}{2})$	−144	any 4 kHz
3 ~ 8	−152	$-152 + (\frac{a-5}{2})$	−142	
8 ~ 11.7	−150	$-150 + (\frac{a-5}{2})$	−140	
12.2 ~ 15.4	−148	$-148 + (\frac{a-5}{2})$	−138	
15.4 ~ 40.5	−115	$-115 + (\frac{a-5}{2})$	−105	any 1 MHz

Note a : Angle of arrival in which the direction to zenith is assumed to be 90°

Graph Showing Allowable Noise Power of Various Systems

Hypothetical Reference Circuit

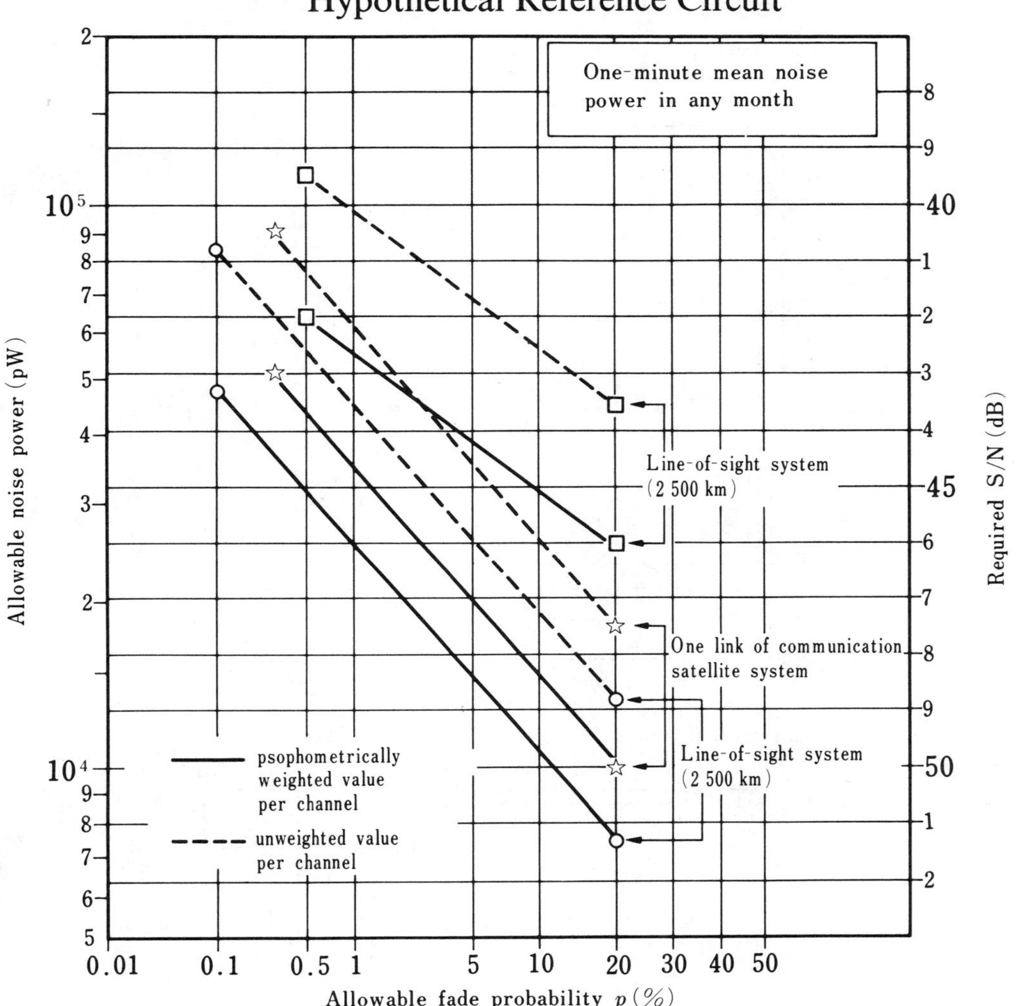

Note 1. The allowable noise power and the required S/N are drawn to the fade probability following the log-normal distribution to give the design criteria on the normal probability chart.

2. The specified values of the allowable noise power are marked by ○, □ and ☆. Satisfying these specified values, no particular restriction is imposed for the intermediate distribution.

Graph of Allowable Noise for Real Line-of-Sight Links

Radio Relay System According to CCIR Rec. 395-1

[A] Requirements I and II

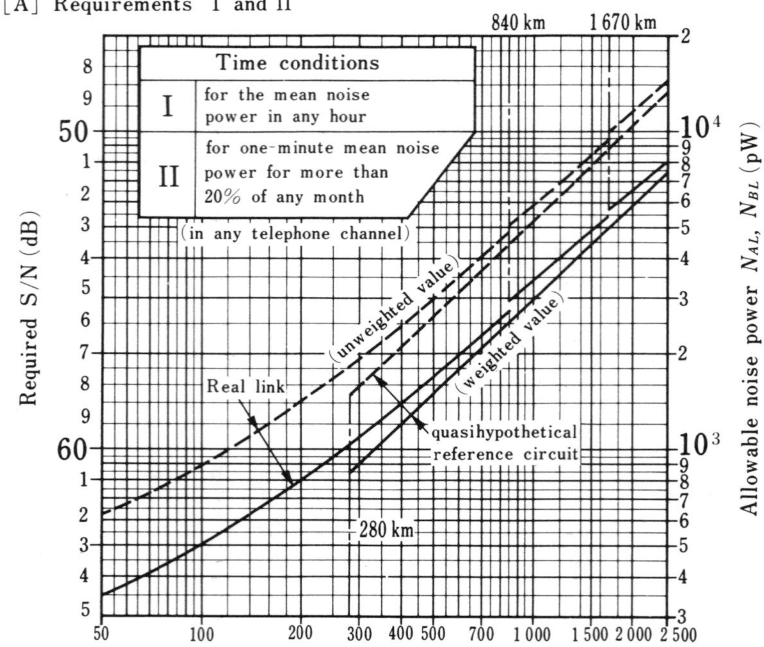

[B] Requirement III (Length of radio-relay link L km)

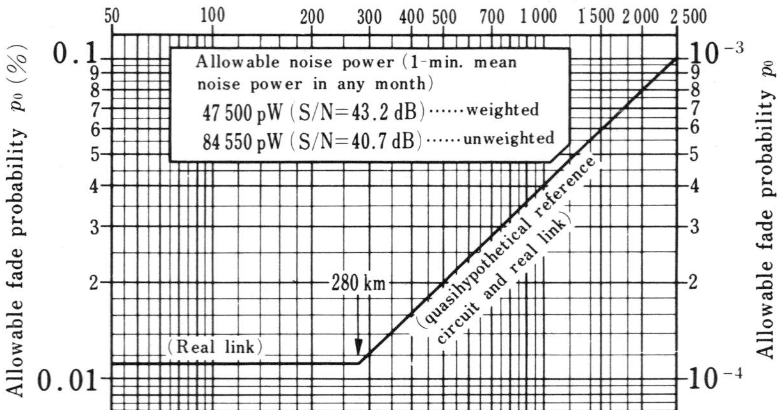

[C] Requirement IV (Allowable outage factor)

Although not specified for real links, the following objectives should be satisfied:

Allowable noise power limit (unweighted)	Allowable outage factor
1 000 000 pW (S/N=30 dB)	$p_C=0.1p_0\%$

Allowable, Designed, and Achieved Noise Power

Line-of-Sight FDM Radio Relay System

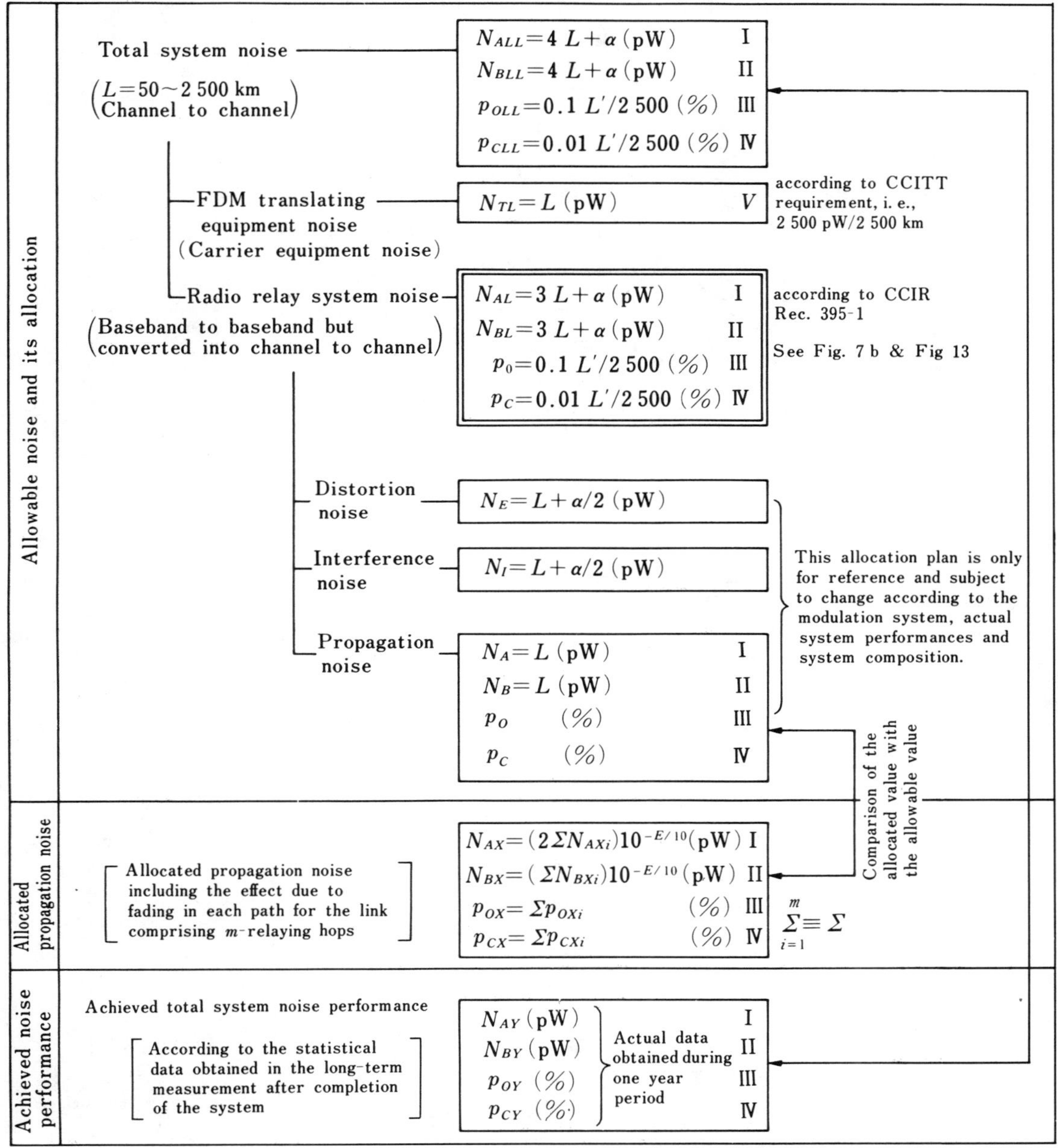

Note : —All the figures of the allowable noise are of weighted value. Therefore multiply 1.78 to obtain the unweighted value.

— p_{OLL}, p_{OY} are the fade probabilities against 50 000 pW, weighted and p_O, p_{OX} are the fade probabilities against 47 500 pW, weighted.

$$\alpha : \begin{cases} 50 \leq L \leq 840 \text{ km} \rightarrow \alpha = 200 \text{ pW} \\ 840 < L \leq 1\,670 \text{ km} \rightarrow \alpha = 400 \text{ pW} \\ 1\,670 < L \leq 2\,500 \text{ km} \rightarrow \alpha = 600 \text{ pW} \end{cases} \text{(weighted value)} \quad \begin{pmatrix} L < 280 \text{ km} \rightarrow L' = 280 \\ L \geqq 280 \text{ km} \rightarrow L' = L \end{pmatrix}$$

Steps for Allocation and Design of Propagation Noise

CCIR/FDM/International Connection System

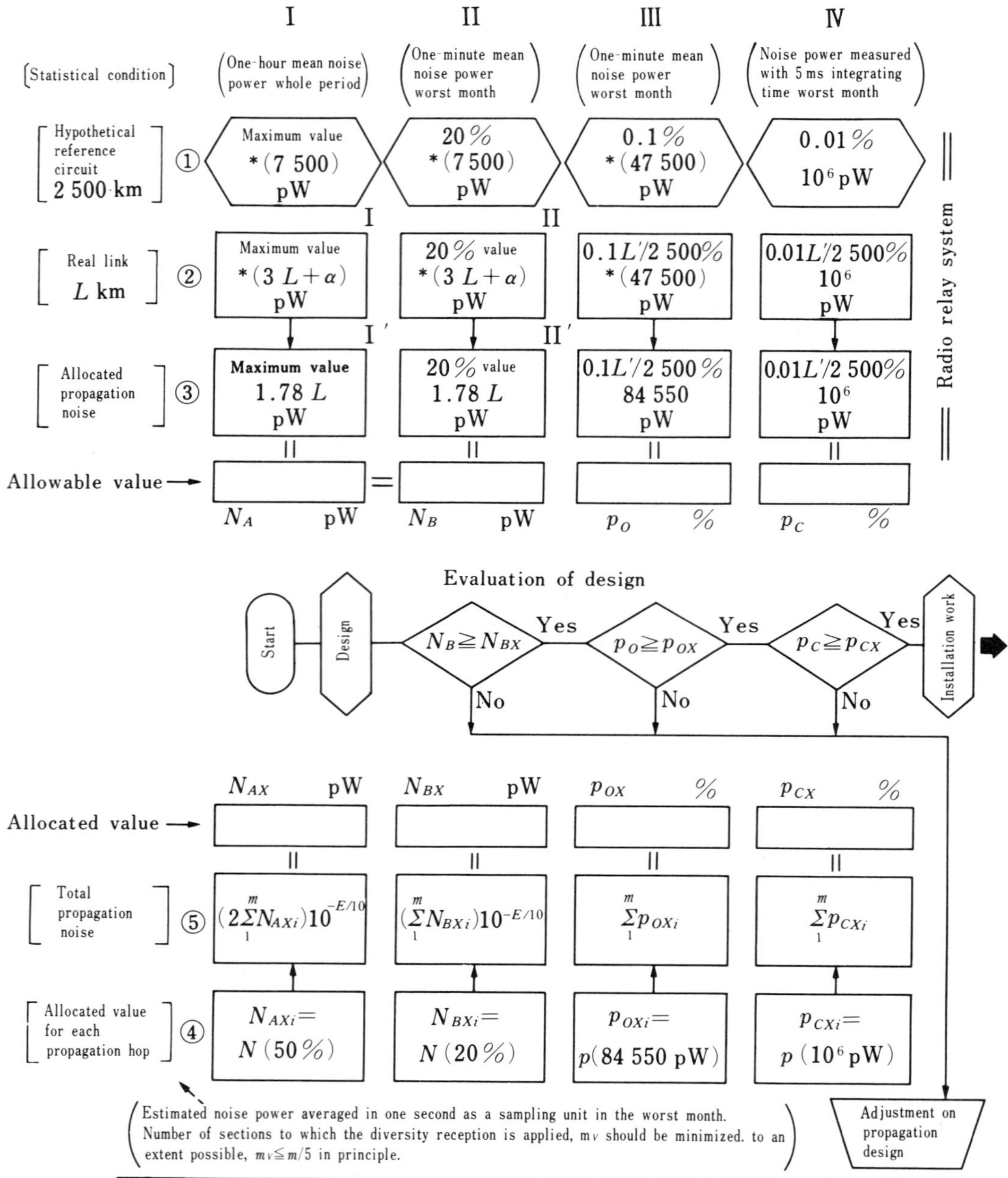

Note α : See Fig. 14 c

L : Baseband section length, $L=50 \sim 2\,500$ km

L' : $L'=280$ km if $L<280$ km and if $L \geqq 280$ km $L'=L$

* Figures in parenthesis show the weighted noise power, figures without asterisks, the unweighted noise power.

Steps for Evaluation and Improvement of Total Noise System

CCIR/FDM/International Connection System

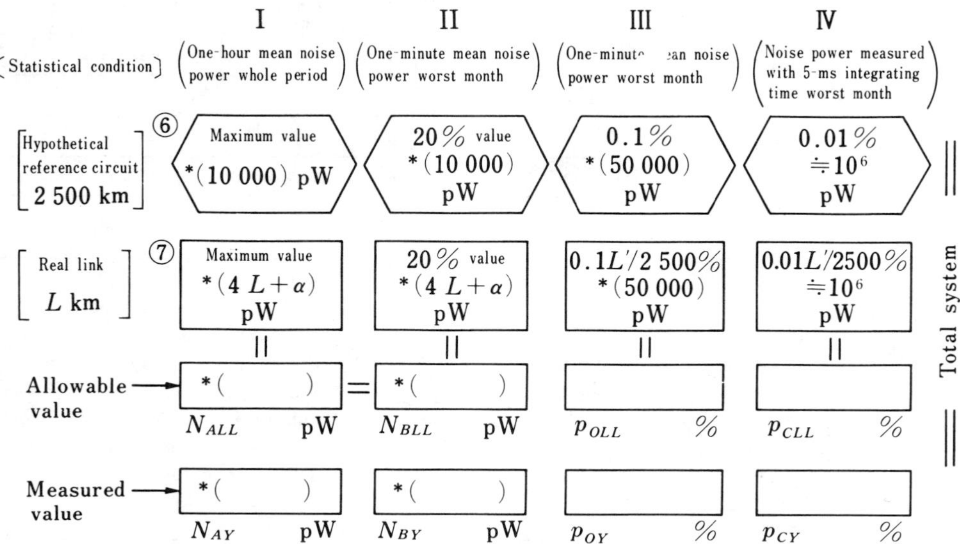

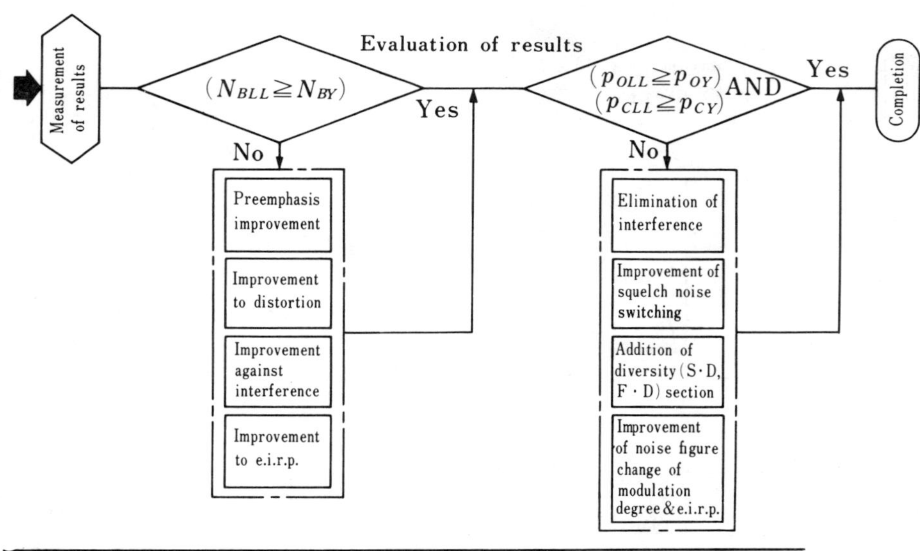

α : L = 50 ~ 840 km, α = 200 pW
L = 840 ~ 1 670 km, α = 400 pW
L = 1 670 ~ 2 500 km, α = 600 pW

} Psophometrically weighted value

Flowchart of Propagation Design for Sections

Line-of-Sight Radio Relay System

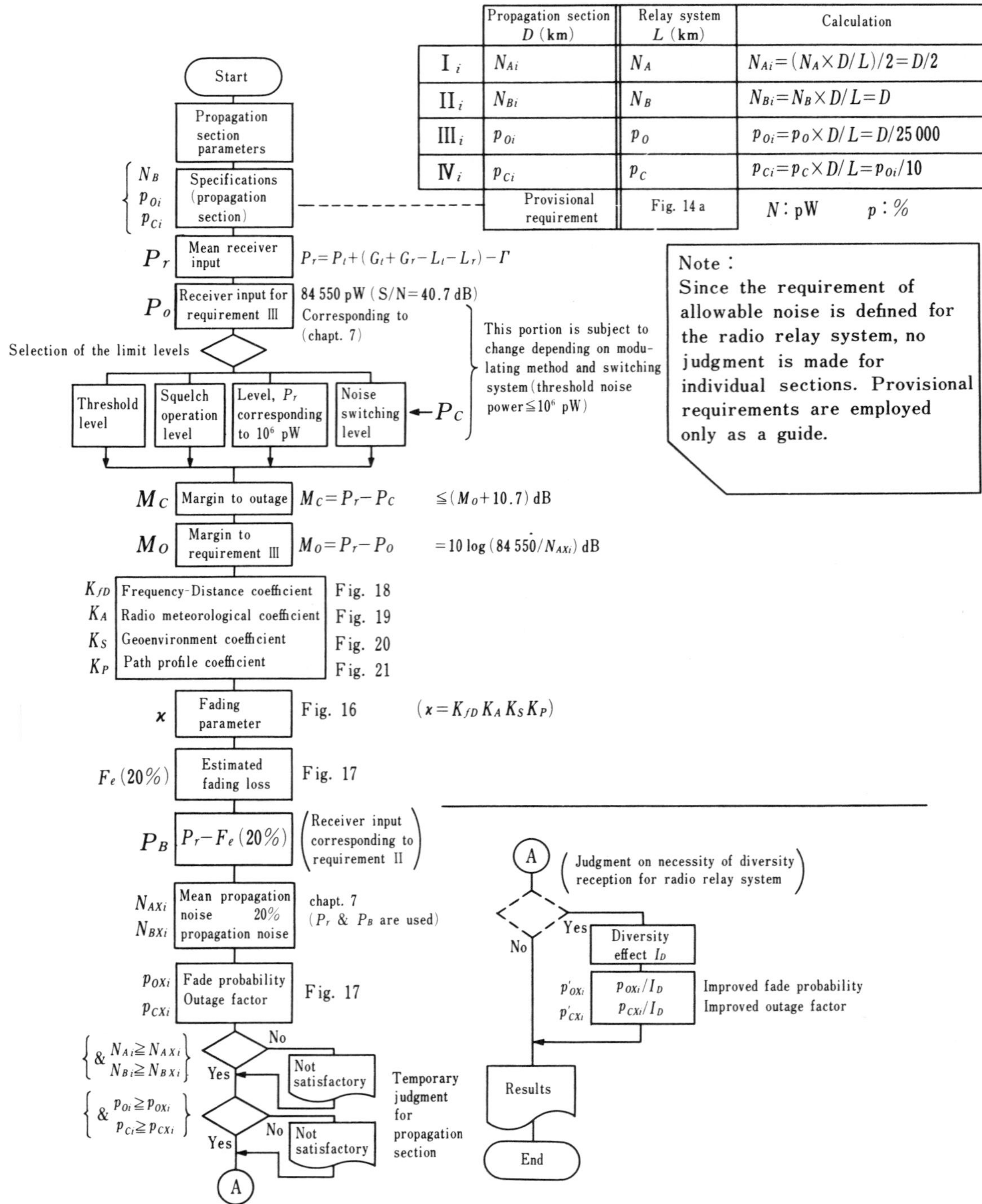

	Propagation section D (km)	Relay system L (km)	Calculation
I_i	N_{Ai}	N_A	$N_{Ai}=(N_A \times D/L)/2=D/2$
II_i	N_{Bi}	N_B	$N_{Bi}=N_B \times D/L=D$
III_i	p_{Oi}	p_O	$p_{Oi}=p_O \times D/L=D/25\,000$
IV_i	p_{Ci}	p_C	$p_{Ci}=p_C \times D/L=p_{Oi}/10$
	Provisional requirement	Fig. 14 a	N: pW p: %

Estimated Loss Due to Fading and Designed Propagation Noise

[A] Estimated loss due to fading, F_e (Fading with a small probability, $p\% < 1\%$)

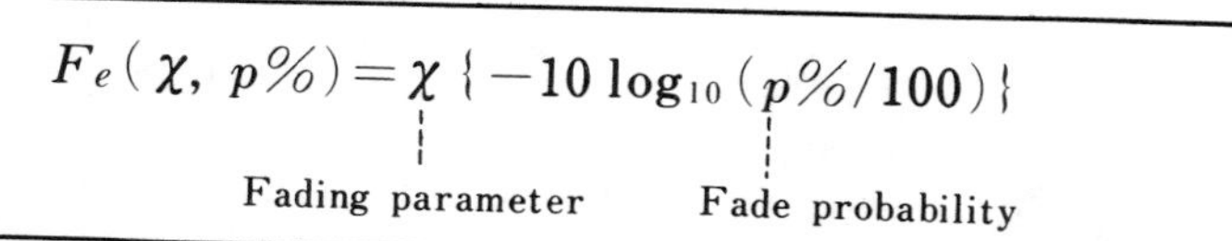

$$F_e(\chi, p\%) = \chi\{-10\log_{10}(p\%/100)\}$$

$$\chi = K_{fD} \cdot K_A \cdot K_S \cdot K_P \leqq 1.5$$

- K_P — Coefficient on path profile
- K_S — Coefficient on surrounding terrain condition
- K_A — Coefficient on atmospheric conditions
- K_{fD} — Coefficient on frequency and distance

$$K_{fD} = g_f g_D$$

$$K_A = a_1 a_2 + 0.05\sqrt{a_1 a_2}$$

$$K_S = S_W + S_P + S_C + S_M$$

$$K_P = (1.04\, b_1 + b_2)\, b_3$$

[B] Designed propagation noise (relay system)

Radio transmission system	Equation	Remark	
Characteristic element of designed propagation noise			
Mean noise power N_{AX} (pW)	$(2\sum_{i=1}^{m} N_{AXi})\, 10^{-E/10}$ — N_{AXi}: Median of one-second mean noise power in the worst month, each section	Maximum mean noise power in any hour E: Preemphasis effect (dB) 0 dB with no preemphasis 3~5 dB with preemphasis	I
20% Noise power N_{BX} (pW)	$(\sum_{i=1}^{m} N_{AXi})\, 10^{-E/10}$ — N_{AXi}: 20% value of one-second mean noise power in the worst month, each section	20% value of one-minute mean noise power in the worst month	II
Fade probability against a specified noise power p_{OX} (%)	$\sum_{i=1}^{m} p_{OXi}$ — p_{OXi}: Time probability that one-second mean noise power exceeds a specified noise power in the worst month, each section	Time probability that one-minute mean noise power exceeds the specified noise power, 84 550 pW (unweighted)	III
Outage probability against a specified noise power p_{CX} (%)	$\sum_{i=1}^{m} p_{CXi}$ — p_{CXi}: Time probability that one-second mean noise power exceeds a specified noise power in the worst month, each section	Time probability that the 5 ms noise power exceeds the specified value, 10^6 pW (unweighted in the worst month)	IV

(1) Noise power and the receiver input for each propagation path are treated using one-second mean value as a sampling unit.
m : number of propagation sections.

(2) Time probability for p_{CXi} may refer to the squelch operating level or the noise switching level corresponding to 10^6 pW of the noise power.

Graph for Fading Estimation (1)

Fade probability p (99.99–1%)

$q = 0.01 \sim 99$ %
$p = 99.99 \sim 1$

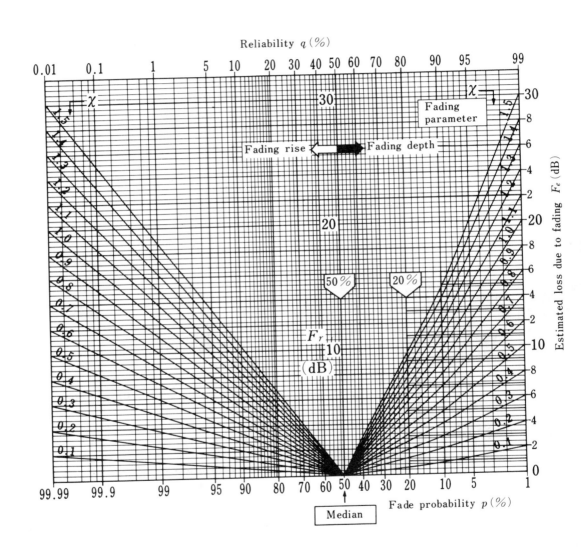

Graph for Fading Estimation (2)

Fade probability p (1–0.003%)

$q = 99 \sim 99.997$ %
$p = 1 \sim 0.003$ %

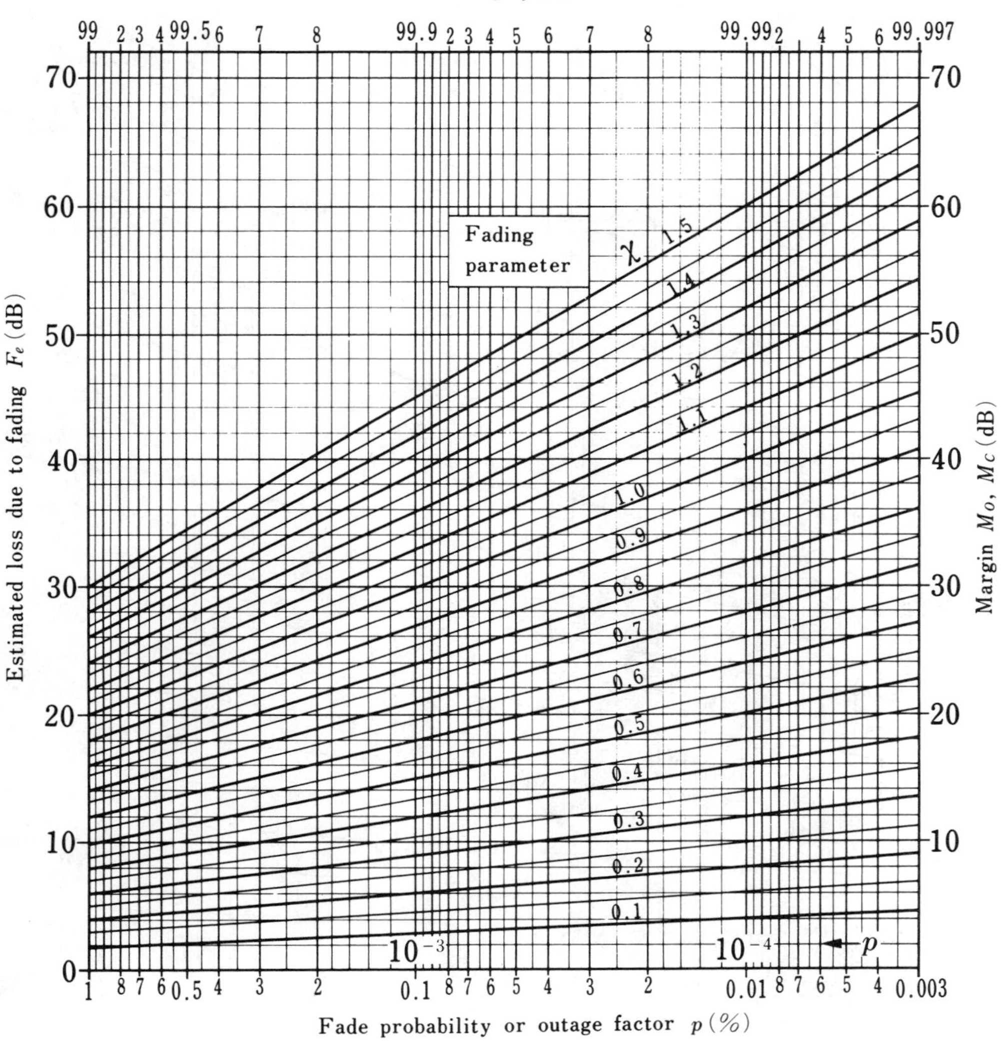

Graph for Fading Estimation (3)

Fade probability p (0.003–0.00001%)

$q = 99.997 \sim 99.99999$ %
$p = 0.003 \sim 0.00001$ %

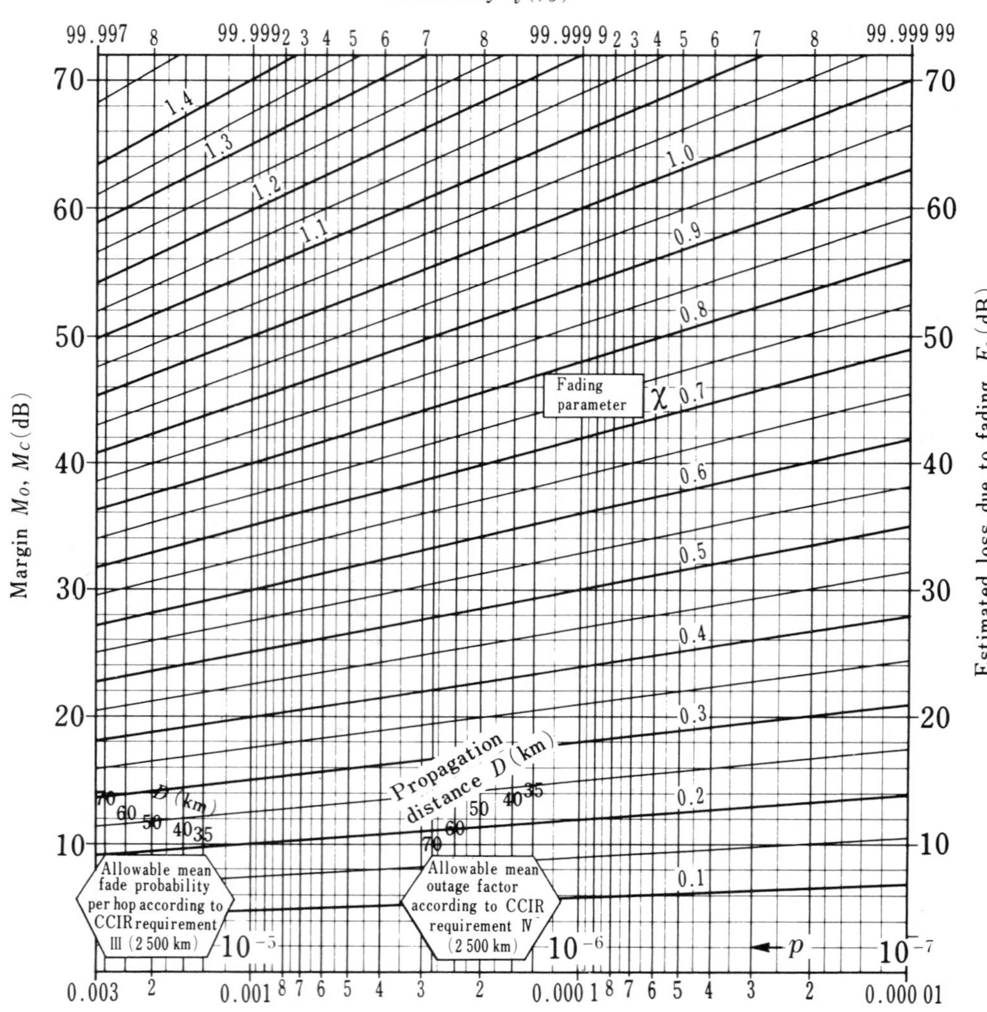

Frequency-distance coefficient K_{fD}

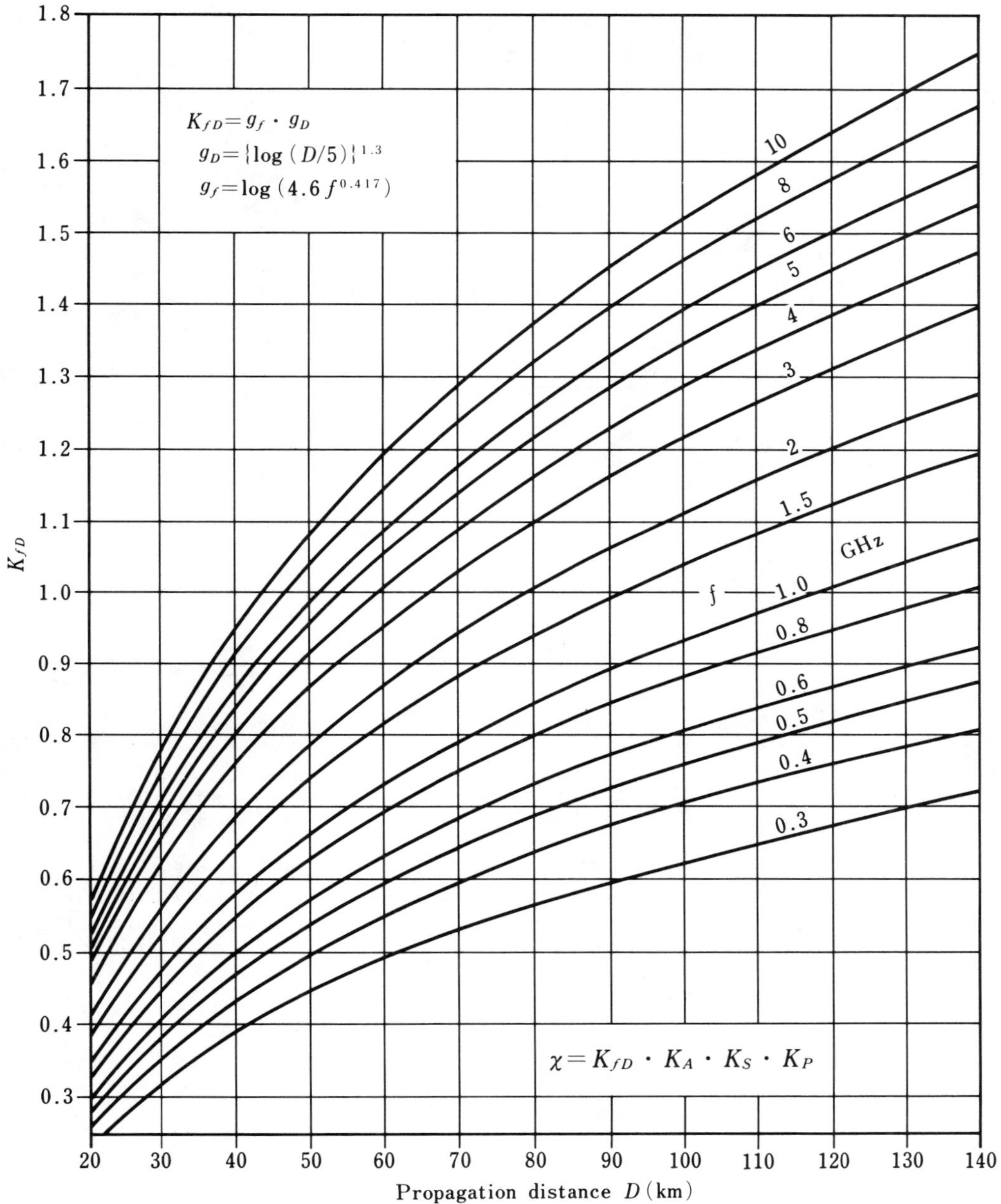

Radio Meteorological Coefficient K_A (1)

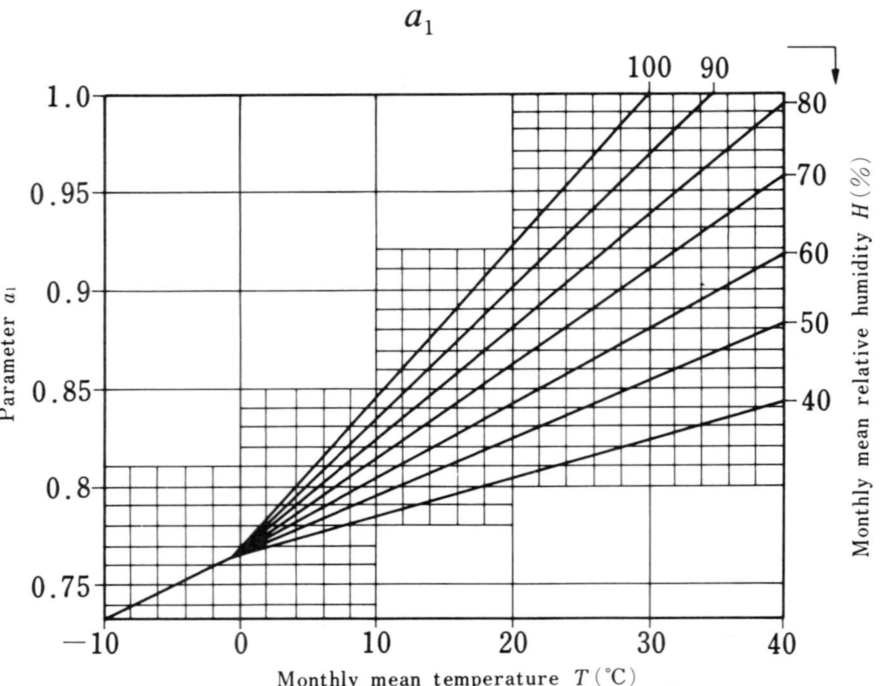

- $a_1 = 0.765 + T\{1.95 + 0.0983(H-40)\}/1\,000$
  where $H=40$ if $H<40$

Note : Figures used as meteorological parameters are the monthly mean values for the same month, for a period of ten years or more if possible, collected from an observatory that is as close as possible to the propagation path.

Radio Meteorological Coefficient K_A (2)

a_2, K_A

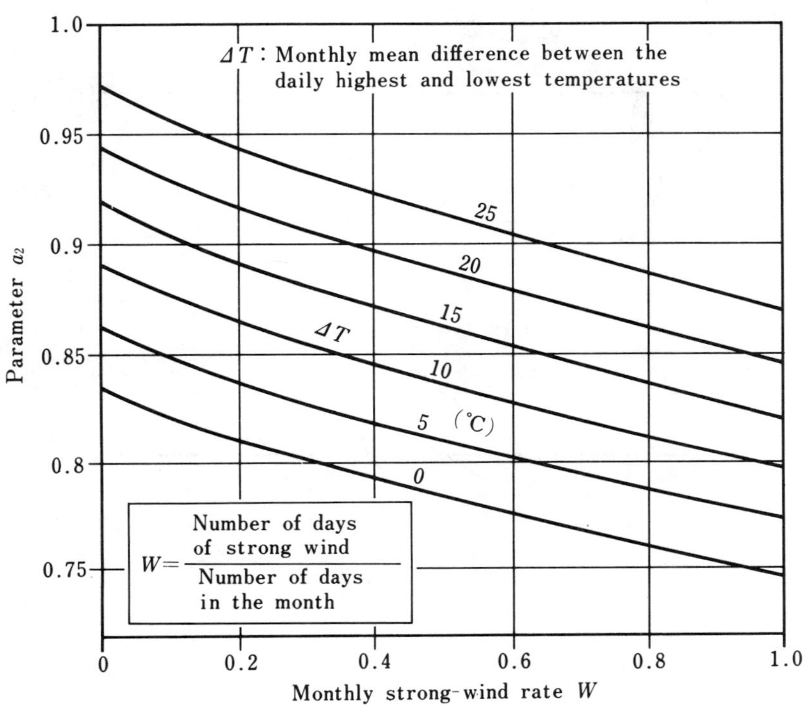

- Strong-wind day : Any day the wind speed is 10 m or greater
- $a_2 = (0.86 + 0.056\,\Delta T)\,(0.97 - W^{0.8}/10)$

$$K_A = a_1 a_2 + 0.05\sqrt{a_1 a_2}$$

Coefficient of Surrounding Terrain Conditions K_S (1)

Classified Surrounding Terrain Map

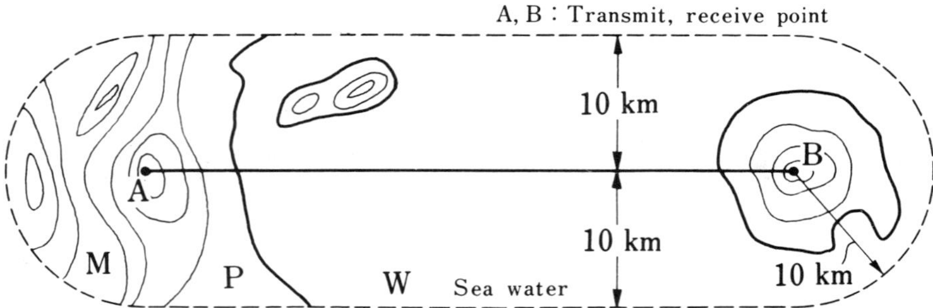

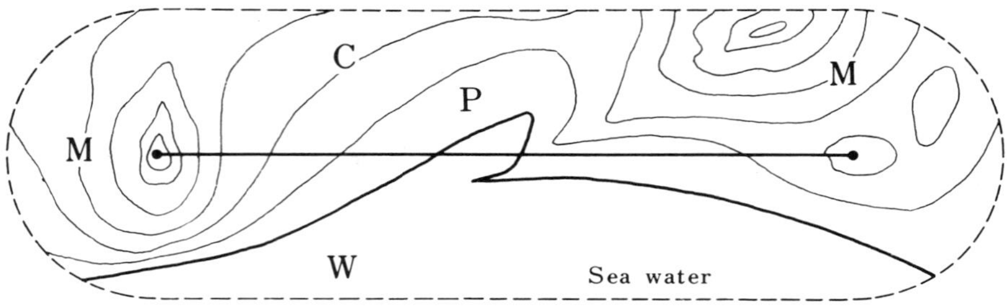

W : Water land (Surfaces of sea, lake, swamp, damp ground, water surface of large river, etc.)

P : Flat land (Flat land such as plain, coastal sandhill, cities, village, desert, snow and ice field, etc.)

C : Sloping land (Skirt of a large mountain, slope such as fan shaped terrain)

M : Mountains, hilly land (Land where undulation can be seen even on 200,000 scale map)

Coefficient of Surrounding Terrain Conditions K_S (2)

How to determine K_S

[1] Topographical study map (Example)

Study unit block □ 2×2 km

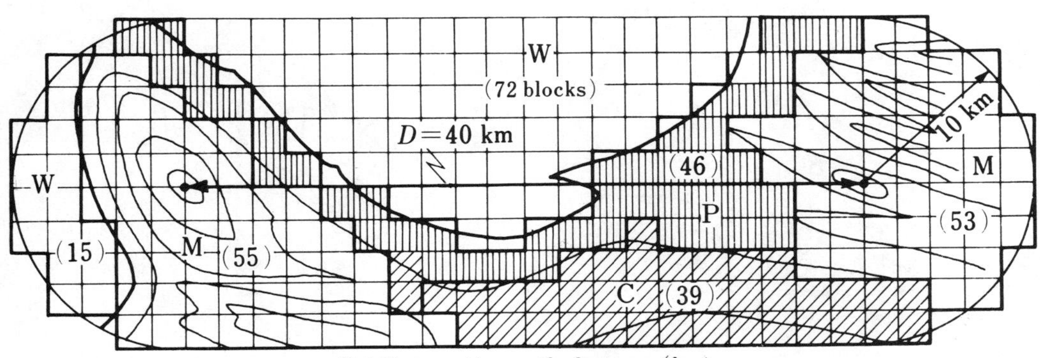

D : Propagation path distance (km)

[2] Calculation of classified terrain ratio

N_O : Total number of blocks		
n_W : Water land	$n_W/N_O = r_W$	
n_P : Flat land	$n_P/N_O = r_P$	
n_C : Slope land	$n_C/N_O = r_C$	
n_M : Mountainous land	$n_M/N_O = r_M$	

Calculation on the map

N_O	280	1.00	r
n_W	87	0.31	r_W
n_P	46	0.16	r_P
n_C	39	0.14	r_C
n_M	108	0.39	r_M

$$K_S = r_W + 0.95\, r_P + 0.9\, r_C + 0.8\, r_M$$

Path Profile Coefficient K_p (1)

Propagation Path without Reflection

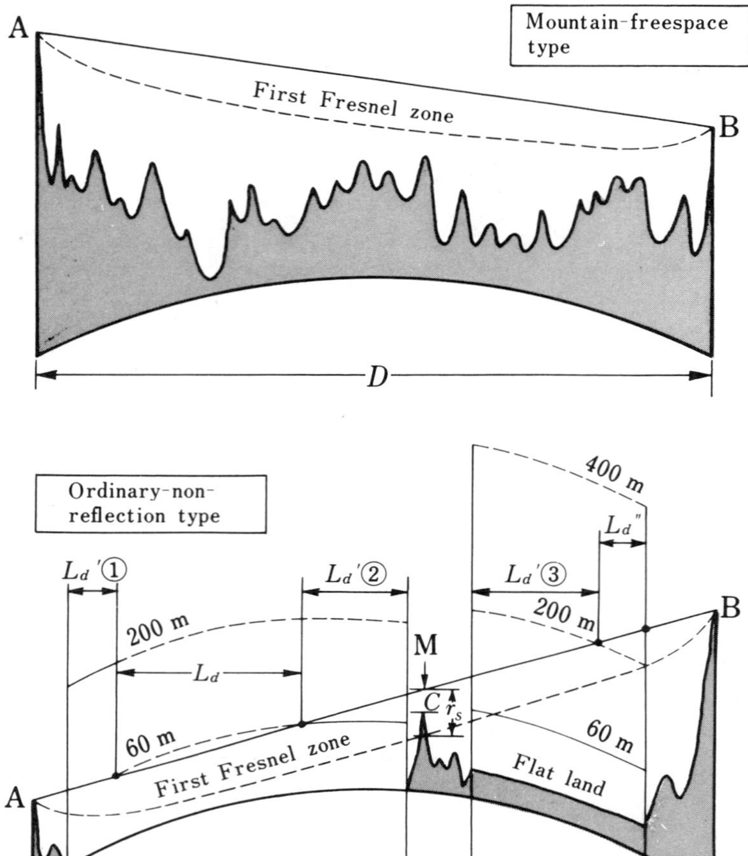

Path profile parameter	Equation	Intermediate parameters	
Propagation echo ratio ρ_θ	<0.1	b_1	Fig. 21 c
Effective low path length $L_e = L_{de}$	$L_d + L_d'/3 + L_d''/5$	b_2	Fig. 21 d
Underlying surface factor T_p	$(d_w + d_p/2)\ D$	b_3	Fig. 21 c

Note : L_d, L_d', L_d'' : Portions of path length where the radio path height above flat land (water, land) for direct wave is 60 m or less, 60 to 200 m, or 200 to 400 m, respectively. d_w : Total length of the water land (sea, lake, river, swamp, paddy field) under the radio path.
d_p : Total length of portions of the ground under the radio path.

Path Profile Coefficient K_p (2)

Propagation Path with Reflection

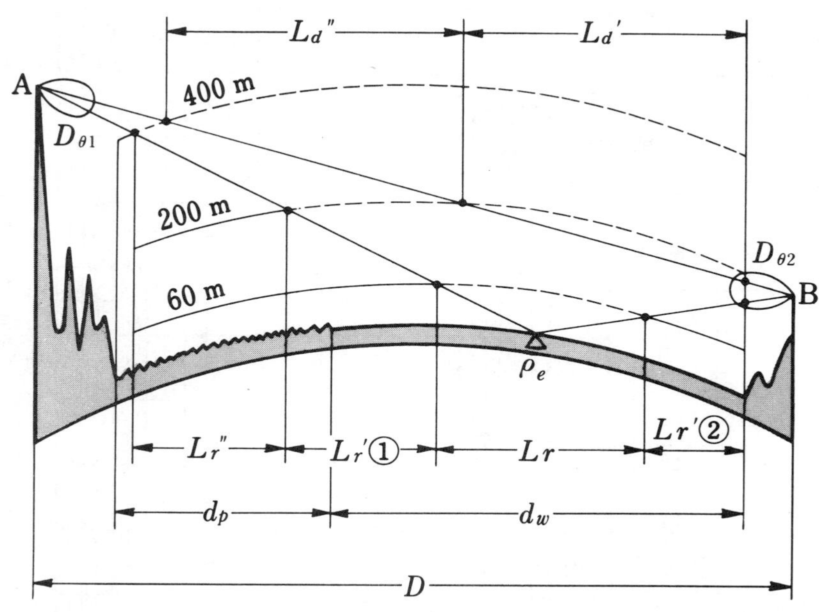

Path profile parameter		equation	Intermediate parameters	
Propagation echo ratio	ρ_θ	$\rho_e \cdot D_{\theta 1} D_{\theta 2}$	b_1	Fig. 21 c
Effective length of the low path	L_e	$L_{de} + L_{re} \cdot \rho_\theta$	b_2	Fig. 21 d
Low-path length of direct ray	L_{de}	$L_d + L_d'/3 + L_d''/5$		——
Low-path length of reflected ray	L_{re}	$L_r + L_r'/3 + L_r''/5$		——
Underlying surface factor	T_p	$(d_w + d_p/2)/D$	b_3	Fig. 21 c

Note : ρ_e : Effective reflection coefficient, $D_{\theta 1}$, $D_{\theta 2}$: Directivities of transmit and receive antennas (Chapter 2) L_d, L_d', L_d'' : Portions of path length where the radio path height above flat land (water, land) for direct wave is 60 m or less, 60 to 200 m, or 200 to 400 m, respectively. d_w : Total length of the water land (sea, lake, river, swamp, paddy field) under the radio path. d_p : Total length of portions of the ground under the radio path.

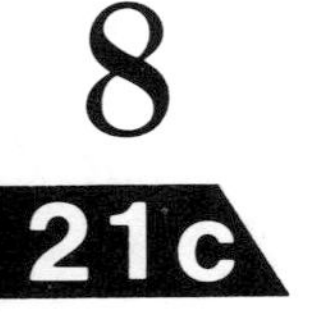

Path Profile Coefficient K_p (3)

b_1, b_3

1.0 0.95 0.9 0.8 0.7

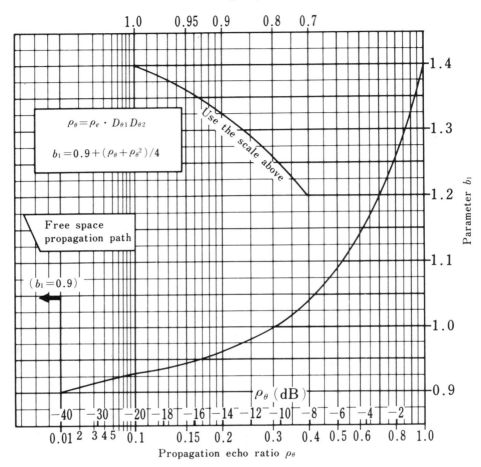

$\rho_\theta = \rho_e \cdot D_{\theta 1} D_{\theta 2}$

$b_1 = 0.9 + (\rho_\theta + \rho_\theta^2)/4$

Use the scale above

Free space propagation path

($b_1 = 0.9$)

ρ_θ (dB)

−40 −30 −20 −18 −16 −14 −12 −10 −8 −6 −4 −2

0.01 2 3 4 5 0.1 0.15 0.2 0.3 0.4 0.5 0.6 0.8 1.0

Propagation echo ratio ρ_θ

Parameter b_1

0.9 1.0 1.1 1.2 1.3 1.4

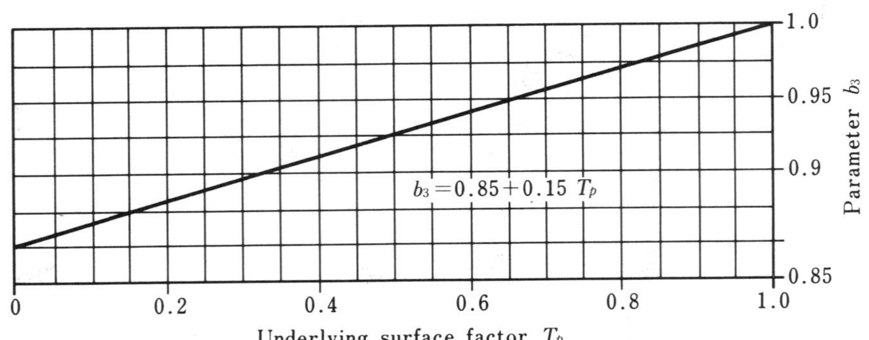

Path Profile Coefficient K_P (4)

b_2, K_P

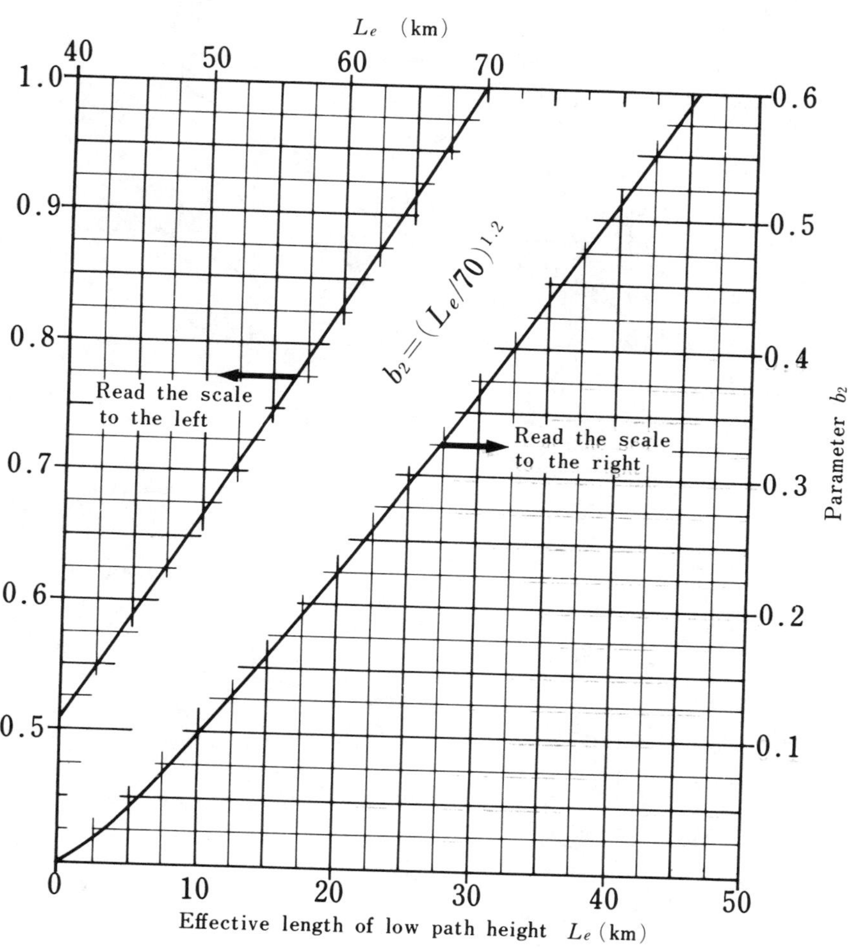

$$K_P=(1.04\,b_1+b_2)\,b_3$$

Alternative Method for Approximation (1)

K_A, K_S

Note : K_{fD} is determined by Fig. 18.

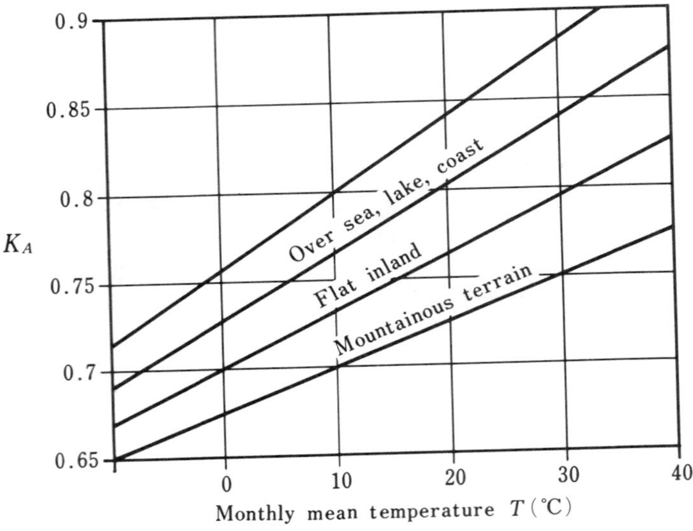

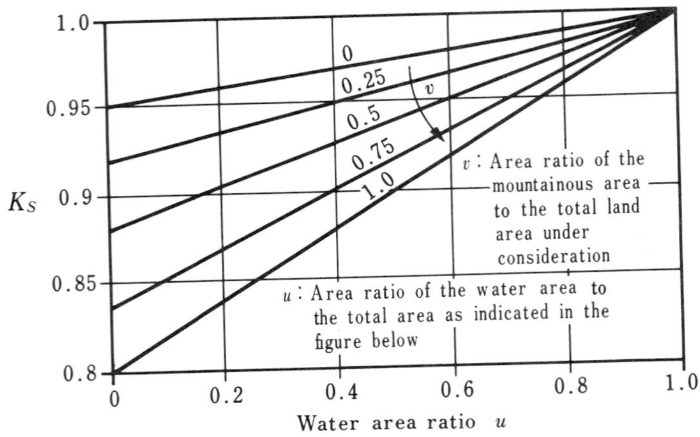

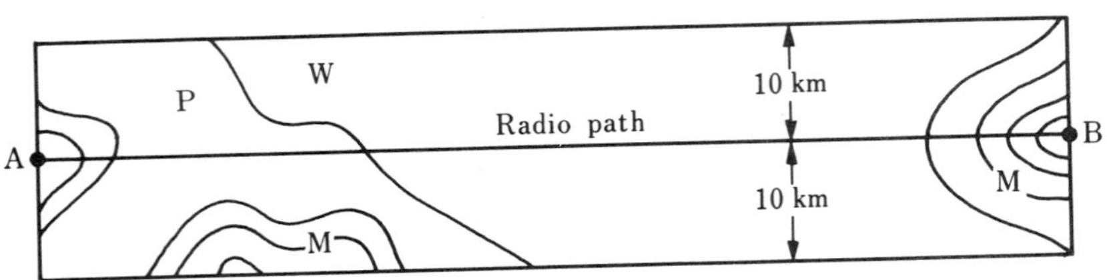

Alternative Method for Approximation (2)

K_P

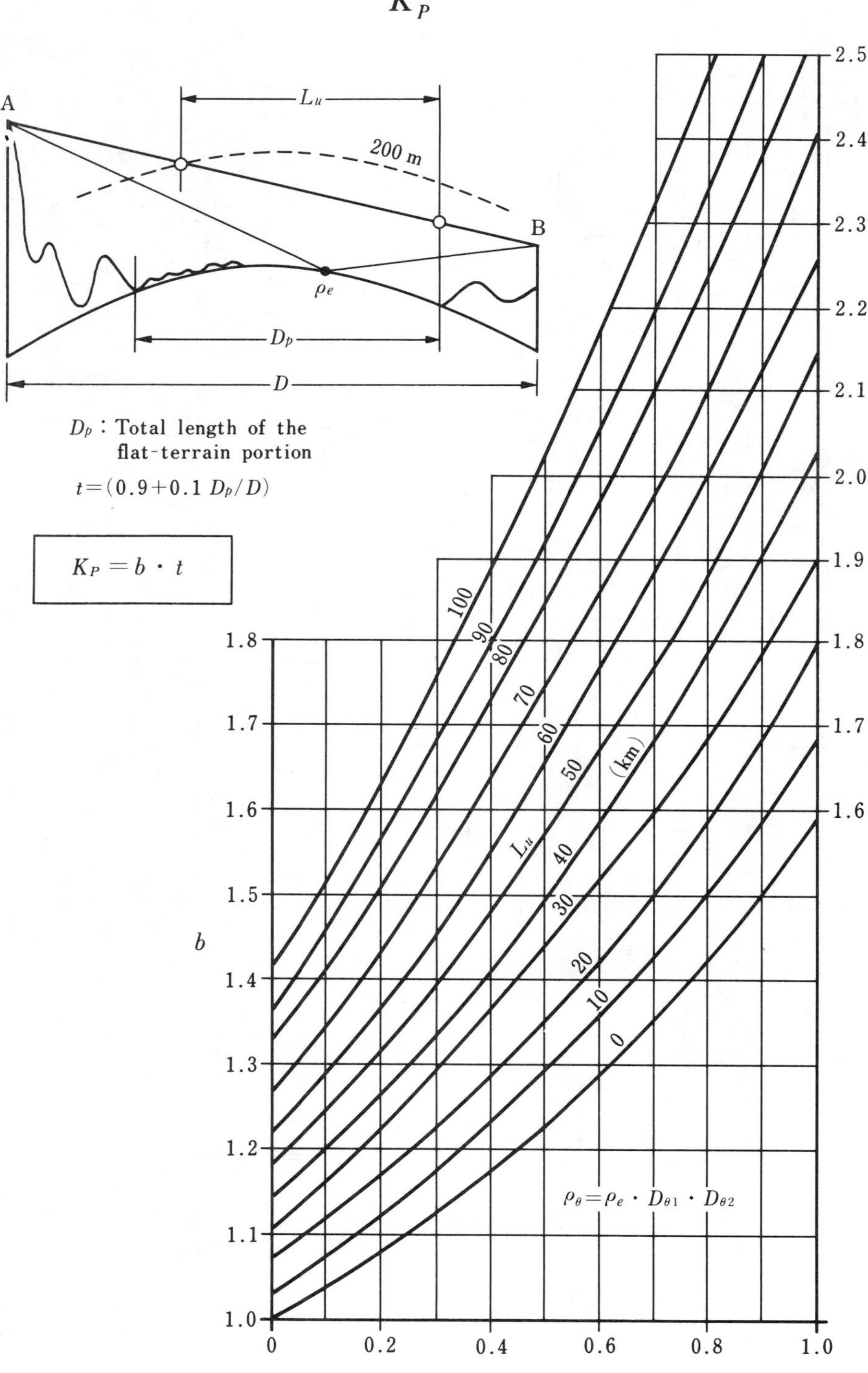

Diversity Effect (FD, SD)

For Reference

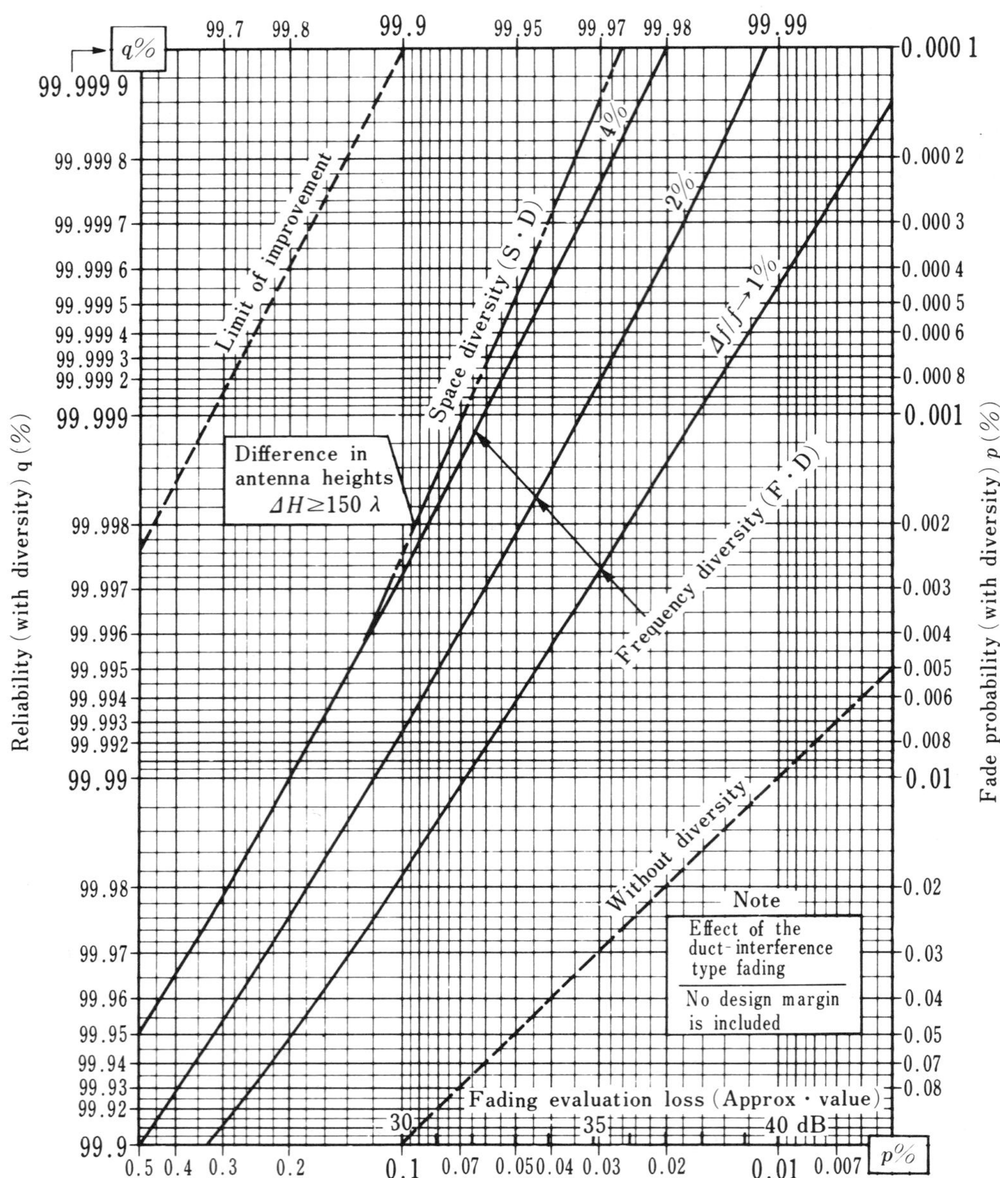

Nomograph for Space Diversity

For Macroscopic Studies

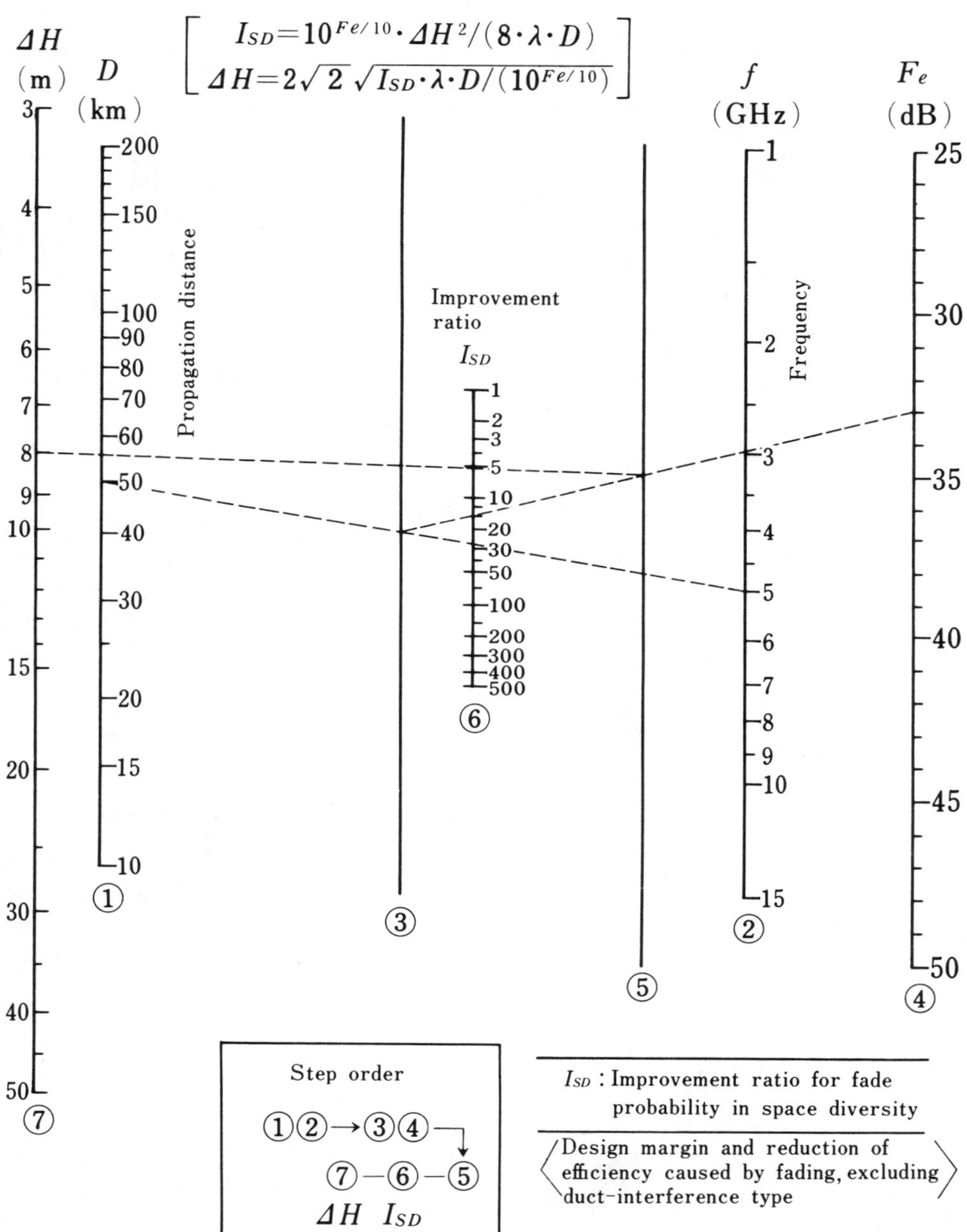

Graph for Space Diversity Design (1)

Equations, Nomograph of λD

[A]

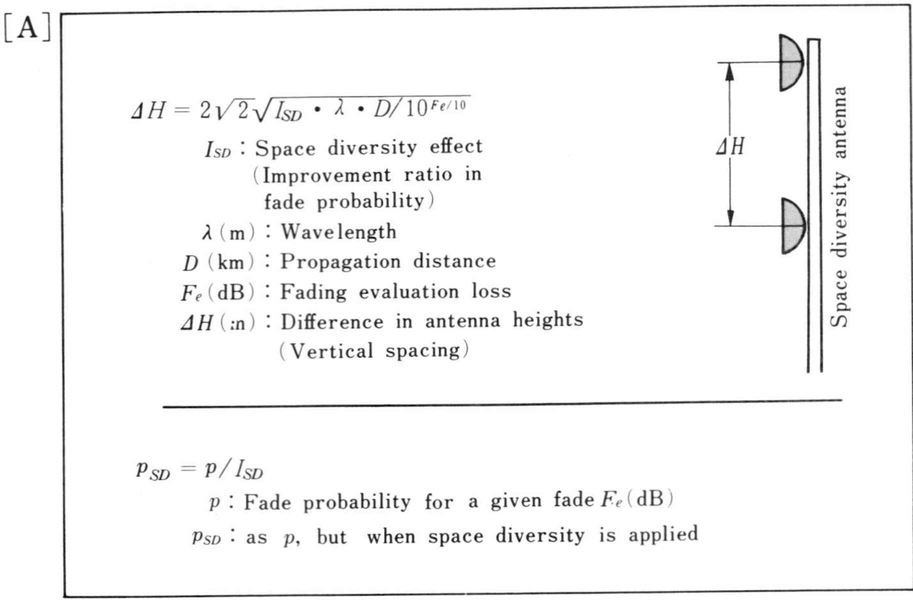

[B] Nomograph of $\lambda \cdot D$

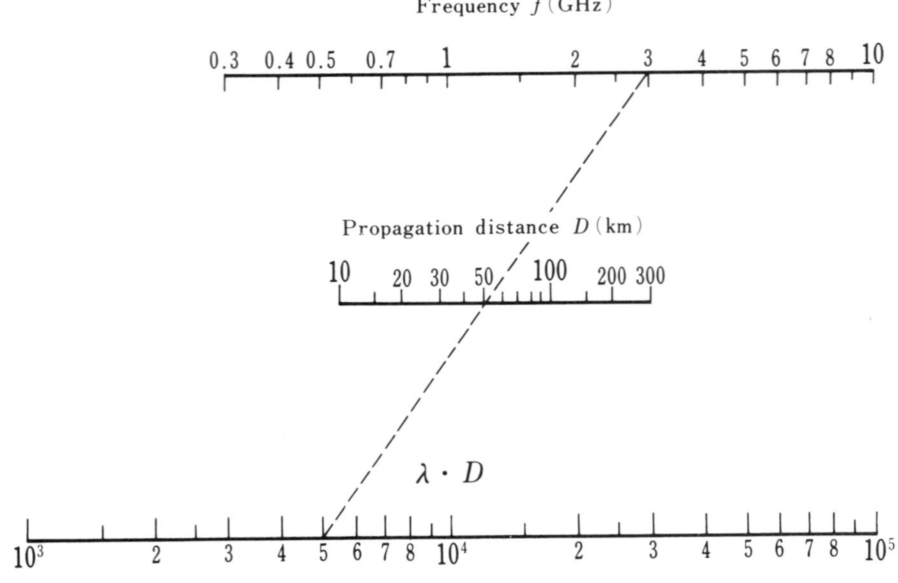

Graph for Space Diversity Design (2)

$I_{SD} = 25, 50$

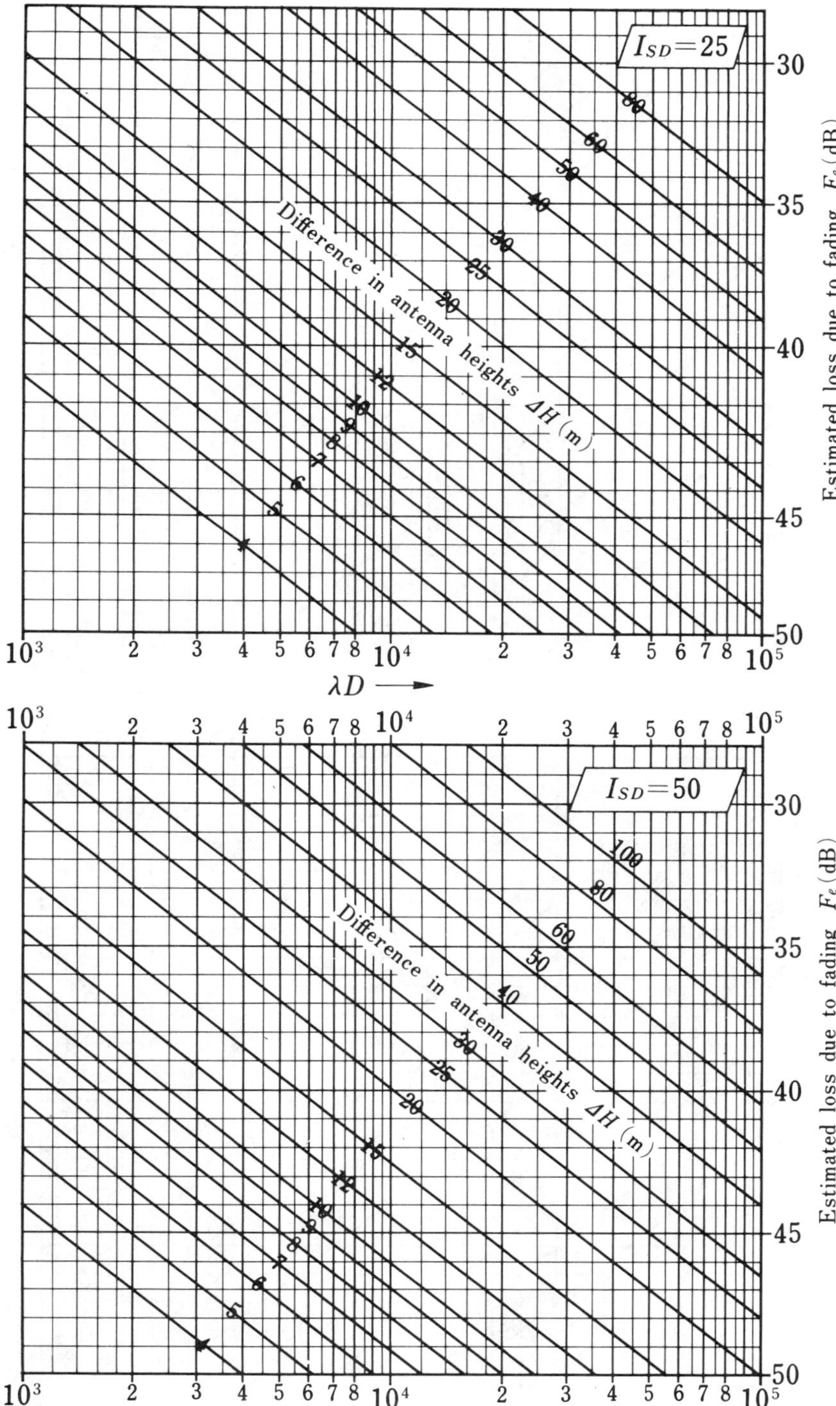

Graph for Space Diversity Design (3)

$I_{SD} = 75, 100$

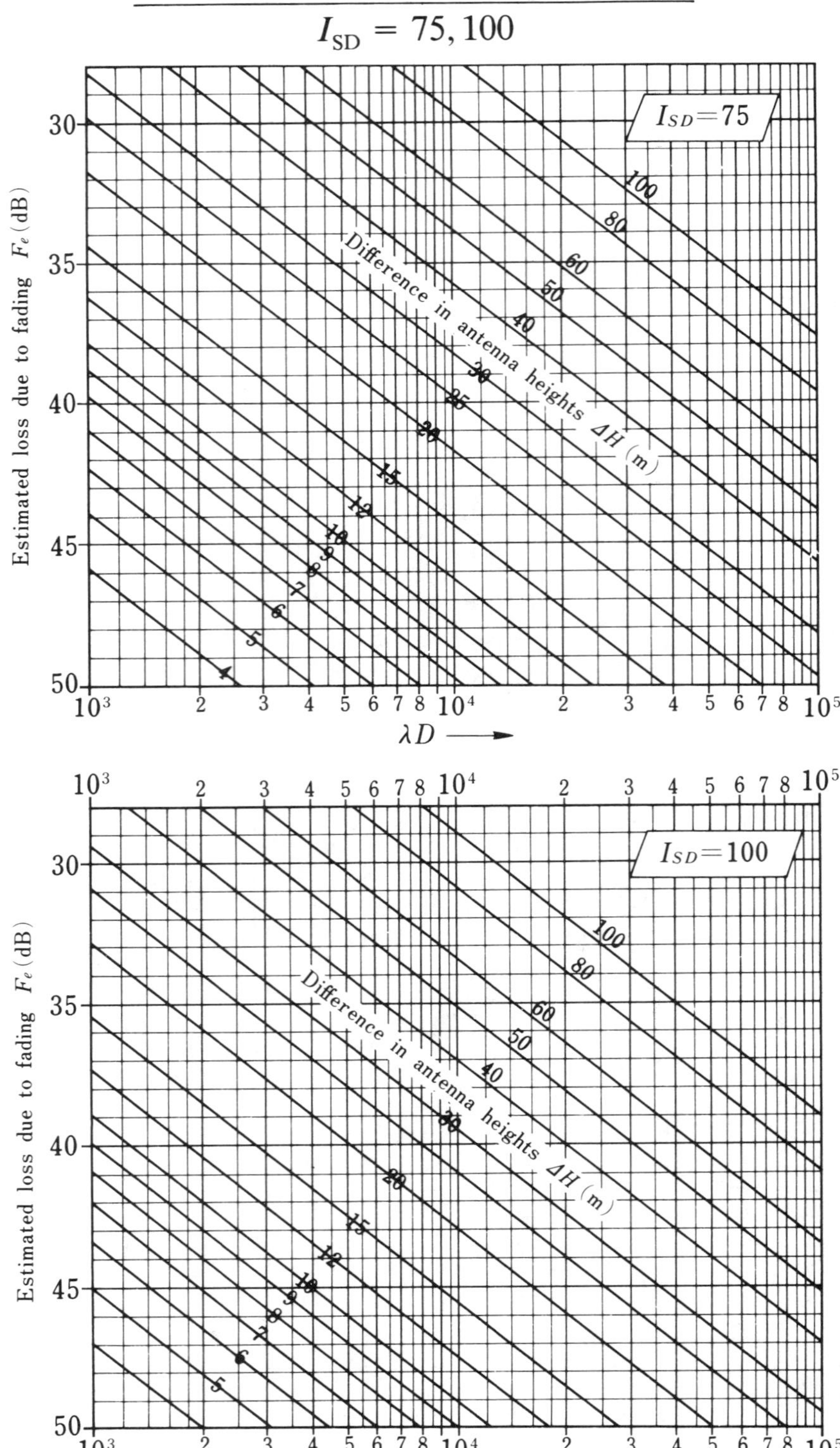

Graph for Space Diversity Design (4)

$I_{SD} = 150, 200$

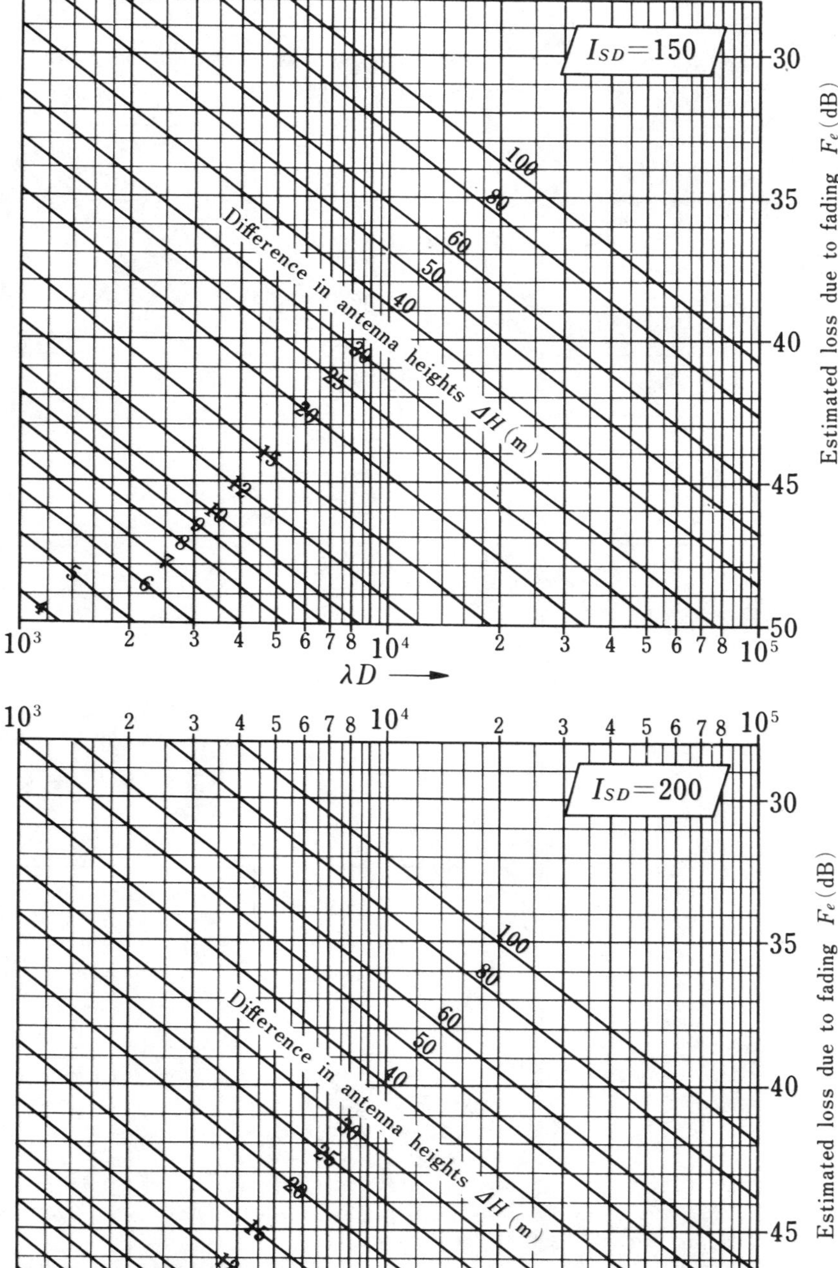

Requirement Applied for Fixed Services in Japan

Required S/N Ratio Reliability, Threshold Margin

(1) Summary of the technical standards in Japan.

(2) The outage factor in this table shows the upper limit and may change according to purposes and conditions to be applied.

Frequency		Service category		Required S/N (dB)	Yearly reliability * q (%)	Threshold margin M_{th} (dB)
MHz	29.7 to 300	Circuit to be connected to the system using 1 GHz or higher		40 (55)*	95 (99.9)*	$0.1 \times D + \{10 \sim 20\}$
		Others		30 (55)*		
	300 to 470	Single channel circuit		30		(for $q=99.5\%$) $0.2 \times D$ (for $q=99.9\%$) $0.2 \times D + 3$
		Multichannel	Public communication	50	99.9	
		Multichannel	Others	30** 40	99.5	
	470 to 1 000	B. C. program relay		55	99.9	
GHz	1 to 10	Public communication		65***	99.9	$\{0.2 \sim 0.3\} \times D$ (for $q=99.5\%$) $\{0.2 \sim 0.3\} \times D+6$ (for $q=99.9\%$).
		Others		55***	99.5	
	more than 10	Public communication B. C. program relay Others		55***	(Outage factor) loss than 0.01	Calculation is made for 10 min precipitation to 0.01% collected at a nearby weather station during July, August and September. (Refer to Chapter 5)
Note				* B.C. program relay; D : Propagation distance (km); ** This figure is applied only when the circuit is not connected to a circuit using 1 GHz or higher frequency.; *** Thermal noise only		

Note : -Different requirements have been specified for microwave public communication ciruit according to CCIR recommendations.

$-M_{th} = P_r(50\%) - P_{th}$ (dB)

$P_r(50\%)$: Median of receiver input

P_{th} : Threshold input level

Margin for Fading (1)

Quick Estimation Graph for Fixed Services

30～300 MHz

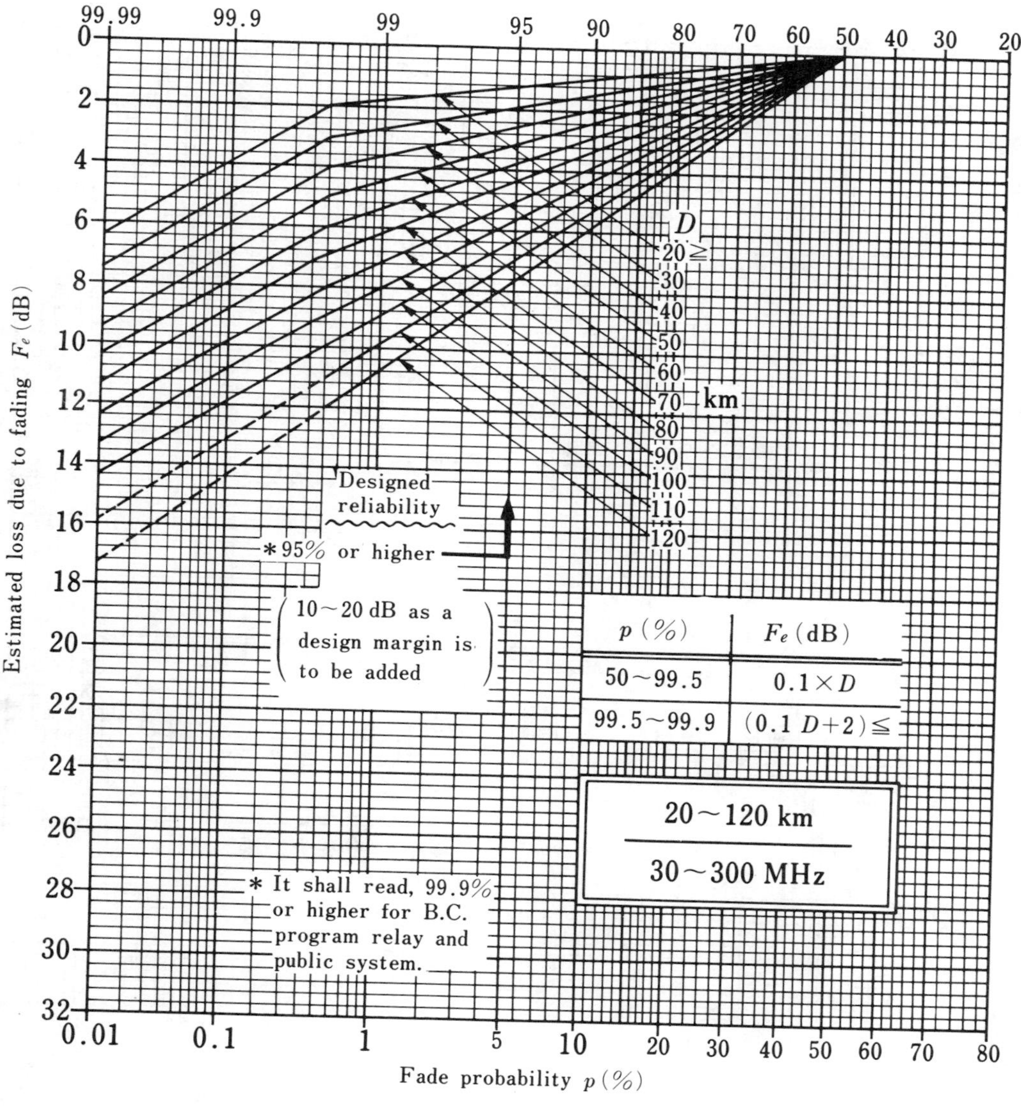

8

26b

Margin for Fading (2)

Quick Estimation Graph for Fixed Services

300 ~ 1 000 MHz

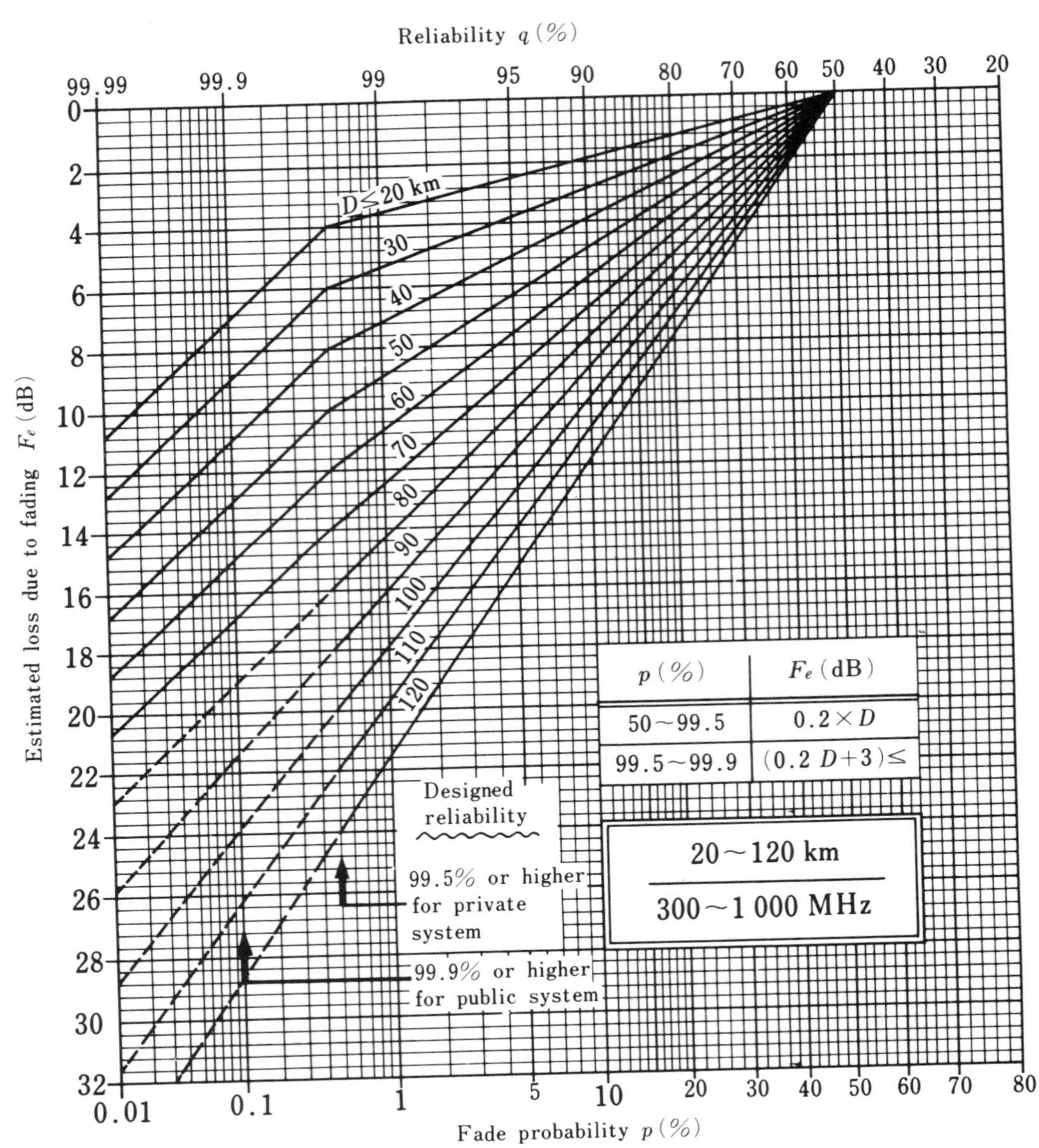

Margin for Fading (3)

Quick Estimation Graph for Fixed Services

1～10 GHz

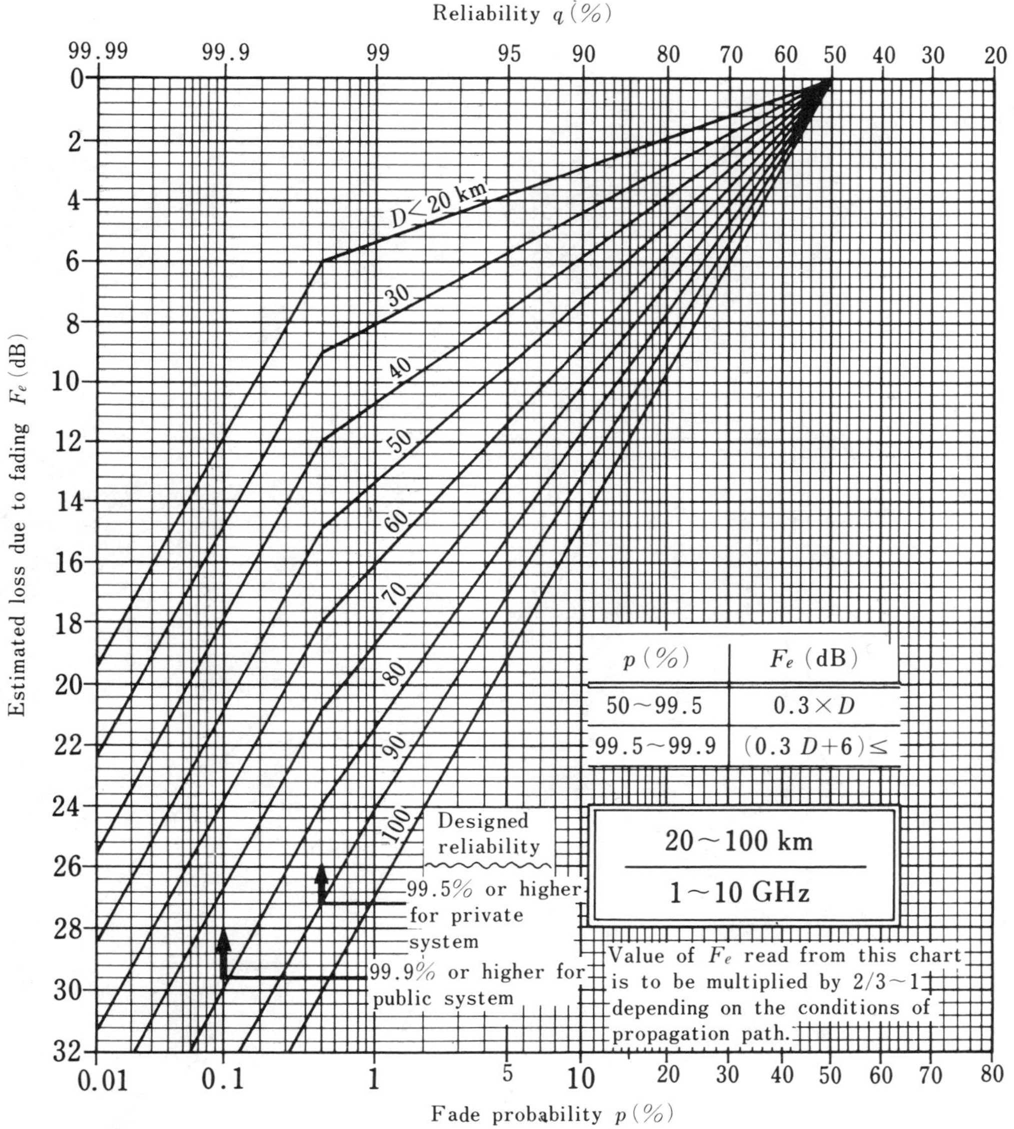

p (%)	F_e (dB)
50～99.5	$0.3 \times D$
99.5～99.9	$(0.3\,D+6) \leq$

Graph for Fading Estimation to Be Used on Initial Planning Stage

Yearly F_e (99.9%)

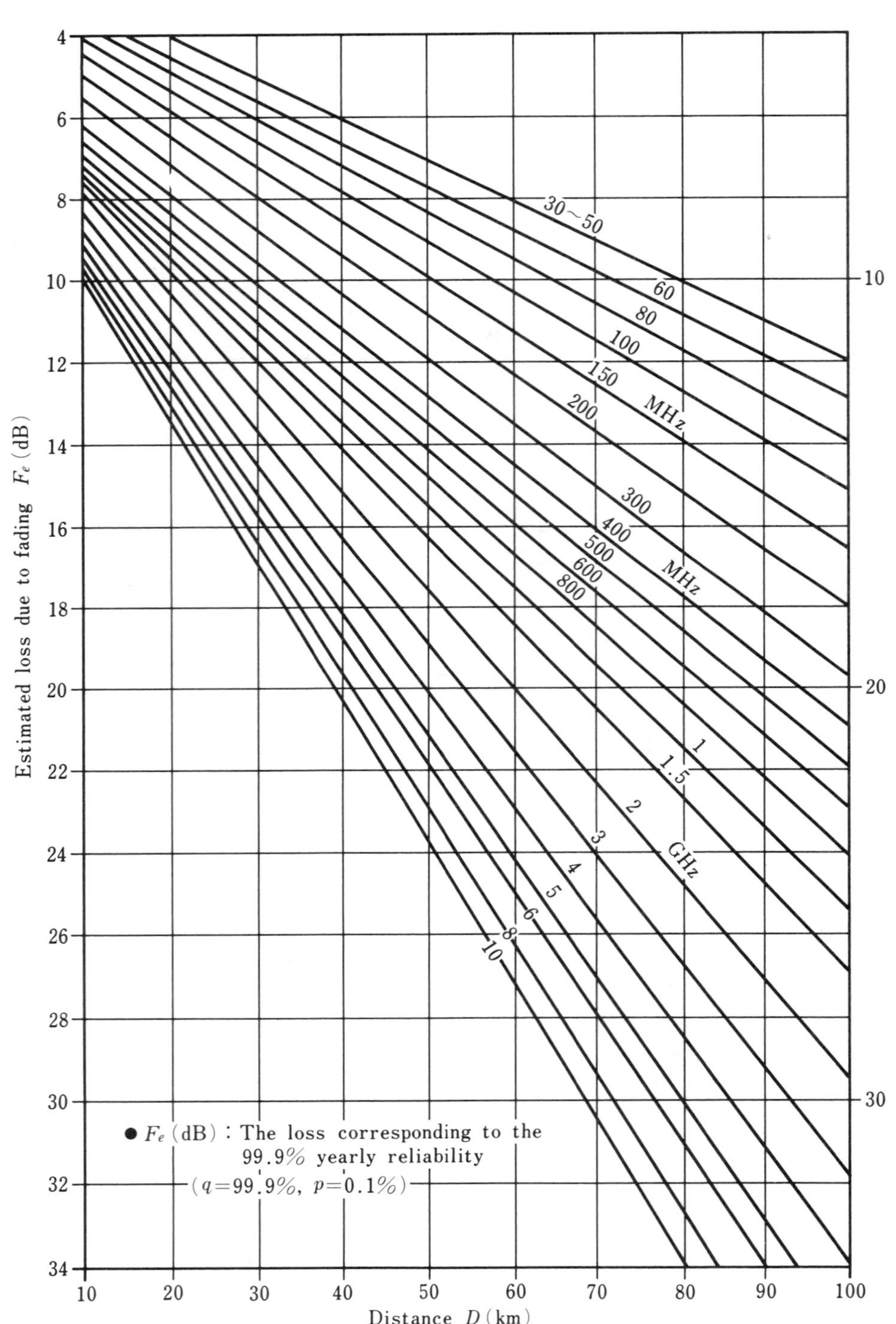

9

ASTRONOMY AND GEOGRAPHY

9.1 ASTRONOMICAL AND GEOGRAPHICAL SURVEY AND RADIO ENGINEERING

Often surveys have to be carried out in the field to determine the location of a radio station and also the geometrical relation between radio paths and various geographical objects.

Such activities as measurement of the distance between two points on the ground, elevation above mean sea level, angles in horizontal and vertical planes, and the preparation of a map by aerial survey photographs and sketches are called geographical surveys. Activities to determine accurate coordinates, bearing angles, etc., of a certain point through celestial bodies such as the sun, moon, planets, and stars are called astronomical surveys. They should be distinguished from astronomical observation or radio astronomy aimed at investigating properties of celestial bodies.

Recently surveys made by geodetic satellites have become more and more important. Public surveys done by a country, a state, or a public organization to locate triangulation points or reference level points for the preparation of maps require high accuracy, large-scale facilities, a large staff, great expenditures, and complicated computation and drafting. However, radio engineering surveys require much less accuracy, and emphasis is on how quickly the task is completed employing simple equipment and instruments. Essential points concerning radio engineering surveys include:

1. Preparation and effective use of maps already published.
2. Use of outstanding objects whose geographical locations are already known, such as triangulation points, reference level points, rocks on top of a

mountain, individual trees, monuments, highway intersections, and milestones.
3. Choice of a suitable celestial body (sun or bright star).
4. Selection of suitable survey equipment and materials which are easy to handle.
5. Determination of a tolerable margin of error to meet the survey purpose.
6. Development of an efficient survey technique and method of computation, and preparation of figures and charts necessary for the survey.

9.2 TOLERABLE ERROR AND CALCULATION METHOD

A. Tolerable Error (in a Simplified Survey)

Let us define a detailed survey as survey work where the accuracy of measurement is 1 s or less for angles, 1 m or less for distance, 1 cm or less for height, and 0.1 s or less for locating a point. The second next accurate survey is called a quasidetailed survey.

A much less detailed survey may be called a simplified survey. The radio engineering survey can be categorized as a simplified survey, and the required criteria for accuracy may be specified as approximately 1 min or less for angles, 1 to 2% of the total or 100 m for distance, 10 m division for height, and 1 to 10 s for surveying to determine a geographical point. More detailed criteria are given as follows:

1 min or less for angles	Equivalent to a maximum error of 3 m at a distance of 10 km
1 to 2% or less for distance	Equivalent to 0.1 to 0.2 dB error for free-space propagation loss
10 m for height	Same error as in the angle measurement, but if it is estimated from the 1:50,000 map, an error of 20 m or less corresponding to the spacing between successive contour lines
1 to 10 s for locating a point in a medium latitude	An error of 1 s corresponding to the error of 0.6 mm along the meridian and 0.5 mm along the parallel on a 1:50,000 map, while 10 s is 1.5 mm along the meridian and 1.25 mm along the parallel on a 1:200,000 map

Accordingly, the survey instrument, that is, a transit compass with an accuracy of 20 to 10 seconds, suffices.

These criteria appear to be easily attainable, but not for an inadequately trained person. Methods of survey and measurement as well as charts have been prepared, keeping in mind the various conditions mentioned in the preceding.

B. Features of Calculation

Most conventional and practical systems of calculating and processing geographical and astronomical surveys have been systematized so as to facilitate the logarithmic summation presupposing the use of:

1. Table of 7- to 8-digit trigonometric functions, inverse trigonometric functions, and their common logarithms.
2. Numerical tables for various corrections.
3. Manual calculation.

The reason for this is that geographical surveying has been conducted based on triangulation, while astronomical surveying is based on spherical trigonometry. Consequently the computation of trigonometric or inverse trigonometric functions cannot be avoided because the computations of various functions require numerical tables of functions, although certain mechanical calculators are available.

The advent of the computer has facilitated the solving of problems that contain a number of mathematical functions, including spherical trigonometry. Hence conventional methods are gradually being modified to up-to-date methods of computation. Also, gradual changes have been made in the equations used for calculation, modifying their original forms or introducing similar, simplified, but more accurate forms.

In this text several equations have been prepared taking into account such circumstances as:

Trigonometric function and inverse trigonometric functions	Calculated directly by pocket calculator
Numerical values for various adjustments	Derived directly from the modified equations for correction as far as possible
Equations to be used for calculation	Equations to obtain the antilogarithm on the basis of the original equation

Attention is given to the difference between calculation equations and conventional equations.

9.3 DETERMINATION OF VARIOUS SURVEY PARAMETERS

A. Position of a Point

The position of a point is determined taking into account its relative position to triangulation points, contour lines, and any other outstanding objects indicated on a map drawn on a scale of 1:50,000 or greater. Then the actual site survey, observation, photographing, and/or measurement are performed.

Finally the position in terms of longitude L, latitude l, and elevation above mean sea level h_s is determined numerically on the map and according to data obtained in the survey.

B. Distance

The approximate distance is obtained by multiplying the actual distance measured on the map by the inverse of the map scale. When a more precise distance is needed, one of these procedures is followed.

1. For a distance of 20 km or more, determine the longitudes and latitudes of both sites as described in Section 9.3A, then apply them to the equations of great-circle distance (Fig. 5).
2. For a distance of approximately 2 to 20 km, make your estimate using the actually measured distance on the map and the map scale.
3. For a distance of 2 km or less, measure the distance on a more detailed map if available, having a scale of 1:25,000 1:10,000, 1:5000, or 1:2500; otherwise conduct an actual field survey (Fig. 7).
4. For a radio path with a large difference between the elevations of both sites, give a correction to the great-circle distance D_g, which is approximately equal to the distance measured on the map, to determine the real distance D_d (Fig. 9).

C. Elevation above Mean Sea Level and Difference in Elevation

An approximate value may be estimated from contour lines on a map. The more accurate value is obtained by measuring a known triangulation point or a known reference level point (Fig. 8). To be considered in the measurement of the elevation difference are corrections due to atmospheric refraction and the earth's curvature.

D. Vertical Angle (Elevation)

The vertical angle can be calculated if the distance between the two points and their elevations above mean sea level are known. If they are not known or are considered to be doubtful, measure them.

E. Direction toward an Adjacent Station

When the direction toward an adjacent station or radio obstacle is to be measured as accurately as possible, the direct observation (for line-of-sight path) or the indirect observation (for transhorizon path) method, in which several intermediate objects between two stations are chosen, is used and the measurements are taken step by step.

To facilitate observation and to prevent erroneous identification of an object, an optical test called mirror test is carried out. In this test, sun rays are reflected by a mirror at the point to be identified.

F. Bearing Angle

When the longitudes and latitudes of both sites are known, the bearing angle measured counterclockwise from the true north toward a point can be estimated at each site using the equations for geographic bearing angle given in Fig. 6. Prior to determining such a bearing angle, the direction of the true north must be determined.

G. Determination of True North (Bearing Angle of Meridian from the North)

When two or more triangulation points are available, the bearing angle from the meridian can be calculated with the distances and the included related angles. No auxiliary ground object is required to find the true north if the astronomical survey of the bearing is able to use the sun or fixed stars, including Polaris. Where less accuracy (more than 1/4 to 1/2°) is allowed, such as in fixing the tower foundation or in the orientation of a radio station building, a general method using the sun is convenient and practical (Fig. 18).

9.4 GEOGRAPHICAL MEASUREMENTS

A. Constants on the Earth

The size and shape of the earth are the bases for related measurement and calculation. Figure 1 shows several constants with their approximate values used in radio engineering. Values given in Fig. 1 are extracted from the *Geodetic Standard System*, 1976 published by IAG, and are the same as those used in radio engineering, as shown in the bottom half of Fig. 1. However, they are slightly different from the conventional figures used in certain countries (Bessel figure in Japan and Hayford figure in the United States).

Regardless of country, elevations above mean sea level indicated on maps are referred to a reference level obtained when the earth is considered a spheroid. (The surface formed by such points is called a geoid.)

For the purpose of this work, the estimated elevation is not necessarily very accurate, that is, the approximate value of elevation and the difference in elevations are of primary concern.

Accordingly the earth's spheroid or the geoid is called simply the earth surface or spherical earth, and the elevation measured therefrom is the elevation above sea level.

Figure 2 gives arc lengths per degree longitude and latitude on the earth surface and is used to determine the distance using the latitude, the difference in latitudes, the longitude, and the difference in longitudes.

B. Map

A map is defined as a flat representation of the whole or a part of the earth according to a certain rule. There are various kinds of maps, for example, topographical maps, transportation maps, and sightseeing maps. The maps used most frequently are the topographical maps and maps prepared from them.

Figure 4 shows various scales of maps selected as suitable. This figure shows how to determine the longitude and the latitude of a point on the map drawn in the transverse Mercator projection system, which is internationally accepted for topographical maps. The accuracy achieved by this method for the 1:50,000 map is within 0.5 to 1″ for latitude and approximately 30 m for distance. However, the accuracy for longitude depends on the latitude of the point of interest, as shown in Fig. 2.

Whether or not the actual point can be pinpointed on the map depends on the topographical conditions at and near the site, the degree of detail and the scale of the map to be used, and also on the field investigation and survey technique.

Figure 5 gives a chart on which the geographical parameters of both the transmitting and the receiving points are to be entered. It also provides a calculation form for the great-circle distance.

The geographic latitude l, obtained from a map, is normally used to indicate the latitude of a position. However, if necessary, the geocentric latitude ϕ, that is, the angle between a line to the center of the earth and the plane of the equator is obtained from Fig. 29.

C. Great-Circle Distance

When the locations of point A (longitude L_1, latitude l_1) and point B (longitude L_2, latitude l_2) are known, the great-circle distance D_g between A and B can be computed by the equation given in Fig. 5.

The great circle is defined as the intersection where the globe surface is cut by a plane passing through the center point of the globe. The great-circle arc connecting these two points gives the shortest path on the surface. The center angle ξ is given in Fig. 5 as $\cos^{-1}[\ ]$.

If $D_g < 20$ km, it is advisable to estimate the distance directly from the map or by the field survey since the estimation error of the longitude and latitude and the computing error may become large.

D. Geographic Azimuth

A rough value of azimuth for the direction $A \to B$ or $B \to A$ is obtained by measuring the clockwise angle included between the meridian passing through point A or B and the direction to the facing station $A \to B$ or $B \to A$ using a protractor on a large-scale map drawn by the transverse Mercator projection system. However, the more accurate value is obtained by computation. Since points A and B are on the sphere (on the ellipsoid to be exact), the relation between β_1, the azimuth A to B, and β_2 the azimuth B to A, cannot be expressed easily like in a flat plane.

The basic equations are given in Fig. 6b. The intermediate parameters β_{01} and β_{02} obtained by the basic equations shall be suitably modified depending upon the relative position of these points and the polarity of the coordinates (L, l), that is, positive for the east, negative for the west, positive for the north, and negative for the south, according to the instructions in the table in Fig. 6b, to determine the azimuths β_1 and β_2.

Figure 6a presents a flowchart of the calculation to which the numbers obtained in Fig. 5 are applied. Since the computation involves a rather troublesome process which is apt to produce a computation error, it is advisable to check whether the calculated results agree approximately with the value measured by a protractor (except for the case of a small-scale map or a map using another system of projection).

Although the calculation of the azimuth on the basis of a quasispheroid such as the shape of the earth can attain better accuracy by making necessary corrections than that on the basis of a sphere, the predominant error lies in determining the coordinates of a point of interest on a map. For example, if the determined position at one end of a 50-km path is in error by 30 m transversally to the path, the corresponding error in the azimuth is 2′, and the error tends to increase as the path length becomes smaller (approximately $10' = 1/6°$ of error against a 10-km path).

Consequently too accurate a presentation of the azimuth for a selected site is considered meaningless; specifically, rounding to one decimal place in degrees such as 235.6° instead of 235°35′24″, is more reasonable.

The same consideration can apply to the measurement of the astronomical azimuth in connection with the position of a celestial body given in terms of declination, right ascension, equation of time, and hour angle. This will justify the use of simplified measurements.

The method of calculation given in Fig. 6*b* is a universal method that can apply to any two points A and B on the earth. In other words, the azimuth can be calculated even for a radio path that connects two points, that is, one in the northern hemisphere and the other in the southern hemisphere, or one in the eastern hemisphere and the other in the western hemisphere. (For example, it can apply to the Tokyo–Rio de Janeiro path, except for the particular conditions given in Fig. 6*b*.)

E. Measurement of Distance

Figure 7*a* gives the measurement procedure and the method of calculation. For measuring a short distance, a tape measure or a stadia can be employed, while triangulation is used to measure the distance of 1 km or more, or the distance over an undulating terrain. As shown in the figure, angles α and β are measured by establishing an additional point. The length of a baseline should be sufficient to minimize the measurement error.

Figure 7*b* illustrates a method of measuring the distance between two remote points P and Q by repeating the above mentioned triangulation in series. In Fig. 7*a* and *b* the distance x to be measured forms one side of the triangle ABC or AQP, and if there is an appreciable level difference in these vertices of the triangle, the calculated figures are adjusted, namely by multiplying the cosine of the elevation angle to obtain the distance on the horizontal plane.

F. Measurement of Level

On a small-scale map the contour lines are so coarse that it is only possible to obtain a rough figure for the elevation above sea level of a selected site or for the difference of elevations. A more accurate measurement of the elevation (on the order of 3 to 15 mm for two points with 2-km distance) or the difference in elevations is carried out through the direct measurement via the level and the leveling staff on the basis of the known level of a benchmark of the existing triangulation point.

On the other hand, the required accuracy in the measurement of elevation (including the ground level plus the tree height) of the points normally encountered is on the order of 1 to 10 and several meters. Therefore the use of a transit compass with an accuracy of 20″ to 1′ for elevation is satisfactory.

Figure 8*a* and *b* illustrates the calculation methods to be used when the elevation of one end of the radio path is known. If surveying has to be made at only one end, additional effects due to atmospheric refraction and to the earth curvature should be taken into account. Figure 10 shows these effects. They are proportional to the square of the distance and should be taken into account when the distance is greater than 4 km. If the measurement can be made at both sites, as in Fig. 8*b*, the terms for the effect at one end and that at the other end cancel each other. Thus when the measurements are made for both directions, the reliability of measurement can be verified by comparing the results obtained by the equation given in Fig. 8*a* with those given in Fig. 8*b*.

9.5 ASTRONOMICAL MEASUREMENTS

A. Definition

The development of man-made satellites and radio astronomy has enhanced remarkably the connection between radio engineering and astronomy. To make the best use of an astronomical phenomenon for measurement without committing any error, comprehension of the true concept of the technical terms in astronomy is essential.

Descriptions of the celestial sphere given in Fig. 11*a* can easily be understood by referring to the illustration in Fig. 11*b*. Figures 12 and 13 present definitions of various terms according to time, and emphasis is placed on differences between mean time and sidereal time as well as on the relationship between the motion of a celestial body and time (including the hour angle).

B. Time and Angles Relating to Celestial Bodies

Figure 13 illustrates the relation between various bases of time (the hour angles of mean sun, apparent sun, and vernal equinox), the hour angles of the ordinary celestial bodies including the sun, and the declination, for reference. The relation between mean time, apparent solar time, and sidereal time appears rather difficult to understand. However, their meaning should be grasped so as to broaden the range of application of astronomical measurements and to perform the measurements correctly.

C. Relationship between Local Sidereal Time and Standard Time

The time commonly used is the standard time adopted uniformly in a certain zone or country. To determine the hour angle of a fixed star, the conversion of observation hours listed in standard time T into local sidereal time at the point of observation θ_t is required. The procedure is as follows.

1. The time of observation (according to standard time) T is converted into the universal time U by

$$U = T - L_s \tag{9.1}$$

where L_s is the longitude of the meridian to define the local standard time (9^h in Japan).
2. Obtain the Greenwich sidereal time θ_0 at $U = 0^h$, referring to a published astronomical ephemeris.
3. Convert the progress of time from the universal time 0^h (i.e., $U = T - L_s$) into the progress of time in sidereal time,

$$(1 + k')(T - L_s) \tag{9.2}$$

where $k' = 0.002\ 737\ 91$.
4. The Greenwich sidereal time at a time of observation θ_g is obtained by adding the value given in step 3 to θ_0,

$$\theta_g = \theta_0 + (1 + k')(T - L_s) \tag{9.3}$$

5. Since the sidereal time at a point of observation is fast by an amount corresponding to the longitude L (positive for the east and negative for the west), the sidereal time at the time and point of observation, that is, the local sidereal time θ_t, is calculated as

$$\boxed{\theta_t = \theta_0 + L = \theta_0 + (1 + k')(T - L_s) + L} \tag{9.4}$$

This equation is the first of the modified equations as given in Fig. 14[3].

The standard time can be determined from the sidereal time by reversing steps 1 to 5 or by solving Eq. (9.4).

Solving Eq. (9.4) for T,

$$T = (\theta_t - \theta_0 - L)/(1 + k') + L_s \tag{9.5}$$

Modifying $1/(1 + k')$ in the right-hand side of Eq. (9.5),

$$(1 - k) = 1/(1 + k') \tag{9.6}$$

where $k = 0.002\ 730\ 43$ and $k' = 0.002\ 737\ 91$. By substituting into Eq. (9.5),

$$\boxed{T = (\theta_t - \theta_0 - L)(1 - k) + L_s} \tag{9.7}$$

This is the second of the modified equations in Fig. 14[3].

D. Computation of Hour Angle (General Case)

It is clearly understood from Figs. 11 and 13 that

1. The local sidereal time at a point of observation θ_t equals the hour angle (clockwise) at the vernal equinox.
2. The celestial body at a right ascension α is positioned at α (counterclockwise) from the vernal equinox.

Accordingly the hour angle (clockwise) of a celestial body h is given by

$$\boxed{h = \theta_t - \alpha} \tag{9.8}$$

Applying θ_t as mentioned in Section 9.5C, Fig. 15 is obtained. This general equation is applicable for the sun, the moon, the planets and the fixed stars. However, the method in the following section can apply to the sun because the equation of time is available in an astronomical ephemeris.

E. Hour Angle and Declination of the Sun

The local mean time is obtained as the sum of the hour angle of the mean sun at a point of observation and 12^h (since the solar meridian passage is obtained as 12^h in the mean time). In other words, the hour angle of the mean sun is obtained by deducting 12^h from the local mean time.

$$\begin{aligned} \text{local mean time} &= \text{standard time } T \\ &\quad + \text{longitude at point of observation } L \\ &\quad - \text{longitude for standard time } L_s \\ &= (T - L_s) + L \\ &= U + L \end{aligned}$$

where U is the universal time.

$$\text{hour angle of mean sun} = (T - L_s) + L - 12^h \tag{9.9}$$

However, the apparent sun instead of the mean (fictitious) sun can be observed.
The equation of time ε is defined as the difference between hour angles for the apparent sun and the mean sun, that is, the difference in declination, and the daily values are given in an astronomical ephemeris.

$$\begin{aligned} \text{equation of time } \varepsilon &= \text{hour angle of apparent sun} \\ &\quad - \text{hour angle of mean sun} \end{aligned} \tag{9.10}$$

If the hour angle of the apparent sun is denoted by $h_\odot$, then using Eqs. (9.9) and (9.10), it is given by

$$\boxed{h_\odot = T - L_s + L + \varepsilon - 12^h + 24^h} \tag{9.11}$$

The last term, 24^h, is applied, as the case may be, too prevent the result from becoming negative. Addition of this term will not produce any error in computed results because $24^h = 360°$.

Correction to Variation of the Equation of Time and the Declination

If the azimuthal measurement requires accuracy better than 1/10°, the corresponding accuracy of the equation of time and the declination must be attained. Usually the astronomical ephemeris indicates values of the equation of time ε_0 and the declination δ_0 for the universal time $U = 0^h$, and may present more detailed information (for every 2 hours, for example).

The appropriate values ε and δ at the time of observation T assuming the daily variations $\Delta\varepsilon$ and $\Delta\delta$, respectively, are given by

$$\boxed{\begin{aligned}\varepsilon &= \varepsilon_0 + \Delta\varepsilon(T - L_s)/24 \\ \delta &= \delta_0 + \Delta\delta(T - L_s)/24\end{aligned}} \tag{9.12}$$

where $(T - L_s) = U$ (universal time) corresponding to the passage of time measured from $U = 0^h$.

If data for every 2 hours are available, use the value nearest to U or obtain the correct value on a proportional allotment for the elapsed time from the time at which the data are available.

Figure 16 gives a calculation for the hour angle and the declination of the sun.

Graphs to Determine Approximate Values of the Equation of Time and the Declination

The required accuracy in determining the orientation of buildings and towers is normally 1 to 3°, and rarely 1/2° or less.

Figure 21 provides the most useful graphs of the equation of time and the declination prepared for simplified surveys. The accuracy in surveying the azimuth is on the order of 1/3 to 1/5° at maximum if the error in determining the time of observation is kept within 10 to 20 s. The day-by-day variations can be found in curves.

F. Surveying of Astronomical Azimuth

Figure 17 shows the relation between various parameters of astronomical surveying and is used with various instructive figures in the preceding pages.

The lower half of Fig. 17 shows the relation between the circumpolar stars and the latitude at a point of observation pertaining to the polar distance P and latitude l. The notation of absolute value is used, for these figures can be applicable to both the southern and the northern hemispheres. In Figure 18*a* and *b* it has been attempted to elucidate the applicable conditions and various problems on an item-by-item basis.

Figures 18*c* to *f* give the equations and the calculation charts for those methods. A sample of calculation is given in the rightmost column of each chart.

Figure 18g shows the Polaris azimuth which can be used conveniently in the northern hemisphere. Although calculating the hour angle is rather troublesome, the Polaris azimuth is easily determined from the table of Fig. 18g using the hour angle and the approximate latitude at the point of observation. The position of Polaris is roughly estimated in the top diagram in Fig. 18g by determining the time of the meridian transit or of the greatest elongation against a specific longitude using an astronomical ephemeris. The hour angle is given by the sum of the difference in longitude and the time difference (using the conversion of 15° to 1^h).

G. Atmospheric Refraction and Dip of Horizon

Atmospheric refraction is the angular difference between the apparent elevation angle and the true elevation angle, produced by light refraction in measurement.

In land surveying the effect of atmospheric refraction is proportional to the square of the distance D^2, as shown in Fig. 10, while in astronomical surveying, it depends principally upon the thickness of the atmospheric layer, measured along the line of sight.

Figure 19 presents the graph prepared from Landau's table showing that atmospheric refraction tends to increase rapidly with decreasing elevation below 5° (attention is paid to the scale).

The larger the atmospheric refraction, the larger is the daily variation. Accordingly, the elevation to be used for measuring a celestial body is preferably more than a minimum of 10°, or possibly more than 30°. However, attention should be paid to the accuracy of the transit compass, which is worsened if the elevation at which the azimuth is to be measured is increased to 70° or more.

The dip of the horizon, as shown in Fig. 20, is a compensation to be made when the elevation is measured with a sextant using the horizon as a reference. However, such a correction is not needed when a transit compass with a weight is employed to determine the plumb line.

H. Comparison of Surveying Methods of Astronomical Longitude and Latitude

1. Figures 22a and b give a comparison of various surveying methods of astronomical longitude and latitude. Among these methods, the single-altitude method and the Polaris altitude method are useful only when either the latitude or the longitude is previously known.

2. The meridian-altitude method seems to be the easiest in both measurement and calculation. However, prior to measurement, the meridian has to be determined. On the other hand, to determine the meridian, either the longitude or the latitude should be known, except for the equal-altitude method.

Consequently, these three methods have several shortcomings in practice, even though they are valuable theoretically.

3. The same-star-at-equal-altitude method is used to determine the longitude without any previous knowledge of the other parameters. If the altitude of a star passing the meridian can be estimated during this measurement, the latitude can be

estimated using the meridian-altitude method. However, measurement with this method may be time consuming.

4. The universal method by two different stars has advantages of imposing fewer restrictions and requiring less time for measurement than the methods mentioned. However, calculation by this method is very complicated, and hence it is not discussed here.

5. The Polaris altitude method is used frequency to determine a more accurate latitude when the longitude is known. The table of corrections required to determine the latitude from the elevation and the hour angle generally is contained in an astronomical ephemeris and if available, it is used for this method.

The method adopted in this Atlas assumes first direct computation by equations for correction and subsequent use of the calculation chart. The described procedures are considered comparatively simple and easy for obtaining the required results.

9.6 BRIGHT FIXED STAR AND ITS APPLICATION

A. Brightness, Apparent Radius, and Symbols of Celestial Bodies

Figure 23 tabulates the maximum luminosity, the apparent radii, and so on, of the sun, the moon, the planets, and fixed stars (only to Sirius in Canis Major, which is the brightest among the fixed stars). No description is given here of the moon, Uranus, Neptune, and Pluto because the moon is very bright and has frequently been used to locate the position of ships on sea. However, the moon produces a very large parallax (the change in the apparent relative orientations of objects when viewed from two different positions, one at a point of measurement and the other at the center of the earth), on the order of 53 to 61′, as well as a large variation of the declination and the right ascension with time.

Therefore the moon, Uranus, Neptune, and Pluto cannot be used unless the astronomical ephemeris contains extremely detailed data (e.g., hourly data). Uranus, Neptune, and Pluto are so dim (M6 to M15), where M denotes magnitude) that their acquisition is almost impossible.

As already noted, the sun is most useful in a simplified measurement because it is the source of daylight. Hence although it is available in the daytime, it produces very large variations of the declination and the equation of time. The other five planets, that is, Mercury, Venus, Mars, Jupiter, and Saturn, are bright. However, they are useful only when the astronomical ephemeris contains the data for the corresponding year, month, and day.

Bright fixed stars are suitable for use in astronomical surveying.

1. For simplified measurement, the data of the declination and the right ascension of that year (not necessarily of that day) are indispensable since there is only an insignificant change in the positions of fixed stars.
2. Figure 24 shows approximate positions of major fixed stars and may be used if the accuracy permits.
3. Since the fixed stars with luminosity M2 or higher number 88 in the entire sky, there are that many stars available for measurement any time during a clear night.

4. A total of 23 stars is presented, including four M2 stars, ten M1 stars, seven M0 stars, and two M-minus stars.

Since the apparent radius of the sun is approximately 16′, it is necessary to determine an appropriate point within the sun for which the measurement is made, such as the center, the upper limb, or the lower limb.

When the point is set at the upper or the lower limb of the sun, corrections by an amount corresponding to the apparent radius must be made. The apparent radius of the moon is approximately as large as that of the sun, as found in the phenomenon of annular eclipse. The apparent radius is 30″ for Venus, 33″ of Jupiter, and less than 10″ for the other planets. Apparent radii of fixed stars are too small to produce an error in measurement.

The symbols used conventionally in an astronomical ephemeris to indicate individually the sun, the moon, and various planets are also illustrated in Fig. 23 together with brief explanations.

B. Major Fixed Stars and Their Positions

Only four stars, that is, Polaris, α Ursae Majoris, α Cassiopeiae, and β Cassiopeiae, among the 23 fixed stars listed are of M2 luminosity; the remainder are of M1 luminosity or more. The reason why these have been chosen is given in the remarks to Fig. 24. The luminosity of a star is determined in such a manner that the dimmest star perceivable by the naked eye is assumed to be of luminosity M6, and the magnitude is reduced by one unit each time the luminosity is increased 2.51 times; for example, an M2 star is 40 times as luminous as an M6 star, while an M1 star is 100 times as luminous.

These 23 stars can all be easily acquired due to their outstanding features, referring to the star-finding chart as given in Fig. 26a. If a simplified measurement suffices, values of right ascension and declination given in the right column of Fig. 24 can be used directly.

As is seen clearly in the declination column, the data on the 23 stars are arranged in the order of declination, beginning with Polaris (N89.1°).

The figure provides data on 14 stars in the N(+) region and 9 stars in the S(−) region, ending with −63° for α Crucis. It is shown that no appropriate star exists in the vicinity of the celestial south pole.

C. Positions of Major Fixed Stars and Their Application

When the chosen stars are concentrated in a certain region of the sky, the measurement may become difficult in the daytime if the sun comes into or near this region.

Figure 25 illustrates the positions of all 23 fixed stars on a graph with the declination on the ordinate and the right ascension on the abscissa.

The sinusoidal wave drawn in the center of the graph shows the eclipse (the apparent annual path of the sun) with the locations of the sun on the first day of each month. Below the main graph is a graph indicating the period unsuitable for measurement due to an excessive approach by the sun. The stars to be observed should be chosen in such a manner that the measurement schedule excludes unsuitable periods.

The main graph shows the suitable time and period for observation, where the sun is sufficiently far away. The graph also shows the right ascension for each suitable month at which the stars to be observed transit the meridian around 20:00 local mean time.

It is clear from these figures that if the measurement is conducted during a period from 2 hours after sunset until 2 hours before sunrise, at least 5 to 10 fixed stars are available for use in any season.

D. Star Finding Chart

All the fixed stars perceivable belong to one of the constallations. However, usually skill and experience are required to determine constellations except for those with outstanding features. Thus it is rather easy to locate a star from relative positions of known bright stars.

Figure 26 presents star finding charts prepared for this purpose. Figure 26*a* and *b* illustrates the celestial north pole and south pole regions, respectively. The horizontal arrows labeled Horizon, show the altitude of the horizon at a point of observation with latitude *l*. This indicates that an observation of stars falling below such a level, which is dependent on the time (hour, day, year), is impossible.

Figures 26*c* to *f*, the star finding charts covering 23^h to 4^h of the declination, show the relative positions of the stars in suitable charts to be chosen according to right ascension and declination.

Vicinity of Celestial North Pole (Fig. 26a)

When the observation point is in the northern hemisphere (N5° or above, in practice), the most useful fixed star is Polaris.

Since Polaris is located within 1° of the polar distance, the azimuth of Polaris viewed from anywhere with latitude between N0° and N30° will not vary more than +1° at any time. It follows that if an error of 1° or more is tolerable in a simplified measurement, the purpose of measurement will be attained using Polaris without making corrections for refinement. How to find Polaris is indicated at the bottom of Fig. 26*a*. If cloud cover makes Polaris difficult to locate, Ursa Major or Cassiopeia, which are easy to find, can be used.

The right ascensions of these two series of stars differ from each other by approximately 10^h, even when one is below the horizon and the other is above the horizon.

Symbols $\alpha, \beta, \gamma, \ldots$ for stars comprising a constellation are given in general in the order of luminosity, but not in all cases. For example, stars forming Ursa Major are given symbols in locational order beginning at the tip of the Big Dipper. The brightest star in Ursa Major is ε (Alioth 1.7 class), while this atlas adopts (Dubhe 2.0 class) for ease of locating it.

Among three bright stars in Cassiopeia, the β star located at one end of the character W and its neighbor, the α star, are also adopted in this atlas. The α star is designated Shedir, while the β star has no proper name. In this study, this star will be designated β Cassiopeiae after the name of the constellation.

Historically most of the bright stars have been given proper names such as Vega, Altair, and Spica. The M1 or brighter stars having academic names only are

α Centauri	0.1 class	Centaurus
β Centauri	0.9 class	
α Crucis	1.1 class	Crux
β Crucis	1.5 class	

Vicinity of Celestial South Pole (Fig. 26b)

No outstanding stars are found in the vicinity of the celestial south pole. Analogous to Ursa Major and Cassiopeia in the north pole, the constellations Centaurus, Crux, Carina, and Eridanus are scattered in an east-to-west direction around the south pole.

Crux, although it can easily be found due to its outstanding geometrical pattern, cannot be used during the period from September to November because of daylight, as shown in Fig. 25. (However, it can be observed below the meridian when the latitude of the observation site $|L|$ is so high that the stars to be observed are circumpolar stars. The same can apply to the following description.)

The objects to observe from September to November are Eridanus, Fomalhaut, and Canopus. Canopus is the second brightest in the whole sky; the third brightest is α Centauri, which is nearest to the sun (4.3 light-years) and may be regarded as the elder brother of the sun because its true luminosity is 1.7 times as great as that of the sun and it comprises a number of planets and satellites.

Pisces Austrinus, Aquila, and Scorpius (Fig. 26c)

Fomalhaut is at 23^h in the far east in the right ascension and $-30°$ in the south in declination. It is easier to find since no such bright stars exist in and around that region.

In May, June, and July, Scorpius is viewed clearly. East of Antares in Scorpius, a series of bright stars are observed, and this constellation will never be missed due to the darkness of the surrounding star field. A large triangle, as shown in the figure, can be viewed on a summer night.

Cygnus, Lyra, and Aquila (Fig. 26d)

Altair, located at the vertex of an acute triangle, passes the meridian at 20^h local mean time in the beginning and the middle of September, as shown in Fig. 25*a*. Around this time, three stars, namely, Deneb, Vega, and Altair, form an acute triangle.

Vega of magnitude 0.1 is approximately two times brighter than Altair of magnitude 0.9. (Vega and Altair rank 4th and 12th, respectively, in the sky.)

Deneb and Vega can be distinguished from each other not only by their brightness, but also by their position, that is, Deneb may be identified since it forms an almost perfect right triangle together with the other three stars, forming a swan's wing.

Virgo, Bootes, Leo, and Scorpius (Fig. 26e)

To the west of Antares in Scorpius glitter Spica (in Virgo) and Arcturus (in Bootes). Further west is found Regulus and in the vicinity of this star is the autumnal equinox. Hence seasons suitable for observation are spring to early summer. (Arcturus is the fifth brightest among all the fixed stars.)

Around Orion (Fig. 26f)

The region surrounding Orion is compared to a beautifully colored flower garden viewed at night in the winter and spring.

A magnificent brilliance is observed between midnight and dawn in autumn. Three stars situated closely together in the trapezoid formed by four stars resemble a glittering pendant.

Sirius in Canis Major of magnitude -1.6 is simply a glitter of a decoration, and it appears to be heroic and worthy of the Chinese designation "wolf in the sky."

The easiest way to find a neighboring star is to search for it through a step-by-step procedure, forming several triangles starting from Orion, as shown in the figure.

Figure 26*f* presents a skyview of the south (the left corresponds to the east). However, it is only part of the sky. Searching for stars in the vast sky should be carried out calmly.

The solar system is located at the edge of the spiral arms of Orion in the galaxy.

E. Simplified Chart of Fixed Stars in the Sky

Figure 27 presents simplified charts of the constellation, and the principal fixed stars centered at the celestial north pole (N) and the celestial south pole (S). These figures have been drawn on the basis of the *Japan Astronomical Ephemeris*, published by the Maritime Safety Agency. The numerical figure attached to each star is useful in consulting the astronomical ephemeris.

9.7 TIME–ANGLE CONVERSION TABLE

Right ascension α, longitude L, and hour angle, as well as solar time, including mean time, apparent time, standard time, universal time, and sidereal time, are expressed in units of hour (h), minute (m), and second (s), while the angles in trigonometric and inverse trigonometric functions are expressed in degree (°) minutes (′), and seconds (″).

Consequently, conversion from time to angle or from angle to time may become necessary. Figure 28 is used conveniently for this purpose. If this table is not available, the relationship

$$1^{\mathrm{h}} \leftrightarrow 15^\circ \qquad \text{or} \qquad 1^\circ \leftrightarrow 4^{\mathrm{m}} \tag{9.13}$$

is used. The conversion is not so difficult since both time and angle use the sexadecimal number system. Conversion of units less than hour or minute is shown at the bottom of Fig. 3.

9.8 RADIO STARS AND APPLICATIONS

The most familiar celestial body from which the received signal wave can be detected is the sun, and the mechanism of radiation taking place over a broad frequency range varies greatly from time to time according to solar phenomena. Compared with the long time period in astronomical observation made during the past millennia, radio astronomy appears to have just started. However, the presence of a great number of radio-wave sources has so far been confirmed, and this elucidates a number of facts.

A. Kinds of Radio-Wave Sources

Sun

The detection of radio waves from the sun is possible from the earth in the frequency band of 15 MHz to 71.4 GHz. Radio waves distributed over the entire frequency band are continuously illuminating the earth. However, the expansion of a sunspot or the explosion of the solar surface will result in the radiation of extremely intense radio waves, a so-called noise storm or burst.

Galactic Radio Waves

In the galactic system to which we belong, the celestial bodies are distributed within a restricted disk-shaped area with a diameter of 100,000 light-years and a thickness of 15,000 light-years.

Consequently the sources of intense radio waves are distributed along the Milky Way, forming nearly a great circle on the celestial sphere. The most intense soure of radio waves is found around the center of the galactic system, that is, near Sagittarius (east of Antares in Scorpius), as shown in Fig. 25*a*.

Other Radio Stars In or Near the Galactic System

Such radio stars are found in a gaseous nebula or in a planetary nebula located slightly away from the Milky Way. Outstanding among these is a radio star called a pulsar, which emits very regular impulsive radio waves with pulse durations ranging from several to several tens of milliseconds and pulse periods of 0.1 to several seconds. The pulsar was first found in 1967, and about 350 pulsars have been discovered so far. However, details of the pulsars have not yet been reported.

Radio Stars Outside the Galactic System

The most typical radio stars are the radio galaxies and the quasars. The name radio galaxy is given to a number of galaxies, excluding the Milky Way, which emit intense radio waves.

Although they appear to be a small stars when observed through an optical telescope, quasars are galaxies or groups of galaxies which show a red shift in the spectrum, and some of which emit intense radio waves.

The emission of radio waves may be attributable to collisions between the galaxies themselves, the large explosions, etc. However, the details of quasar still remain unknown. The number of quasars discovered since 1960 is approximately 300.

B. Application of Radio Stars

Some of the radio stars emit radio-waves of variable intensity, as seen in the variable stars, while other radio stars emit very stable or regular radio waves.

Pulsars are very interesting radio stars and may find application in the future. However, at present, galactic radio waves in the microwave frequency region with stable emission levels are employed as reference measures for the earth stations of communications satellites.

For example, radio stars such as αCas (Shedir), αCyg (Deneb), and αTau (Aldebaran) are chosen because optically they are easy to find, and much progress has been made on these stars in radio astronomy. The positions of these three stars and the method of finding them are given in Figs. 24, 25, 26*a*, 26*d*, and 26*f*. Recently Venus has drawn attention in this respect, based on a great deal of accumulated data.

Chart Index (1), Astronomy and Geography

Chart Index (2), Astronomy and Geography

Shape and Constants of the Earth

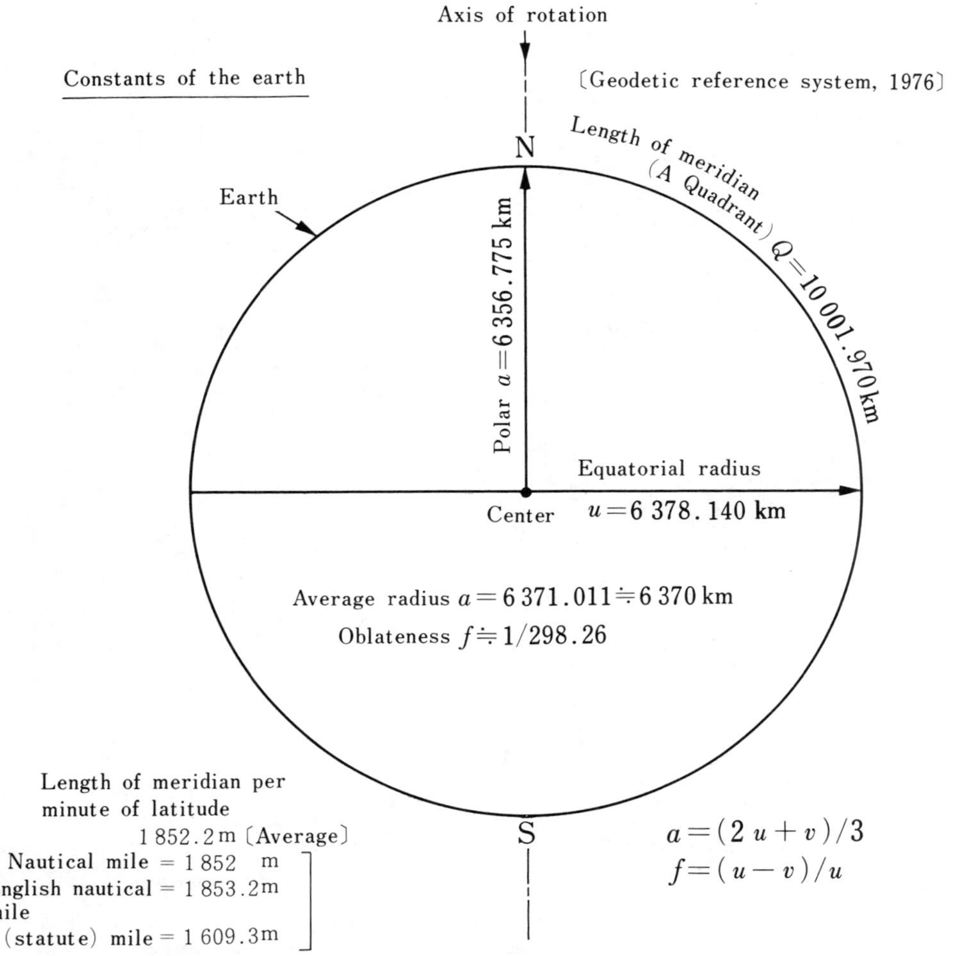

Approximate figures to be used in radio engineering

(1) Earth's radius	a = 6 370 km		Practical figures
(2) Equivalent radius of the earth (K≒4/3)	Ka = 8 500 km		
(3) Length of meridian for a quadrant	Q = 10 000 km		

(4) No distinction is made between the surface of the earth as a spheroid and the geoid.
(5) The undulation of the earth surface, such as mountain, hill and plateau, is not taken into account in the estimation of the earth's radius.

Length Measured along the Arc on the Earth

Length per unit (1°, 1′ or 1″)
of longitude or latitude.

Latitude l (N or S)	Length per unit latitude			Length per unit longitude			Latitude l
	1°	1′	1″	1°	1′	1″	
0°	110.57	1842.9	30.71	111.32	1855.3	30.92	0°
5	110.58	1843.1	.71	110.90	1848.3	.80	5
10	110.61	1843.5	.72	109.64	1827.3	.45	10
15	110.65	1844.2	.73	107.55	1792.5	29.87	15
20	110.70	1845.1	.75	104.65	1744.1	.06	20
25	110.77	1846.2	.77	100.95	1682.5	28.04	25
30	110.85	1847.6	.79	96.49	1608.1	26.80	30
35	110.94	1849.0	.81	91.29	1521.5	25.35	35
40	111.04	1850.6	.84	85.39	1423.2	23.72	40
45	111.13	1852.2	.87	78.85	1314.1	21.90	45
50	111.23	1853.8	.89	71.70	1194.9	19.91	50
55	111.32	1855.4	.92	63.99	1066.6	17.77	55
60	111.41	1856.9	.94	55.80	930.0	15.50	60
65	111.49	1858.2	.97	47.18	786.3	13.10	65
70	111.56	1859.4	.99	38.19	636.4	10.60	70
75	111.62	1860.3	31.00	28.90	481.7	8.02	75
80	111.66	1861.0	.01	19.39	323.2	5.38	80
85	111.69	1861.4	.02	9.74	162.2	2.70	85
90	111.69	1861.6	.02	0.00	0.0	0.00	90
°	km	m	m	km	m	m	°

Length from the equator to each parallel along the meridian

l : Latitude (North or South)

l	Length	l	Length	l	Length	l	Length	l	Length	l	Length
1	110.5	16	1769.6	31	3430.9	46	5096.1	61	6765.5	76	8438.5
2	221.1	17	1880.3	32	3541.8	47	5207.2	62	6876.9	77	8550.2
3	331.7	18	1990.9	33	3652.7	48	5318.4	63	6988.4	78	8661.8
4	442.3	19	2101.6	34	3763.6	49	5429.6	64	7099.8	79	8773.5
5	552.8	20	2212.3	35	3874.6	50	5540.8	65	7211.3	80	8885.1
6	663.4	21	2323.0	36	3985.5	51	5652.1	66	7322.8	81	8996.8
7	774.0	22	2433.8	37	4096.5	52	5763.3	67	7434.3	82	9108.5
8	884.6	23	2544.5	38	4207.5	53	5874.6	68	7545.9	83	9220.1
9	995.2	24	2655.3	39	4318.5	54	5985.9	69	7657.4	84	9331.8
10	1105.8	25	2766.0	40	4429.5	55	6097.2	70	7769.0	85	9443.5
11	1216.4	26	2876.8	41	4540.5	56	6208.5	71	7880.5	86	9555 2
12	1327.0	27	2987.6	42	4651.6	57	6319.9	72	7992.1	87	9666.9
13	1437.7	28	3098.4	43	4762.7	58	6431.3	73	8103.7	88	9778.6
14	1548.3	29	3209.2	44	4873.8	59	6542.6	74	8215.3	89	9890.3
15	1659.0	30	3320.1	45	4984.9	60	6654.1	75	8326.9	90	10002.0
°	km	°	km	°	km	°	km	°	km	°	km

Coordinates of a Point

Longitude & Latitude

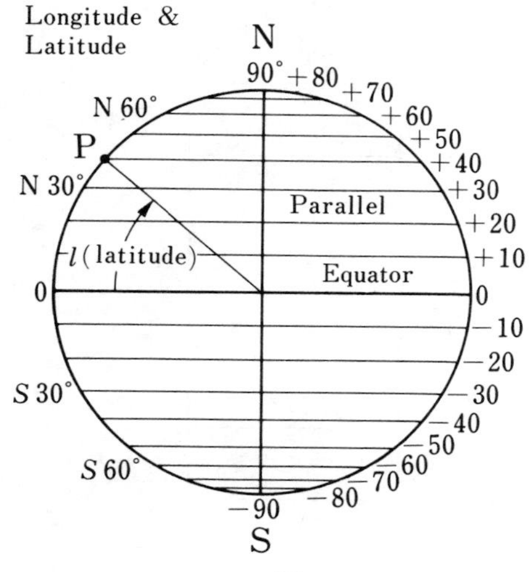

The coordinate of any point on the earth is defined by L or λ longitude and l or φ latitude

The latitude l is expressed with reference to the equator as zero degree as follows. Plus sign for 0° N to 90° N at north pole, minus sign for 0° S to 90° S at south pole

〔Example〕 (N 34° 18′ 20″, + 34° 18′ 20″ / N 34.306°, + 34.306°)

(S 12° 02′ 58″, − 12° 02′ 58″ / S 12.05, − 12.05°)

0 (Greenwich meridian)
(E 30°) 2ʰ — −2ʰ (W 30°)
4ʰ (E 60°) — −4ʰ (W 60°)
L (longitude) N
6ʰ (E 90°) — −6ʰ (W 90°)
8ʰ (E 120°) P — −8ʰ (W 120°)
(E 150°) 10ʰ — −10ʰ (W 150°)
±12ʰ (E W 180°)

The longitude L is expressed with reference to the meridian passing through the royal observatory at Greenwich, England as zero degree as follows.

<Time notation>

Plus sign for eastbound : $0 \sim 12^h$ E

Minus sign for westbound : $0 \sim 12^h$ W

<Angular notation>

Plus sign for eastbound : 0∼180°E

Minus sign for westbound : 0∼180°W

〔Example〕

E $9^h\ 18^m\ 59^s$ = E 139° 44′ 45″

+ $9^h\ 18^m\ 59^s$ = + 139.7458°

Conversion between time and angular notations of a longitude

Time→Angle	$1^h \to 15°$	$1^m \to 15'$ (0.25°)	$1^s \to 15''$ (0.004 167°)
Angle→Time	$1° \to 4^m$	$1' \to 4^s$	$1'' \to 0.067^s$

Map Scale and Application

	Application	Map scale (large←(scale)→small)	Contour spacing
For preliminary study	○ Entire service area ○ Routing plan ○ Macroscopic study on the route	(International) 1/1000 000	—
		1/500 000	200m (100/50)
	○ Spur route ○ Interference study in route branching	1/200 000	100m (50/25)
	○ Macroscopic study on station siting	1/100 000	40m (20/10)
For design	○ Propagation path condition ○ Path profile ○ Field investigation & measurement	1/50 000	20m (10/5)
		1/25 000	10m (5/2.5)
	○ Detailed engineering	1/10 000	5m (2.5/1.5)
		1/5 000	2m (1/0.5)

(1) The large or small reduced scale corresponds to the large or small fraction.

(2) The reduced scale of the map may change depending upon the area comprising a state or local government.

(3) The spacings of contour lines are shown for main curves (middle curve/auxiliary curve). Other than the main curves are omitted according to the topography.

(4) The map projection accepted internationally and used most extensively is transverse Mercator projection.

Method to determine the longitude and latitude of a point.
(Transverse Mercator projection)

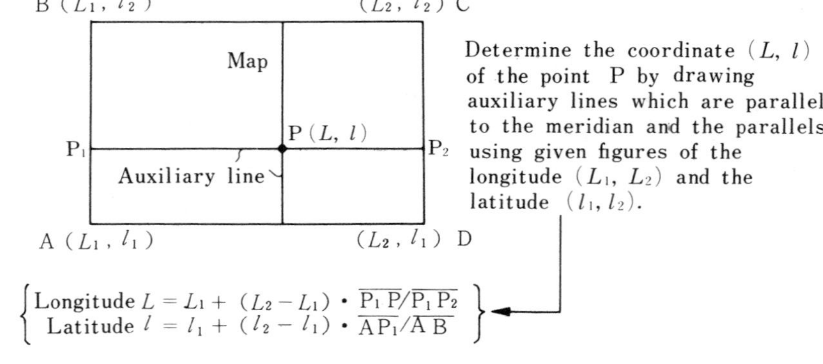

Great-Circle Distance

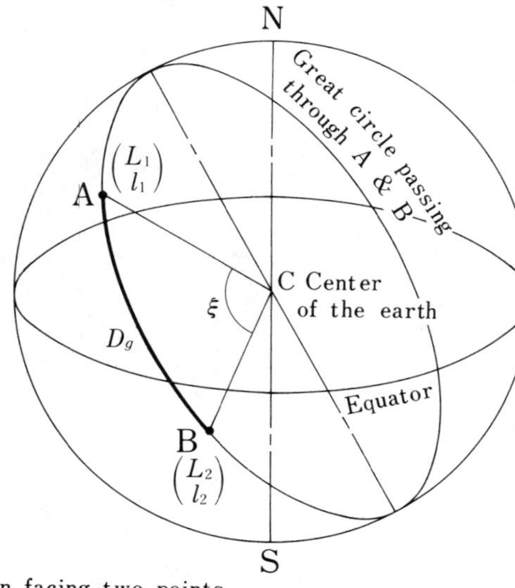

[The intersection (great circle) of the earth and a plane which includes the center of the earth.] Great-circle

[The length of the great circle connects A with B or the shortest distance between A and B on a sphere.] Great-circle distance D_g

[Longitude L (E +, W −); Latitude l (N +, S −)] polarity

Data on facing two points

	Site name · Address (Map)
A	
B	

Longitude / Latitude / Preparatory calculation

			Longitude			Latitude			Preparatory calculation	
A	L_1	+ −	h m s	①				sin l_1	±	⑥
			° ′ ″	②	l_1	+ −	° ′ ″	④	cos l_1 ±	⑦
			. °	③			. °	⑤	tan l_1 ±	⑧
B	L_2	+ −	h m s	①′				sin l_2	±	⑥′
			° ′ ″	②′	l_2	+ −	° ′ ″	④′	cos l_2 ±	⑦′
			. °	③′			. °	⑤′	tan l_2 ±	⑧′

$(L_1 - L_2)$ ± ⑨ — Same magnitude with opposite sign — $\sin(L_1 - L_2)$ ± ⑩

$(L_2 - L_1)$ ± ⑨′ — $\sin(L_2 - L_1)$ ± ⑩′

$\cos(L_1 - L_2) = \cos(L_2 - L_1)$ ± ⑪

Equation of great-circle distance

$$D_g = 6\,370 \times \cos^{-1}[\sin l_1 \sin l_2 + \cos l_1 \cos l_2 \cos(L_1 - L_2)]$$

⑥ × ⑥′ ⑦ × ⑦′ × ⑪

$= 6\,370 \times \cos^{-1}[$ ± ___ + ± ___ $]$

$= 6\,370 \times \cos^{-1}$ ___ (When the results are obtained in degrees, convert them into radians.)

$= 6\,370 \times$ ___

Note:
$1° = 0.017\,453\,2^{rad}$
$1^{rad} = 57.295\,78°$

Symbols ○ Newly obtained figure ◎ Quotation marks

$D_g =$ ___ km (If $D_g < 20$ km, read the distance directly on the map or measure it in the field.)

Great-circle distance

Geographic Azimuth

Calculation Chart

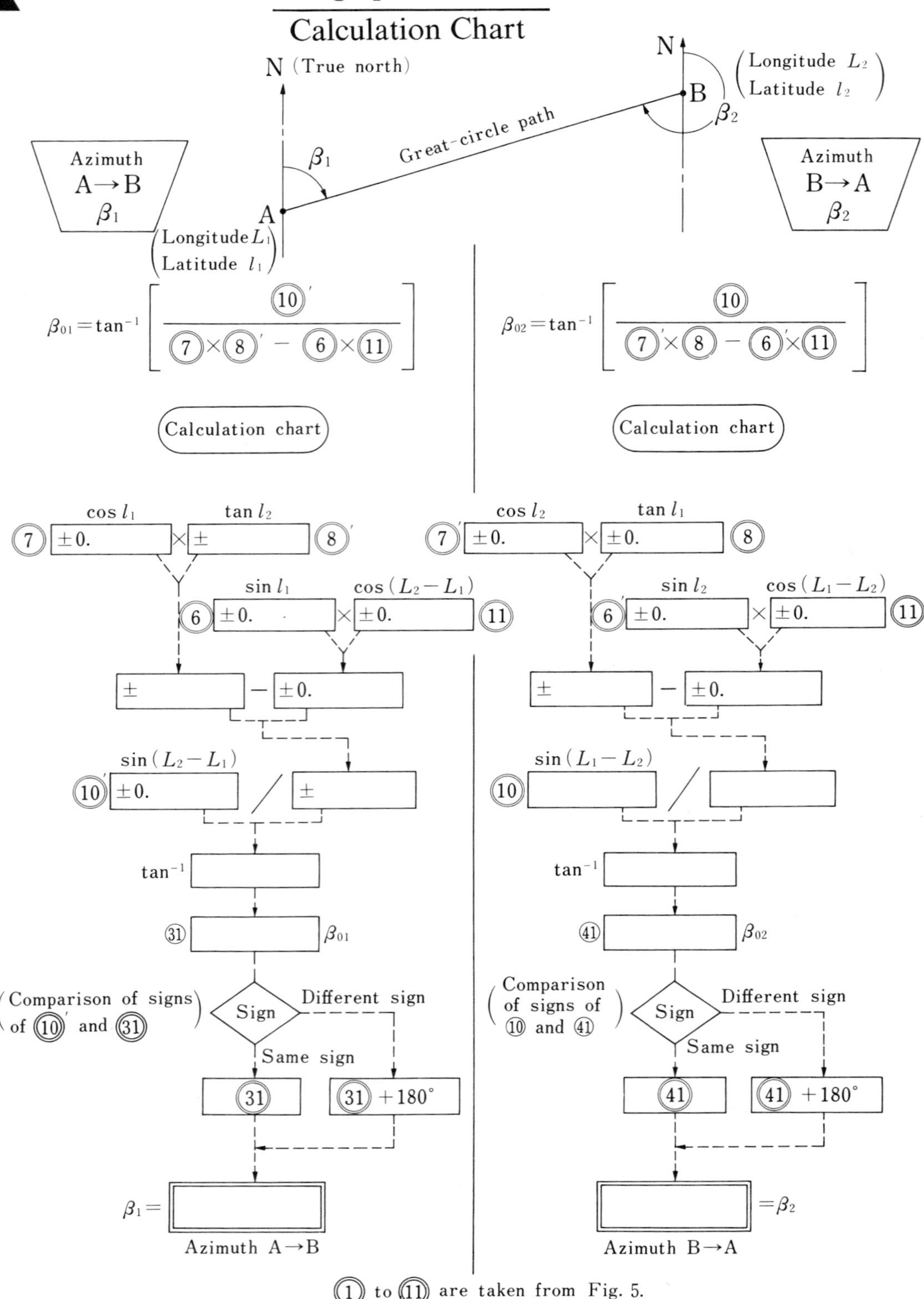

① to ⑪ are taken from Fig. 5.

Geographic Azimuth

Basic Equations and Judgment Table

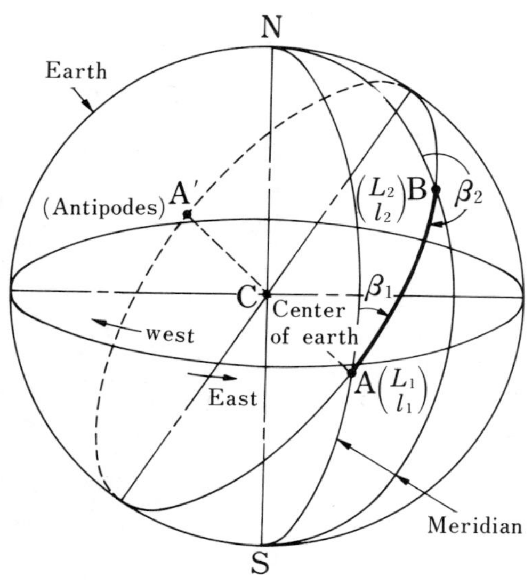

Azimuth

(1) Horizontal angle measured clockwise from the north direction of the meridian to the distant point (0 to 360°).

(2) The angle may be measured clockwise (0 to +180°) or counterclockwise (0 to −180°) from the north direction to the distant point.

Azimuths β_1 (A→B) and β_2 (B→A) are determined from the equations and the judgment tables given below.

Basic equations

$$\beta_{01} = \tan^{-1}\left[\frac{\sin(L_2 - L_1)}{\cos l_1 \tan l_2 - \sin l_1 \cos(L_2 - L_1)}\right] \quad (\text{A}\rightarrow\text{B})$$

$$\beta_{02} = \tan^{-1}\left[\frac{\sin(L_1 - L_2)}{\cos l_2 \tan l_1 - \sin l_2 \cos(L_1 - L_2)}\right] \quad (\text{B}\rightarrow\text{A})$$

Judgment table

General

A→B B→A	$\sin(L_2 - L_1)$ $\sin(L_1 - L_2)$ Ⓜ	β_{01} β_{02} Ⓝ	$\beta_1 =$ $\beta_2 =$	Remark
Comparison of sign	+ (Facing east)	+	β_{01} , β_{02}	If the sign of Ⓜ differs from that of Ⓝ, add 180° to the result of the basic equation.
		−	β_{01} + 180°	
	− (Facing west)	+	β_{02} + 180°	
		−	β_{01} , β_{02}	

Special

Points A and B on the same parallel ($L_1 = L_2$)	Azimuth : 0° or 180°
Points A and B are on the equator ($l_1 = l_2 = 0$)	Azimuth : 90° or 270°
Points A and B are the antipodes of each other	Azimuth : Not fixed

9

7a

Measurement of Distance (1)

Direct Distance

Establish the baseline $\overline{AB}(=r)$, measure the angles α & β, then calculate the distance x.

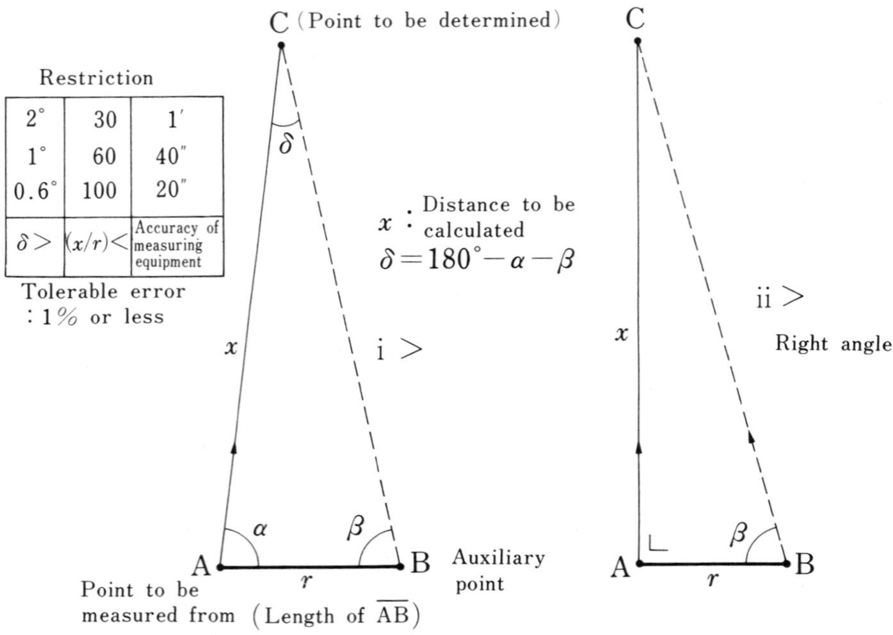

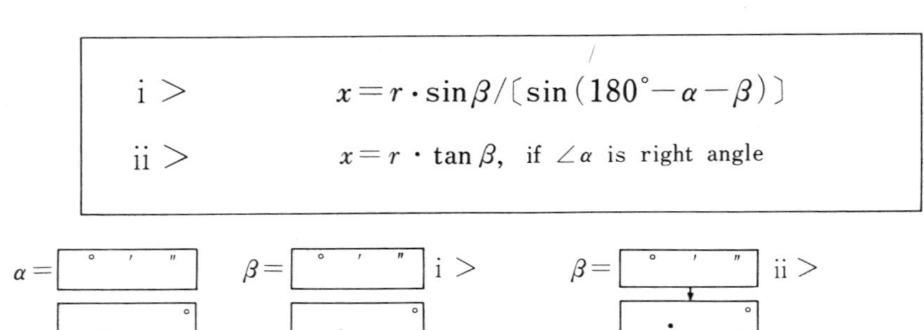

i > $x = r \cdot \sin\beta / [\sin(180° - \alpha - \beta)]$

ii > $x = r \cdot \tan\beta$, if $\angle\alpha$ is right angle

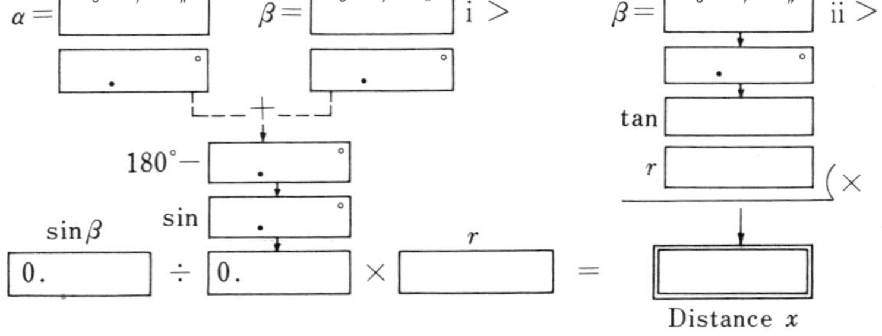

Measurement of Distance (2)

Between Two Distant Points

Establish the baseline $\overline{AB}$ $(=r)$ and measure α_1, α_2, β_1, & β_2 from points A and B, then calculate x.

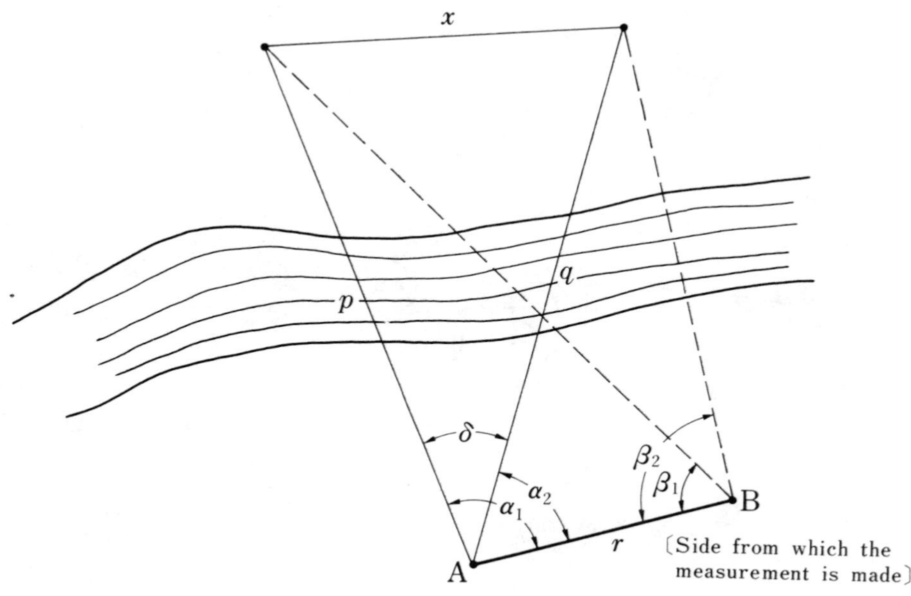

$$p = r \cdot \sin\beta_1 / [\sin(180° - \alpha_1 - \beta_1)]$$

$$q = r \cdot \sin\beta_2 / [\sin(180° - \alpha_2 - \beta_2)]$$

$$\delta = \alpha_1 - \alpha_2$$

$$x = \sqrt{p^2 + q^2 - 2\,pq\cos\delta}$$

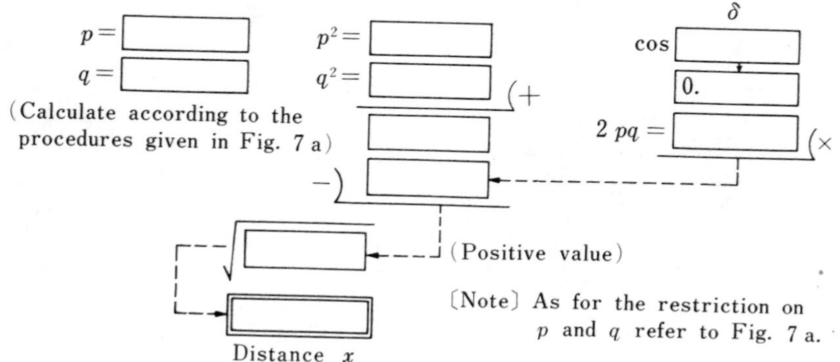

〔Note〕 As for the restriction on p and q refer to Fig. 7 a.

Measurement of Elevation (1)

One-Way Observation

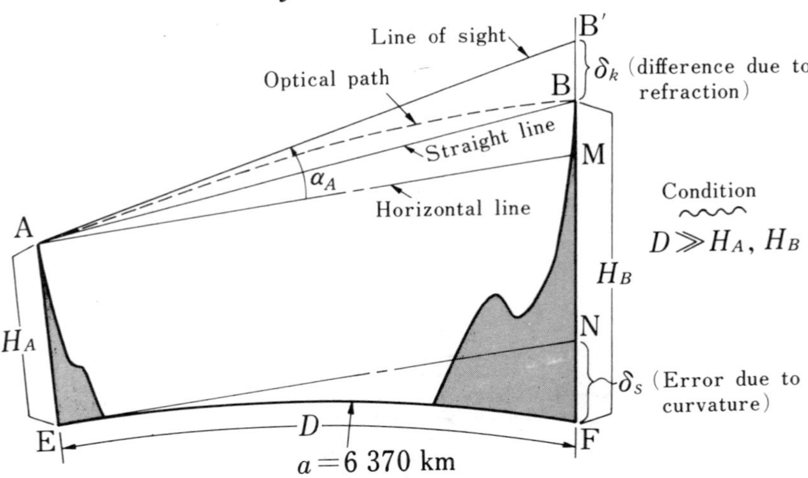

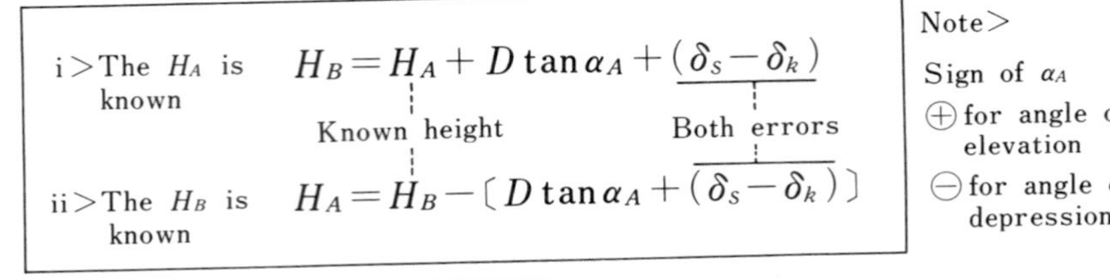

i > The H_A is known $\quad H_B = H_A + D\tan\alpha_A + (\delta_S - \delta_k)$

Known height $\quad$ Both errors

ii > The H_B is known $\quad H_A = H_B - [D\tan\alpha_A + (\delta_S - \delta_k)]$

Note>

Sign of α_A

⊕ for angle of elevation

⊖ for angle of depression

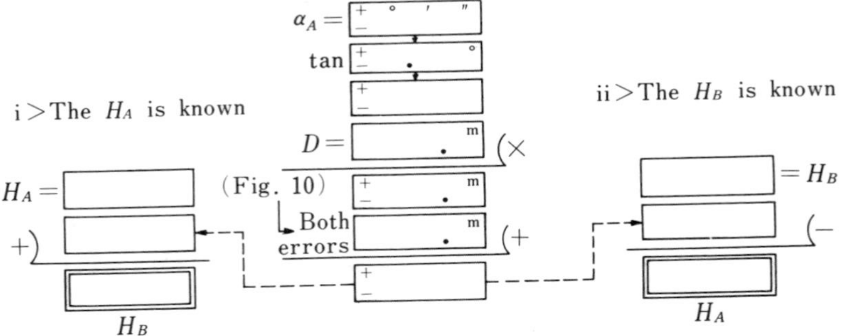

[Remark] As clearly seen on the figure.

$$\overline{MN} = H_A,\ \overline{MB'} \fallingdotseq D\tan\alpha_A$$

$$H_B = \overline{MN} + \delta_s + \overline{MB}$$

$$= H_A + \delta_s + (\overline{MB'} - \delta_k)$$

$$= H_A + D\tan\alpha_A(\delta_s - \delta_k)$$

Measurement of Elevation (2)

Two-Way Observation

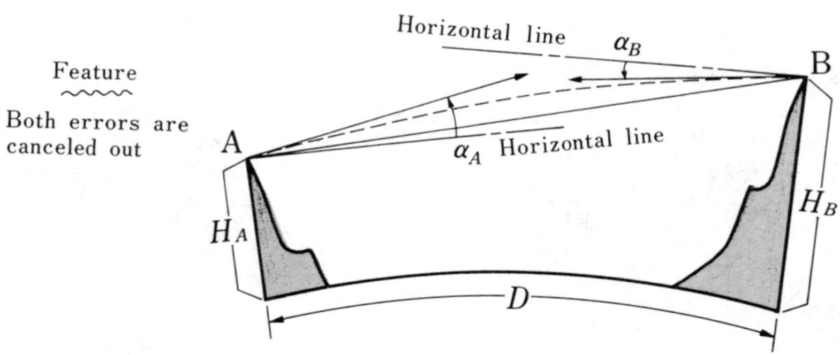

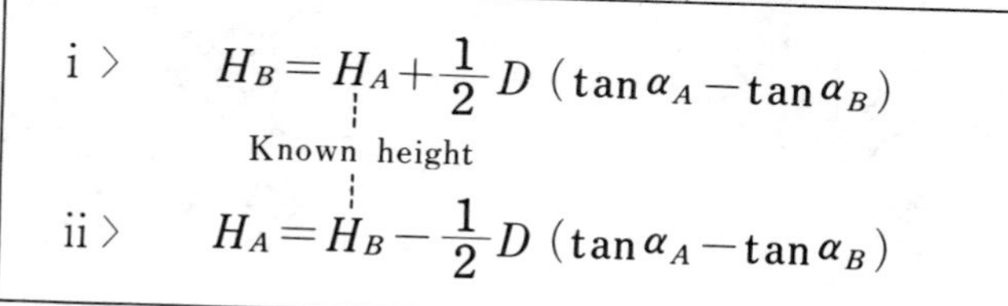

i 〉 $H_B = H_A + \frac{1}{2} D (\tan\alpha_A - \tan\alpha_B)$

Known height

ii 〉 $H_A = H_B - \frac{1}{2} D (\tan\alpha_A - \tan\alpha_B)$

Note:

Sign of α_A and α_B

⊕ for angle of elevation

⊖ for angle of depression

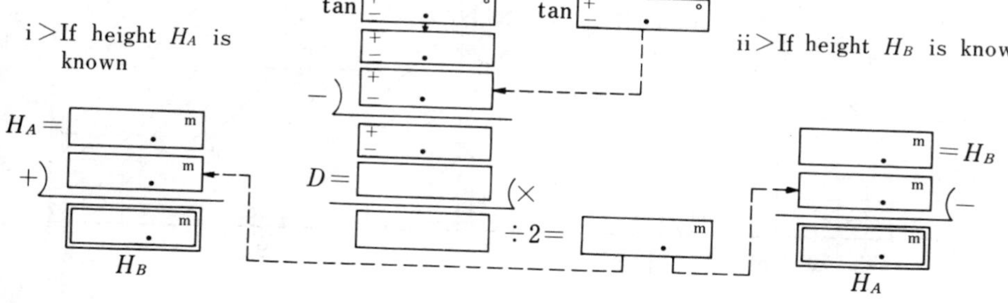

〔Remark〕

<A→B observation> $H_B = H_A + D\tan\alpha_A + (\delta_s - \delta_k)$ ①

<B→A observation> $H_A = H_B + D\tan\alpha_B + (\delta_s - \delta_k)$ ②

Application of the equations in Fig. 8 a

from ② $H_B = H_A - D\tan\alpha_B - (\delta_s - \delta_k)$ ③

(①+③) ÷ 2

$$H_B = H_A + \frac{1}{2} D(\tan\alpha_A - \tan\alpha_B)$$

Distance Increment Due to a Level Difference between Two Points Referred to the Great-Circle Distance

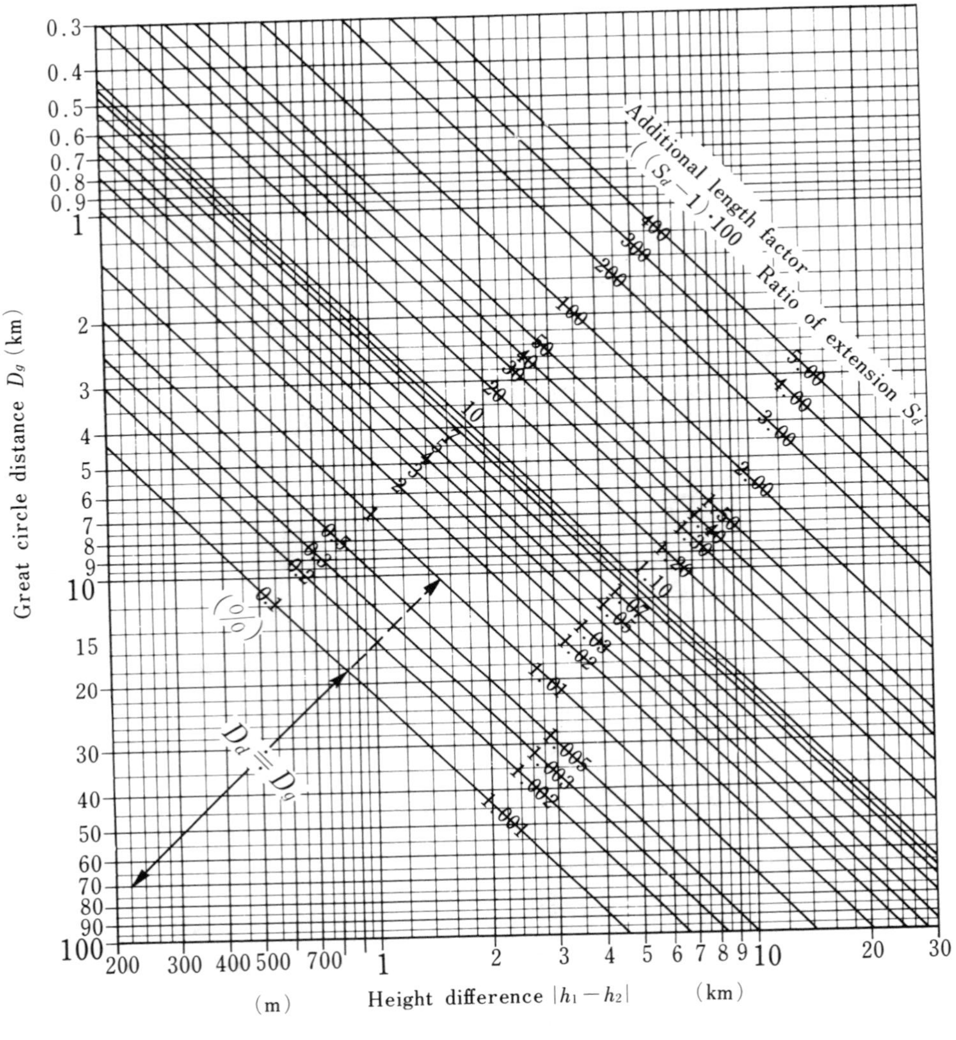

Great circle distance D_g (km)

0.3, 0.4, 0.5, 0.6, 0.7, 0.8, 0.9, 1, 2, 3, 4, 5, 6, 7, 8, 9, 10, 15, 20, 30, 40, 50, 60, 70, 80, 90, 100

Height difference $|h_1 - h_2|$

200, 300, 400, 500, 700 (m); 1, 2, 3, 4, 5, 6, 7, 8, 9, 10, 20, 30 (km)

Additional length factor $((S_d - 1) \cdot 100$ (%))

0.1, 0.2, 0.3, 0.5, 1, 2, 3, 4, 5, 10, 20, 30, 40, 50, 100, 200, 300, 400

Ratio of extension S_d

1.001, 1.002, 1.003, 1.005, 1.01, 1.02, 1.03, 1.05, 1.04, 1.10, 1.20, 1.30, 1.40, 1.50, 2.00, 3.00, 4.00, 5.00

$D_d \fallingdotseq D_g$

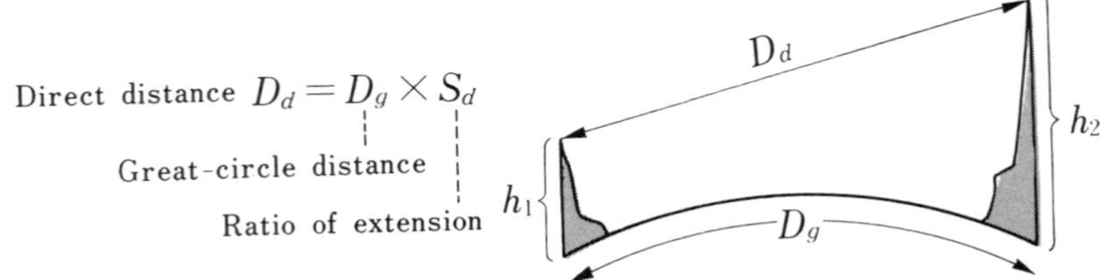

Graph Showing Error Due to Both Earth Curvature and Atmospheric Refraction

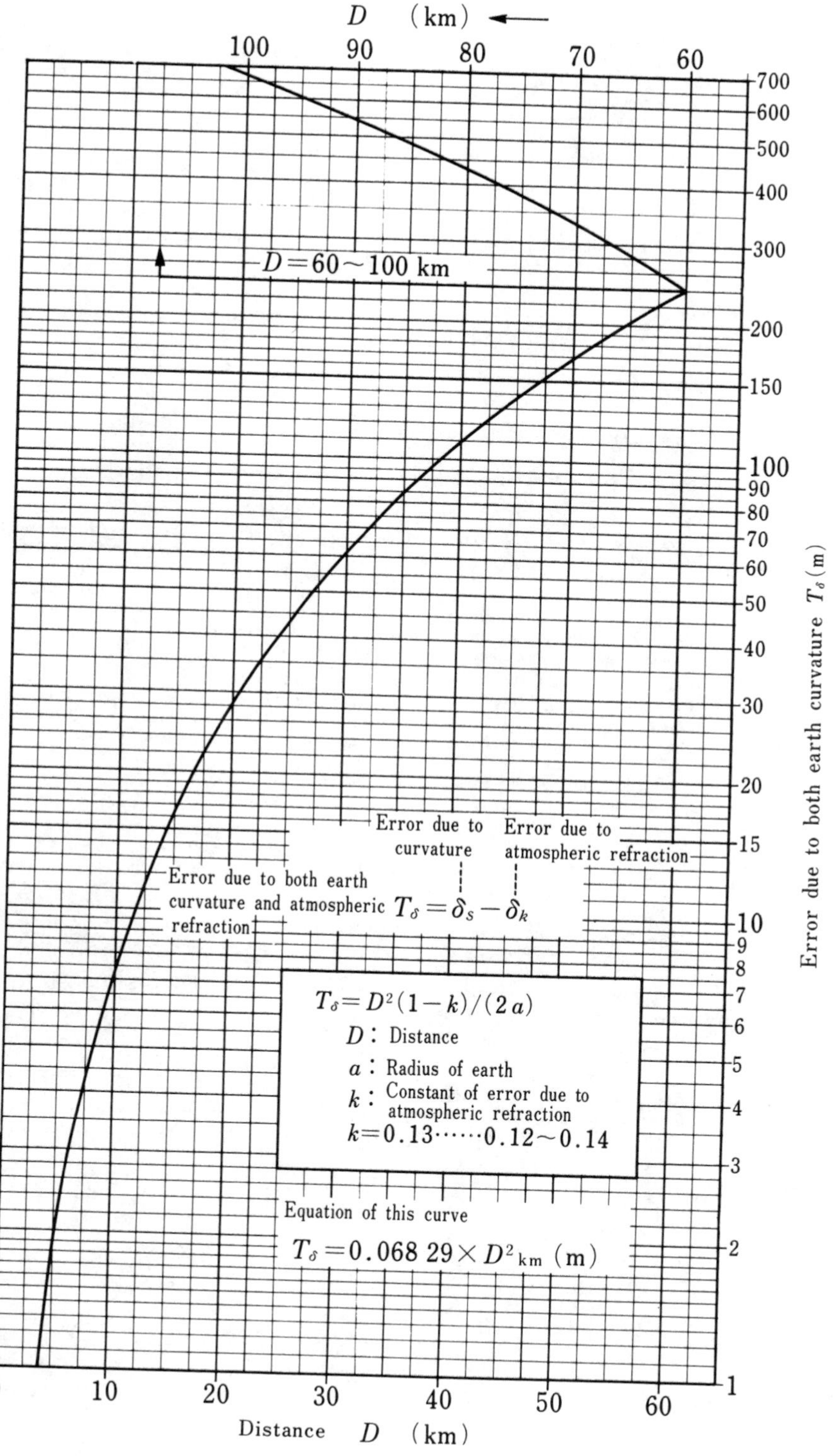

Celestial Sphere (1)

Definitions of Astronomical Terms

a. Celestial sphere — The distances between the earth and all the celestial bodies are extremely great, therefore the difference of the distance can not be recognized by observers on the earth and only the directions are recognized.
In other words, all the celestial bodies appear to be positioned on the inner surface of a tremendously large sphere with its center at the earth, and this sphere seems to be rotating once a day from the east to the west (diurnal movement).
Such an imaginary sphere is called the celestial sphere.

b. Celestial pole — The diurnal movement itself of the celestial sphere is a manifestation of the earth's rotation. Either of two points of intersection of the celestial sphere and the extension of the rotation axis of the earth is called the celestial pole.
These two celestial poles (celestial north pole and celestial south pole) are fixed points on the celestial sphere.

c. Celestial equator — The great circle on the celestial sphere with an equal distance from the celestial north and south poles is called celestial equator. It is equivalent to the intersection of the celestial sphere and the extended plane of the equator of the earth.

d. Zenith — The intersection of the celestial sphere and the line of gravity force at any point of the earth is called the zenith.

e. Hour circle — A great circle passing through both celestial poles is called the hour circle.

f. (Celestial) Meridian — The hour circle which passes through the observer's zenith is called meridian.

g. Ecliptic — Although any fixed star remains at an almost fixed point on the celestial sphere due to its extremely great distance in comparison with the size of the solar system, celestial bodies of the solar system, i.e., the sun, the moon and planets, change their positions daily according to their own movement and the earth's rotation. For example, the sun moves on the celestial sphere 1° per day to the east, rounding the sphere in one year. The apparent annual path of the sun on the celestial sphere is called the ecliptic. (The ecliptic forms an approximate great circle.)

h. Vernal equinox — The point at which the sun moves from the south to the north of the celestial equator is the vernal equinox or first point of Aries, while the point at which the sun moves from the north to the south of the celestial equator is the autumnal equinox or first point of Libra. The former point is used as the coordinate origin to fix the positions on the celestial sphere.
The acute angle between the ecliptic and the celestial equator, approximately 23° 26′, is called the obliquity of the ecliptic and such obliquity produces four seasons, i.e., spring, summer, autumn and winter.

i. Right ascension (α) — The right ascension (α) of a celestial body is defined as the angle included by the hour angle of that celestial body and the hour angle of the first point of Libra expressed in hours at a rate of 24 hours (360°) to the east in the opposite direction to the movement of the celestial bodies.
Similarly to the longitude referred to the Greenwich meridian, the right ascension is measured with reference to the first point of Aries in the celestial sphere.

j. Declination (δ) — As in the case of the latitude on the earth the angle of a celestial body measured from the celestial equator (0°) is the declination (d or δ).
The declination will take a figure 0° to +90° from the celestial equator to the north pole and 0° to −90° from the celestial equator to the south pole.

k. Hour angle (h) — The angle included by the hour circle (meridian) passing through the zenith at a point of observation and the hour circle passing through a celestial body, expressed in hours (at a rate of $24^h/360°$ to the west (forward direction of the celestial body), is called the hour angle (h).

Celestial Sphere (2)

Diagram of Celestial Sphere

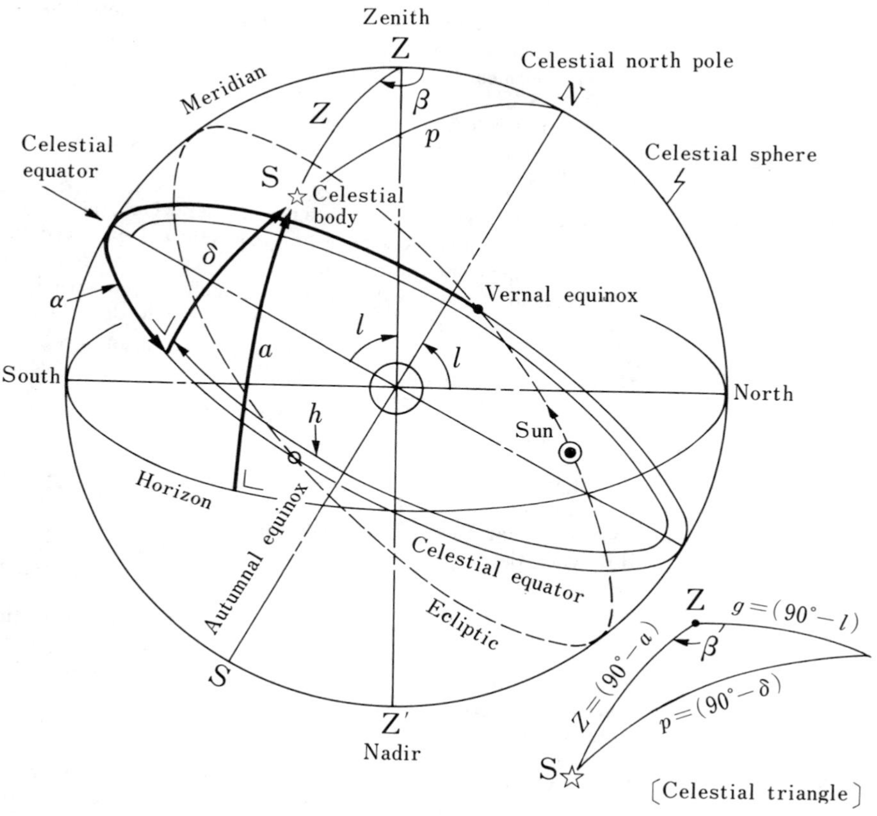

〔Celestial triangle〕

Intrinsic values of celestial bodies (according to astronomical ephemeris)	α : Right ascension	
	δ : Declination	p : Polar distance ($90° - \delta$)
Observation point depending on the time of observation	a : Altitude	Z : Zenith distance ($90° - a$)
	h : Hour angle	

β : Astronomical azimuth — An angle (0~360°) measured from the meridian north clockwise, from the great circle passing through a celestial body and the zenith.

Note : 1) The angle which includes the zenith and the celestial equator equals the latitude of the point of observation or the angle included by the horizon and the celestial pole (the altitude of the celestial pole).

2) The celestial triangle shown in the figure forms a base on which the longitude and the latitude of the point of observation, the celestial azimuth, etc., are measured.

Bases of Time (Sidereal Time, Solar Time)

a. Meridian passage — Meridian passage is defined as the passage of a celestial body across an observer's meridian.

b. Time — The time is measured by the hour angle referring to the culmination of the sun, mean apparent sun or vernal equinox making use of the diurnal movement (actually the rotation of the earth) of the celestial sphere.
If the sun is used, it is called solar time (apparent solar time and mean solar time) while if the vernal equinox is used, it is called sidereal time.

c. Sidereal time — The interval of successive instants (0^h) at which the vernal equinox passes the meridian is defined as one sidereal day, and the time system in which one sidereal day is expressed by 24^h is called sidereal hour. Sidereal hour=Hour angle of the vernal equinox.
Local sidereal time θ_t : Sidereal time in which the culmination of the vernal equinox at an arbitrary location is assumed to be 0^h
Greenwich sidereal time θ_g : The sidereal time in which the culmination of the vernal equinox on the Greenwich meridian (0° of longitude) is assumed to be 0^h.

d. Solar time — Sum of the hour angle of the sun and the time determined by setting the time the sun (apparent sun) passes the meridian at 0^h is called apparent solar time or apparent time.
However, the hour angle of the sun does not proceed at a constant rate because the orbit of the earth takes a form of ellipse and the revolution axis is inclined.
Consequently we normally use mean solar time (or mean time). The time system based on the sun's motion in the year's interval between two successive passages of the sun at the vernal equinox.
This mean solar time may be obtained adding the hour angle of the fictitious mean sun to 12 moon at which hour the mean sun transits the meridian.
Local Apparent Time (L.A.T) : to be used when the time is defined according to the apparent sun and the meridian at the position of an observer.
Local Mean Time (L.M.T) : to be used when the time is defined according to the mean sun and the meridian at the position of an observer.

e. Mean time — Universal Time (*U. T*) : Local mean time at the Greenwich meridian (at 0° of longitude)
Standard Time (*S. T*) : Local mean time defined at a specific meridian (For example 135°(=9^h) of longitude in Japan) and commonly employed in a country or in a region.

Note : Attention should be given to the sidereal time defined by fixing the meridian passage at 0^h and also for the mean time of the mean sun defined by fixing the meridian passage at 12^h.

Time and Angles Relating to Celestial Bodies

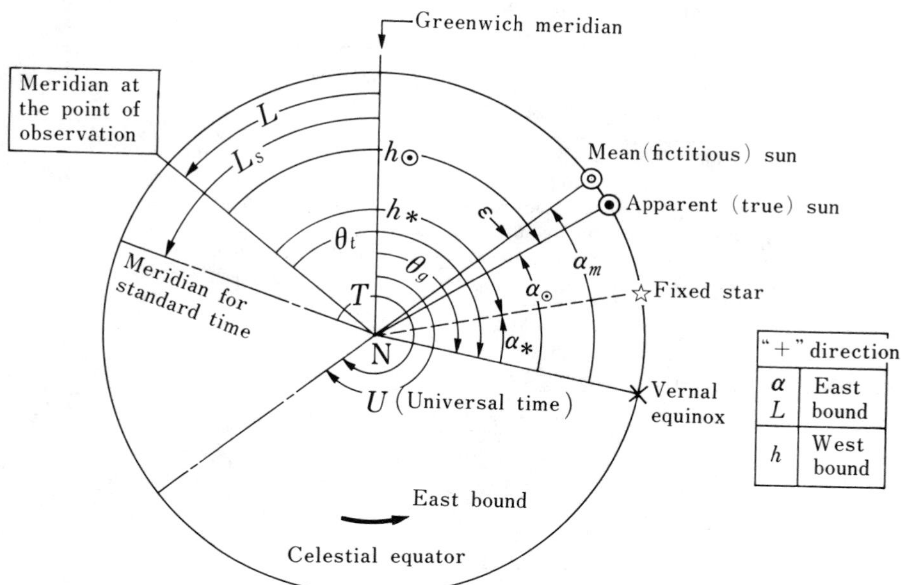

"+" direction	
α, L	East bound
h	West bound

[Basic equations]

Hour angle (Fixed star) $h_* = \theta_g - \alpha_* + L$

(Greenwich sidereal time, Right ascension, Longitude, Universal time)

Hour angle (sun) $h_\odot = \theta_g - \alpha_\odot + L = U + L + \varepsilon - 12^h$

(ε: Equation of time)

Universal time $U = T - L_S$

Equation of time $\varepsilon = \alpha_m - \alpha_\odot$

θ_t : Local sidereal time
θ_g : Greenwich sidereal time
α_* : Right ascension of fixed star
$\alpha_\odot$: Right ascension of apparent sun
α_m : Right ascension of mean sun
L : Longitude
ε : Equation of time
h_* : Hour angle of fixed star
$h_\odot$: Hour angle of apparent sun

T : standard time
L_s : Longitude of meridian for standard time (9^h=E 135° in Japan)

Relation between Local Sidereal Time and Standard Time

[1] Solar year: Time interval between two successive transits of the vernal equinox by the sun.

Solar year	= 365.2422	Mean solar day
	366.2422	Sidereal day

The sun travels eastward by approximately one degree per day. Therefore the meridian transit of the vernal equinox will quicken by approximately 4 minutes every day. This phenomenon is caused by the earth's revolutions ; the mean solar day numbers are less than the Besselian day numbers by one day because the earth makes a revolution once a year in its rotating direction.

[2] Relation between mean solar time and sidereal time

1 mean solar day	= 366.2422/365.2422	
	= 1.002 737 91	Sidereal day
	= $24^h 03^m 56.555^s$	Sidereal time
1 sidereal day	= 365.2422/366.2422	
	= 0.997 269 57	Mean solar day
	= $23^h 56^m 04.091^s$	Mean solar time

[3] Conversion equations of time period

$\Delta\theta$ ······ Time period of sidereal time
Δt ······ Time period of mean solar time

Basic equation

$$\Delta\theta = \Delta t \ (1 + k')$$

k' = 0.002 737 91

$$\Delta t = \Delta\theta \ (1 - k)$$

k = 0.002 730 43

Application

$$\theta_t = \theta_o + (T - L_s)\ (1 + k') + L$$

θ_t: Local sidereal time; $(T - L_s)$: Δt

$$T = (\theta_t - \theta_o - L)\ (1 - k) + Ls$$

T: standard time; $(\theta_t - \theta_o - L)$: $\Delta\theta$

θ_t : Local sidereal time
θ_o : Greenwich sidereal time at $0^h = U$ universal time
L : Longitude
L_S : Longitude at which the standard time is defined
T : Standard time
$(T - L_s)$: Universal time

Calculation of Hour Angle (General Method)

Although this method may be applied to fixed stars, the moon and sun in general, another method as shown in Fig. 16 will be more convenient for the sun.

[1] Basic equation for the hour angle

	Symbol
L_S	Longitude (E)→+
L_S	Longitude (W)→−
α	East wound →+
h	West wound →+

Local sidereal time

Right ascension

$$h=\theta_t-\alpha$$
$$=\theta_0+[1+k'][T-L_S]+L-\alpha$$

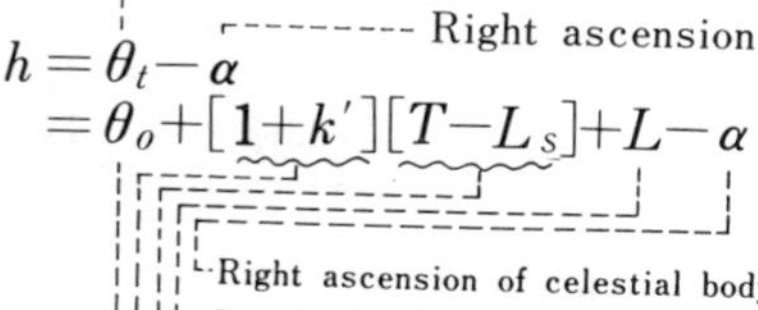

- Right ascension of celestial body
- Longitude of a point where the observation is made.
- Universal time, U
  - T : Time of the observation (standard time).
  - L_S : Meridian longitude to define the standard time.
- Mean hour→Sidereal time conversion conficient
  - $k'=0.002\,737\,91$
- Greenwich sidereal time at $U=0^h$ (to be obtained referring to an astronomical ephemeris).

θ_t = Local sidereal time (Sidereal time against the meridian at the point of observation).

[2] Calculation chart

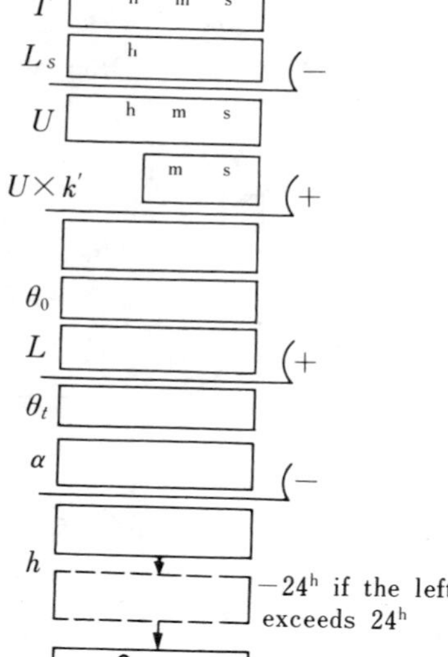

(Example)

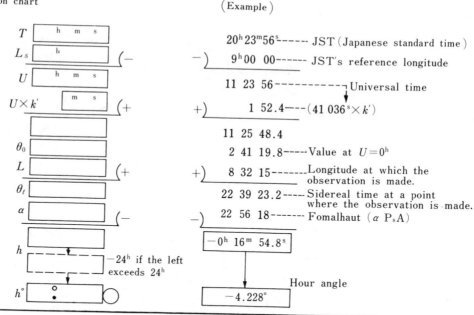

	$20^h23^m56^s$	JST (Japanese standard time)
−)	$9^h00\ 00$	JST's reference longitude
	11 23 56	Universal time
+)	1 52.4	($41\,036^s\times k'$)
	11 25 48.4	
	2 41 19.8	Value at $U=0^h$
+)	8 32 15	Longitude at which the observation is made.
	22 39 23.2	Sidereal time at a point where the observation is made.
−)	22 56 18	Fomalhaut (α PₛA)
	$-0^h\ 16^m\ 54.8^s$	
	$-4.228°$	Hour angle

9

16

Hour Angle and Declination of the Sun

[1] Hour angle of the sun $h_\odot$

General procedure to calculate the hour angle of the sun when the equation of time and the declination (apparent declination) of the day (at 0^h in universal time) of observation is given as below.

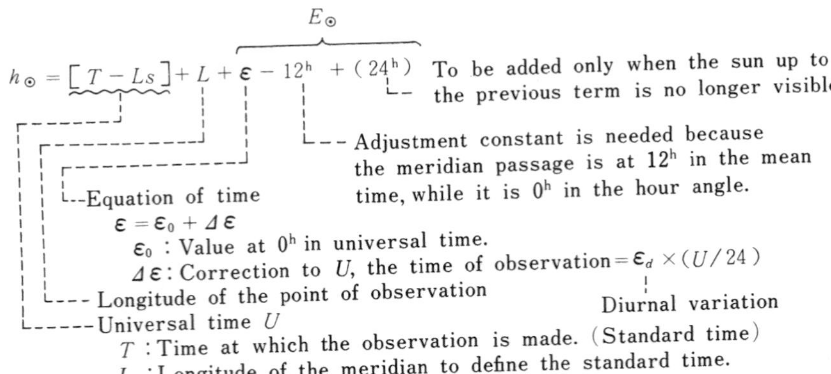

$E_\odot$
$h_\odot = [T - L_S] + L + \varepsilon - 12^h + (24^h)$
To be added only when the sun up to the previous term is no longer visible.
Adjustment constant is needed because the meridian passage is at 12^h in the mean time, while it is 0^h in the hour angle.
Equation of time
$\varepsilon = \varepsilon_0 + \Delta\varepsilon$
ε_0 : Value at 0^h in universal time.
$\Delta\varepsilon$: Correction to U, the time of observation $= \varepsilon_d \times (U/24)$
Diurnal variation
Longitude of the point of observation
Universal time U
T : Time at which the observation is made. (Standard time)
L_S : Longitude of the meridian to define the standard time.

[2] Declination of the sun $\delta_\odot$

$$\delta_\odot = \delta_0 + \Delta\delta$$

- $\Delta\delta$ — Correction to U, the time of observation $= \delta_d \times (U/24.)$ (δ_d: Diurnal variation)
- δ_0 — Value at 0^h in universal time

[3] Calculation chart

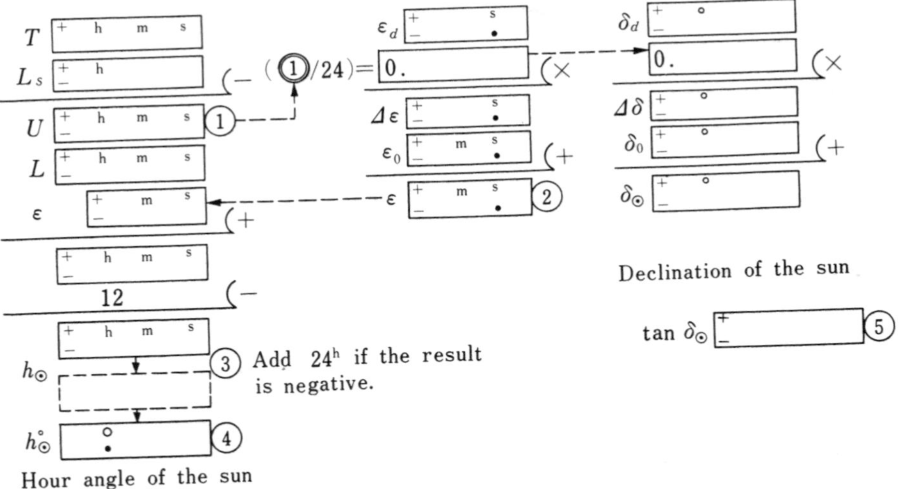

Relation between Various Parameters in Astronomical Measurement

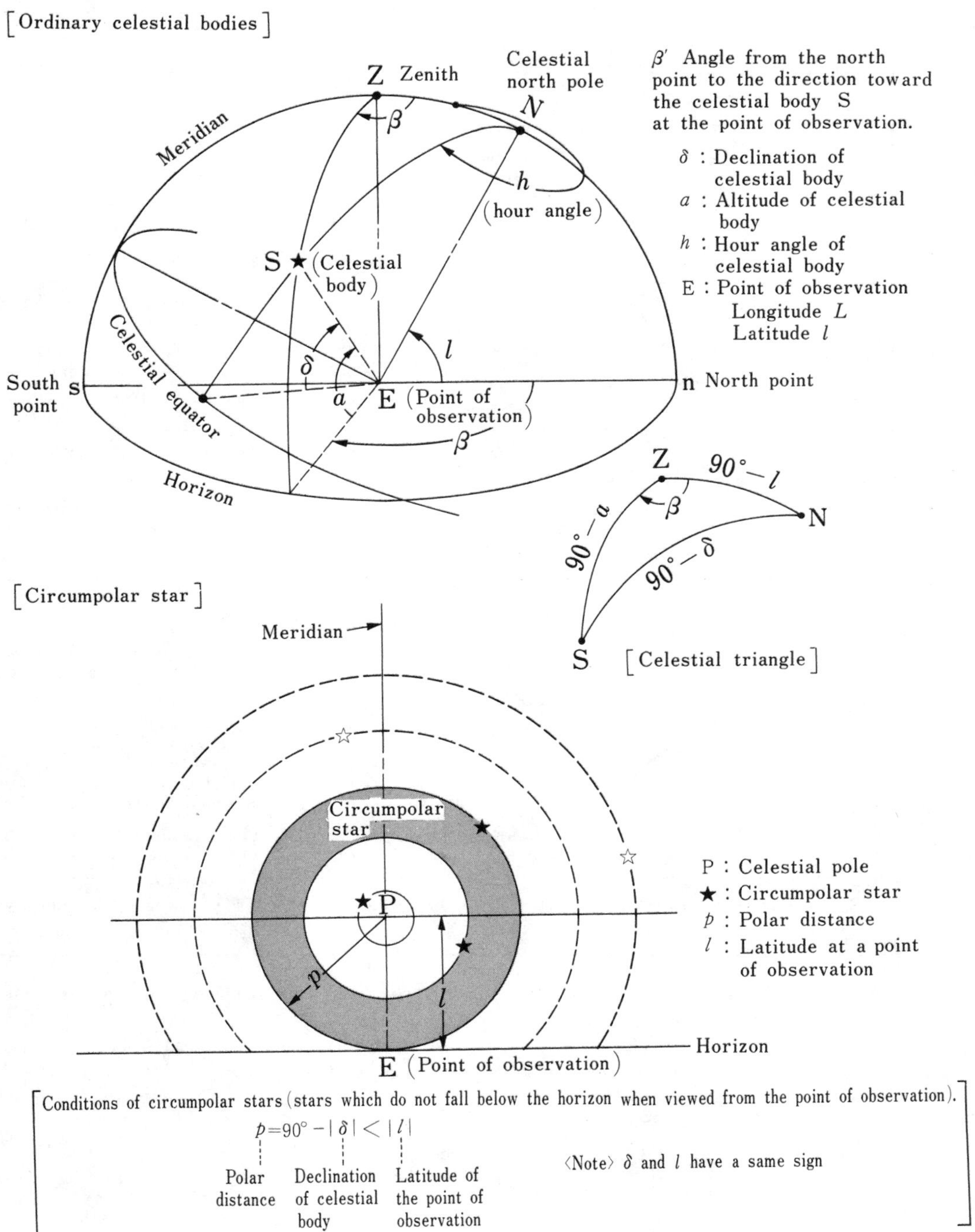
[Ordinary celestial bodies]
Z Zenith
Celestial north pole
N
β'
Meridian
h
(hour angle)
S ★ (Celestial body)
Celestial equator
δ
a
l
South point s
E (Point of observation)
n North point
β
Horizon
β' Angle from the north point to the direction toward the celestial body S at the point of observation.
δ : Declination of celestial body
a : Altitude of celestial body
h : Hour angle of celestial body
E : Point of observation
Longitude L
Latitude l
Z
90°−l
β'
N
90°−a
90°−δ
S
[Celestial triangle]
[Circumpolar star]
Meridian
Circumpolar star
★P
p
l
Horizon
E (Point of observation)
P : Celestial pole
★ : Circumpolar star
p : Polar distance
l : Latitude at a point of observation
Conditions of circumpolar stars (stars which do not fall below the horizon when viewed from the point of observation).
p=90° − | δ | < | l |
Polar distance
Declination of celestial body
Latitude of the point of observation
〈Note〉 δ and l have a same sign

Comparison of Methods to Determine the Astronomical Azimuth (1)

	Item	Summary of method		Advantages and disadvantages
〔A〕	General method (I)	Meridian; β; Point of observation	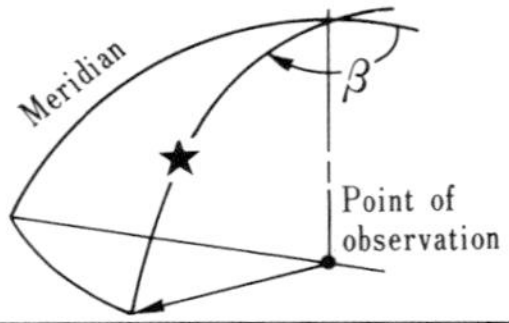 Mark the azimuth of a celestial body on the ground at an arbitrary time and also record the standard time T at which the observation is made.	(1) It is enough to record the observation time. (2) An arbitrary celestial body (sun, various types of fixed stars) may be used any time.
〔B〕	General method (II)	Meridian; β; a_t	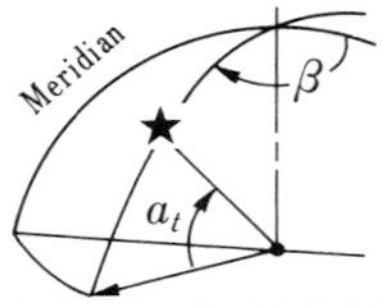 Measure the altitude of a celestial body and mark the azimuth on the ground at an arbitrary time.	(1) Measurement of only the altitude of a celestial body suffices. (2) An arbitrary celestial body can b used at an arbitrary time.
〔C〕	Equal altitude method	b_0; b_3; b_3'; b_2; b_2'; b_1; b_1'; Meridian	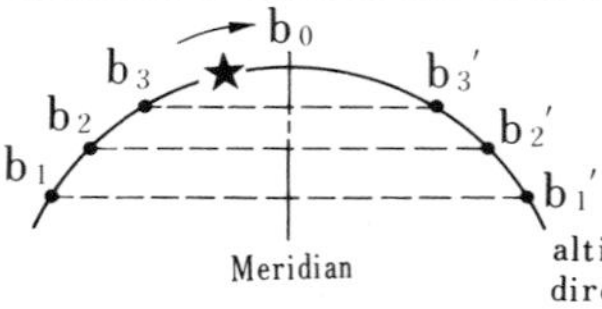 Read the azimuths at two equal altitudes, of celestial bodies, produced, in both sides of the azimuth of the maximum altitude and mark the average direction on the ground.	(1) No complicated computation is required. (2) The measurement takes a considerably long time (2 hours at least)
〔D〕	Culmination method	Meridian	 Compute the time that the celestial body passes through the meridian and mark the direction of the celestial body at such instance, on the ground.	(1) The time of meridian passage is known previously. (2) There is only one opportunity a day to measure a celestrial body (the sun in daytime)
〔E〕	Greatest elongation method	Upper culmination; N; Greatest western elongation; Greatest eastern elongation; β_W; Meridian; β_E	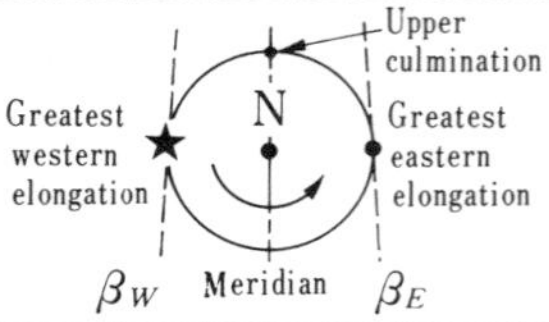 Compute the greatest elongations on the east and on the west, of a circumpolar star from the point of observation and then the direction of the meridian.	(1) This method is suitable for precision measurement since the variation of the azimuth becomes minimum at the greatest elongatic (2) The measurement is possible once a day for a fixed star.
	Remark	The meridian passage by the mean sun occurs at noon or 12^h, while the same by a celestial body (sun or a fixed star) occurs at 0^h of hour angle, h.		

Note : (1) In general the above methods 〔A〕 through 〔D〕 can apply to the sun, moon, planets an fixed stars.

(2) The method 〔E〕 is used mainly for stars (circumpolar stars) where the complementa angle of the declination δ, i.e., the polar distance ($90° - |\delta|$), is less than the latitude at a point of observation.

(3) This atlas deals with the sun and 23 fixed stars.
The moon has been omitted because its large variation in declination requires much time.
The planets can be treated in the same manner as the sun if the astronomical ephemeris is available.

Comparison of Methods to Determine the Astronomical Azimuth (2)

Measurment parameter		Known parameter					Calculation parameter			Basic equations	Item	
		Point of observation		Celestial body								
a_t	b	T	L	l	α	δ	h	β	–			
×	×	○	○	○	○	○	○	○		$\beta=\tan^{-1}\left[\dfrac{\sin h}{\sin l\ \cos h-\cos l\ \tan\delta}\right]$ $\beta=\beta_0$ or $\beta=\beta_0+180$	General method (I)	〔A〕
○	×	× ⊙	×	○	×	○	×	○		$\beta=2\cdot\cos^{-1}\sqrt{\dfrac{\cos\frac{1}{2}(90°-\delta-l-a)\cdot\cos\frac{1}{2}(90°-\delta+l+a)}{\cos l\cdot\cos a}}$ Altitude a = Measured altitude a_t − Atmospheric refraction − (Dip of horizon) + (Parallax)	General method (II)	〔B〕
○	○	× ⊙	×	×	×	× ⊙	×	×	○	Direction of meridian passage $b_o=\dfrac{\sum_{i=1}^{n}(b_i-b_i')}{n}$ (n : The number of a pair of measurements)	Equal altitude method	〔C〕
×	×	○	○	×	○	×	○	×		Fixed star : $T_*=(1-k)\ (\alpha-L-\theta_0)+L_s$ Sun : $T_{\odot}=12-\varepsilon-(L-L_s)$ (ε : Equation of time)	Meridian passage method	〔D〕
×	×	○	○	○	○	○	○	○		Fixed star : $T=(1-k)\ (h+\alpha-L-\theta_0)+L_s$ $\beta=\sin^{-1}[\cos\delta/\cos l]$ $h=\cos^{-1}[\tan l/\tan\delta]$	Greatest elongation	〔E〕
Altitude	Horizontal angle	Time	Longitude	Latitude	Right ascension	Declination	Hour angle	Azimuth	Others	β : Azimuth, T : Time at meridian passage or greatest elongation h : Hour angle a_t : Measured altitude a : True altitude	Remark	

Note : (1) ○ indicates that the parameter is indispensable, × indicates that it is not indispensable, and ⊙ indicates that the parameter is indispensable if the sun is used.

(2) In method 〔A〕, the longitude L and the latitude l of the point of observation, the right ascension and the declination of a celestial body are known, L and α are used for calculation of the hour angle h.

(3) In method 〔B〕 only the latitude l and the declination are to be known and the calculation is rather easy ; however, the true altitude is equal to the measured altitude, at less than the atmospheric refraction.

(4) Parallax to be taken into account in the case of the sun, the moon, and some planets near the earth is an amount of correction to an altitude measured at a point of observation to convert it to the value obtained assuming that the measurements were made from the center of the earth.
The lower the altitude, the greater the parallax; accordingly, the horizontal parallax will be the largest.
Typical figures of the parallax are Sun : 8″ Moon : 53′～61′
Planets : 0～0.5′(Venus) Fixed stars : 0 Therefore it may be igno ed for the fixed stars and also for the sun normally.

(5) For method 〔C〕, if the sun is used some correction is required against the variation with time of the declination and equation of time.

General Method to Determine Astronomical Azimuth (1)

Basic equation

Hour angle — Latitude — Declination

$$\beta_0=\tan^{-1}[\sin h/(\sin l\ \cos h-\cos l\ \tan\delta)]$$

$$=\tan^{-1}[C/(A-B)]$$

Azimuth β: If β_0 and C have the same sign……… $\beta=\beta_0+180°$; If β_0 and C differ in the sign…………… $\beta=\beta_0$

		No. 1	No. 2	No. 3	No. 4	(Example)
	Declination δ	± .	± .	± .	± .	$+/\ominus$ 8.232°
	Hour angle h					58.172°
	Latitude l	± .	± .	± .	± .	$\oplus/-$ 25.220 8°
C	$\sin h$	±0.	±0.	±0.	±0.	$\oplus/-$ 0.849 635
	$\cos h$	±0.	±0.	±0.	±0.	$\oplus/-$ 0.527 371
	$\sin l$	±0.	±0.	±0.	±0.	$\oplus/-$ 0.426 108
	$\cos l$	±0.	±0.	±0.	±0.	$\oplus/-$ 0.904 672
	$\tan\delta$	±	±	±	±	$+/\ominus$ 0.144 672
A	$\sin l\cdot\cos h$	±0.	±0.	±0.	±0.	$\oplus/-$ 0:224 717
B	$\cos l\cdot\tan\delta$	±	±	±	±	$+/\ominus$ 0.130 881
	$A-B$	±	±	±	±	$\oplus/-$ 0.355 598
	$C/(A-B)$	±	±	±	±	$\oplus/-$ 2.389 313
β_0	$\tan^{-1}[C/(A-B)]$	±	±	±	±	$\oplus/-$ 67.289°
						$\beta=\beta_0+180°$
β						$\oplus/-$ 247.289° ≧0
	Day, Month, Year	– –	– –	– –	– –	– –

General Method to Determine Astronomical Azimuth (2)

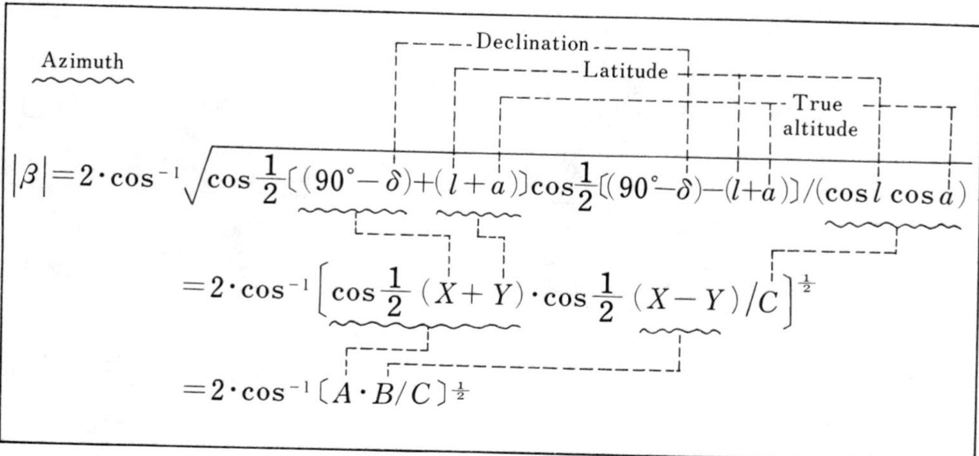

$$|\beta| = 2\cdot\cos^{-1}\sqrt{\cos\frac{1}{2}[(90^\circ-\delta)+(l+a)]\cos\frac{1}{2}[(90^\circ-\delta)-(l+a)]/(\cos l\,\cos a)}$$

$$= 2\cdot\cos^{-1}\left[\cos\frac{1}{2}(X+Y)\cdot\cos\frac{1}{2}(X-Y)/C\right]^{\frac{1}{2}}$$

$$= 2\cdot\cos^{-1}[A\cdot B/C]^{\frac{1}{2}}$$

		No. 1	No. 2	No. 3	No. 4	(Example)
	True altitude a					66.632°
	Latitude l					⊕ 38.392°
	Declination δ					⊕ 17.574°
X	$90^\circ-\delta$					72.426°
Y	$l+a$					105.024°
	$X+Y$					177.450°
	$X-Y$					⊖ 32.598
A	$\cos\frac{1}{2}(X+Y)$					0.022 25
B	$\cos\frac{1}{2}(X-Y)$					0.959 81
	$\cos a$					0.396 64
	$\cos l$					0.783 78
C	$\cos a\cdot\cos l$					0.310 88
	$A\cdot B/C$					0.068 695
	$\sqrt{AB/C}$					0.262 097
	$\cos^{-1}\sqrt{AB/C}$					74.806°
β	$2\cdot\cos^{-1}\sqrt{AB/C}$					149.611°

Day, Month, Year – – – – – – – – – –

Measured altitude a_t					
Atomospheric refraction c_a					
$a=a_t-c_a$					66.632

Culmination Method

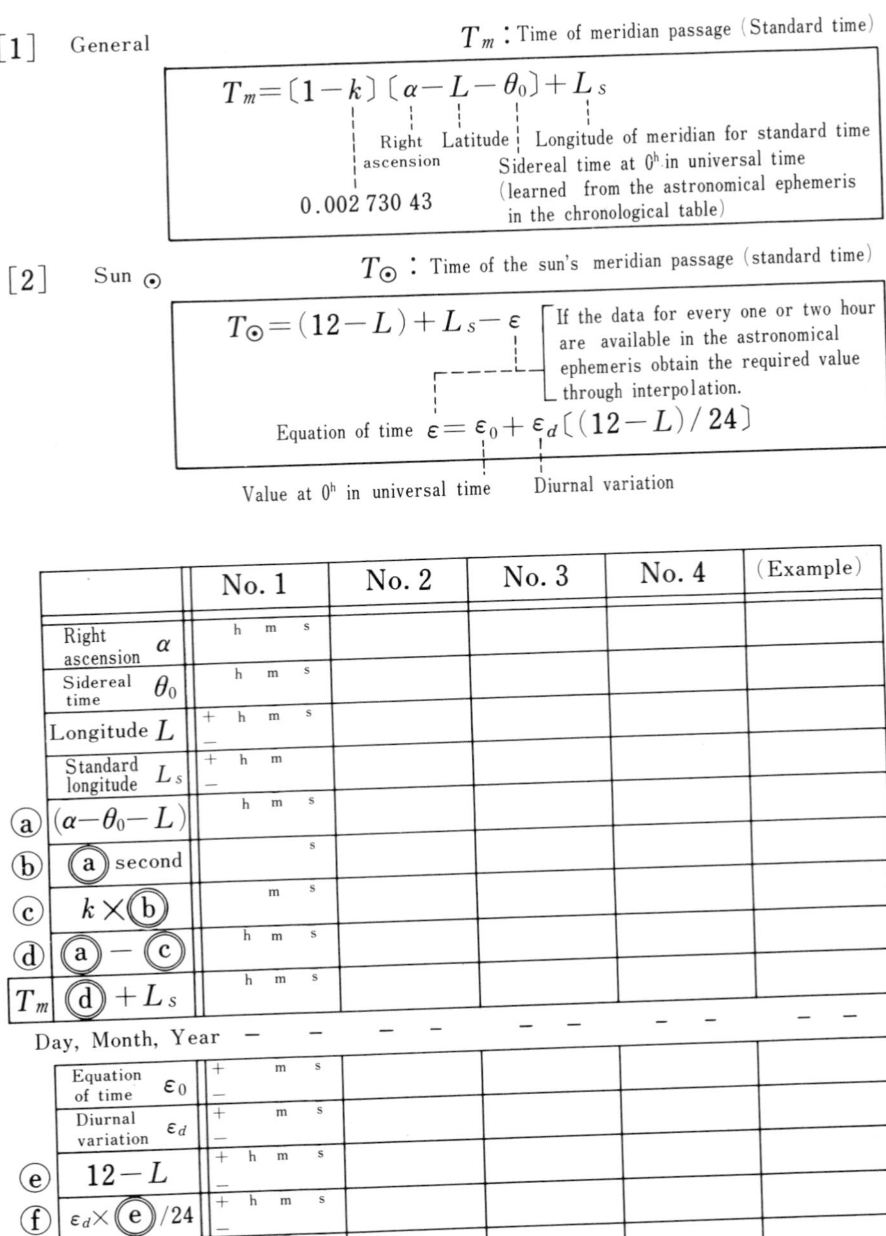

[1] General

T_m : Time of meridian passage (Standard time)

$$T_m = [1-k][\alpha - L - \theta_0] + L_s$$

- k = 0.002 730 43
- α: Right ascension
- L: Latitude
- θ_0: Sidereal time at 0^h in universal time (learned from the astronomical ephemeris in the chronological table)
- L_s: Longitude of meridian for standard time

[2] Sun ⊙

$T_\odot$: Time of the sun's meridian passage (standard time)

$$T_\odot = (12 - L) + L_s - \varepsilon$$

Equation of time $\varepsilon = \varepsilon_0 + \varepsilon_d[(12-L)/24]$

- ε_0: Value at 0^h in universal time
- ε_d: Diurnal variation

If the data for every one or two hour are available in the astronomical ephemeris obtain the required value through interpolation.

		No. 1	No. 2	No. 3	No. 4	(Example)
	Right ascension α	h m s				
	Sidereal time θ_0	h m s				
	Longitude L	+ h m s −				
	Standard longitude L_s	+ h m −				
ⓐ	$(\alpha - \theta_0 - L)$	h m s				
ⓑ	ⓐ second	s				
ⓒ	$k \times$ ⓑ	m s				
ⓓ	ⓐ − ⓒ	h m s				
T_m	ⓓ $+ L_s$	h m s				

Day, Month, Year — — — — — — — — — —

		No. 1	No. 2	No. 3	No. 4	(Example)
	Equation of time ε_0	+ m s −				
	Diurnal variation ε_d	+ m s −				
ⓔ	$12 - L$	+ h m s −				
ⓕ	$\varepsilon_d \times$ ⓔ $/24$	+ h m s −				
ε	$\varepsilon_0 +$ ⓕ	+ h m s −				
$T_\odot$	ⓔ $+ L_s - \varepsilon$	h m s				

Greatest Elongation Method

Azimuth $|\beta| = \sin^{-1}[\cos\delta/\cos l]$

Time (Standard time) $T_n = [1-k][h_0 + \alpha - L - \theta_0] + L_S$

- k = 0.002 730 43
- α: Right ascension
- L: Longitude
- θ_0: Sidereal time at 0^h in universal time
- L_S: Longitude of meridian for standard time

Hour angle $h_0 = \cos^{-1}[\tan l/\tan\delta]$

Altitude $a = \sin^{-1}[\sin l/\sin\delta]$

		No. 1	No. 2	No. 3	No. 4	(Example)
	Right ascension α					
	Declination δ					
	Longitude L					
	Standard longitude L_s					
	Latitude l					
	Sidereal time θ_0					
ⓐ	$\sin\delta$					
ⓑ	$\cos\delta$					
ⓒ	$\tan\delta$					
ⓓ	$\sin l$					
ⓔ	$\cos l$					
ⓕ	$\tan l$					
	Day, Month, Year	– –	– –	– –	– –	– –
h_0	$\cos^{-1}$[ⓕ/ⓒ]					
ⓖ	$(h_0+\alpha-L-\theta_0)$					
ⓗ	ⓖ second					
ⓘ	$k\times$ ⓗ					
ⓙ	ⓖ − ⓘ					
T_n	ⓙ $+ L_S$					
α	$\sin^{-1}$[ⓓ/ⓐ]					
β	$\sin^{-1}$[ⓑ/ⓔ]					

Azimuth of Polaris

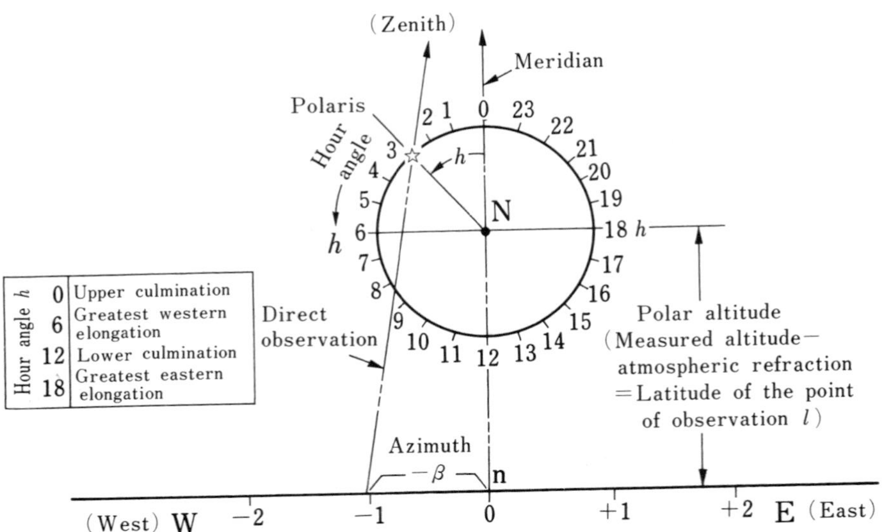

Azimuth to the west (W) Apply − sign to the obtained value

Latitude of the point of observation l \ Hour angle h	0	1	2	3	4	5	6	7	8	9	10	11	12
0°	0	0.2	0.4	0.6	0.8	0.8	0.9	0.8	0.8	0.6	0.4	0.2	0
5°	0	0.2	0.4	0.6	0.8	0.8	0.9	0.8	0.8	0.6	0.4	0.2	0
10°	0	0.2	0.4	0.6	0.8	0.9	0.9	0.8	0.8	0.6	0.4	0.2	0
15°	0	0.2	0.5	0.6	0.8	0.9	0.9	0.9	0.8	0.6	0.4	0.2	0
20°	0	0.2	0.5	0.7	0.8	0.9	0.9	0.9	0.8	0.6	0.5	0.2	0
25°	0	0.2	0.5	0.7	0.8	0.9	1.0	0.9	0.8	0.7	0.5	0.2	0
30°	0	0.3	0.5	0.7	0.9	1.0	1.0	1.0	0.9	0.7	0.5	0.3	0
35°	0	0.3	0.5	0.8	0.9	1.0	1.1	1.0	0.9	0.7	0.5	0.3	0
40°	0	0.3	0.6	0.8	1.0	1.1	1.1	1.1	1.0	0.8	0.6	0.3	0
45°	0	0.3	0.6	0.9	1.1	1.2	1.2	1.2	1.1	0.9	0.6	0.3	0
50°	0	0.4	0.7	1.0	1.2	1.3	1.3	1.3	1.2	0.9	0.7	0.3	0
55°	0	0.4	0.8	1.1	1.3	1.5	1.5	1.5	1.3	1.1	0.7	0.4	0
60°	0	0.5	0.9	1.2	1.5	1.7	1.7	1.7	1.5	1.2	0.8	0.4	0
65°	0	0.5	1.1	1.5	1.8	2.0	2.1	2.0	1.7	1.4	1.0	0.5	0
70°	0	0.7	1.3	1.8	2.2	2.5	2.5	2.4	2.1	1.7	1.2	0.6	0
Hour angle h	24	23	22	21	20	19	18	17	16	15	14	13	12

Azimuth of Polaris (°)

Azimuth to the east (E) Apply + sign to the obtained value

Polaris, general method (Azimuth at an arbitrary time)

(1) This table assumes that the polar distance of Polaris is 0°52′.3.

(2) The table may present an azimuthal error in the vicinity of 0.1° according to the year.

(3) If higher accuracy is required, use the astronomical ephemeris for that day, month and year, and use the equation of azimuth in your calculation.

Atmospheric Refraction

Deviation Due to Light-Ray Refraction Astronomical Surveying

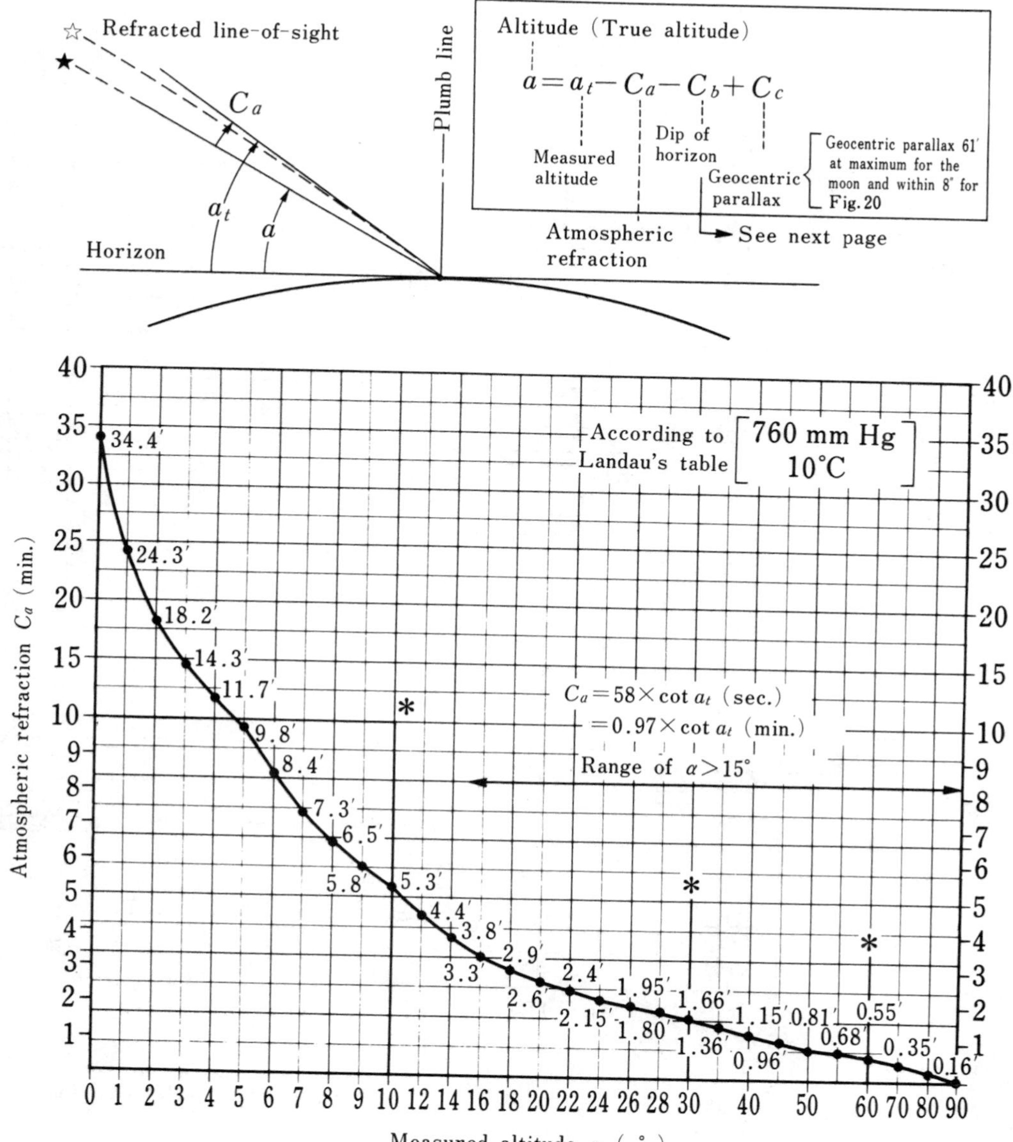

(1) The lower the measured altitude, the greater the influence of the atmospheric refraction. In a simplified measurement the above graph may be used although the refraction is dependent on the meteorological conditions.

(2) Geocentric parallax may be ignored if the moon is excluded from the measurement.

Dip of Horizon

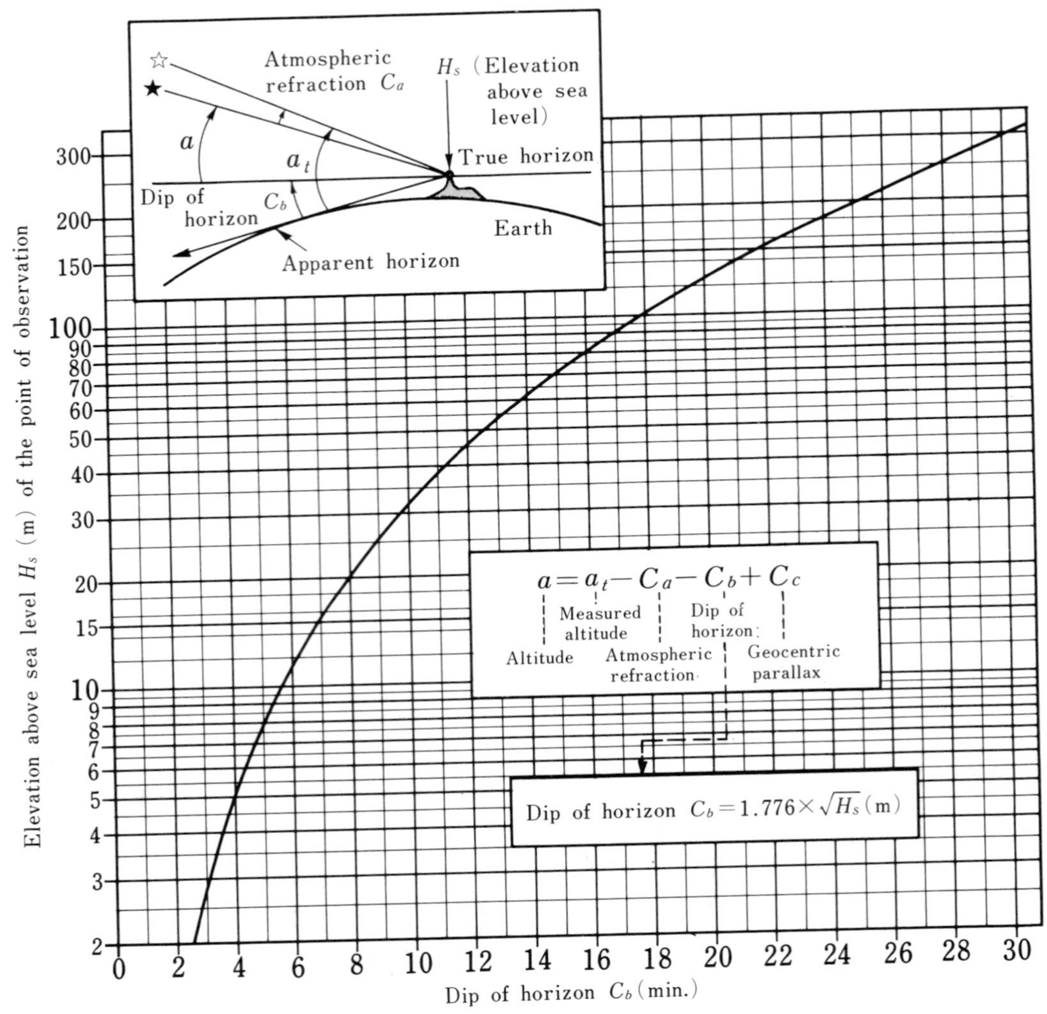

〔Note〕 This graph is used only when the altitude of a celestial body is measured by referring to the apparent horizon (0°) with a sextant, etc. Thus this graph is not necessary when a level is employed.

Apparent Declination and Equation of Time of the Sun (Approximate Value) (1)

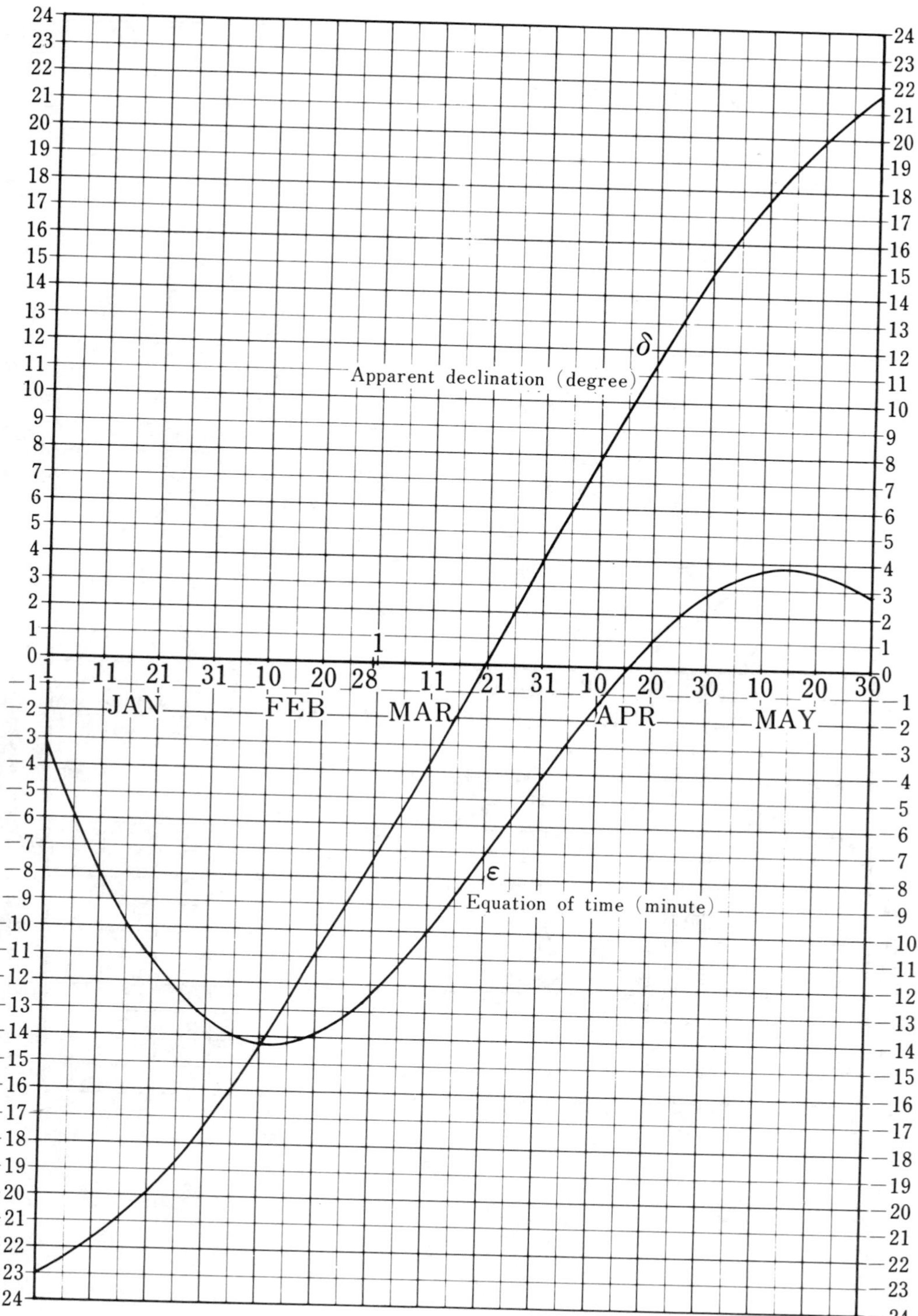

Apparent Declination and Equation of Time of the Sun (Approximate Value) (2)

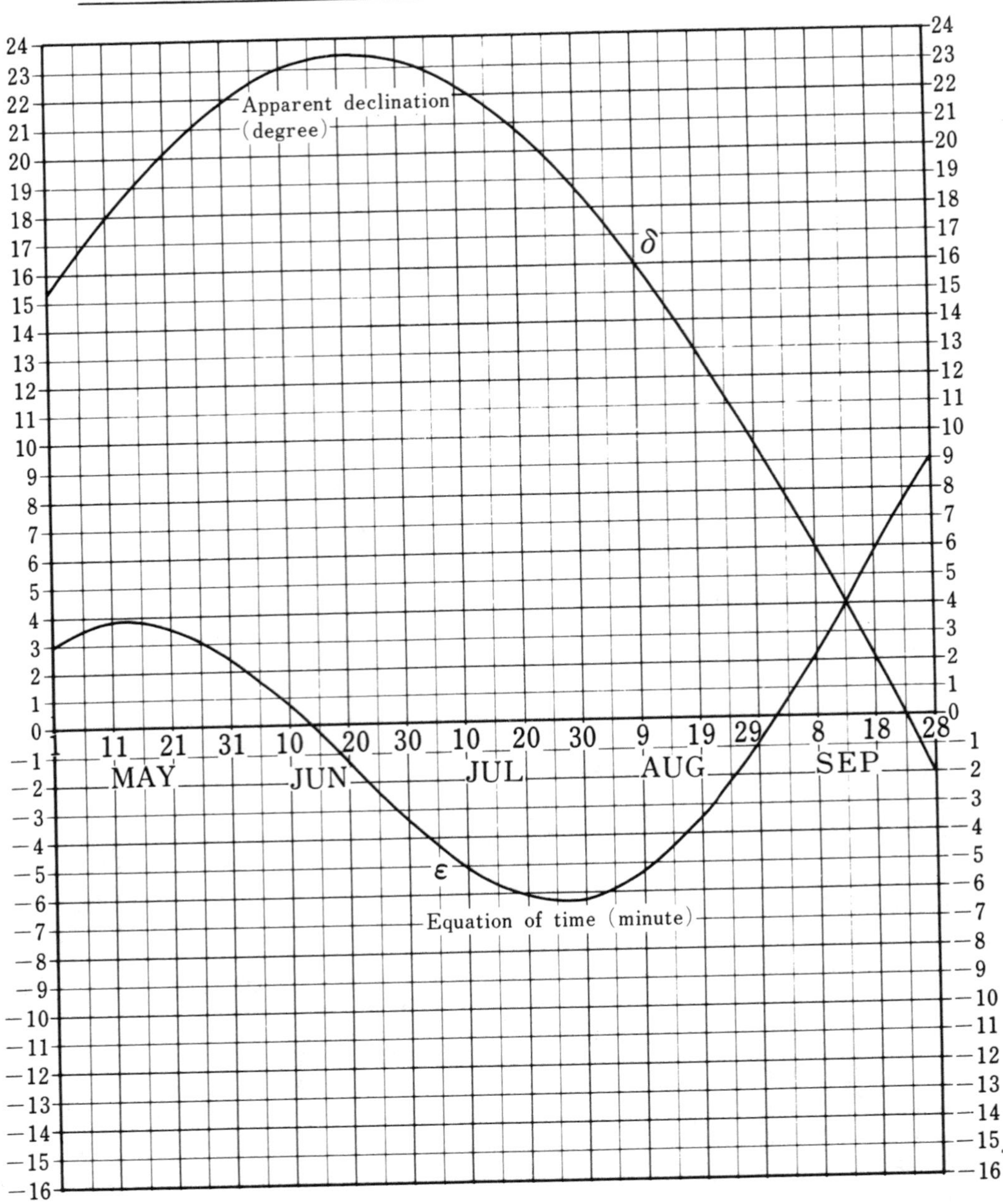

Apparent Declination and Equation of Time of the Sun (Approximate Value) (3)

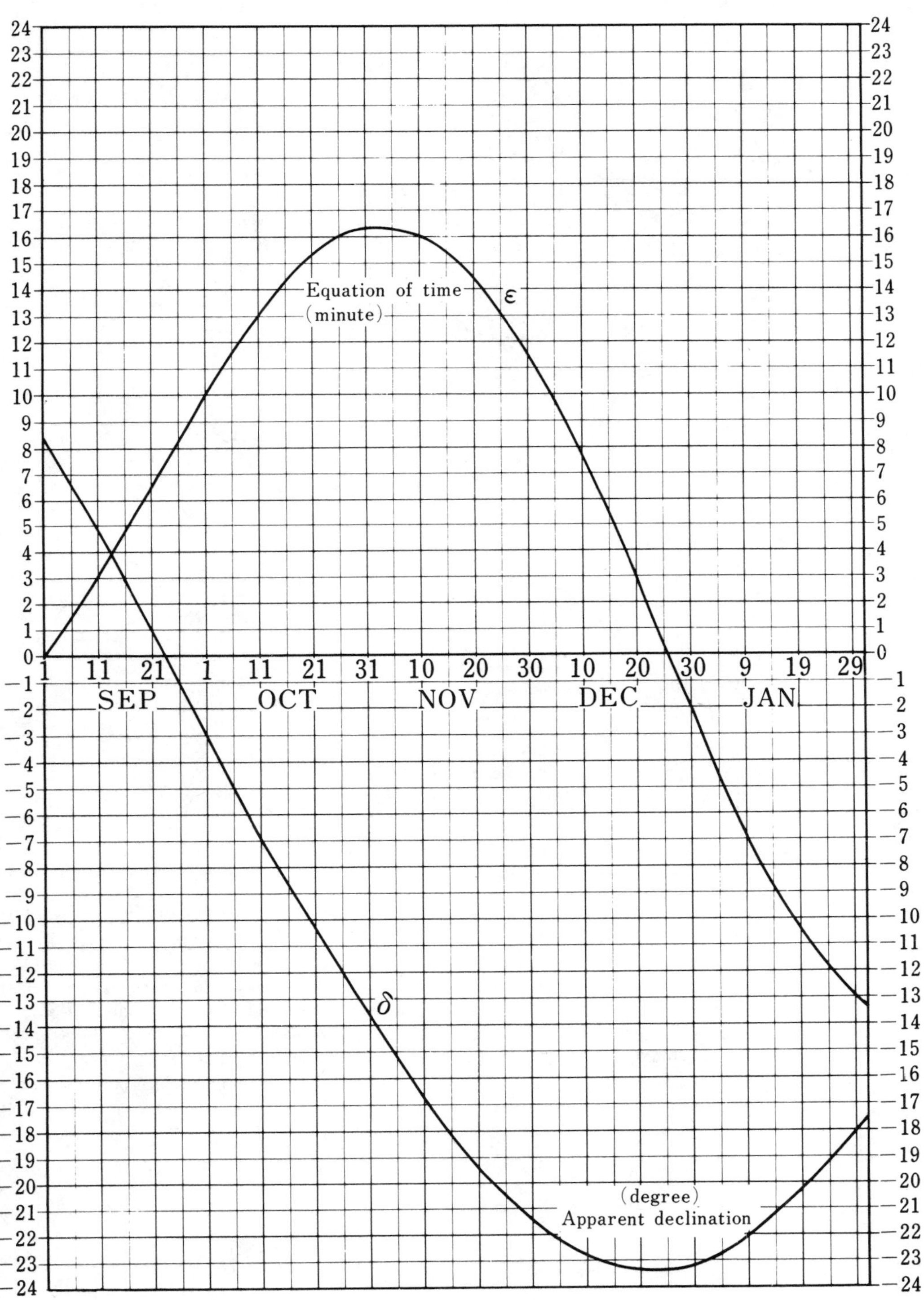

9

22a

Comparison of Measuring Methods of Astronomical Longitude and Latitude (1)

	Item	Summary of method	Advantages and disadvantage
Longitude	Single-altitude method	Measure the celestial altitude a_t at an arbitrary time T.	(1) Measurement can be carried out at any time and the method is simple. (2) The latitude of the observation point should be known previously
Longitude	Same star -at-equal altitude method	Measure when a star passes the same altitude on both sides of the meridian.	(1) Computation is simple. (2) The greater the interval between these two instances, i.e., 2h at least, the better the accuracy. (3) Measurement consumes a rather long time (disadvantage).
Longitude / Latitude	Universal method by two different stars	Measure a pair of altitudes a_{t1} and a_{t2} of two different stars at two different instances.	(1) The longitude and the latitude can be determined simultaneously. (2) Measurement may be accomplished in a short time. (3) Computation is rather complicated.
Latitude	Meridian altitude method	Measure the altitude a_t of the meridian passage of star whose declination is already known.	(1) Computation is easy. (2) Direction of the meridian passage is known previously and accurately. (3) When the time is known, the longitude can be determined.
Latitude	Polaris altitude methed	Measure the altitude of Polaris at an arbitrary time.	(1) Measurement is possible at any time. (2) The longitude is known previously. (3) The method is applicable only in the northern hemisphere.

(1) Highly accurate measurement of longitude and latitude requires the use of special survey equipment and skill.

(2) Determination of the longitude and latitude associated with radio engineering, except for particular cases; the geographic survey will be simple and gives better accuracy than the celestial survey. Close examination of the map compared with the actual topography can determine the longitude and latitude of a point referred to the previously known point (preferably the triangular point).

(3) In the case of a desert, field or snow-ice field with no outstanding known object, the above methods can apply.

Comparison of Measuring Methods of Astronomical Longitude and Latitude (2)

Measuring parameters		Known parameters				Calculating parameter				Basic equations	Item	
		Point of observation		Celestial body								
a_t	T	L	l	α	δ	h	L	l	—			
○	○	×	○	○	○	○	○	×		$h=2\cdot\sin^{-1}\sqrt{\dfrac{\sin\frac{1}{2}(Z+l-\delta)\sin\frac{1}{2}(Z-l+\delta)}{\cos l\cos\delta}}$ $L=(h+\alpha)-(1+k')(T-L_s)-\theta_0$ where $Z=90°-a$	Single-altitude method	Longitude L
×	○	×	×	○	×	×	○	×	θ_g	$\theta_g=\theta_0+(1+k')(T-L_s)$, $\theta_g'=\theta_0+(1+k')(T'-L_s)$] Greenwich sidereal time $L=\alpha-\frac{1}{2}(\theta_g+\theta_g')$	Same-star-at-equal-altitude method	Longitude L
○ a_{t1} a_{t2}	○ T_1 T_2	×	×	○ α_1 α_2	○ δ_1 δ_2	×	○	○	λ, B, P, Q, q	(Omitted due to complexity)	Universal method by two different stars	Longitude L
○	×	×	×	×	○	×	×	○		$l=\delta\pm(90°-a)$ $l=180°-\delta-(90°-a)$	Meridian altitude method	Latitude l
×	○	×	×	○	×	○	○	×		$L=\alpha-\theta_0-(1+k')(T-L_s)$	Meridian altitude method	Longitude L
○	○	○	×	○	○	○	×	○		$l=a-p\cos h+\frac{1}{2}p^2\sin 1''\sin h\tan a$ p : Polar distance	Polaris altitude method	Latitude l
Altitude	Time	Longitude	Latitude	Right ascension	Declination	Hour angle	Longitude	Latitude	Others	True altitude a = Measured altitude a_t − Atmospheric refraction (See Fig. 19) θ_0 : Greenwich sidereal time at 0^h in universal time	Remark	

Single-Altitude Method

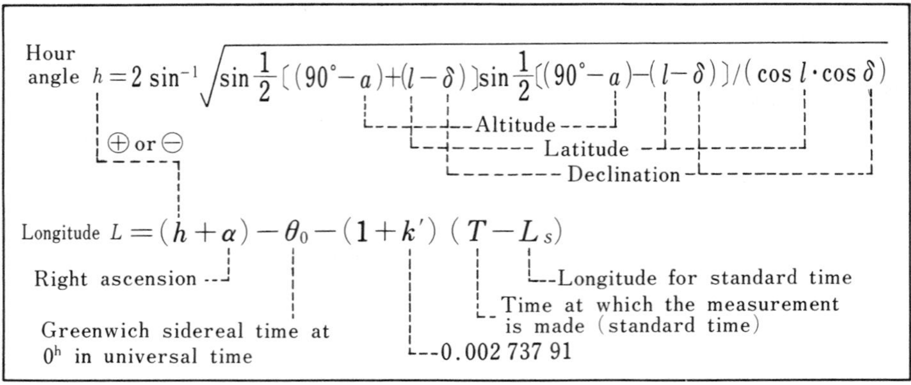

		No. 1	No. 2	No. 3	No. 4	(Example)
	True altitude a					
	Latitude l					
	Declination δ					
	Right ascension α					
Z	$90^\circ - a$					
X	$l - \delta$					
	$Z + X$					
	$Z - X$					
A	$\sin\frac{1}{2}(Z + X)$					
B	$\sin\frac{1}{2}(Z - X)$					
	$\cos l$					
	$\cos\delta$					
C	$\cos l \cdot \cos\delta$					
D	$\sqrt{A \cdot B / C}$					
h	$2\sin^{-1}(D)$					
	Time of measurement T					
	Greenwich sidereal time at 0^h θ_0					
	Standard longitude L_s					
U	Universal time $(T - L_s)$					
	U (second)					
V	$k' \times U$					
W	$U + V$					
E	$h + \alpha - \theta_0$					
L	$E - W$					

Day, Month, Year

(Fig. 19)

	No. 1	No. 2	No. 3	No. 4	(Example)
Measured altitude a_t					
Atmospheric refraction C_a					
$a = a_t - C_a$					

Same-Star-at-Equal-Altitude Method

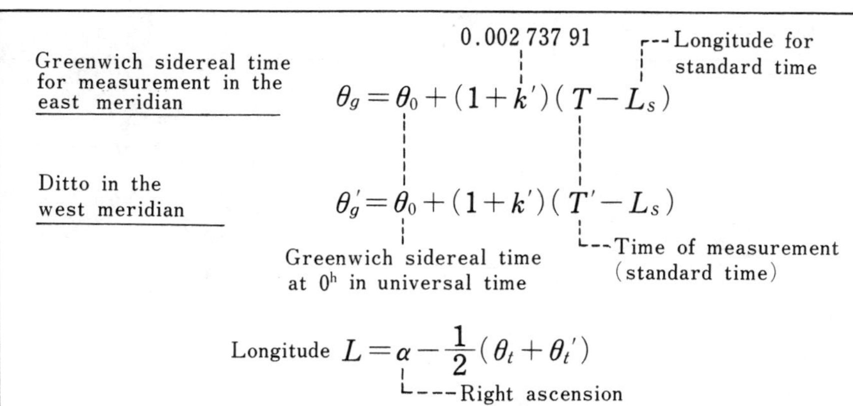

	No. 1	No. 2	No. 3	No. 4	(Example)
Time of measurement T					
Time of measurement T'					
Greenwich sidereal time at 0^h θ_0					
Right ascension α					
Standard longitude L_s					

U	$T - L_s$					
	U second					
V	$k' \times U$					
W	$U + V$					
θ_t	$\theta_0 + W$					

U'	$T' - L_s$					
	U' second					
V'	$k' \times U'$					
W'	$U' + V'$					
θ_t'	$\theta_0 + W'$					

θ_m	$\frac{1}{2}(\theta_g + \theta_g')$					
L	$\alpha - \theta_m$					

Day, Month, Year

9

22e

Meridian-Altitude Method

—Latitude (l)—

<The calculation chart is omitted for simplification>

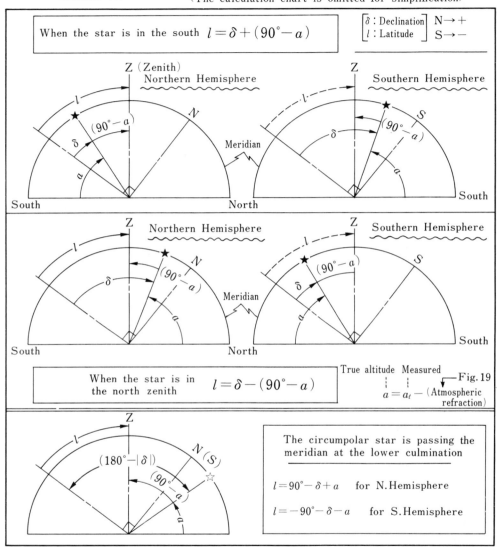

When the star is in the south $l=\delta+(90°-a)$

δ : Declination, l : Latitude — N→+, S→−

Z (Zenith)

Northern Hemisphere — South, North, l, δ, a, $(90°-a)$, N

Meridian

Southern Hemisphere — South, l, δ, a, $(90°-a)$, S

Northern Hemisphere — South, North

Meridian

Southern Hemisphere — South

When the star is in the north zenith $l=\delta-(90°-a)$

True altitude, Measured: $a=a_t-$ (Atmospheric refraction) — Fig. 19

Z — l, $(180°-|\delta|)$, $N(S)$, $(90°-a)$, a

The circumpolar star is passing the meridian at the lower culmination

$l=90°-\delta+a$ for N. Hemisphere

$l=-90°-\delta-a$ for S. Hemisphere

—Longitude (L)—

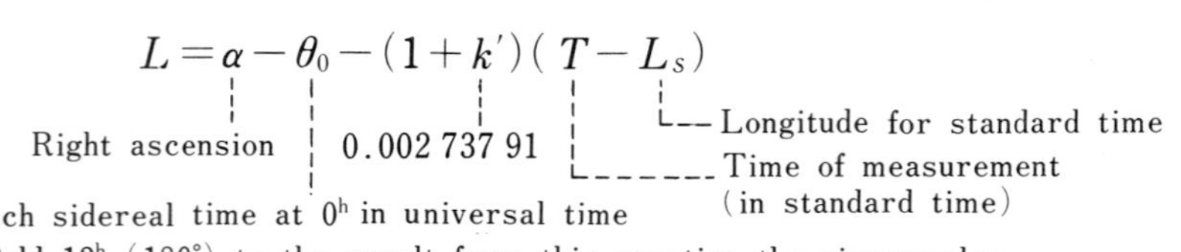

$$L=\alpha-\theta_0-(1+k')(T-L_S)$$

- α: Right ascension
- θ_0: Greenwich sidereal time at 0^h in universal time
- k': 0.002 737 91
- T: Time of measurement (in standard time)
- L_S: Longitude for standard time

Note : Add 12^h (180°) to the result from this equation the circumpolar star is passing the meridian at the lower culmination.

Polaris-Altitude Method

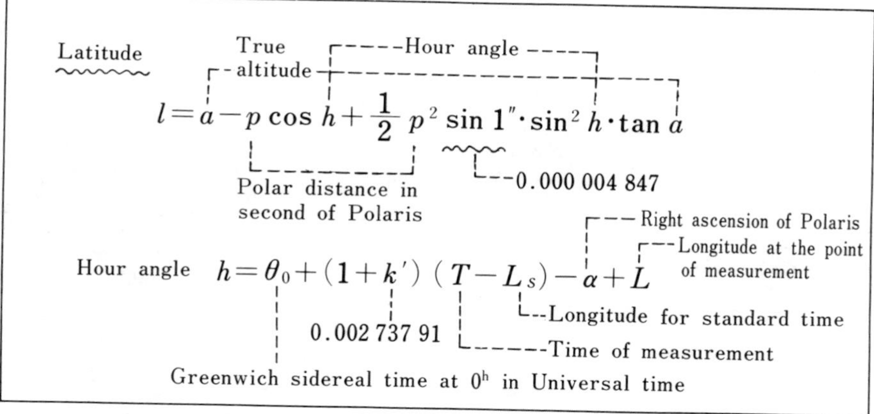

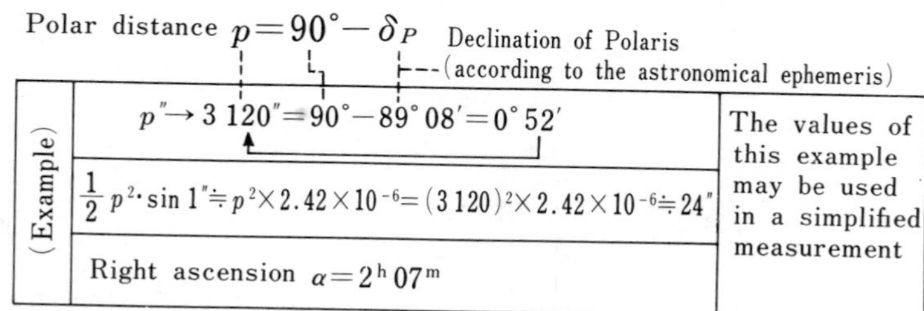

		No. 1	No. 2	No. 3	No. 4	Example
	True altitude a					37.48°
	Right ascension α					$2^h 07^m$
	Polar distance p''					3 120″
	Longitude L					$9^h 23^m 30^s$
	Time of measurement T					$22^h 37^m 46^s$
	Greenwich sidereal time at 0^h θ_0					$17^h 03^m 35^s$
	Longitude for standard time L_s					$9^h 00$
U	Universal time $(T - L_s)$					$13^h 37^m 46^s$
	U sec.					$49\,066^s$
V	$k' \times U$					$134^s \rightarrow 2^m 14^s$
W	$U + V$					$13^h 40^m 0^s$
h	$W + \theta_0 + L - \alpha$					$14^h 00^m 05^s$
	$h°$					210.02°
A	$p \cos h$					−2 702″
B	$\sin^2 h$					0.25
C	$p^2 \times 2 \cdot 42 \times 10^{-6}$					24″
D	$\tan a$					0.766 788
E	$B \times C \times D$					4.6″ → 5″
l	$a - A + E$					38° 13′ 57″

9 23

Luminosities, Apparent Radii, and Symbols of Celestial Bodies

Classification		(Symbols) Name of celestial bodies		Maximum luminosity (Magnitude)	Apparent radius	Mean distance from the sun (×10^8 km) (relative value)	Remark ⟨Origin of the symbol⟩
Sun and moon		☉	Sun	−26.8	15′ 59.6″	0 109.	Ancient symbol to represent gold
Sun and moon		☾	Moon	−12.5	15′ 32.6″	— 0.27	Ancient symbol to represent silver
Planets	Earth planets	☿	Mercury	−1.9	5.5″	0.58 0.38	Messenger, helmet with wings
Planets	Earth planets	♀	Venus	−4.4	30.2″	1.08 0.95	A goddess or mirror
Planets	Earth planets	⊕	Earth	—	—	1.50 1	Symbol of the crusade in the Middle ages
Planets	Earth planets	♂	Mars	−2.8	8.9″	2.28 0.53	God or arrow
Planets	Giant planets	♃	Jupiter	−2.5	23.4″	7.8 11.2	Lightning
Planets	Giant planets	♄	Saturn	−0.4	9.8″	14.3 9.5	Graphically designed K from Kronos
Planets	Giant planets	♅	Uranus	+5.6	1.8″	28.6 3.7	No particular meaning (Discovered in 1781)
Planets	Giant planets	♆	Neptune	+7.9	1.2″	45.0 3.9	The trident of the sea god (Discovered in 1846)
Planets	–	♇	Pluto	+14.9	0.1″	58.8 0.5 ?	P and L are joined (Discovered in 1930)
Fixed star		✳	Star	−1.6 (Sirius)	≒0	(Greater than 4. 3 light-years)	Magnitude minus × 2 M 0× 7 M 1×12 M 2×67

(Uranus, Neptune, Pluto:) Brightness is insufficient for use as an object of astronomical measurement

(1) The solar system consists of 1 fixed star, 9 planets, 32 satellites 10^4 minor planets, 10^{12} comets and others.

(2) The order of (luminosity) magnitude is determined in such a manner that magnitude 6 is given for the darkest star and is reduced by one each time the luminosity is increased 2.5 times. Thus the stellar magnitude 1 is 100 times more luminous than magnitude 6.
(The number of stars visible to the naked eye is approximately 8 500)

Major Fixed Stars and Their Positions

#	Name of star	Abbreviation	Name	Magnitude	Right ascension	Declination
					h m	o
1	α Ursae Minoris	α UMi	Polaris	2.1	2 04	N 89.1
2	α Ursae Majoris	α UMa	Dubhe	2.0	11 02	61.9
3	β Cassiopeiae	β Cas	——	2.4	0 08	59.0
4	α Cassiopeiae	α Cas	Shedir	2.5	0 39	56.4
5	α Aurigae	α Aur	Capella	0.2	5 15	N 46.0
6	α Cygni	α Cyg	Deneb	1.3	20 40	45.2
7	α Lyrae	α Lyr	Vega	0.1	18 36	38.8
8	β Geminorum	β Gem	Pollux	1.2	7 44	28.1
9	α Bootis	α Boo	Arcturus	0.2	14 14	19.3
10	α Tauri	α Tau	Aldebaran	1.1	4 34	16.5
11	α Leonis	α Leo	Regulus	1.3	10 07	12.1
12	α Aquilae	α Aql	Altair	0.9	19 49	8.8
13	α Orionis	α Ori	Betelgeuse	0.9	5 54	7.4
14	α Canis Minoris	α CMi	Procyon	0.5	7 38	N 5.3
15	β Orionis	β Ori	Rigel	0.3	5 13	S 8.2
16	α Virginis	α Vir	Spica	1.2	13 24	11.0
17	α Canis Majoris	α CMa	Sirius	−1.6	6 44	16.7
18	α Scorpii	α Sco	Antares	1.2	16 28	26.4
19	α Pisces Austrinus	α PsA	Fomalhaut	1.3	22 56	29.8
20	α Carinae	α Car	Canopus	−0.9	6 23	52.7
21	α Eridani	α Eri	Achernar	0.6	1 37	57.4
22	α Centauri	α Cen	——	0.0	14 38	60.7
23	α Crucis	α Cru	——	1.1	12 25	62.9

(1) Given are fixed stars of 23 kinds selected from those used frequently in the astronomical survey.

(2) The distribution of magnitudes are 4 stars of magnitude 2, 11 stars of magnitude 1, 6 stars of magnitude 0, and 2 stars of magnitude minus.
Thus all the stars with magnitudes 1 or brighter except two stars (βCen, βCru) are adopted in the given list.

(3) Stars numbered # 1 through # 4 of magnitude 2 are selected for ease of search in the Northern Hemisphere even for beginners and also for extensive use.
Particulars of these stars :
1, Polaris : The declination is approximately 90° →Close to the celestial north pole.
2. The star located at the tip of the Plough.
3. The star located at an end of a series of three bright stars contained in Cassiopeia and forming the letter W. (The right ascension is close to 0^h or close to the right ascension for the vernal equinox.)
4. The star located in the middle of the above three bright stars.

(4) Stars of magnitude 1 or brighter included in the list are 10 in the celestial Northern Hemisphere and 9 in the celestial Southern Hemisphere and these have been selected so that more than a few stars may be observed at any point of the world and all year round.

(5) The star # 22 α Centauri is the fixed star closest (4.3 light-years) to the solar system.

(6) Use the astronomical ephemeris for that year if the more precise location of a fixed star is to be determined.

(7) Radio stars used extensively for satellite communications are :
4 Shedir   # 6 Deneb   # 10 Aldebaran, etc.

Positions of Major Fixed Stars and Applicable Ranges (1)

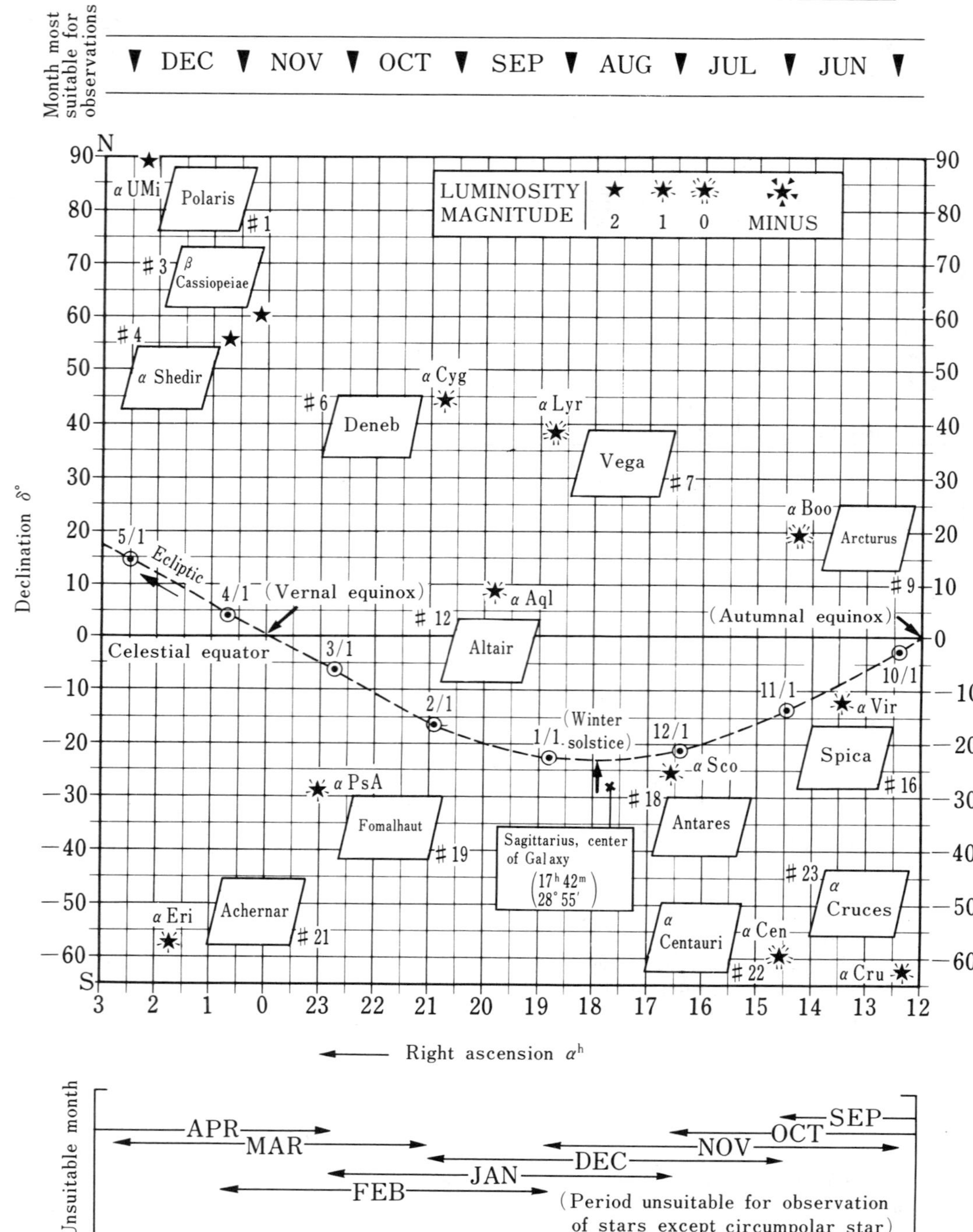

Position of Major Fixed Stars and Applicable Ranges (2)

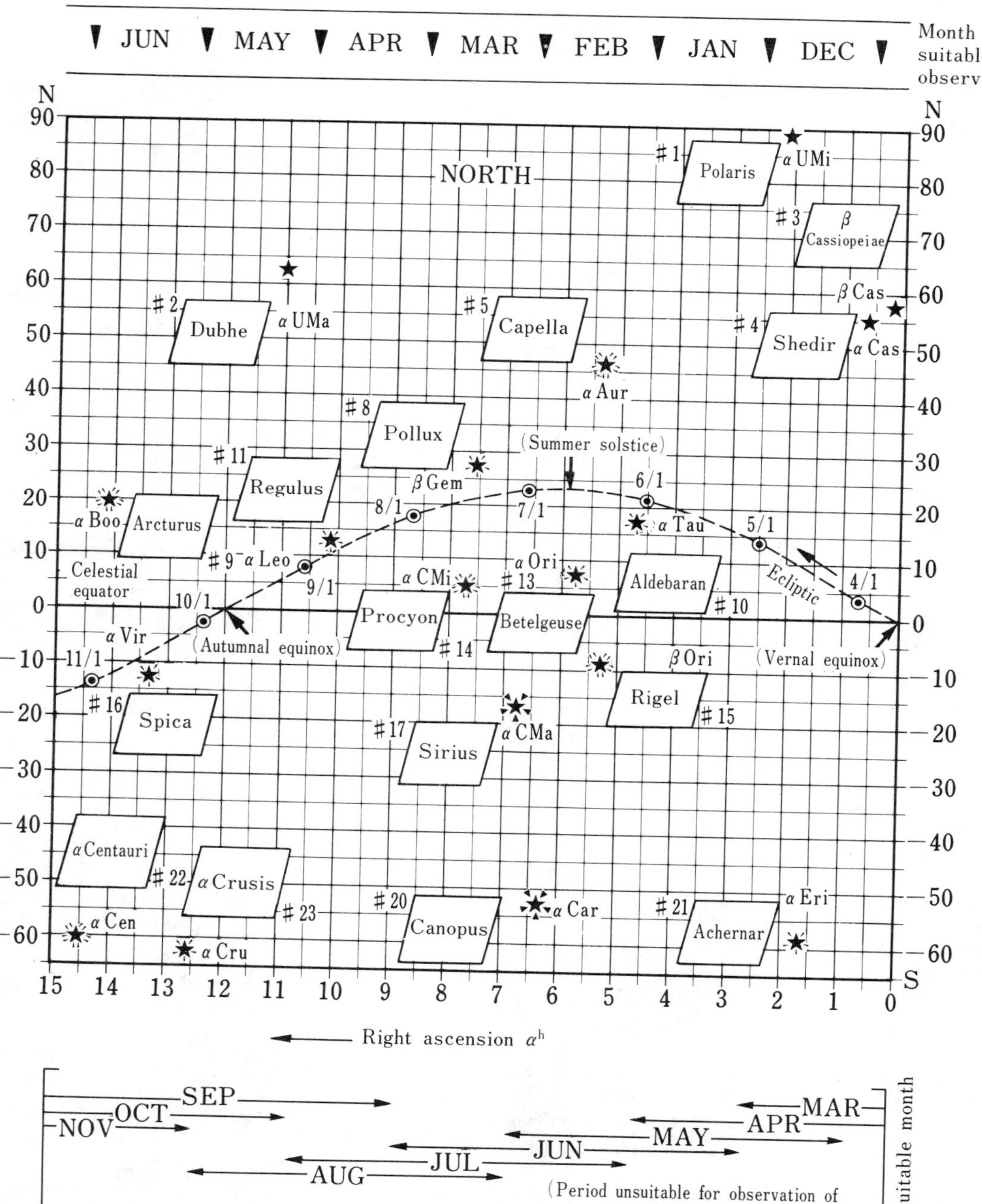

Star Finding Chart (1)

Around the Celestial North Pole

Direction of diurnal rotation

Right ascension

★ Star of magnitude 2
• Star of magnitude 3 or less luminous
( ) Magnitude

URSA MAJOR

η
ζ
Alioth ε (1.7)
β
α #2
Dubhe (2.0)

URSA MINOR

β Kochab (2.2)

N Polaris (2.1)
α #1

CASSIOPEIA

(2.4) #3 β Cassiopeiae
γ
ε
δ
α #4 Shedir (2.5)

10h 11h 12h 13h 14h 15h 16h 17h 18h 19h 20h 21h 22h 23h 0h 1h 2h 3h 4h 5h 6h 7h 8h 9h

80° 70° 60°

N 10° N 15° N 20° N 25° N 30° N 35° N 40° (Horizon)

N 10° N 15° N 20° N 25° N 30° N 35° N 40° (Horizon)

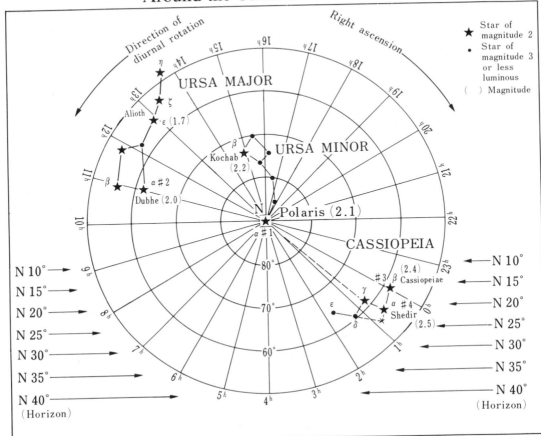

How to find Polaris

(1) Polaris is located approximately on the extension of the line joining the stars β with α (Dubhe) in Ursa Major and separated from the star α by the distance five times that between the stars α and β.

(2) When Ursa Major is below the horizon, find the point × first at the intersection of the ($\beta \rightarrow \alpha$) extension line of Cassiopeia and the ($\varepsilon \rightarrow \delta$) extension line and then find Polaris on the extension of the line passing through × and γ and at a distance five times that between × and γ.
Polaris is located near the second point.

(3) Be careful so as not to mistake Kochab in Ursa Minor for Polaris.

(4) The position of Polaris is verified through checking the relationship, altitude of Polaris $\simeq$ latitude of the point of observation $\pm(0 \sim 1°)$.

(5) The area near α Cassiopeiae (Shedir) is known as an astronomical radio source.
(Utilized as a reference of measurement in satellite communications.)

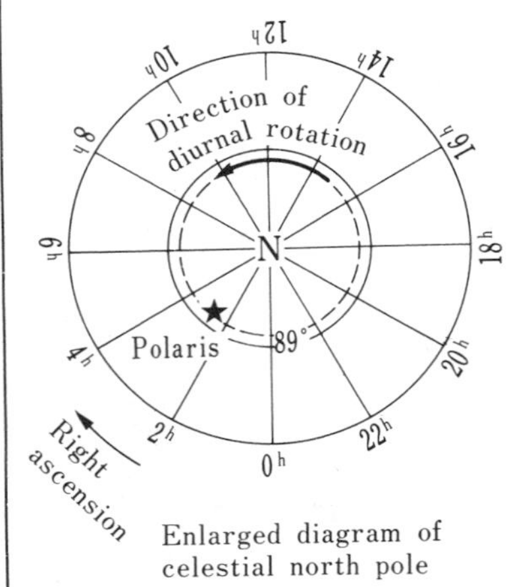

Enlarged diagram of celestial north pole

Star Finding Chart (2)
Around the Celestial South Pole

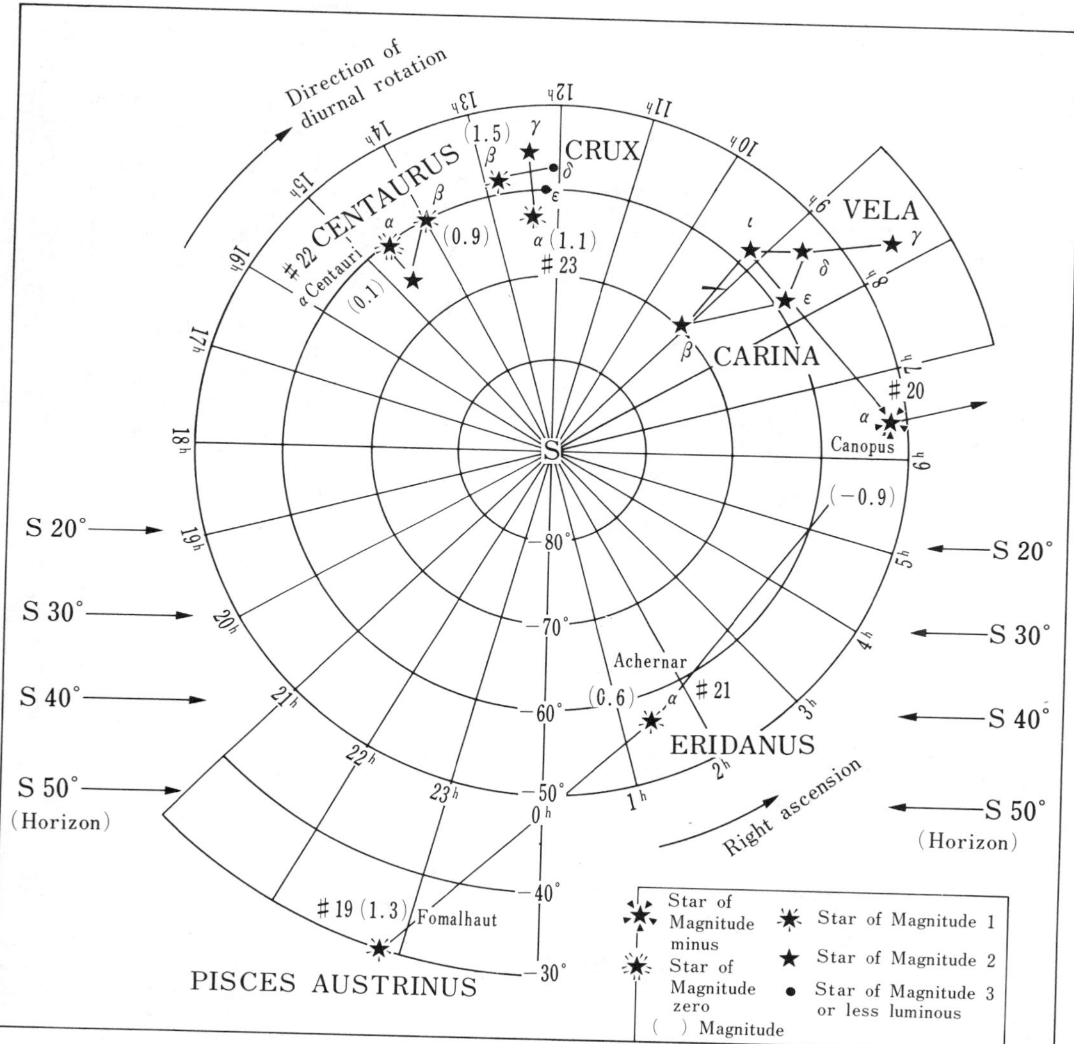

(1) Although the Crux is well known, no outstanding star is found near the south pole.
(2) α Centauri is the fixed star that is nearest (4.3 light-years) to the solar system, and care should be taken not to mistake it for β Centauri shown in the figure.
(3) Canopus is the second most luminous fixed star in the whole sky.

Star Finding Chart (3)

Pisces Austrinus, Aquila, and Scorpius

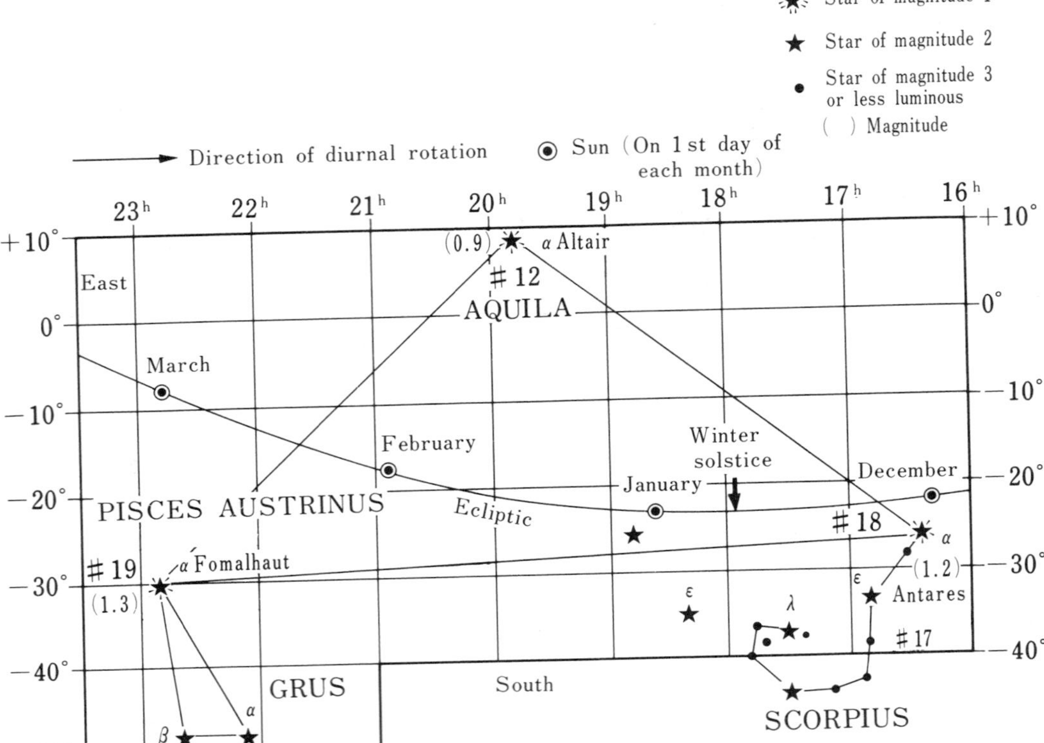

(1) This is a part of the upper view of the southern sky normally presented in star maps.

(2) Many bright stars are found in and around Scorpius, but no outstanding stars in the area between there and α Fomalhaut except for the planets near the ecliptic.

(3) If no luminous star other than the α star is found in Pisces Austrinus, then the location of α Fomalhaut is confirmed through its relation to the α and β stars in Grus.

Star Finding Chart (4)

Cygnus, Lyra, and Aquila

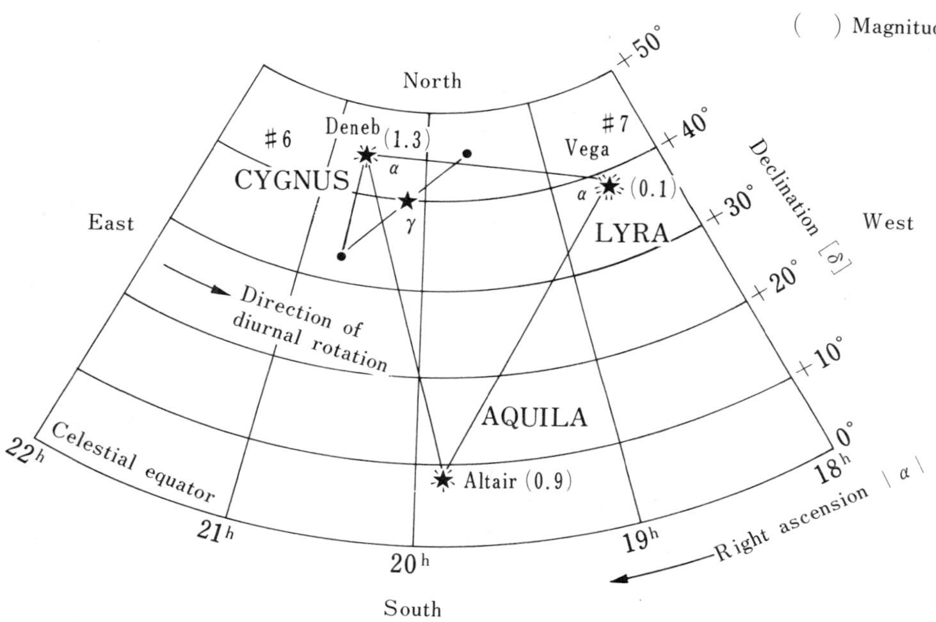

(1) This is a part of the upper view of the southern sky.

(2) On a clear night when the Milky Way is visible, the following stars may be used for reference

Deneb …… located in the middle of the Milky Way
Vega …… outside and to the west of the Milky Way
Altair …… at the eastern edge and in the south of the Milky Way

(3) These three stars form an equilateral triangle, with Altair at the apex and Deneb and Vega at the base, as shown in the figure.

(4) Be careful not to mistake Deneb for Vega or vice versa and bear in mind that the triangle appears to be deformed near the horizon in the east and also in the west. (The same precaution applies to the other star finding chart.)

(5) Deneb is a radio star under study (it is used as a reference point in satellite communications). Recently the presence of a strong source was confirmed, and a black hole was found in the vicinity of the γ star.

Star Finding Chart (5)

Virgo, Bootes, Leo, and Scorpius

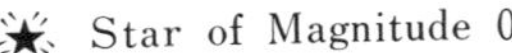

Star of Magnitude 0

Star of Magnitude 1

Star of Magnitude 2

( ) Magnitude

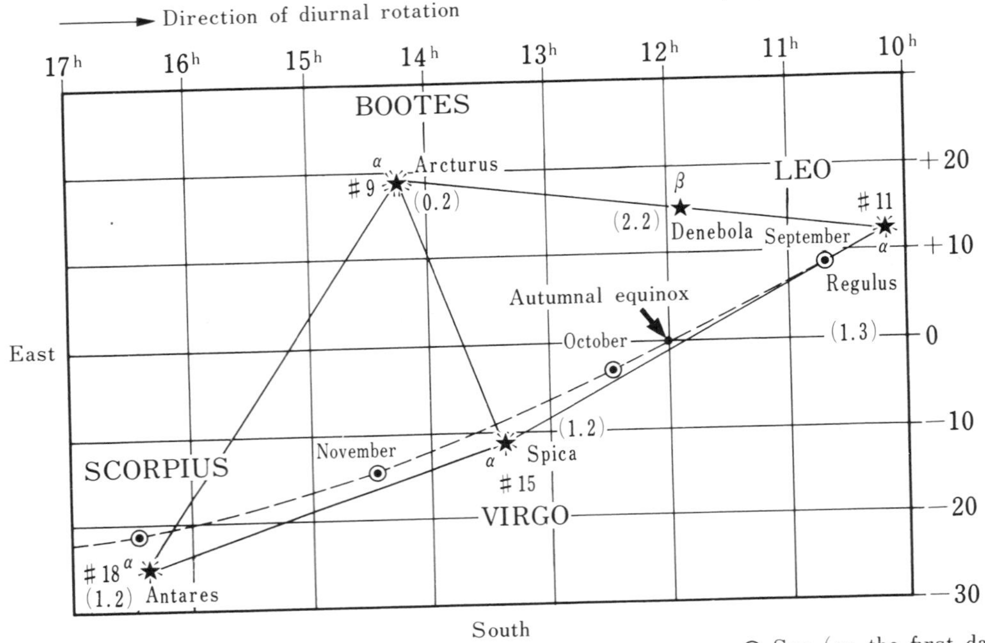

◉ Sun (on the first day of each month)

(1) This is a part of the upper view of the southern sky.

(2) Since the space between Regulus and Antares is approximately 6.5^h, it is probable that one of them sinks below the horizon depending on the time and season.

(3) All these stars are luminous and none of the confusing stars is found in the vicinity; however, it is advisable to find and confirm at least three among these four stars.

(4) The lines connecting Regulus, Spica and Antares fall within the passage of planets, therefore if there is a doubtful star, consult the astronomical ephemeris for that month and year.

Star Finding Chart (6)

View of Area Surrounding Orion

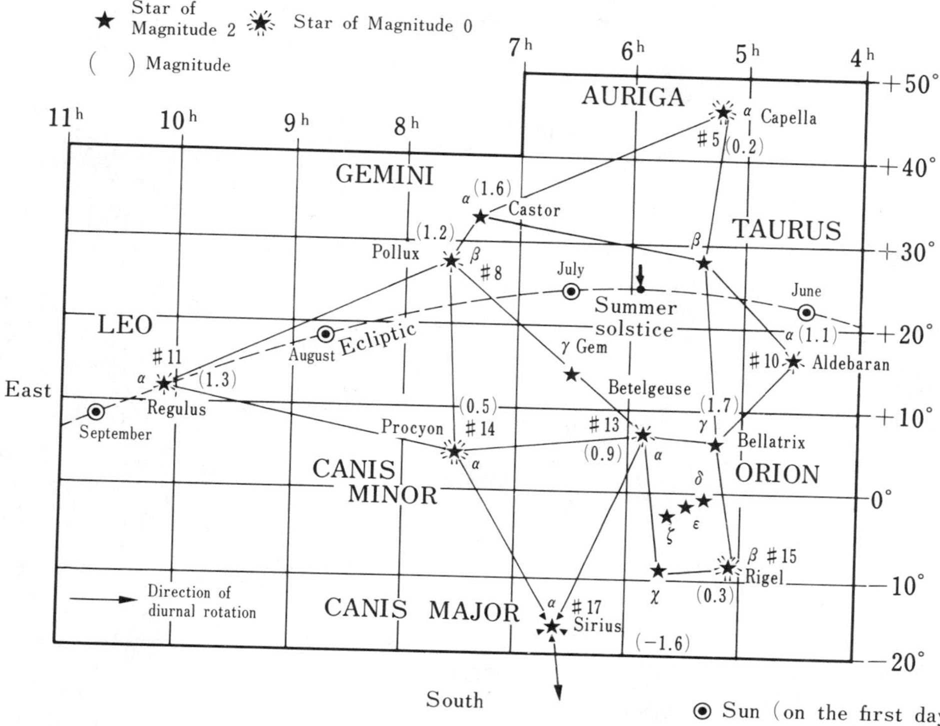

(1) This is a part of the upper view of the southern sky.

(2) The brilliance of Orion may be seen across the celestial equator from October to April except for the north pole region.

(3) Orion may be easily identified through three luminous stars aligned in a line surrounded by four other stars.

(4) Sirius is the most luminous (Mag. −1.6) in the whole sky.

(5) Aldebaran is well known as a radio star used as the measurement standard in satellite communications.

(6) Since the area connecting Aldebaran with Pollox and Regulus overlaps the planet path, if you find any doubtful star in this area consult the astronomical ephemeris.

Simplified Chart of Fixed Stars in the Whole Sky (1)

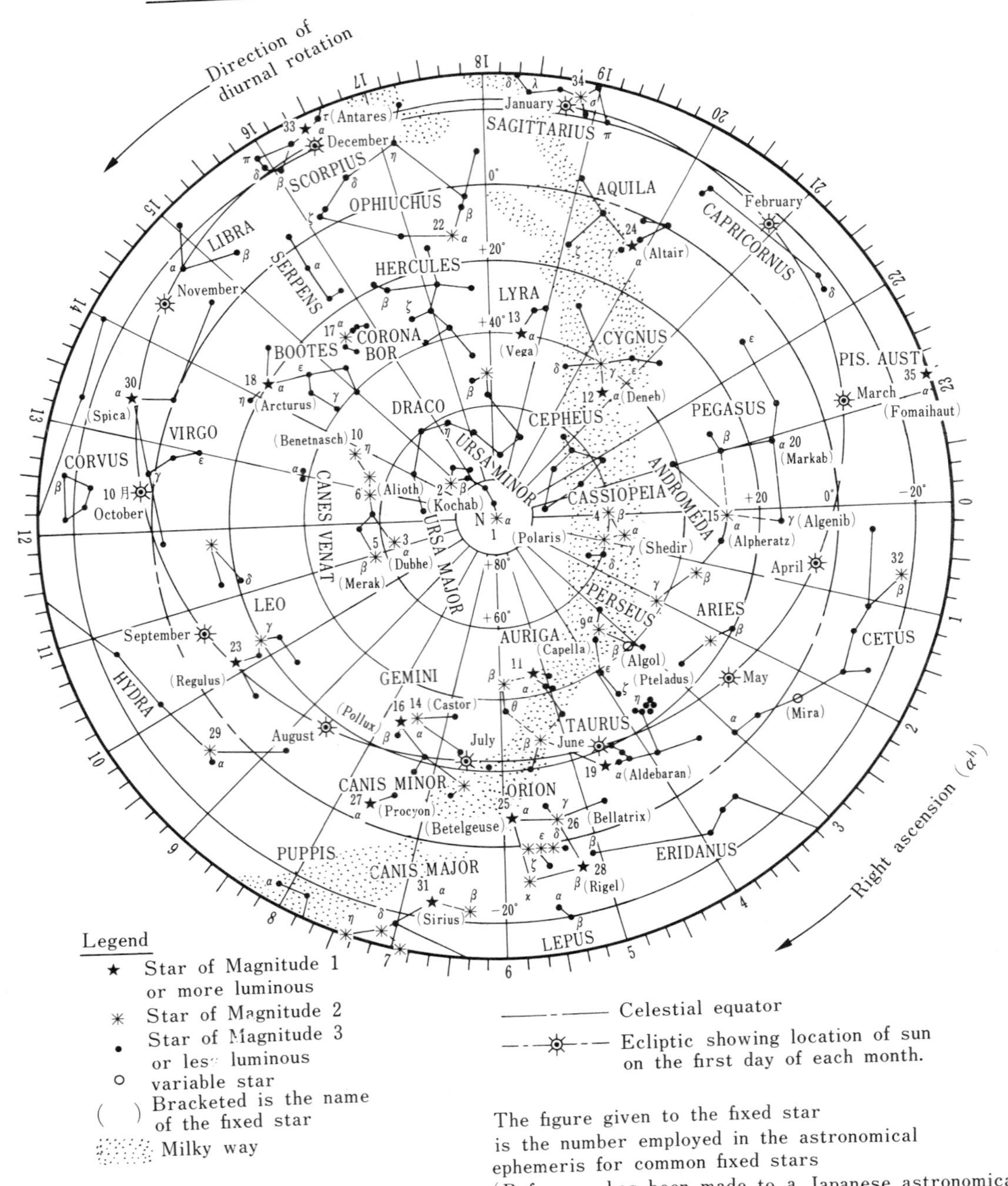

Legend

★ Star of Magnitude 1 or more luminous

✳ Star of Magnitude 2

• Star of Magnitude 3 or less luminous

○ variable star

( ) Bracketed is the name of the fixed star

Milky way

———— Celestial equator

- - -☼- - - Ecliptic showing location of sun on the first day of each month.

The figure given to the fixed star is the number employed in the astronomical ephemeris for common fixed stars

(Reference has been made to a Japanese astronomical ephemeris in preparing showing location of sun chart)

Simplified Chart of Fixed Stars in the Whole Sky (2)

Time–Angle Conversion Table (1)

h	0^h	1^h	2^h	3^h	4^h	5^h	6^h	7^h	8^h	9^h	10^h	11^h	*h*
m	° ′	° ′	° ′	° ′	° ′	° ′	° ′	° ′	° ′	° ′	° ′	° ′	m
0	0 00	15 00	30 00	45 00	60 00	75 00	90 00	105 00	120 00	135 00	150 00	165 00	0
1	15	15	15	15	15	15	15	15	15	15	15	15	1
2	30	30	30	30	30	30	30	30	30	30	30	30	2
3	45	45	45	45	45	45	45	45	45	45	45	45	3
4	1 00	16 00	31 00	46 00	61 00	76 00	91 00	106 00	121 00	136 00	151 00	166 00	4
5	15	15	15	15	15	15	15	15	15	15	15	15	5
6	30	30	30	30	30	30	30	30	30	30	30	30	6
7	45	45	45	45	45	45	45	45	45	45	45	45	7
8	2 00	17 00	32 00	47 00	62 00	77 00	92 00	107 00	122 00	137 00	152 00	167 00	8
9	15	15	15	15	15	15	15	15	15	15	15	15	9
10	30	30	30	30	30	30	30	30	30	30	30	30	10
11	45	45	45	45	45	45	45	45	45	45	45	45	11
12	3 00	18 00	33 00	48 00	63 00	78 00	93 00	108 00	123 00	138 00	153 00	168 00	12
13	15	15	15	15	15	15	15	15	15	15	15	15	13
14	30	30	30	30	30	30	30	30	30	30	30	30	14
15	45	45	45	45	45	45	45	45	45	45	45	45	15
16	4 00	19 00	34 00	49 00	64 00	79 00	94 00	109 00	124 00	139 00	154 00	169 00	16
17	15	15	15	15	15	15	15	15	15	15	15	15	17
18	30	30	30	30	30	30	30	30	30	30	30	30	18
19	45	45	45	45	45	45	45	45	45	45	45	45	19
20	5 00	20 00	35 00	50 00	65 00	80 00	95 00	110 00	125 00	140 00	155 00	170 00	20
21	15	15	15	15	15	15	15	15	15	15	15	15	21
22	30	30	30	30	30	30	30	30	30	30	30	30	22
23	45	45	45	45	45	45	45	45	45	45	45	45	23
24	6 00	21 00	36 00	51 00	66 00	81 00	96 00	111 00	126 00	141 00	156 00	171 00	24
25	15	15	15	15	15	15	15	15	15	15	15	15	25
26	30	30	30	30	30	30	30	30	30	30	30	30	26
27	45	45	45	45	45	45	45	45	45	45	45	45	27
28	7 00	22 00	37 00	52 00	67 00	82 00	97 00	112 00	127 00	142 00	157 00	172 00	28
29	15	15	15	15	15	15	15	15	15	15	15	15	29
30	30	30	30	30	30	30	30	30	30	30	30	30	30
31	45	45	45	45	45	45	45	45	45	45	45	45	31
32	8 00	23 00	38 00	53 00	68 00	83 00	98 00	113 00	128 00	143 00	158 00	173 00	32
33	15	15	15	15	15	15	15	15	15	15	15	15	33
34	30	30	30	30	30	30	30	30	30	30	30	30	34
35	45	45	45	45	45	45	45	45	45	45	45	45	35
36	9 00	24 00	39 00	54 00	69 00	84 00	99 00	114 00	129 00	144 00	159 00	174 00	36
37	15	15	15	15	15	15	15	15	15	15	15	15	37
38	30	30	30	30	30	30	30	30	30	30	30	30	38
39	45	45	45	45	45	45	45	45	45	45	45	45	39
40	10 00	25 00	40 00	55 00	70 00	85 00	100 00	115 00	130 00	145 00	160 00	175 00	40
41	15	15	15	15	15	15	15	15	15	15	15	15	41
42	30	30	30	30	30	30	30	30	30	30	30	30	42
43	45	45	45	45	45	45	45	45	45	45	45	45	43
44	11 00	26 00	41 00	56 00	71 00	86 00	101 00	116 00	131 00	146 00	161 00	176 00	44
45	15	15	15	15	15	15	15	15	15	15	15	15	45
46	30	30	30	30	30	30	30	30	30	30	30	30	46
47	45	45	45	45	45	45	45	45	45	45	45	45	47
48	12 00	27 00	42 00	57 00	72 00	87 00	102 00	117 00	132 00	147 00	162 00	177 00	48
49	15	15	15	15	15	15	15	15	15	15	15	15	49
50	30	30	30	30	30	30	30	30	30	30	30	30	50
51	45	45	45	45	45	45	45	45	45	45	45	45	51
52	13 00	28 00	43 00	58 00	73 00	88 00	103 00	118 00	133 00	148 00	163 00	178 00	52
53	15	15	15	15	15	15	15	15	15	15	15	15	53
54	30	30	30	30	30	30	30	30	30	30	30	30	54
55	45	45	45	45	45	45	45	45	45	45	45	45	55
56	14 00	29 00	44 00	59 00	74 00	89 00	104 00	119 00	134 00	149 00	164 00	179 00	56
57	15	15	15	15	15	15	15	15	15	15	15	15	57
58	30	30	30	30	30	30	30	30	30	30	30	30	58
59	45	45	45	45	45	45	45	45	45	45	45	45	59
h	0^h	1^h	2^h	3^h	4^h	5^h	6^h	7^h	8^h	9^h	10^h	11^h	*h*

h	h°
s	′
0, 1	0
2, 3, 4, 5	1
6, 7, 8, 9	2
10, 11, 12, 13	3
14, 15, 16, 17	4
18, 19, 20, 21	5
22, 23, 24, 25	6
26, 27, 28, 29	7
30, 31, 32, 33	8
34, 35, 36, 37	9
38, 39, 40, 41	10
42, 43, 44, 45	11
46, 47, 48, 49	12
50, 51, 52, 53	13
54, 55, 56, 57	14
58, 59	15
h	h°

Time–Angle Conversion Table (2)

h	12^h	13^h	14^h	15^h	16^h	17^h	18^h	19^h	20^h	21^h	22^h	23^h	h
m	° ′	° ′	° ′	° ′	° ′	° ′	° ′	° ′	° ′	° ′	° ′	° ′	m
0	180 00	195 00	210 00	225 00	240 00	255 00	270 00	285 00	300 00	315 00	330 00	345 00	0
1	15	15	15	15	15	15	15	15	15	15	15	15	1
2	30	30	30	30	30	30	30	30	30	30	30	30	2
3	45	45	45	45	45	45	45	45	45	45	45	45	3
4	181 00	196 00	211 00	226 00	241 00	256 00	271 00	286 00	301 00	316 00	331 00	346 00	4
5	15	15	15	15	15	15	15	15	15	15	15	15	5
6	30	30	30	30	30	30	30	30	30	30	30	30	6
7	45	45	45	45	45	45	45	45	45	45	45	45	7
8	182 00	197 00	212 00	227 00	242 00	257 00	272 00	287 00	302 00	317 00	332 00	347 00	8
9	15	15	15	15	15	15	15	15	15	15	15	15	9
10	30	30	30	30	30	30	30	30	30	30	30	30	10
11	45	45	45	45	45	45	45	45	45	45	45	45	11
12	183 00	198 00	213 00	228 00	243 00	258 00	273 00	288 00	303 00	318 00	333 00	348 00	12
13	15	15	15	15	15	15	15	15	15	15	15	15	13
14	30	30	30	30	30	30	30	30	30	30	30	30	14
15	45	45	45	45	45	45	45	45	45	45	45	45	15
16	184 00	199 00	214 00	229 00	244 00	259 00	274 00	289 00	304 00	319 00	334 00	349 00	16
17	15	15	15	15	15	15	15	15	15	15	15	15	17
18	30	30	30	30	30	30	30	30	30	30	30	30	18
19	45	45	45	45	45	45	45	45	45	45	45	45	19
20	185 00	200 00	215 00	230 00	245 00	260 00	275 00	290 00	305 00	320 00	335 00	350 00	20
21	15	15	15	15	15	15	15	15	15	15	15	15	21
22	30	30	30	30	30	30	30	30	30	30	30	30	22
23	45	45	45	45	45	45	45	45	45	45	45	45	23
24	186 00	201 00	216 00	231 00	246 00	261 00	276 00	291 00	306 00	321 00	336 00	351 00	24
25	15	15	15	15	15	15	15	15	15	15	15	15	25
26	30	30	30	30	30	30	30	30	30	30	30	30	26
27	45	45	45	45	45	45	45	45	45	45	45	45	27
28	187 00	202 00	217 00	232 00	247 00	262 00	277 00	292 00	307 00	322 00	337 00	352 00	28
29	15	15	15	15	15	15	15	15	15	15	15	15	29
30	30	30	30	30	30	30	30	30	30	30	30	30	30
31	45	45	45	45	45	45	45	45	45	45	45	45	31
32	188 00	203 00	218 00	233 00	248 00	263 00	278 00	293 00	308 00	323 00	338 00	353 00	32
33	15	15	15	15	15	15	15	15	15	15	15	15	33
34	30	30	30	30	30	30	30	30	30	30	30	30	34
35	45	45	45	45	45	45	45	45	45	45	45	45	35
36	189 00	204 00	219 00	234 00	249 00	264 00	279 00	294 00	309 00	324 00	339 00	354 00	36
37	15	15	15	15	15	15	15	15	15	15	15	15	37
38	30	30	30	30	30	30	30	30	30	30	30	30	38
39	45	45	45	45	45	45	45	45	45	45	45	45	39
40	190 00	205 00	220 00	235 00	250 00	265 00	280 00	295 00	310 00	325 00	340 00	355 00	40
41	15	15	15	15	15	15	15	15	15	15	15	15	41
42	30	30	30	30	30	30	30	30	30	30	30	30	42
43	45	45	45	45	45	45	45	45	45	45	45	45	43
44	191 00	206 00	221 00	236 00	251 00	266 00	281 00	296 00	311 00	326 00	341 00	356 00	44
45	15	15	15	15	15	15	15	15	15	15	15	15	45
46	30	30	30	30	30	30	30	30	30	30	30	30	46
47	45	45	45	45	45	45	45	45	45	45	45	45	47
48	192 00	207 00	222 00	237 00	252 00	267 00	282 00	297 00	312 00	327 00	342 00	357 00	48
49	15	15	15	15	15	15	15	15	15	15	15	15	49
50	30	30	30	30	30	30	30	30	30	30	30	30	50
51	45	45	45	45	45	45	45	45	45	45	45	45	51
52	193 00	208 00	223 00	238 00	253 00	268 00	283 00	298 00	313 00	328 00	343 00	358 00	52
53	15	15	15	15	15	15	15	15	15	15	15	15	53
54	30	30	30	30	30	30	30	30	30	30	30	30	54
55	45	45	45	45	45	45	45	45	45	45	45	45	55
56	194 00	209 00	224 00	239 00	254 00	269 00	284 00	299 00	314 00	329 00	344 00	359 00	56
57	15	15	15	15	15	15	15	15	15	15	15	15	57
58	30	30	30	30	30	30	30	30	30	30	30	30	58
59	45	45	45	45	45	45	45	45	45	45	45	45	59
h	12^h	13^h	14^h	15^h	16^h	17^h	18^h	19^h	20^h	21^h	22^h	23^h	h

h (s)	h° (′)
0	
	0
1	
2, 3	
	1
4, 5	
6, 7	
	2
8, 9	
10, 11	
	3
12, 13	
14, 15	
	4
16, 17	
18, 19	
	5
20, 21	
22, 23	
	6
24, 25	
26, 27	
	7
28, 29	
30, 31	
	8
32, 33	
34, 35	
	9
36, 37	
38, 39	
	10
40, 41	
42, 43	
	11
44, 45	
46, 47	
	12
48, 49	
50, 51	
	13
52, 53	
54, 55	
	14
56, 57	
58, 59	
	15
h	h°

Correction to Geocentric Latitude

· This graph is used only when the geocentric latitude is to be found.

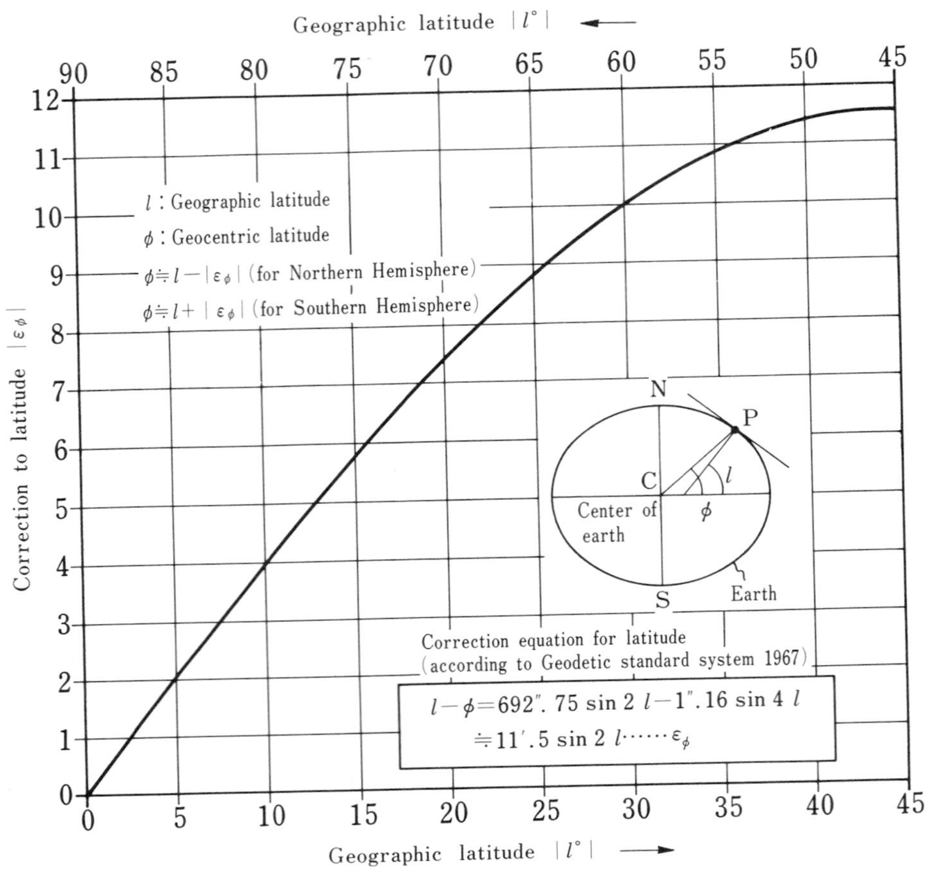

The simplified measurement can assume the following relation.
Latitude : Geographic latitude ≃ Astronomical latitude.

INDEX

Entries in italics refer to chart pages.